W9-CFS-144

Theoretical and Computational Acoustics 2001

Theoretical and Computational Acoustics 2001

Beijing, China   21 – 25 May 2001

Editors

Er-Chang Shang
Scripps Institute of Oceanography, USA

Qihu Li
T F Gao
Chinese Academy of Sciences, China

World Scientific
New Jersey • London • Singapore • Hong Kong

Published by

World Scientific Publishing Co. Pte. Ltd.

P O Box 128, Farrer Road, Singapore 912805

USA office: Suite 1B, 1060 Main Street, River Edge, NJ 07661

UK office: 57 Shelton Street, Covent Garden, London WC2H 9HE

British Library Cataloguing-in-Publication Data
A catalogue record for this book is available from the British Library.

THEORETICAL AND COMPUTATIONAL ACOUSTICS 2001

Copyright © 2002 by World Scientific Publishing Co. Pte. Ltd.

ISBN 981-238-012-4

Printed by FuIsland Offset Printing (S) Pte Ltd, Singapore

PREFACE

The Fifth International Conference on Theoretical and Computational Acoustics (ICTCA) was held on May 21–25, 2001 at the Friendship Hotel in Beijing, China. This conference was a continuation of the Fourth ICTCA held in May, 1999, in Trieste, Italy. The conference was sponsored jointly by the U.S. Naval Undersea Warfare Center (NUWC), the U.S. Office of Naval Research (ONR), the Institute of Acoustics, Chinese Academy of Sciences, the U.S. Naval Research Laboratory (NRL), the NATO SACLANT Undersea Research Center, the Scripps Institute of Oceanography, the Osservatorio Geofisico Sperimentale di Trieste (OGS) Italy, University of Cambridge UK, Yale University, Harvard University, Columbia University, National Taiwan University, University of Petrol Beijing, Bureau of International Cooperation, CAS, China, National Natural Science Foundation of China, Acoustical Society of China, and the Environment Technology Laboratory/NOAA.

The objective of this conference was two-fold: First, it was held to provide a forum for active researchers to discuss state-of-the-art developments and results and their recent contributions in theoretical and computational acoustics and related areas. Second, the conference provided a means to congregate from different branches of science related to acoustics to exchange new ideas and to stimulate new research for the future. Over hundred scholars, scientists, and engineers from 18 countries — representing 64 universities and laboratories — participated in this event. The presented lectures examined various topics on acoustics and related areas ranging from ocean acoustics to seismo-acoustics, aeroacoustics, structural acoustics, and ultrasonics. Particularly, we have two special sessions : one in memorial of Professor David G. Crighton, and one in memorial of Dr. Warren Denner.

The Conference Committee wishes to express deep gratitude to Dr. Ding Lee (NUWC and Yale University) for fulfilling the role of Honorary Chairman of the conference, who is the founder of the ICTCA. This was a great honor for both the Conference Committee and the participants. Special thanks are also extended to our 4 keynote speakers: Dr. Ding Lee (NUWC and Yale University) on *"Parabolic Equation Development in the Twentieth Century"*, Prof. M. J. Buckingham (Scripps Institute of Oceanography) on *"Wave Propagation in Dispersive Marine Sediments: A Memorial to David G. Crighton"*, Prof. Qihu Li (Institute of Acoustics, CAS) on *"The Design Philosophy of Model-based SONAR System"*, and Prof. M. Fink (Lab. Ondes et Acoustique, Universite Paris

7) on *"New Perspertives in Time Reversed Acoustics"*. These excellent keynote lectures gave a very good start to this conference and made a wide open prospect of the future research.

We gratefully acknowledge all the special session organizers and the session chairs, they are: Dr. Ching-Sang Chiu (NPS), Dr. A. Tolstoy (Atolstoy Sciences), Dr. S.F. Wu (Wayne State University), Dr. L. Fishman (NRL), Dr. Y.C. Teng (Columbia University), Prof. Yongguan Mu (University of Petrolem), Prof. M. Pappalardo (University of Rome), Prof. H.L. Zhang (Institute of Acoustics), S. Marburg (Tech. University Dresden), Prof. L. Bjorno (Tech. University Denmark), Prof. A. Hanyga (University of Bergen), Dr. G. Seriani (OGS, Italy), Dr. F.B. Jensen (SACLANT), Prof. G. Rosenhouse (Technion, Israel), Dr. O. Godin (ETL/NOAA), Prof. R.H. Zhang (Institute of Acoustics), Prof. Y.W. Li (State Ocean Admini.), Prof. Y.J. Wang (Nanjing University), Prof. G.Z. Zhu (Tsinghua University), Prof. H.Z. Yang (Tsinghua University), Prof. L.F. Ge (Anhui University), Prof. Ning Wang (Ocean University), Prof. X.H. Chen (University of Petrolem) , Prof. C.F. Chen (National Taiwan University), Prof. Y.Y. Lu (City University, HK).

This was the first time that ICTCA was held in China, our host institute — Institute of Acoustics, CAS — contributed tremendous efforts to make the conference more successful and let the participants to enjoy more Chinese culture, they have arranged the city tour to Great Wall, Forbidden City, Peking Opera entertainment, elegant Chinese food Banquet , and post-conference tour to Xi'an, Lasa, Guilin. Our special thanks go to Ms. Li Ying — the conference secretariat, without her continuous diligent efforts the conference would not be successful.

E.-C. Shang

Qihu Li

T. F. Gao

CONTENTS

Theoretical and Computational Acoustics 2001
E.-C. Shang, Qihu Li and T. F. Gao (Editors)
© 2002 World Scientific Publishing Co.

Parabolic Equation Development in the Twentieth Century

Ding Lee

Department of Computer Science, Yale University, New Haven, CT, USA

Naval Undersea Warfare Center, Newport, RI, USA

Allan D. Pierce

Department of Aerospace and Mechanical Engineering, Boston Unviersity,
Boston, MA, USA

Er-Chang Shang

CIRES, University of Colorado, Boulder, CO, USA

ABSTRACT

In the 20th Century, an important contribution to the modeling of wave propagation
prediction is the Parabolic Equation (PE) approximation method. Five PE review
literatures have been reported. One of them is a comprehensive review of the PE
development in the twentieth century; this paper gives its summary.

1 INTRODUCTION

There have been a few PE review papers reported in open literatures, [1, 2, 3, 4, 5].
Among these literatures, Ref. [4] is quite a comprehensive and complete review of the
PE developments. This paper summarizes the contents of Ref.[4]. Readers who are
interested of all details of individual contributions are suggested to consult Ref.[4].
Some existing issues are outlined.

2 BACKGROUND

After Lenotovich and Fock introduced the PE approximation to solve electromagnetic
problems, it stimulated a lot of interest of applying the PE to solve other scientific

problems. One area is the ocean acoustics. Tappert introduced the PE approximation method to the ocean acoustic community. The PE approximation decomposes an elliptic wave equation into two equations through the choice of an arbitrary constant reference wavenumber. One resulting equation, by means of the far-field elliptic approximation, gives the parabolic wave equation.

3 BASIC THEORETICAL DEVELOPMENT

Parabolic Equation Approximation - The basic PE development begins with the 2-dimensional (range and depth) Helmholtz equation

$$\phi_{rr} + \frac{1}{r}\phi_r + \phi_{zz} + k^2(r, z)\phi = 0, \tag{3.1}$$

where $\phi(r, z)$ is the wave field, $k(r, z) = \frac{\omega}{c(r,z)}$ is the wavenumber, $\omega = 2\pi$ times the source frequency, and $c(r, z)$ is the sound speed. Substituting $\phi(r, z) = u(r, z)v(r)$ into Eq.(3.1) and apply the far-field approximation, $k_0 r \geq 1$, gives

$$u_{rr} + 2ik_0 u_r + u_{zz} + k_0^2(n^2(r, z) - 1)u = 0. \tag{3.2}$$

Standard PE - The first PE introduced by Tappert is regarded as the Standard PE (or narrow-angle PE) which has the expression

$$u_r = \frac{ik_0}{2}(n^2(r, z) - 1)u + \frac{i}{2k_0}\frac{\partial^2 u}{\partial z^2}. \tag{3.3}$$

The Wide-Angle PE - There are problems whose solutions are inaccurate if they are solved by the standard PE due the inappropriate treatment of the angle of propagation. Thus, a wide-angle PE is needed. Discussions have been given in **A Benchmark Test Example**, **Wide-angle Developments**, and **An Estimation of the Propagation Angle**. The size of the angle depends upon the required accuracy. Commonly recognized estimates for the narrow-angle is $\leq 23°$ and for the wide-angle is $\leq 40°$.

The Interface PE - In 1981 Homer Bucker presented an interface problem that cannot be solved by any existing PE model at that time. Lee-McDaniel, then, developed a technique to handle the fluid-fluid interface and incorporated this capability into the IFD model. Then, the IFD solved the Bucker Probelm. Besides the Lee-McDaniel interface development there are 2 theoretical interface models; one by Tappert, one by Gribble.

The Reference Wave Number, k_0 - The selection of k_0 is arbitrary but not unique. Pierce analyzed since the sound speed may change with the range, the k_0 has to be updated after every range step is completed. Pierce introduced the following formula for updating the k_0:

$$k_0^2 = \frac{\int (\frac{\omega}{c})^2 \mid u \mid^2 dz - \int \mid u_z \mid^2 dz}{\int \mid u \mid^2} dz. \tag{3.4}$$

4   ADVANTAGES AND LIMITATIONS

Advantages include: (1) A quick way to solve long-range, low-frequency, and range-dependent problems; (2) Avoiding dealing with the far-end vertical wall boundary condition; and (3) PE Solution is much cheaper to obtain than solving the elliptic equation both in memory and running time savings.

Limitations can be classified in 2 types. Type 1 comes from the mathematical formulation; Type 2 contributed from the solution of the PE.

5   CONTRIBUTIONS PRIOR TO 1984

Mathematical Models include: The Standard PE, A Wide Angle PE, A Pseudo-Partial Differential Equation.

Numerical Solutions include: Split-step Algorithm, A Numerical Ordinary-Differential-Equation Solution, and An Implicit Finite Difference (IFD) Scheme.

Computer Codes include: AESD PE Model, SACLANT PE- PAREQ, and IFD.

The PE Workshop I was held at 1981 directed at reporting of PE progress.

PE Developments for Elastic Solids include developments by Landers-Claerbout, McCoy, Hudson, and Corones.

6   CONTRIBUTIONS FROM 1984 TO 1994

PE contributions in this time interval includes the following:

The PE Workshop II - The second PE Workshop was held at 1993.

Various PE Developments

3-dimensional PE - FOR3D, WRAP, 3DPE-NRL, 3DPE, A Wide-angle 3D PE-SKL, A Wider-angle 3D PE- Hill, A Modifief PE.

Backscattering PE - Two- and 3-Dimensional L1L2 Models, Backscattering from Rough Surfaces, FFRAME, Collins' Development, Brooke and Thomson Two-Way PE.

Elastic PE - Shang-Lee Fluid-Elastic Interface PE, HEPE, Two-Way PE for Elastic Media, SHAPE, Wetton and Brook Development, Thomson's Approximation, Papadakis Model, Elastic PEs-Nagem, PE in Heterogeneous Media.

High-Order PE - St. Mary-Lee Formulation, Knightly's Equation, Estes-Fain Expansion, Lee-Saad-Schultz Expansion, Collins' Equations.

Wide-Angle PE - Gilbert-Lee Development, Greene's Development, Modified Wide-Angle PE, Saad-Lee Wide-Angle Table, Trefethen's Development, A Very Wide-Angle, and Another Very Wide Angle.

Also included are: Stochastic PE, A PE SOlution GENerator, A Range Refraction PE, High-Frequency PE-HYPER, Nonlinear PE, Time Domain PE, Energy-Conserving PE, MOREPE, Adiabatic Mode PE, GF-PE, LOGPE, An Optimal PE-type Wave Equation, Local Parabolic Approximation, U.S.Navy Standard PE,

PE-Related Developments include: A Coupled Ocean-Acoustic Model, Treatment of Interface Conditions, Treatment of Interface Conditions (I: A PE with Density Variations, II: An Irregular Interface Enchancement, and III: Vertical Interface Conditions for Parabolic Approximations), Factorization and Path Integration, Generalized Impedance Boundary Condition, Hybrid Analysis, Range-dependent Duct Boundary Conditions, Alternative Non-reflecting Boundary Conditions, Starting Fields, Further Progress of Reference Wavenumber k_0, and Removing Phase Errors in the PE.

Comparisons With Other Models - Comparisons have been done of PE models against "Normal Mode", "Adiabatic Mode", "Intrinsic Modes", "Mode Coupling", "Ray Theory". Also PE models are compared with one another. In addition, PE models are compared against other non-PE models.

Computational Methods for PE - Finite difference methods, Numerical Ordinary-differential-equation methods, Fast Fourier Transform (FFT), Finite elements methods, Preconditioners, Alternating Direction Implicit (ADI), Adaptive Schemes, Split-step and Combined Algorithms, Energy method, Transformation method, Calculation Frequency method.

Acoustics problems become large and complicated; therefore, supercomputing becomes important. Progress made in supercomputing includes: High Performance PE Solvers, A Parallel 3D PE Solver, Multigrid Methods, A Parallel Algorithm, Parallel

Solution of PE's.

7 Contributions between 1995 and 1999

PE and PE-related contributions during the last five years of the 20th century cover the areas of:

1. **Treatment of Boundary Conditions** which include: PE models with bottom treatment, Non-local Boundary Condition, A Fluid-elastic Interface Model, A New Discrete Boundary Condition, Side-wall Boundaries, and Computational Methods and Numerical Solutions.

2. **Applications** which cover a wide coverage of many areas including: 3-dimensional Problems, Scattering Problems, Travel Time Computation, Global Acoustics, Parametric Array Application, Frontal Effects, Ocean Remote Sensing, Caustics, Current and Current Shear Effects, Downslope Acoustic Propagation Loss, Gulf Stream Eddies, Inversion with PE, Leaky Surface Duct, Mode Propagation in a Wedge, "Oceanographic, Topographic, and Sediment Interactions", Refracting Atmosphere, Shallow Effects, Source and Receiver Motions, Surface Loss, Tomography Irregularities, Simulation by SAFE, Geometrical Theory of Diffraction and the PE, Reciprocity in the Time-domain.

8 REMAINING ISSUES AND PROBLEMS

Remaining issues include: (1) **Is there a group that can use some accurate benchmark problems to evaluate these PE models and reach some kind of recommendations?** (2) **Among available PE models which PE model one should choose for his/her use?** (3) **Should every PE computer model be well-documented?** and (4) **Why NOT use existing accurate PE models for applications?**

Future Needs - No available PE models can handle completely 3D problems with strong interface actions, scattering problems, wave propagation in elastic medium, fluid-elastic interface, backscattering, time domain problems and other related problems.

9 CONCLUSIONS

The PE development covers a variety of categories. Contributions came worldwide and from more than 500 contributors. There are still a number of problems warrant our attention.

(1) More PE contributions are a challenge for ocean scientists and engineers to stimulate their future research; (2) Modellers claim among themselves that one's method is superior to others; as far as computing speed is concerned, this is a true fact but not important in practice because every user is interested in meaningful and accurate results, the speed and amount of work should not be a serious matter; (3) Our experience tells us: **No single existing PE model which is adequate for all applications**; (4) **Is it true that NOT every problem can be solved by the Parabolic Equation Approximation?**

ACKNOWLEDGMENTS

This research was supported jointly by the Department of Computer Science of Yale University, the U.S. Naval Undersea Warfare Center Independent Research project, Boston University and the Environmemtal Technology Laboratory/NOAA, and CIRES, University of Colorado. The authors would like to thank Prof. T.F.Gao and Dr. A.G.Vornovich for their help of a large portion of literature research and for their assistance in valuable technical discussions.

References

[1] Ames, W.F. and D. Lee,"Current Development in the Numerical Treatment of Ocean Acoustic Propagation," *J. Applied Numer. Math.,* **3**(1-2), 1987, 25-47.

[2] Lee,D., "The State-of-the-Art Parabolic Equation Approximation as Applied to Underwater Acoustic Propagation With Discussions on Intensive Computations," Naval Underwater Systems Center TD No.7247, 1984.

[3] Lee, D. and A. D. Pierce, "Parabolic Equation Development in Recent Decade," *J. Comp. Acoust.,* **3**(2), 1995, 95-173.

[4] Lee,D., Allan D. Pierce, and E.C. Shang, "Parabolic Equation Development in the Twentieth Century," *J. Comp. Acoust.,* **8**(4), 2000.

[5] Scully-Power,P.D. and D.Lee (eds), RECENT PROGRESS IN THE DEVELOPMENT AND APPLICATION OF THE PARABOLIC EQUATION, U.S. Naval Underwater Systems Center TR No. 7145, 1984.

Theoretical and Computational Acoustics 2001
E.-C. Shang, Qihu Li and T. F. Gao (Editors)
© 2002 World Scientific Publishing Co.

Wave Propagation in Dispersive Marine Sediments: A Memorial to David G. Crighton

Michael J. Buckhingham

Marine Physical Laboratory, Scripps Institution of Oceanography

University of California, San Diego, 8820 Shellback Way

La Jolla, CA 92093-0238, USA

Also affiliated to: Institute of Sound and Vibration Research, The University

Southampton SO17 1BJ, England

Marine sediments support compressional and shear waves, both of which show weak dispersion and an attenuation that scales essentially as the first power of frequency. Such behavior emerges naturally from a new, linear theory of wave propagation in saturated unconsolidated granular media. The theory is based on the physics of inter-granular shearing, taking account of the microscopically rough surfaces of the mineral grains and the molecularly thin film of pore fluid that separates grains. A stochastic treatment of stress-relaxation occurring at points of contact between asperities, combined with a strain-hardening argument for the behavior of the thin fluid film as shearing progresses, leads to two wave equations, one for compressional and the other for shear waves. From these equations, expressions emerge for the dispersion and attenuation of both types of wave. These theoretical expressions for the phase speeds and attenuations are well behaved at all frequencies and are in agreement with the available experimental evidence.

1. Introduction

A saturated, unconsolidated, granular material such as a sandy marine sediment exhibits interesting wave properties. The grains are unbonded, yet the material is capable of supporting not only a compressional wave but also a shear wave. Both waves show an attenuation that is essentially proportional to frequency[1-4] and both exhibit weak dispersion[5]. The wave speeds vary more or less systematically with grain size and the porosity of the medium[6, 7]. Moreover, the wave speeds and attenuations appear to be correlated, suggesting that they are causally connected through a common physical mechanism.

Over four decades ago, Biot [8, 9] developed a theory of stress waves in a porous, elastic solid containing a compressible, viscous fluid in the pore spaces. In his approach, wave attenuation is attributed to viscous dissipation as pore fluid flows between grains. The Biot theory has been adapted by Stoll[10] and applied to wave propagation in marine sediments. A difficulty with the Biot-Stoll model is that the predicted attenuation has the characteristics of wave propagation in a viscous medium[11] in that it varies as the square of frequency at lower frequencies and then, above some threshold, converts to the square-root of frequency. This type of behavior is clearly illustrated in Fig. 3 of a recent paper on the Biot theory by Williams[12]. It is not possible to approximate satisfactorily the viscous attenuation curve by a straight line over an extended range of frequency.

Since wave attenuation in sediments follows a linear dependence on frequency, it would appear that the primary dissipation mechanism in these materials is not viscosity of the pore fluid. Based on this conclusion, an alternative to the Biot theory of wave propagation in saturated, unconsolidated, granular media has been developed by Buckingham[13]. His treatment is based on the stress induced by inter-granular shearing and the physics of lubricated grain contacts. In this article, the essential new physical ideas underpinning the grain-shearing approach are summarized and the predicted wave properties are shown to compare favorably with the available experimental evidence.

2. Grain-Contact Physics

An ideal unconsolidated sediment is considered in which there are no large scattering centers such as shell fragments, the grains are essentially uniform in size, the pore spaces are free of gas, and the wavelength is considerably longer than the size scale of the grains. The grain contacts are taken to be randomly orientated throughout, allowing the medium to be treated as statistically homogeneous and isotropic, and the boundaries are assumed to be sufficiently far removed for reflections to be negligible. No skeletal mineral frame is included in the analysis, implying that the elastic rigidity modulus of the material to zero. (The elasticity of the mineral grains themselves at points of contact is not neglected). The granular medium is treated as a fluid-like mixture in which the stress-strain relationships are governed by shearing at the grain contacts. It is implicit in these assumptions that only intrinsic attenuation arising from inter-granular interactions is addressed in the model and that high-frequency scattering from individual grains is excluded from the analysis.

The factor which distinguishes a granular material from a suspension is the traction at the grain contacts. In the absence of such traction, the sound speed, c_o, would be given by Wood's equation for a simple two-phase medium [14] with non-interacting components and the attenuation would be zero (assuming that viscosity of the pore fluid and other losses are negligible). Grain-to-grain shearing adds stiffness to the material, which raises c_p, the phase speed of the compressional wave, above c_o, as well as allowing a transverse (shear) wave to propagate with phase speed c_s. Both speeds show weak dispersion, with the wave speed increasing approximately logarithmically with frequency. The shearing also introduces dissipation, which gives rise to an attenuation in both waves that scales essentially as the first power of frequency.

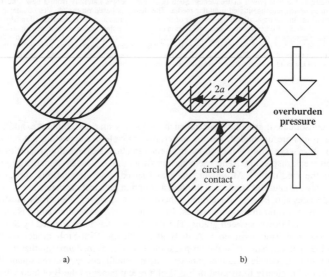

Fig. 1.   Two elastic spheres a) in light contact and b) deformed by
the overburden pressure to create of a circle of contact.

Consider two elastic, spherical grains in light contact (Fig. 1a), a situation which might be encountered immediately beneath the surface of the sediment, where the overburden pressure is negligible. The grains, in this case, remain undistorted. At a finite depth, however, under the influence of the overburden pressure, the grains will be pressed together and will deform. elastically to form a circle of contact [15] of radius a (Fig. 1b). Two types of shearing are possible at the grain contact: translational shearing, due to relative displacement of the grain centers parallel to the plane of the circle of contact (Fig. 2a); and compressive shearing along the radials of the circles of contact, occurring when the grain centers move relative to each other along their connecting line normal to the plane of the circle of contact (Fig. 2b). If the two grains were identical in size and were perfectly smooth, the compressive shearing would be identically zero, just from geometrical considerations. In practice, the grains are likely to have slightly different radii and will be at least microscopically rough, in which case compressive shearing will occur.

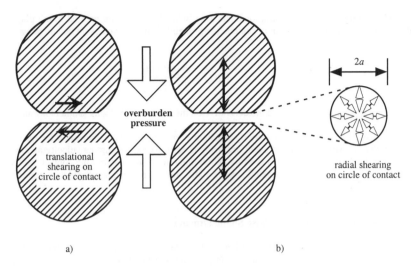

Fig. 2.   a) Translational shearing from tangential strain and
b) compressional shearing, from normal strain, along
radials (open arrows) of the circle of contact .

3.   Random Stick-Slip Stress Relaxation

During the passage of a wave, shearing between the grains is governed by microscopic asperities on the faces of the circles of contact. In fact, the contact is not continuous but occurs discretely at a large number of high points on the circles of radius a. As the grain centers are strained, stress accumulates at the individual points of contact, until eventually one or more fails, slippage occurs and the point-contact relaxes towards a stress-free condition. Since the asperities are unevenly spread, with randomly distributed heights, the total stress, $\sigma(t)$, from shearing over the circle of contact is a stochastic process consisting of a random superposition of individual stick-slip events:

$$\sigma(t) = \sum_{k=1}^{\infty} a_k h(t - t_k) \quad , \tag{1}$$

where $h(t)$, the pulse-shape function from a single relaxation event, must be zero for $t < 0$ to satisfy causality. The kth event occurs at time t_k with amplitude a_k, both of which are taken to be random variables.

The summation on the right of Eq. (1) is known as a random pulse train[16]. Within an elementary volume of the material, which is large compared with the grain size but much smaller than a wavelength, many grain contacts will exist. It follows that the macroscopic wave properties are determined by the average stress over a large number of contacts, which, from Eq. (1), is found to be

$$\overline{\sigma(t)} = \overline{a}\,\overline{K} q(t) \otimes h(t) \quad , \tag{2}$$

where the overbar indicates an ensemble average and the symbol $\otimes$ denotes a temporal convolution. The function $q(t)$ is the probability that a stress-relaxation event is triggered between times t and $t + dt$, $\overline{a}$ is the mean amplitude of the events, and $\overline{K}$ is the mean number of events in an interval of duration T.

4.   The Probability Density Function

If the microscopic stress-relaxation events were Poisson distributed in time, the probability density function, $q(t)$, would be a constant equal to $1/T$. This is not, however, the situation in the granular medium, where the probability of an event occurring must depend on the velocity difference across the plane of contact. Taking the case of translational shearing, with the x-direction parallel to the plane of contact and the z-axis passing through the grain centers, the probability that a given event occurs between time t and $t + dt$ is

$$q(t) = \frac{b}{T} \left| \frac{dv_x(t)}{dz} \right| , \qquad (3)$$

where b is a positive constant and the derivative on the right is the velocity gradient normal to the contact at time t.

Clearly, when there is no velocity gradient across the contact, there can be no shearing and the probability of a stress-relaxation event occurring is zero. By expressing the velocity as $v_x = dx/dt$, it can be seen that the derivative in Eq. (3) is equivalent to the rate of strain across the contact, rather than the strain itself, which is physically reasonable since a static strain represents a steady state condition in which no shearing can occur.

5. Strain-Hardening and the Pulse-Shape Function

The pulse shape function, $h(t)$, characterizes each microscopic event, that is, the sliding of one asperity over another. It can be considered as a material impulse response function (MIRF) in that it represents the time dependence of the stress relaxation that occurs at one microscopic point of contact in response to an impulse of strain rate (or a step-function strain). Thus, to specify $h(t)$, the physics of the individual sliding process must be addressed.

Consider two asperities, A and A', on opposite sides of the circle of contact that are about to undergo sliding (Fig. 3a) in response to a translational strain. Since the material is saturated, a layer of pore fluid will permeate the between the grains, and the asperities themselves will not be in direct contact but will be separated by a molecularly thin film of the same fluid. As sliding progresses, viscous dissipation will occur in this thin fluid film, which gives rise to a drag force opposing the motion that scales with the speed of sliding. A mechanical equivalent of such a force is a dashpot. At the same time, there will be a conservative, elastic force of deformation of the asperities and the fluid film, which may be represented by a Hookean spring. Since they support the same stress, the spring and the dashpot should be connected in series (Fig. 3b), to form what is sometimes known as a Maxwell element[17].

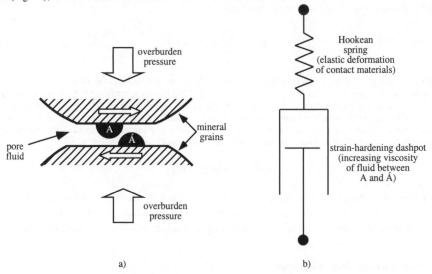

Fig. 3.   a) Asperities A and A' shearing against each other and separated by a thin film of pore fluid (not to scale). b) Equivalent mechanical circuit (Maxwell element) consisting of a series combination of a Hookean spring and non-linear, strain-hardening dashpot.

If we imagine an impulse of strain rate applied across the Maxwell element, the resultant stress in the system is the MIRF, $h(t)$. With constant coefficients for the spring and dashpot, it is easily shown that $h(t)$ decays exponentially in time. Such behavior, however, is not characteristic of granular materials, at least, it does not yield the observed linear scaling of wave attenuation with frequency. This suggests that another physical process is at work, which is postulated to be "strain hardening" originating in the thin fluid film separating the two sliding asperities.

Strain hardening is an increase in the resistance to motion as the sliding progresses, which is represented by a time-dependent coefficient for the dashpot in the Maxwell element. Since the stress in such a dashpot is neither in phase nor in quadrature with the strain, the element is not purely dissipative, as it would be if the coefficient were constant. In fact, the strain-hardening dashpot is non-linear, with cross-coupling between different Fourier components in the stress-strain relationship, and hence the Maxwell element in Fig. 3b does not obey the principle of superposition.

The properties of an aqueous solution confined in a very thin layer (several molecular diameters thick) between solid surfaces are significantly different from those of the bulk material[18]. In particular, the effective viscosity of the thin fluid film may be much higher than the viscosity of the bulk fluid. Moreover, the effective viscosity increases as the film is squeezed harder[19, 20]. This latter phenomenon is exactly the effect represented by the time-dependent coefficient for the dashpot in the Maxwell element.

The time-dependence of the viscous coefficient is not known, but as a first approximation it may reasonably be represented as linear, which is equivalent to truncating its Taylor expansion at the first-order term. Then, the solution for the stress in the Maxwell element is obtained by straightforward integration of a simple non-linear differential equation. The result is the following expression for the MIRF:

$$h(t) = t_o^{-1}\left(1 + \frac{t}{t_o}\right)^{-n} , \tag{4}$$

where the time constant t_o and material exponent n involve the coefficients of the Maxwell element, which themselves represent mechanical properties of the pore fluid and the mineral grains. Eq. (4) is expected to hold, with t_o and n unchanged, for translational and compressive shearing, because the microscopic stress-relaxation occurring at the contact between individual asperities is the same in both cases.

The temporal Fourier transforms of the MIRF in Eq. (4) is

$$H(j\omega) = \left(j\omega t_o\right)^{n-1} e^{j\omega t_o} \Gamma(1-n, j\omega t_o) ,$$

$$\approx \frac{\Gamma(1-n)}{(j\omega t_o)^{1-n}} \tag{5}$$

where ω is angular frequency, $\Gamma(..,..)$ is the complement of the incomplete gamma function, and $\Gamma(..)$ is the gamma function. The approximation in Eq. (5) holds at all frequencies for which $\omega t_o \ll 1$, a condition which appears to be satisfied in unconsolidated sediments, where viscous effects represented by the time constant t_o appear to be negligible.

6. Wave Speeds and Attenuations

When the expressions for the probability density function [Eq.(3)] and the MIRF [Eq. (5)] are substituted into Eq. (2), the average stress for compressional shearing is found to be

$$\sigma_{zz}(t) = \lambda h(t) \otimes \frac{dv_z(t)}{dz} \tag{6}$$

and for translational shearing

$$\sigma_{xz}(t) = \eta h(t) \otimes \frac{dv_x(t)}{dz} , \tag{7}$$

where the overbars have been omitted for brevity and the subscripts xz and zz denote shear stresses in the usual way. All the constants associated with $q(t)$ and $h(t)$ have collapsed into just two scaling factors, λ and η, which are, respectively, compressional and translational stress-relaxation coefficients. Notice that, although the MIRF derives from a non-linear process, the average stresses in the convolutions of Eqs. (6) and (7) are strictly linear in the sense of satisfying superposition.

By extending Eqs. (6) and (7) to three dimensions, the stress tensor relating the stress to the rate of strain at the point (x,y,z) may be specified. On equating the divergence of the stress to the inertial forces in the material and linearizing the result, the first-order Navier-Stokes equation for wave motion in the granular medium is obtained. This, combined with the equation of state, and taking into account conservation of mass, yields a partial differential equation for the particle velocity. Using Helmholtz' theorem[21], it is possible to perform the usual separation into two wave equations, representing compressional and transverse waves in the medium. Even though the shear modulus of the material is identically zero, the equation for shear disturbances is a genuine wave equation representing propagating shear waves. This is a consequence of the stiffness introduced into the material by shearing at the grain contacts.

The speed and attenuation of both types of wave emerge directly from the wave equations. For the compressional wave, the phase speed is given by

$$\frac{1}{c_p} = \frac{1}{c_o} Re\left[1 + \Gamma_p (j\omega T_o)^n + \frac{4}{3}\Gamma_s (j\omega T_o)^n \right]^{-1/2} \tag{8a}$$

and the attenuation is

$$\alpha_p = -\frac{\omega}{c_o} Im\left[1 + \Gamma_p (j\omega T_o)^n + \frac{4}{3}\Gamma_s (j\omega T_o)^n \right]^{-1/2}. \tag{8b}$$

Similarly, for the shear wave, the phase speed is

$$\frac{1}{c_s} = \frac{1}{c_o} Re\left[\Gamma_s (j\omega T_o)^n \right]^{-1/2} = \frac{1}{c_o \sqrt{\Gamma_s}} |\omega T_o|^{-1/2} \cos\left(\frac{n\pi}{4} \right) \tag{9a}$$

and the attenuation is

$$\alpha_s = -\frac{\omega}{c_o} Im\left[\Gamma_s (j\omega T_o)^n \right]^{-1/2} = \frac{|\omega|}{c_o \sqrt{\Gamma_s}} |\omega T_o|^{-n/2} \sin\left(\frac{n\pi}{4} \right) \tag{9b}$$

In these expressions, c_o is the speed of sound from Wood's equation for the equivalent suspension, and T_o is a convenient normalizing parameter, which has been introduced solely to keep terms raised to the fractional power of n dimensionless. Also, a further condensation of parameters has been performed:

$$\Gamma_p = \frac{\lambda}{\rho_o c_o^2 t_o} \left(\frac{t_o}{T_o} \right)^n \Gamma(1-n) \tag{10a}$$

and

$$\Gamma_s = \frac{\eta}{\rho_o c_o^2 t_o} \left(\frac{t_o}{T_o} \right)^n \Gamma(1-n) , \tag{10b}$$

where ρ_o is the density of the two-phase medium. The two dimensionless, stress-relaxation modulii, Γ_p and Γ_s, characterize the average stiffness of the medium due, respectively, to compressive and translational grain-to-grain shearing. The material exponent, n, on the other hand, is a measure of the strain hardening that occurs in a single, discrete sliding event, and is small compared with unity. All three parameters are unknowns that, at present, cannot be estimated from theoretical arguments and must be evaluated from data.

By expanding Eqs. (8) and (9) in Taylor series to first-order in the small parameter n, simple approximations are obtained for the wave speeds and attenuations:

$$c_p \approx c_o \sqrt{1 + \Gamma_p + \frac{4}{3}\Gamma_s \left[1 + \frac{2\beta_p}{\pi}\ln|\omega T_o|\right]} \quad , \tag{11a}$$

$$\alpha_p \approx \frac{n\pi|\omega|\Gamma_p}{4c_o(1+\Gamma_p)^{3/2}} \tag{11b}$$

and

$$c_s \approx c_o \sqrt{\Gamma_s} \left[1 + \frac{2\beta_s}{\pi}\ln|\omega T_o|\right] \tag{12a}$$

$$\alpha_s \approx \frac{n\pi|\omega|}{4c_o\sqrt{\Gamma_s}} \quad . \tag{12b}$$

In these expressions for the wave speeds, the loss tangents are

$$\beta_p \approx \frac{n\pi\Gamma_p}{4(1+\Gamma_p)}\mathrm{sgn}(\omega) \tag{13a}$$

and

$$\beta_s \approx \frac{n\pi}{4}\mathrm{sgn}(\omega) \quad , \tag{13b}$$

where sgn(..) is the signum function. From Eqs. (11b) and (12b), it is evident that both attenuations scale essentially linearly with the frequency, and the phase speeds in Eqs. (11a) and (12a) exhibit weak dispersion, increasing logarithmically with frequency.

The logarithmic form for the dispersion is identical to that derived by several other authors, for instance Futterman[22], for a wave whose attenuation scales as the first power of frequency. An interesting feature of these logarithmic approximations is that they diverge to negative infinity in the limit of low frequency. This unphysical behavior is not a feature of the exact expressions in Eqs. (8a) and (9a), which are well-behaved throughout the entire frequency range. At zero frequency, the exact wave speeds reduce to $c_p = c_o$ and $c_s = 0$; and both attenuations are zero. These are just the properties of the equivalent suspension in which there are no grain-to-grain interactions.

If the frequency dependence of the phase speeds is neglected altogether, by setting the logarithmic terms to zero in Eqs. (11a) and 12a), the resultant zero-order approximations are identical in form to the expressions for the compressional and shear wave speeds in an elastic solid. This could account for the success of Hamilton's[23] empirical elastic model of wave propagation in marine sediments. An essential feature of the elastic model is the skeletal elastic "frame", which Hamilton characterizes by bulk and shear modulii. These two elastic coefficients can be interpreted directly in terms of the stress-relaxation modulii, Γ_p and Γ_s, of the grain-shearing theory.

7. Concluding Remarks

A marine sediment resembles an elastic solid in that it supports a compressional and a transverse wave.

14

Beyond this, however, there is little similarity between the two. In the sediment, at frequencies above a few kHz, available experimental evidence for both types of wave indicates that the attenuation scales as the first power of frequency and the phase speed is weakly dependent on frequency. These wave properties emerge naturally from the linear theory of wave propagation based on inter-granular shearing. According to the model, the material acts as a suspension, with additional stiffness introduced through grain-to-grain interactions. Compressional shearing accounts for dispersion and dissipation in the compressional wave, whilst translational shearing is responsible for supporting a shear wave, as well as characterizing its dispersive and dissipative behavior.

The differences between the inter-granular shearing theory and the Biot theory of wave propagation in a porous elastic medium concern primarily the predicted frequency dependencies of the phase speeds and attenuations. To distinguish between the two theories, measurements of attenuation and dispersion need to be extended to lower frequencies, in the region of 1 kHz or below. Ideally, such measurements should be performed with the source and receiver both buried in the sediment, rather than a source above and a receiver within the medium, in order to avoid the possibility of signal distortion arising from transmission through the water-sediment interface.

Acknowledgement

This work was supported by the Office of Naval Research, Ocean Acoustics Code (Dr. Jeffrey Simmen) under grant number N00014-93-1-0054.

References

1. E.L. Hamilton, "Compressional-wave attenuation in marine sediments," *Geophys.*, **37**, 620-646 (1972).
2. F.A. Bowles, "Observations on attenuation and shear-wave velocity in fine-grained, marine sediments," *J. Acoust. Soc. Am.*, **101**, 3385-3397 (1997).
3. B.A. Brunson and R.K. Johnson, "Laboratory measurements of shear wave attenuation in saturated sand," *J. Acoust. Soc. Am.*, **68**, 1371-1375 (1980).
4. B.A. Brunson, "Shear wave attenuation in unconsolidated laboratory sediments," in *Shear Waves in Marine Sediments*, J.M. Hovem, M.D. Richardson, and R.D. Stoll, Eds. Dordrecht: Kluwer, 1991, pp. 141-147.
5. D.J. Wingham, "The dispersion of sound in sediment," *J. Acoust. Soc. Am.*, **78**, 1757-1760 (1985).
6. M.J. Buckingham, "Theory of acoustic attenuation, dispersion, and pulse propagation in unconsolidated granular materials including marine sediments," *J. Acoust. Soc. Am.*, **102**, 2579-2596 (1997).
7. M.J. Buckingham, "Theory of compressional and shear waves in fluid-like marine sediments," *J. Acoust. Soc. Am.*, **103**, 288-299 (1998).
8. M.A. Biot, "Theory of propagation of elastic waves in a fluid-saturated porous solid: I. Low-frequency range," *J. Acoust. Soc. Am.*, **28**, 168-178 (1956).
9. M.A. Biot, "Theory of propagation of elastic waves in a fluid-saturated porous solid: II. Higher frequency range," *J. Acoust. Soc. Am.*, **28**, 179-191 (1956).
10. R.D. Stoll, *Sediment Acoustics*, Berlin: Springer-Verlag, (1989).
11. M.J. Buckingham, "Acoustic pulse propagation in dispersive media," in *New Perspectives on Problems in Classical and Quantum Physics. Part II. Acoustic Propagation and Scattering - Electromagnetic Scattering*, vol. 2, P.P. Delsanto and A.W. Sáenz, Eds. Amsterdam: Gordon and Breach, 1998, pp. 19-34.
12. K.L. Williams, "An effective density fluid model for acoustic propagation in sediments derived from Biot theory," *J. Acoust. Soc. Am.*, **110**, 2276-2281 (2001).
13. M.J. Buckingham, "Wave propagation, stress relaxation, and grain-to-grain shearing in saturated, unconsolidated marine sediments," *J. Acoust. Soc. Am.*, **108**, 2796-2815 (2000).
14. A.B. Wood, *A Textbook of Sound*, Third ed. London: G. Bell and Sons Ltd., (1964).
15. S.P. Timoshenko and J.N. Goodier, *Theory of Elasticity*, 3rd ed. New York: McGraw-Hill, (1970).
16. M.J. Buckingham, *Noise in Electronic Devices and Systems*, Chichester: Ellis Horwood, (1983).
17. J. Gittus, *Creep, Viscoelasticity and Creep Fracture in Solids*. New York: John Wiley, (1975).
18. S. Granick, "Soft matter in a tight spot," *Phys. Today*, **52**, 26-31 (1999).
19. B. Bhushan, J.N. Israelachvili, and U. Landman, "Nanotribology: friction, wear and lubrication at the atomic scale," *Nature*, **374**, 607-616 (1995).
20. A.L. Demirel and S. Granick, "Glasslike transition of a confined simple fluid," *Phys. Rev. Lett.*, **77**, 2261-2264 (1996).
21. P.M. Morse and H. Feshbach, *Methods of Theoretical Physics: Part 1*, vol. **1**. New York: McGraw-Hill, (1953).
22. W.I. Futterman, "Dispersive body waves," *J. Geophys. Res.*, **67**, 5279-5291 (1962).
23. E.L. Hamilton, "Elastic properties of marine sediments," *J. Geophys. Res.*, **76**, 579-604 (1971).

Theoretical and Computational Acoustics 2001
E.-C. Shang, Qihu Li and T. F. Gao (Editors)
© 2002 World Scientific Publishing Co.

Design Philosophy of Model-Based Sonar System

Qihu Li

Insitute of Acoustics, Academia Sinica

P.O. Box 2712, Beijing 100080

People's Republic of China

ABSTRACT

The performance of sonar system strongly depends on the ocean environment. The real underwater acoustical channel is time/space variant. The system gain of sonar designed based on optimum detection theory often considerably degraded due to the model mismatch. Therefore the robust signal processing, which is less sensitive to the model mismatch, is necessary. The robust signal processor works well in real ocean environment. The model-based sonar system is a system which takes ocean environment as a part of integrate sonar. The design philosophy of this kind of sonar system is described in this paper. The basic idea of robust signal processing and model match concept are described and used to design model-based sonar system.

1.Introduction

Detection and localization are the main task of early sonar system for a long time. With the development of new technology, the submarine become more quiet, the requirement for complicated signal processing technique is a more critical problem for sonar designer. In other hand, in the field of underwater acoustics, research has found that the effect of ocean environment must be taken account in the system design.

The signal transmission channels of underwater sound is complicated, the ambient noise and signal propagation condition vary temporally, spatially and geographically. In the case of shallow water, the acoustic channel become more complicated, because it is boundaried by the rough sea surface and sea bottom.

The traditional sonar system, based on the optimum detection theory has good performance in ideal situation, when the assumption condition are satisfied. But, unfortunately, in real ocean, the performance often severely degraded due to the existance of multipath or frequency spread. It is obviously that we have to find a new schema, which incorporate the ocean environment as a part of integral sonar system, so that the system performance are robust for the model mismatch.

The definition of a model-based sonar processor, according to Camdy [1], is as follows :

A processor that incoporates a mathematical representation of the ocean acoustic propagation and can used to perform various signal processing functions ranging from simple filtering or signal enhancement, to adaptively adjusting model parameters, to localization, to tracking, to sound speed estimation or inversion.

Incorporating a propagation model into signal processing schema was most likely iniated by the work of Hinich, in 1973, who applied it to the problem of source depth estimation [2] . However, as early as 1966, Clay suggested matching the modal function of an acoustic waveguide to estimate source depth in water [3-4].

Traditionally, the optimum detection theory can be considered as the basis of sonar signal processor [5-6], the main criteria of optimum receiver described in literature are maximum likelihood (ML), maximum output signal to noise ration (MSNR) and least mean square error (LMS) criteria. Although there are a quite a few other criteria which derived from above three criteria, but the final solution often can be concluded to the ML, MSNR and LMS criteria. The sonar designer have been noted for a long time that the ocean, particulary the shallow water are far from ideal model of homogeneous and isotropic. Urick, Zhang et al, reported some important results of at sea experiment [7-12]. They show that the time delay and frequency spread in underwater channel often quite serious, sometime will result performance degration of ideal system. For example, the time delay fluctuation in a moderate range (10 – 100 kM) is the order of 10 μs , so that the passive ranging sonar based on the principle of three point array detection will be fail to work in this case. Because, in this range, the target ranging need the accuracy of time delay at least 1 or 2 μs .

A reasonable attempt to solve the model mismatch problem is to design robust signal processor or model-based processor [13-21]. In the mean time some new area in underwater acoustics have been studied, for example, the modal-phase tomography (MPT) and modal horizontal refraction tomography (MHRT) [22], internal soliton [23] etc. It is obvious that, the results of these new topics will finally affect the design philosophy of sonar system. An example is the joint statistical signal detection and estimation theory [24] .

2.Robust Signal Processor

As a kernel part of sonar system, the design of sonar signal processor is based on the optimum recieve method in information theory. According to the traditional signal detection theory, the ambient noise of ocean environment are often described as homogeneous and isotropic. Therefore the optimum detector will be the energy detector. The system gain of such system are, in some condition, proportional to the number of independent element of array and system integration time.

But, in the real ocean, the ambient noise neither homogeneous nor isotropic. Therefore the theoretical value of system gain usually degraded in practical application. This kind of model mismatch not only appear in passive signal detection but also in active signal detection. The typical example is the matched filter. The performance often considerably decrease due to the echo distortion in time domain and frequency domain, i.e. the echo signal is not matched with the original transmited signal.

It is hard to quantatively indicate the model mismatch, but, as we illustrate in Fig.1, the optimum signal processor is often sensitive to the model mismatch due to various factors. That means in the ideal situation, the system gain have a high theoretical value, but once there are some kinds of model mismatch, which is usually unavoidable, the system performance considerablly degrade. In Fig.1 we see the difference between optimum signal processor and robust signal processor. For an optimum signal processor, when the actual ocean environment match the assumption model, the system gain is "optimum". But, in model mismatch situation the system gain decrease. For the robust signal processor, although the peak value of system gain is less than optimum system, but in average, it shows quite good performance than the optimum system. Especially in model mismatch.

It is worth to show that, it is hard to define the exact meaning of model mismatch. It may be taken place in time domain or frequency domain, or both.

The system gain G of a sonar system usually can be separated into two parts, one is space gain G_s and other is time gain G_t , and $G = G_s + G_t$ in dB. The value of G_s come from the array system and G_t come from the time accumulation (integration). Theoretically speaking, $G_s = 10lgN$, where N is the number of array elements, and $G_t = 10lg\sqrt{T/\tau_0}$, where T is the integration time and τ_0 is the correlation radius of background noise. It is seem that both G_s and G_t can be as large as we wish. But in practice, that is not true due to the non-stationary in time domain and non-isotropic, non-homogeneous behavior in space domain of noise field.

Hamson shows that, in early study of Kupman, the maximum system gain G_s is reached at $15--30\lambda$ for vertical array in shallow water [10]. Urick shows that the signal detection in time domain due to the propagation fluctuation make the signal randomly embeded or appeared in the noise background.

The model based robust signal processor can be defined as follows :

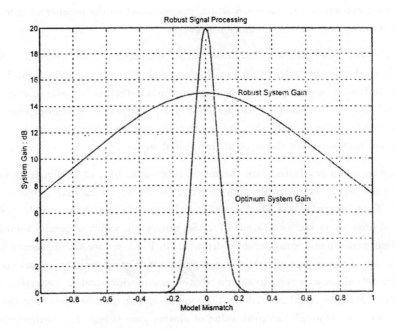

Fig. 1. Comparison of system gain for optimum and
robust signal processor

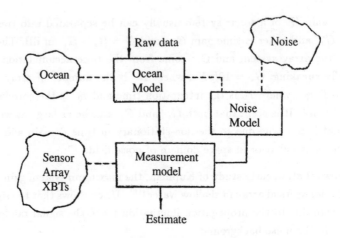

Fig. 2. Model-based sonar signal processing

Suppose H is a possible design space, Q is the set of operating points or models. The metric function $M(h, q)$, $h \in H, q \in Q$. The traditional design is to find $h_0 \in H$ for a specific $q_0 \in Q$ such that

$$M(h_0, q_0) = \min_{h \in H} M(h, q_0) \tag{1}$$

This kind of array processor is actually a matched filter suitable to the environment q_0. If the real environment q_0 varies in some kind of subset, for example $q \in P \subset Q$, it is reasonable to concern $\max_{q \in P} M(h, q)$ for each h. The minmax stratage will be accepted in this procedure. That is

$$\min_{h \in H} \max_{q \in P} M(h, q) \tag{2}$$

The robust signal processor h_R is

$$\min_{h \in H} \max_{q \in P} M(h, q) = \max_{q \in P} M(h_R, q) \tag{3}$$

3.Model based sonar signal processor

The block diagram of model based sonar signal processor is illustrated in Fig.2. The processor get various data from real ocean environment. By combining the ocean model, noise model and measurement model, the signal processor produce the "best" estimate of target parameter, such as range, depth etc.

We have mentioned the definition of model based processor (MBP) in previous section. The advantages of MBP are as follows :

    *recursive

    *statistical, incorporating both noise and parameter uncertainties

    *capable of incorporating non-stationary statistics

    *capable of incorporating both linear/nonlinear space-time varying

    *capable of on-line processing of the measured data at each iteration

    *capable of mornitoring their own performance by testing the residual between the measurement and its prediction value and easily extended to perform adaptively

    *capable of filtering the pressure field as well as simultaneously estimating the model functions

The widely used sound wave model can be categoried into four types, that is :

    A.ray theory

    B.spectral integration

    C.normal mode

    D.parabolic equation

As shown in literature [5], each model have their own application area.

4.Matched Filter Processor (MFP)

MFP is one of MBP used in recent research work. The block diagram of MFP algorithm is shown in Fig.3. As we have shown in section 1, Clay suggested matching the modal function of an acoustic wavequide to estimate source depth. The concept of MFP i.e. compare the measurement results of range and depth by using sound pressure field with the prediction value of modal theory, was first introduced by Bucker [19]. However, matched-field is primary aimed at the localization problem, indeed most estimates implemented by MFP are focused on tracking an estimate of localization parameters.

The computational ocean acoustics often concern to solve the Helmholz equation

$$\nabla^2 p + \frac{\omega^2}{c^2(r,z)} p = \frac{-\delta(r-r_s)\delta(z-z_s)}{r} \tag{4}$$

here $c(r,z)$ is the function of range r and depth z, ω is the angle frequency at source point (r_s, z_s).

In order to completely describe this kind of equation, it is also necessary to add boundary condition and far-field radiated condition. The goal is to solve this equation for the response of the channel to the source, that is to solve for the acoustic pressure $p(r,z)$.

In Fig.3, based on the maximum likelihood criteria, by using the estimate of covariance and the signal plus noise model, we get the optimum matched estimate for the unknown parameter vector in ambienquity plane. As an example, for a vertical array, the covariance matrix of input data can be expressed as

$$K(f_k, t) = \frac{1}{N_t} \sum_{j=1}^{N_t} x(f_k, t_j) x^*(f_k, t_j) \tag{5}$$

where N_t is the point number of FFT, f_k is the frequency.

The correlation of input data value is

$$C(f_k, r, \alpha, t) = \frac{w^*(f_k, r, \alpha) K(f_k, t) w(f_k, r, \alpha)}{Tr[K(f_k, t)]/N} \tag{6}$$

where w is weight vector, r is the search vector (as a function of range, bearing angle, depth), α is the tilt angle of array, N is the number of hydrophones and Tr is the trace of matrix. In the source direction, the MFP output gain in dB is

$$AG = 10lg[C(r_s)] \tag{7}$$

5.DATA Fusion

In order to get an optimum estimate for the paramater, it is necessary to syhthesize the data from various source. That's the goal of data fusion. The data fusion technique is the integration of information from multiple source to produce specific and comprehensive unified data about an entity. In the procedure of data fusion, the environment model considered in the sonar system play an important role.

The basic theorem of data fusion is as follows :

The estimation error of optimum linear data fusion is not greater than that of the individual components.

This fact is true not only for the N independent observation data but also for the dependent data.

Suppose there are N observation data $x_i, i = 1, ..., N$ for estimating a parameter θ .

$$E[x_i] = \theta, \quad Var[x_i] = \sigma_i^2, \quad i = 1, ..., N \tag{8}$$

Let

$$X = [x_1, ..., x_N]^T \tag{9}$$

is the observation data. The weight vector is

$$W = [w_1, ..., w_N]^T \tag{10}$$

Suppose $y = W^T X$ be the linear conbination of X . To find an optimum weight vector, such that

$$I = E[W^T(X - EX)]^2 \tag{11}$$

minimum, subject to the constrain

$$W^T U = 1 \tag{12}$$

where U is the vector of length N , the all element equal to 1.

It can be shown that the solution of above problem is as follows.

$$W_{opt} = \frac{R_{xx}^{-1} U}{U^T R_{xx}^{-1} U} \tag{13}$$

$$I_{min} = I[W_{opt}] = \frac{W_{opt}^T R_{xx} W_{opt}}{U^T R_{xx}^{-1} U} \tag{14}$$

where R_{xx} is the cross correlation matrix of input data X . We can prove that

$$I_{min} \leq Var[x_i] = \sigma_i^2, \quad i = 1, ..., N \tag{15}$$

22

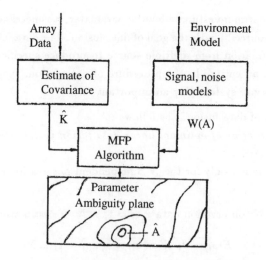

Fig. 3. The block diagram of MFP algorithm

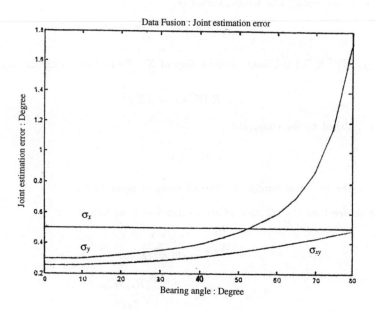

Fig. 4. DATA Fusion : joint estimation error of cyclindrical
array and linear array.

In the case of two dimension, the optimum linear combination of observation x_1 and x_2 should be

$$x_{1,2} = \frac{\sigma_2^2 - r_{12}}{\sigma_1^2 + \sigma_2^2 - 2r_{12}} x_1 + \frac{\sigma_1^2 - r_{12}}{\sigma_1^2 + \sigma_2^2 - 2r_{12}} x_2 \tag{16}$$

$$Var[x_{1,2}] = \frac{\sigma_1^2 \sigma_2^2 - r_{12}^2}{\sigma_1^2 + \sigma_2^2 - 2r_{12}} \tag{16}$$

where r_{12} is the correlation value of x_1 and x_2. It is easy to directely prove that

$$Var[x_{1,2}] \leq Var[x_1], \quad Var[x_{1,2}] \leq Var[x_2] \tag{17}$$

Fig.4 give an example of data fusion. The data fusion for bearing estimation from cylindrical array and linear array. As we see in the plot that the combination of two estimate get much better (uniform estimate error) bearing accuracy than the cylindrical array and linear array. Data fusion should be considered as a technique of model-based sonar system.

REFERENCES

1. J.V.Camdy,"Model based signal processing in the ocean", *IEEE Oceanic Engr. Society News Letter* **Vol.XXV No.3** (2000) 199-205

2. M.J.Hinich,"Maximum likelihood signal processing for a vertical array", *J.Acoust.Soc. Amer.* **Vol.54** (1973) 499-503

3. C.S.Clay,"Use of arrays for acoustic transmission in noisy ocean", *Review of Geophysics,* **Vol.4** (1966) 475-507

4. C.G.Carter and R.E.Robinson,"Ocean effects on time delay estimation requiring adaption", *IEEE J. of Oceanic Engr.* **Vol.18** (1993) 367-378

5. D.J.Edelblute et al., " Criteria for optimum signal detection theory for arrays ", *J. Acoust. Soc.Amer.* **Vol.41 No.1** (1967) 199-205

6. A.A.Winder,"Sonar system technology", *IEEE Trans.* **Vol.SU-22** (1975) 291-332

7. R.J.Urick,"Multipath propagation and its effects on sonar design and performance in the real ocean", *Proc.NATO ASI Study On Underwater Acoustics* 1976

8. Renhe Zhang et al.,"Spatial coherence and time stability of long range sound field in shallow water", *(in Chinese) Acta Acoustica* 1981 No.1 9-18

9. E.Y.Gorodetskay et al.,"Deep water coherence at long ranges theoretical prediction and effects on large array signal processing", *IEEE J. of Oceanic Engr.* **Vol.24 No.2** (1999) 156-171

24

10. R.H.Hamson,"The theoretical gain limitations of a passive vertical line array in shallow water", *J.Acoust.Soc.Amer.* **Vol.68 No.1** (1980) 156-162

11. R.L.Davenport,"The influence of the ocean environment on sonar system design", *EASCON* 1975 *Record* 200-227

12. W.Jobst,"Measurement of the temporal, spatial and frequency stability of an underwater acoustic channel", *J.Acoust.Soc.Amer.* **Vol.63** (1979) 62-69

13. S.A.Kassam and H.V.Poor,"Robust techniques for signal processing : a survey", *Proc. IEEE* **Vol.73** (1985) 21-53

14. D.F.Gingras,"Robust broadband matched field processing", *IEEE J.of Oceanic Engr.* **Vol.18 No.3** (1993) 253-264

15. H.Cox et al.,"Robust adaptive beamforming", *IEEE Trans.* **Vol.ASSP-35** (1987) 1315-1376

16. M.B.Porter,"Acoustic models and sonar systems", *IEEE J. of Oceanic Engr.* **Vol.18** (1993) 425-437

17. P.A.Hazell,"What's the future for ASW in NATO ?", *Sea Technology* **Vol.39 No.11** (1998) 10-17

18. E.J.Sullivin and D.Middeleton,"Estimation and detection issues in matched fiel processing", *IEEE J. of Oceanic Engr.* **Vol.18 No.3** (1993) 156-167

19. C.T.Chen Ed.,"The past, present, and the future of underwater acoustic signal processing", *IEEE Signal Processing",* **Vol.5 No.4** (1998) 21-53

20. H.P.Bucker,"Use of calculated sound fields and matched-field detection to locates sound in shallow water", *J.Acoust.Soc.Amer,* **Vol.59** (1976) 329-337

21. D.Z.Wang and E.C.Shang,*"Underwater Acoustics" In Chinese* **Science Pub** Beijing 1981

22. E.C.Shang et al.,"New Schemes of ocean acoustic tomography", *J. of Computational Acoustics* **Vol.8, No.3** (2000), 459-471

23. J.X.Zhou et al.,"Resonant interaction of sound waves with internal solitons in coastal zone", *J.Acoust.Soc.Amer.* **Vol.90** (1991), 2042-2054

24. D.Middleton and R.Esposito,"Simultaneous optimum detection and estimation of signal in noise", *IEEE Trans. On IT* **Vol.IT-14, No.3** (1968) 434-444

Theoretical and Computational Acoustics 2001
E.-C. Shang, Qihu Li and T. F. Gao (Editors)
© 2002 World Scientific Publishing Co.

Efficient Broadband Signal Simulation in Shallow Water: A Modal Approach

Finn B. Jensen, Peter L. Nielsen and Carlo M. Ferla
SACLANT Undersea Research Centre, Viale San Bartolomeo 400, La Spezia, Italy
E-mail: *jensen@saclantc.nato.int*

Today's minimum requirements for ocean acoustic models are to be able to simulate broadband signal transmissions in 2D varying environments with an acceptable computational effort. Standard approaches comprise ray, normal mode and parabolic equation techniques. In this paper we present a computationally efficient modal approach as implemented in the PROSIM model. There are three key elements to improved efficiency: (1) Local mode calculations for the depth-separated wave equation are done analytically within layers where the sound speed varies linearly in $1/c^2$. This approach was adopted from the ORCA model [E.K. Westwood, C.T. Tindle and N.R. Chapman, JASA 100, 3631-3645 (1996)], which has been shown to be considerably faster than mode models using numerical integration in depth, particularly when the problem involves many hundreds of modes. (2) Range dependence is handled within the adiabatic approximation (no mode coupling), allowing a mode-by-mode spatial redistribution of energy in accordance with the change in local mode properties along the propagation range. (3) Frequency interpolation of local mode properties is performed as in ORCA, i.e. accurate determination of mode wavenumbers and depth functions are required only for a limited number of discrete frequencies (around one in twenty) within the band of interest. Numerical examples show that the PROSIM model in its current configuration is much faster than standard, less optimized models such as C-SNAP and RAM.

1. INTRODUCTION

Broadband models have become indispensable tools for both data analysis and sonar system predictions in ocean acoustics. In particular, these models are used for Monte Carlo studies of acoustic signal fluctuations in the ocean, for tomographic time-domain inversions of ocean structures, for wideband geoacoustic inversions with global search algorithms, for designing underwater acoustic communication systems, and for testing signal processing algorithms in sonar systems in general. For many practical applications, the computational effort involved in using broadband models is still excessive and more efficient solution approaches are continuously being developed. One such model based on normal modes will be described in the following and benchmarked against standard signal models both for accuracy and computational speed. We consider three shallow-water test environments with propagation out to 10 km and a signal bandwidth of 10 –1000 Hz.

2. PROSIM

The broadband acoustic model PROSIM [1] developed at SACLANTCEN over the past few years is essentially a range-dependent version of the ORCA model developed by Westwood *et al.* in the mid 1990s [2]. The design criteria for PROSIM were to perform broadband signal simulations (10 Hz – 10 kHz) in range-dependent shallow-water channels with high computational efficiency. To that end an efficient modal solution involving hundreds of modes was required. Two approaches are commonly applied in the community: (1) Numerical integration of the depth-dependent wave equation resulting in computation times that typically increase quadratically with frequency (the number of modes increases linearly with frequency and so does the required numerical depth discretization). Examples of these types of models are KRAKEN and C-SNAP. (2) Analytical solution of the wave equation in a small number of layers where the sound-speed profile is either constant or varies linearly in $1/c^2$ with depth (Airy function solution). In this case the computation time increases linearly with frequency because the number of modes increases linearly with frequency. The ORCA model is an example of a layer model.

An illustration of the computational performance of ORCA versus KRAKEN on a propagation problem with 15 points in the sound-speed profile is shown in Fig.1. The number of modes computed varies from around 10 at the lowest frequency to around 1000 at the highest frequency. We see that the cross-over point where ORCA becomes more efficient is around 50 modes, and that ORCA is 50 times faster than KRAKEN in computing 1000 modes. Clearly, for acoustic problems involving many hundreds of modes, the layer approach employed in ORCA provides significant savings in CPU time.

Apart from adopting the layer solution approach for PROSIM, we also decided for a fluid environment only, and the evaluation of real-axis modes only. This implies that attenuation is

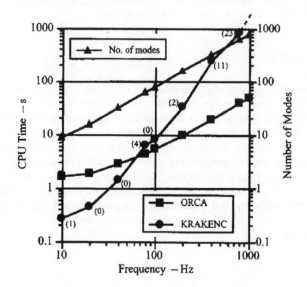

Fig.1 Computational speed versus frequency for two mode models: A layer model (ORCA/PROSIM) versus a numerical integration model (KRAKENC/C-SNAP) [2].

handled as a perturbation, which is an approximation often employed in mode models. The advantage of solving a real-eigenvalue problem is increased speed and a more robust code which ensures that all modes are computed.

Range dependency in PROSIM is handled in the adiabatic approximation, i.e. no cross-mode coupling of energy. This approach works well for weak range dependence and is computationally faster than evaluating the mode coupling coefficients.

Finally, since broadband signal simulations via Fourier synthesis involves computing the acoustic field at many closely-spaced frequencies within the band of interest (often several hundred frequency samples), the concept of frequency interpolation of modal properties was adopted from ORCA. The concept here is that modal eigenvalues and modal depth functions vary smoothly with frequency and that modal information can be obtained with sufficient accuracy by interpolation between computed properties on a coarse grid covering the frequency band of interest. The problem is to build a robust algorithm that ensures accurate modal properties at all frequencies for any sound-speed profile, even when double ducts are present.

The frequency interpolation algorithm implemented in PROSIM [1,3] has been thoroughly checked and has proven to permit a significant speed-up of broadband problems by computing only 1:20 or 1:40 of the required frequency samples and interpolating the remaining information. As a result the PROSIM model should be a factor 20 to 40 faster than other mode models based on brute-force, frequency-by-frequency calculation of the broadband transfer function. The scope of this paper is to demonstrate the computational efficiency of PROSIM compared to less optimized broadband models such as C-SNAP and RAM.

3. C-SNAP

The SACLANTCEN coupled normal-mode model C-SNAP [4] has a modal solver similar to the one used in KRAKEN, i.e. numerical integration is used to solve the depth-separated wave equation. As in PROSIM only real-axis modes are computed and attenuation is handled via a perturbational approach. The range dependency in C-SNAP is dealt with through mode coupling, which should provide more accurate field solutions for strongly range-dependent environments. The approach is to divide the total range into a sequence of range-independent sectors (several tens or several hundreds, depending on the degree of range dependency and the frequency), compute the local mode properties for each sector, compute the acoustic field on a vertical slice at the sector boundary, project this field onto the new mode set in the adjacent sector to determine modal coupling coefficients, re-propagate the field through the next sector, etc. This is not an exact mode coupling procedure since we omit the continuous mode spectrum, which would account for energy propagating into the bottom beyond the critical angle.

Since C-SNAP performs brute-force frequency-by-frequency calculation of the spatial transfer function, this model is expected to be slower than PROSIM. The presence of mode coupling, however, should guarantee more accurate results for range-dependent problems.

4. RAM

The split-step Padé solution of the parabolic wave equation, as implemented in the RAM code [5], is considered the most efficient PE-based technique for solving range-dependent ocean acoustic problems. RAM provides more accurate results than any of the mode models, both because it includes complete coupling among all spectral components, including the continuous mode spectrum, and because losses are handled correctly. The field solution is obtained on a spatial grid (Δr, Δz) which determines both the environmental discretization and the solution accuracy. In essence, a convergence test with decreasing Δr and Δz must be carried out to ensure stable numerical results. Since the required grid size is inversely proportional to frequency, a broadband RAM calculation increases approximately with frequency squared. As for C-SNAP, all frequency samples of the transfer function must be computed.

5. TEST PROBLEMS

To test the accuracy and computational efficiency of the above three broadband models, we have designed a series of simple shallow-water propagation problems as shown in Fig.2. Case 1 is a flat-bottom situation used as a calibration case to see that all models give similar results. Cases 2 and 3 are symmetric upslope/downslope situations where the water depth varies from 200 m at the deep end to 100 m at mid range. This geometry corresponds to a bottom slope of 1.15 deg. The sound-speed profile is downward refracting and consists of 5 input values as shown in the table. This profile is taken to be unchanged along the track. The bottom is a homogeneous fluid halfspace with the properties given in Fig. 2.

We consider a broadband pulse emitted by a source at 100-m depth and calculate the received signal on a hydrophone at 20-m depth and at a range of 10 km. The emitted signal is a Ricker pulse with center frequency of 200 Hz and covering the band 10–450 Hz. Since the received signal at 10 km is found to have a total time dispersion of nearly 1 s, we know that a

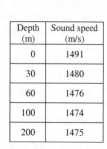

Depth (m)	Sound speed (m/s)
0	1491
30	1480
60	1476
100	1474
200	1475

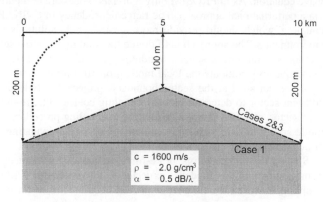

Fig.2 *Shallow water test environments. Case 1 is a constant water depth of 200 m whereas Case 2 is a symmetric upslope/downslope environment. Case 3 has the same geometry as Case 2, but a faster bottom (c = 1800 m/s, ρ = 2.0 g/cm^3, α = 0.1 dB/λ).*

frequency sampling of 1 Hz is required to avoid signal wrap-around in the Fourier transformation ($\Delta f = 1/T$) [6]. Hence 441 frequency samples must be computed to synthesize the received signal at 10 km.

When timing different acoustic models on a particular test problem, it is important to establish uniform convergence criteria for model solutions, i.e. the relative solution accuracy should be the same for all models. We have decided to run each model to a convergence of R = 0.99 for the peak cross-correlation between a super accurate model solution and the one determined to be just accurate enough. The pertinent numerical parameters for each model and for each test problem to obtain stable solutions with R = 0.99 are summarized in Table 1.

Table 1 Test problem summary. NM_{max} is the maximum number of modes computed, NSEG the number of range segments used in the modal computations, NF the number of frequency samples computed, R the peak cross-correlation relative to the RAM result, and CPU is the execution time on an 850 MHz PC.

Test	Model	NM_{max}	NSEG	NF	R	CPU
Case 1 - RI	PROSIM	47	1	441(23)	0.99	1 s
(10-450 Hz)	C-SNAP	47	1	441	0.99	5 s
	RAM	N=4	$\Delta r = 10$ m $\Delta z = 0.25$ m	441	1.00	1200 s
Case 2 - RD	PROSIM	47	100	441(23)	0.78	1 min
(10-450 Hz)	C-SNAP	47	256	441	0.97	15 min
	RAM	N=4	$\Delta r = 10$ m $\Delta z = 0.25$ m	441	1.00	20 min
Case 3 - RD	PROSIM	155	200	1981(50)	0.49	20 min
(10-1000 Hz)	C-SNAP	155	1024	1981	0.95	40 h
	RAM	N=4	$\Delta r = 5$ m $\Delta z = 0.05$ m	1981	1.00	20 h

5.1 Case 1 – RI/LF

This is the flat-bottom calibration case where all three models are expected to provide accurate pulse solutions. In fact, as shown in Table 1, the peak cross-correlation between the PROSIM and RAM results are better than 0.99 and so is the correlation between C-SNAP and RAM. It is also evident from the stacked signal plots in Fig. 3 that the three model solutions are virtually identical.

The calculation times are quite different, however, with the mode solutions being several orders of magnitude faster than the PE solution. Clearly, a range-independent case favors the modal approach since only one mode set is required to compute the acoustic field anywhere in the waveguide. The number of modes is just one at 10 Hz but increases to 47 at 450 Hz. Hence, the C-SNAP model must compute a total of (47-1)/2 x 441 = 10,143 modes to generate a broadband transfer function. PROSIM, on the other hand, uses frequency interpolation in 20 Hz bands and therefore only needs to compute 23 frequency samples. As a result, PROSIM is 5 times faster than C-SNAP for this case.

RAM is intrinsically a range-dependent model that marches the solution out in range on a computational grid (Δr, Δz) which relates to the frequency (inversely proportional) but has

30

TEST CASE 1

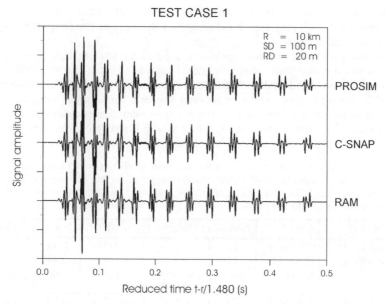

Signal amplitude

R = 10 km
SD = 100 m
RD = 20 m

PROSIM

C-SNAP

RAM

0.0   0.1   0.2   0.3   0.4   0.5

Reduced time t-r/1.480 (s)

Fig. 3 *Comparison of broadband pulse solutions from the three models for Case 1. The source signal is a Ricker pulse with center frequency 200 Hz.*

little dependence on the environmental complexity. Hence, RAM solutions take essentially the same time for range-dependent and range-independent cases, as seen by comparing RAM CPU times for Cases 1 and 2 in Table 1. Accurate RAM solution were obtained with Padé order N=4, using the computational grid size shown in Table 1 for all frequency samples. A potential time saving of a factor 3 could be obtained for this model by using a frequency-dependent computational grid, i.e. a coarser grid at low frequencies and a finer grid at high frequencies. This feature has not yet been implemented in our version of the RAM code.

5.2   Case 2 – RD/LF

This is the upslope/downslope situation with changing bathymetry along the entire track. As mentioned before this will not change the calculation time for RAM, but the mode models will become slower, essentially in proportion to the number of range segments and, hence, additional mode sets, required to obtain a stable solution. For this case with a 1.15 deg bottom slope the adiabatic PROSIM model requires 100 range segments for convergence, whereas the coupled C-SNAP model requires 256 segments. The calculation times (Table 1) are still in favor of PROSIM which is ten times faster than the two other models.

The solution accuracy now becomes an issue since the mode models both treat range-dependence in an approximate fashion. The most accurate solution compared to RAM is seen to be C-SNAP with a peak cross-correlation of 0.97. Less accurate is the adiabatic PROSIM result with a correlation of 0.78. These differences are also evident in the stacked time plots in Fig.4, where we see similar arrival structures in all traces, but with incorrect amplitudes on

TEST CASE 2

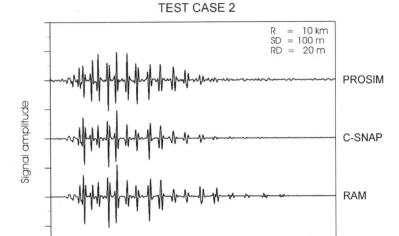

Fig. 4 *Comparison of broadband pulse solutions from the three models for Case 2. The source signal is a Ricker pulse with center frequency 200 Hz.*

some arrivals in the PROSIM solution. This result shows some of the limitations on the adiabatic approximation, which ignores energy coupling between modes.

To investigate this problem in more detail, the various model solutions were also compared at mid-range (5 km) where propagation is entirely upslope. The PROSIM result was here found to be in much better agreement with the RAM reference, indicating that it is the downslope portion which causes problems in the adiabatic solution. This may be explained as being due to the fact that the number of modes present at the apex (5 km) is the total number of modes being used for the whole downslope path, i.e. no extra modes are introduced even though the water depth doubles at 10 km.

5.3   Case 3 – RD/HF

In order to push computations to the limit, we modify Case 2 to have a faster bottom (1800 m/s) and a lower attenuation ($\alpha = 0.1$ dB/λ) which causes more time dispersion due to late, steep-angle ray arrivals. As a consequence, the time window must be increased to 2 s, which, in turn, results in a 0.5 Hz frequency sampling of the transfer function. In addition, we increase the bandwidth to 1000 Hz. These changes result in more modes to be computed (max. 155), more range segments, and several more frequency samples.

As shown in Table 1, PROSIM converges with 200 range segments, and since we still use 20 Hz interpolation bands, only 50 frequencies out of 1981 need to be computed. This case clearly favors the PROSIM interpolation scheme, and the PROSIM execution is now 50 to 100 times faster than the other models. The numerical parameters used for getting convergent answer for C-SNAP and RAM are listed in Table 1.

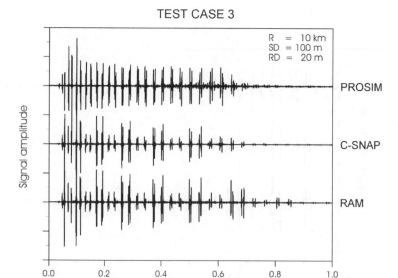

TEST CASE 3

Fig. 5 *Comparison of broadband pulse solutions from the three models for Case 3. The source signal is a Ricker pulse with center frequency 500 Hz.*

The solution accuracy as reflected by the signal cross-correlation coefficients is seen to deteriorate. This was expected since mode coupling becomes more important with increasing frequency. It is again the PROSIM result (R = 0.49) that is least accurate compared to the RAM reference, as clearly evidenced in the signal traces in Fig. 5.

6. SUMMARY AND CONCLUSIONS

Much work has gone into the development of computationally efficient CW propagation models for use in ocean acoustics. These standard techniques, based on ray, mode, wave-number integration and parabolic equation solutions of the wave equation, can be straightforwardly extended to broadband signal simulations via Fourier synthesis of a spectrum of CW solutions. Clearly the computational effort in each model increases with the number of frequency samples required, but, as shown here, not always linearly with the number of frequencies.

Mode models have always been considered optimal for range-independent problems, whereas PE models have become the preferred choice for range-dependent problems. However, when moving to multi-frequency, broadband situations, the picture is not so simple. As shown in this paper, the use of efficient modal solvers combined with frequency interpolation of modal properties across bands of 10-20 Hz can make the modal approach much more efficient than standard PE solutions.

The current implementation in PROSIM of range dependency in the adiabatic approximation is not always sufficiently accurate, and it is suggested that a mode coupling approach be adopted, which would not increase computation times much. Moreover, modal properties could also be interpolated in range, which would lead to a further significant speed-up of mode calculations in range-dependent environments. As to improving the efficiency of broadband PE codes, there is only one obvious way: introduce a frequency-dependent computational grid. This, however, can provide only a factor 3 reduction in computation time for broadband problems. Hence, modes remain the most promising wave theoretic modeling approach for broadband signal simulations in range-dependent shallow water environments.

REFERENCES

[1] F. Bini-Verona, P.L. Nielsen and F.B. Jensen, "PROSIM broadband normal-mode model: A user's guide," SACLANTCEN Report SM-358, 2000.

[2] E.K. Westwood, C.T. Tindle and N.R. Chapman, "A normal-mode model for acousto-elastic ocean environments," J. Acoust. Soc. Amer. **100**, 3631–3645 (1996).

[3] F. Bini-Verona, P.L. Nielsen and F.B. Jensen, "Efficient modeling of broadband propagation in shallow water," *in Proceedings of Fourth European Conference on Underwater Acoustics*, edited by A. Alippi and G.B. Cannelli. CNR-IDAC, Rome (1998), pp. 637-642.

[4] C.M. Ferla, M.B. Porter and F.B. Jensen, "C-SNAP: Coupled SACLANTCEN normal mode propagation model," SACLANTCEN Report SM-274, 1993.

[5] M.D. Collins, "A split-step Padé solution for the parabolic equation method," J. Acoust. Soc. Amer. **93**, 1736–1742 (1993).

[6] F.B. Jensen, W.A. Kuperman, M.B. Porter and H. Schmidt, *Computational Ocean Acoustics* (Springer-Verlag, New York, 2000).

Theoretical and Computational Acoustics 2001
E.-C. Shang, Qihu Li and T. F. Gao (Editors)
© 2002 World Scientific Publishing Co.

Modeling the Effects of Horizontal Refraction and Medium Non-Stationarity in Ocean Acoustics

Oleg A. Godin

CIRES, University of Colorado and NOAA/Environmental Technology Laboratory
Boulder, CO 80305-3328

Mathematical modeling of long-range sound propagation in a 3-D inhomogeneous media with time-dependent parameters is a computationally-intensive problem. In underwater acoustics, the general problem can be significantly simplified by taking into account characteristic properties of the ocean as an acoustic medium. In this paper, we present a simple and efficient analytic technique to approximately reduce 3-D and 4-D problems of underwater sound propagation to better studied 2-D problems. The reduction is made possible by relative slowness of time dependence of environmental parameters and smallness of their horizontal gradients. The horizontal inhomogeneity and non-stationarity of the ocean, although weak, are often non-negligible and will be shown to have important observational consequences. Unlike brute-force numerical approaches, the perturbation theory provides an insight into physics of sound propagation in 3-D inhomogeneous, time-dependent ocean.

1. Introduction

Sound propagation in a horizontally-inhomogeneous ocean is often considered within the *uncoupled azimuth approximation*, also known as the $N \times 2\text{-}D$ approximation. In this approximation one assumes that sound propagates from a point source within a vertical plane, i.e. in a constant azimuthal direction, with negligible scattering between any two planes. One thereby reduces the original 3-D problem to a set of 2-D problems. When combined with the ray-theoretical description of the acoustic field or the "vertical modes - horizontal rays" approach, the $N \times 2\text{-}D$ approximation implies neglecting horizontal refraction, i.e. ray excursions from the vertical source/receiver plane. Eigenrays and ray travel times are then completely determined by depth- and range-dependent sound speed in the vertical plane.

When modeling sound propagation in a time-dependent ocean, the frozen medium approximation is usually invoked. Within this approximation, a time-dependent medium is replaced with a time-independent one parameters of which are "frozen" in time and equal to those of the original medium at some moment, say, when an acoustic signal was emitted or received. The frozen medium approximation can be viewed as a time-domain counterpart of the uncoupled- azimuth approximation, with time-dependence of environmental parameters treated in the former similarly to azimuthal dependence of the parameters in the latter.

In this paper, we develop, in the contexts of adiabatic mode propagation and the ray theory, an analytic approach, which possesses higher accuracy and broader domain of validity than the uncoupled azimuth and frozen medium approximations, while retaining their computational efficiency. By accounting for propagation paths curvature in the horizontal plane, we put forward a simple 2-D description of sound propagation in a 3-D inhomogeneous ocean that is more accurate than the $N \times 2$-D approximation but still requires knowledge of environmental parameters only in the vicinity of the vertical source/receiver plane. We show that excursions of acoustic paths from the vertical source/receiver plane *decrease* adiabatic mode phase and ray travel time regardless of the spatial scale of inhomogeneities causing the horizontal refraction. Further, we introduce a simple technique, the quasi-stationary approximation, that allows one to model the effects of weak time-dependence without actually tracing rays or simulating normal modes in a non-stationary medium. Using this approximation, we address the implications of the medium non-stationarity on acoustic tomography of oceanic currents.

For the stationary case, a more detailed discussion, which includes a systematic derivation of the results outlined below, can be found in Ref. 1.

2. Adiabatic Normal Modes in 3-D and 4-D

Consider an acoustic field at a point with the horizontal coordinates $(x_2, y_2) = r_2$ and the vertical coordinate z_2 due to a point CW source of frequency ω_s at (r_1, z_1). Assuming the unit strength of the source and slow, gradual dependence of environmental parameters on horizontal coordinates, in stationary medium one obtains (Ref. 2, Sec. 7.2)

$$p(r_2, z_2; r_1, z_1) = \sum_n \left(\frac{1}{8\pi D_n} \right)^{1/2} f_n(z_1; r_1) f_n(z_2, r_2) \exp\left(i\theta_n(r_2, t) - \frac{3\pi i}{4} \right), \qquad (1)$$

Here p stands for acoustic pressure, f_n is a normalized shape function of a local mode of the order

$$\theta_n(r_2, t) = \theta_n(r_1, t) + \int_{r_1}^{r_2} q_n \, ds, \quad \theta_n(r_1, t) = -\omega_s t.$$

(2)

n, $q_n = q_n(r)$ is the wave number of the mode, θ_n is its eikonal. Integration in Eq. (2) is along a horizontal eigenray $r = r(\tau, \psi_i)$ (i.e. the horizontal ray that connects the source and receiver), ψ_i is the launch angle of the ray, i.e. the angle the ray makes with Ox coordinate axis at the source, τ is a parameter that defines a point along a given ray. The quantity D_n in (1) is related to a cross sectional area of a modal ray tube and can be calculated as follows:

$$D_n = \frac{\partial(x,y)}{\partial(\tau, \psi_1)} = q_n(r_2) \left[\left(\frac{\partial y}{\partial \psi_1} \right)_\tau \cos\psi_2 - \left(\frac{\partial x}{\partial \psi_1} \right)_\tau \sin\psi_2 \right],$$

(3)

where $\psi_2 = \psi(r_2)$ stands for the ψ value at the receiver. It is assumed that Im q_n ≪ Re q_n and the attenuation due to the imaginary part of the mode wave number is accounted for by the exponential factor in Eq. (1). Therefore, in considering modal ray geometry we assume that $q_n(r)$ is real and non-negative. The rays obey the set of differential equations

$$d r/d\tau = v, \quad dv/d\tau = 0.5\nabla q_n^2,$$

(4)

where $v = \nabla \theta_n$ is horizontal wave vector, $v = q_n$, $v_3 = 0$.

In a waveguide with time-dependent parameters, frequency of each mode $\omega = -\partial\theta/\partial t$ changes as the mode propagates from the source. Mode trajectories also change in time. Differential equations for modal rays become[3]

$$d r/d\tau = v, \quad dt/d\tau = 0.5\partial q_n^2/\partial\omega, \quad dv/d\tau = 0.5\nabla q_n^2, \quad d\omega/d\tau = -0.5\partial q_n^2/\partial t.$$

(5)

Solutions to Eq. (5) are called space-time modal rays because these are curves in three-dimensional space (x, y, t). After the space-time modal rays are found, modal eikonal and amplitude can be expressed in terms of integrals along the eigenrays similar to the stationary case. In particular, the eikonal is given by

$$\theta_n(r_2,t) = \theta_n\left(r_1, t - \int_{r_1}^{r_2} ds\,\frac{\partial q_n}{\partial \omega}\right) + \int_{r_1}^{r_2} ds\left(q_* - \omega\frac{\partial q_n}{\partial \omega}\right), \tag{6}$$

where $\theta_n(r, t)$ is assumed known at $r = r_1$ for a given source and $ds = |d\,r|$ is an increment of arc's length of the modal ray.

3. Effects of Horizontal Inhomogeneities on Mode Phase and Amplitude

Consider weakly inhomogeneous, stationary waveguides where modal wave numbers squared can be represented as

$$q_n^2(r) = k_0^2 + k_0^2 \epsilon\, g(r), \quad |g| \lesssim 1, \quad 0 \le \epsilon \ll 1. \tag{7}$$

The parameter ϵ, which is assumed small, describes deviation of the horizontally-inhomogeneous ocean considered from a layered medium. Dependence of g on mode order is implied but suppressed to simplify notation. When $\epsilon = 0$, the modal eigenray is a straight horizontal line connecting source and receiver. At small ϵ, deviation of the modal rays from the straight lines can be found by using the ray perturbation theory [4] to solve Eq. (4). The perturbation theory leads to the following expressions [1] for the eikonal and the quantity D_n (3):

$$\theta(r_2,t) - \theta(r_1,t) = \int_{x_<}^{x_>} q_n(x,0)\,dx - \int_0^{|x_2 - x_1|}\frac{ds}{2s^2 q_n(x_< + s, 0)}\left(\int_0^s da\, a\,\frac{\partial q_n}{\partial y}(x_< + a, 0)\right)^2 + O(\epsilon^3) \tag{8}$$

$$D_n = q_n(x_1,0)q_n(x_2,0)\left[\int_{x_<}^{x_>}\frac{dx}{q_n(x,0)} - \int_{x_<}^{x_>} dx(x - x_<)(x_> - x)\frac{\partial^2}{\partial y^2}\frac{1}{q_n(x,0)}\right] + O(\epsilon^2). \tag{9}$$

To simplify the notation, here we choose the coordinate system in such a way that $y_1 = y_2 = 0$ and let $x_< = \min\,(x_1, x_2)$, $x_> = \max\,(x_1, x_2)$.

Integration in Eqs. (8) and (9) is along the straight horizontal line between the source and the receiver. Hence, local mode wave numbers q_n need to be calculated only along this line. For a given source and receiver positions, Eqs. (8) and (9) reduce the 3-D problem of sound propagation in horizontally-inhomogeneous ocean to a 2-D problem in the vertical source/receiver plane.

The results (8), (9) are invariant with respect to the interchange of source and receiver positions and, therefore, are consistent with the reciprocity principle. The first term in the RHS of (8) is the eikonal as calculated along the straight modal ray, with the second term giving the correction due to the eigenray curvature. The correction is of second order in ϵ. It is the first cross-range derivative of the propagation constant that determines the phase correction. The derivative can be easily calculated provided that local mode shape functions as well as horizontal environmental gradients are known. An explicit expression for ∇q_n in rather general fluid and fluid/anisotropic solid waveguides can be found in Ref. 5 It is essentially a weighted sum of (i) slopes of the ocean floor and internal interfaces within the ocean bottom, (ii) horizontal gradients of sound speed and density in water, and (iii) horizontal gradients of elastic parameters of the bottom. According to (8), the phase correction is most sensitive to cross-range environmental gradients around the midpoint between source and receiver where ray departure from the Ox axis is at its maximum. As long as horizontal inhomogeneities are weak enough to neglect $O(\epsilon^3)$ and higher-order terms in ϵ, the phase correction is always negative. It means that the uncoupled azimuth approximation which disregards modal rays' curvature is biased towards overestimating mode phase; neglecting horizontal refraction results in a *positive phase bias*.

Positive mode phase errors due to the uncoupled azimuth approximation (up to 180° for the first mode at $f = 50$ Hz and propagation range $R = 420$ km) have been previously reported by Chiu and Ehret [6] in their numerical study of sound propagation through a meandering Gulf Stream current system using fully 3-D mode coupling code. Estimates show that the magnitude of the phase corrections predicted by Eq. (8) and their variation with receiver position are in agreement with the numerical results.

According to (9), effects of cross-range horizontal inhomogeneity on mode amplitude manifest already in terms of the first order in ϵ. It is of interest that mode amplitude depends on the second rather than the first cross-range derivative of the mode phase velocity ω/q_n. Amplitude sensitivity to cross-range variations is highest around the midpoint $x = 0.5(x_1 + x_2)$, where the positive factor $(x - x_<)(x_> - x)$ reaches its maximum. Even when $\partial q_n/\partial y \equiv 0$ at $y = 0$ and eigenrays are confined in the vertical source/receiver plane, mode amplitude - unlike its phase - is sensitive to cross-range variation of environmental parameters. For instance, consider translationally-invariant waveguide with $q_n = q_n(x)$ and cylindrically-symmetric waveguide with the source (or receiver) on the vertical axis of symmetry. According to Eq. (9), mode amplitudes are different in these two media even when parameters of the two waveguides and their cross-range gradients are exactly the same in the vertical source/receiver plane. This conclusion, in particular, and Eq. (9), in general, are in agreement with exact analytic solutions [1] for the translationally-invariant and cylindrically-symmetric waveguides.

4. Ray Travel Time Bias Due to Horizontal Refraction

The perturbation theory[4] can be also applied to quantify the effects of cross-range environmental gradients on conventional 3-D rays. For brevity, we consider horizontal refraction due to sound-speed gradients only. It is assumed, therefore, that if rays are reflected at an interface or a waveguide boundary, these surfaces are horizontal. Assuming horizontal inhomogeneities to be weak, for the ray travel time between a source at $R_1 = (r_1, z_1)$ and a receiver a at $R_2 = (r_2, z_2)$ in stationary ocean, we obtain:

$$T(R_2, R_1) = T_0(R_2, R_1) + \delta\, T_{HR}(R_2, R_1), \quad T_0(R_2, R_1) = \int_{x_1}^{x_2} \frac{dx}{c(x, 0, z_0(x))\, \cos\alpha\,(x, 0, z_0(x))} \quad (10)$$

It is assumed, as before, that $y_1 = y_2 = 0$. Here $c = c(x, y, z)$ is sound speed, $z = z_0(x)$ is the trajectory of eigenray in range-dependent medium with "unperturbed" sound speed $c_0 = c(x, 0, z)$, T_0 and α are the travel time and the grazing angle along the unperturbed eigenray, that is, eigenray in the unperturbed medium. Note that $\cos\alpha$ does not change its sign along the ray and $(x_2 - x_1)\cos\alpha > 0$. The unperturbed eigenray is confined in the vertical source/receiver plane. T_0 is exactly the ray travel time predicted within the uncoupled azimuth approximation. True travel time differs from T_0 due to horizontal refraction, and the correction is given by [1]

$$\delta\, T_{HR} = -\frac{1}{2}\int_{x_<}^{x} \frac{c\, dx}{(x - x_<)^2\, |\cos\alpha|} \left(\int_{x_<}^{x} \frac{\partial c}{\partial y}\, \frac{(x' - x_<)\, dx'}{c^2\, \cos\alpha} \right)^2 + O(\epsilon^3). \quad (11)$$

Here, as in Eq. (10), the integration is along the unperturbed ray. It can be verified that the travel time correction is reciprocal, i.e. $\delta T_{HR}(R_2, R_1) = \delta T_{HR}(R_1, R_2)$. Eikonal (phase) correction is related to the travel time correction by the simple expression $\delta\theta_{HR} = \omega_s\, \delta T_{HR}$.

Corrections to T due to cross-range environmental gradients are negative and of second order in ϵ. Note that sound propagates *faster* than predicted by the uncoupled azimuth approximation despite rays becoming longer because of the horizontal refraction. Equation (11) expresses the travel time bias in terms of cross-range horizontal sound speed gradients integrated along the unperturbed ray. For the particular case of unperturbed medium being uniform ($c_0 = const.$), (11) reduces to the results obtained in [7]. This particular case is thoroughly analyzed in [8] assuming random perturbations in the refraction index.

If the ray theory and adiabatic mode considerations are simultaneously applicable in a waveguide, mode phase corrections due to horizontal refraction are to be consistent with the corrections predicted for 3-D rays corresponding to a given adiabatic normal mode. It can be shown that, indeed, at propagation ranges large compared to ray skip distance, the phase corrections due to horizontal refraction are the same for a mode and for rays with the same grazing angle in a weakly-inhomogeneous waveguide.

5. Modeling Ray Travel Time and Signal Spectrum Variation in the Non-Stationary Ocean

When environmental parameters vary in time, to calculate signal travel time within the ray theory, one has to trace rays and find space-time eigenrays in four-dimensional space (x, y, z, t) which may be very computationally-intensive. The task can be significantly simplified in the case of slow time-dependence of the parameters. Rather than finding space-time eigenrays, in ocean acoustics it is common to rely on the *frozen medium approximation*. Within this approximation, the medium is assumed time-independent ("frozen") during sound transmission but is allowed to change between transmissions. Mathematically, the travel time is given by the integral

$$T_{fr} = \int_{\Gamma_0(t_i)} d\varepsilon / c_0\big(R_0(\varepsilon)\big) \tag{12}$$

along an eigenray Γ_0 in a time-independent medium with sound speed $c_0 = c(R, t_i)$ that coincides with the actual sound speed $c(R, t)$ at some initial moment $t = t_i$. Here ds and $R_0(s)$ are an increment of the arc length along and a point on the ray Γ_0.

We introduce another approximation, to be referred to as *quasi-stationary approximation*:

$$T_{qs} = \int_{\Gamma_0(t_i)} d\varepsilon / c\big(R_0(\varepsilon), t_0(\varepsilon)\big), \quad t_0(\varepsilon) = t_e + \int_0^\varepsilon d\varepsilon_1 / c_e\big(R_0(\varepsilon_1)\big). \tag{13}$$

Here t_e stands for the time when the signal was emitted by the source. Then $t_0(s)$ is the time at which the signal would arrive in a given point on the unperturbed ray Γ_0 if sound propagated in the time-independent medium. So, the difference between the frozen medium and the quasi-stationary approximations is that sound speed variation during acoustic wave propagation is approximately accounted for in the latter. Both the frozen medium and the quasi-stationary approximations can be easily extended to the case of non-stationary ocean with currents by replacing c_0 and c in Eqs. (12) and (13) with the group velocity of sound in moving fluid (Ref. 1, Sec. 5.1).

To evaluate the accuracy of the two approximations, we compare Eqs. (12) and (13) to the exact ray travel time along space-time ray in the case of weak time-dependence of the environmental parameters and currents with velocity $u(R, t)$ small compared to the sound speed. We assume that

$$c(R, t) = c_0(R) + \mu\, \mathcal{C}(R, t), \quad u(R, t) = \mu\, U(R, t), \quad |\mathcal{C}/c_0| \lesssim 1, \quad |U/c_0| \lesssim 1, \quad 0 \le \mu \ll 1 \tag{14}$$

and develop eikonal in powers of the small parameter μ: $\theta(R, t) = \theta_0(R, t) + \mu\theta_1(R, t) + \mu^2 \theta_2(R, t) + \dots$. Dependence of the functions θ and θ_j on the source position and the emission time is implied and suppressed. Assuming that eigenray, Γ_0, and the eikonal function, θ_0, are known in the stationary, motionless case (i.e. at $\mu = 0$), from the eikonal equation (Ref. 1, Sec. 5.1) we obtain

$$\theta_1 = \omega_s \int_{\Gamma_0} \frac{ds}{c_0} \left[\frac{\omega_s}{2}\left(\frac{1}{c^2} - \frac{1}{c_0^2} \right) + \frac{u \cdot \nabla \theta_0}{c_0^2} \right], \tag{15}$$

$$\theta_2 = \int_{\Gamma_0} \frac{ds}{c_0} \left[\frac{1}{2c_0^2}\left(\frac{\partial \theta_1}{\partial t} - u \cdot \nabla \theta_0 \right)^2 - \frac{1}{2}\left(\nabla \theta_1 \right)^2 \right.$$
$$\left. - \omega_s \left(\frac{1}{c^2} - \frac{1}{c_0^2} \right)\left(\frac{\partial \theta_1}{\partial t} - u \cdot \nabla \theta_0 \right) + \omega_s \frac{u \cdot \nabla \theta_1}{c_0^2} \right] \tag{16}$$

Using Eqs. (15) and (16) to calculate corrections to the travel time due to environmental parameters' time dependence, and comparing the result to Eqs. (12) and (13), we find that the difference between the true travel time and prediction of the frozen medium approximation $T - T_{qs} = O(\mu)$. For the quasi-stationary approximation $T - T_{qs} = O(\mu^2)$. These estimates apply to both moving and motionless cases. Hence, the error of the frozen medium approximation is of the same order as the very effect of the medium non-stationarity on the acoustic field. On the contrary, the quasi-stationary approximation correctly reproduces leading-order effects of medium non-stationarity while bypassing the need to trace space-time rays.

The effects of medium non-stationarity on travel time are of particular importance for ocean current tomography, where the *travel time nonreciprocity* (i.e. the difference in time of sound propagation in opposite directions between two acoustic transceivers) is used to measure the velocity of currents.[9] The reciprocity principle ensures that, in a time-independent motionless medium, the travel times are reciprocal. Both sound speed time-dependence and currents contribute to the travel time nonreciprocity. For the purposes of the ocean current tomography, the contribution to the travel time nonreciprocity due to medium motion is a 'signal' while the contribution due to medium non-stationarity is 'noise.' To evaluate feasibility and accuracy of tomographic current inversions, one has to be able to estimate the travel time nonreciprocity due to ocean non-stationarity.

Using the quasi-stationary approximation to calculate ray travel times with accuracy up to terms $O(\mu^2)$, we find for the travel time nonreciprocity:

$$T_+ - T_- = \int_{\Gamma_0} ds \left[\frac{1}{c(R_0(s), t_+)} - \frac{1}{c(R_0(s), t_-)} - \frac{dR_0}{ds} \cdot \frac{u(R_0(s), t_+) + u(R_0(s), t_-)}{c_0^2(R_0(s))} \right], \quad (17)$$

where T_+ is the acoustic travel time from a transceiver A to a transceiver B and T_- is the travel time from a transceiver B to a transceiver A, Γ_0, as before, is the unperturbed eigenray (which is the same for propagation in the opposite directions), and

$$t_+(s) = t_A + \int_0^s \frac{ds'}{c_0}, \quad t_-(s) = t_B + T_0 - \int_0^s \frac{ds'}{c_0} \quad (18)$$

are the times at which sound would arrive in a given point on Γ_0 on its way from the transceiver A or B, respectively, in the unperturbed medium. In Eq. (18), t_A and t_B stand for sound emission times for respective transceivers. It is assumed that the transceivers are at rest. For surface reflected rays, there is an additional contribution to the nonreciprocity due to time-dependence of the ocean surface shape. This contribution is not accounted for in (17).

According to Eq. (17), contribution to the ray travel time nonreciprocity due to medium motion is proportional to the integral over the ray of the current velocity projection on the tangent to the ray. Contribution due to sound speed variation is proportional to the difference of sound speed values at the times when sound travels through a given point in the opposite directions. If one would apply the frozen medium approximation to evaluate the nonreciprocity, the result would be again given by (17) but with t_i substituted for t_+ and t_- in the integrand. Then the contribution to $T_+ - T_-$ due to sound speed time-dependence would be lost entirely. This is consistent with the error of the frozen medium approximation and the nonreciprocity both being of the first order in μ. In the case of stationary media, Eq. (17) reduces to well-known result.[9] In the general case of ocean with weak non-stationarity and slow currents, Eq. (17) allows one to quantify the ray travel time nonreciprocity without tracing rays in moving and/or non-stationary medium.

Another observable consequence of ocean non-stationarity is a variation is acoustic signal spectrum. For an individual narrow-band ray arrival, central frequency changes along the ray in non-stationary medium without noticeable change in the signal spectrum shape. When multipaths are not resolved, change in the spectrum of received signal may be more complex because central frequency shift is generally different for different rays. The sound frequency can be found exactly by tracing space-time rays or approximately, up to terms $O(\mu^2)$, within the quasi-stationary approximation as $\omega = \omega_s (1 - \partial T_{qs} /\partial t)$, where ω_s is frequency at the sound source. Using Eq. (13), we find

$$\omega = \omega_s + \omega_s \int_{r_0}^{r} \frac{ds}{c_0^2} \frac{\partial c}{\partial t}\left(R_0(s), t_0(s)\right). \tag{19}$$

The frequency shift due to sound speed variation with time is proportional to the acoustic wave frequency and, in this respect, is akin Doppler shift.

If sound speed variations occur with frequency Ω and have amplitude δc, then for the change in acoustic frequency one has from Eq. (19) $|\delta\omega| \sim \omega_r R \Omega |\delta c/c^2|$, where R stands for propagation range. Note that at long-range propagation $\delta\omega$ may by far exceed the frequency Ω of the oceanographic process responsible for the sound speed variation.

6. Quasi-Stationary Approximation for Adiabatic Normal Modes

The theory outlined in Sec. 5 for ray arrivals can be extended to modal arrivals. For simplicity, we assume here that medium is motionless. Within the frozen medium approximation, adiabatic mode eikonal (phase) is given by

$$\theta_n^{fr}(r, t) = \int_{r_1}^{r} ds\, q_n\left(\omega_r, r_0(s), t_i\right) - \omega_r t, \tag{20}$$

where integration is along an unperturbed modal ray defined as a modal ray in time-independent waveguide; the initial moment t_i has the same meaning as in Eqs. (12) and (13). In Eq. (20), ω_r stands for the frequency of received signal; $\omega_r = \omega_s$ in the frozen medium approximation. In the quasi-stationary approximation, adiabatic mode eikonal (phase) is given by (cf. (13))

$$\theta_n^{qs}(r, t) = \int_{r_1}^{r} ds\, q_n\left(\omega_r, r_0(s), t_0(s)\right) - \omega_r t, \quad t_0(s) = t - \int_{r_0(s)}^{r} ds_1 \frac{\partial q_n}{\partial \omega}\left(\omega_r, r_0(s_1), t_i\right). \tag{21}$$

Integration in (21) is again along the unperturbed modal ray. In Eq. (21), $t_0(s)$ is the time at which the signal that arrived at r at moment t were at the point $r_0(s)$ if it propagated along the unperturbed ray in the time-independent medium. Note that, in this approximation, to calculate mode phase and frequency $\omega = -\partial\theta/\partial t$ variation it is sufficient to know the mode wavenumber q_n and its derivative $\partial q_n/\partial\omega$ at a single frequency, $\omega=\omega_r$.

To evaluate accuracy of the frozen medium and the quasi-stationary approximations, one has to assume weak time-dependence of the medium parameters, apply a perturbation theory to solve Eq. (5) for space-time modal rays, and compare the exact expression for adiabatic mode eikonal, Eq. (6), to Eqs. (20) and (21). An analysis along these lines shows that, much like the ray arrival case considered in Sec. 5, the quasi-stationary approximation allows one to calculate mode parameters in non-stationary waveguide with accuracy up to the terms $O(\mu^2)$, while the frozen medium approximation differs from the exact result by terms $O(\mu)$ which are of the same order of magnitude as the very acoustic effects of medium non-stationarity.

The quasi-stationary approximation reduces the 4-D problem of acoustic field modeling in three-dimensionally inhomogeneous, time-dependent medium to a 3-D problem of sound propagation in stationary medium. If horizontal inhomogeneities are weak, the results of Secs. 3 and 4 above allow one to further substitute the 3-D problem with a 2-D problem in the vertical source/receiver plane. Moreover, first-order corrections to ray travel time and mode phase due to horizontal refraction and sound speed time-dependence are additive. In other words, in applying the results of Secs. 3-6 to a medium with slow time dependence and weak horizontal inhomogeneity, one can calculate effects of the horizontal refraction neglecting the non-stationarity and effects of the non-stationarity neglecting the horizontal refraction.

7. Significance of 3-D and 4-D Effects

For mode phase perturbation due to horizontal ray curvature we have $|\delta\theta_{HR}| \sim \epsilon^2 k_0 \times R^3/(24 L^2)$ from (8) where R is propagation range and L is the spatial scale of horizontal inhomogeneities. The scale is assumed to be large or comparable to R. In the opposite case of small-scale inhomogeneities, we have $|\delta\theta_{HR}| \sim \epsilon^2 k_0 R^2/L$ from (8). Note rather rapid, nonlinear increase of the phase perturbation with range in both cases. For the uncoupled azimuth approximation to be applicable, it is necessary that $|\delta\theta_{HR}| \ll 1$. From the above estimate of $\delta\theta_{HR}$ we obtain the condition of the approximation validity: $\epsilon^2 k_0 R^3/(24 L^2) \ll 1$ and $\epsilon^2 k_0 R^2/L \ll 1$, for large- and small-scale horizontal inhomogeneities, respectively. Similar estimates can be obtained in the context of geometrical acoustics using Eq. (11).

As an example of a deep-water problem, consider propagation of 250 Hz sound along a 1275 km path in SOFAR channel, as in 1987 Reciprocal Tomography Experiment (RTE87).[9, 10] Assume a depth-independent and range-independent value $\partial c/\partial y = 2 \cdot 10^{-5}$ s^{-1} of the cross-range gradient of the sound speed. The value corresponds to a 2 m/s sound speed change per 100 km and is close to average variation in a north-south direction in the sound speed on SOFAR channel axis. Under these conditions, ray travel-time bias due to the uncoupled azimuth approximation is $|\delta T_{HR}| \sim R^3 (\partial c/\partial y)^2/(24 c^3 \cos^3 \alpha) \sim 10.3$ -11.6 ms depending on the grazing angle of the ray at the waveguide axis [see Eq. (11)]. This is a conservative estimate of the bias because horizontal refraction due to neither mesoscale inhomogeneities nor internal waves has been taken into account.

In Dushaw et al.[10] ray travel-time measurements made during RTE87 have been applied to verify the accuracy of equations used to calculate sound speed in sea water. Measured travel times were compared to travel times predicted for sound-speed fields computed from CTD and XBT data. Uncoupled azimuth approximation was implied, i.e. the ocean was assumed to be range-dependent rather than horizontally-inhomogeneous. It was concluded that Del Grosso's equation for sound speed is substantially more accurate than the Chen and Millero equation. A small correction to Del Grosso's equation of +0.05±0.05 m/s at 4000-m depth was calculated, with 0.05 m/s increase in the sound speed corresponding to 17-ms negative bias in travel time.[10] It is interesting that sign and magnitude of the travel time bias resulting from the apparent error in Del Grosso's equation are similar to the above estimate of travel-time bias due to horizontal refraction.

As another example, consider sound propagation under conditions of the 1993 Pacific Shelf sea trial.[11] The experiment was carried out at a site on the continental shelf southwest of Vancouver Island. Bottom slope and ocean depth were 6° - 8° and H = 300-500 m. The experiment included towing of a source that emitted three CW tones in the band f = 45-72 Hz. Signals were received by a vertical line array with 16 hydrophones. The source was towed along a track with segments roughly along and across isobaths out to a total distance about 10 km from the array. The total number of propagating normal modes and their grazing angles were limited by sound speed in the ocean bottom, which did not exceed 1740 m/s in the upper several hundred meters of the sediment.[12] More than 10 modes contributed considerably to the acoustic field at the array location.

At cross-slope propagation and range R = 7 km we obtain from Eq. (8) the following estimates of phase corrections (in radians) to the predictions of the uncoupled azimuth approximation: 0.004 (n = 2), 0.13 (n = 4), 0.34 (n = 5), 0.77 (n = 6), 1.50 (n = 7), 2.66 (n = 8), 4.38 (n = 9), and 6.84 (n = 10). Mode sensitivity to bottom slope and, hence, to horizontal refraction increases approximately as the fourth power of grazing angle. In the problem considered, significant phase errors occur at grazing angles of 8° and larger. Because of O((1) errors in relative phases of adiabatic modes, the $N{\times}2D$ approximation strongly distorts transmission losses. We see that 3-D effects are clearly important in the field of a point source under conditions of the Pacific Shelf sea trial. Note also that in the shallow-water environment horizontal refraction leads to phase bias of the order 2π at distances of 7 km, while in the above deep-water example such bias occurred at propagation ranges ~ 1,000 km.

Let us now evaluate apparent acoustic Doppler shift due to internal waves in deep ocean conditions of the Moving Ship Tomography (MOST) experiment in the Western Mediterranean Sea (1994).[13] Assume that the internal waves of frequency Ω occupy a layer of vertical extent h

and create a random sound field perturbation with amplitude δc and horizontal correlation length L. From Eq. (19) we obtain then an order-of-magnitude estimate

$$\left\langle \left(\delta f/f_0\right)^2\right\rangle \sim R L \, \Omega^2 \left(\delta c/c^2\right)^2 \left(h/H\right)^2 \tag{22}$$

for the mean value of the relative frequency change squared. Here H is the vertical excursion of the ray. For a refracted -surface reflected ray, H equals depth of its turning point; for purely refracted ray, H equals the difference of depths of the lower and upper turning points, etc. To simulate the conditions MOST experiment, we choose $R = 300$ km, $L = 1$ km, $2\pi/\Omega = 15$ min, $\delta c = 0.5$ m/s, $h = 100$ m, and $H = 2$ km. Then it follows from Eq. (22) that frequency wander due to the internal waves is equivalent to the Doppler shift due to relative source-receiver motion with radial velocity with RMS value about 3 m/s. Such Doppler shifts were indeed measured in the MOST experiment but remained unexplained as the source-receiver motion was monitored to an accuracy much better than 1 m/s. It now appears that the measured frequency shifts were likely due to ocean non-stationarity, more specifically, internal wave activity in the Mediterranean.

Finally, to demonstrate a potential impact of medium non-stationarity on measuring ocean current velocity with reciprocal transmissions, we consider two scenarios: underwater sound propagation on continental shelf in the presence of internal wave solutions under conditions of the Coastal Ocean Probing Experiment [14] (COPE) of 1995 and acoustic probing of an interthermocline lens of Mediterranean water (meddy) in the Atlantic ocean.[15] During COPE, trains of strong internal wave solitons with parallel wave fronts were observed in a shallow water off Northern Oregon with depth about 100 m. Isotherm displacement in the solutions reached 15-25 m leading to sound speed perturbations about $\delta c = 15$ m/s and current velocities up to 0.7 m/s. Horizontal dimensions of an individual soliton were $L = 200$-250 m across its front and soliton phase velocity was 0.8 - 0.9 m/s. Assuming propagation range $R \gg L$ and using Eqs. (17) and (18) to estimate the contributions to travel time nonreciprocity due to sound speed time dependence, δT_c, and due to currents, δT_u, we find $\delta T_u / \delta T_c \sim 0.05 |sin \, \phi|$, where ϕ is the angle between the vertical source/receiver plane and the soliton wave front. Obviously, medium non-stationarity dominates in the travel time nonreciprocity, and inverting the nonreciprocity for the current velocity field appears impossible.

In the other scenario, following, [15] we model an interthermocline lens of Mediterranean water as an eddy with a vertical axis of symmetry and radius 50 km, maximum orbital velocity of rotation 0.5 m/s, vertical extent 1000 m, and sound speed contrast of 10 m/s from surrounding Atlantic waters. The eddy moves horizontally as a whole with drift velocity 0.05 m/s. It is the drift that is responsible for sound speed time dependence in the case of an axially symmetric meddy. Let the meddy intersect the vertical source/receiver plane. Again using Eqs. (17) and (18), we find that typical value of travel time nonreciprocity due to currents for rays crossing the eddy is $\delta T_u \sim 20$ ms while the ratio $\delta T_u / \delta T_c \sim 250/|sin\ \phi|$ where ϕ is the angle between the vertical source/receiver plane and the drift velocity vector. Obviously, contribution of the medium non-stationarity to the nonreciprocity is negligible, and information on water velocity within the eddy can be retrieved from data on travel time nonreciprocity. In a similar manner, it can be demonstrated that the inequality $\delta T_u / \delta T_c$ 1 is also true for much bigger mesoscale eddies.

The above examples show that contributions of ocean non-stationarity to acoustic nonreciprocity may be either dominant or negligible, depending on specific problem.

8. Summary

A technique has been developed to reduce 3-D and 4-D propagation problems to a set of 2-D problems that retains computational efficiency of the uncoupled azimuth and frozen medium approximations but extends their domains of validity and improves accuracy of propagation modeling.

Closed-form applicability conditions are obtained for the $N\times2$-D approximation used in conjunction with the ray or adiabatic mode theory. Corrections to the $N\times2$-D approximation predictions are found to be significant under conditions of some recent experiments in deep and shallow water.

Horizontal refraction *decreases* ray travel time and adiabatic mode phase. The travel time and phase biases are proportional to cross-range environmental gradients squared and increase as third power of range for large-scale inhomogeneities and as second power of range for small-scale (random) inhomogeneities.

Unlike the frozen medium approximation, the quasi-stationary approximation is a sufficiently accurate and efficient approach to modeling low-frequency sound propagation in the time-dependent ocean.

Ocean non-stationarity, particularly internal waves, are likely to be responsible for the anomalous Doppler shifts observed in some tomographic experiments.

Further research is needed to quantify, for specific propagation scenarios, the effect of the cross-range environmental gradients on travel time and amplitude of acoustic arrivals, as well as the acoustic nonreciprocity due to various time-dependent physical processes in the ocean and its implications on current tomography.

Acknowledgments

The author is pleased to acknowledge stimulating discussions with Prof. E. C. Shang. The work reported in this paper was supported, in part, by the US Office of Naval Research.

REFERENCES

1. O. A. Godin, "A 2-D Description of sound propagation in a horizontally-inhomogeneous ocean," to be published in *J. Comput. Acoustics*.

2. L.M. Brekhovskikh and O.A. Godin, *Acoustics of Layered Media. I1: Point Sources and Bounded Beams,* Springer Ser. Wave Phenom., Vol. 10. 2nd, extended edn. (Springer, Berlin, Heidelberg 1999) XIII + 527 p.

3. H. Weinberg and R. Burridge, "Horizontal rays and vertical modes," in *Wave Propagation and Underwater Acoustics*, ed. by J. B. Keller and J. S. Papadakis (Springer, Berlin, Heidelberg 1977) pp. 86-152.

4. Yu. A. Kravtsov and Yu. I. Orlov, *Geometrical Optics of Inhomogeneous Media,* Springer Ser. Wave Phenom., Vol. 6 (Springer, Berlin, Heidelberg 1990). Sec. 2.9.

5. O.A. Godin, "Acoustic mode reciprocity in fluid/solid systems: Implications on environmental sensitivity and horizontal refraction," in *Theoretical and Computational Acoustics '97,* ed. by Y. C. Teng et al. (World Scientific, Singapore, 1999), pp. 59-75.

6. C.-S. Chiu and L.L. Ehret, "Computation of sound propagation in a three-dimensionally varying ocean: A coupled normal mode approach," in *Computational Acoustics*, vol. 1, ed. by D. Lee et al. (Elsevier, North-Holland 1990) pp. 187-202.

7. W. Boyse and J. B. Keller, "Short acoustic, electromagnetic, and elastic waves in random media," *J. Opt. Soc. Am.* **A12**, 380-389 (1995).

8. B. Iooss, Ph. Blanc-Benon, C. Lhuillier, "Statistical moments of travel-times at second order in isotropic and anisotropic random media," *Waves Random Med.* **10**, 381-394 (2000).

9. W. Munk, P. Worcester, and C. Wunsch, *Ocean Acoustic Tomography* (Univ. Press, Cambridge, 1995), Chaps. 3 and 8.

10. B.D. Dushaw, P.F. Worcester, B.D. Cornuelle, and B.M. Howe, "On equations for the speed of sound in seawater," *J. Acoust. Soc. Am.* **93**, 255-275 (1993).

11. N.R. Chapman, M.L. Yeremy, J.M. Ozard, and M.J. Wilmut, "Range dependence of matched field source localization in shallow water," in *Proc. of the Symposium on Shallow Water Undersea Warfare, Halifax, N.S., October 1996,* pp. 115-124.

12. O.A. Godin, N. R. Chapman, M. C. A. Laidlaw, and D. E. Hannay, "Head wave data inversion for geoacoustic parameters of the ocean bottom off Vancouver Island," *J. Acoust. Soc. Am.* **106**, 2540-2551 (1999).

13. O.A. Godin, S. V. Burenkov, D. Yu. Mikhin, S. Ya. Molchanov, V. G. Selivanov, Yu. A. Chepurin, and D. L. Aleynik, "An experiment on dynamic tomography in the Western part of the Mediterranean Sea," *Dokl. Akad. Nauk,* **349**, 398-403 (1996) [In Russian].

14. T. P. Stanton, L. A. Ostrovsky, "Observations of highly nonlinear internal solutions over the continental shelf," *Geophys. Res. Letters,* **25**, 2695-2698 (1998).

15. D. Yu. Mikhin, O. A. Godin, O. Boebel, and W. Zenk, "Simulations of acoustic imprints of meddies in the Iberian Basin: Towards acoustic detection of meddies," *J. Atmos. Ocean Techn.,* **14**, 938-949 (1997).

Theoretical and Computational Acoustics 2001
E.-C. Shang, Qihu Li and T. F. Gao (Editors)
© 2002 World Scientific Publishing Co.

Propagation Modeling in Range-Dependent Shallow Water Scenarios

A. Tolstoy

ATolstoy Sciences, 8610 Battailles Court, Annandale VA 22003

Tel: (703) 978-5801, Fax: (703) 978-1319, E-mail: *atolstoy@ieee.org*

Abstract

Recent investigations of propagation models suggest that models need to be "benchmarked" *each* time they are applied to a different type of scenario. It is important that the user has properly input the various model parameters and checked their validity. Some of this "benchmarking" can be done using the wedge test cases of Jensen and Ferla, '90. Additional testing can involve more complicated scenarios, e.g., multiple sediment layers with range-dependence. Thus, there is a need for a variety of "benchmark" results. These results must often go beyond analytic solutions (such as those for an ideal wedge with point or line sources and perfectly reflecting boundaries). There are very few analytic solutions. *Consistency* checks between different types of models *plus* initial comparisons with carefully generated and accepted benchmark results are highly desirable.

1 Introduction

There has always been interest in the confirmation that a model is producing believable results. This "benchmarking" of models has been a major concern since the first propagation models were developed in the 50s and 60s (Tolstoy and Clay, '66), and often comparisons were made between model predictions and observed data. Unfortunately, comparisons between predicted and observed data are nearly impossible to resolve in the face of discrepancies. Thus, ideally one would like to compare model predictions with *analytic* solutions which have the potential to be known "exactly" (or to within nearly whatever numerical accuracy is desired).

In these days of *shallow water* emphasis, benchmarking has become even more difficult since there can be very complicated environments involved (bottom topography, 3-D effects, multiple sediment layers, elasticity, internal waves). Moreover, there are very few candidate analytic solutions for comparison (Buckingham and Tolstoy, '90). Additionally, with the high interest in *inversion* for the estimation of model inputs (Tolstoy et al., '98), it has become increasingly important to develop fast as well as accurate propagation models (Jensen et al., '94). Yet another motivation for benchmarking is that there are now *many versions* of each propagation model in use, and it can be impossible to know which version of any model one has, and thus, to know what the model truly does and does *not* do correctly.

In any event, "benchmarking" is now interpreted to mean not only comparison with analytic solutions but also comparison with numerical output from generally accepted, highly regarded models (such as from two-way coupled normal modes or an energy conserving high-angle PE). The comparison of a given model output to "benchmark data" (previously generated by an accepted and carefully applied model) can immediately confirm or *not* if another model is working correctly for a given scenario.

2   The ASA Wedge (1990)

In 1990 there was an issue of the Journal of the Acoustical Society of America (JASA **87** (4)) containing a collection of papers devoted to the benchmarking of propagation models in shallow water. Central to many of these papers was a low frequency test case involving a simple wedge (line source) where the bottom boundary was penetrable and had non-zero absorption (Jensen and Ferla, '90; case I subcase 3). This test case and the associated "data" generated for it by means of a two-way coupled normal mode model still comprise an excellent place to start any benchmarking exercise for a range-dependent model.

By comparing energy conserving and non-conserving PE models to those wedge data as well as by comparing adiabatic (a.n.m.) and one-way coupled (c.n.m.) versions of the normal mode code KRAKEN (Porter, '91), we conclude that:

- energy conservation *can be* highly important to a shallow water problem whenever topography is significant – even a low frequencies such as 25 Hz (Fig. 1).

Moreover, we also see that mode coupling (even one-way) versus an adiabatic assumption can be very important (Fig. 2). However,

- we cannot dismiss adiabatic normal modes altogether.

It seems that a.n.m. may *sometimes* do very well but it can be difficult to predict *when*. One might expect that higher frequencies would be less adiabatic. However, we see that for this test case at frequencies of 50, 75, and 100 Hz a.n.m. performs nearly as well as c.n.m. (except near the wedge apex where all propagating modes are past cut-off, Fig. 3).

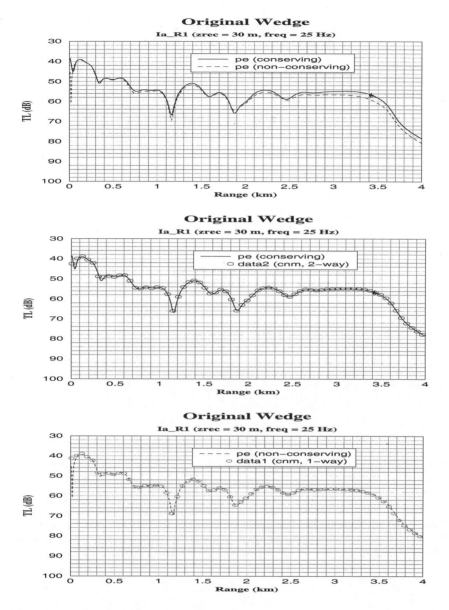

Figure 1: Benchmark Wedge with absorption. Comparisons of (top) energy conserving (EC) PE with non-conserving (NC) PE. Note the higher levels by 2-3 dB for EC near the apex (range of 4 km). (mid) EC with benchmark data assuming the full two-way coupling. Note the essentially perfect agreement. (bottom) NC with benchmark data assuming only one-way coupling. Note the essentially perfect agreement.

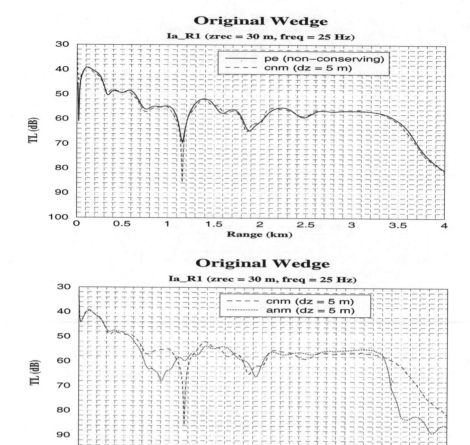

Figure 2: (Top) NC PE versus one-way c.n.m. with crude range increments (dz = 5 m giving range steps of delr = 100 m). Note the excellent agreement except at a strong null near 1.2 km range. (Bottom) a.n.m versus c.n.m. showing strong discrepancies in the neighborhood of 0.75 to 1.25 km range and after 3.2 km range when the last propagating mode is past cut-off.

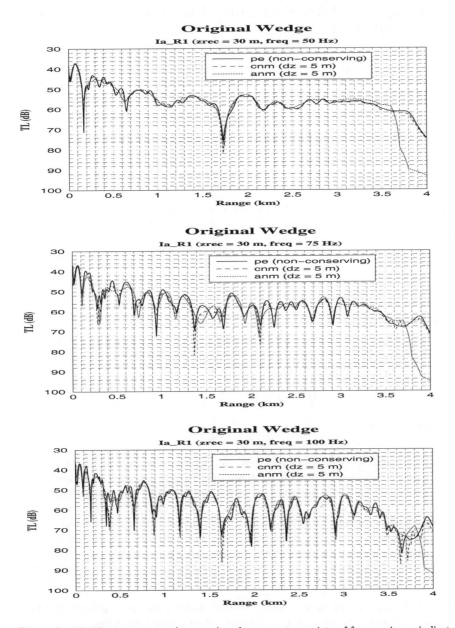

Figure 3: NC PE versus c.n.m. (one-way) and a.n.m. at a variety of frequencies as indicated and higher than 25 Hz. Note the excellent agreements up to the cut-off of the final propagating mode, e.g., beyond 3.7-3.8 km range.

3 Workshop97, SWAM99, Stennis Inversion Workshop '01, Chicago '01 ASA

Since the ASA efforts of 1990 there has been a number of workshops (some have addressed inversion as a primary focus) which have resulted in the generation of "benchmark" data. These include:

- **Workshop97** which generated a family of benchmark data using an FFP code (Jensen et al., '94) for range-*independent* scenarios and was intended for the application of inversion methods. The test cases offered numerous scenarios including multiple sediment layers with bottom sound-speed and density profiles, bottom attenuations, and elasticity.

- **SWAM99** which did *not* offer benchmark data but did suggest a number of test scenarios beyond the wedge for the application of models and which emphasized range-dependent variability including topography, 3-D effects, elasticity, and internal waves. These scenarios were refined for the **Chicago '01 ASA** which offered a subset of 3-D, shelf-break, and internal wave test cases. The intent of both gatherings was to examine the output from a number of community models and to decide if any offered "benchmark" caliber results.

- **Stennis Inversion Workshop '01** which was concerned with inversion for range-dependent environments. Benchmark data were generated using an energy conserving PE and a two-way coupled mode model (Evans, '83) for cases which included a down-slope, a shelf-break, and a flat bottom inclusion.

4 Conclusions

We conclude that benchmarking is still very important in the verification of model accuracy. In general, model developers or users should apply their models to a number of benchmark data sets to be certain that they are using the model correctly, and that it is performing as expected. These data sets include the ASA '90 benchmark wedge with attenuation, some of the Workshop97 range-independent scenarios (depending on whether or not such things as multiple layers or elasticity are of interest), the Stennis Inversion Workshop '01 sets (generally accessible and soon to be written up), and the SWAM99 and Chicago '01 test cases (also soon to be written up with a subset of presentations to qualify as benchmark data).

We also conclude that the failure of adiabatic normal modes is not necessarily predictable. Lower frequencies are not necessarily more adiabatic than higher frequencies.

Finally, we also note that benchmarking now goes beyond comparisons of predictions with analytic solutions. Such efforts should ' ally involve comparisons with different types of models, e.g., PE with high-angle energy conservation capabilities or two-way coupled normal modes. Yet another important test of validity is reciprocity (not explicitly discussed here), i.e., where the source and receiver are interchanged but the final field value at the receiver point should not change.

5 Acknowledgements

The author would like to thank ONR for its continued support.

6 References

1. M. Buckingham and A. Tolstoy (1990), "Line source solution for the wedge with pressure-release boundaries (benchmark problem 1)", *J. Acoust. Soc. Am.* **87** (4), 1511-1513.

2. R.B. Evans (1983), "A coupled mode solution for acoustic propagation in a waveguide with stepwise depth variations of a penetrable bottom", *J. Acoust. Soc. Am.* **74**, 188-195.

3. F.B. Jensen and C.M. Ferla (1990), "Numerical solutions of range-dependent benchmark problems in ocean acoustics", *J. Acoust. Soc. Am.* **87**, 1499-1510.

4. F.B. Jensen, W.A. Kuperman, M.B. Porter and H. Schmidt (1994), *Computational Ocean Acoustics* (Am. Inst. Physics, New York).

5. M.B. Porter (1991), "The KRAKEN normal mode program", SACLANT Undersea Research Centre Report SM-245 (1989), Naval Research Laboratory Mem. Rep. 6920.

6. A. Tolstoy, N.R. Chapman, and G. Brooke (1998), "Workshop97: Benchmarking for geoacoustic inversion in shallow water", *J. Computat. Acoust.* **6** (1&2), 1-28,

7. I. Tolstoy and C.S. Clay (1966), *Ocean Acoustics* (McGraw-Hill, New York).

Theoretical and Computational Acoustics 2001
E.-C. Shang, Qihu Li and T. F. Gao (Editors)
© 2002 World Scientific Publishing Co.

The ASIAEX 2000 Preliminary Experiment in the East China Sea (ECS)

Ching-Sang Chiu and Steven R. Ramp
Naval Postgraduate School, Monterey, CA, USA

James F. Lynch
Woods Hole Oceanographic Institution, Woods Hole, MA, USA

Introduction

In this short paper for the Warren Denner Memorial Session, we would like to discuss three items: 1) Dr. Warren Denner and his visions for the Asian Sea International Acoustic Experiment (ASIAEX) program, 2) a first look at the field program results of the ASIAEX program in general, and 3) a brief computational study based on field hydrography data from the 2000 ECS "preliminary experiment" (as opposed to data from the main ASIAEX field efforts, which were executed in spring of 2001.)

Dr. Warren Denner

Let us begin, appropriately, with Dr. Warren Denner, who was a close colleague and personal friend of the authors. Warren's loves in his professional life were: 1) physical oceanography, 2) ocean acoustics, and 3) people and international relations, which made him an ideal coordinator for an international acoustic and oceanographic science program like ASIAEX. Warren was just as excited by the new oceanography and acoustics that would be learned by going to the ASIAEX sites as by the human interactions and professional collaborations that would occur between the scientists of the USA and some half-dozen important Pacific Rim countries. Warren worked as a "bridge builder," both in science and human relations. Warren died unexpectedly in November of 1999 in Nizhny Novogorod, Russia, building bridges between the USA and Russia on a separate project. Though tragic, Warren's death far from his native California was perhaps the ultimate example of the devotion he had to advancing both science and international relations through science. Warren would have been very pleased – no, overjoyed – to see how well ASIAEX turned out experimentally, and how well the human interactions between the various countries' scientists continue to evolve.

ASIAEX – A brief overview of the 2000-2001 field programs

Before looking at a few computational results based on the ASIAEX ECS preliminary experiment data, we think it might interest the reader to see a brief review of the objectives and accomplishments of the overall ASIAEX field program, which was recently completed (in June 2001). The overall scientific objective of ASIAEX is to study the effects of water-column variability, boundary roughness and sediment structure on sound propagation in a coastal environment. While the study of acoustic volume interaction took place in the South China Sea, the study of acoustic bottom interaction took place in the East China Sea. There were four major components of the ASIAEX field program, which were: 1,2) the spring 2000 preliminary experiments in the East China Sea and South China Sea (SCS) and 3,4) the spring 2001 main experiments in the ECS and SCS.

The April-May 2000 ECS preliminary experiment, centered at roughly 127E, 30N, featured a wide number of measurements. These included: 1) CTD hydrography across the shelfbreak front with the Kuroshio, 2) oceanographic instrument mooring operations, 3) a high resolution Acoustic Doppler Current Profiler (ADCP) survey of the region, 4) seismic profiling operations using both chirp sonar and water gun sources, 5) satellite oceanography (microwave and infra-red imaging), 6) sea-surface temperature from shipboard sensors, and 7) near-surface sediment coring operations. These components all succeeded rather well. The Kuroshio and the non-linear internal wave field were particularly interesting oceanographic features of this area. The sub-bottom geology results are treated in another paper from this conference (Bartek, 2001). We will show acoustics calculations based on a stationary yo-yo CTD time-series and two cross shelf CTD hydrography surveys in the next section.

The April 2000 SCS preliminary experiment, centered around 17N, 122E, featured a very high resolution, Sea Soar hydrographic survey of the physical oceanography of the region, as well as a first glimpse of its rather irregular topography (featuring steep walled canyons and terraces) and a cross-shelf deployment of moored oceanographic sensors. These efforts were also very successful and, aside from not being able to do a high-resolution geology survey there in 2000, the SCS preliminary program completed most of its goals.

The May-June 2001 ECS main experiment (at the same location as the preliminary) was focused on acoustic bottom reverberation issues, and featured the use of both horizontal and vertical receiving arrays, combined with shot and moored array sound sources. Preliminary reports from the project PI's also indicate a good degree of success.

Finally, there was the 2001 main SCS experiment (again at the same site), which the authors all participated in just recently. This was a joint acoustics-geology-physical oceanography effort that extended over three weeks, thus incorporating a full spring-neap tidal cycle. Some of the highlights of this experiment were: 1) the use of a joint vertical/horizontal receiver array to examine acoustic coherence issues, 2) the employment of many acoustic frequencies, covering a band from 50 – 600Hz, 3) forty moorings supplying densely sampled physical oceanography measurements both in space and time, 4) a two week long Sea Soar hydrographic survey, 5) chirp sonar geology data along the two major acoustic paths, and 6) a typhoon transiting the measurement area towards the latter half of the experiment, which was successfully measured by the moored oceanographic and acoustic instruments.

Acoustic Modeling Results

Using hydrographic data obtained in the 2000 ECS preliminary experiment, acoustic model runs to investigate important design questions concerning the bottom and volume acoustic interaction main experiments of 2001 were carried out. The bathymetry of the experimental site and the locations of the hydrographic measurements are shown in Figure 1. The specific questions, which were addressed in the modeling effort, include:

1. Is the water-column variability in northwest quadrant of the ECS survey site robust enough to permit unambiguous observations of the acoustic effects caused by the ocean bottom, if the bottom interaction experiment is to take place there?

2. What are the optimum source depth, receiver range, and frequency in a slope-to-shelf transmission, in the event that the volume interaction experiment is to take place in the vicinity of the shelfbreak inside the ECS survey site?

To explore the answer to the first question, the yo-yo CTD time-series collected on 9 April 2000 and shown in Figure 2 in terms of sound speed was used to estimate the temporal variability of the transmission loss along the water-sediment interface inside the northwest quadrant. Being far away from the Kuroshio Current and with a minimal stratification in Spring, the measured time-series show a very small sound speed variation as anticipated, only 3 m/s over a 4-hour span and was mostly confined in the top layer. However, to a small surprise, the calculated transmission loss (top panel of Figure 3) shows moderate temporal fluctuations of ± 4 to 5 dB along the interface. These calculations were performed for a frequency of 1 kHz and a nominal source depth of 20 m. The transmission loss fluctuations are reducible by local spatial averaging (see bottom panel of Figure 3), i.e., trading resolution for a reduced uncertainty. However, depending on the accuracy and resolution requirements for studying bottom parameters and their acoustic effects, the modeling results indicate that adequate water-column monitoring can be important for a clear separation of bottom and volume effects in the bottom interaction experiment.

The second question was examined using the two cross front CTD sections, obtained on 18 and 24 April 200, respectively (Figure 4). The acoustic wavefields simulated in a slope-to-shelf geometry for a sound source located at the 300-m isobath reveal that 1) the optimum source depth is near the bottom where sound speed reaches its minimum (Figure 5), 2) the optimum frequency band is broad, up to at least 1 kHz, and it peaks at about 200 Hz, and 3) receiver ranges with adequate signal-to-noise ratio well exceeds 70 km. Furthermore, based on the transmission loss calculations using the two CTD sections obtained on the different days, appreciable changes of ± 15 dB due to frontal variability were found.

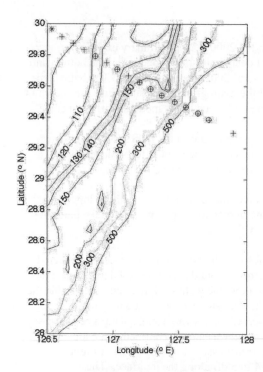

*   Yo-yo (9 April)

+   Section A (18 April)

O   Section B (24 April)

Figure 1. Bathymetry (in meters) of the ASIAEX ECS site and locations of the CTD measurements in the 2000 preliminary experiment.

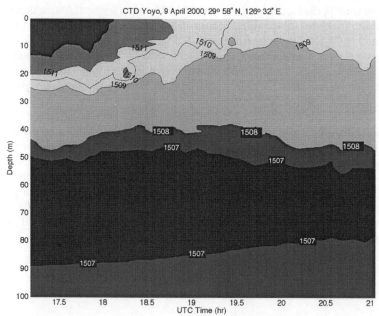

Figure 2. Time series of sound speed profile (m/s) measured by yo-yo CTD at the northwestern corner of the ASIAEX ECS site during the 2000 preliminary experiment.

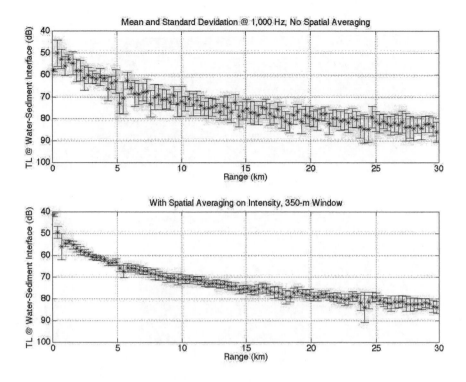

Figure 3. Mean and standard deviation of the modeled transmission loss, without (top panel) and with (bottom panel) local spatial averaging, along the water-sediment interface in the northwestern quadrant of the ECS site. The calculations were based on the measured yo-yo CTD time series and were for a frequency of 1 kHz and a nominal source depth of 20 m.

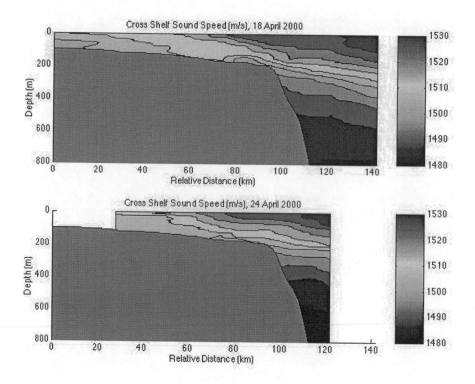

Figure 4. Measured sound speed across the Kuroshio Front. The two sections were measured on different days, 18 April (top) and 24 April (bottom), along the same path during the 2000 preliminary experiment.

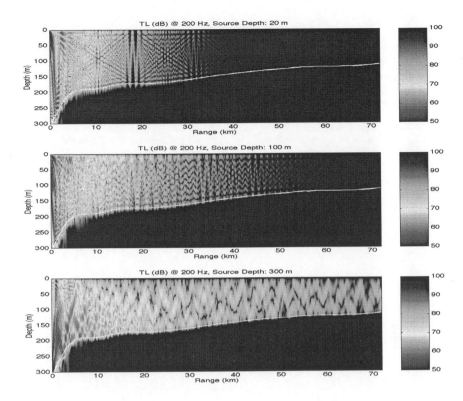

Figure 5. Slope-to-shelf transmission loss at 200 Hz for three different source depths, 20 m (top panel), 100 m (middle panel) and 300 m (bottom panel). Calculated based on a CTD section obtained in the 2000 ECS preliminary experiment, they show a general increase in the distance of sound energy penetration onto the shelf as the depth of the source increases. This dependence on source depth is common to most downward refracting sound channels.

Theoretical and Computational Acoustics 2001
E.-C. Shang, Qihu Li and T. F. Gao (Editors)
© 2002 World Scientific Publishing Co.

Extraction of Modal Back-Scattering Matrix from Reverberation Data in Shallow-Water Waveguide. Part I — Theory

E. C. Shang

MPL, Scripps Inst. Oceanography, UCSD, La Jolla, CA 92093-0238

T. F. Gao

Institute of Acoustics, Chinese Academy of Science, Beijing, China

D. J. Tang

APL. University of Washington, Seattle, WA 98105, USA

Abstract

Extracting the bottom back-scattering information from reverberation data in shallow-water waveguide is an attractive but difficult issue. In previous works, some *a priori* assumption (for instance, the Lambert's law) has been made in order to solve this problem. In this paper, new approaches are proposed. The modal back-scattering matrix can be extracted directly from reverberation data without any *a priori* assumption on scattering coefficient.

1. INTRODUCTION

Reverberation in shallow-water is the main limiting factor for active SONAR operation. It is important to understand the characteristics and the mechanisms of back-scattering in order to establish a predictable model. Scattering, generally, can be attributed to two major mechanisms : (1) roughness of various interface ; (2) volume inhomogeneities of the medium parameters - density and compressibility . Shallow-water reverberation due to roughness [1] and volume inhomogeneities [2] has been developed separately based on perturbation theory [3]. Recently, a unified approach to volume and roughness scattering is proposed [4], which provides the possibility of solving the scattering field due to both of the two mechanisms.

The key component for modeling reverberation in shallow-water is the modal back-scattering matrix Θ_{mn} (defined below by eq.(6)). To extract the Θ_{mn} from the reverberation data has been a challenging topic for a long time. There are, indeed, some work dealing with the extraction of Θ_{mn} from the reverberation data [5-8]. However, the inversions are based on either *empirical* law (Lambert's law) or an *assumption* that Θ_{mn} is separable :

$$\Theta_{mn} = \Theta_m \Theta_n$$

There are many cases where there is marked departure from the Lambert's law [9], and the assumption of the separability of Θ_{mn} [10]. In this paper, we proposed a new approach to extract Θ_{mn} from the reverberation data without any *a priori* assumption on the scattering.

2. REVERBERATION FIELD

In general, the scattered field is described by an integral equation based on Green's theorem. The formal solution of the integral equation can be obtained with the help of an iterative technique in the form of a multiple scattering series [3,4]. In this paper, the first order solution (Born approximation) is considered as the reverberation field. Both the Green's function and the incident field are represented by a sum of normal modes in a stratified waveguide. The incident field is excited by a point source at $(0, z_s)$ and the receiver is placed at $(0, z_r)$. For a limited band pulse, the insonified area is a ring with area $A = 2\pi r_c \Delta r$, where r_c is the range of the center of the insonified ring and Δr is the width of the ring. The scattered (reverberation) field is given by

$$p^s(r_c, z_r ; z_s) = [p_0 / (k_0 r_c)] \sum_m^M \sum_n^M \varphi_m(z_s) \varphi_n(z_r) \, \mathbf{S}_{mn} \int dv_1 \, \eta(R_1) \exp\{i(k_m + k_n)r_1\} \tag{1}$$

where $\varphi_m(z)$ is the modal function, k_m is the complex modal wave number $k_m = \xi_m + i \delta_m$, and η is the random parameter describing the distribution of the medium fluctuation (roughness or volume), $\mathbf{S}_{mn}$ is a scattering kernel relating the incident mode and the scattered mode, its explicit expression depends upon the scattering mechanism, for roughness it can be found in [1], and for volume scattering it can be found in [2].

3. MODAL BACK-SCATTERING MATRIX INVERSION

The back-scattering information inversion is based on the following three components : (1) mode-filtering at the receiving vertical array, (2) changing the point source depth to obtain different incident mode excitation, and (3) using mode stripping in the waveguide as an additional mode filter.

3.1 Mode-filtering of the reverberation feild

When the receiving vertical array is weighted by the j-th mode function, we have

$$p_j^s(r_c ; z_s) = \int p^s(r,z_r;z_s) \, \varphi_j(z_r) dz_r = p_0 /(k_0 r_c) \sum_m^M \varphi_m(z_s) \, \mathbf{S}_{mj} \int dv_1 \eta(R_1) \exp\{i(k_m + k_j)r_1\} \tag{2}$$

The reverberation intensity is

$$I_j = < |p_j^s(r;z_s)|^2 >,$$

$$= (p_0 / k_0 / r_c)^2 \sum_m^M \sum_n^M \varphi_m(z_s) \varphi_n^*(z_s) \, \mathbf{S}_{mj} \mathbf{S}_{nj}^* \, \mathbf{K}_{mjnj} \exp\{i(k_m + k_j - k_n^* - k_j^*)r_c\} \tag{3}$$

Where $\mathbf{K}_{mjnj}$ is the normalized spectrum of the correlation function of random parameter η:

$$\mathbf{K}_{mjnj} = (1/A) \int dv_1 \int dv_2 <\eta(r_1)\eta(r_2)> \exp\{ i(k_m + k_j)r_1 - i(k_n^* + k_j^*)r_2\} \tag{4}$$

Taking only the non-interference term (m=n) as the averaged intensity

$$I_j^{non} (r;z_s) = (I_0 / r_c) \sum_m^M \varphi_m^2(z_s) \mathbf{S}_{mj}^2 \mathbf{K}_{mj} \exp\{-2(\delta_m + \delta_j)r_c\} \tag{5}$$

3.2 Definition of the modal back-scattering matrix Θ_{mn}

In eq.(5), as we can see that the scattering is described by the product of two components - the modal scattering kernel matrix $\mathbf{S}_{mj}^2$ and the spectrum matrix $\mathbf{K}_{mj}$, the combination of them is corresponding to the conventional "back-scattering matrix" which we want to extract from the reverberation data. So, we define it as:

$$\Theta_{mj} = \mathbf{S}_{mj}^2 \mathbf{K}_{mj} \tag{6}$$

Then, the reverberation intensity is

$$I_j^{non} (r;z_s) = (I_0/r_c) \sum_m^M \varphi_m^2(z_s) \Theta_{mj} \exp\{-2(\delta_m + \delta_j)r_c \} \tag{7}$$

It has been noticed that in the Born approximation regime, $\mathbf{S}_{mj}$ is the component contributed from the physical properties of the "secondary sources" and it is more sensitive to modal angle, and $\mathbf{K}_{mj}$ is the component contributed from the random distribution of the secondary sources and it is not sensitive to the modal angle at small grazing angle but more sensitive to frequency, and the "correlation scale" as well as the type of correlation function..

3.3 More accurate time dependence

In eq. (1), the time dependence of the reverberation data was simply described by using the center range r_c of a ring with width Δr. A more accurate time dependence has been discussed in [2] and [11], for a narrow band Gaussian pulse

$$S(t) = \exp[-a^2 (t-t_0)^2] \cos(\omega_0 t) \tag{8}$$

it can be written as (eq.(45) in [2])

$$I_j^{non} (t;z_s) = (2\pi)^3 E \sum_m^M \varphi_m^2 (z_s) \Theta_{mj} A_{mj} \tag{9}$$

where

$$A_{mj} = \exp\{-2(\delta_m + \delta_j)r_{mj} \} / [r_{mj} k_m k_j D_{mj}] \tag{10}$$

$$r_{mj} = (t + t_0) / (\dot{k}_m + \dot{k}_j) \tag{11}$$

$$D_{mj} = \text{sqrt}[\ 1 + (2a^2\ \ddot{k}_m\ r_{mj}\ )^2\ ] \qquad\qquad\qquad (12)$$

$$E = \text{energy of the pulse} \qquad\qquad\qquad (13)$$

$$\dot{k}_m = (\ \partial\ k_m\ /\ \partial\ \omega)_{\omega 0} \qquad\qquad\qquad (14)$$

$$\ddot{k}_m = (\ \partial^2 k_m\ /\ \partial\omega^2)_{\omega 0} \qquad\qquad\qquad (15)$$

The dispersion term D_{mj} is near 1 for not very long distance. For simplicity, we take $D_{mj} \approx 1$, and $r_{mj} \approx r_c = (t + t_0)c_0/2$ in the following discussion of inversion.

3.4 Linear inversion by using different source depth data

Neglecting the dispersion effect, from eq.(9), we have

$$I_j^{non}\ (t;z_s) \approx I_0\ /\ r_c\ \sum_m^{M(r)}\ \varphi_m^2(z_s)\ \Theta_{mj}\exp\{\ -2(\delta_m + \delta_j)r_c\} \qquad\qquad (16)$$

In eq.(16), we have assumed that the unperturbed stratified waveguide is known, which means that Θ_{mj} is the only unknown. For convenience, we define the "effective matrix" Θ_{mj}^{eff} as follows

$$\Theta_{mj}^{eff}\ (r) =\ \Theta_{mj}\exp\{-2(\delta_m + \delta_j)r\} \qquad\qquad\qquad (17)$$

Then, eq.(16) becomes

$$I_j^{non}\ (t;z_s) \approx (\ I_0\ /\ r_c)\ \sum_m^{M(r)}\ \varphi_m^2(z_s)\ \Theta_{mj}^{eff}\ (r_c) \qquad\qquad \text{for } j=1,2,...,M \qquad (18)$$

Eq.(18) is the basic equation for inversion. As we can see from eq.(18), it is a typical linear inverse problem. The unknown number of Θ_{mj}^{eff} is $N = (M \times M)$. If, for each filtered modal reverberation intensity I_j^{non}, we change the source depth z_s M times : z_{s1}, z_{s2}, ..., z_{sM}, then we will have M x M equations for solving the N unknown Θ_{mj}^{eff}. Eq.(18) can be written as the following matrix form :

$$
\begin{matrix}
I_j^{non}\ (t;\ z_{s1}) & \varphi_1^2(z_{s1}) & \varphi_2^2(z_{s1}) & ..... & \varphi_M^2(z_{s1}) & \Theta_{1j}^{eff} \\
\\
I_j^{non}\ (t;\ z_{s2}) & \varphi_1^2(z_{s2}) & \varphi_2^2(z_{s2}) & ...... & \varphi_M^2(z_{s2}) & \Theta_{2j}^{eff} \\
\\
.... & .... & & ... & & \text{For } j=1,2,...,M \quad (19) \\
.... & .... & & ... & & \\
\\
I_j^{non}\ (t;z_{sM}) & \varphi_1^2(z_{sM}) & \varphi_2^2(z_{sM}) & ...... & \varphi_M^2(z_{sM)} & \Theta_{Mj}^{eff}
\end{matrix}
$$

with $=$ between first and second columns.

If the matrix $\Phi^2_{m\,zs}$ is not singular (has a stable inverse matrix $[\Phi^2_{m\,zs}]^{-1}$), then Θ_{mj} can be extracted from data $I_{j\,zs}$ with the help of conventional inversion procedures . A detail discussion on the singularity of the matrix $\Phi_{m\,zs}$ is given in [12].

3.5 Sequential extraction based on mode-stripping

Due to the modal attenuation in A_{mj} , in eq.(9) the higher modes will be stripped out. The upper limit of the summation is determined by an "effective mode number" $-M_{eff}(r)$. Then, eq.(9) becomes

$$I_j^{non} (t;z_s) = (2\pi)^2 \, E \, \sum_m^{M(r)} \varphi_m^2(z_s) \, \Theta_{mj} \, A_{mj} \tag{20}$$

The range dependent effective mode number is determined by the modal attenuation rate. For example, in a Pekeris waveguide, we have [13]

$$M^{eff}(r) = (\, k_0 \, H/\pi) \, \text{sqrt}\{ \, (0.7 \, H) \, / \, (Q \, r) \, \} \tag{21}$$

where H is the water-depth, Q is a parameter of bottom reflection (the slope of the reflection loss at small grazing angle). For a given frequency and bottom, we can estimate the different ranges for different M^{eff} by solving eq.(21), or by the I_j data its self . For example, if we take a Pekeris waveguide with water-depth H=50 m, a sand bottom (Q=0.4), for f=300 Hz, we have

$M^{eff}(r_1) = 1$, at range $r_1 = 35$ km ,
$M^{eff}(r_2) = 2$, at range $r_2 = 9$ km,
$M^{eff}(r_3) = 3$, at range $r_3 = 4$ km,
....

Then, the sequential inversion can be done starting with the lowest order (longest range):

$$I_1^{non}(r_1;z_s) \sim (2\pi)^3 \, E \, \varphi_1^2(z_s) \, \Theta_{11} \, A_{11}$$

$$I_j^{non}(r_2;z_s) \sim (2\pi)^3 \, E \, \sum_m^2 \varphi_m^2(z_s) \, \Theta_{mj} \, A_{mj} \qquad \text{For } j = 1, 2 \tag{22}$$

$$I_j^{non}(r_3;z_s) \sim (2\pi)^3 \, E \, \sum_m^3 \varphi_m^2 (z_s) \, \Theta_{mj} \, A_{mj} \qquad \text{For } j = 1 ,2 , 3$$

....

At range r_n we have (n x n) matrix elements to be determined. However, (n-1)x(n-1) are known from the previous range r_{n-1} , and due to the symmetry of the scattering matrix there are (n-1) more unknown number can be reduced, so the final unknown number at r_n is:

$$n^2 - (n-1)^2 - (n-1) = n$$

We have assumed that the unperturbed waveguide is known, so matrix A_{mj} is known, from eq.(22), we have n equations to determine n unknown Θ_{mj} . Note, since the needed data are only from one fixed source depth z_s , this will be very useful for higher frequency case discussed below.

3.6 Higher frequency case

In the above example, the mode number is small. For higher frequency the mode number could be very large, for example, if f=3000 Hz, then for the same waveguide as above, the mode number will be : M ~ 80, then the matrix elements will be too large ~ 6400 to invert. Actually, the 80 modes cover the same modal grazing angle interval ($0, \theta_{cr}$), where $\theta_{cr} \sim 20^0$. The interval of the modal grazing angle is now about $\Delta\theta_m \sim 0.24^0$. If we do not want to observe the Θ_{mj} matrix with such a small window length as 0.24^0, but for practice interests with a window length of $\Delta\theta_w = 2.4^0$, then the following "window smoothed" processing is proposed.

Firstly, we introduce the symbol "interval group@" -$\{m_c , L\}$ as follows:

$$\{ m_c , L \} = \text{centered at } m_c , \text{ with interval length L}$$

Instead of single mode filtering, we use the following "group filter" at the receiving vertical array :

$$\text{Group filter} = \sum_{j}^{\{jc, L\}} \varphi_j (z_s) \int \varphi_j (z_r) \, dz_r \qquad (23)$$

here $\varphi_j(z_s)$ is a weighting factor to keep the symmetry. By using the group filter given by eq.(23), eq.(2) now becomes

$$P_{\{jc,L\}}^{s} (r;z_s) = \sum_{j}^{\{jc,L\}} \varphi_j(z_s) \int p^s(r, z_r, z_s) \, \varphi_j(z_r) \, dz_r$$

$$\sim \sum_{j}^{\{jc,L\}} \varphi_j(z_s) \sum_{m}^{M} \varphi_m(z_s) \, S_{mj} \int dv_1 \eta(R_1) \exp\{i(k_m + k_j)r_1\} \qquad (24)$$

The intensity due to the non-interference part is

$$I_{\{jc,L\}}^{non}(r_c,z_s) \sim \sum_{j}^{\{jc,L\}} \varphi_j^2(z_s)\exp(-2\delta_j\,r_c) \sum_{m}^{M(rc)} \varphi_m^2(z_s)\exp(-2\delta_m\,r_c\,)\,\Theta_{mj} \qquad (25)$$

$$\sim \sum_{j}^{\{jc,L\}} \varphi_j^2(z_s)\exp(-2\delta_j\,r_c) \sum_{mc}^{M/L} \sum_{m}^{\{mc,L\}} \varphi_m^2(z_s)\exp(-2\delta_m\,r_c)\,\Theta_{mj} \qquad (26)$$

Now, let us define the "window average" as follows

$$Y_{mc}^{ave} \equiv [\,\sum_{m}^{\{mc,L\}} w_m\,Y_m\,] / \sum_{m}^{\{mc,L\}} w_m \qquad (27)$$

By using eq.(27), eq.(26) becomes

$$I_{\{jc,L\}}^{non}(r_c,z_s) \sim G_{jc}(z_s) \sum_{mc}^{M/L} G_{mc}(z_s)\,\Theta_{jc\,mc}^{ave} \qquad (28)$$

where

$$G_{jc}(z_s) = \sum_{j}^{\{jc,L\}} \varphi_j^2(z_s)\exp(-2\delta_j\,r_c\,) \qquad (29)$$

In eq.(28), we have (M/L)x(M/L) unknown **"window averaged"** back-scattering matrix - $\Theta_{jc,mc}^{ave}$. The number of equations in eq.(28) is only (M/L). Due to the $G_{mc}(z_s)$ is composed by a group summation, it seems hard to obtain new equations by changing source depth as in eq.(18). Instead to do the linear inversion, the sequential extraction, discussed in section **3.5**, might be the useful approach. In our case, f= 3000 Hz, $\theta_{cr} = 20^0$, $\Delta\theta_w = 2.4^0$, so (M/L) ~ 8, the group center of modes and the corresponding window-center of grazing angels are :

$$j_c \text{ and } m_c \;=\; 5,\quad 15,\quad 25,\quad 35,\quad 45,\quad 55,\quad 65,\quad 75$$

$$\theta_{mc} \;=\; 1.2^0,\quad 3.6^0,\quad 6.0^0,\quad 8.4^0,\quad 10.8^0,\quad 13.2^0,\quad 15.6^0,\quad 18^0,$$

SUMMARY

(1) Two approaches of inversion the modal back-scattering matrix from reverberation data in shallow-water are proposed. First, the linear inversion by using different source-depth data can be used for lower frequency case, where the effective mode number is not very much large. By setting the source-depths covering over half the water depth gives promising results [12]. It has the capability of retrieving range-dependent Θ_{mn}. Second, "sequential extraction" based on mode-stripping is proposed, which can be used for higher frequency, where the information of modal back-scattering matrix is processed by "window averaged" smoothing.

(2) The proposed approaches do not require any *a priori* assumptions on the mechanism of back-scattering, but requires the knowledge of the unperturbed (or averaged) waveguide - $\varphi_m(z)$, ξ_m, and δ_m. However, the bottom reflection loss related parameter δ_m can be obtained

in the linear inversion for lower frequency case. By using the inverted range-dependent Θ_{mm}^{eff} at two different ranges - $r_{(1)}$ and $r_{(2)}$, then the δ_m is given by

$$\Theta_{mm}^{eff}(r_{(1)}) / \Theta_{mm}^{eff}(r_{(2)}) = \exp\{4\,\delta_m\,(\,r_{(2)} - r_{(1)}\,)\,\} \tag{30}$$

Provided that $S_{mn}^2 K_{mn}$ is not range-dependent.

ACKNOWLEDGMENTS

This work was supported by ONR.

References

[1]. T. F. Gao, "Relation between waveguide and non-waveguide scattering from a rough interface," Acta Acust. **14**, No.2, pp. 126-132, 1989 (in Chinese), also see T. F. Gao and E.C. Shang, J. Acoust. Soc. Am., **94**, No.3, Pt.2, 1786, 1993.

[2]. D. J. Tang, " Shallow-water reverberation due to sediment volume inhomogeneities," (Submitted to IEEE JOE).

[3]. F.G. Bass and I.M. Fuks, *Wave Scattering from Statistical Rough Surface,* Pergamon Press, 1979.

[4]. A. N. Ivakin, "A unified approach to volume and roughness scattering," J. Acoust. Soc. Am. **103**, pp.827-837, 1998.

[5]. D.D. Ellis and P. Gerstoft, " Using inversion technique to extract bottom scattering strength And sound speed from shallow-water reverberation data," Proceedings of 3rd ECUA, Ed. by J. pappadakis, Vol.1, pp.552-562, 1996, Greece.

[6]. V.M. Kurdryashov, "Low-frequency reverberation in shallow-water Arctic Seas," Acoustical Physics, **45**, pp.320-325, 1999.

[7]. G. Jin and Renhe Zhang, "Characteristics of shallow-water reverberation and inversion for Bottom properties" Proceedings of SWAC, Ed. Zhang and Zhou, pp.303-308, 1997.

[8]. Ji-Xun Zhou and Xue-zhen Zhang, " Shallow-water acoustic reverberation and small grazing Angle bottom scattering," Proceedings of SWAC, Ed. Zhang and Zhou, pp.315-322, 1997.

[9]. P.D. Mourd and D.R. Jackson," A model/data comparison for low frequency bottom back-scatter," J. Acoust. Soc. Am., **94**, pp.344-358, 1993.

[10]. F. Li, J. Liu, and R. Zhang, " Shallow-water reverberation and the bottom scattering mechanism," Proceedings of 5th ECUA, pp.1135-1140, Ed. M. Zakharia, 2000, Lyon, France.

[11]. K. LePage, " Bottom reverberation in shallow water: Coherent properties as a function of Bandwidth, waveguide characteristics, and scatterer distribution," J. Acoust. Soc. Am., **106**, Pp.3240-3254, 1999

[12]. T.F. Gao and E.C. Shang," A discussion on the singularity of the matrix in reverberation inversion" (to be published)

[13]. D.Z. Wang and E.C. Shang, *Underwater Acoustics,*(in Chinese), p.214,Science Press 1981.

Theoretical and Computational Acoustics 2001
E.-C. Shang, Qihu Li and T. F. Gao (Editors)
© 2002 World Scientific Publishing Co.

Analysis of the Split-Step Fourier Algorithm for the Solution of Parabolic Wave Equations

F. D. Tappert

Division of Applied Marine Physics, University of Miami, RSMAS, Miami, FL 33149

K. B. Smith

Department of Physics, Naval Postgraduate School, Monterey CA 93943

M. A. Wolfson

Department of Meteorology, Penn State University, Univesity Part, PA 16802

ABSTRACT

The Split-Step Fourier (SSF) algorithm is an explicit, unconditionally stable, and efficient method for solving a class of forward propagation wave equations in ocean acoustics that are obtained by making parabolic approximations. A general motivation of the SSF algorithm is given in terms of the concepts of semigroups, operator splitting, path integrals, Trotter's product formula, and the Baker-Campbell-Hausdorff expansion. Then an explicit numerical scheme for the SSF algorithm is introduced for the boundary conditions and input data appropriate to the underwater acoustics problem, and it is shown that this scheme is consistent, unconditionally stable, and convergent. A numerical example of high frequency (30 kHz) pulse propagation in shallow water is included.

INTRODUCTION

Before about 1970, all known ocean acoustic models were based on the assumption that the lateral inhomogeneities of the ocean could be neglected along a given track or radial at fixed bearing between an acoustic source and receiver. Such "range independent" models, whether ray-based or mode-based, failed to account for certain observed acoustic propagation phenomena, such as the effects of bathymetric variations and the effects of ocean fronts and eddies. Thus the severe approximation that the ocean is perfectly stratified and variables can be separated in the acoustic wave equation had become untenable more than twenty years ago, and a need arose for "range dependent" acoustic models.

Attention was focused on propagation at a fixed bearing or azimuth because estimates show that the effects of azimuthal coupling, or "out-of-plane" effects, are normally small. Even range independent acoustic models use different ocean environments to predict propagation at different bearings, and thus have a "quasi-three-dimensional" capability. Range dependent acoustic models of course have this same capability.

The problem of developing range dependent ocean acoustic models is formidable because the propagation is unconstrained—there is no Snell's invariant or other constant of the motion that allows the solution of the wave equation to be represented in integral form. The solution must be marched out in range at fixed bearing from the source (or receiver, using reciprocity) to the variable field point. Whether the marching is done with rays, coupled modes, or directly with the acoustic field, the computations are numerically intensive. Thus robust algorithms are required that are accurate, stable and efficient.

The "parabolic equation method" was introduced by Leontovich and Fock in the 1940's to theoretically analyze long range radio wave propagation in the troposphere. Much later, this method was developed into a numerical model and applied to the ocean acoustic propagation problem.[1, 2, 3, 4, 5, 6] This PE/SSF (Parabolic Equation/Split-Step Fourier) numerical model was the first range dependent full-wave ocean acoustic model. Direct descendents of this model are now used operationally throughout the world by the U.S. Navy.[7, 8, 9]

The "split-step Fourier" algorithm,[3, 10] called the SSF algorithm in the following, has a parallel and convoluted history. It combines "spectral" methods[11, 12] for solving constant coefficient partial differential equations and "splitting" or "fractional step" methods[13, 14] for solving variable coefficient equations. The SSF algorithm was first developed for solving nonlinear wave equations. Applications of the SSF algorithm to nonlinear wave equations are given in Refs. ([15],[16],[17]).

The remainder of this paper is organized as follows. Sect. I gives the ocean acoustic forward propagation wave equation, and a class of parabolic wave equations is obtained. Boundary conditions that give a properly posed initial value problem are stated, and the semigroup of "propagators" is defined. Then in Sect. II the SSF algorithm is derived and analyzed, and unconditional stability and second order convergence are established. Practical advice concerning selection of mesh sizes in depth and range are provided. Sect. III contains a numerical example of the application of the PE/SSF model to high frequency pulse propagation in shallow water.

I. PARABOLIC WAVE EQUATION

The starting point of the underwater acoustics problem is the forward propagation wave equation[18] at fixed bearing and one frequency, also known as the one-way Helmholtz equation,

$$ik_0^{-1}\frac{\partial\Psi}{\partial x} = \hat{H}(x)\Psi. \tag{1}$$

Here x is the range variable and the range dependent pseudo-differential operator $\hat{H}(x)$ that acts on the depth variable z is given by[18]

$$\hat{H}(x) = -\sqrt{n^2(z,x) - \hat{p}^2}, \tag{2}$$

and the "momentum" operator is

$$\hat{p} = -ik_0^{-1}\partial_z. \tag{3}$$

Here c_0 is the reference sound speed, $k_0 = \omega/c_0$ is the reference wavenumber, $n(z,x) = c_0/c(z,x)$ is the acoustic index of refraction, $c(z,x)$ is the range dependent sound speed profile at fixed bearing, and the symbol ∂_z in the operator $\hat{p}$ denotes the partial derivative with respect to depth z. In this article, operators are adorned with "hats." The complex-valued full-wave acoustic pressure P at frequency $f = \omega/2\pi$ and at fixed bearing is given by the operator WKBJ expression[18]

$$P(f,z,x) = P_0(f)[-rk_0\hat{H}(x)]^{-1/2}\Psi(z,x), \tag{4}$$

where $P_0(f)$ is the Fourier frequency representation of the transmitted signal time series. The time domain analytic signal at the receiver is obtained by Fourier synthesis as[19]

$$\tilde{P}(t,z,x) = \int_0^\infty P(f,z,x)\exp(-i2\pi ft)df. \tag{5}$$

The one-way Helmholtz equation neglects backscattering, and the derivation using the operator WKBJ approximation is based on the assumption that $\hat{H}(x)$ varies slowly in range on the scale of the acoustic wavelength, or

$$\varepsilon_x = \left\| \frac{1}{k_0} \frac{\partial \ln \hat{H}}{\partial x} \right\| \ll 1. \tag{6}$$

The smallness of ε_x is important to the efficiency of the SSF algorithm for it allows a relatively large numerical range step to be used. If $\hat{H}(x)$ contains a component that varies rapidly in x on the scale of the acoustic wavelength or less, then backscattering should be explicitly included by means of the two-way PE model.[18] Another way the forward (WKBJ in range) approximation can fail is if $\hat{H}(x) = 0$, which would be a turning point in range where waves propagate vertically. Due to the large bottom absorption of steeply propagating waves, this unlikely event is not known to occur in underwater acoustics.

The operator $\hat{K}(x) = -k_0\hat{H}(x)$ may be interpreted physically as the horizontal component of wavenumber in direction x. A basic result is the conservation relation,[18, 20]

$$\int |\Psi|^2 dz = \text{const}, \tag{7}$$

which follows from the self-adjoint property of $\hat{H}(x)$. A physical interpretation[20] is that this constant is proportional to the outgoing radial component of acoustical power W_x in the small constant bearing interval Δb, since Eq.(4) gives

$$\int |\Psi|^2 dz \propto x\Delta b \int P^* \hat{K}(x) P dz \propto W_x. \tag{8}$$

The phrase "parabolic approximation" refers to the procedure of replacing the exact operator $\hat{H}(x)$ with an approximate operator $\hat{H}_{PE}(x)$ and replacing the exact forward propagation wave equation with an approximate wave equation,[18]

$$ik_0^{-1} \frac{\partial \Psi_{PE}}{\partial x} = \hat{H}_{PE}(x)\Psi_{PE}. \tag{9}$$

The primary reason for introducing such approximations is to take advantage of the efficient SSF algorithm. The general conditions that $\hat{H}_{PE}(x)$ must satisfy are that it be self-adjoint, yielding a conservation relation, and that it obey the principle of reciprocity.[19] The fact that parabolic equation models, unlike some coupled mode models, have never had difficulty with power conservation is worth stressing.

In this article, the following class of parabolic approximations is considered:

$$\hat{H}_{PE}(x) = -1 + \hat{T} + \hat{U}(x), \tag{10}$$

where the "kinetic energy" operator $\hat{T}$ depends only on $\hat{p}$ and is a constant coefficient (with respect to z) differential or pseudo-differential operator, and the "potential energy" operator $\hat{U}(x)$ is a multiplication operator (with respect to z). For example, the "standard" (STD) parabolic approximation is the following:

$$\hat{T} = \hat{T}_{STD} = \hat{p}^2/2, \tag{11}$$

$$\hat{U}(x) = \hat{U}_{\mathrm{STD}}(x) = [1 - n^2(z,x)]/2 - i\alpha(z,x)/k_0. \tag{12}$$

Here $\alpha(z,x)$, the non-negative attenuation coefficient of sound waves, has been added because it is essential for the validity of this parabolic approximation and because it is necessary to include attenuation, especially in marine sediments, for accurate predictions of acoustic wavefields in the ocean. A second example is the wide-angle Thomson-Chapman[21] approximation:

$$\hat{T} = \hat{T}_{\mathrm{TC}} = \hat{p}^2/[1 + \sqrt{1 - \hat{p}^2}], \tag{13}$$

$$\hat{U}(x) = \hat{U}_{\mathrm{TC}}(x) = 1 - n(z,x) - i\alpha(z,x)/k_0. \tag{14}$$

The recently introduced c_0-insensitive parabolic approximation,[20] that is fully second order accurate, provides a third example. It uses the same form of Hamiltonian operator as Thomson-Chapman after the depth coordinate and sound speed profile have been mapped by the "tilde" transformation.[18, 20]

Let $\Psi_{\mathrm{PE}} = \psi \exp(ik_0 x)$. Then Eq.(9) becomes the so-called parabolic wave equation,

$$ik_0^{-1}\frac{\partial \psi}{\partial x} = \hat{H}(x)\psi, \tag{15}$$

where the approximate Hamiltonian has been redefined as

$$\hat{H}(x) = \hat{T} + \hat{U}(x). \tag{16}$$

As mentioned above, this operator must vary slowly in range, and the measure of slowness is

$$\varepsilon_x = \left\| \frac{1}{k_0}\frac{\partial \hat{U}}{\partial x} \right\| /h \ll 1, \tag{17}$$

where h is an estimate of the magnitude (norm) of $\hat{H}$. At short ranges, the steep angles that are transmitted by an omnidirectional source give $h = O(1)$. At longer ranges, these steep angles are "stripped away" by lossy bottom interactions leading to $h = O(\bar{\theta}^2) = O(u) \ll 1$, where $\bar{\theta} \ll 1$ is an estimate of the rms grazing angle, $u \ll 1$ is an estimate of the variation of $U(z,x)$ in depth, and approximate equipartition of kinetic and potential energy yields $\bar{\theta}^2 \sim u$. With these estimates, in the next section the SSF algorithm for solving Eq.(15) as an initial value problem is developed and analyzed.

The ocean volume is the region in depth, $\eta_s(x) \le z \le z_b(x)$, where $\eta_s(x)$ is one realization of the random displacement of the sea surface from its mean value $z = 0$, and $z_b(x) = \bar{z}_b(x) + \eta_b(x)$, where $\bar{z}_b(x)$ is given by detailed nautical charts and $\eta_b(x)$ is one realization of the random displacement of the seafloor from its mean value $z = \bar{z}_b(x)$. The acoustic source is inside the ocean at range $x = 0$ and the acoustic receiver is also inside the ocean at range $x = x_{\mathrm{max}}$. The boundary condition at the sea surface is the "pressure release" condition, $\psi(\eta_s(x), x) = 0$. The sea floor at $z = z_b(x)$ is penetrable and cannot be handled as a boundary condition. Thus the seabed $z \ge z_b(x)$ is described as a fluid with its own physical acoustic properties, and the computational domain is extended to great depths where attenuation of the sound field causes $\lim_{z \to \infty} \psi(z,x) = 0$. The initial condition is prescribed as $\psi(z,0) = \psi_0(z)$.

The Dirichlet boundary condition at $z = \eta_s(x)$ is treated by the method of images by extending the domain to $z < \eta_s(x)$, and requiring that the solution of Eq.(15) satisfy the condition of odd symmetry about the surface at every range x,

$$\psi(-z + \eta_s(x), x) = -\psi(z - \eta_s(x), x). \tag{18}$$

Although the SSF algorithm has been extended to this rough surface scattering case,[22] in this article only the much simpler case of $\eta_s(x) = 0$ is considered. The sea surface is then at $z = 0$, and the potential U is extended to negative z by even symmetry, $U(-z, x) = U(z, x)$. The initial condition is extended by odd symmetry, $\psi_0(-z) = -\psi_0(z)$. It is then easy to show that the solution of Eq.(15), extended to the domain $-\infty < z < \infty$, has odd symmetry for all x, i.e., $\psi(-z, x) = -\psi(z, x)$. It is assumed that with these conditions, Eq.(15) is a properly posed initial value problem with a unique solution.

Let $\hat{G}(x, x')$ be the operator that propagates the solution of Eq.(15) from range x' to the range x, $x \geq x'$:

$$\psi(x) = \hat{G}(x, x')\psi(x'). \tag{19}$$

This "propagator" satisfies the operator equation,[23]

$$ik_0^{-1}\frac{\partial \hat{G}}{\partial x} = \hat{H}(x)\hat{G}, \tag{20}$$

with the initial condition at range $x = x'$ given by $\hat{G}(x', x') = \hat{I}$, where $\hat{I}$ is the identity operator. When these linear operators are represented as integrals, the kernel is called the Green's function:

$$\psi(z, x) = \int_{-\infty}^{\infty} G(z, x; z', x')\psi(z', x')dz'. \tag{21}$$

These operators form a semigroup with the property

$$\hat{G}(x, x') = \hat{G}(x, x'')\hat{G}(x'', x'), \tag{22}$$

where $x \geq x'' \geq x'$. In quantum mechanics, this relation is called the "composition" rule.[23]

II. SPLIT-STEP FOURIER ALGORITHM

The total range, x_{max}, is divided into M equal parts with range step $\Delta x = x_{max}/M$. The mesh in range is uniform: $x_m = m\Delta x$, for $m = 0, 1, \ldots, M$. The solution of Eq.(15) at range x_m is denoted by the vector (in depth) ψ_m, for $m = 0, 1, \ldots, M$. The solution at range x_m is related to the solution at the previous range by the operator (infinite dimensional matrix), $\hat{G}_m(\Delta x)$. Thus

$$\psi_m = \hat{G}_m(\Delta x)\psi_{m-1}, \tag{23}$$

for $m = 1, 2, \ldots, M$, where $\hat{G}_m(\Delta x)$ is given according to Eq.(19) by

$$\hat{G}_m(\Delta x) = \hat{G}(x_m, x_{m-1}). \tag{24}$$

Then the desired solution at the total range may be obtained in terms of the initial condition ψ_0 as the product,

$$\psi(x_{max}) = \psi_M = \prod_{m=1}^{M} \hat{G}_m(\Delta x)\psi_0. \tag{25}$$

If volume loss (the imaginary part of $\hat{U}(x)$) is neglected, then the operator $\hat{G}_m(\Delta x)$ is unitary, and the L_2 norm of the solution is preserved,

$$\|\psi\|^2 = \int |\psi(x_M)|^2 dz = \int |\psi(x_0)|^2 dz, \tag{26}$$

which physically expresses the conservation of power. If loss is included, then $\hat{G}_m(\Delta x)$ is a "contraction" operator and the norm of the solution is decreased.

The numerical analysis problem is to devise an efficient algorithm for approximating $\hat{G}_m(\Delta x)$ for small Δx, to estimate the accuracy of the approximation, to show stability, and to show convergence of the product in Eq.(25) as $\Delta x \to 0$. Also a good approximation will maintain the unitary property of $\hat{G}_m(\Delta x)$ when losses are set to zero. The SSF algorithm provides an explicit unitary approximation to $\hat{G}_m(\Delta x)$ that allows recursive advancement of the solution in accordance with Eq.(23) with the computations being done "in place," thereby saving core storage. It is further shown below that the SSF algorithm is unconditionally stable and pointwise convergent for smooth $U(z, x)$.

An integral equation for $\hat{G}_m(\Delta x)$ that may be obtained from Eq.(20) is the following:

$$\hat{G}_m(\Delta x) = \hat{I} - ik_0 \int_{x_{m-1}}^{x_m} \hat{H}(x)\hat{G}(x, x_{m-1}) dx. \tag{27}$$

Iteration of this equation and formal summation yields the formula[23]

$$\hat{G}_m(\Delta x) = \hat{D} \exp(-ik_0 \Delta x \overline{H}_m), \tag{28}$$

where the average Hamiltonian is

$$\overline{H}_m = \int_{x_{m-1}}^{x_m} \hat{H}(x) dx / \Delta x, \tag{29}$$

and $\hat{D}$ is the Dyson operator that disentangles the power series expansion of $\exp(-ik_0\Delta x \overline{H}_m)$ into range-ordered form. This complication arises from the range dependence of $\hat{H}(x)$. Application of the midpoint rule and the integral mean value theorem to Eq.(29) yields

$$\overline{H}_m = \hat{H}_m + O(k_0^2 \Delta x^2 h \varepsilon_x^2), \tag{30}$$

where ε_x is defined in Eq.(17), and here we define

$$\hat{H}_m = \hat{T} + \hat{U}_m, \tag{31}$$

and the potential operator $\hat{U}_m$ is evaluated at the midpoint in range,

$$\hat{U}_m = \hat{U}(x_{m-1} + \Delta x/2). \tag{32}$$

The endpoint (trapezoidal) rule would do just as well as the midpoint rule, and has been used in the past.

Next, it can be shown that the effect of the ordering operator $\hat{D}$ begins at third order in Δx for small Δx. An estimate gives

$$\hat{G}_m(\Delta x) = \hat{D} \exp(-ik_0\Delta x \overline{H}_m) = \exp(-ik_0\Delta x \overline{H}_m)[1 + O(\delta_x)], \tag{33}$$

where

$$\delta_x = k_0^3 \Delta x^3 h^2 \varepsilon_x^2. \tag{34}$$

Combining this with Eq.(30) then shows that the the estimated remainder of both of these approximations is the same, and we obtain

$$\hat{G}_m(\Delta x) = \hat{E}_m(\Delta x)[1 + O(\delta_x)], \tag{35}$$

where

$$\hat{E}_m(\Delta x) = \exp(-ik_0 \Delta x \hat{H}_m) = \exp\left(-ik_0 \Delta x (\hat{T} + \hat{U}_m)\right). \tag{36}$$

The fact that the remainder term, $O(\delta_x)$, is third order in Δx is one order better than is needed to show convergence. For a range independent problem ($\varepsilon_x = 0$), we have $\hat{G}_m(\Delta x) = \hat{E}_m(\Delta x)$ and no approximation has been made. As stressed in the Introduction, however, the range independent problem has no practical importance in modern ocean acoustics, and the PE/SSF model was designed from the beginning to crunch the hard problem of range dependent propagation.

We now turn to the central subject of operator splitting. In Eq.(36) the operator $\hat{T}$ is diagonal in momentum (Fourier) space and the operator $\hat{U}_m$ is diagonal in coordinate (depth) space. The main idea of the SSF algorithm is to apply each operator in the space where it is diagonal, and to maintain the operators in exponential form yielding exact unitarity and infinite order accuracy. Since $\hat{T}$ does not in general commute with $\hat{U}_m$, an approximation is necessary. The Baker-Campbell-Hausdorff formula[25, 26] may be used to show that

$$\exp\left(-ik_0 \Delta x (\hat{T} + \hat{U}_m)\right) = \exp(-ik_0 \Delta x \hat{T}) \exp(-ik_0 \Delta x \hat{U}_m)[1 + O(k_0^2 \Delta x^2 C)], \tag{37}$$

where the commutator $\hat{C}$ is defined by $\hat{C} = \hat{T}\hat{U}_m - \hat{U}_m\hat{T}$, and its magnitude may be estimated as $C = O(h\varepsilon_z)$, where again h is the estimated magnitude of $\hat{H}_m = \hat{T} + \hat{U}_m$, and ε_z is a measure of how rapidly $\hat{U}_m$ varies in the depth variable z:

$$\varepsilon_z = \left\| \frac{1}{k_0} \frac{\partial \ln \hat{H}_m}{\partial z} \right\| \approx \left\| \frac{1}{k_0} \frac{\partial U}{\partial z} \right\| / h. \tag{38}$$

Within the water column, ε_z is naturally small compared to unity and this makes C quite small even for steep angles, $h = O(1)$. It is worth noting that for the isovelocity problem, $\partial U / \partial z = 0$, then $C = 0$ and the approximation given in Eq.(37) is exact. In either case, the range step Δx can be chosen so large that the solution changes by a large amount in one step. This is not so for finite difference algorithms in range, and thus the SSF algorithm is much more efficient for an important class of ocean acoustic problems.

At the water-sediment interface, however, $U(z, x)$ may have a jump discontinuity. This requires remedial action in order to make the SSF algorithm efficient for problems that involve sound reflection from the sea floor. One of the "trade secrets" of the PE/SSF model that was published in Ref. [24] is that jumps in the potential function $U(z, x)$ are smoothed in depth by a "mixing" function over the distance of a wavelength or so. This smoothing procedure, if done carefully, does not alter the physical properties of partial reflection and transmission at penetrable boundaries, and can speed up computer runs with the PE/SSF model by a factor of up to ten. The examples published in Refs. [4, 18] were done with this smoothing technique. With smoothing in depth, we obtain $\varepsilon_z = O(1)$, at most. Then to make the SSF error per step small, the range step Δx should be selected to satisfy

$$k_0 \Delta x \overset{<}{\sim} 1/\sqrt{h}. \tag{39}$$

For small grazing angles, $h \ll 1$, the range step can be much larger than the acoustic wavelength; for steep angles, $h = O(1)$, then the range step is the same order as the acoustic wavelength.

A symmetric version[3] of the Baker-Campbell-Hausdorff formula, believed to be novel in this context, yields the second order approximation

$$\hat{E}_m(\Delta x) = \hat{g}_m(\Delta x)[1 + O(k_0^3 \Delta x^3 C^2)], \tag{40}$$

where

$$\hat{g}_m(\Delta x) = \exp(-ik_0 \Delta x \hat{T}/2) \exp(-ik_0 \Delta x \hat{U}_m) \exp(-ik_0 \Delta x \hat{T}/2). \tag{41}$$

Using the above estimate of C, we obtain

$$\hat{E}_m(\Delta x) = \hat{g}_m(\Delta x)[1 + O(\delta_z)], \tag{42}$$

where

$$\delta_z = k_0^3 \Delta x^3 h^2 \varepsilon_z^2. \tag{43}$$

Combining this with Eq.(35), we obtain

$$\hat{G}_m(\Delta x) = \hat{g}_m(\Delta x)[1 + O(\delta)], \tag{44}$$

where

$$\delta = \delta_x + \delta_z = k_0^3 \Delta x^3 h^2 (\varepsilon_x^2 + \varepsilon_z^2). \tag{45}$$

The SSF algorithm uses the approximation $\hat{G}_m(\Delta x) \approx \hat{g}_m(\Delta x)$, and the estimated error per range step is $O(\delta)$. With this approximation, the solution is advanced in range according to

$$\psi_m(\text{SSF}) = \hat{g}_m(\Delta x)\psi_{m-1}(\text{SSF}). \tag{46}$$

To estimate the global error after M range steps, the SSF algorithm yields

$$\psi_M(\text{SSF}) = \prod_{m=1}^{M} \hat{g}_m(\Delta x)\psi_0. \tag{47}$$

The exact solution is given by Eq.(23), and with the estimate in Eq.(44) it follows that

$$\psi_M = \prod_{m=1}^{M} \hat{g}_m(\Delta x)[1 + O(\delta)]\psi_0 = \psi_M(\text{SSF})[1 + O(M\delta)]. \tag{48}$$

Since $M = x_{\max}/\Delta x$, we obtain the pointwise estimate

$$\psi_M - \psi_M(\text{SSF}) = O(\delta_g)\psi_M(\text{SSF}), \tag{49}$$

where the global error due to discrete range step Δx is

$$\delta_g = x_{\max} k_0^3 \Delta x^2 h^2 (\varepsilon_x^2 + \varepsilon_z^2). \tag{50}$$

Letting $\Delta x \to 0$ for fixed $x_{\max}$, we obtain second order convergence of $\psi_M(\text{SSF})$ to ψ_M. A mathematical proof of convergence to first order is the content of the "Trotter product formula."[25, 27, 28] The SSF algorithm may also be viewed as a numerical implementation of a Feynman path integral, as discussed by Nelson.[29]

Implementation of the SSF algorithm does not use the operator $\hat{g}_m(\Delta x)$ as in Eq.(47), but instead uses the equivalent formula

$$\psi_M(\text{SSF}) = \exp(ik_0 \Delta x \hat{T}/2) \prod_{m=1}^{M} \hat{R}_m(\Delta x) \exp(-ik_0 \Delta x \hat{T}/2)\psi_0, \tag{51}$$

where $\hat{R}_m(\Delta x)$ is the asymmetric operator indicated in Eq.(37):

$$\hat{R}_m(\Delta x) = \exp(-ik_0 \Delta x \hat{T}) \exp(-ik_0 \Delta x \hat{U}_m). \tag{52}$$

The solution is computed recursively according to

$$\psi'_m(\text{SSF}) = \hat{R}_m(\Delta x)\psi'_{m-1}(\text{SSF}), \quad m = 1, 2, \ldots, M, \tag{53}$$

where $\psi'_0(\text{SSF}) = \exp(-ik_0 \Delta x \hat{T}/2)\psi_0$. Then the desired solution is

$$\psi_M(\text{SSF}) = \exp(ik_0 \Delta x \hat{T}/2)\psi'_M(\text{SSF}). \tag{54}$$

Although the SSF algorithm updates the solution in range using the operator in Eq.(52) that is only first order accurate, the final solution actually has second order accuracy at almost no cost.

The depth variable z is discretized by first truncating to $z_{\max}$ and then introducing N mesh points equally spaced in the interval $-z_{\max} \leq z \leq z_{\max}$. The mesh points are at z_j and the mesh spacing is $\Delta z = 2z_{\max}/N$. Then the SSF algorithm given in Eq.(53) becomes

$$\psi'_{j,m}(\text{SSF}) = \sum_{j'=1}^{N} R_{jj',m}(\Delta x)\psi'_{j',m-1}(\text{SSF}), \tag{55}$$

where the $N \times N$ matrix at range m, $R_{jj',m}(\Delta x)$, is the discrete representation of the operator $\hat{R}_m(\Delta x)$ specified in Eq.(52). Let $U_{j,m} = U(z_j, x_{m-1} + \Delta x/2)$ be the discrete representation of the multiplication operator $\hat{U}_m$ defined in Eq.(32). Then $V_{j,m}(\Delta x) = \exp(-ik_0 \Delta x U_{j,m})$ is the diagonal matrix, or N component vector, that is the discrete representation of the operator $\exp(-ik_0 \Delta x \hat{U}_m)$. If the "mixing length" L is the scale length in depth over which the function $U(z, x)$ is smooth, then $\Delta z \ll L$ makes $V_{j,m}(\Delta x)$ an infinite order approximation to the corresponding operator.

From Eq.(52) we then obtain

$$R_{jj',m}(\Delta x) = T_{jj'}(\Delta x)V_{j',m}(\Delta x), \tag{56}$$

where $T_{jj'}(\Delta x)$ is the discrete representation of the operator $\exp(-ik_0\Delta x \hat{T})$. Since $\hat{T}$ is a constant coefficient differential operator, it is diagonal in Fourier, or "momentum," space. Let F_{kj} be the unitary $N \times N$ matrix that performs an N point discrete Fourier transform (DFT) from z_j to $p_k = k\Delta p$, $k = -N/2+1, \ldots, N/2$, where $\Delta p = \pi/k_0 z_{max}$. The maximum grazing angle that can be represented on the mesh with given Δz is therefore $\sin(\theta_{max}) = p_{max} = \pi/k_0\Delta z$. In momentum space, $\hat{T}$ is the N-dimensional vector T_k, $k = -N/2 + 1, \ldots, N/2$. Thus $T_{jj'}(\Delta x)$ is the unitary $N \times N$ matrix,

$$T_{jj'}(\Delta x) = \sum_{k=-N/2+1}^{N/2} F_{jk}^{-1} \exp(-ik_0\Delta x T_k) F_{kj'}, \tag{57}$$

that has norm $\|T(\Delta x)\| = 1$. This is an infinite order approximation to the corresponding operator because each angle that is sampled is propagated "exactly." If θ_{max} is the maximum grazing angle in the problem, then the above relation shows that the depth mesh should be chosen to satisfy

$$\Delta z \le \pi/k_0 \sin(\theta_{max}). \tag{58}$$

This also follows from the Nyquist "sampling theorem."

Thus the matrix $R_m(\Delta x)$, defined by Eq.(56), that advances the SSF solution one step in range according to Eq.(53) satisfies

$$\|R_m(\Delta x)\| \le 1, \tag{59}$$

equality holding if losses are neglected. This shows that $\|\psi_m(\text{SSF})\| \le \|\psi_{m-1}(\text{SSF})\|$. It follows that the SSF algorithm is unconditionally stable, for any Δx and Δz. Convergence of the SSF algorithm then follows from the Lax equivalence theorem[11, 30] that is stated in Ref. [31] as "The combination of consistency and stability is equivalent to convergence."

Although this proof of convergence of the SSF algorithm is reassuring, in practice one wants conditions on Δx and Δz that provide adequate accuracy of the approximate solution $\psi_M(\text{SSF})$ without testing for convergence. We recommend use of Eq.(39) for Δx and Eq.(58) for Δz, with reasonable estimates of h and θ_{max}. These estimates can be computed internally as the run proceeds, and then the mesh sizes Δx and Δz are dynamically adapted as a function of range.[32, 33]

III. NUMERICAL EXAMPLE

A difficult modeling problem in ocean acoustics is high frequency pulse propagation in a range dependent shallow water environment with a relatively hard reflecting bottom, such as a sandy seabed. Fig. 1 shows the solution to this kind of problem obtained with a PE/SSF model called the UMPE (University of Miami Parabolic Equation) model.[20, 34] An omnidirectional source at depth 80 m transmits a pulse at center frequency 30 kHz and bandwidth 5 kHz, giving about 0.2 ms pulse length. This signal is received at range $x_{max} = 750$ m as a sequence of short pulses due to multipath propagation conditions as shown in the wavefront plots in Fig. 1.

The water depth is 120 m, which is about 2 400 wavelengths at $\lambda_0 = 5$ cm corresponding to the center frequency. The final range, $x_{max} = 750$ m, is about 15 000 wavelengths. The

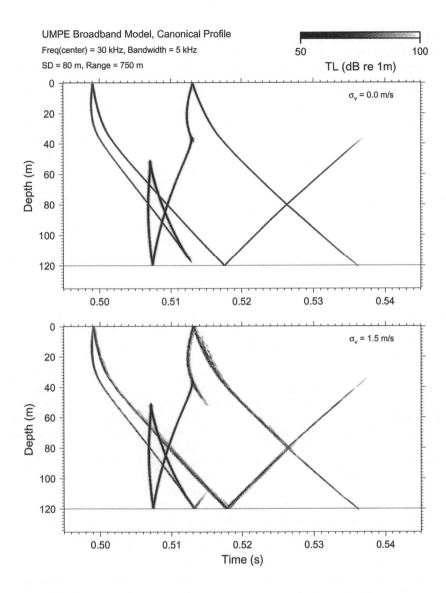

Figure 1: Wavefront plots of high frequency propagation in shallow water with a flat sandy bottom at the depth of 120 m. The upper panel shows range independent propagation, and the lower panel shows the effects of range dependent propagation through microscale turbulence in the ocean mixed layer.

computational depth is $z_{max} = 180$ m, and the sediment thickness is 15 m, ending at depth 135 m where a fully absorbing substrate is located.

The mean sound speed profile is a canonical downward refracting summertime profile, with an isothermal layer (called the "mixed" layer by oceanographers) having temperature $T_s = 23\,°C$ between the surface at $z = 0$ and about 25 m, a thermocline centered at $z = 35$ m that extends from about 25 m to 45 m with a smooth transition to another isothermal layer having temperature $T_b = 7\,°C$ below the thermocline. Range dependence is modeled by microscale turbulence in the mixed layer near the surface (extending downward to about 40 m) having vertical wavelength 4 m, horizontal wavelength 20 m, and rms sound speed fluctuation of $\sigma_v = 1.5$ m/s. Setting $\sigma_v = 0.0$ gives a range independent problem.

The 15 m thick sediment layer is modeled as follows. The sound speed at the top of the sediment layer is 1.08 times the sound speed at the bottom of the water column, yielding a critical grazing angle at the water-sediment interface of $\theta_c = 22.1°$. The density of the sediment layer is constant at 1.4 times water density.[18, 24] The sound speed in the sediment increases linearly with depth at the rate 1.2 m/s/m. The volume absorption coefficient of the sediment is constant at 0.04 dB/kHz-m, which yields 1.2 dB/m at the center frequency of 30 kHz. This attenuation, although small, is sufficient to make head wave propagation quite negligible. These physical properties of the marine sediment layer are typical of acoustically hard, sandy types of shallow water sediments.

Two computer runs are made ($\sigma_v = 0.0$ and $\sigma_v = 1.5$ m/s) with $N = 2^{14} = 16\,384$ points in z giving $\Delta z = 2.2$ cm. Since this is less than half a wavelength ($\lambda_0 = 5$ cm), the maximum grazing angle allowed by this mesh is greater than 90°. The total number of range steps is $M = 1\,000$, giving $\Delta x = 75$ cm $\approx 15\lambda_0$. This is adequate for this problem because bottom interactions strip away most of the energy having grazing angles greater than the critical angle, $\theta_c \approx 22°$, at the final range $x_{max} = M\Delta x = 750$ m. The computational bandwidth is $W = 10$ kHz, with a Hann window in frequency, and $N_f = 512$ equally spaced frequencies are used. This yields a time window having duration $N_f/W = 51.2$ ms, and with zero padding by a factor of two the time increment is $\Delta t = 0.5/W = 0.05$ ms. The CPU time required for each run is about 8 hr on a desktop workstation (DEC Alpha).

The result of the range independent run ($\sigma_v = 0.0$) is displayed in the upper panel of Fig. 1. There are two surface reflections and two bottom reflections, with a third bottom interaction that is very lossy due to the incident grazing angle being larger than the critical angle. A classical "triplication" is seen between 0.507 s and 0.513 s below 50 m that is due to downward refraction of direct path waves by the thermocline. A later, nascent triplication is seen near time 0.513 s at depth about 40 m that is due to downward refraction by the thermocline of upgoing waves that are reflected once from the bottom.

The result of the range dependent run ($\sigma_v = 1.5$) is displayed in the lower panel of Fig. 1. Multiple small angle forward scattering by the microscale turbulence in the mixed layer above about 35 m causes spreads in depth and time of those waves that have traversed the mixed layer. The earlier triplication now extends to the seafloor where it is reflected. The later, nascent triplication is now extended in depth by more than 10 m. The time spread of some portions of the wavefront exceeds 1 ms which is five times the transmitted pulse length. This computed effect has obvious implications for high frequency acoustic imaging in the ocean. Chaos theory says that this time spread proceeds by the process of "stretching and folding" of the wavefront. No doubt there are a large number of "micro-triplications" in the spread portions of the wavefront that are unresolved by the 5 kHz bandwidth. This phenomenon is called "wave chaos," a finite frequency manifestation of "ray chaos" that is a well-known phenomenon in ocean acoustics.

ACKNOWLEDGMENTS

This research was partially supported by the Office of Naval Research, Code 3210A. An ocean acoustic PE/SSF model, called UMPE, Version 1.3, is publicly available by anonymous ftp. For instructions on downloading the source code and users guide,[34] send email to the second author at 'kevin@physics.nps.navy.mil'. An upgraded model, Version 2.0, will soon be made available.

References

[1] F. D. Tappert and R. H. Hardin, in "A Synopsis of the AESD Workshop on Acoustic Propagation Modeling by Non-Ray Techniques," C. W. Spofford, AESD Tech. Note **73-05**, Arlington, VA, 1973.

[2] F. D. Tappert, "Parabolic equation method in underwater acoustics," J. Acoust. Soc. Am. Suppl. **55**, S34 (A) (1974).

[3] F. D. Tappert and R. H. Hardin, "Computer simulation of long-range ocean acoustic propagation using the parabolic equation method," in *Proc. 8th Intern. Cong. on Acoustics* (Goldcrest, London, 1974), Vol. 2, p. 452.

[4] F. D. Tappert, "Selected applications of the parabolic-equation method in underwater acoustics," in *International Workshop on Low-Frequency Propagation and Noise*, Woods Hole, Massachusetts, 14–19 October, 1974 (Department of the Navy, Washington, D. C.,1977), Vol. 1, pp. 155–194.

[5] F. D. Tappert, "The parabolic approximation method," Workshop on Wave Propagation and Underwater Acoustics, Mystic, Connecticut, 19–21 November 1974.

[6] S. M. Flatté and F. D. Tappert, "Calculation of the effect of internal waves on oceanic sound transmission," J. Acoust. Soc. Am. **58**, 1151–1159 (1975).

[7] J. A. Davis, D. White, and R. C. Cavanagh, Editors, "NORDA parabolic equation workshop," Report **TN-143**, Naval Ocean Research and Development Activity, SSC, MS, 1982).

[8] F. D. Tappert, "Computation intensive propagation modeling using the PESOGEN system," J. Acoust. Soc. Am. Suppl. **76**, S37 (A) (1984).

[9] S. A. Chin-Bing, D. B. King, J. A. Davis, and R. B. Evans, *PE WORKSHOP II: Proceedings of the Second Parabolic Equation Workshop*, (Naval Research Laboratory, Stennis Space Center, 1993).

[10] R. H. Hardin and F. D. Tappert, "Applications of the split-step Fourier method to the numerical solution of nonlinear and variable coefficient wave equations," SIAM Rev. **15**, 423 (1973).

[11] R. D. Richtmyer and K. W. Morton, *Difference Methods for Initial-Value Problems*, second edition (Interscience Publishers, New York, 1967).

[12] S. A. Orszag, "Comparison of pseudospectral and spectral approximations," Stud. in Appl. Math. **51**, 253–259 (1972).

[13] G. Strang, "Approximating semigroups and the consistency of difference schemes," Proc. Amer. Math. Soc. **20**, 1–7 (1969).

[14] G. I. Marchuk, *Methods of Numerical Mathematics*, translated by J. Ružička (Springer-Verlag, New York, 1975).

[15] F. D. Tappert and C. N. Judice, "Recurrence of nonlinear ion acoustic waves," Phys. Rev. Lett. **29**, 1308–1311 (1972).

[16] A. Hasegawa and F. D. Tappert, "Transmission of stationary nonlinear optical pulses in dispersive dielectric fibers. I. Anomalous dispersion," Appl. Phys. Lett. **23**, 142–144 (1973).

[17] A. Hasegawa and F. D. Tappert, "Transmission of stationary nonlinear optical pulses in dispersive dielectric fibers. II. Normal dispersion," Appl. Phys. Lett. **23**, 171–172 (1973).

[18] F. D. Tappert, "The parabolic approximation method," in *Wave Propagation and Underwater Acoustics*, Lecture Notes in Physics, Vol. 70, edited by J. B. Keller and J. S. Papadakis (Springer-Verlag, New York, 1977), Chap. V, pp. 224–287.

[19] L. Nghiem-Phu and F. D. Tappert, "Modeling of reciprocity in the time domain using the parabolic equation method," J. Acoust. Soc. Am. **78**, 164–171 (1985).

[20] F. D. Tappert, J. L. Spiesberger, and L. Boden, "New full-wave approximation for ocean acoustic travel time predictions," J. Acoust. Soc. Am. **97**, 2771–2782 (1995).

[21] D. J. Thomson and N. R. Chapman, "A wide angle split step algorithm for the parabolic equation," J. Acoust. Soc. Am. **74**, 1848–1854 (1983).

[22] F. D. Tappert and L. Nghiem-Phu, "A new split-step Fourier algorithm," J. Acoust. Soc. Am. Suppl. **77**, S101 (A) (1985).

[23] J. J. Sakurai, *Modern Quantum Mechanics* (Benjamin-Cummings, Menlo Park, California, 1985).

[24] F. D. Tappert and L. Nghiem-Phu, "Modeling of pulse response functions of bottom interacting sound using the parabolic equation method," in *Ocean Seismo-Acoustics*, Ed. T. Akal and J. M. Berkson (Plenum Pub. Corp., New York, 1986), pp. 129-137.

[25] B. Simon, *Functional Integration and Quantum Physics* (Academic Press, New York, 1979).

[26] V. S. Varadarjan, *Lie Groups, Lie Algebras, and their Representations* (Springer-Verlag, New York, 1984), Sect. 2.15.

[27] H. F. Trotter, "Approximation of semigroups of operators," Pacific J. Math. **8**, 887–919 (1958).

[28] H. F. Trotter, "On the product of semigroups of operators," Proc. Amer. Math. Soc. **10**, 545–551 (1959).

[29] E. Nelson, "Feynman integrals and the Schrödinger equation," J. Math. Phys. **5**, 332–343 (1964).

[30] P. D. Lax and R. D. Richtmeyer, "Survey of the stability of linear finite difference equations," Comm. Pure Appl. Math. **9**, 267–293 (1956).

[31] G. Strang, *Introduction to Applied Mathematics* (Wellesley-Cambridge Press, Wellesley, Massachusetts, 1986), pp. 574–576.

[32] R. N. Buchal and F. D. Tappert, "A Variable Range Step in the Split-Step Fourier Algorithm," AESD Tech. Memo, November, 1975.

[33] F. D. Tappert, "High angle PE and travel time modifications," SYNTEK Engineering and Computer Systems, Inc., Rockville, MD 20852, Progress Report on Project 501M54, September, 1988.

[34] Kevin B. Smith and Frederick D. Tappert, "UMPE: The University of Miami Parabolic Equation Model, Version 1.0," MPL Technical Memorandum **432**, May 1993.

Theoretical and Computational Acoustics 2001
E.-C. Shang, Qihu Li and T. F. Gao (Editors)
© 2002 World Scientific Publishing Co.

Recent Progresses on Shallow Water Reverbervation

Renhe Zhang and Fenghua Li

National Laboratory of Acoustics, Chinese Academy of Science, Beijing 100080, China

In shallow water, the reverberation is often the most severe limiting factor in relation to the use of active sonar systems for

target detection, and it is of great interest to underwater acousticians. In this paper, some recent progresses on the

experiment and theory of shallow water reverberation are presented. The data were obtained from several experiments at

different sites in different seasons. Some of the progresses include: (1) Ray-Mode reverberation model, which can be used

to calculate the reverberation intensity; (2) Coherent reverberation model, which was developed to explain the observed

oscillation phenomenon of the reverberation loss; (3) The bottom inversion from the reverberation; (4) Theoretical and

experimental results of the spatial correlation of the reverberation, (5) Experimental results of the direction of the

reverberation, etc.

Nomenclature:

ω	frequency
$D_i(\alpha_m, \alpha_n)$	scattering directivity of the ith bottom scattering element
N	bottom scattering element number per unit area
$<>$	statistical average
$k(z)$	wavenumber, $k(z) = \omega / c(z)$
v_m	complex eigenvalue of normal mode, i.e., $v_m = \mu_m + i\beta_m$,
β_m	mode attenuation
S_m	cycle-distance of the eigenray
α_m	grazing angle of the mth normal mode, which is determined by $k(h)\cos\alpha_m = \mu_m$
$\sigma(\alpha_m, \alpha_n)$	bottom scattering coefficient, $\sigma(\alpha_m, \alpha_n) = N\langle D_i(\alpha_m, \alpha_n)\rangle$
Lambert scattering model	$\sigma(\alpha_m, \alpha_n) = \mu \sin\alpha_m \sin\alpha_n$
superscript star (*)	complex conjugate.
E_0	energy flux density of the transmitting signal at unit distance
p_i	sound pressure scattered from the ith bottom scattering element
τ	time duration of source signal
$p(r, z_s, z)$	reverberation time series
z_r	depth of the receiver
z_s	depth of the source
r	range between the receiver and the scatterer,
h	water depth

d	the slope of $k^2(z)$, $d = dk^2(z)/dz$
$q_m(z)$	a factor derived from WKBZ[1] theory, $q_m(z) = \left[0.875d^{2/3} + k^2(z) - \mu_m^2\right]^{-1/4}$
θ_m	determined by the surface reflection phase, Ref.[3]
$\Psi(z)$	phase intergerate in the waveguide, $\Psi_m(z) = \int_{\zeta_m}^z \sqrt{k^2(y) - v_m^2}\, dy + \theta_m$
$\Theta_m(z,h)$	$\Theta_m(z,h) = \left(\sqrt{\dfrac{8\pi}{r}} \dfrac{\sqrt{\mu_m}\, q_m(z) q_m(h)}{S_m} \sin[\Psi_m(z)] e^{-\beta_n r}\right) e^{i[\varphi_m(h)]}$
$T_{m,n}(r,\varphi)$	a function of the mode coupling coefficient that couples the field of nth mode with the field of mth mode in the direction of φ in a range-dependent environment
$\phi_m(z_s)$	eigenfunction at the position $(0, z_s)$
$\phi_n(r,z)$	eigenfunction at the position (r,z)
V	bottom reflection coefficient
Q	$-\ln\lvert V(\alpha)\rvert = Q\alpha$ for $0 \le \alpha \le \alpha_c$.

1. Introduction

The predominant background interference of the active sonar in shallow water is often bottom reverberation. In recent years, the shallow water reverberation has been paid more and more attentions. In this paper, some recent progresses on the shallow water reverberation were presented.

In the past years, several reverberation models based on ray [1], PE [2], and normal mode [3,4] theories have been developed. Those models mainly treat the propagation from the source to scatterers and from scatterers to the receiver. The reverberation model based on normal mode, combined with ray-mode analogies, is one of the most practical models. The basic idea of this Ray-mode reverberation model was first introduced by Bucker and Morris, further developed by Zhang and Jin [3], and reviewed by Ellis [4]. The numerical calculations and theoretical analyses in this paper are mainly based on this Ray-mode reverberation model. The brief discussions about this reverberation model were discussed in section 2.

Two extensions of the Ray-mode Reverberation were developed in this paper. One is to extend the Ray-mode Reverberation theory to coherent reverberation model [5]. The primary object of this theoretical work is to explain the observed oscillation phenomenon of the reverberation intensity in '96 China-US Yellow Sea Experiment, which cannot be explained by the incoherent ray-mode reverberation theory.

Another approach is to extend the Ray-mode reverberation theory to Nx2-D reverberation model. The Ray-mode can provide good prediction to many experimental data in range-independent shallow water, however, many ocean environment have considerable space variability. Full simulation of reverberation in the range-dependent environment involves not only the local scattering processes, but also the accurate propagation model. Several 2D reverberation codes have been developed in last few years. In this paper, a Nx2-D reverberation model was developed by combining the coupled-mode parabolic-equation [6] (CMPE) propagation code for a range-dependent environment and the Ray-mode reverberation model. Some numerical and experimental results of the reverberation directionality were also presented.

Besides the developments of the reverberation model, other approaches about reverberation, including spatial correlation, mode filtering, Geoacoustic inversion, were also presented in this paper. Measurement of correlation of reverberation [7,8,9] at spatially separated points can provide useful information for the active sonar. In this paper, based on the ray-mode reverberation theory, a concise analytical expression was derived. The experimental data show that this expression can predict the experimental data well.

Mode filtering is an important approach for the research of mode propagation and scattering in shallow water. Some papers [10,11] about mode filtering from sound propagation have been published in the last few decades. However, few results about the mode filtering from reverberation have been presented. In this paper, a novel mode filtering method from reverberation was introduced to estimate the normal mode attenuation and bottom scattering index at small grazing angles. The results show that the bottom attenuations obtained from measured reverberation data and propagation data are in good agreement with each other.

The bottom parameters often play important roles in the shallow water sound propagation. However, direct measurement of these parameters is usually difficult, particularly for lower frequencies and low grazing angles. The Geoacoustic properties of the seabed are of interest to underwater acousticians for many years. Many inversion methods[12-23] have been developed. The inversion method from reverberation discussed in this paper has the advantages of: 1) The experiment can be easy conducted; 2) The bottom scattering coefficients can be obtained at the same time.

2. Ray-Mode Reverberation Model [3]

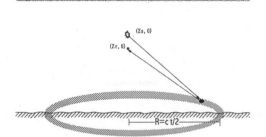

Fig. 1. Schematic illustrating reverberation scattered from the bottom.

Figure 1 illustrates the reverberation generated by the bottom scattering. The reverberation pressure at the receiver $(z_r, 0)$ is often calculated by the summation of the sound pressures scattered from the bottom scattering elements over the area insonified by the incident pulse,

$$p(t) = \sum_i p_i \,,$$

(2.1)

Omitting the cross terms, the received reverberation intensity can be written as

$$I(t) = p(t)p^*(t)$$
$$= \sum_i |p_i|^2$$

(2.2)

The backscattering pressure p_i can be expressed as [3,24]:

$$p_i = \sum_m \sum_n \Theta_m(z_s, h) e^{i\mu_m r} \Theta_n(z_r, h) e^{i\mu_n r} D_i(\alpha_m, \alpha_n),$$ (2.3)

The incoherent summation can be written as,

$$I_{inc} = E_0 \pi r c \sum_m \sum_n |\Theta_m(z_s, z_b)|^2 |\Theta_n(z_r, z_b)|^2 \sigma(\alpha_m, \alpha_n),$$ (2.4)

The reverberation loss is defined as,

$$RL_{inc} = 10 \log_{10} \frac{E_0}{I_{inc}}$$

$$= -10 \log_{10} \left[\pi r c \sum_m \sum_n |\Theta_m(z_s, z_b)|^2 |\Theta_n(z_r, z_b)|^2 \sigma(\alpha_m, \alpha_n) \right]$$ (2.5)

The joint China-U.S. Yellow Sea '96 Experiment was conducted in the middle of Yellow Sea. Summer conditions in Yellow Sea produce an approximate three-layer sound-speed profile, with a near-linear thermocline connecting a warmer surface isovelocity layer to a cooler isovelocity bottom layer (Figure 2). The explosives were denoted at depths of 7m and 50m, respectively. A vertical line array was used to receive the reverberation signals.

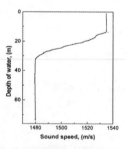

Fig. 2. A sound speed profile during '96 Yellow Sea Experiment.

The experimental reverberation losses for several frequencies and source depths in the Yellow Sea Experiment are compared in figure 3 with the numerical calculations from Eq (2.5). The depth of receiver in figure 2 is 50m that is outside of the thermocline. In the Figure, the solid lines are numerical calculations, and the dotted lines are the experimental data. The bottom parameters are from Reference [6], and the Lambert scattering model is used. It can be seen from the Figures that the numerical results and experimental data are in good agreements.

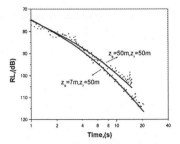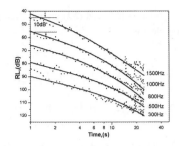

Fig. 3. Comparisons of the experimental data and the numerical predictions. The central frequency for the left figure
is 1000Hz. The depths of the source and receiver in the right figure are 7m and 50m, respectively.

3. Coherent Reverberation Model [5]

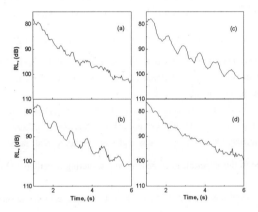

Fig. 4. The reverberation loss versus time for the explosive source denoted at 7m. The central frequency is 2000Hz.
The receiver depths are 6m, 18m, 20m and 50m for (a), (b), (c) and (d), respectively.

The primary objective of the development of coherent reverberation model is to explain the observed
oscillation phenomenon of reverberation loss in the experiment as shown in Fig. 4. In figure 4 are shown the
reverberation losses versus time. The source depth is 7m. The hydrophone depths for figures (a), (b), (c), and (d)
were 6m, 18m, 20m, and 50m, respectively. The central frequency of the measured data was 2000Hz, and a
1/3oct-frequency averaging had been made on the data. It can be seen from the figure 4 that for the hydrophones at
18m and 20m that were located within the thermocline, the received reverberation losses had a regular oscillation.
This phenomenon has not been observed before and cannot be explained by the Ray-mode reverberation theory
presented in Sec. 2.

Combining equations (2.2) and (2.3) gives the reverberation intensity at the receiver scattered from the ith bottom scattering element,

$$I_i = p_i p_i^*$$

$$= \sum_m \sum_n |\Theta_m(z_s,z_b)|^2 |\Theta_n(z_r,z_b)|^2 |D_i(\alpha_m,\alpha_n)|^2 +$$

$$\sum_m \sum_n \sum_{\substack{n' \\ (n' \neq n)}} |\Theta_m(z_s,z_b)|^2 \Theta_n(z_r,z_b)\Theta_{n'}^*(z_r,z_b)D_i(\alpha_m,\alpha_n)D_i^*(\alpha_m,\alpha_{n'})e^{i(\mu_n - \mu_{n'})r} +$$

$$\sum_n \sum_m \sum_{\substack{m' \\ (m' \neq m)}} |\Theta_n(z_r,z_b)|^2 \Theta_m(z_s,z_b)\Theta_{m'}^*(z_s,z_b)D_i(\alpha_m,\alpha_n)D_i^*(\alpha_{m'},\alpha_n)e^{i(\mu_m - \mu_{m'})r} +$$

$$\sum_m \sum_{m'} \sum_n \sum_{\substack{n' \\ (m' \neq m; n' \neq n)}} \Theta_m(z_s,z_b)\Theta_{m'}^*(z_s,z_b)\Theta_n(z_r,z_b)\Theta_{n'}^*(z_r,z_b)D_i(\alpha_m,\alpha_n)D_i^*(\alpha_{m'},\alpha_{n'}) \times \quad (3.1)$$

$$e^{i(\mu_m + \mu_n - \mu_{m'} - \mu_{n'})r}$$

Numerical calculations show that in the shallow water with a thermocline, when the source is above or below the thermocline, the 3rd and 4th terms in equation (3.1) can be neglected after an averaging over frequency due to interference of great number of cross terms with factors $e^{i(\mu_m - \mu_{m'})r}$. If the receiver is not within the thermocline, the 2nd term in equation (3.1) can be also neglected due to the same reason. If the receiver is within the thermocline, only few scattering normal modes have important effects on the reverberation, and the modal interference can not be neglected, which will be discussed in the following text.

When receiver is within the thermocline, only few modes have large amplitudes. Those modes are often called effective normal mode, and the grazing angles of those normal modes are called effective grazing angle. When the receiver is within the thermocline, the effective grazing angles are within a small range of angle, which is about from 12 degree to 15 degree in this experiment. In general, the scattering directivity $D_i(\alpha_m,\alpha_n)$ is complicate and difficult to measure. However, it is reasonable to assume that for each scattering element, the $D_i(\alpha_m,\alpha_n)$ is almost a constant in such a small range of the effective scattering grazing angle, that is

$$D_i(\alpha_m,\alpha_n) \approx D_i(\alpha_m,\alpha_c) , \qquad (3.2)$$

where α_c is the median of effective scattering grazing angle.

Combining equations (3.1) and (3.2), the coherent reverberation intensity [5] and the coherent reverberation loss which include the effect of the modal interference can be written as:

$$I_{coh} = E_0 \pi r c \left[\begin{array}{l} \sum_m \sum_n |\Theta_m(z_s,z_b)|^2 |\Theta_n(z_r,z_b)|^2 \sigma(\alpha_m,\alpha_n) + \\ \sum_m \sum_n \sum_{\substack{n' \\ (n' \neq n)}} |\Theta_m(z_s,z_b)|^2 \Theta_n(z_r,z_b)\Theta_{n'}^*(z_r,z_b)\sigma(\alpha_m,\alpha_c)e^{i(\mu_n - \mu_{n'})r} \end{array} \right] \qquad (3.3)$$

$$RL_{coh} = 10\log_{10}\frac{E_0}{I_{coh}}$$

$$= -10\log_{10}\left\{\pi r c\left[\begin{array}{c}\sum_m\sum_n\left|\Theta_m(z_s,z_b)\right|^2\left|\Theta_n(z_r,z_b)\right|^2\sigma(\alpha_m,\alpha_n)+\\\sum_m\sum_n\sum_{\substack{n'\\(n'\neq n)}}\left|\Theta_m(z_s,z_b)\right|^2\Theta_n(z_r,z_b)\Theta_{n'}{}^*(z_r,z_b)\sigma(\alpha_m,\alpha_c)e^{i(\mu_n-\mu_{n'})r}\end{array}\right]\right\} \tag{3.4}$$

In figure 5 are given the comparisons of the coherent reverberation loss, incoherent reverberation loss and the measured data. In the Figure, the solid lines are the experimental reverberation data, the dashed lines are the incoherent reverberation losses calculated from equation (2.5), and the dotted lines are the coherent reverberation losses calculated from equation (3.4). The bottom parameters are from reference [23]. The receiver depths for figures (a) and (b) in figure 5 are 18m and 20m, respectively. It can be seen from the Figures that the numerical results from the coherent reverberation model can predict the oscillation phenomenon and give better agreement with the measured data. It also should be pointed out that the oscillation phenomenon of the reverberation in figure 5 is due to the interference of the normal modes. The detailed physical explanation of this oscillation phenomenon was presented in reference [5].

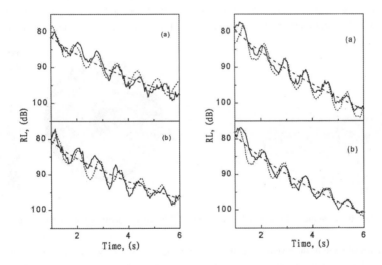

Fig. 5. The comparisons of the calculated coherent reverberation losses (dotted line), incoherent (dashed line) reverberation losses and the measured data (solid line). The receiver depths are respectively 18m and 20m for (a) and (b). The center frequency is 1000Hz, the source depth is 50m (Left), and. The center frequency is 2000Hz. The source depth is 7m. (Right)

4. Coupled Mode-PE Nx2-D Reverberation Model

The pressure $p(r, z_s, z)$ at position (r, z) due to an acoustic source located at position $(0, z_0)$ in a range-dependent shallow water can be written as:

$$p(r, z_s, z) = \sum_n \sum_m \phi_m(z_0) T_{m,n}(r, \varphi) \phi_n(r, z) \tag{4.1}$$

In this paper, the $T_{m,n}(r, \varphi)$ is calculated by the coupled-mode parabolic-equation (CMPE) method[6]. It has been discussed in Reference [6] that CMPE has fast computing speed and high accuracy.

Decomposing the nth normal mode at position (r, z) into the up-doing and down-going plane waveform[3]:

$$\phi_n(r, z) = Up(r)e^{if(z)} + Down(r)e^{-if(z)} \tag{4.2}$$

For a narrowband signal pulse of duration τ, the reverberation time series from φ direction $R(t, \varphi)$ is:

$$R(t, \varphi) = \int_t^{t+\tau} \sum_m \sum_n \sum_i \sum_j \left\{ \begin{matrix} \phi_m(z,)T_{m,n}(r, \varphi)Down_n(r)D_{n,i}(r, \varphi) \\ \times Up_i(r)T_{i,j}(r, \varphi)\phi_j(z,) \end{matrix} \right\} 2\pi ct dt d\varphi \tag{4.3}$$

The reverberation time series $R(t)$ from all direction is:

$$R(t) = \int_0^{2\pi} R(t, \varphi) d\varphi \tag{4.4}$$

The coupled mode-PE Nx2-D reverberation model was used to calculate the reverberation directionality in a shallow water shown in Fig. 6. The environment consists of an ideal slope with the water depth increased linearly from 65 to 135m. The source and the receiver were located at (50m,0km,0km) and (20m,0km,0km) respectively.

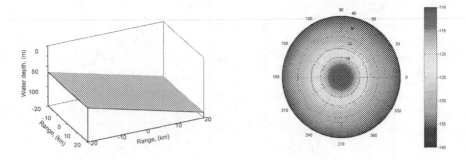

Fig. 6. Geometry for numerical example 1.   Fig. 7. The directionarity of the reverberation in a slop.

The numerical result of the reverberation from diffirent directions was shown in Fig. 7. The Lambert bottom scattering model was used. It indicated from Fig. 7 that the received reverberation is asymmetric. The revererbation level from the direction with deeper water is samller.

In Fig 8 was shown the received beam power in dB vs reverberation time by a horizontal line array in a shallow water reverberation experiment. The omnidirectional source and receivers are located ar depths 39m and 10m.

It is indicated from Fig 8 that there is a bright spot in the $40°$ direction at the reveberation time of 21s. A reasonable explanation of this phenomenon is due to a seamount. The configuration of the bottom surface from the map was shown in the figure 10, where a seamount was existed. The numerical prediction with Lambert scattering model of the received beam power was shown in Fig. 9. The comparison between the numerical and experimental data show some correspondences.

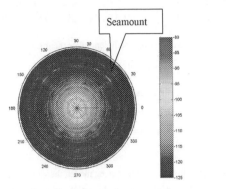

Fig. 8. Received Beam power in dB vs reverbearion time by a horizontal line array. Frequency is 800Hz.

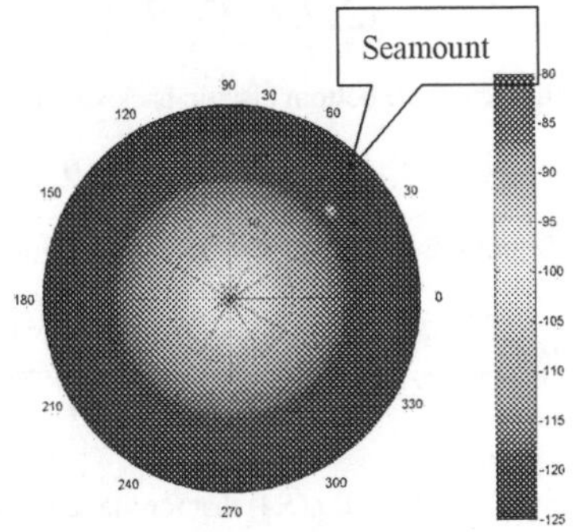

Fig. 9. Numerical prediction of the directionarity of reverberation.

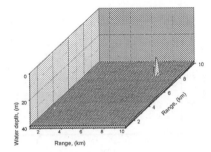

Fig. 10. Configuration of the bottom surface.

5. Spatial Correlation Theory Of Reverberation

The vertical correlation of the reverberation received at two spatially separated points $(z_{r1},0)$ and $(z_{r2},0)$ is defined as:

$$C(z_s,z_{r1},z_{r2};t_0) = \frac{\int_{t_0}^{t_0+\Delta t} p(z_s,z_{r1},t)p(z_s,z_{r2},t)dt}{\left[\int_{t_0}^{t_0+\Delta t}|p(z_s,z_{r1},t)|^2 dt \int_{t_0}^{t_0+\Delta t}|p(z_s,z_{r2},t)|^2 dt\right]^{1/2}} . \tag{5.1}$$

On the basis of ray-mode theory, omitting the cross terms, one gets the vertical correlation as:

$$C(z_s, z_{r1}, z_{r2}; t_0) = \frac{\sum_m \sum_n |\Theta_m(z_s, h)|^2 \Theta_n(z_{r1}, h) \Theta_n^*(z_{r2}, h) \sqrt{\sigma(\theta_m, \theta_m) \sigma(\theta_n, \theta_n)}}{\sqrt{\sum_m \sum_n |\Theta_m(z_s, h)|^2 |\Theta_n(z_{r1}, h)|^2 \sigma(\theta_m, \theta_n)} \cdot \sqrt{\sum_m \sum_n |\Theta_m(z_s, h)|^2 |\Theta_n(z_{r2}, h)|^2 \sigma(\theta_m, \theta_n)}} \quad (5.2)$$

For the separable bottom bistatic-backscattering model, the scattering coefficient takes the form:

$$\sigma(\theta_m, \theta_n) = \sqrt{\sigma(\theta_m, \theta_m)} \sqrt{\sigma(\theta_n, \theta_n)} \quad (5.3)$$

Substituting Eq. (5.3) into Eq. (5.2), the vertical correlation can be expressed as

$$C(z_s, z_{r1}, z_{r2}; t_0) = \frac{\sum_n \Theta_n(z_{r1}, h) \Theta_n^*(z_{r2}, h) \sqrt{\sigma(\theta_n, \theta_n)}}{\sqrt{\sum_n |\Theta_n(z_{r1}, h)|^2 \sqrt{\sigma(\theta_n, \theta_n)}} \sqrt{\sum_n |\Theta_n(z_{r2}, h)|^2 \sqrt{\sigma(\theta_n, \theta_n)}}} . \quad (5.4)$$

It can be seen from Eq.(5.4) that for the separable bottom bistatic-backscattering model, the vertical correlation is independent on the source depth. Fig. 11 shows the comparisons of the vertical correlations for different source depths. The sound speed profile is shown in Fig. 2. The depths of the two vertical separated receivers are 60m and 62m, respectively. It is shown in the figure that the vertical correlation of reverberation is independent on the source.

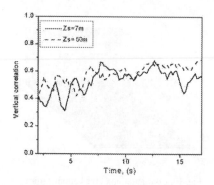

Fig. 11. Comparison of vertical correlation for different source depths. Frequency is 500Hz. The depths of the two receivers are 60m and 62m. The source depths are 7m (solid line) and 50m (dashed line), respectively.

In an isovelocity shallow water, the vertical correlation can be approximated as:

$$C_v = \frac{1}{\left(1 + \frac{k^2 \Delta z^2 h}{2Qr}\right)^{(n+1)}} \quad (5.5)$$

where Δz is the receiver separation.

In Figs. 12 and 13 are shown the comparisons of the experimental reverberation and the theoretical prediction from Eq. (5.5). In Fig 12 is shown the vertical correlation vs. reverberation time. The rough curve is the experimental data, and the smooth curve is the theoretical prediction. In Fig. 13 is shown the vertical correlation vs.

receiver separation. The dots are the experimental data, and the smooth curve is the theoretical prediction. The frequency is 500Hz. It is shown from the figure that the analytical expression and provide good prediction to the experimental data.

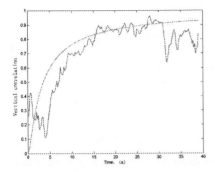

Fig. 12. vertical correlation vs. reverberation time.

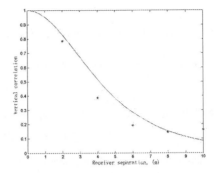

Fig. 13. vertical correlation vs. receiver separation.

6. Mode Filtering From Propagation And Reverberation Data

In a range-independent shallow water, from Eq. (2.3), the reverberation time series $R(z,t)$ is[1]:

$$R(z,t) = \sum_m \sum_n \sqrt{\frac{8\pi}{r}} \frac{\sqrt{\mu_m} q_m(z) q_m(z_b)}{S_m} \sin \varphi_m(z_s) e^{-\beta_m r + i \mu_m r} \qquad (6.1)$$
$$\times \sqrt{\frac{8\pi}{r}} \frac{\sqrt{\mu_n} q_n(z) q_n(z_b)}{S_n} \sin \varphi_n(z_r) e^{-\beta_n r + i \mu_n r} D_i(\alpha_m, \alpha_n)$$

The mode filtering from reverberation can be written as:

$$\sum_{z_r} \sin \varphi_i(z_r) p(z) = \sum_m \sqrt{\frac{8\pi}{r}} \frac{\sqrt{\mu_m} q_m(z) q_m(z_b)}{S_m} \sin \varphi_m(z_s) e^{-\beta_m r + i \mu_m r} \qquad (6.2)$$
$$\times \sqrt{\frac{8\pi}{r}} \frac{\sqrt{\mu_i} q_i(z) q_i(z_b)}{S_i} e^{-\beta_i r + i \mu_i r} \sum_{z_r} \sin^2 \varphi_i(z_r) \wp_{m,i}$$

On the basis of the separable bottom scattering model, one has:

$$\Delta RL_{\text{mode2-mode1}}$$
$$= 10 \log_{10}(I_{rev-\text{mode2}}) - 10 \log_{10}(I_{rev-\text{mode1}}) \qquad (6.3)$$
$$= 10 \log_{10} \left[\left| \sum_{z_r} \sin \varphi_2(z_r) p(z) \right|^2 \right] - 10 \log_{10} \left[\left| \sum_{z_r} \sin \varphi_1(z_r) p(z) \right|^2 \right]$$
$$= 10 \log_{10} \sqrt{\sigma(\alpha_2, \alpha_2)} - 10 \log_{10} \sqrt{\sigma(\alpha_1, \alpha_1)} - 20[\log_{10}(e)](\beta_2 - \beta_1) r$$
$$= A + Br$$

It is shown in Eq. (6.3) that A is a function of bottom scattering model, and B is a function of mode attenuation. In an isovelocity shallow water, it can be assumed that:

$$\alpha_2 \approx 2\alpha_1, \quad \ln|V(\alpha)| \approx Q\alpha, \quad S_i \approx \frac{2h}{\tan(\alpha_i)} \qquad (6.4)$$

From BDRM[2] theory, the mode attenuation can be expressed as:

$$\beta_i = \frac{\ln|V(\alpha_i)|}{S_i} \tag{6.5}$$

Substitute Eqs. (6.4,6.5) into Eq. (6.3),

$$\beta_1 = -\frac{B}{60\log_{10}(e)} \tag{6.6}$$

If the bottom scattering model can be written as the following expressing,

$$\sigma(\alpha_1,\alpha_2) = \mu \sin^{n/2} \alpha_1 \sin^{n/2} \alpha_2 \tag{6.7}$$

where n is the bottom scattering index. Substitute Eq. (6.7) into Eq. (6.3), one has

$$n = \frac{A}{5\log_{10}(2)} \tag{6.8}$$

Mode filtering from propagation can be expressed as:

$$\sum_{z_r} \sin\varphi_i(z_r)p(z) = \sqrt{\frac{8\pi}{r}} \frac{\sqrt{\mu_i}\,q_i(z)q_i(z_b)}{S_i} e^{-\beta_i r + i\mu_i r} \sum_{z_r} \sin^2\varphi_i(z_r) \tag{6.9}$$

In this section, the mode filtering method was used to the measured reverberation and propagation data. The data were collected in an isovelocity shallow water by a 16-elment vertical line array, which spans the whole water column. The water depth is 40m.

In Fig. 14 was shown $\Delta RL_{mode2-model}$ defined in Eq. (6.3) vs. reverberation time at frequency of 400Hz. The line in the figure is the linear fitting of the experimental results. A and B are shown in table 1. The mode attenuation and bottom scattering index derived from Eqs. (6.6) and (6.8) were also shown in table 1.

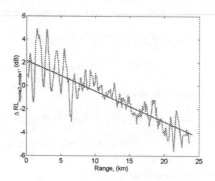

Fig. 14. $\Delta RL_{mode2-model}$ defined in Equation s. reverberation time at the frequency of 400Hz.

Table 1. Results from reverberation mode-filtering

Frequency(Hz)	A	B(dB/km)	β_1 (dB/km)	n
400	2.172	-0.26	0.010	1.44
500	2.775	-0.34	0.013	1.84

In Fig. 15 was shown the waveform of normal mode 1 and 2 by using mode filtering method from pulse waveform propagation. The central frequency is 400Hz, and the range is 19km. In Fig. 16 was shown the amplitude of the mode 1 in unit of dB vs. propagation range.

The comparison of the mode attenaution obtained from reverberation and proapagtion was shown in Table 2. It is shown in Table 2 that the results from the reverberation and proapgation are in good agreement.

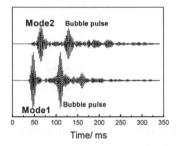

Fig. 14. The waveform normal mode 1 and 2 by using the mode filtering technology from pulse waveform propagation.

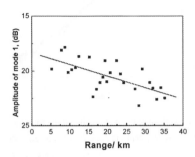

Fig. 15. The amplitude of the normal mode 1 in unit of dB vs. propagation range.

Table 2. Comparison of the bottom attenuation from reverberation and propagation

Frequency(Hz)	β_1 (dB/km) from reverberation	β_1 (dB/km) from propagation
400	0.010	0.012
500	0.013	0.011

7. Geo-Acoustic Inversion From Reverberation

The reverberation inversion method is to obtain the acoustic properties of the bottom which minimums the cost function:

$$E(\Omega) = \sqrt{\sum_i [RL(t_i,\Omega) - RL(t_i)]^2} \times \sqrt{\sum_i [C(t_i,\Omega) - C(t_i)]^2} \qquad (7.1)$$

where $RL(t_i,\Omega)$ and $RL(t_i)$ are the numerical and measured reverberation losses, and $C(t_i,\Omega)$ and $C(t_i)$ are the numerical and measured vertical correlation of the reverberation. The advantages of the reverberation inversion are: 1) The experiment can be easy conducted; 2) The bottom scattering coefficients can be obtained at the same time.

The measured and the numerical reverberation losses and vertical correlations of the reverberation are shown in Fig.16. Frequency is 1000Hz. Source depth is 25m. And the receiver depth of the RL is 26m, the two receiver depths of the vertical correlation are 24m and 26m. The agreement between the measured and calculated waveforms is good. The inverted bottom reflection losses and bottom scattering coefficient of different frequencies are shown in Fig. 17.

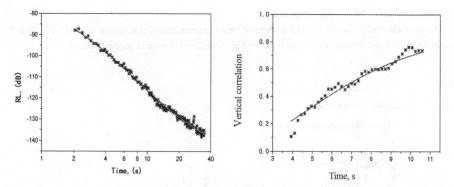

Fig.16 Comparison of the numerical and measured RL(left) and reverberation vertical correlation(right). Frequency is 1000Hz. Source depth is 25m. And the receiver depth of the RL is 26m, the two receiver depths of the vertical correlation are 24m and 26m.

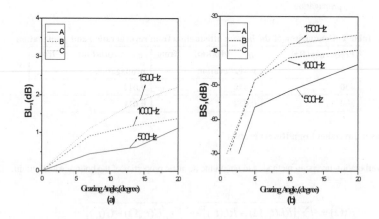

Fig. 17. The inverted bottom reflection losses and bottom scattering coefficient of different frequencies.

8. SUMMARIES

1. A ray-mode reverberation model was introduced in this paper. The comparisons between the numerical results and experimental data show that the ray-mode reverberation model can provide good prediction to the experimental data.

2. Based on the coherence of the normal modes, a coherent reverberation model was discussed to explain an observed oscillation phenomenon of the reverberation level.

3. A Nx2-D reverberation model based on the coupled ray mode theory was developed. Some numerical examples

was also discussed.

4. A concise analytical expression of the vertical correlation of reverberation for the isovelocity shallow water was derived. The experimental data show that this expression can predict the experimental data well.

5. A mode filter method for reverberation was presented. The mode attenuation and bottom scattering coefficient were derived from the mode-filter results. The mode attenuations derived from reverberation are good agreement with that from propagation

6. A Geo-acoustic inversion method from reverberation was shown in this paper. The results show that the inverted bottom properties can predict the experimental data well.

REFERENCES

1. C. Wu "Calculation of shallow water reverberation intensity based on ray theory", *Acta Acustica* 4, 114-119. (1979)

2. M.D. Collins, G.J. Orris and W.A.Kuperman "Reverberation modeling with two-way parabolic equation", 1993 in *Ocean Reverberation* (D.D. Ellis, J.R. Preston, and H.G Urban, editor), 119-224. Netherlands: Kluwer Academic Publishers.

3. R. Zhang and G Jin "Normal-mode theory of average reverberation intensity in shallow water.", *Journal of Sound and Vibration* 119(2), 215-223, (1987)

4. D.D. Ellis "A shallow-water normal-mode reverberation model", *J.Acoust.Soc.Am.* 97(5), 2804-2814, (1995)

5. F. Li, G Jin, R.Zhang, "The coherent reverberation theory and the oscillation phenomenon of reverberation in shallow water", *Science in China,* 30, 560-566, (2000)

6. Z. Peng, F. Li, "The coupled mode-PE theory based on WKBZ theory", *Science in China,* 31, 165-172(2001)

7. K. B. Smith, J. Acoust. Soc. Am. 60, 305-210 (1976)

8. J. Zhou, Chinese Phys., 1, 494-504, (1981)

9. F. Li, J, Liu, "the vertical correlation of the shallow water bottom reverberation", Proceeding of 7th WESTPRAC, Kumamoto, Japan, 2000

10. R. Zhang, "The analysis of the single mode in shallow water", *Acta Acustica* 73-85, (1980)

11. W. Hou, C. Wu, R. Zhang, "A simplified method for identification and extraction of normal mode," Chinese Jouranl of Acoustics, 6, 43-51, (1987)

12. P.R.Shaw and J.A.Orcutt, "Waveform inversion of seismic refraction data and application to young pacific crust,"Geophys.J.R. Astron. Soc. 82 375-414 (1985)

13. N.M. Carbone, G.B.Deane, and M.J.Buckingham, "Estimating the compressional and shear wave speeds of a shallow water seabed from the vetical coherence of ambient noise in the water column," J.Acoust.Soc.Am. 103(2), 801-813 (1998)

14. T.C. Yang and T.Yates "Improving the sensitivity of full-field geoacoustic inversion for estimating bottom sound speed," Full Field Inversion Methods in Ocean and Seismo-Acoustics, editor: O.Diachok, A.caiti, P.Gerstoft, and H.Schmidt, Kluwer Acasemic Publishers

15. Clinton Siedenburg, Norm Lehtomaki, and Juan Arvelo "Iterative Full-Field Inversion Using Simulated Annealing,"

16. S.M. Jesus, "A sensitivity study for full-field inversion of Geoacoustic data with a towed array in shallow water," Full filed inversion methods in ocean and seismo-Acoustics, Kluwer Academic publishers, , 109-114, 1995

17. A.Basu and L.N.Frazer, "Rapid determination of the critical temperature in simulated annealing inversion,"

Science 249,1409--1412 (1990).

18. Mrinal K.Sen and Paul L.Stoffa, "Nonlinear one-dimensional seismic waveform inversion using simulated annealing," Geophysics 56, 1624-1638 (1991).

19. C.E.Lindsay and N.R.Chapman, "Matched field inversion for Geoacoustic modal Parameter Using Adaptive Simulating Annealing," IEEE Journal of Oceanic Engineering 18,224-231 (1993)

20. M.D.Collins and W.A.Kuperman, "Nonlinear inversion for ocean-bottom properties," J.Acoust.Soc.Am. 92,2770--2783 (1992).

21. W.Cary and C.H.Chapman, "Automatic 1_D inversion of marine seismic refraction data," Geophys. J. 93, 527-546 (1988)

22. Subramaniam D. Rajan, "Determination of Geoacoustic parameters of the ocean bottom-data requirements," J.Acoust.Soc.Am. 92, 2126-2140 (1992)

23. F.Li and R.Zhang "The Bottom Speed and Attenuation Inverted by Waveform and Transmission Loss", *Acta Acustica* 25, 297-302, (2000)

24. R.Zhang and F.Li , "Beam-Displacement Ray-Mode Theory of Sound Propagation in Shallow Water.", *Science in China* 42, 739-749, (1999)

Theoretical and Computational Acoustics 2001
E.-C. Shang, Qihu Li and T. F. Gao (Editors)
© 2002 World Scientific Publishing Co.

On Acoustic Tomography Scheme for Reconstruction of Hydrophysical Parameters for Marine Environment

V. A. Akulichev, V. P. Dzyuba, P. V. Gladkov, S. I. Kamenev, Yu. N. Morgunov
Pacific Oceanological Institute, Far Eastern Branch, Russian Academy of Sciences,
43 Baltiyskaya Street, Vladivostok 690041, Russia
E-mail: *pacific@online.marine.su*

1. Introduction

Modern acoustic tomography is a powerful tool for hydrophysical measurements of the characteristic of the sea medium inhomogeneties and currents. As a rule the problem inverse of acoustic tomography is solved by means of the travel times of the rays and/or the models corresponding the acoustical sounding signal [1-3]. Often we can't measure these times as accurate as we need. Especially, it is right for the shallow sea and straits where the intense hydrodynamic processes usually are in existence. Modern acoustic tomography is powerfull tool for hydrophysical measurements of characteristic of the sea medium inhomogeneities and currents. As a rule the problem inverse of acoustic tomography solves by means of arrival times of the rays and the modes which corresponds the acoustical sounding signal. Unfortunately, often, we can't determine these times with correct necessary us especially for small distance, a shallow sea and straits where can be intensity hydrodynamic processes. The results of acoustic monitoring of the medium sea showed what the cross-correlation function between transmitted and received sounding signals $K_{ss_0}\left(\vec{r},\vec{r}_0,t,t_0\right)$ is very sensitive hydrophysical parameteres of the medium sea [4]. The numerical modeling of $K_{ss_0}\left(\vec{r},\vec{r}_0,t,t_0\right)$ showed what the propagation times of both the rays and the modes were not being determined by times corresponding maximums of the function $K_{ss_0}\left(\vec{r},\vec{r}_0,t,t_0\right)$ for small distances between the source and the receiver (10 – 100 kilometers), as shown in Fig. 1. Therefore, we are need of methods not using the travel times but which capable of decide of the inverse problem.

In the article the possibility using of the cross-correlation function $K_{ss_0}\left(\vec{r},\vec{r}_0,t,t_0\right)$ to obtain the sound speed field directly are considered.

2. Theory

Let source is situated at point $\vec{r}_0$ ($\vec{r}_0$ is radius-vector) and radiates the acoustic signal $S_0(t)$. This source produce at a point $\vec{r}$ acoustic field $S(t,\vec{r})$. Using the Green's function of the sound channel we obtain

$$S(t,\vec{r}) = \frac{1}{2\pi} \int_{-\infty}^{\infty} S_0(\omega) \cdot G(\omega, \vec{r}, \vec{r}_0) \cdot e^{i\omega t}\, d\omega,$$

where - $S_0(\omega) = \int_{-\infty}^{\infty} S_0(t) \cdot e^{-i\omega t}\, dt$ is Fourier transform of $S_0(t, \vec{r}_0)$ with respect to the time argument.

The Green's function must satisfy following equation

$$\left[\Delta + \frac{\omega^2}{c^2(\vec{r})} \right] \cdot G(\omega, \vec{r}, \vec{r}_0) = -4\pi \cdot \delta(\vec{r} - \vec{r}_0),$$

and corresponding condition on the boundary of the waveguide.

In the ray approach we have

$$G(\omega, \vec{r}, \vec{r}_0) = \sum_{j=1}^{N} A_j(\vec{r}) \cdot e^{-i\omega t_j(\vec{r})},$$

where - $t_j(\vec{r})$ and $A_j(\vec{r})$ are amplitude and travel time of the signal along the j-th ray, N is number the rays entering into the point $\vec{r}$.

Using above expression we can written the cross-correlation function $K_{ss_0}(\vec{r}, \vec{r}_0, t, t_0)$ as

$$K_{ss_0}(\vec{r}, \vec{r}_0, t, t_0) = \frac{1}{(2\pi)^2} \int_{-\infty}^{\infty} |S_0(\omega)|^2 \cdot \sum_{j=1}^{N} A_j(\vec{r}) \cdot e^{i\omega\left[t - t_j(\vec{r}) - t_0\right]}. \tag{1}$$

Assuming the spectral density of $S_0(t, \vec{r}_0)$ to have following form

$$|S_0(\omega)|^2 = \begin{cases} B, & for \quad \omega + \Delta\omega \geq \omega \geq \omega - \Delta\omega \\ 0, & for \ other \ \omega \end{cases}. \tag{2}$$

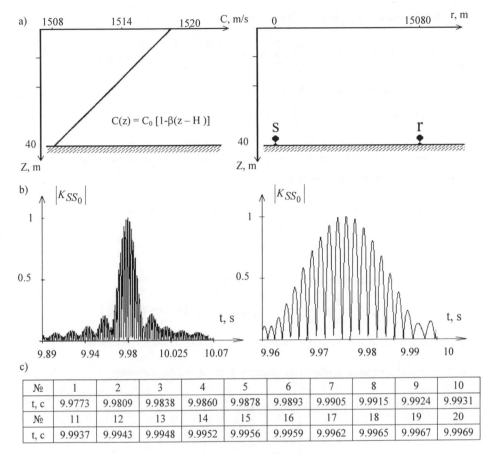

Fig. 1. Sound speed profile and location both the receiver and the source (a);
Module of the cross-correlation function $K_{SS_0} (\vec{r}, \vec{r}_0, t, t_0)$ (b);
Table: number of the ray and his arrival time into the receiver (c).

We find real part $K_{ss_0}(\vec{r}, \vec{r}_0, t, t_0)$ in the form

$$\text{Re K}_{ss_0}(\vec{r}, \vec{r}_0, t, t_0) = \frac{2B \Delta\omega}{(2\pi)^2} \sum_{j=1}^{N} A_j(\vec{r}) \frac{\sin \Delta\omega(t - t_j - t_0)}{\Delta\omega(t - t_j - t_0)} \cos \omega_0(t - t_j - t_0) , \quad (3)$$

and it is function what we measure. For the monochromatic signal of frequence ϖ_0 we have

$$\text{Re K}_{ss_0}(\vec{r}, \vec{r}_0, t, t_0) = \frac{B}{(2\pi)^2} \sum_{j=1}^{N} A_j(\vec{r}) \cos \omega_0(t - t_j - t_0) . \quad (4)$$

In the mode approach the Green's function form determinates by propertys of the waveguide directly. However, can be shown that $K_{ss_0}(\vec{r}, \vec{r}_0, t, t_0)$ have same structure as for the ray approach.

The Green's function can be written

$$G(\vec{r}, \vec{r}_0, t, t_0) = \frac{1}{(2\pi)^2} \int\int_{-\infty}^{\infty} T\Phi(\omega, \vec{\gamma}, z, z_0) \exp(-i\vec{\gamma}(\vec{\rho} - \vec{\rho}_0)) d\vec{\gamma} , \quad (5)$$

where - $\vec{\gamma} = \{\gamma_x(\omega); \gamma_y(\omega)\}$ is a horisontal component the wave vector of the signal propagating in the waveguide.

The function (mode) $\Phi(\omega, \vec{\gamma}, z, z_0)$ must satisfy following the equation

$$\left[\frac{d^2}{dz^2} + \frac{\omega^2}{c^2(z)} - \gamma^2 \right] \Phi(\omega, \vec{\gamma}, z, z_0) = -4\pi \delta(z - z_0)$$

The vector $\vec{\rho}$ and $\vec{\rho}_0$ are both horisontal component of the vectors $\vec{r}$ and $\vec{r}_0$. The operate T is defined as

$T = \sum_{j=1}^{N} \delta(\vec{\gamma} - \vec{\gamma}_j)$ for discrete portion of the spectrum $\vec{\gamma}$ and $T = I$ for continues

values $\vec{\gamma}$. The vector $\vec{\gamma}$ can be written a Teylor's series

$$\vec{\gamma} = \left[\vec{\gamma}(\omega_0) + \frac{(\omega - \omega_0)}{\upsilon^2} \vec{\upsilon} + \dots \right],$$

where - $\vec{\upsilon} = \dfrac{d\omega}{d\vec{\gamma}}\bigg|_{\omega = \omega_0}$ is horisontal component of the group velosity of the mode $\Phi(\omega, \vec{\gamma}, z, z_0)$. Using

two first members the series and assuming what modes don't dependence from ω near ω_0 we have

$$G(\vec{r}, \vec{r}_0, t, t_0) = \frac{1}{(2\pi)^2} \int\int_{-\infty}^{\infty} T\Phi(\omega_0, \vec{\gamma}, z, z_0) e^{-i(\omega - \omega_0)t(\omega_0) + i\gamma(\omega_0)(\vec{\rho} - \vec{\rho}_0)} d\vec{\gamma} , \quad (6)$$

where - $t(\omega_0) = \dfrac{(\vec{\rho} - \vec{\rho}_0)}{\upsilon^2} \vec{\upsilon}$.

Using (1),(2) and (6) we obtain for discrete spectrum $\vec{\gamma}$ that

$$\text{Re}K_{ss_0}(\vec{r}, \vec{r}_0, t, t_0) = \frac{2B\Delta\omega}{(2\pi)^2} \sum_{j=1}^{N} \frac{\sin\Delta\omega(t - t_j - t_0)}{\Delta\omega(t - t_j - t_0)} \text{Re}\left[A_j(\vec{r}) e^{-i\omega_0(t - t_0)} \right], \quad (7)$$

where - $A_j(\vec{r}) = \dfrac{1}{(2\pi)^2} \Phi\Big(\omega_0, \vec{\gamma}_j(\omega_0), z, z_0\Big) e^{-i\vec{\gamma}_j(\omega_0)(\vec{\rho}-\vec{\rho}_0)}$.

Making limits of (7) for $\Delta\omega \to 0$ we obtain

$$\mathrm{Re}\, K_{ss_0}\Big(\vec{r}, \vec{r}_0, t, t_0\Big) = \frac{2B}{(2\pi)^2} \sum_{j=1}^{N} \mathrm{Re}\Big[A_j(\vec{r}) e^{i\omega_0(t-\tau_j-t_0)} \Big].$$
(8)

where - τ_j is the travel time j-th mode is measured by the phase velosity of the mode.

Examination the expression (3), (4), (7) and (8) we can see that being discussed correlation function is determined only by spectral density of the signal and properties of waveguide. The travel times both of the rays and the modes can be determined under following condition $\Delta t = t_{j+1} - t_j > \dfrac{\pi}{\Delta\omega} = \dfrac{1}{2\Delta f}$.

This type of dependence $K_{ss}(\vec{r}, \vec{r}_0, t, t_0)$ of both the signal spectrum and waveguide properties permit to use the cross correlation function directly to solve of the problem inverse in acoustic tomography of the ocean. As minimum, we see two ways to decision this problem. First, we can obtain the travel time and/or parameters of the channel sound. Second, we may solve the problem inverse directly defining of the sound speed $c(\vec{r})$ as function of coordinates. Take in to account written up we can permit following main stages of the scheme to decision of the problem inverse.

I. Choice a modeling sound speed field $c(\vec{r})$ and calculate the modeling cross-correlation function

$\quad K^M{}_{ss_0}\Big(\vec{r}, \vec{r}_0, t, t_0\Big)$.

II. Construct a mistake function and formulate task mathematical.
1. It is the task of mathematical programming to determine of the travel times or/and waveguide parameters.
2. It is the task of variational calculus to determine of the sound speed field as function of position.
III. Solve of the formulated problem for necessary us values.

The varying of the signal spectrum we can obtain equation systems for parameters of $c(\vec{r})$ if need.

3. Numerical modeling

We have made the numerical modeling experiment corresponded to real conditions shelf of the Japan sea. The numerical experiment relevant to actual conditions of the shallow sea was conducted. Were used a different sound speed profiles. Distance between the source and receiver was about 15 kms. The source and receiver placed near to bottom. The form of the bottom corresponded to a slant with depths 27 m at the source and 76 m at the receiver, or flat bottom with depth of 40 meters, both at the source, and at the receiver.

The mathematical task was formulated as a task of mathematical programming for definition of the relative vertical gradient of the sound speed, depths of the rupture of the sound speed profile, minimum and maximum quantities of the sound speed. The reflection coefficient from bottom $m = 0.9$. As a discrepancy can be the following residual function

$$\eta = \int_{t_1}^{t_2} \Big| K_{SS_0}^{Ex} - K_{SS_0}^{M} \Big|^2 dt .$$

The ray approach was used, as it allowed to obtain an analytical solution for many used profiles. The outcomes of simulation have shown that the inverse task has a unique solution for required parameters which reduce in an absolute minimum a residual function. The cross-correlation function K_{ss_0} has appeared is very sensitive to a variations of a sound speed profile, as shown in Fig. 2. Let consider the following sound speed profile, which is shown in Fig. 1, corresponding to real conditions of the shallow sea

112

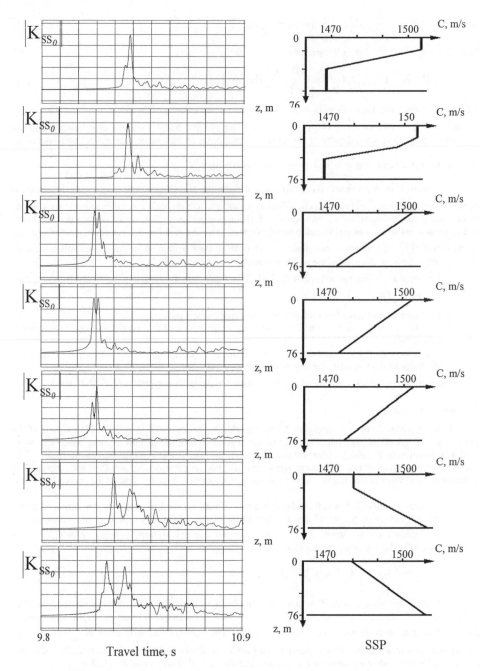

Fig. 2. Modelling sound speed profiles (right) and corresponding them models of the cross-correlation function $K_{SS_0}(\vec{r},\vec{r}_0,t,t_0)$ (left).

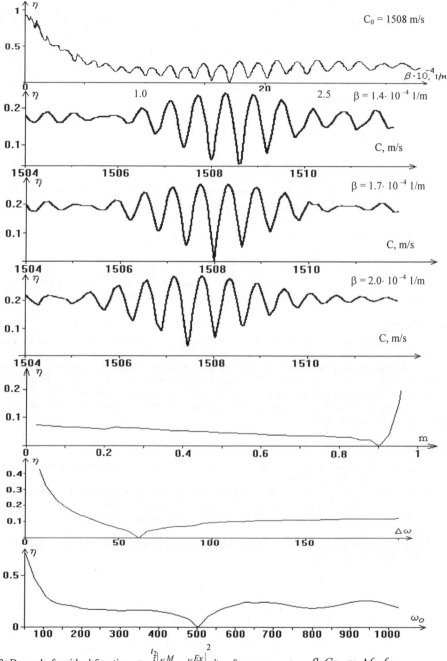

Fig. 3. Depend of residual function $\eta = \int\limits_{t_1}^{t_2} \left| K^{M}_{ss_0} - K^{Ex}_{ss_0} \right|^2 dt$ from parameters $\beta, C_0, m, \Delta f, f_0$.

$$c(z) = c_0 \left[1 - \beta(H - z) \right],$$

where H is depth of the sea. This profile have minimum sound speed near bottom equal $c_0 = 1508 \frac{m}{s}$ and $\beta = 1.7 \cdot 10^{-4} \frac{1}{M}$. The receiver and source both was disposed near bottom on distance 15080 meters between them. Transmitted signal had the rectangular density spectral width 30 Hz and central frequence 250 Hz. Mathematical task formulated as the mathematical programming problem for the obtain values β and c_0. The mistake function had following form

$$\eta(\beta, c_0) = \int_{t_1}^{t_2} \left| K_{SS_0}^{Ex} (\vec{r}, \vec{r}_0, t) - K_{SS_0}^{M} (\vec{r}, \vec{r}_0, t) \right|^2 dt , \tag{9}$$

where the travel time satisfied conditions $t_1 \le t_i(r) \le t_2$. We used the ray approach allowing the correct theoretical solution on the analitical form for travel times.

The resalts of the numerical modeling showed what problem of the inverse has singl-valued solution corresponding $c_0 = 1508 \frac{m}{s}$ and $\beta = 1.7 \cdot 10^{-4} \frac{1}{m}$. What is interesting, if one of values β or c_0 differ from $1508 \frac{m}{s}$ or $1.7 \cdot 10^{-4} \frac{1}{m}$ but function $\eta(\beta, c_0)$ still had minimum for c_0 or β equal true of the second of them, as shown in Fig. 3.

The theoretical predictions are being based on the fact that cross-correlation function contains the information about medium ocean in both propagation times of the ray or modes and their amplitudes. Important, what Green's function is single for many cases of boundary conditions and speed sound profiles.

4. Conclusions

It is possible to using cross-correlation function between transmitted and received sounding signals to obtain the sound speed field directly.

The task to determine both of the travel times rays and modes and their numbers, or waveguide parameters may be formulated as the task of mathematical programming. The task to determine of the sound speed field as function from position may be formulated as the task of variational calculus.

The travel times both of the rays and the modes and their numbers can be determined reliably under following condition $\Delta \omega \cdot \Delta t \ge \pi$.

References

1. W. Munk., C. Wunsch *Ocean acoustic tomography: a scheme for large scale monitoring*. Deep-Sea Res. 1979. Vol. 26A. pp. 123-161.
2. A.B. Baggeroer, W.A. Kuperman, P.N. Mikhalevsky *An overview of matched-field methods in ocean acoustics* IEEE J. Of Ocean Engineering. 1993. Vol. 18. N. 4. pp. 401-424.
3. W. Munk, P. Worchester, C. *Wunsch Ocean acoustic toomography*. Cambridge: Cambridge University Press. 1995.
4. Yu.N. Morgunov and etc. *Acoustic tomography for monitiring of the Japan Sea.* Proc. Of WESTPRA VII. 2000. Pp. 1159-1164.

Theoretical and Computational Acoustics 2001
E.-C. Shang, Qihu Li and T. F. Gao (Editors)
© 2002 World Scientific Publishing Co.

Acoustic Tomography of Water Hydrophysical Structure in the Japan Sea

Victor A. Akulichev
Pacific Oceanological Institute (POI), Far Eastern Branch, Russian Academy of Sciences,
43 Baltiyskaya Street, Vladivostok 690041, Russia, E-mail: *akulich@marine.febras.ru*

Vladimir V. Bezotvetnykh
POI, 43 Baltiyskaya Street, Vladivostok 690041, Russia

Vladimir P. Dzyuba
POI, 43 Baltiyskaya Street, Vladivostok 690041, Russia

Sergey I. Kamenev
POI, 43 Baltiyskaya Street, Vladivostok 690041, Russia
E-mail: *pacific@online.marine.su*

Evgeny V. Kuz'min
POI, 43 Baltiyskaya Street, Vladivostok 690041, Russia

Yury N. Morgunov
POI, 43 Baltiyskaya Street, Vladivostok 690041, Russia
E-mail: *morgunov@poi.dvo.ru*

Anatoly V. Nuzhdenko
POI, 43 Baltiyskaya Street, Vladivostok 690041, Russia

In September-October of 1999 under the JESAEX (Japan/East Sea Acoustic Experiment) project, in the coastal zone of the Japan Sea experimental studies were performed on the acoustic monitoring of water temperature variations [1]. Using multiplex phase-manipulated signals comprising M-codes with the center frequencies 250, 366, 406 and 604 Hz, the water medium on stationary traces was insonified [2-4]. During May-November 2000, a series of experiments were conducted with the aim of improving the instrument support for monitoring acoustic propagation using a two-way propagation technique of measuring ocean flows. The second part of our paper is devoted to monitoring a field of temperature. Experimental and theoretical cross-correlation functions among transmitting and receiving signals are used in one reconstruction method for identification of oceanographic parameters.

1. Introduction

The two-way propagation technique involved transmitting a sound in opposite directions along the same path. This was done to monitor the hydrophysical processes on the Japan Sea shelf on a stationary trace in October 2000 (Fig.1, 2). To implement the above method, two identical electromagnetic-type sources S1 and S2 were developed with special purpose dampers intended for broadening the passband. They projected multiplex phase manipulated signals of M-code, with a symbol selection of 511 and carrier frequency of 250 Hz. These signals were transmitted every minute for several hours. Using the cable, source S1 was placed at a distance of 200 m from the shore and at a depth of 25 m (the bottom depth was 27 m). Autonomous source S2 was placed at a distance of 15 km offshore and at a depth of 75 m (the bottom depth was 76 m). The signals from receivers R1 and R2 were transmitted through radio communication channels to the coastal post. Receivers R1 and R2 were each fixed 20 cm from the center of their respective sources.

The hydrological conditions on the shelf, in the arrangement area of receiving and transmitting systems were characterized by striking 24-hourly variations of temperature and salinity with the secondary exhibitions of half-24-hourly peaks (Fig.3). With a full tide, at a depth of 16m and higher there occurred to be the replacement of warm near-shore waters by cold and more salt sea waters. In that case the temperature fluctuations in lower layers reached 10° C.

The figure 7 shows the 24-hourly variations of cross-correlation function of acoustic signals received by the hydrophone at point R1 and monitoring hydrophone R2 in the reception point. As seen, the largest propagation velocity of the signal principal energy corresponds to the times 6.52 and 18.52, thus falling in the same time intervals with tides, that is with the coming out the shelf colder and saltier water It is observed, that velocity is great at 6.52 a.m. - at the moment of the main back tidal (24-hourly phase). The latest incomings of the principal energy of signal is noted at 22.52 and 14.52. In this time colder and saltier water is coming in the shelf.

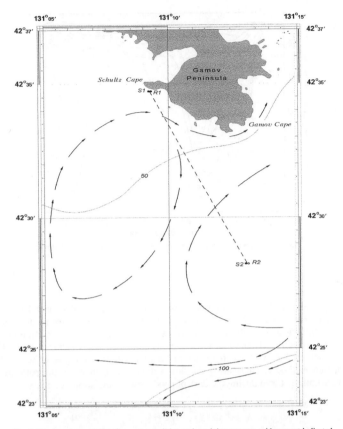

Fig. 1. The arrangement of stationary transmitting and receiving systems, with current indicated

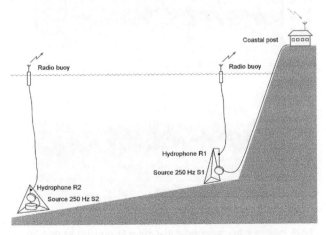

Fig. 2. A set of acoustical systems on the shelf near Gamov Peninsula

118

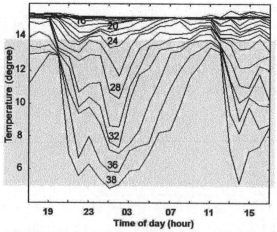

Fig. 3. Dependence of water temperature in °C from time of day at different depths in *m*

2. Experimental Results

There was obtained cross-correlation function of the received and transmitted signals that were propagated in opposite directions. The figure 4 shows variations in this function. We can see the temporal parameters are very similar for an hour and a half. The differences in travel times of the basic groups of rays, from which the water flow velocity between the transceivers is computed, are visualized in the figure 5. Here cross-correlation functions for two fixed moments of time are shown. Examples of successful use of the two-way transmission method for measuring the flow velocity component on the path are found in the results of 12-hour sounding, taken during the days of good weather and strong currents.

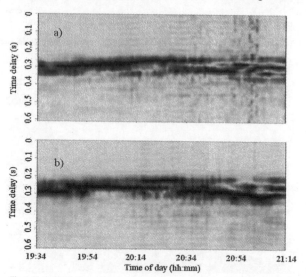

Fig. 4. Variations in the cross-correlation function over 2 hours: (a) S1-R2, (b) R2-S1

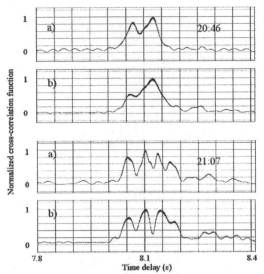

Fig. 5. Two fragments of the function from figure 4: (a) S1-R2, (b) S2-R1

The plot of computed velocity component of flow directed along the path, and the rise of tide for this time interval, are shown in the figure 6. As illustrated by this figure, the velocity component of flow fluctuates mainly from 0.4 m/s to -1.0 m/s. Moreover, the quantitative measurements of flow velocity observed approximately agreed with calculations.

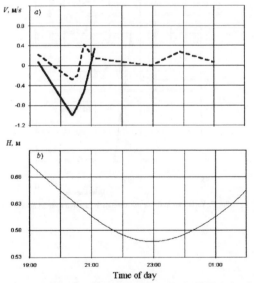

Fig. 6. Velocity component of flow directed along the trace (a) (a solid line – the whole waveguide, a dashed line – the near bottom layer), and the height of the tide (b)

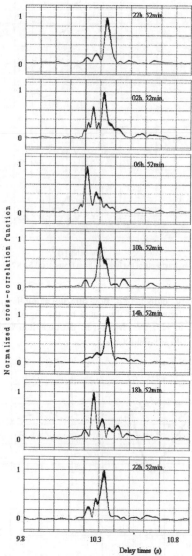

Fig. 7. Variations in the cross-correlation function over 24 hours

3. Modeling cross-correlation function

This part shows result of the using cross-correlation function between transmitted and received sound signals to decide problem of inverse [5]. We can see experimental cross-correlation function on the left (Fig. 8a), on the right you can see modeling cross-correlation function (Fig. 8b), which corresponds to this sound speed profile (Fig. 8c). We can see good coincidence modeling and experiment cross-correlation function. Sound speed profile, obtained within solving inverse problem, agreement with natural measurements.

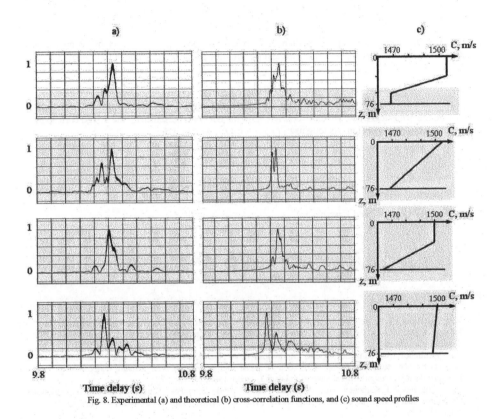

Fig. 8. Experimental (a) and theoretical (b) cross-correlation functions, and (c) sound speed profiles

4. Conclusion

In conclusion it should be said that the sound sources of electromagnetic type, especially developed for transmitting multiplex phase-manipulated signals, provide an opportunity to conduct acoustic sensing the Japan Sea water area. The sources have small mass and size characteristics, and are simple in fabrication and operation. The technology of setting sources on the ground has been tested in cable and autonomous variants.

Measured at different travels, the waveguide pulsed characteristics-time relationships that were obtained while sounding water medium by M-code signals during many 24-hours, correlate with the variations of shelf water temperature, associated with the internal waves. Quality of the obtained our acoustic data shows good promise for further investigations to aimed at improving the methods and instruments intended for tomography of the Japan Sea water structure and dynamics. Experimental results the presented enable us to improve and supplement the technical and methodological base to carry out tomography investigations in the Sea of Japan [1-4].

The theoretical predictions are based on the fact what cross-correlation function contains the information about medium ocean as for propagation times of the ray or modes and their amplitudes [5]. In summary the cross-correlation function can be very useful for directly to solve of the inverse problem in acoustic tomography.

Appendix

To calculate the averaged group delay time of individual normal waves (or rays) by using the phase function for an acoustical field, we require an expression for the propagation time of the signals:

$$t_n^{\pm} = \int_0^r \frac{dr}{c_n(r) \pm V(r)} , \qquad (A.1)$$

where $c_n(r)$ is the group velocity of the nth normal wave (nth ray), $V(r)$ is the projection of the current velocity on the acoustical trace, which is dependent on the horizontal distance r, the $\pm$ sign corresponding to the direction of transmission. As long as $c_n(r) \gg V(r)$, we can write:

$$\Delta t = t_n^- - t_n^+ \approx 2 \int_0^r \frac{V(r) dr}{c_n^2(r)} , \qquad (A.2)$$

$$t_n^+ + t_n^- \approx 2 \int_0^r \frac{dr}{c_n(r)} . \qquad (A.3)$$

For the average values c_n and V we have:

$$V = \Delta t \cdot c_n^2 / 2r , \qquad (A.4)$$

$$c_n = 2r / (t_n^+ + t_n^-) . \qquad (A.5)$$

References

1. V.A. Akulichev, V.V. Bezotvetnykh, S.I. Kamenev, E.V. Kuz'min, Yu.N. Morgunov, A.V. Nuzhdenko, "The Japan/East Sea Acoustic Experiment (JESAEX) Proect: Acoustic tomography for coastal areas," *Proc. Institute of Acoustics*, Southampton, UK, 2001, **23**, Part 2, pp. 315-320.
2. V.A. Akulichev, S.I. Kamenev, Yu.N. Morgunov, A.V. Nuzhdenko and S.I. Penkin, "Instrumentation and methods for acoustical monitoring of the Sea of Japan shelf," *Proc. of International Symposium on Acoustic Tomography and Acoustic Thermometry*, Yokosuka, Japan, 1999, pp. 188-192.
3. V.A. Akulichev, S.I. Kamenev, Yu.N. Morgunov, A.V. Nuzhdenko and S.I. Penkin, "A set of acoustical systems for environment monitoring on the oceanic shelf," *Proc. of the IEEE Conference Oceans'99*, Seattle, USA, 1999, **2**, pp. 630-636.
4. V.A. Akulichev, V.V. Bezotvetnykh, S.I. Kamenev, E.V. Kuz'min, Yu.N. Morgunov, A.V. Nuzhdenko and S.I. Penkin, "Acoustic tomography for monitoring of the Japan Sea," *Proc. of the Seventh Western Pacific Regional Acoustics Conference*, Kumamoto, Japan, 2000, **2**, pp. 1159-1163.
5. V.A. Akulichev, V.P. Dzyuba, P.V. Gladkov, S.I. Kamenev, Yu.N. Morgunov, "On acoustic tomography scheme for reconstruction of hydrophysical parameters for marine environment," *in present Proceedings*.

Theoretical and Computational Acoustics 2001
E.-C. Shang, Qihu Li and T. F. Gao (Editors)
© 2002 World Scientific Publishing Co.

Anomalous Sound Propagation Due to Bottom Roughness

Fenghua Li, Renhe Zhang, Zhenglin Li, Zhaohui Peng
National Laboratory of Acoustics, Institute of Acoustics, Chinese Academy of Sciences,
Beijing 100080, China

This paper presents an investigation of the effect of bottom roughness on the sound propagation in shallow water with a thermocline. The data obtained from several shallow water transmission loss experiments in summer indicate that there is an anomalous transmission loss at 2 to 4 kHz when the source and the receiver are both above the thermocline. With an assumption of bottom roughness, a theory based on the coupled mode model is presented to explain this phenomenon. The relationships between the anomalous loss and the frequency, the rms interface height of the bottom, the interface correlation length are also given out. Those relations may be able to be used to estimate the seafloor roughness.

1. Introduction

The studies about the backscattering due to sea-sediment roughness have been published in many papers [1,2,3]. In general, however, the discussions of the forward scattering due to sea-sediment roughness compared to that of the backscattering were ignored. A forward scattering experimental phenomenon was published in Ref. [4] by Jin, Zhang, etc.. In the shallow water waveguide bounded below by a lossy bottom, the continuously redistribution of energy between normal modes due to the various scattering processes (such as soliton [5]) will cause, increased overall loss and a lower mean field. Some scattering processes will not affect the sound propagation significantly in the isovelocity shallow water, but in the shallow water with thermocline, some scattering processes can cause anomalous sound propagation in some special conditions.

In the shallow water with a thermocline, when the source and the receiver are both above the thermocline, only very few effective normal modes [6] have great contributions to the sound propagation. The energies scattered from those effective modes to other modes due to bottom roughness will decay very rapidly in range due the large mode attenuation or small eigenfunction. This mechanism may cause extra transmission loss, which is called anomalous TL in this paper.`

In this paper, an observed anomalous TL phenomenon in '96 China-US Yellow Sea Experiment was given out, and preliminary physical explanations and discussion were offered. A concise analytical expression was derived to predict the anomalous TL.

2. Observed Anomalous Sound Propagation

In Figure 1 was shown the sound speed profile measured in the '96 China-US Yellow Sea Experiment, which was conducted in the middle of Yellow Sea in 1996. It is shown from the figure that summer conditions in the shallow water produce an approximate three-layer sound speed profile, with a near-linear thermocline connecting a warmer surface isovelocity layer to a cooler isovelocity bottom layer.

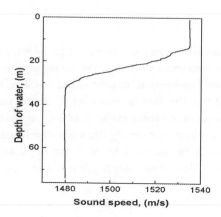

FIG. 1. Sound speed profile from '96 China-US Yellow Sea Experiment.

In Figure 2 was shown the measured transmission loss and the numerical results. The frequency is 2000Hz. The source depth is 7m. The difference between the left figure and the right figure is the receiver depth. The receiver depth in the left figure is 60m, which is below the thermocline. The receiver depth in the right figure is 8m, which is above the thermocline. The dots are the experimental data, and the solid curves are the numerical calculation by using Beam-Displacement Ray-Mode [6] theory based on the flat bottom assumption. The bottom parameters used by the numerical calculation are from Ref. [7] (extrapolation from the low frequency Geoacoustic inversion results). It is shown from the figure that the numerical calculation is in good agreement with the experimental data in the left figure; however, the experimental transmission loss data is larger than numerical results in the right figure. The difference between the experimental data and the numerical results is about 20dB at 13km. The numerical calculation cannot provide good explanations to both the experimental data in the left figure and right figure just by adjusting the bottom parameters. This kind of phenomenon that the measured TL data were larger than the numerical predictions at frequency higher than 2000Hz when the source and the receiver are both above the thermocline was also observed in other experiments. A reasonable physical explanation of this phenomenon is that such as internal wave, or bottom roughness make the range-independent model is not suitable for the case of receiver and source both above the thermocline, but it is suitable to predict the sound propagation when the source and the receiver are not both above the thermocline. The range-dependent model should be used. There is no data indicate there exist significant internal wave during the experiment. In the following section a theoretical explanation of the anomalous transmission loss by the coupled mode theory based on the assumption of rough bottom is presented.

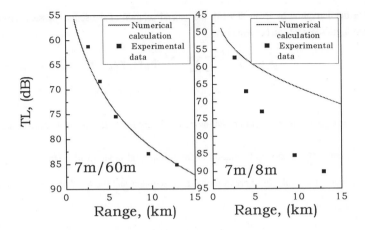

FIG. 2. Comparison of the Measured TLs and numerical results (Flat bottom assumption).

3. The theoretical derivation of the anomalous sound propagation due to bottom roughness

Geometry of a shallow water with rough bottom was shown in Fig. 3. Several roughness spectrums have been developed to describe the bottom roughness. As a preliminary explanation, the Gaussian roughness spectrum was used in this paper. The Guassian spectrum can be expressed as [1],

$$W(K) = \frac{\sigma^2 L}{2\sqrt{\pi}} e^{-K^2 L^2 / 4} \tag{3.1}$$

where σ is the rms interface height, L is the interface correlation length.

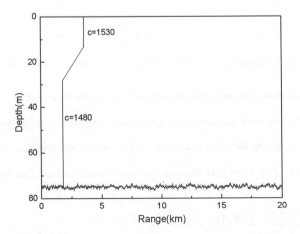

FIG. 3. A shallow water with rough bottom.

126

Following the derivation by Evan [8,9,10], the general solution of the sound pressure (single scattering approximation) in jth segment can be written as,

$$p^j(r,z) = \frac{1}{r}\sum_{l1}\sum_{l2}\cdots\sum_{lj}\Phi_{l1}(z_s)\Phi_{lj}(z)C_{l1,l2}C_{l2,l3}C_{l3,l4}\cdots$$
$$\times e^{i(\mu_{l1}r_1 + \mu_{l2}(r_2 - r_1) + \cdots) - (\beta_{l1}r_1 + \beta_{l2}(r_2 - r_1) + \cdots)} \tag{3.2}$$

where $\Phi_{lj}(z)$ is eigenfunction of ljth normal mode in jth segment, μ_{lj} is eigenvalue of the ljth normal mode in jth segment, β_{lj} is mode attenuation of the ljth normal mode in jth segment, $C_{lj,l(j+1)}$ is mode coupling coefficient between the ljth normal mode in jth segment and the $l(j+1)$th normal mode in $(j+1)$th segment. $C_{lj,l(j+1)}$ is

$$C_{l_1,l_2} = \frac{\int \frac{\Phi_{l1}^j(z)\Phi_{l2}^{j+1}(z)}{\sqrt{\rho_j \rho_{j+1}}}dz}{\int \frac{\Phi_{l2}^{j+1}(z)\Phi_{l2}^{j+1}(z)}{\rho_{j+1}}dz} \quad \text{and} \quad \int \frac{\Phi_{l1}^j(z)\Phi_{l1}^j(z)}{\rho_j}dz = 1 \tag{3.3}$$

Omitting the cross term, the incoherent sound intensity can be expressed as,

$$I = p^j(r,z)p^j(r,z)^*$$
$$= \frac{1}{r}\sum_{l1}\sum_{l2}\cdots\sum_{lj}\Phi_{l1}(z_s)^2\Phi_{lj}(z)^2 C_{l1,l2}^2 C_{l2,l3}^2 \cdots$$
$$\times e^{-2\beta_{l1}r_1 - 2\beta_{l2}(r_2 - r_1) - \cdots} \tag{3.4}$$

In shallow water with thermocline as shown in Fig. 1, a critical grazing angle α_c is defined as,

$$\alpha_c = \arccos\left(\frac{c_1}{c_0}\right) \tag{3.5}$$

where c_1 and c_0 are the sound speed in the lower layer and upper layer shown as Fig. 1, respectively.

Numerical calculations show that those normal modes with grazing angle larger than α_c has very large mode attenuation, and will not be important to the sound field; those normal modes with grazing angle smaller than α_c has very small mode amplitude when the source and the receiver both above the thermocline, and have very small contribution to the sound field. From the view of this point, the incoherent sound intensity can be approximated as

$$I \approx \frac{1}{r}\sum_{l1}\sum_{l2}\cdots\sum_{lj}\Phi_{l1}(z_s)^2\Phi_{lj}(z)^2 C_{l1,l2}^2 C_{l2,l3}^2 \cdots$$
$$\times e^{-2\beta_{l1}r_1 - 2\beta_{l2}(r_2 - r_1) - \cdots} \qquad l1 = l2 = l3 = \cdots = lc \tag{3.6}$$

Denoting $\dfrac{\Phi_{lc}^{j+1}}{\sqrt{\rho_{j+1}}}$ as

$$S_k^{j+1} = \frac{\Phi_{lc}^{j+1}}{\sqrt{\rho_{j+1}}} \tag{3.7}$$

Using perturbational approximation, S_{lc}^{j+1} can be approximated as,

$$S_k^{j+1} \approx S_k^j + \frac{dS_k^j}{dh} dh + \frac{1}{2} \frac{d^2 S_k^j}{dh^2} (dh)^2 + \cdots \tag{3.8}$$

where dh is the water depth difference between the jth segment and (j+1)th segment.

Substituting Eqs. (3. 7), (3.8) into Eq. (3.3), one has,

$$C_{lc,lc} \approx \frac{\displaystyle\int \left[S_{lc}^j S_{lc}^j + S_{lc}^j \frac{dS_{lc}^j}{dh} dh + \frac{1}{2} S_{lc}^j \frac{d^2 S_{lc}^j}{dh^2} dh^2 \right] dz}{\sqrt{\displaystyle\int \left[S_{lc}^j S_{lc}^j + 2 S_{lc}^j \frac{dS_{lc}^j}{dh} dh + S_{lc}^j \frac{d^2 S_{lc}^j}{dh^2} dh^2 \right] dz}} \tag{3.9}$$

$$\approx e^{-\frac{1}{2} X dh^2}$$

where $X = \displaystyle\int \left[\frac{dS_{lc}^j}{dh} \frac{dS_{lc}^j}{dh} - \frac{dS_{lc}^j}{dh} S_{lc}^j \right] dz$.

Substitute Eq. (3.9) into Eq. (3.6), with the Guassian roughness spectrum, define ΔTL as the difference between the TL with rough bottom and TL with flat bottom, one has,

$$\Delta TL = TL_{rough} - TL_{flat}$$

$$\approx 20[\log e] X \frac{\sigma^2 r}{L} \tag{3.10}$$

It is indicated from Eq. (3.10) that ΔTL is in proportion to range, and σ^2, $1/L$.

4. Numerical Results

In this section, some numerical calculations about ΔTL caused due to the bottom roughness were presented, the numerical results were compared to the theoretical prediction from Eq. (3.10). The theoretical expression was also used to explain the experimental data.

The numerical results in this section are calculated by RAM [11]. The bottom roughness satisfied the Gaussian spectrum. The sound speed profile is shown in Fig. 3. The bottom properties are 1587m/s, 1.85g/cm^3, 1.0dB/wavelength, and the frequency is 2000Hz. The source and receiver depths are both 7m. The definition of ΔTL is shown in Eq. (3.10). In Fig. 4 is shown the ΔTL vs. interface correlation length L. The dotted are the numerical results from RAM, and solid curves are the theoretical calculation from Eq. (3.10). It is shown from the figure the theoretical prediction fits the numerical results well.

In Fig. 5 is shown the ΔTL vs. interface rms height σ. The dotted are the numerical results from RAM, and solid curves are the theoretical calculation from Eq. (3.10). It is shown from the figure that from small interface rms height, the theoretical prediction fits the numerical results well. When the interface height is large, the approximation in Eq. (10) will not be satisfied.

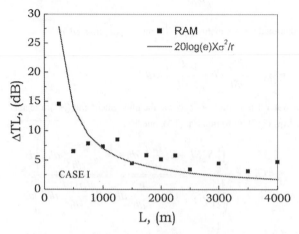

Fig. 4. ΔTL vs. interface correlation length L. The dotted are the numerical results from RAM, and solid curves are the theoretical calculation from Eq. (3.10)

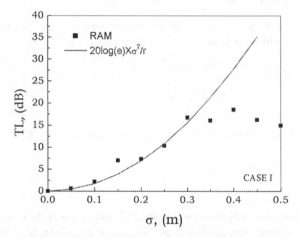

Fig. 5. ΔTL vs. interface rms height σ. The dotted are the numerical results from RAM, and solid curves are the theoretical calculation from Eq. (3.10)

In Fig. 6 is shown the comparison of measured TL and the calculation TL. The dots are the experimental data, and the solid curves are numerical results calculated by BDRM with the flat bottom assumption as shown in Fig. 2. The dashed curve is the numerical results with bottom roughness, which is $TL_{dash} = TL_{solid} + \Delta TL$, and $\dfrac{\sigma^2}{L} = 0.0002$. It is from the figure that based on the bottom roughness assumption, the numerical

prediction fits the experimental data well, and the bottom roughness can explain the observed anomalous TL.

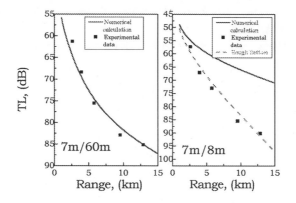

Fig. 6. the comparison of the measured TL and the calculation TL.

5. Summary

In this paper, an anomalous sound propagation phenomenon in shallow water when the source and the receiver are both above the thermocline was explained based on the hypothesis of the bottom roughness. A concise analytical expression was derived to predict the Anomalous TL. The comparisons of numerical calculations with the analytical express show that bottom roughness may cause anomalous TL, which is in

proportion to range, and σ^2, $1/L$. The analytical expression was also used to explain the theoretical data.

References

[1]. Jackson, D.R., and Briggs, K.B. "High-frequency botoom backscattering: Roughness versus sediment volume scattering," J.Acoust.Soc.Am. 92, 962-977, (1992).

[2]. Eric, I. Thorsos, "The validity of the Kirchhoff approximation for rough surface scattering using a Guassian roughness spectrum," J.Acoust.Soc.Am., 83:78-92, 1988.

[3]. Anatoliy N. Ivakin, "A unified approach to volume and roughness scattering," J.Acoust.Soc.Am., 103:827-837, 1998

[4]. G Jin, R. Zhang, and W. Hou, "A novel phenomenon in shallow water propagation—bottom scattering in the long-range propagation," 21, 905-911, 1996

[5]. J. X. Zhou, X. Z. Zhang and P. H. Rogers, ``Resonant Internation of Sound Wave with Internal Solitons in the Coastal Zone," J. Acoust. Soc. Am. 90, 2042--2054, (1991)

[6]. R. Zhang and F. Li, ``Beam-Displacement Ray-Mode Theory of Sound Propagation in Shallow Water," Science in China, 42, 739--749 (1999).

[7]. F. Li and R. Zhang, ``The Bottom Speed and Attenuation Inverted by the Waveform and Transmission Loss," Acta Acustica, 25, 297--302 (2000).

[8]. R. B. Evans, ``A Coupled Mode Solution for Acoustic Propagation in a Wave-Guide with Stepwise Depth Variations of a Penetrable Bottom," J. Acoust. Soc. Am. 74, 1414--1419 (1986).

[9]. M. B. Porter, F. B. Jensen and C. M. Ferla, ``The Problem of Energy Conservation in One-Way Models,"   J. Acoust. Soc. Am. 89, 1058-1067 (1991).

[10] F. B. Jensen, W. A. Kuperman, M. B. Porter and H. Schmidt, Computational Ocean Acoustics, American Institute of Physics, New York, (1993).

[11]. M. D. Collins, "A split-step pade solution for the parabolic equation method," J. Acoust. Soc. Am. 93(4), 1736-1742. (1993)

Theoretical and Computational Acoustics 2001
E.-C. Shang, Qihu Li and T. F. Gao (Editors)
© 2002 World Scientific Publishing Co.

The Application of Hydroacoustics on Ocean Observation

Yunwu Li

The Institute of Ocean Technology, State Oceanic Administration, 60 Xianyang Road,
Tianjin, 300111, China

Abstract

Acoustical method is the only method of ocean remote observation. Compared with other
waves, the absorption less of acoustic wave by sea water is very lower. The IOC
(Intergovernmental Oceanographic Commission) has carried out the GOOS(Global Ocean
Observation System) program. Some parameters of ocean environment can be measured by
means of the inversion of the acoustic field theory. In this paper the following applications are
discussed.

a) Profiling of the current with Doppler or correlation technique.

b) Inversion of sea surface wave direction and/or wind field by means of bottom-
mounted acoustical device.

c) Measurement the sound velocity, attenuation, temperature, salinity of sea water and
the density and particle diameter spectrum of suspended sediment in sea water, by
means of standing wave resonance of sweep-frequency ultrasonic wave as well as
acousto-optical diffraction.

d) Detection the detail topography with SAS (Synthetic Aperture Sonar).

e) The progress of underwater information transmission using the state-of-the-art
acoustic field theory.

1. Acoustical Method is the only Method of Remote Ocean Observation.

Beside the methods based on mechanics, electro-magnetics, optics, chemistry
and biology, the acoustical method has been applied on observation of ocean
environmental elements. Each of them has its own privilege. But the
absorption loss of acoustical wave propagated in sea water is less than thousandth
of that of electro-magnetical or optical wave. So that the acoustical wave
becomes the only one which is able to transmit in sea water, and at the same time
the acoustical method is the only method of remote ocean observation. The
much lower velocity and frequency of acoustical wave compared with electro-

magnetical and optical ones restricts the resolution and efficiency of observation. The acoustical wave can merely carry much less information. Meantime the inhomogeneity and variability of sea water and its boundary makes the acoustical field very complex, and ocean observation by means of acoustical method extremely difficult. After 70's last century, the following successes in R&D have helped to realize acoustical ocean observation.

1.1 The Development of Acoustical Field Theory in Hydro-acoustics

Based on the results in hydro-acoustical theoretical research, the theoretical models and experimental data of vertical stratified inhomogeneity of sea water medium, the effect of horizontal inhomogeneity on acoustical field, the effect of sea surface and bottom boundaries, the scattering of bubbles and particles suspended in sea water, the scattering and reflection of stratified inhomogeneity of sea bottom sediment etc. , it is able to inverse the ocean environmental parameters from the variations of receiving acoustical signals.

1.2 The Development of the Design of the Underwater Acoustical Systems

The development of the materials, design and transducer array techniques allows the realization of underwater acoustical system with low frequency, broad band, high power. Then the long distance acoustical detection is able to carry out by means of broad band, complex signal.

1.3 The Development of Hydro-acoustics Hi-tech

Along with the application of advanced electronics, computer techniques, DSP techniques in underwater acoustical engineering it is realizable to use the acoustical method to observe the ocean environmental parameters and detect the sea bottom.

In order to exploit sustainably the ocean resources, protect the ocean environment foresightedly and forecast the atmospheric/oceanic variation and disaster timely, the observation of ocean parameters has been an very important task in 21 century — the oceanic century. The IOC(Intergovernmental Oceanographic Commission)has carried out the GOOS(Global Ocean Observation System)Program. In this program the coastal countries ought to set out local ocean observation system and monitor the ocean environmental parameters simultaneously conformed to the standard.

In the past five years, Chinese Government has promoted a so-called 818 Subject of 863 Hi-Tech Program — Ocean Monitoring Subject headed by

Academician Qihu Li. And Chinese scientists and engineers have researched and developed a series of techniques and devices for ocean observation. Among them, the observation of a lot of parameters of ocean environment has been realized by means of the inversion of the acoustic field. To resume the detail of the subject is not my work. I'd like to introduce only some aspects of the development as following.

2   Profiling of the Current with Doppler or Correlation Technique

Since 1970's the IOT has explored the application technique of Doppler effect of echo signal to measure current. After effort in 20 years, this technique has matured. Various products of ADCP have been available abroad and in China. Among them, different designs have been selected to meet the demand of different observation object. Generally, they are classified in the measuring system for stratified current in deep sea and that for local current in shallow water or for flow between two points(particularly, the system of measuring current velocity and direction in narrow space of experimental water pool). The ADCP used to detect farther distance transmits broad band coded pulse signals of lower frequency. Then the pseudonoise signals bring rich information. Therefore the statistical calculation can be decreased greatly and the observation accuracy can be improved obviously. A central vertical-oriented beam has been added to the so-called Janus array to profile the spatial distribution of suspended particle concentration. Strenuous labor of sampling and weighing method can be avoided and continuous remote profiling can be carried out during the ship is sailing. Bottom-mounted ADCP can profile the current near sea bottom. Some ADCP systems adopt phase-controlled array as transmitter and receiver to increase detection depth and improve performances. But the precision ADCP used to detect short distance often applies focusing structure, and chooses pulse-to-pulse interferential method to process the received signals.

Scientists in the Institute of Acoustics and IOT developed the ACCP successfully. They designed an inhomogeneous array and statistically processing software. It is expected to produce market available device in the near future.

3. Inversion of Sea Surface Wave and/or Wind

A bottom-mounted acoustical system including an acoustical Doppler profiler and an ambient noise receiver was used to receive the Doppler shift of back-scattering signal and ambient noise both from sea surface and produced by fluctuation of rough air-water boundary which is blowed by wind near the surface. Provided these signals are correlated with the surface condition, we can inverse the wind and wave above the receiving system from both Doppler shift and noise intensity.

4. Deposition Monitoring System

Monitoring the deposition process is very important to offshore engineering and environmental protection. Scientists in IOT and other institutions integrated ultrasonic resonance cavity, acousto-optical diffraction device, Doppler profiler, automatic mechanical suspended particles sampler as a monitoring system to observe the deposition process. Prof Proni of AOML, NOAA invited Chinese colleagues to collaborate with this system in the Miami Coral Reef Protection project. From the Doppler shift of back-scattering by the suspended particles it is suitable to review the deposition process. Sweeping frequency ultra sound produces standing wave resonance in a tube-shape resonance cavity filled with sea water. The velocity of sea water can be calculated accurately from the interval between two successive resonance frequencies. At the same time the acoustical attenuation, temperature and salinity of sea water can be deduced by means of acousto-optical diffraction.

5. Acoustical Detection of Sea Bottom

The acoustical detection of sea bottom (topography, profile, sediment layer, object above sea bottom or buried in sediment) is very effective.

In deep ocean, deep-towed system integrated with echo sounder, multibeam echo sounder, side-scan sonar, bottom profiler is often used to detect the deep ocean floor and multimetalic nodules. In recent years ultrashort baseline positioning system has been adopted to resolve the precision positioning of towed fish. The detectors carried by the fish move so close the sea bottom that the resolution and accuracy of detection is satisfying.

The artificial nervous network or other statistical technique with reasonably selected characteristics of sediment (e.g. porosity, roughness) and that of reflection and scattering signal (e.g. duration of pulse tail, attenuation velocity) have been used to discriminate the type of the sediment.

In SAS (synthetic aperture sonar) sophisticated DSP technique increases equivalently the aperture of the array, then enhances the resolution and accuracy of detection.

6. Underwater Information Transmission

Acoustical communication is again the only means of transmission of measured data and control command. Because the information volume transmitted by acoustical communication is limited, and the acoustical field is very complex, particularly there is multi-path transmission, over a long period of time the acoustical communication has not developed rapidly.

In recent years the development of bottom-mounted instruments, mooring buoy system, submerged buoy system and underwater vehicle have required to transmit data from deep ocean to sea surface or between underwater platforms. In the meantime, the development in acoustical field theory, state-of-the-art coded pulse and matched field technique enhances the effective and reliability of acoustical communication. As a result, acoustical communication has turned into an useful tool to ocean environment observation and offshore engineering.

Theoretical and Computational Acoustics 2001
E.-C. Shang, Qihu Li and T. F. Gao (Editors)
© 2002 World Scientific Publishing Co.

On Some Approximations for Nonstationary Problem of the Acoustical Pulse Scattering by a Random Medium

Oleg E. Gulin

Harbin Engineering University, Harbin 150001, China, *gulin98@sina.com*

Il'ichev's Pacific Oceanological Institute of RAS, Vladivostok 690041, Russia, *gulin2001@mail.ru*

Igor O. Yaroshchuk

Il'ichev's Pacific Oceanological Institute of RAS, Vladivostok 690041, Russia, *yaroshchuk@lycos.com*

Under the exact wave formulation in the spatial-time domain statistical boundary value problem of the scattering of sound pulses incident on the randomly fluctuating layered medium is considered. For the solving of this problem the analytical-numerical approach has been developed by us earlier and results of a statistical simulation for different durations of the original incident pulses and various thicknesses of a random medium layer were presented. Analysis was carried out for the statistical moments, correlation functions and power spectral densities of the backscattered wave field. Comparison with the results of an approximate asymptotical analysis, carried out earlier by another authors for this problem, finds out the number of differences. In this paper we examine both the results of the exact statistical simulation and the approximate analytical ones and as a generalization propose some rather simple approximations for the description of the statistical moments of the backscattered field in the region of nonstationarity. Studying of the considered problem has both the fundamental significance for theoretical acoustics and the practical application for the interpretation of data associated with the time pulse probing of the ocean water column and the sediments of a bottom.

1. Introduction

This paper is devoted to some generalizations of numerical and analytical results concerning the studying of nonstationary problem of the normal incidence of pulses and signals on the halfspace of a random non dissipative medium. In our previous papers on the problem of pulses scattering by the fluctuating medium [1-3] the approach with the help of statistical simulation has been developed, that has allowed to obtain the exact pattern of backscattering processes by the investigation of the field statistical moments as the function of a time. A rather small number of papers is known in which this problem is considered similarly to our studying that is directly in the spatial-time domain and in the framework of a wave approach [4-7] is uniquely adequate for the investigation of a backscattering. Analytical results obtained in these papers are completely based on the asymptotic analysis of the scattering processes at the long times of an observation. Our theoretical-numerical generalizations in the proposed paper use both known asymptotic expressions and results of the statistical simulation. Owing to that we have been successful in the obtaining of the simple analytical dependencies well approximating the exact behaviour of the backscattered field statistical moments in the regions where the asymptotic formulas are incorrect.

2. Setting Up the Problem, Numerical Scheme and Main Parameters

For the convenience of readers below we shortly give the problem mathematical formulation and some accompanying notes which in the complete form not one time were considered in the previous papers (see, for example, [1-3]). The pulse $\varphi[t + (z - L)/c_0]$ incidence on a layer of inhomogeneous medium $L_0 < z < L$ in some instant $t = +0$ is described by the initial-boundary-value problem for the wave equation. In the case of the normal incidence of a pulse from the right side homogeneous halfspace it may be written in the following form for a wave field inside the medium

$$\partial_z^2 U(z, t) - c^{-2}(z)(\partial_t^2 U) = 0 , \tag{2.1}$$

$$[\partial_z U(z, t)]|_{z=L} + c_0^{-1}[\partial_t U(L, t)] = 2 c_0^{-1} \partial \varphi(t)/\partial t , \tag{2.2a}$$

$$[\partial_z U(z, t)]|_{z=L_0} - c_0^{-1}[\partial_t U(L_0, t)] = 0 . \tag{2.2b}$$

Here c_0 is the sound speed value in the homogeneous halfspace $z > L$, from which the incident pulse arrives. Fluctuations of the sound speed $c(z)$ inside the random medium provide the backscattered field $r(t) = U(L, t) - \varphi(t)$ in the right hand side of the layer $z > L$ at $t > 0$. This field is the object of our investigation. In the case of incidence on the random medium layer of the pulse having a form of the Green function of a one-dimensional free space $\varphi(t) = \theta[t + (z - L)/ c_0]$ (θ is the Heaviside unit function) the backscattered field $r(t) = \theta(t)R_L(t)$ for each realization of $c(z)$ obeys to the following integro-differential equation with the initial condition, that may be derived by the imbedding method [1]

$$[(\partial_L) + 2c^{-1}(L)(\partial_t)]R_L(t) = (2c_0)^{-1} [1 - c_0^2/2c^2(L)] \int_0^t [\partial_t R_L(t - \xi)][\partial_\xi R_L(\xi)] \, d\xi , \quad R_{L_0}(t) = 0 . \tag{2.3}$$

For an arbitrary realization of a random sound speed profile the effective scheme of a computation has been proposed by us for this equation solving. It is constructed on the basis of the analytical-numerical method allowing to obtain the solution of (2.3) in the explicit form [1,8]. For the stochastic problem solving, when the random function $c(z)$ is specified by its ensemble of realizations, we use the method of statistical simulation has been developed earlier in application to the investigation of steady-state problems [9]. It implies the approximation of a random function $c(z)$ by the to-limitary process, constructing by the uniform separation of the entire medium layer into the set of N elementary sublayers Δz_k , within every of which $c_k = $ const , and these values are given by the random number generator. If $\Delta z_k \to 0$ then this to-limitary process reduces to the original one. Mathematical aspects of the correctness of such approximation may be found in [1].

Random fluctuations of a sound speed inside the medium over the depth z , we specify in the form $c(z) = c_0 [1 + \varepsilon(z)]$, where $\varepsilon(z)$ is the Gaussian Markov's process having the next characteristics

$$<\varepsilon> = 0 , \quad <\varepsilon(z)\varepsilon(z')> = \sigma_\varepsilon^2 \exp(-|z - z'|/l) . \tag{2.4}$$

The following situation is of interest, when the intensity of fluctuations $\sigma_\varepsilon^2 \ll 1$, while the correlation radius l is the least spatial parameter for the problem at hand. Though we do not introduce the dissipation, nevertheless in the spatial-time representation it is valid to consider the medium as the random halfspace implying such times of the observation for which the presence of the left boundary is not essential because of the scattering pulse can't find time to reach it. Owing to the statistical simulation we have obtained earlier [2,3] the exact dependencies for the behaviour of the backscattered field statistical moments in the case of the normal incidence on the fluctuating medium of the rectangular pulses having different durations as well as of the narrow band signal. Some of these dependencies for the average intensity of the backscattered field $<R^2(\tau)>$ as well as for the fourth moments $<R^4(\tau)>$ are presented in figures 1-6, where τ is the dimensionless time. We came to the conclusion that for calculations it is convenient to make the reference to the diffusion coefficient D at some based carrier frequency Ω. This coefficient is known appears as the important parameter when we analyze the wave statistical problems under the steady-state formulation [6,7], where they satisfy to the boundary value problem for the stochastic Helmholtz equation. In terms of physics the diffusion coefficient characterizes the some spatial scale, during which the accumulating effects of the random inhomogeneities of $c(z)$ in statistical sense become the essential. For fluctuations specified in the form (2.4), $D(\omega) = 2\sigma_\varepsilon^2 \omega^2 l/(c_0^2 + 4\omega^2 l^2)$, where ω is an arbitrary frequency of harmonic components of the incident pulse. Thus in the process of statistical simulation the temporal coordinate t everywhere was normalized by the temporal scale of diffusion $T_\Omega^{-1} = D(\Omega)c_0$ accordingly to $\tau = tT_\Omega^{-1}$. By similar way the transition to other variables having the character of the dimensionless time was realized (further they are marked by the notation «~»). Final formula for the statistical simulation by such dimensionless variables has the following form [2]

$$R_N(\tau)=K_N\theta(\tau)+(1+K_N)\left[\sum_{i=1}^{N-1} a_{N,i}\,\theta(\tau-2i\Delta) + \sum_{i=0}^{N-1} b_{N,i}\,\theta(\tau-2\Delta(i+1))\right]-\sum_{i=1}^{N} a_{N+1,i}\,R_N(\tau-2\Delta i) \quad,$$

where coefficients $a_{N,i}$, $b_{N,i}$, K_N delineate the reflection processes on the interfaces of the elementary sublayers $\Delta_k = \Delta$ and they are found from the simple recurrent relations [1,2]. Statistical characteristics of the backscattered field are determined from here by the usual procedure of the ensemble averaging. It is necessary to notice that in the most of cases for the reliable statistics the averaging over 1000 realizations is a rather sufficient for the problem at hand.

3. General results

As it was indicated above for the problem at hand some theoretical results of the asymptotic analysis are known [4-7]. They were obtained at the large observation times $t \to +\infty$ and are based on the statistical investigation of the steady-state boundary value problem for the Helmholtz equation. In general form one can easy write the solution of nonstationary problem (2.1)-(2.2) by virtue of the Fourier transform. Such a way for the intensity of wave field in some point inside the inhomogeneous medium

$$I(z, L, t) = U^2(z, L, t) = (2\pi)^{-1} \int_{-\infty}^{\infty} d\omega \int_{-\infty}^{\infty} d\psi\, I_{\omega,\psi}(z, L)\, \varphi(\omega+\psi/2)\, \varphi^*(\omega-\psi/2)\, \exp(-i\psi t) . \quad (3.1)$$

Here $\varphi(\omega)$ is the Fourier transform of the time pulse $\varphi(t)$, $U_\omega(z, L)$ is the solution of the steady-state boundary value problem for the Helmholtz equation correspondent to the original nonstationary problem (2.1)-(2.2), and $I_{\omega,\psi}(z, L) = U_{\omega+\psi/2}(z, L) U^*_{\omega-\psi/2}(z, L)$ is the two-frequency analog of the field intensity of monochromatic plane waves. It proves to be that on the basis of the representation (3.1) and known relations for the intensity and power flux of a steady-state problem due to the method of analytical prolongation of the steady-state problem solution into the complex plane by the stochastic parameter $\beta = (\delta/Dc_0)$ (δ is the dissipation that is necessary for the steady-state problem analysis), such as $\lim_{\delta \to 0} \beta = (0 - i\psi)/Dc_0$, a formula for the field second moment (average intensity) $< I(z, L, t)>$ can be derived [7]. In particular for the average intensity of the field at the boundary, when a pulse $\varphi(t)$ is incident on it, the following expression takes place

$$< I\,(\text{L},\, t)> \; = c_0\pi^{-1} \int\limits_{-\infty}^{\infty} d\omega\, |\varphi(\omega)|^2\, D(\omega)[2 + D(\omega)c_0 t]^{-2} \; . \tag{3.2}$$

This formula is valid for the large t , and also for the kind of pulses satisfying to certain conditions of smoothness and spatial-time localization, which correspondingly bound the form of their spectra $\varphi(\omega)$. For instance, the next integrals should exist [7]:

$$< I\,(\text{L},\, \infty)> \; = c_0(2\pi)^{-1} \int\limits_{-\infty}^{\infty} d\omega\, |\varphi(\omega)|^2\, D(\omega) \; ,$$

$$E(\infty) \; = \int\limits_{-\infty}^{L} dz < I\,(z,\infty)> \; = c_0(2\pi)^{-1} \int\limits_{-\infty}^{\infty} d\omega\, |\varphi(\omega)|^2 \; . \tag{3.3}$$

From here it follows that towards high frequencies a spectrum $|\varphi(\omega)|$ must decay not slower than according to the power law $|\varphi(\omega)| \sim \omega^{-\alpha}$ with the power index $\alpha > 1.5$. For example, the incident pulse in the form of the Green function $\theta(t)$ as well as the delta-pulse are not satisfied to these conditions and integrals (3.3) delineating the average intensity of a field and the total energy within the halfspace of a random medium will be divergent. In paper [7] is mentioned that for pulses without the high frequency content, characterizing only by one parameter – is their duration η , from (3.2) one can obtain the asymptotical law $< I\,(\text{L},\, t)> \sim t^{-3/2}$. Strictly speaking, for any rectangular pulse of duration η it is incorrect to seek the asymptotic of the backscattered field by the formula (3.2) since the restrictions on its applicability region. Nevertheless further we abstract ourselves from these restrictions since we'll be interested in the possibility of a qualitative understanding how the asymptotic behaviour of the average intensity varies and by what it is caused. Let us consider the rectangular pulse of a unit amplitude and η in length : $\theta(t) - \theta(t - \eta)$. Its spectrum is $\varphi(\omega) = 2\,\omega^{-1}\sin(\omega\,\eta/2)\exp(i\omega\eta/2)$ and formally substituting such spectrum in (3.2) we obtain

$$< I\,(\text{L},\, t)> \; = 2^{-1}(a/t)^{1/2}\,\{1 - \exp\,[-2\,\eta\,(at)^{-1/2}\,] - 2\,\eta\,(at)^{-1/2}\exp\,[-2\,\eta\,(at)^{-1/2}\,]\}\;, \tag{3.4}$$

$a = \sigma_\varepsilon^2\,l/c_0$. In this expression (3.4) one can see the parameter $2\,\eta\,(at)^{-1/2}$, or by the introduced dimensionless variables $\chi = 2^{3/2}\,\tilde{\Omega}\,\tilde{\eta}/\tau^{1/2}$, $\tilde{\eta} = \eta T_\Omega^{-1}$. In order to formula (3.2) could delineate the backscattered field it is necessary that $\tau > \tilde{\eta}$. In this case, if even $\tau \gg \tilde{\eta}$ the parameter χ may have both the small $\chi \ll 1$ and the large $\chi \gg 1$ values. Values $\chi \to 0$ correspond to the asymptotical condition $\tau \to +\infty$, so expanding the exponent in (3.4) into the series up to the third term we obtain $<I (L, t)> = a\eta^2 (at)^{-3/2}$, i. e. the law is $\sim t^{-3/2}$, that corresponds to the assertions of the papers [5,7] though in this case of the rectangular incident pulse, as it was indicated, applicability conditions of the formula (3.2) are broken. Situation is of the most interest and which is not considered by the authors of papers [5,7] takes place, when $\tau \gg \tilde{\eta}$, while $\chi \gg 1$. In this case from (3.4) the asymptotic $\sim t^{-1/2}$ follows. It is easy to convinced that such situation can be typical for the pulses of not very short duration. Thus the analysis of the formula (3.4) for rectangular pulses of the duration $\tilde{\eta}$ without the content shows that the behaviour of the backscattered field conditionally can be delineated by the several regions. The first region is characterized by the times $\tilde{\eta} \ll \tau \ll 8\tilde{\Omega}^2\tilde{\eta}^2$. Inside this region for the second statistical moment qualitatively the power law $<R^2(\tau)> \sim \tau^{-0.5}$ is valid. While τ increases the power index α also increases from the value $\alpha = 0.5$ up to $\alpha = 1.5$ if $\tau \gg 8\tilde{\Omega}^2\tilde{\eta}^2$ and χ becomes much lesser than unity. It is seen from here that the asymptotic condition $\tau \to +\infty$ under which the analytical formula (3.2) is rigorously valid is equivalent to the condition $\tau \gg 8\tilde{\Omega}^2\tilde{\eta}^2$. Certainly, during the analysis of the short pulses, $\tilde{\eta} \ll 1$, we imply the description of the region $\tau > 1$, characterizing the multiple scattering effects.

Presented above a rather not strict reasons on the analysis of formulas (3.2), (3.4) for rectangular pulses is easy to verify by virtue of the comparison with the exact dependencies obtained owing to the statistical simulation. For calculations the following values of the parameters have been specified: $\sigma_\varepsilon^2 = 0.025$, $\tilde{\Omega} = 100$, $l = 0.01 T_\Omega c_0$. In fig. 1 results of the simulation for the second statistical moment of the backscattered field are presented in the case of the incidence on the random medium of the model delta-pulse, defined as $\delta(t) = \lim_{\eta \to 0} [\theta(t) - \theta(t - \eta)]\eta^{-1}$, with the calculation value $\tilde{\eta} = 0.01$. For such model the parameter $\tilde{\Omega}^2\tilde{\eta}^2 = 1$ and qualitative estimations yield the values $\alpha < 1.5$ if $\tau \sim 8$ and $\alpha = 1.5$ if $\tau \gg 8$. In fig. 1 the monotonous approximating curve that obeys to power law $A\tau^{-1.4}$ is plotted. It is easy to see, that during the observation time interval, $20 < \tau < 80$, the second statistical moment of the backscattered field is well approximated by the discovered law $\sim \tau^{-1.4}$. Recently [3] we have presented the approximation for the smaller observation times and during the interval $20 < \tau < 40$ the law of decay was $\sim \tau^{-1.3}$. Difference between power indexes for approximations and theoretical dependencies testifies that under the simulation we have considered times of the observation, which though are rather large but not enough yet to obtain the so far theoretical asymptotic. This fact is supported by the following example of the incident pulse with a duration in 10 times greater than the previous one, $\tilde{\eta} = 0.1$ (see fig. 2). In this case the parameter $\tilde{\Omega}^2\tilde{\eta}^2 = 100$ and for all region of the observation we obtain inequality $\tau \ll 8\tilde{\Omega}^2\tilde{\eta}^2$. As a result the power index α should be remarkable different from the value $\alpha = 1.5$. Indeed, from fig. 2 one can see that the law $<R^2(\tau)> \sim \tau^{-1.1}$ is a good approximation for the average intensity of the backscattered field over the interval $40 < \tau < 80$. For the interval $20 < \tau < 40$ the approximation $<R^2(\tau)> \sim \tau^{-0.9}$ has been obtained earlier [3]. For this example it is obviously that values $\alpha \sim 1.5$ will be reached at very large observation times τ.

Let us turn attention to the description of the backscattered field behaviour for the case of the Green function, when a pulse of the form of the Heaviside θ - function is incident on the random medium. In terms of physics the disturbance in such form as the Green function is the model of a pulse with very large duration that is longer than interval of the observation. It proves to be that for the second statistical moment $<R^2(\tau)>$ over the all interval of times of the observation ($\tau > 1$) the approximation $0.005\ \tau^{-0.5}$ is valid with the high accuracy, that is shown in fig. 3. In this case formula (3.2) is quietly unsuitable for the estimation since the applicability conditions of this formula are not realized, integrals for $<I(L,\infty)>$ and $E(\infty)$ are divergent, whereas the duration of θ - pulse exceeds all the others temporal scales of the problem. In the absence of a dissipation the energy yielding inside the medium by the pulse in this case continuously increases and correspondingly rises during the time the energy of the backscattered field ($\sim \tau^{0.5}$), which is formed under the increase of τ by the lower and lower frequency components of the initial pulse spectrum $|\varphi(\omega)|^2 = \omega^{-2}$. In this case it is not right to say about the spatial-time localization of the pulse as it takes place in the above considered situation, when $\tau >> \tilde{\eta}$ [7]. However for θ - pulse we can testify the fact of the enough fast transition from nonstationary regime to quasi-stationary one for $\tau > 15\text{-}20$ [2], which is described by the values of moments $<R(\tau)> \approx -0.01$, $<R^2(\tau)> \approx 0.005\ \tau^{-0.5}$. We see that in this limiting case of $\tilde{\eta} \to +\infty$ the power index value $\alpha = 0.5$ has coincided with its value of the short pulse asymptotic when $\tilde{\eta} << \tau << 8\tilde{\Omega}^2\tilde{\eta}^2$. It has happened due to the slower decay law that is associated with the scattering of low frequency components of the pulse spectrum, which in this case provide the general contribution in the backscattered field whereas, for example, in the case of delta-pulse the contribution of low frequency components is comparatively small.

It is interesting to analyze the case when a quasi-monochromatic signal is incident on the random medium. We specify such signal in the form $\theta(\tau)\cos(\tilde{\omega}\tau)$, modeling the incidence of monochromatic wave with some frequency ω. From the steady-state stochastic problem investigation is known [7] that incident plane monochromatic wave should be totally reflected by the halfspace of a random medium. The basis for such reflection is a phenomenon of the stochastic parametric resonance. In our case of the incident quasi-monochromatic signal the statistical simulation confirms this outcome. During the time the gradual increase of the average intensity of the backscattered field takes place. The lower a current frequency the slower such increase. Owing to the great duration of the signal (theoretically is infinite) the formula (3.2) is not suitable for the asymptotical analysis and the formal substitution into it the spectrum of signal reduces to the incorrect result. However at this the examine of a structure of the obtaining integrals allows to guess the probable form of the approximate law. The function $\pi^{-1}\ \text{arctg}\ [\ f(\omega)\ t\ ]$, or in dimensionless variables $\pi^{-1}\ \text{arctg}\ [\ f(\tilde{\omega})\tau\ ]$, provides the approximation to exact results of statistical simulation that is shown in fig. 4 for the signal current frequency $\tilde{\omega} = 10$. It is seen the good coincidence almost for the all time interval. Deviations are not more than several percents and they lie within the limits of the calculation accuracy for the statistics. Function $f(\tilde{\omega})$ characterizes the spectrum of sound speed fluctuations and the co-relation between the diffusion coefficient $D(\tilde{\omega})$ and the basic one $D(\tilde{\Omega})$. In [7] is indicated, that in the case of incident on the random halfspace of the pulse with high frequency content the asymptotic behaviour of average intensity of the backscattered field from the formula (3.2) $\sim \tau^{-2}$, i. e. the decay must be faster than for all rectangular pulses considered above. But there is implied the situation is quietly different from our case. Such asymptotic description is valid for the great times after the back front or the end of the coda of incident pulse whereas in our case as it was already mentioned the signal is very prolong and during its length the formula (3.2) and its outcomes on backscattering are unsuitable.

Finally we have studied the behaviour of the forth moments describing fluctuations of the intensity of the pulse backscattering field in terms of the obtaining approximations for them. As the example in figures

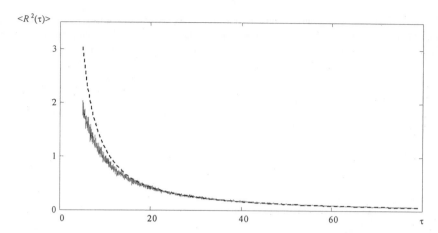

Fig. 1. Second statistical moment of the backscattered field as a function of the observation time in the case of the incident delta-pulse. Oscillating curve is the result of statistical simulation, monotonous dashed curve is the approximation $28.7\,\tau^{-1.4}$.

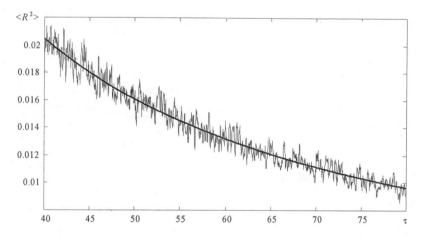

Fig. 2. Second statistical moment of the backscattered field as a function of the observation time in the case of the incidence of the pulse with duration $\tilde{\eta} = 0.1$. Monotonous black curve is the approximation $1.13\,\tau^{-1.1}$ to results of the statistical simulation.

5, 6 one can see the comparison between statistical simulation results and approximations for the cases of delta- and θ- pulses. Earlier there was indicated in [4,5] on the Gaussian character of the probability density function for the backscattered field. We also have verified this reason (see, for example, [1,3]) and came to the conclusion that it is really valid with the sufficient accuracy at least in the case of the short

144

incident pulses. If to suppose the Gaussian form for the probability density, then it is obvious that the forth field moments reduce to the combination of second and first moments by virtue of the formula $<R^4> = 3(<R^2>)^2 - 2(<R>)^4 \approx 3(<R^2>)^2$, since the second term is much lesser respectively the first one for all

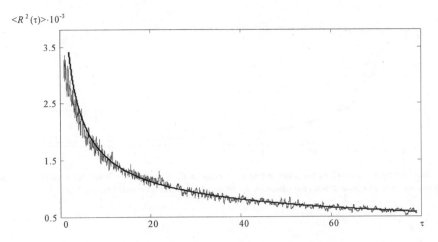

Fig. 3. Second statistical moment of the backscattered field as a function of the observation time in the case of incident θ - pulse. Monotonous black curve is the approximation $0.005 \, \tau^{-0.5}$ to results of the statistical simulation.

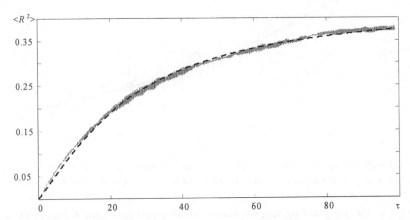

Fig. 4. Second statistical moment of the backscattered field as a function of the observation time in the case of the incidence of the quasi-monochromatic signal $\theta(\tau) \cos(\tilde{\omega}\tau)$, $\tilde{\omega} = 10$. Oscillating curve is the result of statistical simulation, monotonous dashed curve is the approximation $\pi^{-1} \arctan(0.039\tau)$.

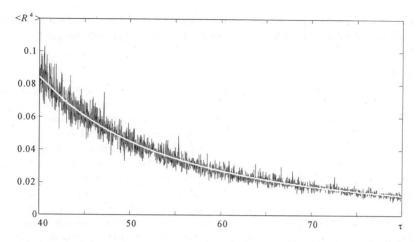

Fig. 5. Forth statistical moment of the backscattered field as a function of the observation time in the case of the incidence of the delta-pulse. Monotonous white curve is the approximation $2600\,\tau^{-2.8}$.

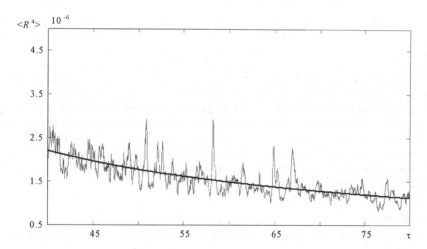

Fig. 6. Forth statistical moment of the backscattered field as a function of the observation time in the case of the incidence of the θ- pulse. Monotonous black curve is the approximation $0.00008\,\tau^{-0.98}$.

considered cases [2]. From figures one can see the good correspondence to this relation for approximation curves. It is the additional side argument supporting the Gaussian character of the probability density for the backscattered field in the region of a multiple scattering and besides it is obtained that the character of the probability distribution is independent on the duration of a pulse incident on the random medium.

4. Conclusion

Thus in the presented paper we have shown that the behaviour of the average intensity (energy in the point) of the backscattered field for the problem of pulse incidence on the halfspace of the random medium, whose fluctuations are specified by the Gaussian Markov's process is well approximated by the simple temporal dependencies over the general part of the observation time interval for which earlier the simulations have been carried out. Analytical expressions, that are known from the investigations of this problem indicate on the power character of the asymptotic laws for the second moment of the backscattered field. But frequently they are not suitable since have the applicability region bounded by the special pulse waveform (describing by the smooth functions, that have no short fronts and so without the high frequency content in the case of a broad-band signal) but what is more remarkable – they describe very large times of the observation. For example, even in the case of the delta-pulse model both the approximation given by us and the asymptotic due to the analytical formula (3.2) are of the power type but with the different power indexes. Such difference becomes more essential for the pulses of greater duration. It indicates on the fact that asymptotical laws following from the theory are valid only for so great observation times when there is no energy remains in the backscattered field from the initial one of the pulse incident on the random medium. This fact we have remarked in previous papers [2,3], here it was exhibited by the results of the calculations. Since the asymptotics are very far the verification of them owing to the statistical simulation is difficult because of the unreasonable large volume of calculations. But in our opinion it has no any physical sense since can't provide the new results for the scattering process description. From this point of view the asymptotics mainly are of the academic interest since they are rather poor delineate the real pattern of the pulse scattering. From the other hand the asymptotical formulas similar to (6) are useful as we have seen, because owing to them we got the possibility to better understand the structure of the obtaining solution and to find the good approximations for statistical moments of the backscattered field, that have been demonstrated in this paper. Finally, the studying of the forth moments of the backscattered field has allowed to obtain approximation dependencies, which additionally testify to the Gaussian law of the probability distribution for the field in the region of multiple scattering. This fact is clear from the physical reasons and proves to be valid for pulses both of the short and great duration, i. e. it is independent on the length of the pulse.

Acknowledgements

These investigations were supported by Foundation for University Key Teacher by the Ministry of Education of China.

References

1. O. E. Gulin and I. O. Yaroshchuk , *Russ. Izv. Vyssh. Uchebn. Zaved., Radiofiz.* **42** 383 (1999).
2. O. E. Gulin and I. O. Yaroshchuk , *Acoustical Physics* **45** 704 (1999).
3. O. E. Gulin and I. O. Yaroshchuk , *Proc. UK Inst. of Acoustics* **21** (9), 1999, pp. 111-116.
4. R. Burridge, G. Papanicolaou and B. White , *SIAM J. Appl. Math.* **47** 146 (1987).
5. R. Burridge, G. Papanicolaou, P. Sheng and B. White , *SIAM J. Appl. Math.* **49** 582 (1989).
6. M. A. Guzev and V. I. Klyatskin , *Waves in Random Media* **1** 7 (1991).
7. V. I. Klyatskin and A. I. Saichev , *Russ. Usp. Fiz. Nauk* **162** 161 (1992).
8. O. E. Gulin and V. V. Temchenko , *Comp. Math. and Math. Phys.* **37** 487 (1997).
9. I. O. Yaroshchuk , *Russ. Zh. Vychisl. Matem. i Matem. Fiz.* **24** 1748 (1984).

Theoretical and Computational Acoustics 2001
E.-C. Shang, Qihu Li and T. F. Gao (Editors)
© 2002 World Scientific Publishing Co.

Numerical and Imitative Simulation of Sound Signal and Noise Propagation in Shallow Sea

Desen Yang[1], Igor O. Yaroshchuk[2], Oleg E. Gulin[1,2]

[1]Harbin Engineering University, Harbin 150001, China, *euipc@public.hr.hl.cn*

[2]Il'ichev's Pacific Oceanological Institute of RAS, Vladivostok 690041, Russia, *yaroshchuk@lycos.com*

Abstract. On the basis of the exact wave approach and under the using of the imbedding method ideas the correspondent software was developed and calculations of acoustical fields in the shallow sea from monochromatic source, radiating in the middle frequencies 500 – 1000 Hz, were carried out. We investigate the influence of real factors presenting in the coastal ocean area on the possibilities of a reception of useful signal from the source. Namely these are the antiwaveguide sound speed depth stratification, having the features of the thermocline origin, the presence of a bottom, having some complex impedance structure, finally the presence of a complex field of the underwater acoustical noise. For the last factor the possibilities of imitative simulation are considered. Sound field simulation is carried out on an example of the scalar and vectorial power characteristic behaviour, namely these are the acoustical energy (scalar intensity) and the sound power flux density (vector). The considered models and calculation results can be useful for the prognosis of results of experimental measurements in a coastal ocean area.

Introduction

Quality of measurements and statistical estimation of scalar and vector hydroacoustical signals are appreciably determined by a successful choice of statistical field models of signals and noises. A main way of the development of such field models are the complex both experimental and analytical investigations of processes of hydroacoustical fields formation.

The problem of development of statistical models of hydroacoustical fields is especially urgent and at the same time is difficult for conditions of the shallow sea, where the process of field formation essentially depends on the reflecting properties of a sea bottom.

The overwhelming majority of analytical investigations of underwater acoustical fields in the shallow sea is based on the simulation of the scalar characteristic of a field that is the sound pressure (see, for example, references in [1]). So the developed numerical methods are oriented mainly to calculation of scalar fields. Possibilities to use these methods for the simulation of vector characteristics of fields are rather limited.

In the present paper we propose some original approaches to solving of the problem of signal and noise fields formation in the shallow sea.

Setting up the problem and govern equations

We shall assume further that a water layer $[0,H]$ of the ocean environment is stratified, the bottom represents a homogeneous layer [-h,0] of liquid sediments, and that the underwater acoustical noise field is excited by fluctuations of atmospheric pressure. Then the boundary value problem on generation of acoustic waves by fluctuations of atmospheric pressure $p_a(t, r)$ (here t is the temporal, and $r = \{\rho, z\}$ are the spatial, coordinates) for spatial-time spectral components is reduced to solving of the set of imbedding equations for the water layer $z = [0,H]$ (see, for example, [2,3]):

$$\frac{d\phi_H}{dH} = -q^2(H) - \phi_H^2, \quad \phi|_{H=0} = \phi_0 \quad ,$$

$$\frac{\partial\psi(z,H)}{\partial H} = -\phi_H \psi(z,H), \quad \psi(z,z) = 1 \ . \tag{1}$$

Here $q^2(z) = (k^2(z) - æ^2)$ is the square of vertical wave number, $k^2(z) = \omega^2/c^2(z)$, $æ = |æ|$ is the horizontal wave number, $\omega = \omega_0(1 - i\gamma)$ is the sound frequency, where ω_0 is its real part and where γ is the small medium dissipation. Parameter $c^2(z)$ describes the sound speed layered stratification inside the ocean medium, having certain mean value c_0. Initial condition to the first equation of the set (1) (that is the Riccati equation) is determined via the bottom impedance of the sediment layer [-h,0] $\sim \phi_0^{-1}$ and for the large values h it has the following form:

$\phi_0 = ik_0\alpha_1\sqrt{\alpha_2^2(1-i\gamma_1)^2 - æ^2/k_0^2}$, where $k_0 = \omega_0/c_0$, while the parameters α_1 and α_2 represent correspondingly the ratios of the water density and sound speed to the same ones of the liquid bottom $\alpha_1 = \rho_0/\rho_1$, $\alpha_2 = c_0/c_1$, and where γ_1 is the bottom wave dissipation.

Spatial-time spectral amplitude of a pressure of the point monochromatic source, located in the point z_0 inside the water column, obeys to the following imbedding equation:

$$\frac{\partial p(\omega,æ,z,z_0,H)}{\partial H} = -q^2(H)\phi_{z_0} \psi(z_0,H) \psi(z,H),$$

$$p(\omega,æ,z,z_0,H)\big|_{H=\max(z,z_0)} = \begin{cases} \psi(z,z_0), & z_0 \geq z, \\ \psi(z_0,z), & z_0 \leq z, \end{cases} \tag{2}$$

Vertical component of an oscillatory velocity w , normalized by the water environment wave resistance $\rho_0 c_0$ is equal to

$$w(\omega,æ, z, z_0, H) = (-i/k_0)\phi_z p(\omega,æ, z, z_0, H) \ . \tag{3}$$

Further we assume that the field of noise sources is statistically stationary and uniform and is described by virtue of a spatial-time spectral density (further, spectrum) $S_a(\omega,æ)$. Then spectral components of the acoustical field inside the water layer is entirely determined via the spectrum $S_a(\omega,æ)$ and quadratures from the function $\phi_z(\omega,æ)$. For instance, spectral amplitudes of the pressure and oscillatory velocity are determined by the expressions

$$p(\omega,æ; z) = p_a(\omega,æ)\psi(z,H) \ , \quad w(\omega,æ; z) = -i\phi_z p_a(\omega,æ)\psi(z,H)/k_0 \ . \tag{4}$$

The spatial-time spectra of the pressure and average vertical power flux (the real part of mutual p-w spectrum) take the following form:

$$S_{pp}(\omega,æ; z) = S_a(\omega,æ)|\psi(z,H)|^2 \ ,$$

$$S_{pw}(\omega,æ; z) = k_0^{-1} S_a(\omega,æ)|\psi(z,H)|^2 \operatorname{Re}\{-i\phi_z\} \ .$$

The field of monochromatic source is determined by the integration over the plane waves

$$f(\omega,\rho,z,z_0,H) = 2\pi \int_0^\infty d\text{æ}\,\text{æ}\,f(\omega, \quad z,z_0,H)\,J_0(\text{æ}\rho) \quad,\quad f = \{p,w\}\,, \tag{5}$$

and elements of a temporal spectral tensor of a noise field are determined by the following expressions:

$$S_{pp}(\omega,z) = 2\pi \int_0^\infty d\text{æ}\,\text{æ}\,S_a(\omega, \quad |\psi(z,H)|^2\,,$$

$$\tag{6}$$

$$S_{pw}(\omega,z) = 2\pi k_0^{-1} \int_0^\infty d\text{æ}\,\text{æ}\,S_a(\omega, \quad \mathrm{Re}\,(-i\phi_z)|\psi(z,H)|^2$$

Integrals (5) and (6) essentially depend on the mode structure of wave fields. The dispersing equation for considered boundary conditions: $p(H) = 0$ on the surface and $p(0) + \Omega_0\,w(0) = 0$ at the bottom, - has the form $\Omega_H(\omega,\mu) = 0$. Here for the convenience we use the notation $\mu = \text{æ}^2$, while the function $\Omega_z = -ik_0\phi_z^{-1}$ is the impedance of the layer $[-h, z]$, and it satisfies to the equation

$$\frac{\partial \Omega_z(\omega,\mu)}{\partial z} = -ik_0 + ik_0^{-1}q^2(z)\,\Omega_z^2(\omega,\mu), \quad \Omega_z(\omega,\mu)\big|_{z=0} = \Omega_0(\omega,\mu) \ . \tag{7}$$

Numerical schemes

The field of monochromatic source (5) as known can be calculated by the various ways. A rather widely spread and in a lot of cases the economic method is a summation of normal waves (modes). The main problem for it is the necessity of the effective and exact calculation of eigenvalues of the homogeneous boundary value problem for linear acoustic equations $\mu^{(n)}$ (here n is the number of an eigenvalue).

We shall assume that $\mu^{(n)}$ is the function of a frequency and water layer surface position $\mu^{(n)} = \mu^{(n)}(\omega,H)$. Introduce also the notation $\Theta_H(\omega,\mu) = \partial\Omega_H(\omega,\mu)/\partial\mu$. Then, for the function $\mu^{(n)}(\omega,H)$ we obtain the equation

$$\partial \mu^{(n)}(\omega,H)/\partial H = -\Theta_H^{-1}[\omega,\mu^{(n)}(\omega,H)]\,\partial\,\Omega_H[\omega,\mu^{(n)}(\omega,H)]/\partial H \quad, \tag{8}$$

where the function Θ_H itself is determined by the following equation

$$\partial\Theta_H/\partial H = -ik_0^{-1}(\Omega_H^2 - 2q^2\Omega_H\,\Theta_H)\,,\ \Theta_H\big|_{H=0} = \partial\Omega_0/\partial\mu \quad. \tag{9}$$

Taking into account the expressions for normal waves via eigenfunctions

$$w(\omega,\mu^{(n)},z,H) = \exp\left\{ik_0^{-1}\int_z^H d\zeta\,q^2(\zeta)\,\Omega_\zeta\right\}, \tag{10}$$

$$p(\omega,\mu^{(n)},z,H) = -\Omega_z\,w(\omega,\mu^{(n)},z,H)$$

and also equations (9), (7), the equation (8) for $\mu^{(n)}(\omega,H)$ may be rewritten in the form

$$\frac{\partial\mu^{(n)}}{\partial H} = \frac{ik_0}{\partial\Omega_0/\partial\mu - ik_0^{-1}\int_0^H d\zeta\,p^2(\omega,\mu^{(n)},\zeta,H)} \ . \tag{11}$$

Initial condition to (11) will be the meaning of eigenvalue in a thin homogeneous layer

$\mu^{(n)}(\omega,H)\Big|_{H\to 0} = \mu_0^{(n)}$. Calculating eigenvalues (11) and functions (10), one can represent the

point source field in the form of the sum of normal waves. However in many important practical cases it is necessary to take into account for calculations the continuous part of the problem spectrum, involving in the integral representation (5). For such situation the entire scalar-vector source field in the water layer one can calculate directly by the formula (5). Taking into consideration that the integrand has the singularities in the points $\mu = \mu^{(n)}$, we split the integration area on æ so that the points $Re(\mu^{(n)})$ would be in the nodal points of a grid $\{æ_j\}$ ($j = 1, N$). And between nodal points ($æ_j$, $æ_{j+1}$) we replace the Green function f to the linear interpolation $f = a_j + b_j\,æ$, here a_j and b_j are the coefficients of such interpolation. Then the integral (5) will be rewritten as a series

$$f(\omega,\rho,z,z_0,H) \approx \sum_j \left\{ a_j \frac{1}{\rho}\left[æ\,J_1(æ\rho)\right]_j^{j+1} + b_j \frac{1}{\rho^2}\left[æ\,J_0(æ\rho)\right]_j^{j+1} + b_j \frac{1}{\rho}\left[æ^2\,J_1(æ\rho)\right]_j^{j+1} \right\}. \quad (12)$$

$J_{0,1}$ is the Bessel function of the zero or the first order. In formula (12) we omitted the terms which are not essential both for the near and for the far wave field. Notice that formula (12) is convenient for calculations and it allows easily to determine envelopes of the various scalar-vector characteristics of an acoustical field.

Similarly to formula (12) it is possible to obtain a numerical scheme for calculations of the temporal spectral tensor of a noise field (6). However it is implied as more prospective the calculation of a noise field from the surface by the method of randomization of the noise source spectra (that is the imitative simulation) [4]. This approach, firstly, allows to obtain not only the second statistical moments of a field but the higher ones also and to calculate the phase characteristics of fields, which are necessary for studying of vector values. Secondly, random realizations of a noise field are possible to use effectively for both the development of statistical noise wave models and the creation of various methods for signal estimation on a background of noises.

Let us consider arbitrary splitting of the wave vector space $\mathbf{K} = \mathbf{R}^2$ ($æ \in \mathbf{K}$) on M not intersected areas $\mathbf{K}_1, ..., \mathbf{K}_M$ ($\mathbf{K}_1 \cap \mathbf{K}_m = \varnothing$). Then an arbitrary realization of the surface source field takes the form

$$p_a(\omega,\rho) = \sum_{m=1}^{M} \chi_m^{1\backslash 2}(\omega)\eta_m\,exp\{i\,æ_m\,\rho\}$$

$$\chi_m(\omega) = \int_{K_m} d^2\,æ\,S_a(\omega,æ),$$

$$P_m(\omega,æ) = \chi_m^{-1}(\omega)\,S_a(\omega,æ), \quad (13)$$

where $æ \in \mathbf{K}_m$ and random vectors $æ_m$ are distributed according to the probability densities P_m , while η_m are the random independent Gaussian values having zero mean value and unit variance. Random field of the surface sources at $z = H$, obtained by such way, is the statistically uniform and its spatial spectrum is equal to $S_a(\omega, æ)$.

Using formulas (13) it is easy to obtain random realizations of fields of the pressure and oscillatory velocity vertical and horizontal components inside the water layer

$$p_a(\omega, \rho, z) = 4\pi^2 \sum_{m=1}^{M} \chi_m^{1\backslash 2}(\omega)\eta_m \, exp\{i\,æ_m\,\rho\}\psi(z, H) \ ,$$

$$w(\omega, \rho, z) = 4\pi^2(-i/k_0) \sum_{m=1}^{M} \chi_m^{1\backslash 2}(\omega)\eta_m \, exp\{iæ_m\rho\}\phi_z\,\psi(z, H) \ ,$$

$$\mathbf{u}(\omega, \rho, z) = 4\pi^2 \sum_{m=1}^{M} \chi_m^{1\backslash 2}(\omega)\eta_m \, exp\{iæ_m\rho\}(æ_m/k_0)\,\psi(z, H) \ , \tag{14}$$

where $\mathbf{u} = [u_1\ u_2]^T$ is the horizontal velocity vector.

Formulas (13) and (14) represent in general the scheme of imitative simulation of the surface noise field. Notice, that formulas (14) are the exact solution of the linear acoustic equations for the sources of the form (13).

During the noise field simulation the space K of wave numbers was divided into the concentric circles

$$K_m = \{\, æ_m \leq |æ| \leq æ_{m+1}\, \}, m = (1, M) \ . \tag{15}$$

At the same time this splitting (15) was fulfilled by the such way that the value M would be much greater than number of normal waves and the points of splitting (15) without fail were hitted in all values $Re(\mu^{(n)})$. It has allowed to take relatively not large values M, and to carry out the calculations for the desired time. Nevertheless notice, that enough small-scale and uniform separation of the wave number space (15) $(M \to \infty)$ reduces to the same results. In the latter case it is not necessary to solve the dispersing equation (11), but calculations require the greater time.

Results of the simulation

For the simulation the typical hydrological conditions, characterizing for the Yellow sea in summer period were taken [5]: the sea depth $H = 40$m, the sound speed inside the sediment layer $c_1 = 1583$ m/s, the density $\rho_1 = 1.85$ g/sm^3. The bottom attenuation was taken according to the empiric-formula $0.37\ f^{1.84}$ [dB/m kHz] [5] while the attenuation of the water layer was taken as $(0.00165 + 0.0263\ f^{1.45})$ [dB/m kHz] [1]. Sound speed profile for the water layer was considered for the case of the expressed thermocline at the depth 20 m, while the sound speed near the surface was 1540 m/s , and at the bottom – 1500 m/s . These data in general reflect the underwater sound propagation conditions during the experiment carried out in summer months in coastal zone of the Yellow sea. Simulation of wave fields was carried out according to expressions (12)-(14).

In figs.1-3 results of calculations for scalar intensity (energy in the point) and vertical power flux depth distributions are presented for the local monochromatic source placed on the depth 10 m and the frequencies 500 Hz, 750 Hz and 1000 Hz. In all figures except figs.4 along the vertical axis is the depth in meters from the sea surface [0] towards the bottom [-40]. The main features of the scalar intensity distributions, is seen, demonstrate the underwater pressure field structure yielding by the modes which remain at the considered distances. For example, at the distance 1000 m from the source and for the frequency 500 Hz the structure of the field is formed by the first 6 modes. When the distance increases the quantity of such modes decreases. The number of these modes which consist of the waves having the small grazing angles (respectively the bottom) are smaller than the angle of a total reflection are quietly determined by the bottom impedance, mainly by the value $\alpha_2 = c_0/c_1$. So on the depth structure of the intensity we can make the conclusion respectively the approximate value c_1 in sediments. Along the horizontal direction the average intensity law of decay is very well approximated by the curve $A\rho^\beta$ where β exceeds the cylindrical law power index $\beta = 1$ for the frequencies about 500 Hz and becomes much greater for higher frequencies of 750 and 1000 Hz. These features are illustrated in figs. 4.

As for the vertical power flux it is formed mainly by the higher modes with the large grazing angles on which the bottom absorption influences especially significantly. In comparison with the level of intensity its level is rather small and has more expressed value in the near source field. Vertical power flux fastly decreases with a distance as well as with the frequency. In figures one can see relatively expressed values of the power flux only for 500 Hz. Pay attention on the power flux sign variations which characterize the different direction of a power arrival.

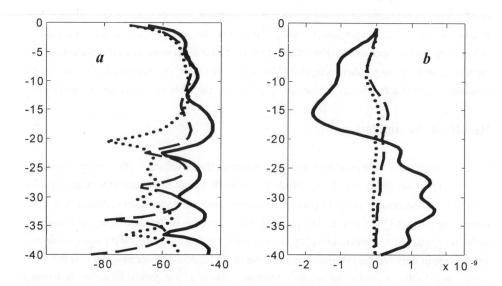

Fig. 1. *a*: Scalar intensity depth dependencies in dBs reduced to 1 m from the source level, frequency is 500 Hz. Solid curve corresponds to the distance of 1000 m from the source, dashed curve – to 2000 m, dotted curve – to 3000 m. *b*: Vertical power flux depth dependencies in linear scale reduced to 1 m also in equivalent intensity units. Solid curve corresponds to the distance of 1000 m from the source, dashed curve – to 2000 m, dotted curve – to 3000 m.

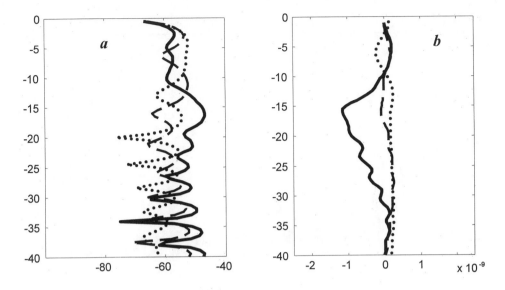

Fig. 2. Dependencies are similar to ones of fig. 1, but frequency is 750 Hz.

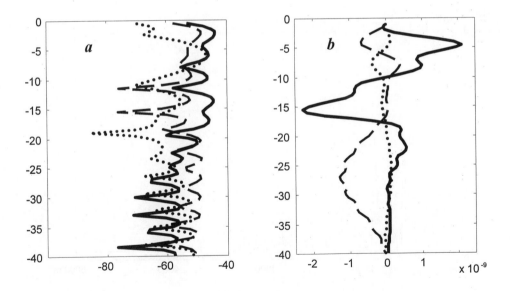

Fig. 3. Dependencies are similar to ones of fig. 1 and 2, but frequency is 1000 Hz.

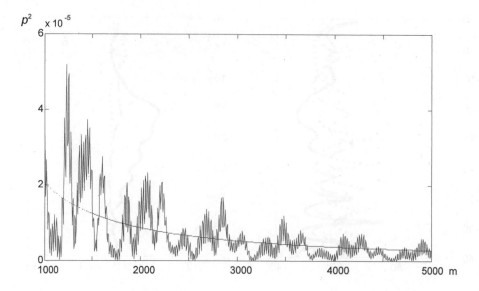

Fig. 4*a*. Scalar intensity as a function of the horizontal distance from the source for $z = 10$ m. Frequency is 500 Hz. Monotonous curve is the approximation law $\sim \rho^{-1.22}$.

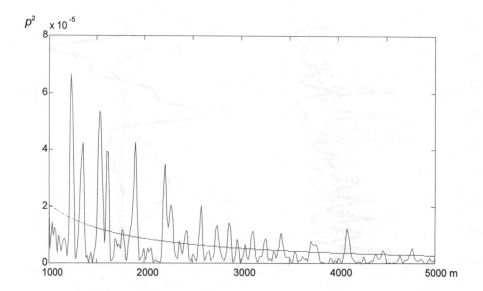

Fig. 4*b*. Scalar intensity as a function of the horizontal distance from the source for $z = 10$ m. Frequency is 1000 Hz. Monotonous curve is the approximation law $\sim \rho^{-1.25}$.

In figs.5 results of imitative simulation of the noise field are presented for the case of surface sources having the spectrum density taken in the Gaussian form $S_a(\omega, æ) =$ $= \sigma^2 l^2 S_a(\omega) \ exp\{- æ^2 l^2/2\}$. For the convenience in figures we used the notation <p²> for the temporal spectrum $S_{pp}(\omega, z)$ and <pw> for the spectrum $S_{pw}(\omega, z)$ (see (6)), also for the total velocity <V²> instead of analog $S_{VV}(\omega, z)$. Results are given as the illustration of the described above method of a simulation (see formulas (13), (14)). Imitative simulation was carried out for the specified shallow sea bottom conditions in the absence of water column stratification. The last fact is not principle since for the utilized underwater noise field model the influence of a thermocline is very little in comparison with the influence of the impedance bottom. From the figures one can see very close dependencies at high frequencies (higher than 500 Hz) both for the intensity and vertical power flux. The difference is observed under the conditions of a small quantity of excited modes in the layer at the low frequencies (100 Hz). For the intensity the amplification of a noise field inside the water column takes place as it must be while on the contrary for the vertical power flux the decrease of the level is observed.

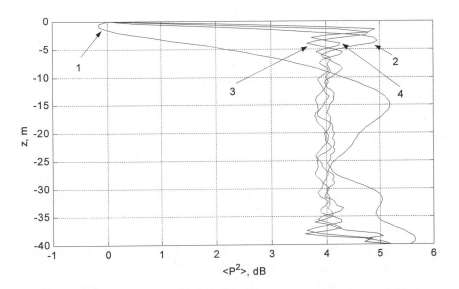

Fig. 5a. Average scalar intensity (potential energy) as a function of the depth in dBs relatively its value on the surface. Curves: 1 is 100 Hz, 2 is 500 Hz, 3 is 750 Hz, 4 is 1000Hz.

156

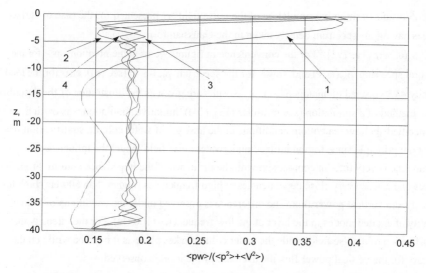

Fig. 5*b*. Average vertical power flux as a function of the depth in units of energy and normalized by the total energy. Curves: 1 is 100 Hz, 2 is 500 Hz, 3 is 750 Hz, 4 is 1000Hz. $\langle V^2 \rangle = \langle w^2 \rangle + \langle u^2 \rangle$.

Acknowledgements

These investigations were supported by Foundation for University Key Teacher by the Ministry of Education of China.

References.

1. Katsnelson B.G. and Petnikov V.G. Shallow water acoustic. – *M.: Nauka*, 1997. –191p.
2. Gulin O.E. and Klyatskin V.I. (1993) Generation of low frequency acoustical noise in the layered ocean by surface sources. *Natur. Phys. Sources of Underwater Sound*, B.R.Kerman (ed)., 247-253.
3. Gulin O.E. and Yaroshchuk I.O. (2001) Modeling of dynamic noise scattering in layered fluctuating ocean. *Proceedings of UK Inst. of Acoust*. **23**.
4. Shvyrev A.N. and Yaroshchuk I.O. (1999) Statistical modeling of ocean wind-generated noise field. *2-nd Internat. Workshop IWAET'99. (China)*. 78-87.
5. Zhou Ji-xun "Normal mode measurements and remote sensing of sea-bottom sound velocity and attenuation in shallow water. (1985) *JASA*. Vol. **78**. No.3. 1003-1009.

Theoretical and Computational Acoustics 2001
E.-C. Shang, Qihu Li and T. F. Gao (Editors)
© 2002 World Scientific Publishing Co.

Measurement of Tidal Vortices by the Coastal Acoustic Tomography

Arata Kaneko
Graduate School of Engineering, Hiroshima University,
Higashi-Hiroshima 739-8527, Japan

Jae-Hun Park
Frontier Observational Research System for Global Change,
Japan Marine Science and Technology Center, Yokosuka 237-0061, Japan

Noriaki Gohda and Haruhiko Yamaoka
Graduate School of Engineering, Hiroshima University,
Higashi-Hiroshima 739-8527, Japan

Hong Zheng
SEA Corporation, Ichikawa 272-0127, Japan

Tadashi Takano
AAS Institute, Asia Air Survey Co., Ltd, Atsugi 243-0016, Japan

The time evolution of tidal vortices generated in a small channel of the Seto Inland Sea, Japan was measured by the coastal acoustic tomography system (CATS) composed of five moored-type acoustic stations during March 2-3, 1999. The vortex fields were not well reconstructed by the conventional damped least squares method because of the failure of sound transmission for part of the station pairs. The reconstruction of vortex fields were significantly improved by assimilating the station-to-station differential travel time data into a barotropic ocean model, applying the ensemble Kalman filter technique.

1. Introduction

In the last two decades, the ocean acoustic tomography (OAT) had a continuous development as a powerful tool to measure four-dimensional structures of temperature and current velocity in the deep ocean where the underwater sound channel exists[1-3]. However, application of the OAT to the shallow sea is still at a standing phase in spite of its high potential ability as an environmental monitoring tool in the shallow sea characterized by strong variability due to complicated shorelines and bottom topographies. Until now, a few attempts to the shallow-sea acoustic tomography were done for measuring the structure of temperature and current velocity in the vertical section[4-5], but its extension to the horizontal-section measurement still remains as a target because of the limited number of the acoustic station smaller than three.

Since 1994, big effort has been devoted by Hiroshima University to construct multiple sets of the coastal acoustic tomography system (CATS) with an aim of velocity field measurement. The pilot experiments were successfully completed by using two sets of shipboard-type CATSs and it was confirmed that the accuracy of velocity measurement was good enough to increase the number of CATSs[6-7]. After that, the CATS took a model change from the shipboard to moored (self-contained) type, and five sets of the moored-type CATSs were constructed at the beginning of 1999. Part of the data obtained in the first moored-type CATS experiment is reported as a rapid correspondence publication by Park and Kaneko[8]. The full result of the first field experiment is here presented together with a detailed explanation of the methods of data analysis.

2. Experiment

Five moored-type CATSs were deployed during March 2-3, 1999 at the stations S1-S5 of about 10 m depth near the shore, surrounding the Neko-Seto Channel of the Seto Inland Sea, Japan (Fig.1). In this channel, a pair of tidal vortices is generated by the eastward flowing tidal jet, flushing out from a narrow inlet on the western side. The vortices are translated and diminished as the direction of tidal current changes. The main purpose of this experiment is to get a continuous mapping for all processes of initiation, growth, translation and decay of the vortices. A trench with depths about 100 m extends toward the NE from the inlet along the path of the tidal jet. The clockwise/counterclockwise vortex is formed on a shallower bank on the right/left side of the trench after the phase of maximum eastward current.

The mooring configuration is sketched in Fig.2. All electronic units such as card PC, amplifier, A/D converter, DSP (digital signal processing) and memory are stored together with batteries inside the pressure housing supported by a surface buoy.

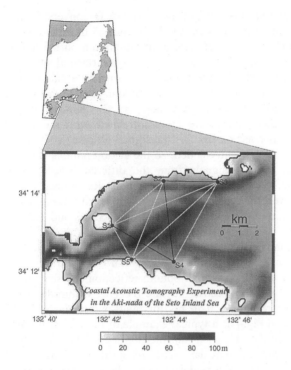

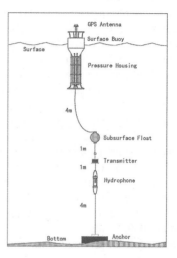

Fig.1 Location map of the observation region. S1-S5 are stations at which CATSs are deployed. The white and black lines indicate the successful and failed sound transmissions, respectively.

Fig.2 Sketch of the CATS mooring deployed at the 10m depth of S1-S5

Station-to-station timings for sound transmission, receiving and A/D conversion are coherently synchronized by clock signals obtained via a GPS antenna. A transmitter and hydrophone are mounted at 6 m and 5 m, respectively, from the bottom on the lower portion of the mooring line lifted up from the bottom by a subsurface float. The transmitter and hydrophone are connected with the pressure housing by a cable. The carrier of frequency 5.5 kHz modulated by the pseudo random signal, the Gold sequence of 10th order[9], is transmitted from all the stations every five minutes at the same timing. Received signals at each station are cross-correlated with the Gold sequence used in the transmission to increase remarkably the signal-to-noise (S/N) ratio. The Gold sequence for each station is differently coded, so the station number for the received signals can be identified even if signals from the different stations are overlapped. Only the cross-correlated data are stored in the memory. Reciprocal sound transmissions between the stations were successfully performed for seven station pairs. No data were acquired for the remaining three station pairs (S1-S4, S2-S3 and S2-S4) because of the unexpected troubles.

A weak surface duct is formed in the near surface layer to guide sound propagation in the upper 20 m as seen in the mean vertical profiles of T, S and C. There are five eigenrays which propagate in the surface duct without bottom reflection and scattering (Fig.3). These direct rays are not resolvable because they are received within the time range of 0.09 ms less than the time resolution 0.56 ms for multi-path arrivals. Data obtained at the station S2 are shown in Fig.4 as a typical example of the waterfall plot, in which the time plot of the received data for each transmission are stacked from the bottom to top with increasing time of the day. The received data after the cross-correlation make a sequence of correlation peaks, showing the successive arrival of rays. The first arrival peak is usually the biggest one, followed by a few smaller peaks. Data analysis is done for the first and biggest arrival peak corresponding to the direct ray paths. A small number of data with the first arrival peak not the biggest are extracted from the data analysis because attention is paid on the direct ray paths.

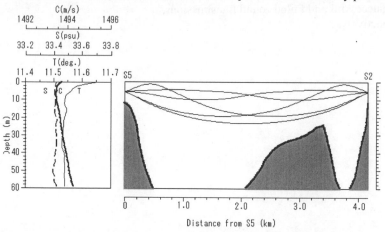

Fig.3 Ray diagram for five eigenrays without surface and bottom scatterings. The mean vertical profiles of T, S and C are drawn on the left side of the figure.

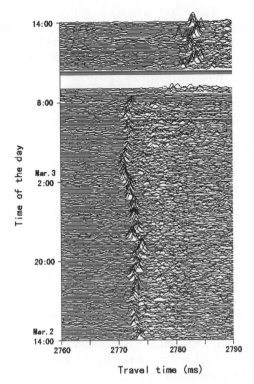

Fig.4 Waterfall plot of the cross-correlation function obtained at S2.

3. Methods of Data Analysis
3.1 Basic equations for the acoustic tomography

The travel times $t^{\pm}$ of sound propagating between the stations S1 and S2 along the ray paths $\Gamma^{\pm}$ may be formulated by

$$t^{\pm} = \oint_{\Gamma^{\pm}} \frac{ds}{C_0(z) + \delta C(x, y, z) + \mathbf{u}(x, y, z) \cdot \mathbf{n}} \tag{3.1}$$

where C_0 denotes the mean sound speed field and δC the sound speed field deviated from the mean. The $\mathbf{u}$ is the velocity vector, $\mathbf{n}$ the unit vector tangential to the ray and ds the arc length measured along the ray. When putting the ray path for C_0 into Γ_0, assuming $C_0 \gg \delta C, |\mathbf{u}|$ and taking a subtraction of t^+ and t^-, Eq.(3.1) serves to get the differential travel time

$$\Delta t = \frac{1}{2}\left(t^+ - t^-\right) \approx -\oint_{\Gamma_0} \frac{\mathbf{u} \cdot \mathbf{n}}{C_0^{\ 2}} ds = -\int_0^R \frac{\left(u + v \tan\phi\right)}{C_0^{\ 2}} dx \tag{3.2}$$

where (u, v) denotes the velocity components on the horizontal plane (x, y) and ϕ the angle between the eastward x-axis and the ray path. Note that the vertical component of velocity (w) is here neglected as assumed usually in the oceanographic problem.

The vertical shear of u and v become very small when water is well homogenized by the vertical mixing due to strong tidal currents and vortex formation }Under this condition, velocity fields can be well approximated by the depth-averaged values. Useful information on velocity fields may still be extracted from the depth-averaged velocities even in the oceanographic environment characterized with large vertical shear.

3.2 Damped least squares method

We shall here introduce the stream function   to describe the horizontal velocity field (u, v):

$$u = -\partial \Psi / \partial y, \quad v = \partial \Psi / \partial x \tag{3.3}$$

and expand it into the Fourier series like

$$\Psi(x,y) = \sum_{k=0}^{N_x} \sum_{l=0}^{N_y} \left\{ A_{k,l} \cos 2\pi \left(\frac{kx}{L_x} + \frac{ly}{L_y} \right) + B_{k,l} \sin 2\pi \left(\frac{kx}{L_x} + \frac{ly}{L_y} \right) \right\} = \sum_{j=1}^{(N_x+1)(N_y+1)} D_j Q_j(x,y) \tag{3.4}$$

where the coefficient and function vectors are expressed by

$$\mathbf{D} = \{D_j\} = \left[A_{00}, B_{00}, A_{01}, B_{01}, \cdots\cdots\cdots, A_{N_x N_y}, B_{N_x N_y} \right] \tag{3.5}$$

$$\mathbf{Q}(x,y) = \{Q_j\} = \left[1, 0, \cos\frac{2\pi y}{L_y}, \sin\frac{2\pi y}{L_y}, \cdots\cdots\cdots, \cos 2\pi(\frac{N_x x}{L_x} + \frac{N_y y}{L_y}), \sin 2\pi(\frac{N_x x}{L_x} + \frac{N_y y}{L_y}) \right] \tag{3.6}$$

, respectively. Both N_x and N_y are taken as 2 in this study.

Two-dimensional vector field **u** may generally be decomposed into two parts of solenoidal and nonrotational components:

$$\mathbf{u} = \mathbf{u}_\Psi + \mathbf{u}_\Phi \tag{3.7}$$

The solenoidal component $\mathbf{u}_\Psi$ satisfies the continuity equation and permits us to introduce the stream function. On the other hand, the nonrotational component $\mathbf{u}_\Phi$ describes the divergent flow generated by the sea-surface changes and bottom topography. When based on the projection slice theorem[1], the nonrotational component can not be measured by the sound transmission experiment in which travel times for individual ray paths are acquired as data. It should be noted that the acoustic tomography can measure selectively the non-divergent component of velocity fields. Substituting Eqs. (3.3) and (3.4) into Eq. (3.2), we obtain

$$\Delta t_i = \sum_{j=1}^{(N_x+1)(N_y+1)} D_j \int_0^{R_i} \frac{\frac{\partial}{\partial y} Q_j - \tan\phi_i \frac{\partial}{\partial x} Q_j}{C_0^{\,2}} dx \tag{3.8}$$

When using the matrix notation and adding the differential travel time error **e**, Eq. (3.8) is rewritten

$$\mathbf{y} = \mathbf{E}\mathbf{x} + \mathbf{e} \tag{3.9}$$

where

$$y = [\Delta t_1, \Delta t_2, \cdots\cdots, \Delta t_i, \cdots\cdots, \Delta t_M]$$

$$E_{ij} = \int_0^{R_i} \frac{\dfrac{\partial}{\partial y} Q_j - \tan\phi_i \dfrac{\partial}{\partial x} Q_j}{C_0^{\,2}} dx \tag{3.10}$$

$$x = D^T$$

$$e = [e_1, e_2, \cdots\cdots, e_i, \cdots\cdots, e_M]$$

We shall introduce the objective function (J)

$$J = e^T e + \alpha^2 x^T x = (y - Ex)^T (y - Ex) + \alpha^2 x^T x \tag{3.11}$$

where   is the weighting factor. The expected solution $\hat{x}$ is determined to minimize J by taking a derivative of J with respect to x as

$$\hat{x} = (E^T E + \alpha^2 I)^{-1} E^T y \tag{3.12}$$

The expected error $\hat{e}$ is determined as follows:

$$\hat{e} = y - E\hat{x} = \left\{ I - E(E^T E + \alpha^2 I)^{-1} E^T \right\} y \tag{3.13}$$

The square of the solution ($\hat{x}^T \hat{x}$) is plotted against the square of the error ($\hat{e}^T \hat{e}$), changing   as a parameter. The optimum   is determined at a point of maximum curvature on the L-shaped curve drawn. At this point, $\hat{x}^T \hat{x}$ is minimized in combination with $\hat{e}^T \hat{e}$. The process of determining optimum   is performed for fixed y to get an optimum solution whenever travel time data are obtained. It is certified through the computer simulation of the coastal acoustic tomography that velocity fields reconstructed by using the optimum value of   have good cross-correlation properties with the prescribed model velocity fields[10].

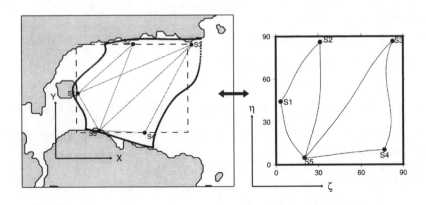

Fig.5 Transformation of the tomography domain by the orthogonal curvilinear coordinates. The physical domain (left panel) surrounded by the thick solid lines is converted into the rectangular computational domain (right panel).

As shown in Fig.5, the physical domain with the complicated shoreline is converted into the computational domain of a rectangular shape by using the orthogonal curvilinear coordinates (ξ, ς). As a result, the transmission lines are changed from the straight to curved ones. By this coordinate conversion, the boundary condition of no normal current at the shoreline is easy to be satisfied. The current fields reconstructed by the damped least squares method can generally not satisfy the non-divergent (no normal) flow condition at the shorelines due to errors. As a result, the nonrotational velocity component u occurs near the shoreline. The reconstructed current field is here modified to satisfy the non-divergent flow condition not only in the interior but also at the northern and southern shorelines, using u obtained by solving the Laplace equation[11].

3.3 Data assimilation

The differential travel time data y_t obtained at time t in the tomography experiment are related to the state vector x_t by the following observation equation

$$y_t = E_t x_t + e_t \qquad (3.14)$$

where the state vector is composed of current velocities in the observation domain and e_t is the travel time error. The data assimilation is an advanced technique in which the model results are corrected to match optimally with the observation data. When the Kalman filter technique is applied, the data assimilation is done by giving a correction based on the observation data to the forecasted result of the state vector x_t^f:

$$x_t^a = x_t^f + K_t \left(y_t - E_t x_t^f \right) \qquad (3.15)$$

where x_t^a is the assimilated result of the state vector[12]. Note that all the forecasted values with the suffix (f) come from the model. The operator matrix K_t is called the Kalman gain and the innovation matrix $\left(y_t - E_t x_t^f \right)$ is put the travel time differences between the observation and model; y_t is the observation data and $E_t x_t^f$ is determined through the ray transmission simulation in the ocean circulation model.

When minimizing the following covariance matrix

$$P_t^a = \left\langle \left(x_t^a - x_t^f \right) \left(x_t^a - x_t^f \right)^T \right\rangle \qquad (3.16)$$

the Kalman gain K is determined in the form

$$K_t = P_t^f E_t^T \left(E_t P_t^f E_t^T + R_t \right)^{-1} \qquad (3.17)$$

where $P_t^f = \left\langle \left(x_t^f - x_t \right) \left(x_t^f - x_t \right)^T \right\rangle$ is the error covariance matrix for the forecasted state vector, $R_t = \left\langle e_t e_t^T \right\rangle$ is the error covariance matrix for the observation data and < > denotes the ensemble mean.

After the data assimilation, the time evolution of velocity fields is performed by the nonlinear ocean circulation model

$$x_{i,t+1}^f = F \left(x_{i,t}^a, w_t + q_{i,t} \right) \qquad (3.18)$$

obeying the ensemble Kalman filter technique[13] where w_t and $q_{i,t}$ are the external forces and the model errors, respectively. The subscript (i) is used for the member of ensemble. The state vector at time (t+1) is estimated as an ensemble mean of all members like

$$\overline{\mathbf{x}}_{t+1}{}^{f} = \frac{1}{N_e} \sum_{i=1}^{N_e} \mathbf{x}_{i,t+1}{}^{f} \tag{3. 19}$$

The error covariance matrix for the forecasted state vector is expressed by

$$\mathbf{P}_{t+1}{}^{f} = \mathbf{S}_{t+1}{}^{f} \left[\mathbf{S}_{t+1}{}^{f} \right]^{T} \tag{3. 20}$$

where the i-th column of $\mathbf{S}_{t+1}$ is given by the following vectors

$$\mathbf{s}_{i,t+1}{}^{f} = \frac{1}{\sqrt{N_e - 1}} \left(\mathbf{x}_{i,t+1}{}^{f} - \overline{\mathbf{x}}_{t+1}{}^{f} \right) \tag{3. 21}$$

The error covariance matrix for the observation data $\mathbf{R}_{t+1} = \mathbf{R} = \langle \mathbf{e}\mathbf{e}^{T} \rangle$ is put time-invariant and only the diagonal components are considered on the assumption of no cross-correlation between station-to-station errors.

For simplicity, the computational steps for data assimilation are here summarized. The model we apply is the Princeton Ocean Model (POM) with the coordinate fitting to the bottom topography[14]. Only the external (barotropic) mode is here considered to describe the horizontal distribution of the depth-averaged velocity. The non-slip boundary conditions are imposed at the northern and southern shores (Fig.6). The tidal sea-level changes are given as external forces at the western and eastern open boundaries, considering the nearest tidal gauge station data and the random disturbances superimposed on them. The range of the sea level change is put as 3.5 m/1.5 m ¶ random errors for the western/eastern open boundary. The random errors have the Gaussian distribution with a standard deviation of 0.8 m. The model is spin up for 100 different open boundary conditions for a day without using any observation data. The time step and grid size of the model are set to 1 s and 200 m x 200 m, respectively. From the second day, the travel time data are assimilated into the ocean model every five minutes after the estimate of the ensemble mean $\overline{\mathbf{x}}$ and the error covariance matrix

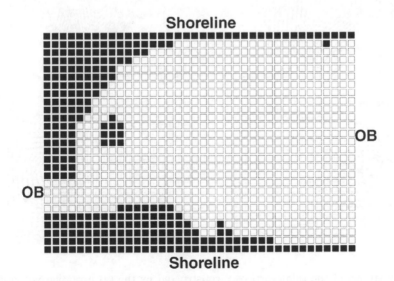

Fig.6 Computational grids and boundary conditions. Marks OB indicate the open boundaries at the eastern and western edges.

166

$\mathbf{P}^f$ from the 100 different model results. The diagonal components of $\mathbf{R}$ are determined to be the square of travel time difference errors corresponding to the constant velocity error of 5 cm/s. Note that the differential travel time errors are increased with decreasing station-to-station distances. Here the data assimilation can be completed. Then the ocean circulation model evolves the velocity fields until the acquisition of the next observation data at which data assimilation is possible again. The cycle of the data assimilation and the model time evolution is repeated for the second and third days.

4. Results

Current structures reconstructed at 23:30 of March 2, 1999 by four kinds of methods; (a) inversion without the boundary condition, (b) inversion with the boundary condition, (c) ocean model and (d) data assimilation are shown in Fig.7 with the vector plots. The snap shots were taken at the tidal phase about three hours later than the phase of maximum eastward current. In all the figures, a pair of vortices formed in nearly east-to-west arrangements while the size and magnitude of the vortices were quite different among individual cases. Unnatural currents across the northern and southern shorelines are visible in Fig. 7(a). There was another unnatural aspect that the strong southeast currents over 50cm/s occurred in the southeast of the small island located on the western side of the observation region. The cross-shore currents at the northern and southern shores are diminished in Fig.7(b) in which the boundary conditions of no normal currents are considered at the shores, but the strong currents southeast of the island still exist. As seen in Fig.7(c), a weak vortex pair with a smaller size is generated

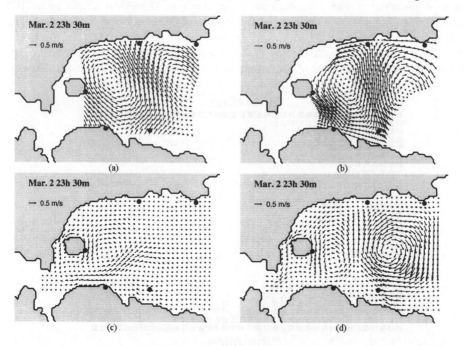

Fig.7 Comparison of the tidal vortices reconstructed by the (a) inversion without the boundary condition, (b) inversion with the boundary condition, (c) ocean model and (d) data assimilation. The results at 23:30 of March 2, 1999 are presented.

in the model domain. The weak vortices are remarkably strengthened by data assimilation (Fig.7(d)). When compared with the inversion results, the currents south of the small island are very weak in both the results of modeling and data assimilation. This implies that the result of data assimilation is close to the condition of slack water (no currents) expected in this tidal phase.

The hourly maps of horizontal current fields reconstructed by data assimilation are shown in Fig.8 with the vector plots during 14:30 of March 2 to 13:30 of March 3. According to the nearest tidal gauge data, the eastward current occurred during 17:00-23:00 of March 2 and 5:30-11:30 of March 3, and the westward current prevailed in the

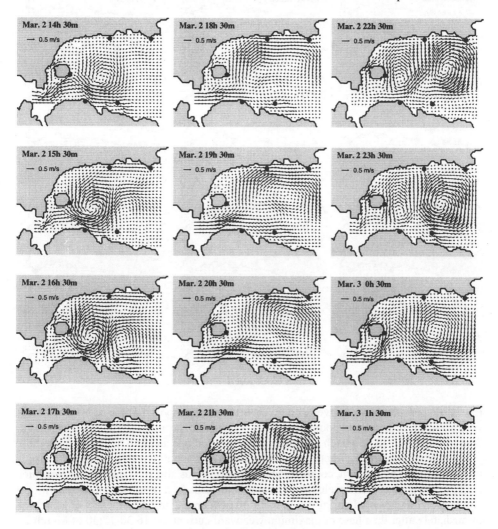

Fig.8 Hourly plot of the tidal current fields for the day reconstructed by the data assimilation. The shipboard ADCP data for the depth-averaged values in the upper 20 m are superimposed on the reconstructed currents with thick, bright arrows whenever the ADCP observations are carried out.

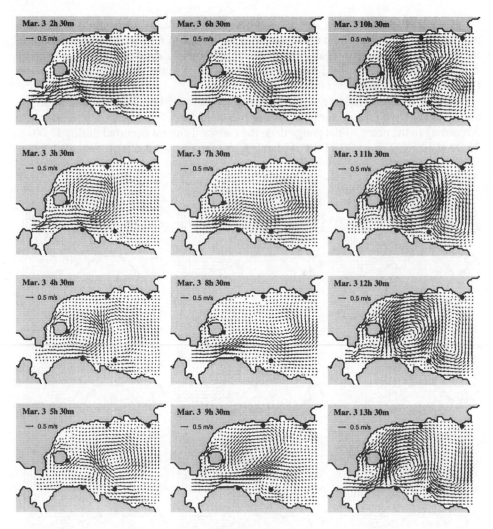

Fig.8 Continued

remaining periods. A well-developed tidal-vortex pair composed of the right, clockwise vortex and the left, counterclockwise vortex was visible during 21:30 to 23:30 of March 2 and 10:30 to 12:30 of March 3. These periods started one and half hour later than the phase of maximum eastward current and continued for about two hours over the phase of slack water. The western, counterclockwise vortex was absorbed into the narrow inlet at the western edge with developing westward currents. As a result, only the eastern, clockwise vortex existed during 14:30 to 15:30 of March 2 and 2:30 to 3:30 of March 3 immediately after the phase of maximum westward current. At 16:30-17:30 of March 2 and 4:30-5:30 of March 3 following the tidal phase with one clockwise vortex, a new vortex pair formed with a rotation opposite to that mentioned above. This vortex pair is suddenly diminished in the subsequent slack water.

5. Summary and Discussion

The coastal acoustic tomography system (CATS) composed of five moored-type acoustic stations is applied to measure, at a 5-minute interval, tidal vortices induced in the small channel of the inland sea. The data are analyzed by two kinds of methods; the inversion based on the damped least squares method and the assimilation of the tomography data into the ocean model by use of the ensemble Kalman filter technique. The depth-averaged current fields are poorly reconstructed by the inversion method because of the partial failure of data acquisition due to unexpected troubles. The poorly reconstructed current fields are significantly improved by applying the data assimilation technique. The validity of the assimilation result is also confirmed in comparison with the shipboard ADCP data acquired during the tomography experiment[8]. As a result, all the processes of initiation, growth, translation and decay of the tidal vortices are well understood through the continuous mapping of vortex fields. It is, thus, proposed that the ensemble Kalman filter may be the best technique of data assimilation suitable for coastal acoustic tomography in the shallow sea dominated by strongly nonlinear tidal currents.

Acknowledgments

We thank Drs. Y. Takasugi and X-H. Zhu for the kindest supply of the CTD data. This program is partly supported by the Japan Science and Technology Corporation.

References

1. W. Munk, P. F. Worcestor and C. Wunsch, Ocean Acoustic Tomography (Cambridge Univ. Press, Cambridge, 1995).
2. F. Gaillard, Y. Desaubies, U. Send and F. Schott, "A four-dimensional analysis of the thermal structure in the Gulf of Lion", J. Geophys. Res. **102** (1997), 12515-12537.
3. G. Yuan, I. Nakano, H. Fujimori, T. Nakamura, T. Kamoshida and A. Kaya, "Tomographic measurements of the Kuroshio Extension meander and its associated eddies", Geophys. Res. Lett. **26** (1) (1999), 79-82.
4. D. B. Chester, P. Malanotte-Rizzoli, "Acoustic tomography in the Straits of Florida", J. Geophys. Res., **96** (1991), 7023-7048.
5. C. S. Chiu, J. H. Miller and J. F. Lynch, "Inverse technique for coastal acoustic tomography", in Theoretical and Computational Acoustics Vol.2, eds. D. Lee and M. H. Schultz (World Scientific, Singapore, 1994), pp. 917-931.
6. H. Zheng, N. Gohda, H. Noguchi, T. Ito, H. Yamaoka, T. Tamura, Y. Takasugi and A. Kaneko, "Reciprocal sound transmission experiment for current measurement in the Seto Inland Sea", J. Oceanogr., **53** (1997), 117-127.
7. H. Zheng, H. Yamaoka, N. Gohda, H. Noguchi and A. Kaneko, "Design of the acoustic tomography system for velocity measurement with an application to the coastal sea", J. Acoust. Soc. Jpn. (E), **19** (3) (1998), 199-210.
8. J-H. Park and A. Kaneko, "Assimilation of coastal acoustic tomography data into a barotropic ocean model", Geophys. Res. Letters, **27** (20) (2000), 3373-3376.
9. M. K. Simon, J. K. Omura, R. A. Scholtz and B. K. Levitt, Spread Spectrum Communications Handbook (McGraw-Hill, New York, 1994).
10. J-H. Park and A. Kaneko, "Computer simulation of the coastal acoustic tomography by a two-dimensional vortex model", J. Oceanogr., **57** (2001), in press.
11. H. Sielschott, "Measurement of horizontal flow in a large scale furnace using acoustic vector tomography", Flow Meas. Instrum., **8** (1997), 191-197.
12. R. E. Kalman, "A new approach to linear filtering and prediction problems", Trans. ASME, Series D, J. Basic. Engr., **82** (1960), 35-45.
13. G. Evensen, "Sequential data assimilation with a nonlinear quasi-geostrophic model using Monte Carlo methods to forecast error statistics", J. Geophys. Res., **99** (1994), 10143-10162.
14. A. F. Blumberg and G. L. Meller., "A description of a three-dimensional coastal ocean circulation model", Three-dimensional coastal ocean models, Coastal Estuarine Sci., **4** (1987), 1-16.

Theoretical and Computational Acoustics 2001
E.-C. Shang, Qihu Li and T. F. Gao (Editors)

Acoustic Tomography for Monitoring Ocean Cold Water Mass

Peng Linhui, Wang Ning, Qiu Xiaofang, Er Chang Shang*
Ocean University of Qingdao, Qingdao, 266003

*CIRES, University of Colorado/NOAA, ETL, Boulder,
Colorado 80303-3328, USA

Abstract

Ocean cold water mass is an oceanographic phenomena that oceanographers concerned many years. Method of monitoring the cold water mass is a problem to be solved urgently. A new tomography approach, modal wave-number tomography (**MWNT**), is proposed for monitoring the cold water mass and numerical simulation is carried out for Yellow sea cold water mass. The perturbation theory is used to invert the coefficients of empirical orthogonal functions of sound speed profile (SSP) from the local modal wave number perturbation. The numerical simulation shows that the modal wave number tomography can invert average SSP, in particular, is of potential to monitoring range dependent SSP structure.

Keywords: monitoring, tomography, wave-number, Yellow sea

I. Introduction

Cold water mass is one of the importance phenomenon in the near sea of China. Yellow sea cold water mass have been followed with interest by oceanographer for many years[1] since it exists in the whole Yellow Sea all year. Method for monitoring the cold water mass is a problem to be solved urgently. The exist of the cold water mass arouse variety of the ocean sound speed profile, so the method of acoustic tomography can be used to monitoring the cold water mass. Modal wave-number inversion of seabed developed by Rajan et al.[2] is succeed in obtaining bottom geoacoustic parameters as a function of depth in shallow water. The method was extended to accommodate a weakly range-dependent environment[3]. It is shown how one can obtain the SSP perturbation through the perturbation in the modal wave number with respect to the background modal wave number perturbation. The modal wave number can be picked up from beam-formed out of a horizontal towed. Here this method is extended to obtained sound speed profile of water column for weakly range-dependent environment. In this paper, numerical simulation of the modal wave number tomography for Yellow Sea cold water mass have been done. There are region of weakly range-dependent and region of intense rang-dependent either in the cold water mass area. Numerical simulations are carried out for the two kind cases. **EOF** of sound speed profile is used in the work so integral equation of perturbation inverse can be solved easily. In the numerical simulation PE calculation code **FOR3D**[4] is used for calculate sound pressure, then wave number decomposition is carried out to pick up the modal wave number. The modal wave number of the background field is obtained by **KRAKEN** program[5].

In Sec.II of this paper, the background material of Yellow Sea cold water mass is described. Then the method of modal wave-number tomography for sound speed profile of water column is presented. This consists of the perturbation inversion equation, **EOF** expand of sound speed profile and modal wave number decomposing of sound field. At last, results of numerical simulation are given and analyzed.

II.   Background material of Yellow Sea cold water mass

The Yellow Sea cold water mass lies in the center area of South Yellow Sea and North Yellow Sea.   The vertical stratification of it is most obvious in summer.   Water temperature of the upper layer can be 24 C~27 C high, the salinity is about 31.0, in the deep district, salinity is about 32.0~32.5 and water temperature is 6 C low, contract with high temperature of upper layer water. So it is called "Yellow Sea cold water mass".   There exist cold centers of the cold water mass in South Yellow Sea and North Yellow Sea, respectively.   The average location of North Yellow Sea over many years is about $38°14'N$ 、 $122°12'E$. Taking the section of Darian-chengshan, the average temperature distribution over many years is shown in Fig.1[1]. The sound speed distribution is calculated by Eq.(1)[6] from Fig.1, and the density is taken as constant in this area.   The numerical simulation is done for sound speed and it can be converted to temperature by :

$$c = 1449.2 + 4.6T - 0.055T^2 + (1.34 - 0.010T)(S - 35) + 0.016z \qquad (1)$$

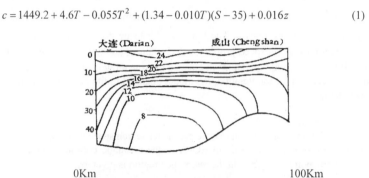

0Km   100Km
Fig1. Temperature distribution of Yellow Sea cold water mass

III. Perturbation inverse of modal wave-number

The sound speed variation in water will cause variation of local modal wave-number.   So the relation of them can be used to determine sound speed variation from the variation of the modal wave-number. We start out with some "background" problem that approximates the problem of interest.   The real environment is different from the background can be consider as a perturbation over the background.   Perturbation theory provides the means for determining the difference of sound speed profile $\Delta c(z)$ between the background and the true model.

(1). Perturbation inverse of sound speed profile

Consider a wave-guide of stratified horizontally with a constant density profile. For background problem with a sound speed profile $c_0(z)$, the spatial part of pressure $p(r,z)$ (assume with harmonic time dependence $e^{-j\omega t}$) satisfies the homogeneous Helmhotz equation

$$\nabla^2 p + k^{(0)^2} p = 0 \qquad (2)$$

At place where is far from the source, neglecting the contribution of continuous spectrum, solution of the equation can be expressed as

$$p(r,z) = \sum_m A_m(r)Z_m^{(0)}(z) \tag{3}$$

$Z_m^{(0)}(z)$ is the mth vertical normal mode for the background satisfying surface condition, bottom condition and equation

$$\left(\frac{d^2}{dz^2} + k^{(0)^2}(z)\right)Z_m^{(0)}(z) = k_m^{(0)^2}Z_m^{(0)}(z) \tag{4}$$

where, $k^{(0)}(z) = \omega/c_0(z)$, $k_m^{(0)}$ is the mth eigenvalue for the background problem.

For small perturbation sound speed profile with $c(z) = c_0(z) + \Delta c(z)$, the wave number $k(z) = \omega/c(z)$ is different from wave number $k^{(0)}(z)$ of the background. The modal wave number perturbation is given by

$$\Delta k_m^{(1)} = \frac{1}{k_m^{(0)}} \int_0^\infty \rho_0^{-1} \left|Z_m^{(0)}(z)\right|^2 k^{(0)^2}(z)\frac{\Delta c(z)}{c_0(z)} dz \tag{5}$$

where $\Delta k_m = k_m - k_m^0$. This is the equation of modal wave-number perturbation inversion. Where, quantities of the background are known. The term $\Delta k_m^{(1)}$ can be obtained by experiment. Solving Eq.(5) the perturbation of sound speed profile $\Delta c(z)$ is gained, and the sound speed profile can be obtained further. If the background model is not close to the true model, an iterative procedure can be employed where, at each step of the iteration, the model obtained in the previous iteration is used as the background model.

(2). Solving the integral equation

In order to solve the integral equation (5), the parameter search space had to be reduced. Here the sound speed profile is expressed as a sum of EOFs. In the rectangle coordinate system, the sound speed $c(x, y, z_i)$ at any point (x, y, z_i) in the researched region is given by

$$c(x, y, z_i) = \bar{c}(z_i) + \sum_{k=1}^K a_k(x, y)f_k(z_i) \tag{6}$$

where $\bar{c}(z_i)$ is the average sound speed of all of sound speed samples in the region at z_i depth. $f_k(z_i)$ is a eigenvector of the covariance matrix of SSP sample matrix and $a_k(x, y)$ is the coefficient of $f_k(z_i)$. Substituting Eq.(6) into Eq.(5) and discreting Eq.(5), we change question of solving a integral equation into the question of solving the algebraic equation.

(3). Decomposing of the modal wave-number

In Eq.(5), $\Delta k_m^{(1)}(z)$ is difference between the background wave number and the true modal wave number. The wave number of the true field can be obtained from beam-formed output of horizontal towed. Here we decompose the modal wave-number spectrum from the numerical simulating sound field. The points corresponding to the spectrum peaks are the modal wave numbers. For the horizontal stratified wave guide wave number spectrum is defined as[3]

$$g(k_r; z, z_0) = \int_0^\infty p(r, z, z_0)J_0(k_r r)rdr$$

where, k_r is the horizontal component of the total wave number $k(z) = \omega/c(z)$, r is the horizontal range between the source and the receiver. z, z_0 are depths of the receiver and the source respectively. J_0 is the zeroth-order Bessel function, g is the Green's function corresponds to the problem of range-dependent, satisfying

$$\left[\frac{d^2}{dz^2} + \left(k^2(z) - k_r^2\right)\right] g(k_r; z, z_0) = -2\delta(z - z_0)$$

δ is the Dirac delta functional. At far field, wave number can be obtained by translation of approximating zeroth-order Bessel function as

$$g(k_r; z, z_0) \sim \frac{e^{i\frac{\pi}{4}}}{\sqrt{2\pi k_r}} \int_0^{\infty} p(r, z, z_0) \sqrt{r} e^{-ik_r r} dr \qquad (7)$$

(4). Range-dependent case

In range-dependent case, the Bessel transform relation in Eq.(7) is no longer rigorously valid, since the modal wave number and the eigenfunction is different at different place as sound speed profile or other parameter are different. For environment with weakly range-dependent, one can discredited the integral aera into a finite number of segments whose properties are range-independent[3]. Assume a segment $r1 < r < r2$, the modal wave number can be estimate as

$$g(k_r; z, z_0) \sim \frac{e^{i\frac{\pi}{4}}}{\sqrt{2\pi k_r}} \int_{r1}^{r2} p(r, z, z_0) \sqrt{r} e^{-ik_r r} dr \qquad (8)$$

The modal wave-numbers obtained from eq.(8) do not exactly correspond to the local eigenvalues because the eigenvalues do not remain constant over range covered by the integration. Different integral length and distribution of sound speed in this area affect precision of the estimation.

IV.   Numerical simulation

(1) Adiabatic Test

Considering the ocean environment as **Fig.1**.   It can be seen from the figure that the variety of sound speed in cold water mass region is different along horizontal range.   At the edge of the water mass, the sound speed profile varies quickly with range.   At the center region of the water mass, the sound speed profile varies slowly with range.   Taking tow kind of regions for the numerical simulation, the one segment is r=0~15Km whose sound speed profile varies quickly with range, and the second segment is r=25~45Km which sound speed profile varies slowly with range.   In the numerical simulation, depth of the sound source is $z_0 = 10$m.   The source frequency is f=300Hz. Depth of the receiver is h=40m.   Acoustic pressure is calculated by **PE** calculation Code **For3D**.

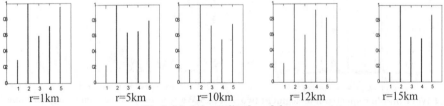

r=1km   r=5km   r=10km   r=12km   r=15km

Fig.2. Modal amplitudes of the segment one
(Mode Coupling segment)

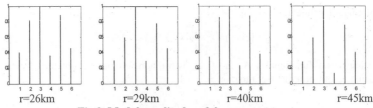

| r=26km | r=29km | r=40km | r=45km |

Fig.3. Modal amplitudes of the segment two
(Adiabatic segment)

At first, adiabatic test is carried out.   The sound pressure of the segment one is expanded as a sum of the local modal of r=5Km section.   The amplitudes of normal modals at different section are show in **Fig.2**.   The sound pressure of the segment two is expanded as the local mold of r=38.3Km section.   The amplitudes of normal mold at different section in the second segment are show in **Fig.3**. It can be seen from **Fig.2** that modal amplitudes vary in evidence with range in the first segment. Where amplitude of the third normal mold do not vary with range in substance.   Amplitudes of the forth and the fifth normal mold vary with range obviously.   This is shown that acoustic propagation in this segment cannot be considered as adiabatic.   In **Fig.3**, it can be seen that there are no obvious variation of normal mode amplitude in the second segment.   So acoustic propagation in this segment is adiabatic in evidence.

At the groundwork of adiabatic test, inverse of sound speed profile is carried out.   The modal wave number of the background is calculated by the normal mode program **KRAKEN** code, where sound speed profile at r=0.0m is taken as the background of segment one, and sound speed profile at r=25Km is taken as the background of segment two.   Calculated value of the background modal wave number are line in **Table.1**. The modal wave number spectrum of 25~45Km segment is shown in **Fig.4**. The modal wave numbers obtained from the spectrum is line in **Table.2**.

Table1. modal wave number of the background

	Kr1	Kr2	Kr3	Kr4	Kr5	Kr6
R=0Km	1.25050203	1.24259068	1.23472743	1.22526656	1.21320533	
R=25Km	1.26891804	1.26006598	1.24918498	1.23656950	1.22389568	1.20969952

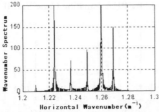

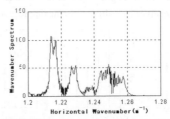

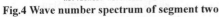

| **Fig.4 Wave number spectrum of segment two** | **Fig.5 Wave number spectrum of segment one** |

Table.2. The modal wave numbers of segment two

Kr1	Kr2	Kr3	Kr4	Kr5	Kr6
1.26885327	1.25983509	1.24915877	1.23678908	1.22398440	1.20982475

For the segment r=0~15Km, integral range is taken as 0.001~10Km, the modal wave number obtained is shown

in Fig.5. It can be seen that the wave number spectrum is intermixed in this segment. Distinct wave number cannot be distinguished. This is result of nonadiabatic propagation. So, discrete this segment into four portion further: D1(0~4Km), D2(4~8Km), D3(8~12Km) and D4(12~15Km). In these new portions, varation of environment parameters is slow relatively, approximating to range-independent. The wave number spectra of these portions are shown as Fig.6, and the corresponding modal wave number is line in Tab.3

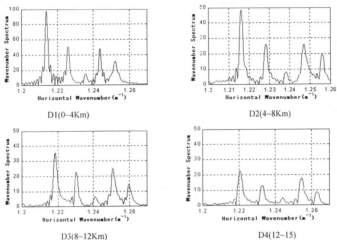

D1(0~4Km)   D2(4~8Km)

D3(8~12Km)   D4(12~15)

Fig.6 Wave number spectrum of segment one

Tab.3. The modal wave numbers of segment two

	Kr1	Kr2	Kr3	Kr4	Kr5
D1(0~4Km)	1.25204500	1.24359250	1.23549875	1.22590875	1.21390375
D2(4~8Km)	1.25581802	1.24699037	1.23809273	1.22817147	1.21604549
D3(8~12Km)	1.25972878	1.25088363	1.24099737	1.23038459	1.21845144
D4(12~15Km)	1.26289618	1.25446258	1.24387102	1.23261456	1.22058823

(2) Inversion of SSP

Samples of sound speed profile for **EOF** are taken from the region of Yellow Sea could water mass. The parameters and results above are used to inverse sound speed profiles. The sound speed profile inversed for segment r=25~45Km is shown in **Fig.7**. Where, **Fig.7(a)** is comparison of inversed profile with the average profile of this segment. **Fig.7(b)** is comparison of inversed profile with the sound speed profiled of section r=35Km. The parameter of section r=25Km is used as the background parameter for this region. The parameter of section r=0Km is used as the background parameter for inverse in region r=0~15Km. The comparisons of inversed profiled of portion D1, D2, D3, D4 with the profile of section r=8Km which is the center of the region are shown in Fig.8. Fig.9 are comparisons of inversed profiled of portion D1, D2, D3, D4 with the average profile of segment 0~15Km. Fig.10 are comparisons of inversed profiled of portion D1, D2, D3, D4 with the average profiles of these portion respectively.

It can be seen from result of **Fig.7** that for segment two which sound profile varies weakly with range inversed sound profile is coincidence with the average profile and profile of center section of this segment either. Inversed sound speed profile of potion D1, D2, D3, D4 of segment two are not perfectly coincidence with the average profile of segment one or center section profile of the segment. But Inversed sound speed profile of potion D2, D3 of the segment are coincidence basically with the average profile of segment one or center section profile of the segment as shown in Fig.8 and Fig.9. In Fig.10 it is clear that inverse is succeed for one potion, because these potions are basically rang-independent. We have conclusion that method of modal wave number tomography can be used to inverse sound speed profile of water column. The inversed profile is coincidence with true profile for the case of range-independent and weakly range-dependent. For case of range-dependent, inversed profile obtained from above method is differ from the true profile. In order to pick up modal wave number, a finite horizontal range is needed to get wave number spectrum. When environment parameter vary intensely in the range, one cannot obtain the wave numbers exactly, and inversed profile is not presice. So the length of range for wave number decomposing should be select carefully.

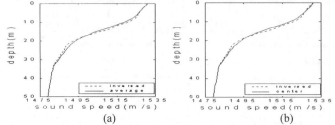

(a)   (b)

Fig.7. The sound speed profile inverted for segment r=25~45Km

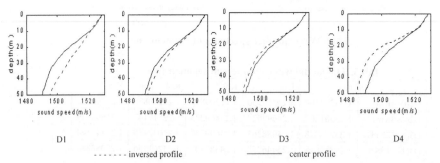

D1   D2   D3   D4

- - - - - - inversed profile   —————— center profile

Fig.8. Comparisons of inversed profiled of portion D1, D2, D3, D4 with the profile of section r=8Km

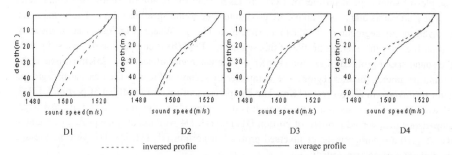

D1   D2   D3   D4

- - - - - - inversed profile   —————— average profile

Fig.9. Comparisons of inversed profiled with the profile the average profile of segment 0~15Km

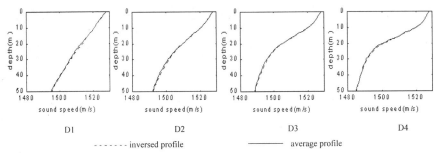

D1   D2   D3   D4

- - - - - - inversed profile   ———— average profile

Fig.10. Comparisons of inversed profiled with of average profiles

V.   Conclusion

The method of modal wave number tomography is extended to obtain sound speed profiles of water column. Numerical simulation of modal wave number for monitoring Yellow Sea could water mass has been carried out. It can be seen from the result that modal wave number tomography is succeed in monitoring the cold water mass. The further work refer to ocean experiment is expected. For environment with weakly range-dependent, inversed sound speed profile is coincidence with the average profile of the range. For environment which parameters vary with range quickly, we can discredite the area into a finite number of segments whose properties are range-independent or weakly range-dependent. Then modal wave number tomography can be used . We are now developing a new modal wave number-mode coupling tomography method for monitoring strong range dependent environment.

Reference

[1]Yuan ye li " Circulation of Yellow Sea cold Water Mass I . Thermal and circulation structure in the central region"。 Ocean and limnology，Vol10，No3，1979. inChinese.

[2]Subramaniam D. Rajan, James F. Lynch and George V. Frisk, "Perturbative inversion methods for obtaining bottom geoacoustic parameters in shallow water",JASA,**82**(3),998-1017(1987)

[3] George V. Frisk, James F. Lynch and Subramaniam D. Rajan, "Determination of compressional wave speed profiles using modal inverse techniques in range-dependent environment", JASA, **86**(5),1926-`939(1989)

[4]Lee.D., M.H.Schultz. Numerical Ocean Acoustic Propagation in Three Dimensions. [M] World Scientific.Singapore,1995

[5]M.B.Porter, The KRAKEN Normal Mode Program, NRL/MR/5120-92-692.

[6]Wang De Zhao & Shang Er Chang, Underwater Acoustics。Scientific Publication China Co.1981,in Chinese.

V. Conclusion

Reference

Theoretical and Computational Acoustics 2001
E.-C. Shang, Qihu Li and T. F. Gao (Editors)
© 2002 World Scientific Publishing Co.

Mode Coupling Inversions — I. Perturbation Theory

Wang Ning*, Er Chang Shang
Information Science and Technology College
Ocean University of Qingdao, P. R. China, 5-Yushan Rd, Qingdao, 266003

Abstract:

Coupled-WKB perturbation and generalized perturbation relations for scattering matrices in arbitrary background are derived. The relations can be considered as a linear integral mapping between model perturbation and scattering data perturbation. Sensitivity kernel is introduced to investigate the functional dependence of the scattering data on the model parameters. Numerical simulation shows that for strong range dependent waveguide, the sensitivity kernels for the Coupled-WKB and the general perturbations behavior completely different. It means that the usual perturbation tomography inversion has its limit for the inversion of strong range dependent model since the Frechet derivative of the scattering matrix with respect to the model parameter has remarkable nonlinear property.

PACS numbers: 43.30Bp

I. INTRODUCTION

ABNORMALLY LARGE acoustic mode coupling transmission attenuation due to the acoustic interaction with naturally occurring internal solitary wave trains, shelfbreak oceanfront was found experimentally [1--5]. However, the current acoustic tomography methods based on the adiabatic modal phase or travel time, are difficult to handle the mode-coupling problem [6-8]. There have been very few studies on the mode coupling tomography inversion in the ocean acoustic community. To include mode coupling in costal acoustic tomography, Chiu, Miller and Lynch developed a broadband forward coupled-mode approach to estimate arrival structure due to mode coupling [9]. In the geophysics community, several tries to incorporat mode coupling into diffration tomography have been made. Li & Tanimoto improved the adiabatic inversion by considering the mode coupling based on the Born approximation [10]. Marquering & Snieder proposed an inverse method based on a scalar exponent approximation [11]. The inversion incorporating mode coupling has two main advantages over the adiabatic inversion. The first is that including the mode coupling neglected in the adiabatic approximation, is absolutely needed to strong range dependent tomography. The second is that the mode coupling may provide more information than that only using modal phase or travel time perturbation. There are, however disadvantages that a larger model parameter set is needed to parameterize sufficiently realistic structure, this makes the inversion highly non-unique, and the inversion may be strong nonlinear. An accepted way reducing parameter set is to make eigenvector analysis or singular value decomposition based on perturbation relation. A perturbation theory directly provides Frechet derivatives with respect to model perturbation, is an indication of sensitivity. It is expected that the inversion in a strong range dependent environment may depends on the background, and analysis of the sensitivity to key environmental parameter is important for the inversion. The purpose of this paper is to develop this perturbation theory, in particular, mode coupling perturbation theory in arbitary range dependent background. An integral equation approach is developed in **Sec. II**, and generalized perturbation formula of forward and backward scattering matrices for arbitrary background is derived. In **Sec. III**, we derive perturbation formula from Evans's stepwise coupled mode theory, and investigate the relationships between the discrete form in this Section and the continuous forms given in **Sec. II**, respectively.

II. INTEGRAL EQUATION APPROACH

For clarity, we restrict our discussion to 2D - wave propagation problem with the coordinate system such that the z-axis lies along the depth direction, and x-axis points the range direction with origin at the source position. In this section, we shall consider an integral equation approach for mode coupling perturbation theory.

2-1: Born Approximation

At first, we start to derive simple Born approximation. The background model is a range independent layered waveguide. An inhomogeneous region lies in the interval **(0,L)** as shown in **Fig. 1**. The govern equation is (a harmonic time dependence exp (-iωt) is assumed)

$$\rho(x,z)[\frac{\partial}{\partial x}\frac{1}{\rho(x,z)}\frac{\partial}{\partial x}+\frac{\partial}{\partial z}\frac{1}{\rho(x,z)}\frac{\partial}{\partial z}]p(x,z)+\frac{\omega^2}{c(x,z)^2}p(x,z)=0, \qquad (1)$$

where $p(\omega;x,z)$ is the acoustic pressure, c (x, z) and ρ(x, z) are the medium sound speed and density as functions of depth z and range x. Assuming that the medium perturbation is small, i.e.

$$\rho(x,z) = \rho_0(z)+\rho_1(x,z) \text{ and } \frac{\rho_1}{\rho_0} <<1, \quad (2)$$

$$\cdot \ c(x,z) = c_0(z)+c_1(x,z) \text{ and } \frac{c_1}{c_0} <<1. \quad (3)$$

In (2)-(3), c_1(x,z) and ρ_1(x,z) denote the perturbation quantities, and c_0(z) and ρ_0(z) are the background medium parameters. Under this assumption, **(1)** can be rewritten as

$$\rho_0(z)[\frac{\partial}{\partial x}\frac{1}{\rho_0(z)}\frac{\partial}{\partial x}+\frac{\partial}{\partial z}\frac{1}{\rho_0(z)}\frac{\partial}{\partial z}]p(x,z)+\frac{\omega^2}{c_0(z)^2}p(x,z)+\Delta(x,z)p(x,z)=0, \quad (4)$$

where the differential operator $\Delta(x,z)$ is defined by

$$\Delta(x,z)=-[\frac{\partial}{\partial x}(\frac{\rho_1}{\rho_0})]\frac{\partial}{\partial x}-[\frac{\partial}{\partial z}(\frac{\rho_1}{\rho_0})]\frac{\partial}{\partial z}-2k_0^2\left(\frac{c_1(x,z)}{c_0}\right). \quad (5)$$

Without loss of generality, throughout the paper we neglect the variation of density, i.e. $\rho_1=0$.

Neglecting the contribution from continuous spectrum, the wave function can be written as

$$p(x,z)=\sum_{n=1}^{\infty}A_n(x)\phi_n(z), \quad (6)$$

where $\phi_n(z)$ ($n=1,2\ldots$) are the vertical modal eigenfunctions satisfying

$$\{\rho_0\frac{d}{dz}\frac{1}{\rho_0}\frac{d}{dz}+k_0^2(z)-k_n^2\}\phi_n(z)=0, \quad n=1,2\ldots N \quad (7)$$

and the orthogonal relations

$$<\phi_m,\phi_n>\equiv\int\frac{\varphi_m(z)\varphi_n(z)}{\rho(z)}dz=\delta_{mn}. \quad (8)$$

The modal coefficients $A_n(x)$ are range dependent, and describe the mode coupling. Using the orthogonal relation (8) we have

$$\frac{d^2}{dx^2}A_m(x)+\sum_{n=1}^{N}[k_m^2\delta_{mn}+V_{mn}(x)]A_n(x)=0, \quad m=1,2,\ldots,N. \quad (9)$$

The matrix potential (MP) $V(x)$ in (9) is determined by

$$V_{mn}(x)\equiv<\phi_m,\Delta\phi_n>. \quad (10)$$

In principal, (9) is an infinite dimensional ordinary differential equation (ODE.) System. However, it is known that for a fixed frequency only finite numbers (assuming N_m) of modes contribut remarkably at a large distance.

In the first approximation, neglecting the off-diagonal elements of the MP, (7) is reduced to

$$\frac{d^2}{dx^2}A_n(x)+[k_n^2+V_{nn}(x)]A_n(x)=0, \quad n=1,2,\ldots, \quad (11)$$

(11) is a decoupled equation; its solution gives the adiabatic-mode approximation of (9). As (9) has exactly the same form as the coupled equation of the one-dimensional many S-channel quantum scattering problem, we adapt below an analogy of the formulation used in the quantum problem. At first we define a specific matrix solution Jost solution used in the quantum physics [12-13]. The solution are determined by the following ($N_m \times N_m$) matrix-valued integral equation

$$A_J(K,x)=\exp(iKx)+\int_x^L\frac{\sin K(x-x')}{K}V(x')A_J(K,x')dx', \quad x\in(0,L) \quad (12)$$

$$A_J(K,x)\rightarrow\exp(iKx), \quad as \quad x\rightarrow L.$$

$$\exp(iKx)\equiv diag(e^{ik_1x},e^{ik_2x},\ldots,e^{ik_Nx}), \quad (13)$$

and

$$\frac{Sin[K(x-x')]}{K}=Diag\{\frac{\sin[k_1(x-x')]}{k_1},\ldots,\frac{\sin[k_N(x-x')]}{k_N}\}, \quad (14)$$

are diagonal matrices. As K appears in the form K^2 in (9), then $A_J(-K,x)$ is the solution of (9) if $A_J(K,x)$ is. Defining the Wronskian matrix

$$W[A_1,A_2]\equiv A_1^T\frac{d}{dx}A_2-\frac{d}{dx}A_1^TA_2,$$

it satisfies

$$W[A_J(K,x),A_J(-K,x)]=const. , \quad x\in(0,L). \quad (15)$$

$$W[A_1,A_2]^T=-W[A_2,A_1], \quad W[A_1,A_1]=0. \quad (16)$$

The superscript T here denotes the matrix transpose. It is clear that $A_J(-K,x)$ and $A_J(K,x)$ consist a fundamental solution pair of (9).

We next investigate the relation between the Jost solution and the physical scattering solution. Physically, the matrix valued physical scattering solution is required to satisfy the following asymptotic form

$$A_{phys}(x,K) = \begin{cases} e^{iKx} + e^{-iKx}S_b(K,L,J); & x \longrightarrow 0 \\ e^{iKx}S_f(K,L,J); & x \longrightarrow L \end{cases} \quad (17)$$

In (17), the N×N matrices $S_f(K,L,J)$ and $S_b(K,L,J)$ define the Jost forward scattering matrix (JFSM) and Jost backward scattering matrices (JBSM), respectively. To distinguish to the (physical) scattering matrices defined in the next section, we use an addtional argument J in the above definition. The physical solution consists from the superposition of incidence and reflection waves as x→0 and is pure outgoing as x→L. (17) means directly

$$A_{phys}(x,K) = A_J(x,K)S_f(K,L,J). \quad (18)$$

This equation gives a relation between the physical and the Jost solution. Using (17) and (18), we have the asymptotic form for the Jost solution

$$A_J(K,x) = e^{iKx}S_f(K,L,J)^{-1} + e^{-iKx}S_b(K,L)S_f(K,L,J)^{-1}, \text{ as x→0} \quad (19)$$

Taking x→0 in (12) and comparing with (19) obtains the following integral representations for the JFSM

$$S_f(K,L,J)^{-1} = I + \int_0^L \frac{\exp(-iKx)}{2iK}V(x)A_J(x,k)dx, \quad (20)$$

and the BFSM

$$S_b(K,L,J)S_f(K,L,J)^{-1} = -\int_0^L \frac{\exp(iKx)}{2iK}V(x)A_J(x,k)dx. \quad (21)$$

For an experimental setup in which receivers (vertical line array) locate at depths (x=L, z=z_J), (j=1,2,M1), the pressure of the jth receiver generated by a point source at (x=0,z=z_s) is given by

$$p_j(L,z_j) = \sum_{n,m}^N \varphi_m(z_j)\exp(ik_mL)[S_f(K,L,J)]_{mn}\varphi_n(z_s) \quad j=1,2,...,M1, \quad (22)$$

and the amplitude for the nth mode is given by

$$q_n = \sum_m^N \exp(ik_mL)[S_f(K,L,J)]_{mn}\varphi_n(z_s) \quad n=1,2,...,N. \quad (23)$$

From (20) and (21), the Born approximation for the scattering matrices and modal travel time perturbation can be easily derived; the result is summarized in Table I.

2-2: Distorted Born Approximation

Formulae presented in Table I are useful for both the direct and the inverse problem, but these can only be used to the range independent background. We next derive a general perturbation formula for arbitrary background. Instead of (9) we start from the following equation

$$\frac{d^2}{dx^2}A(x) + (K^2 + V_0(x) + \Delta V(x))A(x) = 0, \quad (24)$$

$V_0(x)$ is a range dependent potential for the background. Assuming that the Jost solution with perturbation $V_0(x)$ has been solved, it can be shown (Appendix A) that the perturbation formulae for the JFSM and the JBSM are given by

$$\Delta S_f(K,L,J) = -\frac{1}{2iK}\int_{-\infty}^{+\infty}A^0_J(-K,y)^T\Delta V(y)A^0_J(K,y)dyS^0_f(K,L,J), \quad (25)$$

and

$$\Delta S_b(K,L,J) = -S^0_f(K,L,J)^{-1}\frac{1}{2iK}\int_{-\infty}^{+\infty}A^0_J(K,y)^T\Delta V(y)A^0_J(K,y)dyS^0_f(K,L,J). \quad (26)$$

In (25) and (26), the quantities with superscript 0 indicate that for the background. With the help of (17), (25) can be rewritten to

$$\Delta S_f(K,L,J) = -\frac{1}{2iK}[S^0_f(-K,L,J)^T]^{-1}\int_{-\infty}^{+\infty}A^0_p(-K,y)^T\Delta V(y)A^0_p(K,y)dy. \quad (27)$$

When the sound propagation with $V_0(x)$ is weak mode coupling, the physical solution for $V_0(x)$ can be approximated by the adiabatic approximation, then

$$\Delta S_f(K,L,J) = -\frac{1}{2iK}[S^0_f(-K,L,J)^T]^{-1}\int_{-\infty}^{+\infty}[\exp-i\int^{x'}K(s)ds]\Delta V(x')[\exp-i\int^{x'}K(s)ds]dx', \quad (28)$$

where $K(s)$ is local eigenvalue matrix. Taking $S^0_f(-K,L,J) = Id$ in (28), we have a **WKBJ-Born** approximation. It is conventional to rewrite (27)-(28) into the following forms

$$[S^0_f(-K,L,J)^T][2iK]\Delta S_f(K,L,J) = 2k_0^2\int_0^{+L}\int dz A^0_p(-K,x)^T\Psi(z)\frac{\delta c(x,z)}{c_0(z)}A^0_p(K,x)dx, \quad (29)$$

and

$$[S^0_f(-K,L,J)^T][2iK]\Delta S_f(K,L,J) = 2k_0^2\int_0^{+L}\int dz \exp[-\int_0^x iK(s)ds]\Psi(z)\frac{\delta c(x,z)}{c_0(z)}\exp[\int_0^x iK(s)ds]dx. \quad (30)$$

In (29) and (30), the matrix $\Psi(z)$ is defined by

$$[\Psi(z)]_{mn} \equiv \frac{\varphi^0_m(z)\varphi^0_n(z)}{\rho(z)}, \quad (31)$$

and the integral kernels are named as the sensitivity kernels. A discrete form for (25) is derived in **Appendix-B**, it is useful for numerical consideration.

III. STEPWISE COUPLED MODE APPROACH

In the last section, we derived perturbation formulae for the **JSM**, but the derivation is not physically straightforward. In this section, we derive the perturbation formulae by dividing the range axis into a number of segments and approximating the field as range independent within the each segment as shown in **Fig.2**.

Assuming that the region is divided into N segments each with range length h_j ($j=1,2....,N_h$). Neglecting the contribution from continuous spectrum，the sound pressure in the jth horizontal segment can be expressed as

$$p^j(r,z) = \sum_{m=1}^M[a^j_m\exp(ik^j_m(x-x_j)) + b^j_m\exp(-ik^j_m(x-x_j))]\varphi^j_m, \quad j=1,2...N_h \quad (32)$$

where (a^j_m,b^j_m) are the modal coefficients, and k^j_m (m=1,2,3,...; j=1,2,...,N$_h$) are the local eigenvalues of normal modes in the jth segment. In (32), subsript – denotes the quantities at the left side of an interface and + denotes that at the right side, respectively. By the continuity of pressure and normal velocity obtains [14-15]

$$\begin{bmatrix} b_+^{j+1} \\ a_+^{j+1} \end{bmatrix} = \begin{bmatrix} M^J_1 & M^J_2 \\ M^J_3 & M^J_4 \end{bmatrix}\begin{bmatrix} b_-^j \\ a_-^j \end{bmatrix}. \quad (33)$$

In (33) the coefficient matrix is called interface matrix (**IM**) in this paper, its submatrices are given by

$$M_1 = M_4 = \frac{1}{2}(\bar{C}_i + \hat{C}_i);$$
$$\quad (34)$$
$$M_2 = M_3 = \frac{1}{2}(\bar{C}_i - \hat{C}_i)$$

and

$$[\bar{C}_i]_{mn} = <\varphi^{i+1}_m,\varphi^i_n>;$$
$$\quad (35)$$
$$[\hat{C}_i]_{mn} = \frac{k^i_n}{k^{i+1}_m}<\varphi^{i+1}_m,\varphi^i_n>$$

Assuming the income wave in the left segment is given, and we require that the solution is purely outgoing in the right segment, i.e. $b^{j+1}=0$. Solving for the backscattered amplitudes finds

$$b_-^{\,j} = -(M^{\,j}{}_1)^{-1} M^{\,j}{}_2 a_-^{\,j} , \quad \textbf{(36)}$$

and

$$a_+^{\,j} = [M^{\,j}{}_4 - M^{\,j}{}_3 (M^{\,j}{}_1)^{-1} M^{\,j}{}_2] a_-^{\,j} . \quad \textbf{(37)}$$

for forward scattered amplitudes. The coefficient matrices in **(36)** and **(37)** are the backward and the forward scattering matrices due to the interface, respectively. Under the single scatter approximation, **(36)** and **(37)** are reduced to

$$b_-^{\,j} = -M^{\,j}{}_2 \equiv S^{\,j}{}_b a_-^{\,j} , \quad \textbf{(38)}$$

and

$$a_+^{\,j} = [M^{\,j}{}_4] a^{\,j} - = S^{\,j}{}_f a_-^{\,j} . \quad \textbf{(39)}$$

It is physically straightforward that under the one-way scattering approximation, i.e. neglecting the coupling between the forward and backward scattering components, the total forward scattered and the backscattered matrices are given by

$$S_f(K,L) = \prod_{j=1}^{N_h} [P_j S^{\,j}{}_f(K)] , \quad \textbf{(40)}$$

and

$$S_b(K,L) = \sum_{j=1}^{N_h} [\prod_{k=1}^{j} P_k] S^{\,j}{}_b [\prod_{k=1}^{j} P_k] , \quad \textbf{(41)}$$

respectively.

$$P_j \equiv \exp(iK^{\,j} h_j) , \quad \textbf{(42)}$$

is the phase displacement matrix in the jth segment. It is clear that the physical forward scattering matrices is just the matrix-valued physical scattering solution defined in the last section, so related with **JFSM** by the following relation (see (18))

$$S_f(K,L) = \exp(i \sum K^{\,j} h_j) S_f(K,L,J)$$

Letting

$$S^{\,j}{}_f = (S^{\,j}{}_f)_{dia} + (S^{\,j}{}_f)_{off} , \quad \textbf{(43)}$$

the off diagonal part of the total forward scattering matrix can be rewritten to

$$S_f(K,L)_{off} = \sum_{j=0}^{N_h} [\prod_{k=j}^{N_h} P^k][S^{\,j}{}_f]_{off} [\prod_{k=1}^{j} P^k] + o(\Delta^2) . \quad \textbf{(44)}$$

In the derivation of **(44)**, we used the fact

$$(S^{\,j}{}_f - Id)_{dia} * (S^{\,j}{}_f)_{off} \approx o(\Delta^2) , \quad \textbf{(45)}$$

Id denotes the identity matrix.

From **(44)** we see that the off diagonal part of the total forward scattering matrix is superposition of the forward scattered waves from the different interfaces. For example, the component from the jth interface is

$$[\prod_{k=j+1}^{N_h} P^k][S^{\,j}{}_f]_{off} [\prod_{k=1}^{j} P^k] . \quad \textbf{(46)}$$

In **(46)**, the term $[\prod_{k=1}^{j} P^k]$ represents the phase shift from the 1st to the jth segments, $[S^{\,j}{}_f]_{off}$ is the forward scattering matrix at the jth interface, and $[\prod_{k=j+1}^{N_h} P^k]$ is the phase shift from the j+1th to the N_hth segment.

Assuming there is only small change of the medium properties with respect to the background in each segment, we can write the modal equation in the following form

$$\{\rho_0 \frac{d}{dz}\frac{1}{\rho_0}\frac{d}{dz} + k_0^2(z) - k^j{}_m{}^2 + \Delta^j\}\phi^j{}_m(z) = 0, \quad x \in \text{jth segment} \quad \textbf{(47)}$$

where Δ^j is the perturbation with respect to the background environment. Using the standard perturbation theory, the following relations are obtained

$$\varphi^{j+1}{}_m(x + h_{j+1}) = \varphi^0{}_m(x) + \sum_{i \neq m} \frac{<\varphi^0{}_m, \Delta^{j+1}\varphi^0{}_i>}{k^0{}_m{}^2 - k^0{}_i{}^2}\varphi^0{}_i + o(\Delta_{j+1});$$

$$\varphi^j{}_n(x + h_j) = \varphi^0{}_n(x) + \sum_{i \neq n} \frac{<\varphi^0{}_n, \Delta^j\varphi^0{}_i>}{k^0{}_n{}^2 - k^0{}_i{}^2}\varphi^0{}_i + o(\Delta_j);$$

$$\quad \textbf{(48)}$$

where we use Landau's symbol $o(\Delta)$ to express a high order small quantity.

Substituting **(48)** into **(35)** yields

$$[S^j{}_f]_{mn} = \frac{1}{2}\frac{\{<\varphi^0{}_m,(\Delta^j - \Delta^{j-1})\varphi^0{}_n> + \frac{k^0{}_n}{k^0{}_m}<\varphi^0{}_m,(\Delta^j - \Delta^{j-1})\varphi^0{}_n>\}}{(k^0{}_m{}^2 - k^0{}_n{}^2)} + o(\Delta), \quad m \neq n; \quad \textbf{(49)}$$

and

$$[S^j{}_b]_{mn} = \frac{1}{2}\frac{\{<\varphi^0{}_m,(\Delta^j - \Delta^{j-1})\varphi^0{}_n> - \frac{k^0{}_n}{k^0{}_m}<\varphi^0{}_m,(\Delta^j - \Delta^{j-1})\varphi^0{}_n>\}}{(k^0{}_m{}^2 - k^0{}_n{}^2)} + o(\Delta), \quad m \neq n. \quad \textbf{(50)}$$

Inserting **(49)-(50)** into **(45)** and **(41)** yields

$$S_f(K,L)_{mn} = \frac{1}{2}\exp(ik^0{}_m L)[\sum_j^{N_h} \frac{1}{k^0{}_m}\exp[-i(k^0{}_m - k^0{}_n)\sum_{k=1}^j h_k]\frac{<\varphi^0{}_m,(\Delta^j - \Delta^{j-1})\varphi^0{}_n>}{(k^0{}_m - k^0{}_n)}]$$

$$\to \frac{1}{2}\exp(ik^0{}_m L)[\int_0^L \frac{1}{k^0{}_m}\exp[-i(k^0{}_m - k^0{}_n)x]\frac{<\varphi^0{}_m,(\frac{d}{dx}\Delta)\varphi^0{}_n>}{(k^0{}_m - k^0{}_n)}dx], \text{ as } h_j \to 0 \quad m \neq n$$

$$\quad , \quad \textbf{(51)}$$

and

$$S_b(K,L)_{mn} = \frac{1}{2}\sum_j^{N_h} \frac{1}{k^0{}_m}\exp[-i(k^0{}_m + k^0{}_n)\sum_{k=1}^j h_k]\frac{<\varphi^0{}_m,(\Delta^j - \Delta^{j-1})\varphi^0{}_n>}{(k^0{}_m + k^0{}_n)}$$

$$\to \frac{1}{2}\int_0^L \frac{1}{k^0{}_m}\exp[-i(k^0{}_m + k^0{}_n)x]\frac{<\varphi^0{}_m,(\frac{d}{dx}\Delta)\varphi^0{}_n>}{(k^0{}_m + k^0{}_n)}dx, \quad h_j \to 0, m \neq n$$

$$\quad . \quad \textbf{(52)}$$

Recalling $\Delta(x) = 0$, $x \notin [0, L]$, it can be shown by the partial integration that **(51)** and **(52)** are equivalent to the results in **Table-1**.

For range-dependent background, **(51)-(52)** are modified to

$$S_f(K,L)_{mn} = \frac{1}{2}\exp(i\sum k^j{}_m h_j)[\sum_i^{N_h} \frac{1}{k^j{}_m}\exp[-i\sum_{k=1}^j h_k(k^k{}_m - k^k{}_n)]\frac{<\varphi^j{}_m,(\Delta^{j+1} - \Delta^j)\varphi^j{}_n>}{(k^j{}_m - k^j{}_n)}] \quad m \neq n,$$

$$\quad \textbf{(53)}$$

and

$$S_b(K,L)_{mn} = \frac{1}{2}\sum_j^{N_h} \frac{1}{k^j{}_m}\exp[-i\sum_{k=1}^j h_k(k^k{}_m + k^k{}_n)]\frac{<\varphi^j{}_m,(\Delta^{j+1} - \Delta^j)\varphi^j{}_n>}{(k^j{}_m + k^j{}_n)}, \quad m \neq n. \quad \textbf{(54)}$$

(53) is equivalent to the relation **(30)**.

III. CONCLUGING REMARKS

The general perturbation theory in arbitrary background were developed in this paper. The results presented here are useful to the sensitivity analysis in future tomographic inversion. The results are also can be used as the theoretical framework for linearized inversion.

In a separate paper, we shall consider a perturbation mode coupling tomography based on the result given in this paper. It will be shown that the method can be used successfully to image small model perturbation.

ACKNOWLEDGMENT
We are grateful to Professor Ji-Xun Zhou providing the paper list on acoustic interaction with internal wave.

References:

[1] Ji-xun Zhou, Xue-zhen Zhang and P. Rogers, " Resonance interaction of sound waves with internal solitons in coastal zone, ", J.Acoust.Soc.Am. **90**(1991),2042-2054.

[2] Ji-xun Zhou, Xue-zhen Zhang, "Anomalous Sound Propagation in Shallow Water due to Internal Wave Solitons," IEEE Proc. OCEANS'93, 1, 87-92 (1993).

[3] A. N. Rutenko, " Experimental Study of the effect of internal waves on the frequency Interference Satructure of the Sound field in shallow Sea," Acoustic Physics, **46(2)**(2000), 207-213.

[4] E.C.Shang and Y.Y.Wang, " The impact of mesoscale oceanic structure on global-scale acoustic propagation," in Theoretical and Computational Acoustics, ed. Ding Lee, et.al, (World Scientific Publishing Co.Sigapore, 1996), pp.409-431.

[5] E.C.Shang and Y.Y.Wang, " The Frontal effects on Long-Range Acoustic Propagation in the North Pacific Ocean," in Theoretical and Computational Acoustics, ed. Y.C.Teng, et.al, (World Scientific Publishing Co.Sigapore, 1999), pp.475-486.

[6] W.Munk, P. Worcester and C. Wunsch, Ocean Acoustic Tomography, Cambridge University Press, 1995.

[7] E.C.Shang, " Ocean acoustic tomography based on adiabatic mode theory," J.Acoust.Soc.Am. **85**(1989),1531-1537.

[8] V.V. Gocharov and A.G.Voronovich, " An experiment on matched-field acoustic tomography with continuous wave Signal in Norway Sea ," J.Acoust.Soc.Am. **93**(1993),1873-1881.

[9] C.S. Chiu, J.H.Miller, and J.F.Lynch, " Forward coupled mode propagation modeling for coastal acoustic tomography," J.Acoust.Soc.Am. **99**(1996),793-802.

[10] Li, X, D and Tanimoto, T, " Waveforms of long-period body waves in a slightly aspherical Earth model," Geophys. J. Int. Vol. 112, pp.92-102 (1993).

[11] H. Marquering and R. Snieder, " Surface-wave mode coupling for efficient forward modeling and inversion of body-wave phases," Geophys, J. R. astr. Soc., **120** (1995), 186-208.

[12] K. Chadan and P.C. Sabatier: Inverse Problems in Quantum Scattering Theory (Springer-Verlage, New York, 1989 Second Edition).

[13] Ning Wang and Sadayuki Ueha: " Inverse Scattering of a Medium which Supports M-types of wave", Journal of The Physical Society of Japan, **61**(2) (1992),pp.455-461.

[14]R.B.Evans, " A Coupled mode solution for acoustic propagation in a wave-guide with stepwise depth variations of a penerable bottom," J.Acoust.Soc.Am. **74**(1983),188-195.

[15] F.B.Jensen, W.A.Kuperman, M.B.Porter and H.Schmidt, Computational Ocean Acoustics, Springer-Verlage, New York, 1992.

Appendix A: Derivation of (29)-(30)

To derive relations (29) and (30), we construct a range-dependent Green function satisfying the following integral equation

$$G_1(K;x,y) = -\theta(y-x)\frac{\sin K(x-y)}{iK} + \int_x^{+\infty}\frac{\sin K(x-y)}{iK}V(y)G_1(K;y,y)dy \qquad , \text{ (A-1)}$$

and the boundary conditions

$$G_1(K;x,y) = 0, \quad x > y, \qquad \text{(A-2)}$$

$$G_1(K;x,y)\big|_{x=y} = 0 \qquad \text{(A-3)}$$

$$\frac{\partial}{\partial x}G_1(K;x,y)\big|_{x=y-\varepsilon} = 1, \text{ as } \varepsilon \to 0, \quad \text{(A-4)}$$

where $\Theta(x)$ is the Heaviside function. Using the basic solutions $A^0{}_J(\pm K,x)$, this Green's function can be written as

$$G^1(K;x,y) = \theta(y-x)\{A^0{}_J(K,x)C(K,y) + A^0{}_J(-K,x)D(K,y)\}, \text{ (A-5)}$$

where coefficient matrices $C(K,y)$ and $D(K,y)$ are determined by the boundary conditions

$$0 = A^0{}_J(K,y)C(K,y) + A^0{}_J(-K,y)D(K,y),$$

$$1 = \frac{\partial}{\partial y} A^0{}_J(K,y)C(K,y) + \frac{\partial}{\partial y} A^0{}_J(-K,y)D(K,y) \quad \textbf{(A-6)}$$

Using the relation

$$W[A^0{}_J(K), A^0{}_J(-K)] = -2iK$$

the following result can be easily shown

$$C(K,y) = \frac{1}{2iK} A^0{}_J{}^T(-K,y),$$

$$D(K,y) = \frac{-1}{2iK} A^0{}_J{}^T(K,y) \quad , \quad \textbf{(A-7)}$$

i.e.

$$G^1(K;x,y) = -\Theta(y-x)\{A^0{}_J(K,x)\frac{1}{2iK}A^0{}_J{}^T(-K,y) - A^0{}_J(-K,x)\frac{1}{2iK}A^0{}_J{}^T(K,y)\}. \quad \textbf{(A-8)}$$

Now the perturbed Jost function is given by

$$A^1{}_J(x,k) = A^0{}_J(x,k) - \int_x^{+\infty} G^1(K;x,x)\Delta V(x')A^1{}_J(x',k)dx', \quad \textbf{(A-9)}$$

then

$$\frac{1}{T^1(K)} = \frac{1}{T^0(K)} + \frac{1}{T^0(K)}\frac{1}{2iK}\int_0^{+\infty} A^0{}_J(-K,y)^T\Delta V(y)A^1{}_J(K,y)dy. \quad \textbf{(A-10)}$$

for transmission coefficient and

$$R'(K)\frac{1}{T^0(K)} = R^0(K)\frac{1}{T^0(K)} - \frac{1}{T^0(K)}\frac{1}{2iK}\int_{-\infty}^{+\infty} A^0{}_J(K,y)^T\Delta V(y)A^1{}_J(K,y)dy. \quad \textbf{(A-11)}$$

(29) and **(30)** are obtained by approximating $A^1{}_J(K,x)$ by $A^0{}_J(K,x)$.

Appendix B:

In this section, we derive the discrete form of **(27)**. Assuming that the inhomogeneous region is divided into N segments each with range length dx_j $(j=1,2....,N_h)$, we have

$$\Delta S_f(K,L) = \frac{-1}{2iK}[S^0{}_f(-K,L)^T]^{-1}\{\sum_{i=1}^{N}[A^0{}_p(-K,x_i)]^T\Delta V(i)A^0{}_p(K,x_i)dx\}. \quad \textbf{(B-1)}$$

From **(18)**, we have

$$A^0{}_p(K,L) = \exp[\sum_{i=1}^{N} K^i dx_i]S^0{}_f(K,L). \quad \textbf{(B-2)}$$

From (40), the matrix-valued physical scattering solution has the following recursive relation

$$A_p(K,L) = A_p(K,x_j \to L)A_p(K,0 \to x_j). \quad \textbf{(B-3)}$$

Insertting (B-2) and (B-3) into (B-1) yields

$$\Delta S_f(K,L) = \frac{-1}{2iK}[S^0{}_f(-K,L)^T]^{-1}\{\sum_{i=1}^{N}[A^0{}_p(-K,x_i)]^T\Delta V(i)A^0{}_p(K,x_i)dx\}. \quad \textbf{(B-4)}$$

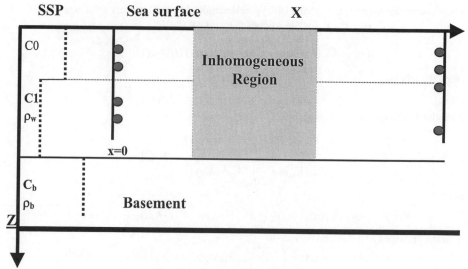

Figure 1. The model environment

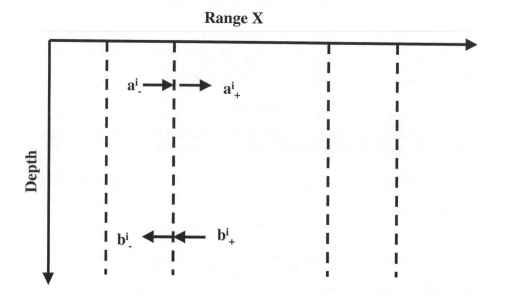

Figure 2 Range Dependent Environment

Table –1

	Scattering Matrix	Travel-time Perturbation
Adiabatic Approx.	$S_{nn} = \delta_{nn} + i \iint \frac{\Delta c}{c(z)^3 k_n} \varphi_n^{\ 2} dzdx; \quad S_{nm} = 0 \quad if\ m \neq n,$	$\Delta t_n = \frac{\partial}{\partial \omega} \{ \iint \frac{\Delta c}{c(z)^3 k_n} \varphi_n^{\ 2} dzdx \}$
Born Approx.	$S_{nn} = \delta_{nn} + i \iint \frac{\Delta c}{c(z)^3 k_n} \varphi_n^{\ 2} \exp[i(k_m - k_n)x] dzdx$	$; \Delta t_{nn} = \frac{\partial}{\partial \omega} \{ \iint \frac{\Delta c}{c(z)^3 k_n} \varphi_n^{\ 2} \exp[i(k_m - k_n)x] dzdx \}$

Theoretical and Computational Acoustics 2001
E.-C. Shang, Qihu Li and T. F. Gao (Editors)
© 2002 World Scientific Publishing Co.

Modeling Uncertainties in the Prediction of the Acoustic Wavefield in a Shelfbeak Environment

P. F. J. Lermusiaux[1], C.-S. Chiu[2], A. R. Robinson[1]

[1]Harvard University, Cambridge, MA 02138, USA

[2]Naval Postgrauate School, Monterey, CA 93943, USA

Abstract

The uncertainties in the predicted acoustic wavefield associated with the transmission of low-frequency sound from the continental slope, through the shelfbreak front, onto the continental shelf are examined. The locale and sensor geometry being investigated is that of the New England continental shelfbreak with a moored low-frequency sound source on the slope. Our method of investigation employs computational fluid mechanics coupled with computational acoustics. The coupled methodology for uncertainty estimation is that of Error Subspace Statistical Estimation. Specifically, based on observed oceanographic data during the 1996 Shelfbreak Primer Experiment, the Harvard University primitive-equation ocean model is initialized with many realizations of physical fields and then integrated to produce many realizations of a five-day regional forecast of the sound speed field. In doing so, the initial physical realizations are obtained by perturbing the physical initial conditions in statistical accord with a realistic error subspace. The different forecast realizations of the sound speed field are then fed into a Naval Postgraduate School coupled-mode sound propagation model to produce realizations of the predicted acoustic wavefield in a vertical plane across the shelfbreak frontal zone. The combined ocean and acoustic results from this Monte Carlo simulation study provide insights into the relations between the uncertainties in the ocean and acoustic estimates. The modeled uncertainties in the transmission loss estimate and their relations to the error statistics in the ocean estimate are discussed.

1. Introduction

Data-assimilative, high-resolution ocean physical models, when coupled to accurate acoustic propagation models, usually improve the prediction of the acoustic wavefield associated with the transmission of low-frequency underwater sound. Ideally, a comprehensive prediction should include a characterization of the reliability or uncertainty of forecast quantities. This allows the correct interpretation or processing of these quantities in a scientific or tactical application. Uncertainties in both the ocean and acoustic estimates arise from imperfect data, imperfect models and environmental variability not explicitly known. In a comprehensive coupled ocean physics and acoustic prediction system, the forecast of uncertainties involves the attribution of errors within each of the physical (oceanographic) and acoustical components, and the transfer of these error estimates or probabilities through the coupled system. In this paper, we outline an approach for carrying out such coupled predictions, focusing more on the transfer than on the attribution of errors. The approach is exemplified by a short hindcast of the large-mesoscale physics and transmission loss in a slope-to-shelf sound propagation experiment in the Middle Atlantic Bight (MAB) shelfbreak region, off the east coast of the United States, during the summer of 1996.

The basic concepts of the coupled physical-acoustical prediction approach presented here build on recent advances made in error estimation and data assimilation research in meteorology and oceanography (Ehrendorfer, 1997; Robinson *et al.*, 1998; Lermusiaux, 1999a), which in turn have their roots in classical estimation and control theory. To control and minimize errors in the prediction, dynamical models and data are optimally combined through data assimilation as a function of their respective error statistics or uncertainties. In numerical simulations, an issue that arises relates to the large dimension of the statistical properties of the coupled physical-acoustical fields. Efficient reduction and representation of uncertainties are necessary. To do so, the Error Subspace Statistical Estimation (ESSE) approach (Lermusiaux, 1999a,b; Lermusiaux and Robinson, 1999) is employed. This scheme is based on a reduction of the evolving error statistics to their dominant components or subspace. Presently, these statistics are measured by a variance or least-squares criterion (Tarantola, 1987; Robinson et al, 1998): a subspace is then characterized by the dominant components of the eigen-decomposition of a covariance matrix (subspace definitions that focus on higher-order moments can also be used). In our coupled prediction, a main task is thus to estimate and track the evolving uncertainty subspace of the coupled ocean physics and acoustic variables.

In what follows, Section 2 overviews the dynamics and sources of physical-acoustical variabilities in the MAB. Section 3 describes the data, models and methods employed. Section 4 consists of the results and Section 5 of the conclusions.

2. The Middle Atlantic Bight Variability and Uncertainty

The MAB shelfbreak marks a dramatic change, not only in water depth, but also in the dynamics of the waters that lie on either side (Beardsley and Boicourt, 1981; Colosi *et al.*, 2001). The shelf is about 100 km wide, extending from Cape Hatteras to Canada. The shelfbreak, which refers to the first rapid change in depth that occurs between the coastal and deep ocean, is near the 100-m isobath. The main oceanographic feature in the MAB is a mesoscale front of temperature, salinity and hence sound speed, separating the shelf and slope water masses. Located near the shelfbreak, this front is usually tilted in the opposite direction of the bottom slope. The shelf-water to the north is cold and fresh while the slope-water to the south is warm and salty. In the summer, the surface layers (top 10 to 30 m) are stratified (warmer and lighter) which has a tendency to reduce the influence of the atmosphere on the front below.

In the MAB region, atmospheric fluxes, buoyancy-induced pressure gradients, tides, river run-offs and shelf-slope water exchanges combine to generate a rich variety of physical phenomena. The shelf-water variability is often driven by wind forcings and tides. In the slope region, mesoscale variability is significant, especially that induced by Gulf Stream rings and meanders. Near the shelfbreak front, instabilities occur. Shelfbreak eddies, meanders, and internal waves, bores and solitons (SWARM Group, 1995; Colosi *et. al.*, 2001) are frequently observed, leading to complex, energetic variability on multiple time and space scales.

Variability and uncertainty of an estimate are inherently related. The portion of the variability, oceanic and acoustic, that is estimated with errors contributes to uncertainty. For example, variability that is totally unresolved is pure uncertainty. Mathematically, uncertainty can be defined by the *probability density function (PDF) of the error in the estimate*. Errors here refer to differences between the true and estimated fields. In a prediction, both errors in the initial data (conditions) and errors in the models and boundary conditions impact the accuracy of the forecast. Predicted uncertainties thus contain the integrated effects of the initial error and of the errors introduced continuously during the model integration.

3. Data, Models and Uncertainty Forecast Methodology

Data: During July and August of 1996, data were collected in the MAB south of New England, as part of the ONR Shelfbreak PRIMER Experiment (Lynch *et al.*, 1997). The main objective was to study the influence of oceanographic variability on the propagation of sound from the slope to the shelf. Intensive measurements were carried out in a 45 km by 30 km domain between the 85 m and 500 m isobaths. The measurements consisted of temperature, salinity, velocity, chlorophyll, bioluminescence and acoustic transmissions. Some additional wide-coverage data were obtained from outside that domain, including atmospheric fluxes and satellite surface temperature and height. To initialize physical fields and their uncertainties, these synoptic but also historical data are utilized as well as a feature model for the shelfbreak front (Lermusiaux, 1999a).

Ocean Physics Model: The physical variables are temperature, salinity, velocity and pressure. For this first coupled ocean physics-acoustic uncertainty study, only large mesoscales are considered (small mesoscales are not resolved by the 9 km grid resolution). Physical fields are initialized for August 1 in a domain (Fig. 1) of 320.26km by 355.29km centered on 39.86 N, 70.06 W. A simulated 5-day forecast is issued for August 5. The dynamical evolution is computed by the numerical ocean model of the Harvard Ocean Prediction System (HOPS) (e.g. Robinson, 1996 and 1999). Atmospheric fluxes based on buoy data are imposed at the surface. The model parameters and boundary condition schemes were calibrated based on data and sensitivity studies.

Acoustic Model: The acoustic model used in this study is that of Chiu, 1994, and Chiu *et al.*, 1995 and 1996. It is based on the physics of coupled normal modes. The basic formulation of the model involves decomposing the acoustic pressure into slowly-varying complex envelopes that modulate (mode by mode) analytic, rapidly-varying, adiabatic-mode solutions. Given sound speed, density, attenuation rate and bathymetry as a function of space, the acoustic solution is thus obtained by integrating a coupled set of differential equations governing these complex modal envelopes. Model output contains sound pressure, transmission loss, and travel time, phase and amplitude of the individual modes. One, several or all of these acoustic variables can be included in the joint ocean-acoustic state space for the coupled ocean-acoustic prediction. For simplicity, our discussion will focus on the prediction of the transmission loss and its uncertainty along an actual Shelfbreak-PRIMER acoustic path. Figure 1 shows the geometry of such a path and its relation to the HOPS model domain.

Coupled Uncertainty Forecasts: As discussed in Sec. 1, our approach is the ESSE methodology. For the MAB case study presented here, the principal components constituting the error subspace of the ocean physical state and their coefficients are first initialized for August 1 combining data and dynamics, following Lermusiaux *et al.*, 2000. To account for nonlinearities, this initial uncertainty is forecast using an ensemble of Monte-Carlo prediction realizations obtained as follows. The initial physical oceanographic state is first perturbed using random combinations of the initial error principal components. For each of these perturbed initial conditions, the nonlinear ocean dynamical model is then integrated for 5 days. These Monte-Carlo integrations are

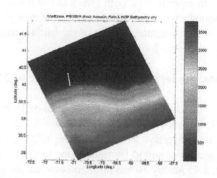

Figure 1. *Geometry of the Shelfbreak PRIMER western acoustic path superposed on the HOPS model bathymetry.*

carried out in parallel until the size of the ensemble is large enough to describe most of the error variance in the forecast. This is assessed by a convergence criterion. For the scales considered, an ESSE estimate of the error covariance for the cross-slope sound-speed field along the MAB transmission path was obtained after 80 integrations. With these 80 forecast realizations of the sound speed field, 80 Monte-Carlo realizations of the acoustic wavefield are then computed to estimate the uncertainty in the predicted acoustic wavefield and its linkages to uncertainties in the ocean forecast. Note that larger ensembles of ocean forecast realizations (from 100 to 267 samples) have been computed for the region, with and without stochastic error forcings. The stochastic forcings aimed to account for the uncertainties associated with smaller-scale ocean processes including sub-mesoscale eddies and internal tides. In what follows, only forecasts of mesoscale uncertainties, i.e., corresponding to the case of no stochastic forcings, are presented.

4. Results

4.1 Physical Oceanography

The 5-day forecast of the large-mesoscale physical fields and their uncertainty estimates (based on 100 realizations) are illustrated by Fig 2. The focus is on the critical coupling variable for our study, sound speed (computed here using the UNESCO 1983 polynomial, based on the local pressure and forecast temperature and salinity). The surface sound speed map (Fig 2a) indicates the location the shelfbreak front. Note also three primary water masses: the Gulf of Maine water southeast of Cape Cod (lowest sound-speed), the shelf water to the north and slope water to the south. A meander develops in the western side of the domain. A slope-water eddy and a shelf-water eddy are starting to form upstream and downstream of this meander, respectively. The surface error standard deviation (Fig. 2b) relates to this mesoscale variability. It is largest along the front, especially downstream of the meander. Note also some streak patterns due to surface wind effects (to the southeast). The relatively large standard deviation at the inflow of the shelfbreak jet is due to uncertainties in the position and strength of the front,

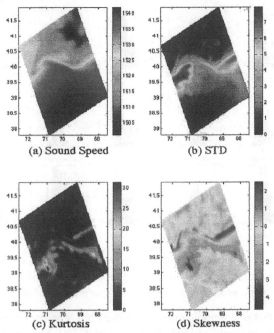

Figure 2. Surface values of the 5-day forecast sound speed and its error statistics: the ensemble mean, error standard deviation and higher-order statistics on the top model level are shown.

and also to uncertainties brought upon by the open-boundary condition. Higher-order error statistics (Fig. 2c-d) as well as error correlation functions (not shown) can also be computed by ESSE. If forecast errors were Gaussian, their skewness (3rd/2nd moment, Fig. 2d) would be null. An interesting result is that, on average, this skewness changes sign at the front. It is estimated to be positive on the shelf and negative on the slope. Its extrema corresponds to negative values on the slope because the day before the forecast time some westerly winds occurred and because of internal dynamics. If errors were Gaussian, the kurtosis (4th/2nd moment, Fig. 2c) would be 3. On average, it is here forecast to be maximum near the skewness extrema. The kurtosis extrema is near the region where a small mesoscale shelf-water eddy (mainly subsurface) is forming. Being able to compute such high-order statistics is important to characterize the statistical shapes of uncertainties. It has significant consequences in scientific, operational and also technical (e.g. data assimilation schemes) applications.

To illustrate the PDF of the errors in the physical estimates, a set of local sound speed error PDFs (based on 100 realizations) are plotted on Fig.3. Each of these histograms corresponds to a surface numerical grid point in the ocean physical dynamical model, going along a straight line across the shelfbreak from south (offshore) to north (along the US coast). At each of the 15 locations, the x-axis consists of 21 equally spaced bins in deviations (m/sec) with respect to the ensemble mean (the 0). The y-axis

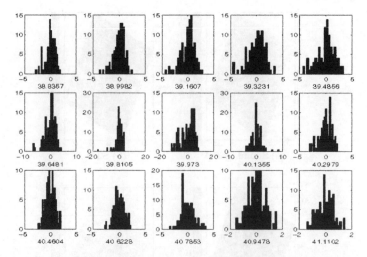

Figure 3. Forecast sound-speed error PDFs. Histograms of surface values (top model level) of the forecast sound speed error, at 15 locations in an across-shelf direction (top left is above slope near 39N, 69E; bottom right is at the shelf, near Nantucket).

consists of the number of ensemble members in each bin. The front location is close to the histogram number 8 (middle of the 15 plots, at 39.973N). The forecast results show a maximum variance near the front (see 20 m/sec extremes) and a lowest variance near Nantucket (see 2 m/sec extremes). The negative and positive skewness identified on Fig. 2 are again clearly visible just to the south and north of front, respectively. Note also that the error PDF of the sound speed is steepest on each side of the front and close to Gaussian away from the front except near the edges of strong eddies or meanders where high shear occur.

4.2 Acoustics

As discussed in the previous sections, based on observed oceanographic data during the 1996 Shelfbreak Primer Experiment, HOPS was initialized with perturbed physical oceanographic fields that are in statistical accord with a realistic error subspace and then integrated to produce 80 realizations of a regional forecast of the sound-speed field. One of these realizations of the sound-speed field along the transmission path is shown in Fig. 4. The different forecast realizations of the sound speed were then fed into a coupled-mode sound propagation model to produce realizations of the TL prediction for a low-frequency transmission from the slope, across the shelfbreak, onto the shelf. Specifically, the transmission frequency was 400 Hz and the sound source was located near the bottom at the 300-m isobath on the slope. Six of the 80 different realizations of the transmission loss (TL) prediction are shown in Fig. 5. They show that the structure in the spatial distribution of the acoustic energy is quite different from one realization to another, even in the shelf region where the ocean variance is minimal. The TL structure on the shelf is largely determined by what happens to the acoustic energy prior to entering the shelf. With the large sound-speed variances over the slope, the initial distribution of the acoustic energy over the set of acoustic normal modes (i.e., modal excitation) near the source range, as well as the redistribution of modal energy along the slope due to mode coupling, are different for the different ocean realizations. This results in different TL structures on the shelf (Fig. 5).

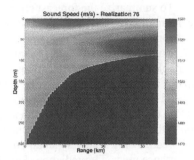

Figure 4. *A realization of the sound-speed forecast along the transmission path.*

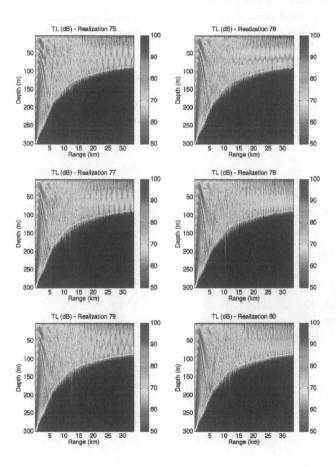

Figure 5. *Six different realizations of the TL prediction*

To summarize the forecast error statistics (uncertainty) for the TL prediction and its relation to the uncertainty in the ocean forecast, we show the error standard deviations of the sound-speed forecast in Fig. 6, the error standard deviations of the TL prediction in Fig. 7, and the corresponding histograms (error PDF estimates) for the sound speed and TL variables at two different locations, shelfbreak and shelf, in Fig. 8, respectively (all computed from 80 realizations). The error standard deviations of sound speed (Fig. 6) calculated from the 80 ocean realizations show that large uncertainties are confined in the top layers (from 0 to 50m depth, with a maximum around 30m depth) over the slope region at the frontal zone. Over the shelf, for the 5-day period considered, the HOPS model predicts only relatively small error variances at the mesoscale. Accordingly, the error variance in the TL prediction (Fig. 7) is small near the source below the top layers., but as the acoustic energy reaches these top layers where large sound speed error variances are confined, the error variances in TL increase. Note that the uncertainty in the TL does not grow in range over the shelf where sound speed uncertainties are relatively small. The complexity and inhomogeneity of the predicted error statistics in this slope-to-shelf transmission in the MAB are further revealed in the PDF estimates shown in Fig. 8. In particular, note the transformation of the PDF shape as uncertainties are transferred from the ocean (sound speed) estimate to the acoustic (TL) estimate. Because the sound pressure field, from which TL is computed, is composed of multiple acoustic modes, an in-depth understanding of the linkage between the error statistics of TL and sound speed behooves a careful analysis on the behavior of the errors in the amplitude and phase of each acoustic mode. This modal error analysis is being carried out at present.

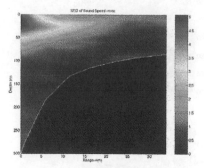

Figure 6. *Error standard deviation estimate of sound-speed forecast.*

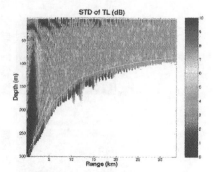

Figure 7. *Error Standard deviation estimate of TL prediction.*

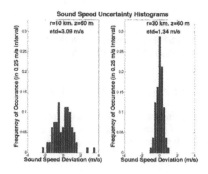 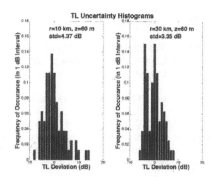

Figure 8. The histograms (PDF estimates) of the sound speed and TL uncertainties at two different locations (shelfbreak and shelf).

5. CONCLUSIONS

A methodology for the modeling and prediction of coupled ocean physics and acoustic uncertainties was outlined and exemplified. It is based on Error Subspace Statistical Estimation (ESSE). The focus was on the transfer of ocean physical uncertainties and their impact on the acoustical fields and uncertainties. The example considered consisted of the prediction of uncertainties for the large-mesoscale ocean physics and transmission loss in a slope-to-shelf sound propagation across the shelfbreak front in the Middle Atlantic Bight (MAB) region.

The results reveal the strong influence of the oceanic variability on the coupled uncertainties. They also illustrate possible modifications of the properties of error PDFs in their transfer from the ocean physics to the acoustic. Future challenges include the careful and comprehensive attribution of all error sources in such coupled predictions and their multiscale transfer through the coupled system. Scientific progress and error reduction should arise form the study and understanding of these error sources and transfers. Important feedbacks involve the coupled physical-acoustical data assimilation and the joint improvement of physical and acoustical data sampling schemes based on coupled error forecasts and quantitative adaptive sampling.

Acknowledgements

We are grateful to the Office of Naval Research for support under grants N00014--00--1--0771, N00014-01-1-0771, N00014-95-1-0371, N00014-97-1-0239 and N0001401WR20335. This is a contribution of the UNcertainties and Interdisciplinary Transfers through the End-to-End System (UNITES) team.

References

Beardsley, R.C.and W.C. Boicourt, 1981: On estuarine and continental-shelf circulation in the Middle Atlantic Bight. Evolution of Physical Oceanography. Scientific Surveys in Honor of Henry Stommel, edited by B. Warren and G. Wunsch, MIT Press, Cambridge, MA.

Chiu, C.-S., "Downslope Modal Energy Conversion," J. Acoust. Soc. Am., **95**(3), 1654-1657, 1994.

Chiu, C.-S., J.H. Miller, W.W. Denner, and J.F. Lynch, "A Three-Dimensional, Broadband, Coupled Normal-Mode Sound Propagation Modeling Approach," in Full Field Inversion Methods in Ocean and Seismic Acoustics (O. Diachok, A. Caiti, P. Gerstoft and H. Schmidt editors), Kluwer Academic Publishers, 57-62, 1995.

Chiu, C.-S., J.H. Miller and J.F. Lynch, "Forward coupled-mode propagation modeling for coastal acoustic tomography," J. Acoust. Soc. Am., **99**(2), 793-802, 1996.

Colosi J.A., J.F. Lynch, R.C. Beardsley, G. Gawarkiewicz, C.-S. Chiu and A. Scotti, "Observations of nonlinear internal waves on the New England continental shelf during summer shelfbreak PRIMER, J. . Geophys. Res., **106**(C5), 9587-9601, 2001.

Ehrendorfer, M. 1997. Predicting the uncertainty of numerical weather forecasts: a review. *Meteorologische Zeitschrift*, **6** (4) 147--183.

Lermusiaux, P.F.J. 1999a. Data assimilation via error subspace statistical estimation, Part II: Middle Atlantic Bight shelfbreak front simulations and ESSE validation. *Mon. Weather Rev.* **127**(7), 1408-1432.

Lermusiaux, P.F.J. 1999b. Estimation and study of mesoscale variability in the Strait of Sicily. *Dyn. Of Atmos.Oceans*, Special issue in honor of Professor A. R. Robinson, **29**, 255-303.

Lermusiaux, P.F.J. and A.R. Robinson, 1999. Data assimilation via error subspace statistical estimation, Part I: theory and schemes. *Mon. Weather Rev.* **127**(7), 1385--1407.

Lermusiaux, P.F.J., D.G. Anderson and C.J. Lozano, 2000. On the mapping of multivariate geophysical fields: error and variability subspace estimates. *Q.J.R. Meteorol. Soc.*, April B, 1387-1430.

Lynch, J.F., G.G. Gawarkiewicz, C.-S. Chiu, R. Pickart, J.H. Miller, K.B. Smith, A. Robinson, K. Brink, R. Beardsley, B. Sperry, and G. Potty (1997), "Shelfbreak PRIMER - An integrated acoustic and oceanographic field study in the mid-Atlantic Bight," ," in *Shallow-Water Acoustics* (R. Zhang and J. Zhou Editors), China Ocean Press, 205-212, 1997.

Robinson, A.R., 1996. Physical processes, field estimation and an approach to interdisciplinary ocean modeling. *Earth-Science Rev.*, **40**, 3-54.

Robinson, A.R., P.F.J. Lermusiaux and N.Q. Sloan, III, 1998. Data Assimilation, in *The Sea: The Global Coastal Ocean I*, Processes and Methods (K.H. Brink and A.R. Robinson, Eds.), Volume 10, John Wiley and Sons, New York, NY.

Robinson, A.R., 1999. Forecasting and simulating coastal ocean processes and variabilities with the Harvard Ocean Prediction System, in ``Coastal Ocean Prediction'', (C.N.K. Mooers, Ed.), AGU Coastal and Estuarine Studies Series, 77--100.

SWARM Group (J. Apel, M. Badiey, C.-S. Chiu, S. Finnette, R. Headrick, J. Kemp, J. Lynch, A. Newhall, M. Orr, B. Pasewark, D. Tielbuerger, A. Turgut, K. von der Heydt, S. Wolf, "An overview of the 1995 SWARM shallow water internal wave acoustic scattering experiment," IEEE J. Oceanic Engineering, **22**(3), 465-500, 1997.

Tarantola, A., 1987: *Inverse Problem Theory. Methods for Data Fitting and Model Parameter Estimation.* Elsevier Science Publishers, Amsterdam, The Netherlands.

Theoretical and Computational Acoustics 2001
E.-C. Shang, Qihu Li and T. F. Gao (Editors)
© 2002 World Scientific Publishing Co.

The Effects of Internal Waves on Broadband Source Localization in Shallow Water

Zhenglin Li, Jin Yan

National Laboratory of Acoustics, Chinese Academy of Sciences, Beijing, China

Internal waves are the primary source of ocean variations in shallow waters. The ability of passive source localization may be degraded by mismatch between forward modals and data because of the activities of internal waves. In this paper, the effects of Garrett-Munk and solitary internal waves on broadband coherent matched-field processing (MFP) and incoherent MFP are investigated. The true data fields are simulated with the presence of internal waves. The replica fields are calculated using the average sound profile (without internal waves). It is shown that both coherent MFP and incoherent MFP give the correct source location if there are Garrett-Munk internal waves only. For the case that there are solitons or both Garrett-Munk internal waves and solitons, incoherent MFP shows a complex ambiguity surface structure and fails to localize the source correctly. However, coherent MFP demonstrates consistent true source localization and small sidelobes.

1 Introduction

In shallow waters, internal waves are the primary source of ocean variations. The activities of the internal wave make the temperature of the sea-water varying with the time and space, and then effect the sound propagation in the sea [1, 2, 3, 4]. Matched-field source localization can be degraded by environment mismatch caused by the internal waves [5]. So, it is significant to understand fully the effects of internal waves on the matched-field source localization. In these years, the interest of the underwater acoustics community about MFP has shifted from deep-water to shallow-water, and from narrow-band MFP to broadband fashion [5, 6, 7]. In this paper, the effects of Garrett-Munk and solitary internal waves on broadband coherent frequency domain matched-field processing (MFP) and incoherent MFP are investigated.

2 The Acoustic environment and internal waves model

The average temperature and salinity profiles of the seawater we used for theoretical study about the effects of internal waves on MFP are listed in Fig.1[4]. The corresponding mean sound speed profile is shown in Fig.2(a), and the buoyance frequency shown in Fig.2(b). The depth of sea-water is 75m. There exists a thermocline from 11m to 35m. The bottom is assumed to be a flat bottom with a constant sound speed of 1580m/s, a density of 1.8g/cm^3 and an attenuation coefficient of 0.23 dB/λ.

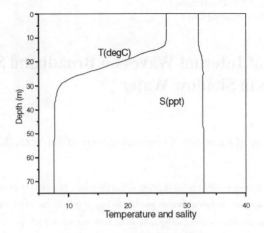

Fig.1 The average TS profiles from YS'96

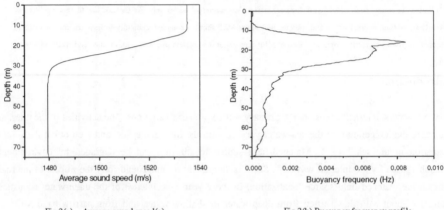

Fig.2(a)   Average sound speed(a)   Fig.2(b) Buoyancy frequency profile

The sound-speed perturbation caused by internal waves is related to the displacement of the water mass by[2]

$$\delta c(r,z,t) = \int_z^{z+\eta(r,z,t)} \overline{c}(z')G(z')N^2(z')dz' \qquad (2.1)$$

where $\eta(r,z,t)$ is the internal wave displacement fields, $\overline{c}(z)$ is the mean sound-speed profile, $N(z)$ is the buoyancy frequency, and $G(z)$ is a function of the derivatives of the salinity and temperature profiles as a function of depth. $\eta(r,z,t)$ can be written in the form

$\eta(r,z,t) = \eta_D(r,z,t) + \eta_S(r,z,t)$, where $\eta_D(r,z,t)$ represents the diffuse component and $\eta_S(r,z,t)$ represents the soliton contribution.

The internal wave displacement can be expressed as a sum of discrete normal modes

$$\eta_D(r,z,t) = \frac{1}{2\pi} \int_{\omega_c}^{N_{\max}} \sum_j F(\omega, j) W(\omega, j, z) \exp[i(k_{hj}r - \omega t)] d\omega \qquad (2.2)$$

where k_{hj} is the horizontal wavenumber of the jth mode, W is the vertical mode function of the internal waves, ω_c is the Coriolis frequency: $\omega_c = 2\Omega \sin(\text{latitude})$ with Ω being the Earth's angular speed, and $N_{\max}$ is the maximal buoyancy frequency. $F(\omega, j)$ represents the amplitude of the jth mode internal wave which is a zero mean, complex Gaussian random variable with a spectral representation

$$\left\langle |F(\omega, j)|^2 \right\rangle = E_0 M (j^2 + j_*^2)^{-p/2} \omega^q \qquad (2.3)$$

where $<\ldots>$ represents the ensemble average, and E_0 is the average energy density in J/cm² unit.

The characteristic wave number j_* and the spectral power law exponent p are empirically determined. The normalization factor M satisfies $1/M = \sum_{j=1}^{\infty} (j^2 + j_*^2)^{-p/2}$, q is the spectral slop. In calculation, we take $E_0 = 90 \text{J/cm}^2$, $p = 3$, $j_* = 1$ and $q = -3/2$ [4].

The displacement caused by soliton packets consisting of M solitons can be written as

$$\eta_S(r,z,t) = W(\omega, j = 1, z) \sum_{m=1}^{M} \Lambda_m \text{sec h}^2 \left(\frac{r_m - V_m t}{\Delta_m} \right) \qquad (2.4)$$

where $W(\omega, j = 1, z)$ is the first order vertical mode function as in Eq.(2.2) , Λ_m is the amplitude factor of individual soliton, and Δ_m is the characteristic width of the soliton and can be expressed as $\Delta_m = \sqrt{12\beta/(\alpha\Lambda_m)}$ with α and β being the nonlinear and dispersion parameters of KdV equation [2]. V_m is the amplitude-dependent wave speed and related to the linear speed v by

$$V_m = v + \alpha \Lambda_m / 3.$$

A single realization of the sound-speed distribution including the deterministic profile (the same as Fig.2(a)), a Garrett-Munk contribution and soliton packets, is illustrated in Fig.4. Sound-speed variation associated with a soliton packet consisting of six solitons is visible in the figure from 500m to 3500m. The solitons are generated about twenty kilometers away from the left edge of the figure and traveling to

the left. Where $\alpha = 0.035$, $\beta = 100$, $v = 0.6 \text{m/s}$, $\Lambda_m = 10e^{-0.3(m-1)}$ ($m = 1 \cdots 6$), and

$t = 7.5$ h.

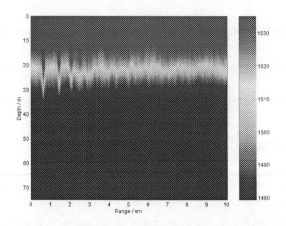

Fig 4 Single realization of the sound-speed distribution consisting of

a deterministic profile, a Garrett-Munk contribution and soliton packets.

3 Broadband Matched field processing (MFP)

Broadband MFP can be done by coherent or incoherent summation of the narrow-band range-depth ambiguity surfaces. Coherent frequency domain MFP can be written as[5, 6]

$$B_{MF}(r,z) = \frac{\left| \sum_{i=1}^{M_f} \sum_{j=1}^{N} p_j^{\text{rplc}}(f_i)^* p_j^{\text{data}}(f_i) \right|^2}{\left[\sum_{i=1}^{M_f} \sum_{j=1}^{N} \left| p_j^{\text{rplc}}(f_i) \right|^2 \right] \left[\sum_{i=1}^{M_f} \sum_{j=1}^{N} \left| p_j^{\text{data}}(f_i) \right|^2 \right]} \tag{3.1}$$

where j is the hydrophone index, r, z represent the range and depth of the replica source respectively, i is the frequency index. If the spectral correlation is squared before it is summed over frequency in Eq.(3.1), one has the incoherent MFP.

4 Simulation results

To study the effects of internal waves on broadband source localization in shallow water, we simulated source localizations in the presence of Garrett-Munk linear internal waves only, solitary internal waves only and both kinds of internal waves. Broadband sound propagation is modeled using RAM[6] PE program. Frequency band is from 400Hz to 600Hz in 4Hz steps. The replica fields are calculated using the average sound profile for a range spanning from 100m to 10km in 100-m steps and a depth covering 1m to 70m in 1m steps. The true data fields are simulated with the presence of internal waves, and the source is located at 50m in depth, and 14 elements receiver array at range 5km and depth covering 5m to 70m in 5m steps.

Fig.5(a) is the incoherent MFP ambiguity function with the presence of linear internal wave only. We can see from this figure that because of the linear internal waves, the correlation coefficient drops to less than 0.5. but it gives the correct source location. Comparing with incoherent MFP result, the coherent MFP gives the correct source location also, and has smaller sidelobes(Fig.5(b)).

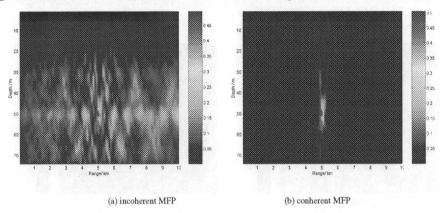

(a) incoherent MFP　　　　　　　　(b) conherent MFP

Fig.5 MFP range-depth ambiguity surfaces with presence of linear internal waves

If there are solitary internal waves only between the source and receivers, the incoherent MFP fails to localize the source correctly and gives a complex ambiguity surface structure. The coherent MFP gets the true source localization (see Fig.6).

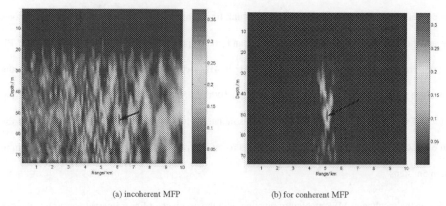

(a) incoherent MFP   (b) for conherent MFP

Fig.6 MFP range-depth ambiguity surfaces with presence of solitary internal waves only

For the case that there are both linear and solitary internal waves, incoherent MFP shows a more complex ambiguity surface structure. Through ten times of simulation, we find that MFP source location appears to be randomly due to the random internal waves. But for the coherent MFP, it could consistently give the correct source location roughly.

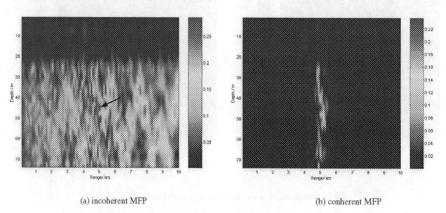

(a) incoherent MFP   (b) conherent MFP

Fig.7 MFP range-depth ambiguity surfaces with both linear and solitary internal waves

5 Conclusions

Presence of the internal waves degrades the performance of MFP. It is shown that both coherent MFP and incoherent MFP give the correct source location if there are linear internal waves only. For the case that there are solitons or both linear internal waves and solitons, incoherent MFP shows a complex ambiguity surface structure. However, coherent MFP consistently gives the true source localization roughly.

In this paper, we did not take care of the source spectrum. In fact, when we use coherent MFP as Eq.(3.1) for real experiment data, the source spectrum must be considered. For this case, one could use only parts of the cross-spectral density matrix[7] or use one of hydrophone signal as reference signal to normalize the source phase spectrum[8].

References

1. S. Flatte, R. Dashen, W. Munk *et al*. *Sound transmission through a fluctuating ocean* (Cambridge: Cambridge Univ. Press, 1979).

2. D. Tielburger, S. Finette, S. Wolf, "Acoustic propagation through an internal wave field in a shallow water waveguide", *J. Acoust. Soc. Am.,* **101**(2) (1997), 789-808.

3. J. X. Zhou, X. Z. Zhang, P. H. Rogers, "Resonant interaction of sound waves with internal solitons in the coastal zone", *J. Acoust. Soc. Am.,* **90**(5) (1991), 2042-2054.

4. YAN Jin, ZHANG Renhe, ZHOU Shihong, SHA Liewei, "Characteristics of the internal waves and their effects on the sound transmission in the Midst of the Yellow Sea", *Chinese Journal of Acoustics*, **18**(1) (1999), 47-54.

5. Yoo K, Yang T C, "Broadband source localization in shallow water in the presence of internal waves", *J. Acoust. Soc. Am.,* **106**(6) 1999; 3255-3269.

6. Collins M D, "A split-step Pade solution for parabolic equation method", *J. Acoust. Soc. Am.*, **93**(4) (1993), 1736-1742.

7. Westwood E K, "Broadband matched-field source localization", *J. Acoust. Soc. Am.,* **91**(5) (1992), 2777-2789.

8. Michalopoulou Z –H, "Robust multi-tonal matched-field inversion: A coherent approach", *J. Acoust. Soc. Am.*, **104**(1) (1998), 163-170.

Theoretical and Computational Acoustics 2001
E.-C. Shang, Qihu Li and T. F. Gao (Editors)
© 2002 World Scientific Publishing Co.

The Coupled-Mode Parabolic-Equation Method Based on the WKBZ Theory

Zhaohui Peng, Fenghua Li and Renhe Zhang
Institute of Acoustics, Chinese Academy of Sciences, P.O. Box 2712, Beijing,
China, 10080

On the base of the generalized phase integral (WKBZ) theory, the coupled-mode parabolic-equation (CMPE) method is studied in the range-dependent waveguides. The
CMPE solution is a hybrid model expressed in terms of the normal modes and mode coefficients, which uses a PE approach in a radial direction and normal modes in depth direction. Considering the contribution of the imaginary part of the eigenvalues, a fast and
accurate approach for eigenvalues-finding of local normal modes is presented. The
complex eigenvalues are solved from a transcendental equation in real plane. This approach is conveniently applied to other normal mode model. Combining the improved
WKBZ theory and CMPE theory, a new range-dependent propagation model is studied.
Examples are presented to illustrate the accuracy and efficiency of CMPE.

INTRODUCTION

On problems of sound propagation in range-independent and gradually range-dependent waveguides, there are many models to resolve, such as normal mode theory and adiabatic mode approximations [1], parabolic equation (PE) approaches [2], ray methods [3], etc. A lot of codes have been developed. The generalized phase integral (WKBZ) approximation [4] and the beam- displacement ray-mode (BDRM) theory [5] provide accurate prediction of sound field in horizontally stratified water efficiently. However, on range-dependent problems, the horizontal change of environment must be taken into consideration. The sound field calculated from adiabatic mode approximations has large errors. Although has high accuracy, the coupled modes theory [6] need very long computer time. The parabolic equation (PE) method, which has become the most popular wave-theory technique for solving range-dependent propagation problems, can only provide solutions of far field, and need very long computer time for high frequency and very long range problems.

Combined with the coupled mode theory and parabolic equation method, a new model is developed by Abawi, Kuperman and Collins [7], that is the coupled-mode parabolic-equation method. The coupled-mode parabolic-equation solution is expressed in terms of the normal modes and mode coefficients, which satisfy coupled horizontal wave equations and can be solved with PE method. It is practical to apply the coupled-mode parabolic-equation to solve large-scale problems and possibly even global scale problems at low frequency.

On the progress of resolving sound propagation problems with the coupled-mode parabolic-equation, the computation of local modes and coupled coefficients is one of the most difficult aspects, and takes most computer time. Therefore, an efficient algorithm of the computation of local modes and coupled coefficients is the key to improve the efficiency of the coupled-mode parabolic-equation. Considering the contribution of the imaginary part of the eigenvalues, a fast and accurate approach for eigenvalues-finding of local normal modes is presented. This approach is conveniently applied to other normal mode model. And applying this approach, which can calculate the eigenvalues of very high-order number, the near sound field can be well studied. On the combination of the improved WKBZ theory and the coupled-mode parabolic-equation theory, a new range-dependent propagation model (CMPE) is presented.

The theory of CMPE is presented in Sec. 1. In Sec. 2, the approach to compute the local modes and eigenfunctions is derived. Examples are presented in Sec. 3 to illustrate the accuracy and efficiency of CMPE.

1. Coupled-Mode Parabolic-Equation [7]

We work in cylindrical coordinates, where z is the depth below the pressure-release ocean surface, r is the range from a harmonious point source at $z = z_0$. The acoustic pressure $p(r, z)$ satisfies the reduced wave equation

$$\rho \nabla \cdot \left(\frac{1}{\rho} \nabla p \right) + k^2 p = 0 \tag{1}$$

where, $\rho(r, z)$ is the density, and $k(r, z) = \omega / c(r, z)$ is the wave number. $c(r, z)$ is the sound velocity, ω is the circular frequency.

Using the method of Ref.7, the series solution of Eq.(1) can be obtained,

$$p(r, z, \theta) = r^{-\frac{1}{2}} \sum_{n=1}^{\infty} [k_n(r, \theta)]^{-\frac{1}{2}} u_n(r, \theta) \phi_n(z; r, \theta) \tag{2}$$

where the local eigenvalue k_n and normal mode $\phi_n(z; r, \theta)$ satisfy

$$\frac{\partial^2 \phi_n(z; r, \theta)}{\partial z^2} + \left[\tilde{k}^2 - k_n^2 \right] \phi_n(z; r, \theta) = 0 \tag{3}$$

and

$$\int_0^\infty \phi_n \phi_m dz = \delta_{nm} \tag{4}$$

where $\tilde{k}^2 = k^2 + \frac{1}{2} \left[\frac{1}{\rho} \nabla^2 \rho - \frac{3}{2\rho^2} (\nabla \rho)^2 \right]$.

The u_n of Eq.(2) is the mode coefficient satisfied

$$\frac{\partial \vec{u}}{\partial r} = -A_r \vec{u} + i \left(\frac{1}{r^2} \frac{\partial^2}{\partial \theta^2} + K^2 \right)^{1/2} \vec{u} \tag{5}$$

where $\vec{u} = [u_0 \quad u_1 \quad u_2 \quad \cdots \quad u_M]^T$ is the mode coefficients vector. The diagonal matrix K^2 is the eigenvalues matrix. The coupling coefficients matrix A_r is defined by

$$A_{r,i,j} = \int \phi_i \frac{\partial \phi_j}{\partial r} dz .$$

Eq.(5) can be solved with splitting method.

2. Calculation of the eigenvalues and normal modes

One of the most difficult aspects of normal mode computation is finding the roots of the characteristic equation [8]. There are various methods to solve the characteristic equation; all have their virtue and shortcoming. In this section a new efficient method for the eigenvalues finding is presented.

The characteristic equation is [11]

$$2\int_{\xi_l}^{\eta_l}\sqrt{k^2-k_l^2}\,dz + \varphi_s(k_l) + \varphi_b(k_l) - i\ln\left|V_s(k_l)V_b(k_l)\right| = 2l\pi \qquad (l=0,1,2,\cdots) \quad (6)$$

where, l is the number of modes, η_l and ξ_l are the depth of upper and lower turning points, respectively. $k_l = R_l + iI_l$ is the complex eigenvalue, R_l is the horizontal wave number, I_l is the attenuation, i is the complex unit that satisfies $i^2=-1$, k is the wave number. V_s and V_b are the reflection coefficients of the plane wave from the surface and bottom, respectively. φ_s and φ_b are the reflection phases from surface and bottom, respectively.

We define $k_l^2 = (R_l + iI_l)^2 = a + i2b$

where, $a = R_l^2 - I_l^2$ and $b = R_l I_l$. It should be noticed that a will be negative when the imaginary part of eigenvalue is larger than the real part.

The difference between the square of the wave number and eigenvalues is $k^2 - k_l^2 = k^2 - a + i2b$. By assuming the imaginary part is very small, that is $2b \ll \left|k^2 - a\right|$, one obtains

$$\sqrt{k^2 - k_l^2} \approx \sqrt{k^2 - a} - \frac{ib}{\sqrt{k^2 - a}}$$

where $\sqrt{k^2 - k_l^2}$ is a complex function. In order to calculate the higher-order modes, the Pekeris branch cut is selected in this paper.

Using the same method with Ref.11, one obtains

$$\ln\left|V_s(k_l)V_b(k_l)\right| \approx \ln\left|V_s(a)V_b(a)\right| + 2b\delta(a) + i[\varphi_s(a) + \varphi_b(a)] \qquad (10)$$

where, $\delta(a) = \dfrac{\partial}{\partial a}[\varphi_s(a) + \varphi_b(a)]$.

Substituting Eq.(9) and Eq.(10) into Eq.(6) gives

$$2\int_{\xi_l}^{\eta_l} \sqrt{k^2 - a}\,dz + \varphi_s(a) + \varphi_b(a) = 2l\pi \,, \tag{11}$$

and

$$\int_{\xi_l}^{\eta_l} \frac{2b}{\sqrt{k^2 - a}}\,dz + \ln\left|V_s(a)V_b(a)\right| + 2b\delta(a) = 0 \tag{12}$$

Eq.(11) and Eq.(12) are two equations which can be solved in real plane. One can easily solve the transcendental equation (11) for a by using an iterative method, and then substituting the value of a into Eq.(12), one obtains b,

$$b = a\beta \tag{13}$$

where $\beta = -\dfrac{\ln\left|V_s V_b\right|}{\displaystyle\int_{\xi_l}^{\eta_l} \frac{2a}{\sqrt{k^2 - a}}\,dz + 2a\delta(a)}$. $\tag{14}$

Substituting a and b into Eq.(7), one obtains

$$R_l = \sqrt{\frac{a + |a|\sqrt{1 + 4\beta^2}}{2}}\,, \tag{15}$$

and

$$I_l = a\beta/R_l \tag{16}$$

Now, it is easy to obtain numerical solutions. During the whole procedure, there is only one transcendental equation in real plane need to be solved. By this way, the calculation speed can be great improved.

When $4\beta^2 \ll 1$, Eq. (15) and Eq.(16) are reduced to

$$R_l^2 = a\,, \tag{17}$$

and

$$I_l = R_l\beta = -\frac{\ln\left|V_s V_b\right|}{S_l(R_l) + \delta_l(R_l)} \tag{18}$$

where, $S_l = \displaystyle\int_{\xi_l}^{\eta_l} \frac{R_l}{\sqrt{k^2 - R_l^2}}\,dz$ is the mode span, $\delta_l = \dfrac{\partial}{\partial R_l}\left[\varphi_s + \varphi_b\right]$ is the beam displacement of surface and bottom. Eq.(18) is the formula Ref.11 first gave for calculating the mode attenuation.

In this paper, the generalized phase integral (WKBZ) theory [4] is used to calculate the normal modes,

$$\varphi_l = \sqrt{\frac{4k_l}{S_l}} \times \begin{cases} \dfrac{\exp\left(-\displaystyle\int_z^{\eta_l}\sqrt{k_l^2-k^2(y;r,\theta)}\,dy-\gamma\right)}{\left\{Bb^{4/3}-Db^{2/3}\left[k^2(z;r,\theta)-k_l^2\right]+16\left[k^2(z;r,\theta)-k_l^2\right]^2\right\}^{1/8}} & 0<z<\eta_l \\[6mm] \dfrac{\sin\left(-\displaystyle\int_{\eta_l}^z\sqrt{k^2(y;r,\theta)-k_l^2}\,dy+\dfrac{\pi}{2}-\dfrac{\varphi_s}{2}\right)}{\left\{Bb^{4/3}-Db^{2/3}\left[k^2(z;r,\theta)-k_l^2\right]+\left[k^2(z;r,\theta)-k_l^2\right]^2\right\}^{1/8}} & \eta_l\le z\le\xi_l \\[6mm] \dfrac{(-1)^l\exp\left(-\displaystyle\int_{\xi_l}^z\sqrt{k_l^2-k^2(y;r,\theta)}\,dy-\gamma\right)}{\left\{Bb^{4/3}-Db^{2/3}\left[k^2(z;r,\theta)-k_l^2\right]+16\left[k^2(z;r,\theta)-k_l^2\right]^2\right\}^{1/8}} & \xi_l<z\le H \end{cases}$$

where, η_l and ξ_l are the depth of upper and lower turning points, respectively. φ_s is the mode phase of reflection from surface, H is the sea depth, $B=2.152$, $D=1.619$, $\gamma=-\ln\cos(\varphi_s/2)$, $b=\left|\partial k^2/\partial z\right|$, S_l is the mode span.

3. Examples

3.1 Calculation of eigenvalues

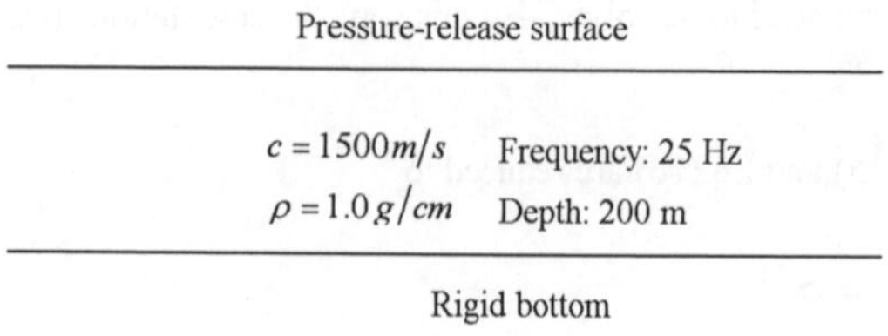

Pressure-release surface

$c=1500m/s$	Frequency: 25 Hz
$\rho=1.0g/cm$	Depth: 200 m

Rigid bottom

Fig.1 Homogenous water volume with
pressure-release surface and rigid bottom

Two examples are illustrated in this section. The example A is a typical problem of homogenous water volume with pressure-release surface and rigid bottom as shown in Fig.1. The eigenvalues of example A can be obtained by an analysis formula. Comparing the numerical solution with the analysis solution, one can see the precision of CMPE. The example B is a problem with typical sound velocity profile as shown in Fig.3. Comparing the numerical results between CMPE and KRAKENC model, one can see the efficiency of CMPE.

The numerical eigenvalues and the analytical eigenvalues are listed in table 1 and shown in Fig.2, the analytical eigenvalues of the example A is given by

$$k_m=\sqrt{\left(\frac{\omega}{c}\right)^2-\left[\left(m-\frac{1}{2}\right)\frac{\pi}{H}\right]^2} \qquad m=1,2,\cdots$$

Table 1   The numerical eigenvalues and the analytical eigenvalues of example A (1/m)

Number	Numerical eigenvalues		Analytical eigenvalues	
	Real part	Imaginary part	Real part	Imaginary part
1	0.10442481	0.00000000	0.10442481	0.00000000
2	0.10203461	0.00000000	0.10203461	0.00000000
3	0.09707781	0.00000000	0.09707781	0.00000000
4	0.08912722	0.00000000	0.08912722	0.00000000
20	0.00000000	0.28784840	0.00000000	0.28784838
21	0.00000000	0.30450994	0.00000000	0.30450994
22	0.00000000	0.32107536	0.00000000	0.32107535
88	0.00000000	1.37045166	0.00000000	1.37045169
89	0.00000000	1.38620489	0.00000000	1.38620484
90	0.00000000	1.40195711	0.00000000	1.40195715

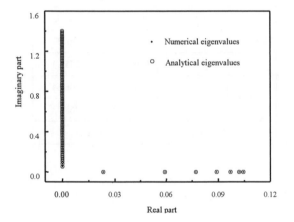

Fig.2 Comparisons between the numerical eigenvalues
and the analytical eigenvalues. Dots denote the numerical
eigenvalues and circles denote the analytical eigenvalues.
(unit: m^{-1})

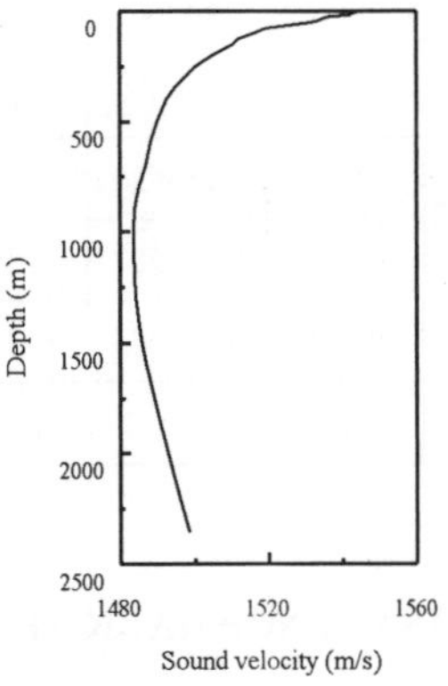

Fig.3 A typical sound velocity
profile of example B

Total 90 eigenvalues are calculated for example A. The first seven eigenvalues are real, the other are pure imaginaries. Fig.2 and Table 1 indicate that the numerical method has high precision.

Example B involves a typical sound velocity profile given in Fig.3. The water depth is 2350 meters, and the depth of sound channel axis is 1024.81 m where the sound velocity is 1483.5 m/s. The water sound velocity is 1498.18 m/s on surface and 1591.1 m/s on bottom. The density ratio between bottom and water is 1.7670, while the bottom attenuation is 0.15dB/(m·KHz). The frequency is 100Hz. Three approaches are used to calculate the eigenvalues of example B: CMPE, the generalized phase integral (WKBZ) theory [4] and KrakenC [9]. Total 1000 modes are calculated, some of them are given in table 2. The — denotes that the mode can't be calculated. The time of WKBZ is the computing time of the first 300 modes. Table2 indicate that CMPE, WKBZ and Krakenc have the same precision, while CMPE and WKBZ is more efficient than KrakenC.

Table 2   Eigenvalues of example B calculated by three approaches. (unit: 1/m)

Num-ber	CMPE		WKBZ		KrakenC	
	Real part $(10^{-3}m^{-1})$	Imaginary part $(10^{-3}m^{-1})$	Real part $(10^{-3}m^{-1})$	Imaginary part $(10^{-3}m^{-1})$	Real part $(10^{-3}m^{-1})$	Imaginary part $(10^{-3}m^{-1})$
1	423.439	0.00000	423.439	0.00000	423.437	0.00000
2	423.306	0.00000	423.306	0.00000	423.304	0.00000
3	423.153	0.00000	423.153	0.00000	423.157	0.00000
100	399.493	0.06455	399.493	0.06455	399.494	0.06419
200	325.984	0.18116	325.984	0.18116	325.985	0.18113
298	138.683	0.71791	138.682	0.71792	138.684	0.71785
299	134.789	0.74159	134.787	0.74160	134.790	0.74153
300	130.765	0.76744	130.762	0.76745	130.766	0.76737
998	0.285620	1265.24	—	—	0.285619	1265.24
999	0.285592	1266.64	—	—	0.285591	1266.64
1000	0.285565	1268.05	—	—	0.285564	1268.05
Time	2.42seconds		0.49 second		233.11 seconds	

3.2 ASA BENCHMARK RESULTS

A benchmark problem [12] on sound propagation in wedge-shaped ocean is solved in this section. The schematic of environmental scenario is given in Fig.5. The initial water depth is 200m decreasing linearly to 0 m at a range of 4 km giving a wedge angle of about 2.86° . The point source is

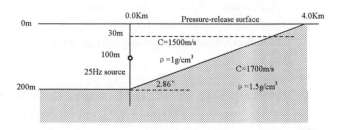

Fig.5 Schematic of the wedge problem

placed at 100m and the receiver depth is 30m. The water sound velocity is 1500 m/s, and the bottom velocity is 1700 m/s. the density ratio between bottom and water is 1.5 and the bottom attenuation is 0.5 dB/λ. In Fig.6, the numerical solutions of transmission loss from CMPE and COUPLE [13] are compared. The frequencies are 25 and 100Hz in Fig.6(a) and Fig.6(b), respectively.

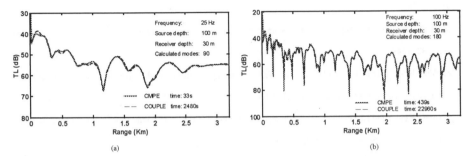

(a)   (b)

Fig.6 Comparisons of transmission loss from CMPE and COUPLE at 30m for wedge-shaped problem involving a point source at 100m. The solid curves are calculated by the CMPE. The dashed curves are calculated by the COUPLE. (a) The frequency is 25Hz and 80 modes are calculated. (b) The frequency is 100Hz and 180 modes are calculated.

In Fig.6, the results are in good agreement with each other. It takes 33 seconds to resolve the wedge-shaped problem using CMPE and 2480 seconds using COUPLE for 25Hz point source. And it takes 439 seconds to resolve the wedge-shaped problem using CMPE and 22960 seconds using COUPLE for 100Hz point source. One can draw a conclusion that the CMPE is much faster than COUPLE.

4 CONCLUSIONS

On the base of the generalized phase integral (WKBZ) theory, a new approach of coupled-mode parabolic-equation (CMPE) method is studied in the range-dependent waveguides. Results have indicated that the eigenvalues-finding algorithm is efficient and high precision, and can calculate very high-order modes. It only needs simple change to apply this algorithm in WKBZ theory. CMPE is efficient and high precision for range-dependent waveguides, and can be easily extended to 3-D problems.

5 ACKNOWLEDGMENTS

The work was supported by the National Natural Science Foundation of China.(Grant No. 10074070)

REFERENCE

1. ZHANG Rehe, LIU Hong, HE Yi, and, V. A. Akulichev. WKBZ Adiabatic Mode Approach to Sound Propagation in Gradually Range-Dependent Channels, Ata Acoustica (in Chinese), 1994,19(6): 408

2. D. Lee and A. D. Pierce. Parabolic Equation Development in Recent Decade, *J. Comp. Acoust.* 1995, **3**(2): 95-173

3. M. B. Porter and H. P. Bucker. Gaussian beam tracing for computing ocean acoustic fields, *J. Acoust. Soc. Am.* 1987, **82**: 1349-1359

4. ZHANG Rehe, LIU Hong, and HE Yi, The WKBZ mode approach to sound propagation in range-independent ocean channels, Chinese Jour. of Acous. (in Chinese), 1994, **13**(1): 1-12

5. ZHANG Renhe and LI Fenghua, Beam-displacement ray-mode theory of sound propagation in shallow water. Science in China (Series A), 1999, 42(7): 740-749

6. R. B. Evans, A coupled mode solution for acoustic propagation in a waveguide with stepwise depth variations of a penetrable bottom, *J. Acoust. Soc. Am.* 1983, **74**(1): 188-195

7. A. T. Abawi, W. A. Kuperman and M. D. Collins, The coupled mode parabolic equation, *J. Acoust. Soc. Am.* 1997, **102**(1): 233-238

8. F. B. Jensen, W. A. Kuperman, M. B. Porter etc, Computational Ocean Acoustics, New York： AIP Press, 1993

9. M. B. Porter. The KRAKEN Normal Mode Program. NRL/MR/5120-92-6920, Washington D.C.: Naval Research Laboratory, 1992

10. J. F. Miller and S. N. Wolf. Modal Acoustic Transmission Loss (MOATL): A Transmission-loss Computer Program Using a Normal-Mode Model of the Acoustic Field in the Ocean, NRL Report 8249, 1980

11. R. Zhang and Z. Lu. Attenuation and group velocity of normal mode in shallow water, Journal of Sound and Vibration, 1989, **128**: 121-130

12. F. B. Jensen and C. M. Perla. Numerical solution of range-dependent benchmark problems in ocean acoustics, *J. Acoust. Soc. Am.* 1990, **87**(4): 1499-1510

13. R. B. Evans. A coupled mode solution for acoustic propagation in a waveguide with stepwise depth variations of a penetrable bottom, *J. Acoust. Soc. Am.* 1983, **74**(1): 188-195

14. R. B. Evans. The decoupling of stepwise coupled modes, *J. Acoust. Soc. Am.* 1986, **80**(5): 1414-1418

Theoretical and Computational Acoustics 2001
E.-C. Shang, Qihu Li and T. F. Gao (Editors)
© 2002 World Scientific Publishing Co.

The Use of 4D Seismic for Ocean Acoustic Tomography

Accaino F., Böhm G., Dal Moro G., Madrussani G., Rossi G. and Vesnaver A.

Istituto Nazionale di Oceanografia e di Geofisica Sperimentale, Trieste, Italy

Abstract

Usually, in the conventional processing of marine data, the velocity variations of the sea water are not considered because their effects have no influence on the imaging reconstruction, especially at large depths; but in the time-lapse monitoring of hydrocarbon reservoirs, the velocity time-variations in the seawater should be measured and compensated.
We show here the application of a 3-D tomographic tool on a 4-D real data set and the results of the depth imaging by considering or not the velocity variations in the sea water in the time-lapse analysis. The results here presented evidence also how the huge number of time-lapse surveys in various parts of the world may be of great importance for oceanography, since can be a way of monitoring temperature time and space variations within the oceans.

4-D analysis

The basic idea of time-lapse monitoring is that, in a producing hydrocarbon reservoir, the Earth response does not change in time, except at the reservoir itself. Here, variations are expected both in the seismic velocities, due to the different rock saturation, and in the interfaces among the common reservoir fluids, i.e. gas, oil and water.

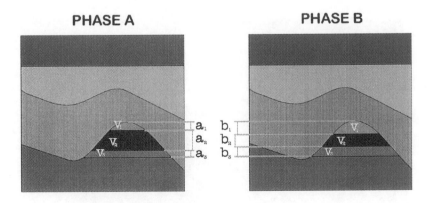

Figure 1 - Reservoir changing in two different time instants.

The parameters which can influence the seismic data recorded at different times are: the interface positions defining the gas-oil-water contacts; the velocity of P and S waves in the reservoir; the signature of the sources and the recording acquisition geometry.

Another important variable, which is independent upon the reservoir exploitation but which must be considered in the time-lapse case, is the velocity in the sea water, whose variation in time could significantly distort the time-lapse analysis results.

Sea water velocity and its variations in time and depth

If we consider the Mackenzie (1981) equation to determine the sound speed (v) in the sea water as a function of temperature (T), salinity (s) and depth (z):

$$v(T,s,z) = 1448.96 + 4.591T - 0.05304T^2 + 0.0002374T^3 + (1.340 - 0.01025T)(s - 35) +$$
$$0.0160z + 1.675 \cdot 10^{-7} z^2 + 7.139 \cdot 10^{-13} Tz^3$$

we observe that temperature is the parameter which plays the major rule, as shown in Figure 2, where two graphical representations of this formula are presented, keeping respectively constant salinity (on the left) and depth (on the right). In both graphics it is evident the major influence of the temperature variation with respect to the other parameters.

It is known that sea temperature changes both in depth and time. Time variations depend upon the climatic seasonal variation (e.g. Rajan and Frisk, 1992). In some cases a significant variation of the sound speed in shallow water could be detected also from one day to another, due to sudden changes of the weather conditions (air temperature, wind). As regards as the temperature depth variations, the oceans have a well-mixed surface layer, where water temperature is relatively constant, below which the *thermocline* is present, a layer where temperature changes very rapidly with depth. Depth and thickness of thermocline vary with the season (thermocline is shallower and well defined during the summer, deeper and less developed during the winter) and can be very different from one year to another.

For all these reasons an accurate measure of the sea water velocity must be detected for a time-lapse analysis, to remove its effects from the analysis results.

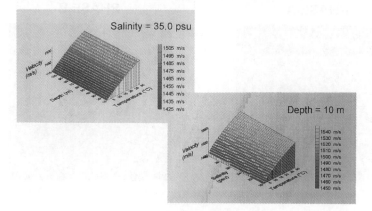

Figure 2 - Variation of sound speed in water with respect of temperature, salinity and depth.

Thermocline detection

To detect the depth and thickness of the thermocline, the oceanographers use a vertical streamer to record direct waves and obtain a ray disposition which follows the layer bedding (Figure 3A). Usually this acquisition is not available in conventional seismic. In this latter case, the thermocline thickness and depth can be detected by using the travel-time inversion of three different waves: *direct, reflected and refracted* (Figure 3B). However the ray paths obtained by these waves (in particular the reflected ones) is not optimal to resolve the vertical velocity variation in the sea. This fact could generate linear dependent elements in the tomographic matrix and, consequently, the presence of the null space in the system (Vesnaver, 1995).

A   B

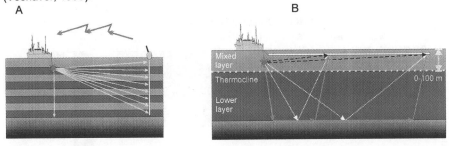

Figure 3 - Optimal acquisition for detecting vertical velocity variation (A). Ray paths distribution for the wave types in conventional seismic (A).

To avoid this drawback and dealing in some case with reflected and refracted arrivals only, in the particular real case here presented we applied the method of the staggered grids in the tomographic procedure (Vesnaver and Böhm, 2000) to detect the correct position and thickness of the thermocline.

Figure 4 shows a simple scheme of the staggered grid approach: we average different results of the tomographic inversion, obtained using a very rough, but robust, discretization (left) to achieve a final grid with a much finer resolution (right) than the starting models, but not too much affected by null space influence.

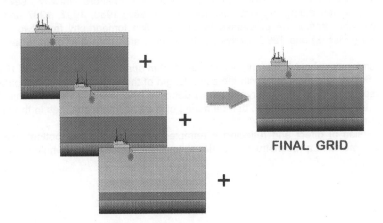

Figure 4 - Staggered grid method for detecting thermocline depth and thickness.

Figure 5 displays a vertical section showing the result of the tomographic inversion from a single 2-D line extracted from one of the real 3-D data sets we analyse in this paper. The water layer velocity field is obtained by the joint inversion of the reflected and refracted arrivals using the staggered grid method.

In the upper section the final discretization of the model is shown, to point out the vertical finer resolution resulting from the staggered grids. The interpolated values (below) divide the profile in two zones with different mean velocity and show the position in depth of the vertical velocity change (thermocline), in agreement with the thermocline average depth observed (e.g. Lenhart et al., 1995).

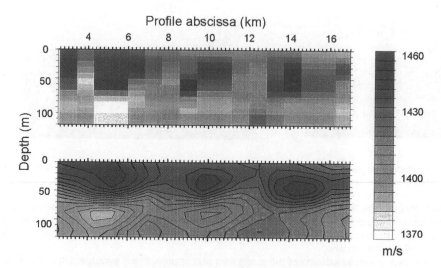

Figure 5 - Thermocline detection with staggered grids method.

A case history

At a producing reservoir in the North Sea (Oseberg field, offshore Norway), several 3D surveys were collected during the last decades, i.e. in 1982, 1989, 1992, 1997 and 1999. Figure 6 shows the profiles we processed by 3D reflection tomography from two of these vintages: 1989 (crosses) and 1992 (circles); both surveys are composed of irregular, nearly parallel lines.

In 1989 two streamers about 3 km long and located 100 m apart were used, while in 1992 two boats trailed five streamers in total with the same length of those used in 1989 but only 75 m distant. Sensor spacing in 1989 survey (25 m) doubles 1992's one. Moreover, due to the presence of strong irregular currents, the feathering of the streamers changes in the different areas and from line to line.

The joint inversion of direct and reflected + refracted waves of the sea bottom, using the staggered grid, provides a reconstruction of the water-layer velocity in the whole investigated area for the two different vintages (Figure 7). The velocity fields obtained points out the differences between the sound speed in the two time intervals.

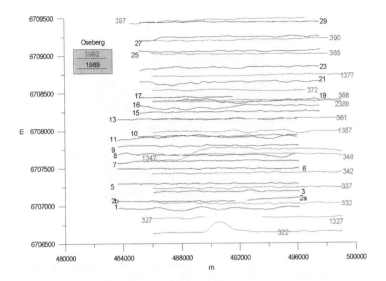

Figure 6 - Plane view of the whole acquisition in two different years.

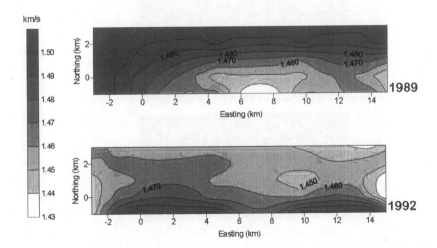

Figure 7 - Sound speed of the uppermost part of the sea water from the joint inversion of direct, reflected and refracted arrivals (plane view).

Migrated lines

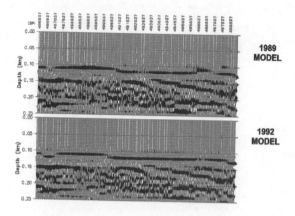

Figure 8 - Depth migrated section of the very shallow layers below sea-bottom of the 1992 data set. In the upper image the sea-water velocity relative to 1989 survey is used.

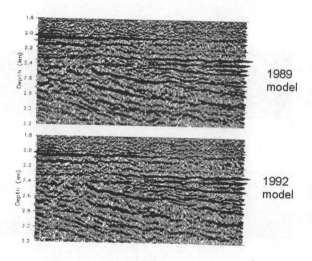

Figure 9 - Depth migrated section of the reservoir area of the 1992 data set. In the upper image the sea-water velocity relative to 1989 survey is used.

The velocity field obtained from a tomographic inversion is commonly used as input for a pre-stack depth imaging of the data (e.g. Vesnaver et al., 1999). The better the velocity, the more focused is the image obtained, so that migration and focusing is a commonly used test of the reliability of the velocity field. To show how the even small differences in sea-water velocity may be of importance in the process, the depth migration of the 1992 data set has been performed using the same velocity field, but in the water layer, where both velocity (1989 and 1992) were used. Figure 8 and 9 show the results, both for the very shallow sediments below the sea-bottom (Figure 8) and for the deeper part near the reservoir (Figure 9). The differences are subtle, but in general it may be observed that the continuity of the reflectors is affected when the wrong sea-water velocity is used, also at a depth of about 2 km.

Conclusions

The time-lapse tomographic inversion of the North-Sea data set, and their successive depth migration showed how it may be of importance to compensate the effects of sea-water time and space variations in such a case. The very small velocity variations due to the changed relationships among water, gas and oil in the reservoir could otherwise be masked by changes due to the different times of the acquisition. The joint inversion of direct, reflect and refracted arrivals give enough information on water velocity distribution, even in lack of a vertical streamer. Moreover, to reduce the possible influence of null space, the staggered grid method may be adopted, to obtain a robust and detailed map of the thermocline in the area under study. With these premises, ocean studies could have great advantage from the huge number of time-lapse surveys that in many parts of the world are performed in the last decades, since, almost without additional efforts, important information about ocean temperature changes may be obtained from them.

References

Lenhart, H.J., Radach, G., Backhaus, J.O. and Pohlmann T., 1995. *Simulation of the North Sea circulation, its variability and its implementation as hydrodynamical forcing in ERSEM.* Neth. J. Sea Res., 33: 271-299.

MacKenzie, K.V., 1981. *Nine-term equation for sound speed in the Ocean*, J. Acoust. Soc. Am., 70: 807-812.

Rajan S.D and Frisk, G.V., 1992. *Seasonal variations of the sediment compressional wave speed profile in the Gulf of Mexico.* J. Acoust. Soc. Am., 91: 127-135.

Vesnaver, A., 1995. *Null space reduction in the linearized tomographic inversion.* In:O. Diachok et al., Ed., Full field inversion methods in ocean and seismo-acoustics: Kluwer, 139-145.

Vesnaver A., Böhm G., Madrussani G., Petersen S. e Rossi G.; 1999. *Tomographic imaging by reflected and refracted arrivals at the North Sea.* Geophysics, 64, 1852-1862.

Vesnaver, A. and Böhm G., 2000. *Staggered or adapted grids for seismic tomography?* The Leading Edge, 19: 944-950.

Theoretical and Computational Acoustics 2001
E.-C. Shang, Qihu Li and T. F. Gao (Editors)

The Gain Limitation of Dense Spacing Linear Array in the Presence of High Wave-Number Noise

Shan Bingyi, Wang Qing, Wang Zhongkang
State Key Laboratory of Ocean Acoustics, Hangzhou Applied Acoustic Research Institute,
Hangzhou 310012, China, Box: 1249

Abstract: The gain limitation of dense spacing linear array in high or low wave-number noise field is discussed in this paper. The gain expression is derived, and the computer simulation is carried out, which compares the processing results obtained from the towing trial. The theoretical model is consistent with the experimental results, so it can be used to predict the actual array gain. When high wave-number noise predominates, about 30dB gain limitation can be achieved with the aperture of about two wavelengths, which far exceeds the conventional theoretic gain; while low wave-number noise will lead to gain degradation.
Key words: gain limitation, dense-spacing array, sound speed ratio, beam-shift

1. Introduction

Research on array gain is a classic subject, which people are greatly concerned about in radar and sonar applications. In isotropic homogenous electromagnetic or sound field, where noise and signal propagate at the same speed, array gain has been studied and conclusions for different array can been seen in texts. For example, uniform linear array gain $G_l = 10\log(2L/\lambda)$ and planar array gain $G_p = 10\log(4\pi S/\lambda^2)$ are well known, where λ is the signal wavelength, L and S is the length and area of the array respectively. The array gain under the directive ocean noise field caused by wind has also been studied early. And Elena etc. has studied large-array gain in long-range and deepwater environments[1] and pointed out that because the reduction of signal coherence caused by sound scattering leads to gain degradation. But they didn't consider the effect of high wave-number noise. In practical situations, the case of gain loss often appears, but sometimes there is the case that gain exceed the relative aperture[2,3]. This is neither super-gain nor measure error, this is just caused by different sound field condition.

In this paper, the gain limitation of dense spacing linear array is discussed, and the gain expression is derived under the condition that there exist not only ambient ocean noise but also high wave-number noise such as mechanical vibration noise and flow noise. Furthermore, the computer simulation is carried out for representative cases using the special data obtained from the towing trial, which shows that theoretical predictions are in good agreement with experimental results. The gain limitation strongly depends on the structure of noise field: for the isotropic noise field with sound speed c in water, the classical gain expression is correct; but it is not suitable when sound field contains the noise component whose speed is different from sound signal's speed c. In the noise field that low wave-number noise predominates, gain limitation is smaller than the relative aperture calculated by signal wavelength; when high wave-number noise predominates, the gain limitation can exceed the corresponding value mentioned above.

2. Background noise model

In the working circumstance for the modern sonar array such as towed-array and flank array, etc., the radiant noise from the towing ship is an important noise source besides the ocean noise. It can be regarded as the interference with strong directivity coming from end-fire, which propagates with the sound speed c. In addition, the pressure fluctuation noise of turbulent boundary layer and mechanical vibration noise are also a class of important noise sources, which propagate with wave speed v different from the sound speed c. Within the frequency band of interest under normal towing speed, the spectrum level of flow noise is about 20dB higher than that of ambient ocean noise, and mechanical vibration noise about 20dB higher than flow noise.

Besides distinctly different correlation radii, another characteristic of various noise components is that they have different space wave-numbers, or say they have different wave propagation speeds. Suppose various noise components satisfy the superposition theorem, each of them can be expanded as the sum of many plane waves, and suppose the array is very fine, the field is axisymmetric, and the spectrum W_q of each noise component is different, then the total noise pressure output of the mth hydrophone is:

$$p_m(t) = \sum_{q=1}^{Q} \int_{\omega} \int_{\theta} A_q W_q D_q(\omega_q,\theta) B_q(\omega_q,\theta) e^{-j(\omega_q t - k_q x_m)} \sin\theta d\theta d\omega \qquad (1)$$

where $k_q = \omega_q \cos\theta_q / C_q$ represents the space wave-number of the qth noise component, A_q, ω_q and C_q represent the amplitude, angular frequency and wave propagation speed, respectively. $B_q(\theta,\omega)$ is the space-frequency response function of the sound field, $D_q(\theta,\omega)$ is the space-frequency response function of the limitative size hydrophone corresponding to the qth noise component. It should be emphasized that D_q is not only determined by the features of the hydrophone, but also is closely related with the wave-number's characteristics of the noise field. For line element, there is

$$D_q(\omega,\theta) = \frac{\sin(k_q l/2)}{k_q l/2} \qquad (2)$$

A linear hydrophone of limitative size l may be regarded as a point source in isotropic ocean noise, but it has very strong directivity in the pressure fluctuation noise field of turbulent boundary layer, because their wave-numbers differ by two orders of magnitude.

3. Gain of linear array

A uniform linear array of M sensors with spacing d is placed into a mixed noise field which consists of Q noise components with different wave-number k_q, if the beamforming time delay is $\tau = d\cos\theta_0/c$ and control wave-number is $k_0 = \omega\cos\theta_0/c$, then the output pressure $p(t)$ of linear array with M elements is

$$
\begin{aligned}
p(t) &= \sum_{m=1}^{M} p_m(t) \\
&= \sum_{m=1}^{M}\sum_{q=1}^{Q} \iint A_q W_q D_q(\omega,\theta) B_q(\omega,\theta) e^{-j[\omega t + (k_0-k_q)x_m]} \sin\theta d\theta d\omega
\end{aligned} \qquad (3)
$$

For brevity, we assume that there are only two independent noise components in sound field and their wave-numbers are k_n and k_v, respectively. When the time factor $e^{-j\omega t}$ is ignored, (3) can be expressed as

$$p(\Delta\omega) = p_n(\omega) + p_v(\omega) \qquad (4)$$

where

$$p_n(\Delta\omega) = A_n \sum_{m=1}^{M} \iint_0^\pi W_n B_n(\omega,\theta) D_n(\omega,\theta) e^{-jmd(k_n-k_0)} \sin\theta d\theta d\omega \qquad (5)$$

$$p_v(\Delta\omega) = A_v \sum_{m=1}^{M} \iint_0^\pi W_v B_v(\omega,\theta) D_v(\omega,\theta) e^{-jmd(k_v-k_0)} \sin\theta d\theta d\omega \qquad (6)$$

Defining $\sigma_c = c/v$ as the propagation speed ratio of two noise components and exchanging the symbols of sum and integral, the output power of two noise components can be expressed respectively, by

$$P_n(\Delta\omega) = (\frac{A_n}{M})^2 \iint_0^\pi [W_n B_n(\omega,\theta) D_n(\omega,\theta) \frac{\sin Mx}{\sin x}]^2 \sin\theta d\theta d\omega \qquad (7)$$

$$P_v(\Delta\omega) = (\frac{A_v}{M})^2 \iint_0^\pi [W_v B_v(\omega,\theta) D_v(\omega,\theta) \frac{\sin My}{\sin y}]^2 \sin\theta d\theta d\omega \qquad (8)$$

where

$$x = \frac{\omega d}{2c}(\cos\theta - \cos\theta_0) \qquad (9)$$

$$y = \frac{\omega d}{2c}(\sigma_c \cos\theta - \cos\theta_0) \qquad (10)$$

Furthermore, defining $\sigma_p = A_v/A_n$ as the amplitude ratio of two noise components, we can deduct the directivity gain of array G which is the ratio of the noise power received by point hydrophone and M elements linear array as follows

$$G\ (\Delta\omega,\sigma_p,\sigma_c) = \frac{M^2 \int\int_0^\pi (B_n^2 W_n^2 + \sigma_p^2 W_v^2)\sin\theta d\theta d\omega}{\int\int_0^\pi [(W_n B_n D_n \frac{\sin Mx}{\sin x})^2 + (\sigma_p B_v W_v D_v \frac{\sin My}{\sin y})^2]\sin\theta d\theta d\omega} \quad (\ 11\ )$$

Discussion:

(i) As σ_p approaches zero, namely high wave-number noise is very weak and ocean noise predominates, then

$$G(\omega,\sigma_c,\sigma_p) = \frac{M^2 \int\int B_n^2 W_n^2 \sin\theta d\theta d\omega}{\int\int_\omega (W_n B_n D_n \frac{\sin Mx}{\sin x})^2 \sin\theta d\theta d\omega} \quad (\ 12\ )$$

For the hydrophone of normal size, we can regard $D_n(\omega,\theta)=1$ approximately. If $\theta_0 = \pi/2$, it can be proved that G is the conventional gain G_0 of general linear array.

$$G_0(\omega) = 2L/\lambda \quad (\ 13\ )$$

where $L = (M-1)d$ is the length of array, $d \ll \lambda$ and λ is the sound wave-length.

(ii) As $\sigma_p \gg 1, \sigma_c \gg 1$, the high wave-number noise predominates now, then G can be expressed as

$$G(\omega,\sigma_c,\sigma_p) = \frac{2M^2 \int W_v^2 d\omega}{\int\int_0^\pi (W_v D_v \frac{\sin My}{\sin y})^2 \sin\theta d\theta d\omega} \quad (\ 14\ )$$

In general case, $G \gg G_0$. The position of the ith order maximum value of space response is

$$\theta_i = \arccos[\frac{1}{\sigma_c}(\cos\theta_0 \pm \frac{i\lambda}{d})]\ ,\quad i=0,\ 1,\ 2\cdots\cdots \quad (\ 15\ )$$

When $\sigma_c \gg 1$, in spite of how to control the beam, the principal(0th order) maximum value is at nearby $\pi/2$ all the time and not at the direction of control beam, which can result in beam-shift.

(iii) As $\sigma_p \gg 1, \sigma_c \gg 1$ and $d \to 0, M \to \infty$, gain can reach its maximum value, that is

$$G_{max} \approx \sigma_c G_0 = 2L/\lambda_v \quad (\ 16\ )$$

where λ_v is the wave-length of high wave-number noise.

(iv) As $\sigma_p \gg 1, \sigma_c < 1$, low wave-number noise predominates, the grid lobe in space response will not occur in general, but the position of principle maximum value will be shifted, i.e., $\theta_i \neq \theta_0$ and $G < G_0$. For example, for the bending wave in steel plate, $\sigma_c \approx 0.3$, and then the mechanical vibration noise incidenting from end-fire will produce a maximum value near $110°$ and cause gain decrement.

4. Simulation

The computer simulation has been carried out for a uniform linear array with $M = 16, L = 3m, \Delta f = 10Hz \sim 1250Hz$. We assume that there are general noise p_n with directivity distribution $B_n \neq 1$ and high wave-number noise in sound field. The former propagates with sound speed c, the latter propagates with wave speed v. We have also compared the simulation results in the case $B_n = 1$ for the range of σ_P from -20dB to 50dB and σ_c from -10dB to 70dB. Fig.1 shows the directivity gain G of the array as the function of σ_P with σ_c as parameter indicated above curves.

For velocity ratio $\sigma_c = -10$dB, which corresponds to low wave-number noise, G decreases with the increase of σ_P. The bending wave in steel plate applies to this case. For $\sigma_c = 10$, which corresponds to mechanical vibration noise of towed linear array, G increases slightly with the increase of σ_P. For $\sigma_c = 40$dB, it corresponds to pressure fluctuation noise of turbulent boundary layer at normal towing speed. When $\sigma_P > 40$dB, whatever σ_c is, G gets closer to a saturation value which varies with different σ_c. As $\sigma_c \geq 60$ and $\sigma_P \geq 50$, the saturation value is approximate to the gain limitation, while $\sigma_P < 10$dB, regardless of σ_c, G always comes to a asymptotic value—the conventional gain G_0.

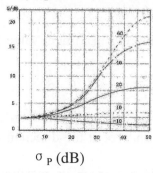

σ_P (dB)

Fig.1 The relations of gain G and σ_P in isotropic noise filed

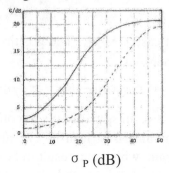

σ_P (dB)

Fig.2 The effect of sound field configuration on G

Fig.2 shows the effect of sound field configuration with $\sigma_c = 60$dB caused by cardioid distributed noise(solid line) and isotropic noise(dash line) respectively. It is seen that the gain in the directivity noise field is higher and it reaches its saturation value with σ_p about 10dB ahead of that in isotropic noise. Except for special explanation, we refer to general noise as directivity noise and non-acoustic noise as isotropic noise in the following text, which is in accord with the actual situations.

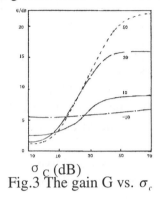

σ_c (dB)

Fig.3 The gain G vs. σ_c

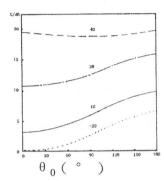

θ_0 (°)

Fig.4 The gain G vs. θ_0

Fig.3 shows G versus propagation speed ratio σ_c for different noise components with corresponding σ_p values above curves. When $\sigma_c > 20$ dB, the gain G increases with the increase of σ_p and σ_c; and adversely it decreases as the decrease of σ_c if $\sigma_c < 10dB$ (the larger σ_p,the faster G drops). When $\sigma_p < 0$ dB, i.e. the noise propagating with speed c predominates, G goes to a fixed value and basically doesn't change with σ_c as expected.

Fig.4 shows G versus beam control direction θ_0. The gain increases with the increase of θ_0 when $\sigma_p \le 20dB$ .It doesn't change with θ_0 basically when $\sigma_p \ge 40dB$, and the higher σ_p is , the larger is the gain G under the same condition. Here, we assume σ_c=46dB.

Gain G versus hydrophone's length l is shown in Fig.5. The gain does not change obviously for the simulated 16 elements uniform linear array, when the length of hydrophone increases from 1cm up to continuous and $\sigma_c \le 30dB$. Only when $\sigma_c >> 20dB, \sigma_p \ge 40dB$, the gain can be raised obviously with the increase of l. In another words, the increase of l has effects on G only when the pressure fluctuation noise of turbulent boundary layer is very strong. In normal conditions, it has no obvious effect on suppressing high wave-number noises, this is out of accord with the result of Knight[4]. It is only the sub-array processing that is the effective method to suppress these noises.

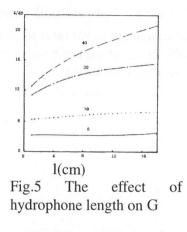

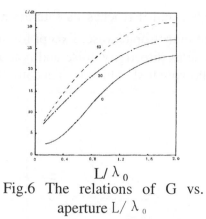

Fig.5 The effect of hydrophone length on G

Fig.6 The relations of G vs. aperture L/λ_0

Directivity gain G versus relative aperture L/λ_0 for fixed size linear array is shown in Fig.6, where λ_0 is the wave-length of signal corresponding to the center frequency. The gain increases with the increase of L/λ_0 when σ_P is constant, and it also increases with the increase of σ_P when L/λ_0 is constant. Because it already approaches the gain limitation at $\sigma_P \geq 40dB$, the increase of G can be ignored if σ_P increases any more. As $\sigma_P \geq 0$, $G \gg G_0 = 10\log(2L/\lambda)$, i.e. as long as there is high wave-number noise in sound field, the actual gain of dense spacing linear array is much greater than the conventional theoretic value G_0 as well as the logarithm of the sensor numbers. The increased amount depends on both parameters σ_c and σ_P. The changing trend of gain G is in a similar way for different θ_0, but the difference of their magnitudes is larger.

5. Experimental verification

In order to verify the theoretical model for gain presented in the paper, the towing experiment in a lake has been carried out for 12 elements linear array, and the experimental results are compared with that of computer simulation. The experimental equipment and conditions are described in reference[3]. The main noise sources in the experiment consist of two parts: the radiant noise and wake noise from towing ship with sound velocity c, and the flow noise caused by pressure fluctuation of turbulent boundary layer and structure vibrating, which is high wave-number noise. The spectrum level of background noise in the lake is much lower. The amplitude ratio of noise components σ_P can be calculated from the output of hydrophone using the difference of coherence for every noise component. The Fig.7 shows the inverse of amplitude ratio for correlative and uncorrelated noise σ_P^{-1} versus frequency under a typical condition of this experiment. It can be seen from the Fig.7 that the amplitude ratio $\sigma_p \approx 0db$ in this case for 100Hz<f<1kHz.

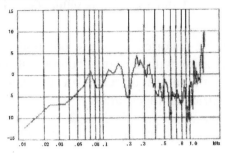

Fig 7   The relative spectrum level σ_p^{-1} of flow noise to towed ship noise

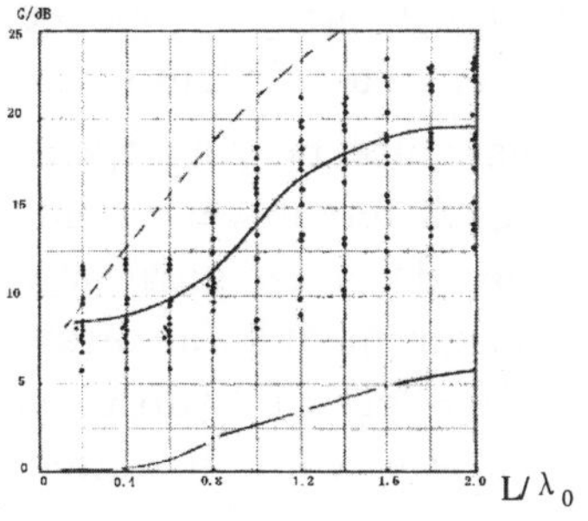

Fig.8   The comparison of the theoretical and the experimental values of the gain G vs. relative aperture

Under the condition showed in Fig.7, the comparison of the theoretical and experimental values of gain G for a linear array with 12 elements is shown in Fig.8. In it the dash line corresponds to the calculated values from the equation (11) with $\sigma_p = 0dB$, and the data point " " corresponds to the values measured for every voyage, while the solid line to an average value for 14 voyages. For convenience of comparison, the classical theoretical gain G_0 in equation (13) is also indicated by dot-dash line. It can be seen from Fig.8 that the vast majority of experimental values of gain G are between the gain limitation (dash line) and the classical value G_0 in the presence of high wave-number noise, and the gain G is 10dB to 25dB higher than G_0 under the condition indicated in Fig.7, which is a tremendous value and is of great importance in engineering practice. Because practical hydrophone is regarded as a point hydrophone in data processing despite of its size, the experimental value of gain is estimated slightly lower but still much larger than the classical theoretical value G_0 yet.

6. Conclusion

Based on the theoretical analysis, computer simulation and the experimental verification, the following conclusions can be drawn:

(1) The actual gain of dense spacing linear array can far exceed the classical theoretical gain G_0 in the presence of high wave-number noise. The classical theoretical value is approximately suitable only when $\sigma_p < 10dB$. The practical gain of an array can be predicted by the theory developed in the paper.

(2) The gain limitation of an array strongly depends on the sound filed structure and array configuration as well as its processing method. It

can either far exceed the classical theoretical value G_0, or much lower than it in the presence of low wave-number noise. The gain of practical array is between the G_0 and the maximum value G_{max} in general cases. Here $Gmax \approx \sigma_c G_0$.

(3) The noise bearing spectrum can be distorted by noise components with different wave-number for an array with electrical compensation. This deserves great attention in the noise directing or the noise measurement by an array with electrical compensation.

(4) The effect of hydrophone length on the gain is less in general as $\sigma_p < 40dB$, while the effect of beam control direction θ_0 on the gain is obvious. Only if $\sigma_p \geq 40dB$ and $\sigma_c \geq 40dB$, the gain is obviously increased by raising the length of hydrophone.

Acknowledgements

The authors would like to acknowledge Prof. Zheng Zhaoning and Prof. Wang Yanlin for their suggestions on the subject and thanks are also given to Lou Jufang for her help on the translation of the paper.

Reference

[1]Elena Y.G, etc. "Deep-Water Acoustic Coherence at Long Ranges",IEEE Journal of Oceanic Engineering. vol.24 (1999) No.2,156-171

[2]Shan Bingyi,Ma Chen,Zheng Zhen Yu,"Noise Suppression of a Hydrophone Group in the Mixed Noise Field" ,Proceeding of 2nd inter. workshop on Acoustical Engineering and Technology, Harbin, China,1999, 373-383

[3]Shan Bingyi ,Zheng Guolun, Zou Tengrong, "Fractional Beamforming of Dense Spacing Array for Time-frequency Mix", Chinese Journal of Acoustics, vol.19 (2000) No.2, 149-158

[4]Knight A., "Flow noise calculation for expended hydrophones in fluid and solid-filled towed array", JASA 100(1) 1996, 245-251

[5]S P Beerens, etc. "Flow noise analysis of towed sonar array", Proceedings of UDT, 1999, 392-397

Theoretical and Computational Acoustics 2001
E.-C. Shang, Qihu Li and T. F. Gao (Editors)
© 2002 World Scientific Publishing Co.

The Passive Acoustical Method of Monitoring and Measuring Sea Surface Wind Speeds

Jianheng Lin, Daoqing Chang, Xuejun Li, Guojian Jiang
Qingdao Lab of Institute of Acoustics, the Chinese Academy of Sciences,
Qingdao, China

Abstract: Ocean ambient noise is highly correlated with surface wind speed. The relationship between ambient noise and wind speed is given based on the measurements made in recent years. It is shown that ocean ambient noise levels (NSL) are linear with the logarithm of surface wind speeds at usual wind speeds: $SPL = a + b * \log_{10} V$. The coefficients, a and b, depend not only on measurement positions and frequencies, but also on wind speeds. Their values determined at a frequency usually adapt only in the wind speed range from 1.5 to 18m/s (corresponding to the Pushurl wind level of 1 to 6). Our study shows that the coefficients of the previous expression must be re-determined in the cases when the wind level is higher than 15m/s and when the wind speed is lower than 1.1m/s (corresponding to 0-1 sea state). The expression provides the basis for estimation of wind speed by ambient noise within ±10% of the measured value in wind speed range of 0.1-22.6m/s when the coefficients are determined piecewise. Results of surface wind speed estimation by ocean ambient noise are given. The best estimation frequency band is 2kHz-5kHz.

1. Introduction

Many measurements and analysis [1-22] show that different noise sources contribute much to the ambient noise in different frequency bands. Wind-generated ambient noise and distant ship noise are generally thought to be the most important sources in 10Hz-20kHz band, while in the band of 10kHz -- 20kHz, wind-generated ambient noise plays an important role.

It is noticeable that ocean ambient noise is highly correlated with surface wind speed. If an accurate relationship between ocean ambient noise and wind speed is given, the sea surface wind speed can be estimated accurately from the measured noise data. The monitoring of the ocean in a storm is possible by adopting the automatically monitoring systems.

Knusden [1] analyzed the data of ocean ambient noise during World War II in 1948, and provided the famous Knusden Spectrum parameterized by wind forces or sea states. Since then people found that wind speeds are better correlated with ocean ambient noise than sea states.

In 1962, Wenz [2] obtained the famous Wenz spectrum curve based on many statistical data of ocean ambient noise and pointed out that the relationship between ambient noise and wind speed is similar in the whole frequency range between 1 Hz and 10kHz. It was demonstrated by Piggot [4] in 1964 that in shallow sea (less than 40 meters deep), ocean ambient noise level depends on the logarithm of surface wind speed linearly. In 1972, Crouch and Bur [5] further found a similar linear relationship in deep sea (to depth of 5000m). Rossby suggested an unbiased estimation of wind speed and wind stress in a wide sea by the ambient noise measurements in 1973. In the same year, Shaw et al [9] calculated surface wind speeds from ambient noise data obtained at the ocean bottom and successfully comparing them with shipboard measurements.

If the observed spectrum level of ambient noise is expressed in SPL in dB re $1 \mu Pa^2 Hz^{-1}$, the relationship between SPL and the logarithm of the observed surface wind speed appears to be linear and satisfies:

$$20 * \log_{10} V = a * SPL + b \qquad (1)$$

Where V is wind speed, the base of the logarithm is 10. The coefficients a and b should take different values in different seas and receiving depths, which can be determined by the least squares fit of the ambient noise levels response to wind speeds. Shaw suggested $a=1.01$, $b=30.4$, which enables us to estimate wind speeds from ambient noise data within 5 knots for wind speeds greater than 5 knots (about 2.5m/s). In the end of 1996, R.L.Gordon et al published their results: the relation between wind speed and SPL also satisfies Equation (1), and the values of $a=0.84$, $b=0.217$ are given at 4.0kHz, where V is the surface wind speed in meters per second.

The Qingdao Lab of Institute of Acoustics, the Chinese Academy of Sciences fitted the measured data to the logarithm of the observed surface wind speeds, the relationship satisfies [22]:

$$SPL = A(f) + B(f) * \log_{10} V \qquad (2)$$

Where f is frequency in Hz, SPL is the noise spectrum level in decibels re $2*10^{-14}$ dyne/centimeter at frequency f. At $f=1kHz$, $A(f)=20.9$, $B(f)=19.3$, where V is wind speed in meter/second.

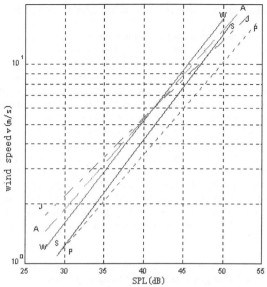

Fig.1 The relationship between *SPL* (14.5kHz)and the log of the wind
speed published in the literature. Lines are marked by author
or deployment: W is Wenz, S is Shaw et al.[1978], J is JASIN
deployment, P is Pacific deployment, and A is Atlantic deployment.

Fig.1 displays the relationship between *SPL* (at *14.5kHz*) and the logarithm of wind speed obtained respectively by *Wenz* et al in the literature[14].

By far, ocean surface wind speed has been successfully estimated from ambient noise over the range from 2.5m/s to 18m/s. Reports on the range of high wind speeds are few, and the estimating precision is low.

2.   The system of monitoring and measuring surface wind speed by ambient noise in Qingdao sea

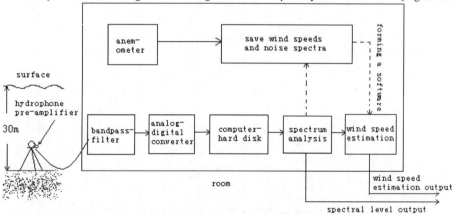

Fig.2. A block diagram of the principle of estimating surface wind speed by measuring ambient noise

Fig.2 demonstrates the principle of obtaining surface wind speed by measuring ocean ambient noise. The ambient sound field was monitored using a low noise bottom-mounted hydrophone. The unit has a flat response from 1kHz-50kHz and a sensitivity of $172dB/volt/1\mu Pa$. The hydrophones were fixed on an iron shelf of 28 m deep, and the signal cable led to the bank lab shack where the data were filtered, converted from analog to digital, sampled and recorded on computer hard disk directly. The sampling and recording of the digital data of ambient noise are achieved both automatically and manually. If an auto-recorded manner is adopted, the record length and the time interval between two adjacent records can be set at will.

In order to estimate surface wind speed by ocean ambient noise, the following steps were taken:

1) At first, we measured ambient noise for a wide range of wind speed using the monitoring and measuring system shown in Fig.2, and enough records were obtained.

2) Second, spectral analysis was made for each record. The wind speed recorded by anemoscope at that time is also given. Dividing all of the wind speeds into many scopes, and averaging the spectrum levels in the same wind speed scope, we can obtain an average spectrum curve of ambient noise corresponding to this wind speed scope (or an average wind speed in this scope)

3) Third, calculating the spectrum level of ambient noise for other wind speed (or average wind speed) scopes in the same way, finally we can obtain all the ambient noise curve for all of the wind scopes.

4) Fourth, the curve of ambient noise spectrum level against the logarithm of wind speed is plotted.

5) Choosing a group of frequencies in the frequency band from several hundred Hz to 20kHz to calculate the fitting parameters, we can determine the best estimating frequency band (or frequency), and write the software of estimating the surface wind speed by ambient noise spectrum level.

6) An ambient noise record at any wind speed is given, we can calculate the ambient noise spectrum, and estimate its corresponding wind speed by the software, and then compare the estimated value with the measured one.

3. The results of estimating the wind speed by ambient noise in Qingdao Sea

3.1 The spectrum curve of wind-generated ambient noise at Qingdao Sea

Between July 1999 and February 2000, we measured the ambient noise at Qingdao Sea by using a recoverable ambient noise receiving system, and estimated wind speeds from ambient noise. Figure 3 shows the spectrum levels of ambient noise (in dB re $1\mu Pa^2 Hz^{-1}$) at Qingdao Sea during a south-wind period. This figure is parameterized by wind speeds. The highest wind speed is 22m/s (corresponding to wind level 8). Fig.3 also shows the spectrum level curve during a north-wind period. This curve have a similar spectrum level, curve shape and curve slope with that of south wind period at low wind speed, while the spectrum levels for south wind are higher than those for north wind at high wind speed. For medium wind speed, the spectrum level for south wind is about 1-1.5dB higher than for north wind. This difference reaches 2dB for high wind speed.

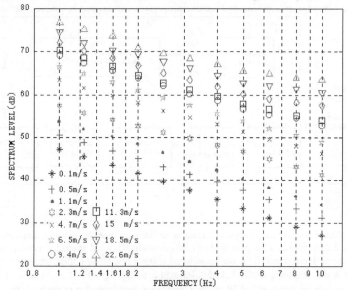

Fig.3. Ambient noise spectrum in Qingdao Sea for south wind

Since both the shapes and the slopes of the spectra have similar characteristics between north wind and south wind over the observed wind speeds, the effects of the surf slapping on the bank are little and its effect on the estimation of absolute value of wind speed is negligible.

3.2 The relationship between wind-generated ambient noise and surface wind speed

Fig.4 shows the relationship between wind-generated ambient noise and wind speed at Qingdao Sea during a south wind period between 1kHz to 10kHz.

The eleven curves display two characteristics as follows:

1) The spectrum level is not strictly linear with the logarithm of wind speed over the whole observed wind speed range (from 0.1m/s to 22.6m/s), especially when the wind speed increases from very low to medium wind speed, the corresponding ambient noise spectrum level increases rapidly with the wind speed.

2) There is a different linear relationship between ambient noise spectrum level and the logarithm of wind speed at different frequencies, which shows that there exists a best frequency band or frequency to estimate the surface wind speed.

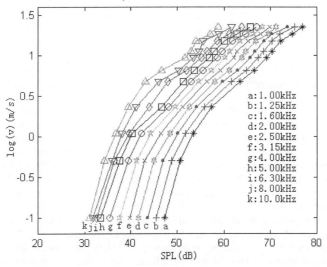

Fig.4. The relationship between wind-generated ambient noise and surface wind speed in Qingdao sea.

3.3 Estimating the surface wind speed by three-section fitting method

In order to improve the precision of estimating surface wind speed by ambient noise level, we adopted a three-segment fitting method, dividing the whole wind speed range into three scopes: low scope from 0.1m/s to 1.1m/s, medium scope from 1.1m/s to 15m/s, and high scope from 15m/s to 22.6m/s. We fit the spectrum level of ambient noise response to the observed wind data for each scope.

The relationship between the ambient noise spectrum level SPL and the logarithm of wind speed $logV$ appears to be linear and satisfies:

$$SPL = a + b * \log_{10} V \quad (3)$$

Where V is the sea surface wind speed in meters per second, and SPL is the noise spectrum level in decibels, the values of a and b for three scopes are given in Table 3.1.

Table 3.1 Coefficients a and b for three-segment fitting method at $f=3.15kHz$

a,b	Low (0.1-1.1m/s)	Medium (1.1-15.0m/s)	High (15.0-22.6m/s)
a	43.95	43.58	31.00
b	6.39	16.86	27.52

3.4 The best frequency (frequency band) for estimating wind speed

For a given wind speed scope, such as the medium scope (1.1-15m/s), the linear relationship between spectrum level and the logarithm of wind speed in different frequency is different. We can select a best fitting curve at different frequencies on which the estimating errors of wind speeds are the least, and the corresponding frequency is just the best one for wind speed estimation. Apparently, the more linear the relation is, the higher precision of estimation can be obtained. For medium wind speed scope, we calculated the relation between the noise spectrum level and the logarithm of wind speed at eleven frequencies over the frequency band from 1kHz to 10kHz, and obtained eleven straight lines by the method of the least square errors, then calculated the corresponding correlation coefficients. Values of the fitting and correlation coefficients at each frequency are given in Table 3.2.

Table 3.2 The fitting coefficients of a and b, and the correlation coefficient of r
at eleven frequencies for medium wind speed scope

coeff. \ f(kHz)	1.0	1.25	1.6	2.0	2.5	3.15	4.0	5.0	6.3	8.0	10.0
a	52.24	50.62	48.92	47.35	45.47	43.58	41.73	39.84	38.34	36.84	34.90
b	17.03	16.85	16.78	16.49	16.62	16.86	17.07	17.26	17.43	17.62	18.20
r	0.996	0.997	0.997	0.998	0.999	0.999	0.999	0.999	0.994	0.996	0.993

The data in Table 3.2 demonstrates that the correlation and linearity between the ambient noise level and the logarithm of wind speed are better at several frequencies: 2.5kHz, 3.15kHz, 4.0kHz, 5.0kHz, and the inversion precisions are quite high. It should be noticed that the errors of estimating wind speeds comes not only from the fitting precisions, but also from statistical average calculation. For example, the spectrum level of ambient noise corresponding to the average wind speed 2.3m/s is the average result of all the spectrum levels corresponding to all the wind speeds in the wind speed scope from 1.8 to 2.5m/s. Apparently, if the wind speed scope from which the average wind speed of 2.3m/s calculated extends from 1.8m/s – 2.5m/s to 1.5m/s-3.0m/s, the statistical precision of the average noise spectrum level at 2.3m/s deceases and its square increases, even though the number of the records averaged increases. To ensure the statistical precision of spectrum level curves against wind speed, the wind speed scope from which the average value calculated should be within $(1\pm10\%)$ of the labeled wind speed. The scope should not be too narrow so that a high accuracy can be ensured.

We noticed that when frequency is larger than 10kHz, the slope of the fitting line becomes distinctly small although there still exists the linear relation between ambient noise spectrum level and the logarithm of wind speed. The ambient noise spectrum level is less sensitive to the change of surface wind speed. In order to improve the estimation, from both the correlation and sensitivity between ambient noise and wind speed, we concluded that the best frequency band to estimate the surface wind speed from ambient noise is between 2kHz and 5kHz at Qingdao Sea, and 3.15kHz is the best frequency.

3.5 The comparison of wind speed between the estimated value and measured value .

The ambient noise spectrum levels at 3.15kHz are selected to estimate the surface wind speed. First the appropriate straight line should be selected to estimate the surface wind speed according to Equation (4), and the value of estimating coefficients of a and b (shown in Table 3.2) should be determined, then the estimated values of wind speeds at $f=3.15\,kHz$ are obtained and shown in Table 3.3.

$a= 43.9506$, $b = 6.3921$ when $SPL \leqslant 44.3dB$
$a= 43.5800$, $b = 16.8630$ when $44.3dB < SPL < 63.4dB$
$a= 31.0037$, $b = 27.5161$ when $SPL \geqslant 63.4dB$

(4)

Table 3.3   The comparison between the estimated value and measured value of wind speed.

SPLs(dB) at f=3.15kHz	Estimated value (m/s)	Measured value (m/s)	Estimating Errors
40.5	0.31	0.3	3.3%
42.1	0.56	0.6	-6.7%
46.4	1.47	1.4	5.0%
53.8	4.04	4.3	-6.0%
56.4	5.76	6.0	-4.0%

58.3	7.46	6.9	8.2%
59.5	8.79	8.2	7.2%
60.5	10.08	10.8	-6.7%
62.1	12.54	11.5	9.0%
62.9	13.99	13.6	2.8%
64.5	16.31	17.0	-4.1%
66.8	19.77	20.1	-1.6%

Table 3.3 shows that the estimating precision is greatly improved by using the multi-section fitting wind speed method, and the estimated wind speed is within ±10% of the measured speed. This is the most marked characteristic of our research. Another important result is that we have determined a best frequency to estimate wind speed, which also helps to improve the estimating precision.

If we made a two-section fitting for the relationship curve of ambient noise spectrum level against the logarithm of wind speed at $f=3.15kHz$, we obtain the coefficients as follows :

$\quad$ *First paragraph:* $\qquad$ *when SPL $\leqslant$ 44.3dB, $\quad a = 43.6795, \quad b = 6.2656$*

$\quad$ *Second paragraph:* $\qquad$ *when SPL>44.3dB, $\quad a = 43.6104, \quad b = 17.8204$*

$$(5)$$

When we adopt the two-section method to estimate the surface wind speed, the inversion errors are increased drastically as shown in Table 3.4 when SPLs are higher than 44.3dB, where fives times the error is larger than 10%, and the highest reaches 17.9%.

We fit the relation curve of noise spectrum level against the logarithm of wind speed only to one straight line, the coefficients a and b are given as follows:

$\quad\quad a = 44.9377$

$\quad\quad b = 13.9001$ $\qquad\qquad$ (6)

and the relative estimating error is bigger than *20%* in this case.

Table3.4　The comparison between the estimated value and measured value of wind speed.

SPLs(dB) at f=3.15kHz	Estimated value （m/s）	Measured value （m/s）	Estimating errors
40.5	0.31	0.3	3.3%
42.1	0.56	0.6	-6.7%
46.4	1.43	1.4	2.4%
53.8	3.73	4.3	-13.2%
56.4	5.22	6.0	-13.0%
58.3	6.67	6.9	-3.3%
59.5	7.79	8.2	-5.0%
60.5	8.86	10.8	-17.9%
62.1	10.90	11.5	-5.2%
62.9	12.09	13.6	-11.1%
64.5	14.87	17.0	-12.6%
66.8	20.00	20.1	-1.00%

Above data show that the multi-section fitting method notably improves the precision of estimating the surface wind speed by ambient noise.

4.Discussion

1) About multi-segment fitting method of estimating surface wind speed

Three-segment linear fitting wind speed method has remarkable advantages. It improves the estimating precision greatly, especially for very low and high wind speed situations. In Qingdao sea, when *SPL* at *f=3.15kHz* is lower than *44.3dB*, the corresponding wind speed belongs to low wind speed scope, while when *SPL* at *f=3.15kHz* is higher than *63.4dB*, the corresponding wind speed belongs to high wind speed scope. These are two limens for determining a appropriate wind speed scope.

It should be pointed out that the two limens at different sea are different, which should be determined based on analyzing the measured data of ambient noise for a long time. That is to say that "learning" is requested for the value of a and b.

2) The effect of precipitation on the estimating of wind speed.

It is well known that the spectra of rainfalls in the frequency band 1kHz-10kHz are similar to white noise. Apparently, the wind speed while companied by rain cannot be accurately estimated by the ambient noise.

Generally speaking, a wind process is continuous, and the instantaneous value of wind speed is fluctuant. However, the rainfall is often intermittent. Once sea area is chosen, the speed of wind companied by intermittent rain still can be estimated approximately by the ambient noise data when the rain stops, and in the meantime. We can compare the noise spectrum level data between rain and rain-stop, and then estimate wind speed with rain by subtracting the contribution of precipitation for different rainfall. The decreasing slope of spectrum curve of rain noise in the frequency band 1kHz-10kHz is generally $(-1 \sim -3)dB/octave$. If this slope of spectrum curve is observed, it means that rain appears on the sea surface, and we cannot simply estimate the wind speed by spectrum level at that time.

3) The effect of slapping-bank surf on estimating surf wind speed.

It is reported that stronger sound can be generated by surf slapping on the bank, and the spectrum has a decreasing slope of −5dB/octave.

The shape of the Qingdao bank in the east of Qingdao city is mostly like a "bow". The back of the bow points to the measurement position, and the receiving point is more than 6 kilometers away from the bank. The sound generated by surf slapping on the bank is emanative to the receiving hydrophone, rather than concentrated. Thus, the shape of the bank decreases the effect on ambient noise measurement greatly and the slapping-bank surfs contribute little to the wind-generated ambient noise, and their sound can be neglected at usual wind speed.

We notice that the sound generated by slapping-bank surfs is much lower for north winds than for south winds even they have same speeds. With the same wind speed 15m/s at south wind and north wind, we observed the difference of sound levels is within 2dB in frequency band 100Hz-20kHz at the same position, which shows that the sound generated by surfs slapping on the bank during south wind period in Qingdao sea is not high as imagined before. The effect of slapping-bank surf on the measurements of ambient noise in the experimental position can be neglected.

It seems that in Qingdao sea, the difference of 1-1.5dB of ambient noise sound level in frequency 1kHz-10kHz between south wind and north wind is not always caused by slapping-bank surf, which is a subject for further study.

4   Conclusions

The method of estimating surface wind speed from ambient noise is playing an important role in underwater passive monitoring ocean by sound, and is being paid more and more attention in many countries. This project supported by Chinese Ocean 863 Plan provides the basis for building an underwater passive sound monitoring and measuring system in China. We successfully estimated surface wind speed by ambient noise in Qingdao Sea, and obtained the following results:

1) The method of multi-segment linear fitting the relation of ambient noise spectrum level against the logarithm of wind speed greatly improves the precision of estimating surface wind speed, which is one of the most noticeable characteristics of our research.

2) Another noticeable characteristic is that we have determined the best frequency band for the estimation in which the estimating error of wind speed is the least. The best frequency band in Qingdao Sea is between 2.0kHz and 5.0kHz.

3) The estimating error of surface wind speed

The wind speed can be estimated within $\pm 10\%$ of the wind speed by the method suggested by us.

4) The effect of the surf on the estimating of wind speed is discussed. Qingdao Sea is wide, and the slapping-bank surf has little effect on the ambient noise for the advantageous bank shape and position. The analysis of observed ambient noise data for high wind speed situations shows that the

surf which slaps on the bank contributes little to the ambient noise at the receiving position at south wind. The reason for the slight difference of the spectrum level between south wind and north wind is expected to study in the future, and it seems that this difference most probably comes from the topography of the sea bottom.

5) The effect of precipitation on the ambient noise received by hydrophone is analyzed. The effective estimating wind speed method during wind and rain together has not been provided in this paper, while the wind speed can be estimated during the rain gap. If we observe the ambient noise data without stop during continuous rain, intermittent rain, and just before and after rain at same area, we are sure to take the rain noise out of the measured ambient noise data, and estimate the surface wind speed by ambient noise during rain and wind.

Not only wind speed but also other parameters such as precipitation can be monitored and measured automatically by ambient noise measurements; Otherwise, the dynamic change of fish resource and the status of zoology of the fish breeding field can also be monitored by similar method.

Acknowledgments. This work was partially supported by Ocean 863 Youth Fund Program under contract 818-Q-14. Helpful discussions were held with Prof. Hao LongSheng, Xie BaoXing of Institute of Acoustics, the Chinese Academy of Sciences regarding the field observation and data sampling.

Reference:

[1]Knudsen V.O.,R.S.Alford and J.W.Emling, "Underwater ambient noise", Journal of Marine Research, 7, 410-429(1948).

[2]Wenz G.M., "Acoustic ambient noise in the ocean: spectra and sources", Journal of the Acoustical Society of America,34,1936-1952(1962).

[3]Wenz,G.M., "Review of underwater acoustics research : Noise", Journal of the Acoustical Society of America, 51, 1010-1024(1971).

[4]Piggot C.L., "Ambient Sea Noise at low frequencies in shallow water of the scotian Shelf", Journal of the Acoustical Society of America, 36,2152-2163(1964).

[5]Crouch W.R. and P.J.Burt ,"The logarithmic dependence of surface generated ambient-sea-noise spectrum level on wind speed", Journal of the Acoustical Society of America,51,1066-1072(1972).

[6]Eckart C.,"The theory of noise in continuous media", JASA, 1953, 25, No.2,195-199.

[7]Urick R.J.,"Correlative properties of ambient at Bermuda", JASA, 1966,40,No.5,1108-1109.

[8]Urick R.J,"Ambient Noise in the Sea", Peninsula,Los Altos,CA,1986.

[9]P.T.Shaw, "On the estimation of oceanic wind speed and stress from ambient noise measurements", Deep Sea Research ,Vol.25, 1973, pp.1225 to 1233.

[10]L.Ding and D.M.Farmer, "On the dipole acoustic source level of breaking waves", JASA.96(1994),3036-3044.

[11]R.L.Gordon et al.,"Ocean Abient Sound Instrument System", Memorial University of Newfoundland,1996 Zedel@weejordy.physics.mun.ca.

[12]D.M.Farmer ,"Observations of High Frequency Ambient Sound Generated by Wind", B.R.Kerman(ed.),Sea Surface Sound,403-415 1988 by Kluwer Academic Publishers.

[13]David D.Lemon,"Acoustic Measurements of Wind Speed and Precipitation Over a Continental Shelf", Jouranl of Geophysical Research, 89(C3),pp. 3462-3472,May 20,1984.

[14]David L.Evans," Oceanic Winds Measured From the Seafloor", Journal of Geophysical Research, 89(C3),pp.3457-3461,May 20,1984.

[15]David Shonting," On the Spectra of Wind Generated Sound in The Ocean", B.R.Kerman(ed.),Sea Surface Sound, 417-427 1988 by Kluwer Academic Publishers.

[16]Duncan B. Ross," Observations of Oceanic Whitecaps and Their Relation to Remote Measurements of Surface Wind Speed", Journal of geophysical research 79(3),January 20,1974 .

[17]Svein Vagle ,"An Evaluation of the WOTAN Technique of Inferring Oceanic Winds from Underwater Ambient Sound",Journal of atmospheric and oceanic technology Vol.7 .

[18]A.S.Burgess," Wind-generated surface noise source levels in deep water east of Australia", J.Acoust.Soc. Am.73(1), January 1983 .

[19]Sarah J.Bass ,"Ambient Noise in the Natural Surf Zone:Wave-Breakin Frequencies",JOURNAL OF OCEAN ENGINEERING 22 (3),.JULY 1997 .

[20]Josette Paquin Fabre, "Noise Source Level Density Due to Surf—Part II: Duck",NC IEEE JOURNAL OF OCEANIC ENGINEERING 22 (3),JULY 1997.

[21]Jeffrey A.Nystuen, "The Underwater sound generated by heavy rainfall", J.Acoust.Soc.Am. 93(6),June 1993 .

[22]张殿云等，中科院声学所北海站科技档案。

Theoretical and Computational Acoustics 2001
E.-C. Shang, Qihu Li and T. F. Gao (Editors)
© 2002 World Scientific Publishing Co.

The Estimation Method for Self Noise of Underwater Fluid-Coming Structure

Peng Linhui, Lu Jianhui
Ocean University of Qingdao, Qingdao, 266003

Abstract: Method of statistical energy analysis is used to analyze self-noise of underwater fluid-coming structure excited by turbulent boundary layer pressure fluctuation. The estimation formula for the self-noise is obtained.

Key word: statistical energy analysis, self-noise

1. INTRODUCTION

It is a important problem that self-noise of underwater fluid-coming structure can be predicted. The traditional mode approach is applied to the presage of low frequency sound-vibration environment. However, in practice it is difficult to presage the sound-vibration results for the engineering problem of wide-band high frequency. Statistical energy analysis (SEA) approach, which has been developed since 1960s, can overcome the difficulty confronted the mode analysis approach and thus provide a favorable tool for dealing with wide-band high

frequency kinetic problem in complex systems [1]. It is specially succeede in avigation technology.

But the method is used less for underwater problem. Here, SEA approach is used to analyses self-noise of underwater fluid-coming structure excited by turbulent boundary layer pressure fluctuation and the corresponding engineering estimation formula is obtained, which is appropriate for case of high frequency.

2. SELF-NOISE OF UNDERWATER FLUID-COMING STRUCTURE

(1). Analysis mode

Suppose the underwater fluid-coming structure is a water-drop shaped shell. When sailing with high speed, surface of the shell is excited by turbulent boundary layer pressure fluctuation to vibrate and to produce self-noise in the shell. Taking the fluid-coming structure as ellipsoid shell,

ρ_s is material density of the shell, c_s is longitudinal wave sound speed, R_2 and R_1 are semi

major and semi minor axes of the shell respectively, h is thickness of the shell, ρ_a and c_a are the

density and sound speed of liquid, which fill and surround the shell, fluid-coming velocity is U.

The above physical model consists of three parts: turbulent boundary layer pressure fluctuation, ellipsoid shell and sound field in the shell. When the shell sailing with high speed, turbulent boundary layer pressure fluctuation excites vibration of the shell to radiate noise, and transmits noise to the shell as a sound source directly. Here, turbulent boundary layer pressure fluctuation and hydrodynamic noise environment are regarded as input power sources of SEA sub-system. The approximate case is considered as turbulent boundary layer pressure fluctuation acts on the whole ellipsoid shell uniformly. Since the problem to be solved is the sound field in the shell, we only take account of the shell sound vibration modal group, which can radiate sound pressure effectively and is considered as sub-system 1. Noise field in the shell is taken as sub-system 2.

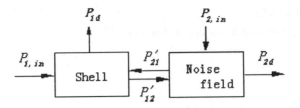

Fig.1. SEA model

Assuming $P_{i,in}$ is the internal input power for the ith sub-system, P'_{ji} the one-way power flow form the ith sub-system to the jth sub-system, and P_{id} the power loss of the ith sub-system. η_i is the internal loss factor of the ith sub-system, η_{ij} is the coupling loss factor. The transmitting paths of power flow in the system are illustrated in figure 1.

(2). Loss factors

For most engineering structure, the structure damping loss factor are far less than 0.1, so it can be neglected, ig,
$$\eta_1 = 0 \tag{1}$$

The internal loss factor of sound field sub-system can be obtained from reverberation time and expressed as
$$\eta_2 = \frac{0.04 c_a \alpha S}{f V_a} \tag{2}$$

where S and V_a are the surface area and the volume of the sound field space. α is average sound-absorbing coefficient of the wall.

Suppose the area of the connecting interface between the sub-sound field and the sub-structure is A_p, η_{12} is coupling loss factor from the structure sub-system to the pressure field sub-system respectively. η_{12} can be obtained as

$$\eta_{12} = \frac{\rho_a c_a}{\omega \rho_s h} \sigma_{rad} \tag{3}$$

Where σ_{rad} denotes the radiation radio of the structure.

According to the reciprocity principle of SEA, η_{21} can be obtained as

$$\eta_{21} = \frac{n_s}{n_a} \eta_{12} \tag{4}$$

Where, n_s is the mode density of ellipsoid shell. n_a is the mode density of the 2th sub-system (sound field in the shell).

The mode density of ellipsoid shell is shown to be [2]

$$
n_s(f) = \begin{cases}
0 & f < f_{r2} < f_{r1} \\[3ex]
\dfrac{\sqrt{2}}{\pi} \dfrac{f^{\frac{3}{2}} F(\frac{\pi}{2},\zeta)}{(f_{r1}^2 - f^2)^{\frac{1}{2}}(f_{r1}^2 - f_{r2}^2)^{\frac{1}{2}}} n_p(f) & f_{r2} < f < f_{r1} \\[4ex]
\dfrac{2}{\pi} \dfrac{f^2 F(\frac{\pi}{2},\frac{1}{\zeta})}{(f^2 - f_{r1}^2)^{\frac{1}{2}}(f + f_{r1})^{\frac{1}{2}}(f - f_{r2})^{\frac{1}{2}}} n_p(f) & f_{r2} < f_{r1} < f
\end{cases}
\tag{5}
$$

where, f_{r1} and f_{r2} is ring frequency of the shell: $f_{r1} = \dfrac{c_s}{2\pi R_1}$, $f_{r2} = \dfrac{c_s}{2\pi R_2}$

$n_p(f)$ is the mode density of the flat plate which has the same surface area and thickness with the ellipsoid shell. It can be written as $n_p(f) = \dfrac{\sqrt{3} A_p}{h c_s}$, here A_p、c_s are area and longitudinal wave sound speed of the flat plate.

$$
\zeta = \frac{(f + f_{r1})^{\frac{1}{2}}(f - f_{r2})^{\frac{1}{2}}}{[2f(f_{r1} - f_{r2})]^{\frac{1}{2}}} \qquad F(\frac{\pi}{2},\zeta) = \int_0^{\frac{\pi}{2}} (1 - \zeta^2 \sin^2 x)^{-\frac{1}{2}} dx
$$

The mode density of the 2th sub-system (sound field in the shell) is described as [1]:

$$
n_a(\omega) = \frac{\omega^2 V_a}{2\pi^2 c_a^3}
\tag{6}
$$

here V_a is the volume of the sound field in the shell, c_a is the sound speed of liquid in the sound field.

(3). Input power

The pressure on the fluid-coming structure shell is the sum of turbulent boundary layer pressure fluctuation (blocked pressure) and the radiation sound pressure of the shell vibration. The radiation sound pressure is ignored here because it is far less than turbulent boundary layer pressure fluctuation. Turbulent boundary layer pressure fluctuation is considered as an ideal pressure source of surface distributed, and radiation damping and structure damping are considered in loss factors.

The input power of 1st sub-system can be obtained as

$$P_{1,in} = \frac{2\pi^2 c_a^2 n_s}{\omega^2 \rho_s h} \sigma_{rad} p_a^2 \qquad (7)$$

where p_a is mean square value of turbulent boundary layer fluctuation

$$p_a = \sigma \rho_a U^2. \qquad (8)$$

σ is a constant. U is the sailing speed.

As the peseudosound, the radiation efficiency of turbulent boundary layer pressure fluctuation is very low. So we can neglect the sound power $P_{2,in}$, which directly transmits to the shell, i.e. $P_{2,in} = 0$.

(4). Sound pressure level of the self-noise

The balance equation of power flow is

$$\begin{cases} P_{1,in} = P_{1d} + P'_{12} - P'_{21} = \omega \eta_1 E_1 + \omega \eta_{12} E_1 - \omega \eta_{21} E_2 \\ P_{2,in} = P_{2d} + P'_{21} - P'_{12} = \omega \eta_2 E_2 + \omega \eta_{21} E_2 - \omega \eta_{12} E_1 \end{cases} \qquad (9)$$

The sound pressure level in the shell is $\qquad SPL = 10\log \dfrac{E_2 \rho_a c_a^2}{V_a P_{ref}^2} \qquad (10)$

From equation (1)~(10), sound pressure level of the underwater fluid-coming structure self-noise can be derived. For high frequency case, we can take into account some approximate relation.

$$\sigma_{rad} \approx 1, \qquad n_s \approx \frac{2}{\pi} \frac{\sqrt{3} A_p}{h c_s} F(\frac{\pi}{2}, \frac{1}{\zeta}),$$

$$F(\frac{\pi}{2}, \frac{1}{\zeta}) \approx \frac{\pi}{2}(1 + \frac{1}{4\zeta^2}) = \frac{\pi}{2} + \frac{c_s(R_2 - R_1)}{8 f R_1 R_2}$$

We can finally obtain the self-noise level of underwater fluid-coming structure as

$$SPL \approx 20\log \sigma \rho_a U^2 + 10\log \frac{\rho_a c_a}{\rho_s c_s} - 20\log f + 10\log \frac{c_a^2}{h^2 \alpha}$$

$$+10\log \left[\frac{\pi}{2} + \frac{\pi}{4} \frac{\lambda_s(R_2 - R_1)}{R_1 R_2} \right] + 139.4 (dB) \qquad (11)$$

where $\lambda_s = \dfrac{c_s}{f}$.

The equation (11) can be used to estimation the self-noise of underwater fluid-coming structure excited by turbulent boundary layer pressure fluctuation. The relation is appropriate for case of high frequency and low structure loss. The first term in the equation is relation of the mean square value of the turbulent boundary layer fluctuation with the self-noise. It can be seen that the relation is linear. The second team is relation of the self-noise with sound impedance ratio of the fluid with the structure. It is shown that enhancing the structure sound impedance can reduce the self-noise level. The third team shows that the sound level reduces with frequency. The forth team is relation of the noise with thickness and inner surface sound absorbing coefficient of the shell. The fifth team is relations of the noise and the structural shape.

3.CONCULTION

Suppose turbulent boundary layer pressure fluctuation distributes uniformly on underwater fluid-coming structure. We use SEA approach to obtain the estimation formula for self-noise of fluid-coming structure excited by turbulent boundary layer pressure fluctuation. According to the engineering reality, the formula is simplified. Damping loss in the shell and directed transmation of boundary layer peseudosound are neglect in the analysis. The formula provides the concrete of self-noise of fluid-coming structure with the mean square value of turbulent boundary layer pressure fluctuation, fluid-medium, acoustic impedance radio of structure material, frequency, thickness of structure shell, sound absorption coefficient of inner surface as well as the shape of fluid-coming structure.

REFERANCE

[1]   Lyon R.H.. Statistical Energy Analysis of Dynamical Systems: Theory and Applications. The MIT Press, Cambridge, Massachusetts, and London, England, 1975

[2]   Yiao Deyuan, Wang Qicheng. Theory and Applications of Statistical Energy Analysis. Beijing Science and Technology University Press, 1995

[3]   M.K.Bull.Wall-prssure fluctuations Beneath Turbulent Boundary Layer: Some Reflecions on Forty years of Research. Journal of Sound and Vibration, 190(3),299-315

Theoretical and Computational Acoustics 2001
E.-C. Shang, Qihu Li and T. F. Gao (Editors)
© 2002 World Scientific Publishing Co.

Improvement of the Bottom Boundary Condition Treatment in the 3-D PE

Liu Jinzhong, Li Qi
Ocean University of Qingdao, China

Many methods have been offered in the treatment of the interface conditions of the three-dimensional parabolic wave equation. There are mainly two improved methods in handling irregular boundary interface: the stair approximation and the slope approximation. The accuracy of each method has been shown in this paper by numerical results.

KEYWORDS: parabolic equation, finite difference, interface condition

I . INTRODUCTION

The Parabolic Equation (PE) approximation method was first introduced into the field of underwater acoustics in 1977 by Tappert [1]. In the last decades, numerous contributions have been made in the enhancement of PE method by the great development of the computing science and the computer science. Now. The PE method has been shown to be a useful tool for solving realistic problems involving long-range sound propagation. Method of parabolic approximation is an efficient numerical analysis method of the sound field, especially for range dependent problems and strong-coupled problems. But there are also some imperfect areas within the PE method.

It is important to handle the interface condition correctly in the PE method when a higher precision is need. It is simple to solve the problems of the horizontal interface. But the real ocean bottom is complicated ,and some even involve hills. There are energy-conserving problems [2] for the range-dependent conditions when we handle the irregular interface using standard PE. We often handle the irregular interface using two methods: the stair approximation and the slope approximation. With stair approximation, both of the interface conditions, continuity of pressure and continuity of the normal component of particle velocity, should be enforced across the runs (horizontal parts) of the steps and the rises (vertical parts) of the steps. In standard PE the later conditions is not satisfied, so the energy is not conserved and an error is introduced. We can obtain a high precision with the slope approximation if the interface is handled properly. But the mesh size (in θ direction) should be small in three-dimensional conditions when we compute the field numerically using the finite difference methods. The computation speed is slow and it's difficult to make program because that the PE finite difference form nearby the interface should be rewritten if the interface changed.

Analysis [2] has shown if u is replaced by the scaled $u^* = u / \sqrt{\rho c}$ in the PE algorithm, then, a good approximation to the true energy-conserving condition is realized. The ocean bottom interface conditions were handled by this method in the stair approximation. The numerical computation has shown a higher accuracy of this method.

II .The slope approximation method[6]

In this chapter, we handle the bottom boundary conditions with the methods suggested by reference [6] for 3-D parabolic equation.

The interface is handled by the sloped ocean bottom boundary conditions in this method, as shown in figure 3

The wave equation in inhomogeneous media

$$\rho \nabla (\frac{1}{\rho} \nabla p) + k_0^2 n^2 (r,z,\theta) p = 0 \tag{1}$$

Where p is the acoustic pressure, k_0 is the reference wavenumber and $n(r,z,\theta)$ is the index of refraction.

Substitute

$$p(r,z,\theta) = u(r,z,\theta) H_0^1 (k_0 r) \tag{2}$$

into Eq.(1) we obtain the parabolic equation in inhomogeneous media

$$\frac{\partial u}{\partial r} = (-ik_0 + ik_0 \sqrt{1 + X^+ + Y^+}) u \tag{3}$$

Where

$$X^+ = n^2(r,\theta,z) - 1 + \rho \frac{\partial}{\partial z} \left(\frac{1}{\rho} \frac{\partial u}{\partial z} \right)$$, $\rho = \rho(r,z,\theta)$

$$Y^+ = \frac{1}{k_0^2 r^2} \rho \frac{\partial}{\partial \theta} (\frac{1}{\rho} \frac{\partial u}{\partial \theta})$$

The treatment of the ocean bottom interface condition is to obtain the finite difference form of the operator X^+, Y^+. Due to the discontinuity of ρ on the interface boundary, the handling of

$\frac{\partial}{\partial z} (\frac{1}{\rho} \frac{\partial u}{\partial z})$ and $\frac{\partial}{\partial \theta} (\frac{1}{\rho} \frac{\partial u}{\partial \theta})$ must satisfy the interface condition on the bottom boundary. A finite difference technique, suggested by Varga, is applied to treat the interface boundary to fulfill this requirement [3].

$$\int_{z_{m-\frac{1}{2}}}^{z_{m+\frac{1}{2}}} \frac{\partial}{\partial z} (\frac{1}{\rho} \frac{\partial u}{\partial z}) dz = \left(\frac{1}{\rho} \frac{\partial u}{\partial z}\right)_{m+\frac{1}{2}} - \left(\frac{1}{\rho} \frac{\partial u}{\partial z}\right)_{m-\frac{1}{2}} + \left(\frac{1}{\rho} \frac{\partial u}{\partial z}\right)_m^- - \left(\frac{1}{\rho} \frac{\partial u}{\partial z}\right)_m^+ \tag{4}$$

Hence

$$\frac{\partial}{\partial z} (\frac{1}{\rho} \frac{\partial u}{\partial z}) = \frac{1}{\Delta z^2} (\frac{1}{\rho_{m+\frac{1}{2},l}} u_{m+1,l} + \frac{1}{\rho_{m-\frac{1}{2},l}} u_{m-1,l}) - \frac{1}{\Delta z^2} (\frac{1}{\rho_{m+\frac{1}{2},l}} + \frac{1}{\rho_{m-\frac{1}{2},l}}) u_{m,l} + \frac{1}{\Delta z} [\frac{1}{\rho_1} \frac{\partial u_1}{\partial z} - \frac{1}{\rho_2} \frac{\partial u_2}{\partial z}] \tag{5}$$

The following form can be obtained by the same method

$$\frac{\partial}{\partial \theta} (\frac{1}{\rho} \frac{\partial u}{\partial \theta}) = \frac{1}{\Delta \theta^2} (\frac{1}{\rho_{m,l+\frac{1}{2}}} u_{m,l+1} + \frac{1}{\rho_{m,l-\frac{1}{2}}} u_{m,l-1}) - \frac{1}{\Delta \theta^2} (\frac{1}{\rho_{m,l+\frac{1}{2}}} + \frac{1}{\rho_{m,l-\frac{1}{2}}}) u_{m,l} + \frac{1}{\Delta \theta} [\frac{1}{\rho_1} \frac{\partial u_1}{\partial \theta} - \frac{1}{\rho_2} \frac{\partial u_2}{\partial \theta}] \tag{6}$$

The key is to obtain the form of the operator $\frac{1}{\rho_1} \frac{\partial u_1}{\partial z} - \frac{1}{\rho_2} \frac{\partial u_2}{\partial z}$, $\frac{1}{\rho_1} \frac{\partial u_1}{\partial \theta} - \frac{1}{\rho_2} \frac{\partial u_2}{\partial \theta}$ on the boundary.

Now, we show how to treat the interface condition by figure 4

Let n, m, l indicate the discretization index in the r-direction, z-direction and θ-direction. Let s, h, q indicate the r-increment, the z-increment and the θ-increment. We suggest the wave field at range r=ns has been known and the field at range r=(n+1) s is to be computed. Suppose the density nearby both sides of the bottom boundary change slowly. The following equation can be obtained using the Helmholtz equation in the homogeneous media.

$$\frac{\partial u}{\partial r} = (-ik_0 + ik_0\sqrt{1 + X + Y})u \tag{7}$$

Where $X = n^2(r,\theta,z) - 1 + \frac{1}{k_0^2}\frac{\partial^2 u}{\partial z^2}$, $Y = \frac{1}{k_0^2 r^2}\frac{\partial^2 u}{\partial \theta^2}$

Make an approximation to the operator $\sqrt{1 + X + Y}$

$$\sqrt{1 + X + Y} \approx 1 + \frac{1}{2}X - \frac{1}{8}X^2 + \frac{1}{2}Y \tag{8}$$

Hence $\quad u_r = ik_0[-1 + (1 + \frac{1}{2}X - \frac{1}{8}X^2 + \frac{1}{2}Y)]u \tag{9}$

Neglect the effect of $\frac{1}{8k_0^4}(\frac{\partial^2 u_1}{\partial z^2})^2$

We obtain: $\quad \frac{\partial u}{\partial r} = au + b\frac{\partial^2 u}{\partial z^2} + c\frac{\partial^2 u}{\partial \theta^2} \tag{10}$

Where

$$a = \frac{ik_0}{2}[n^2(r,\theta,z) - 1]\{1 - \frac{1}{4}[n^2(r,\theta,z) - 1]\}, b = \frac{i}{2k_0}\{1 - \frac{1}{2}[n^2(r,\theta,z) - 1]\}, c = \frac{i}{2k_0 r^2}$$

Hence the PE form is obtained in the water (media 1) nearby the bottom boundary become

$$\frac{\partial u_1}{\partial r} = a_1 u_1 + b_1\frac{\partial^2 u_1}{\partial z^2} + c_1\frac{\partial^2 u_1}{\partial \theta^2} \tag{11}$$

And the PE form in the bottom (media 2) nearby the boundary become

$$\frac{\partial u_2}{\partial r} = a_2 u_2 + b_2\frac{\partial^2 u_2}{\partial z^2} + c_2\frac{\partial^2 u_2}{\partial \theta^2} \tag{12}$$

Both of the interface conditions, continuity of pressure and continuity of the normal component of particle velocity, should be enforced across the steps .We can obtain the following equations

$$p_1 = p_2 \tag{13}$$

$$\frac{1}{\rho_1}\nabla p_1 \cdot \vec{h} = \frac{1}{\rho_2}\nabla p_2 \cdot \vec{h} \tag{14}$$

Substituting Eq. (13,14) into Eq. (2), we obtain the following bottom boundary interface conditions for the parabolic wave equation.

$$u_1 = u_2$$

$$\left(\frac{1}{\rho_1}\frac{\partial u_1}{\partial r} - \frac{1}{\rho_2}\frac{\partial u_2}{\partial r}\right)n_r + \left(\frac{1}{\rho_1}\frac{\partial u_1}{\partial \theta} - \frac{1}{\rho_2}\frac{\partial u_2}{\partial \theta}\right)n_\theta / r + \left(\frac{1}{\rho_1}\frac{\partial u_1}{\partial z} - \frac{1}{\rho_2}\frac{\partial u_2}{\partial z}\right)n_z \tag{15}$$

$$= \left(ik_0 - \frac{1}{2r}\right)\left(\frac{1}{\rho_2} - \frac{1}{\rho_1}\right)n_r u$$

Using a Taylor series expansion for $u_{m-1,l}^{n+2}$ upon $u_{m,l}^{n+1}$, we find

$$(u_{m-1,l}^{n+2})_1 = u_1 + s\frac{\partial u_1}{\partial r} + \frac{s^2}{2}\frac{\partial^2 u_1}{\partial r^2} - h\frac{\partial u_1}{\partial z} + \frac{h^2}{2}\frac{\partial^2 u_1}{\partial z^2}$$

$$(u_{m-1,l}^{n+2})_2 = u_1 + s\frac{\partial u_2}{\partial r} + \frac{s^2}{2}\frac{\partial^2 u_2}{\partial r^2} - h\frac{\partial u_2}{\partial z} + \frac{h^2}{2}\frac{\partial^2 u_2}{\partial z^2}$$

We obtain Eq. (16) by using the following three equations

$$(u_{m-1,l}^{n+2})_1 = (u_{m-1,l}^{n+2})_2$$

$$u_{m+1,l}^n = u_1 - s\frac{\partial u_1}{\partial r} + \frac{s^2}{2}\frac{\partial^2 u_1}{\partial r^2} + h\frac{\partial u_1}{\partial z} + \frac{h^2}{2}\frac{\partial^2 u_1}{\partial z^2}$$

$$u_{m+1,l}^n = u_1 - s\frac{\partial u_2}{\partial r} + \frac{s^2}{2}\frac{\partial^2 u_2}{\partial r^2} + h\frac{\partial u_2}{\partial z} + \frac{h^2}{2}\frac{\partial^2 u_2}{\partial z^2}$$

$$s\frac{\partial u_1}{\partial r} - h\frac{\partial u_1}{\partial z} = s\frac{\partial u_2}{\partial r} - h\frac{\partial u_2}{\partial z} \tag{16}$$

We can obtain the following equation by the same method

$$s\frac{\partial u_1}{\partial r} + \frac{q}{r}\frac{\partial u_1}{\partial \theta} = s\frac{\partial u_2}{\partial r} + \frac{q}{r}\frac{\partial u_2}{\partial \theta} \tag{17}$$

Using a Taylor series expansion for $u_{m,l}^{n+1}$ upon $u_{m,l}^n$ and the difference method, we find

$$u_m^n = u_1 - s\frac{\partial u_1}{\partial r} + \frac{s^2}{2}\frac{\partial^2 u_1}{\partial r^2}$$

$$(\frac{\partial u_1}{\partial r})_{m,l}^n = \frac{\partial u_1}{\partial r} - s\frac{\partial^2 u_1}{\partial r^2} = a_{m,l}^n u_{m,l}^n + [b_{m,l}^n \frac{u_{m+1,l}^n - 2u_{m,l}^n + u_{m-1,l}^n}{h^2} + c_{m,l}^n \frac{u_{m,l+1}^n - 2u_{m,l}^n + u_{m,l-1}^n}{q^2}]$$

We obtain

$$u_1 - \frac{s}{2}\frac{\partial u_1}{\partial r} = (1 + a_{m,l}^n s)u_{m,l}^n + \frac{s}{2}[b_{m,l}^n \frac{u_{m+1,l}^n - 2u_{m,l}^n + u_{m-1,l}^n}{h^2} + c_{m,l}^n \frac{u_{m,l+1}^n - 2u_{m,l}^n + u_{m,l-1}^n}{q^2}] \tag{18}$$

Using the Taylor series expansion we can also obtain the following four equations

$$u_{m-1,1}^{n+1} = u_1 - h\frac{\partial u_1}{\partial z} - \frac{h^2}{2}\frac{\partial^2 u_1}{\partial z^2} \quad u_{m,l+1}^{n+1} = u_1 + q\frac{\partial u_1}{r\partial\theta} + \frac{q^2}{2r^2}\frac{\partial^2 u_1}{\partial\theta^2}$$

$$u_{m+1,1}^{n+1} = u_1 + h\frac{\partial u_2}{\partial z} - \frac{h^2}{2}\frac{\partial^2 u_2}{\partial z^2} \quad u_{m,l-1}^{n+1} = u_1 - q\frac{\partial u_2}{r\partial\theta} + \frac{q^2}{2r^2}\frac{\partial^2 u_2}{\partial\theta^2}$$

By eliminate the two-order items we obtain

$$b_1\frac{u_{m-1,l}^{n+1}}{h^2} + c_1 r^2\frac{u_{m,l+1}^{n+1}}{q^2} - (\frac{b_1}{h^2} + \frac{r^2 c_1}{q^2})u_1 = -\frac{a_1}{2}u_1 + \frac{1}{2}\frac{\partial u_1}{\partial r} - \frac{b_1}{h}\frac{\partial u_1}{\partial z} + \frac{rc_1}{q}\frac{\partial u_1}{\partial\theta} \qquad (19)$$

$$-b_2\frac{u_{m+1,l}^{n+1}}{h^2} + c_2 r^2\frac{u_{m,l-1}^{n+1}}{q^2} + (\frac{b_2}{h^2} + \frac{r^2 c_2}{q^2})u_2 = -\frac{a_1}{2}u_2 + \frac{1}{2}\frac{\partial u_2}{\partial r} - \frac{b_2}{h}\frac{\partial u_2}{\partial z} - \frac{rc_2}{q}\frac{\partial u_2}{\partial\theta} \qquad (20)$$

By solving the equations group constituted by equations from (15) to (20) we can obtain the form of

variables $\dfrac{\partial u_1}{\partial z}$, $\dfrac{\partial u_1}{\partial r}$, $\dfrac{\partial u_2}{\partial z}$, $\dfrac{\partial u_2}{\partial r}$, $\dfrac{\partial u_1}{\partial\theta}$, $\dfrac{\partial u_2}{\partial\theta}$. Hence we know the form of the operator

$\dfrac{1}{\rho_1}\dfrac{\partial u_1}{\partial z} - \dfrac{1}{\rho_2}\dfrac{\partial u_2}{\partial z}$, $\dfrac{1}{\rho_1}\dfrac{\partial u_1}{\partial\theta} - \dfrac{1}{\rho_2}\dfrac{\partial u_2}{\partial\theta}$ on the interface boundary.

A NUMERICAL EXAMPLE

We test the validity of the interface treatment with an example of the Acoustical Society of America (ASA) Benchmark Wedge as shown in figure 5, where $\Delta z = 0.5, \Delta r = 5, \Delta\theta = 1$.

The base solution is computed by the image method [5]

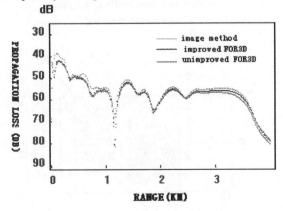

Figure1 Three-dimensional sound propagation in a wedgelike ocean Azimuth angle 0^0

Depth of receiver 30m

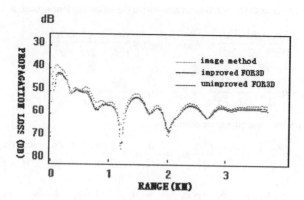

Figure 2 Three-dimensional sound propagation in a wedgelike ocean Azimuth angle 0^0

Depth of receiver 30m

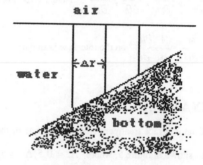

Figure 3 the bottom interface in the slope

approximation method

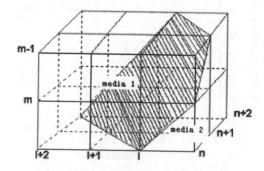

Figure 4 the three-dimensional bottom interface

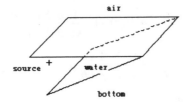

Figure 5 three-dimensional Benchmark Wedge

the density is 1.0 in water,1.5 in   bottom .

the sound velocity is 1500m/s in water, 1700m/s in bottom

the absorption coefficient is 0 in water, 0.5dB/wave length

the source is 200m depth, 25Hz in frequency

III.The stair approximation and energy-conserving method

Both of the interface conditions, continuity of pressure and continuity of the normal component of particle velocity, should be enforced across the runs (horizontal parts) of the steps and the rises (vertical parts) of the steps in the stair approximation method, as shown by figure 8.   It is complicated to fulfill both of the interface conditions across the steps.

Analysis [2] has shown if u is replaced by the scaled $u^* = u/\sqrt{\rho c}$ in the PE algorithm, then, a good approximation to the true energy-conserving condition is realized. We applied this method in the FOR3D model

The three-dimensional Parabolic Equation may be written as the form:

$$\frac{\partial^2 u}{\partial r^2} + 2ik_0 \frac{\partial u}{\partial r} + k_0^2 (X + Y)u = 0 \tag{21}$$

If the field at (r, z) has been given the field at $(r + \Delta r, z)$ has the following form using the methods described in Ref.(3)

$$u(r + \Delta r, \theta, z) = (\frac{1+PX}{1+\overline{P}X})(\frac{1+QY}{1+\overline{Q}Y})u(r,z) \tag{22}$$

where $X = n^2(r,z) - 1 + \frac{1}{k_0^2} \rho \frac{\partial}{\partial z}\left(\frac{1}{\rho}\frac{\partial}{\partial z}\right)$, $Y = \frac{1}{k_0^2 r^2}\frac{\partial}{\partial \theta}(\frac{1}{\rho}\frac{\partial u}{\partial \theta})$

$P = 1/4 + \delta/4$,$\overline{P}$ is the complex conjugate of P , $\delta = -ik_0 \Delta r$

We write $\mu = \sqrt{\rho(\theta,z)c(\theta,z)}$ and $u^* = u/\mu$

Hence the formula (22) becomes

$$u^*(r + \Delta r, z) = (\frac{1+PX^*}{1-\overline{P}X^*})(\frac{1+QY^*}{1-\overline{Q}Y^*})u^*(r,z) \tag{23}$$

Where

$$X^* = n^2(r,z) - 1 + \frac{1}{k_0^2} \frac{\rho}{\mu} \frac{\partial}{\partial z} \left(\frac{1}{\rho} \frac{\partial}{\partial z} \mu \right)$$

$$Y^* = \frac{1}{k_0^2 r^2} \frac{1}{\mu} \frac{\partial}{\partial \theta} (\frac{1}{\rho} \frac{\partial u}{\partial \theta} \mu)$$

The formula(3) can be rewritten as

$$u^{j+1} = [1 + (-\frac{\delta}{4} + \frac{1}{4})X^*]^{-1} [1 + (\frac{\delta}{4} + \frac{1}{4})X^*][1 + \frac{\delta}{4}Y^*][1 - \frac{\delta}{4}Y^*]^{-1} u^j \quad (24)$$

Hence

$$[1 + (-\frac{\delta}{4} + \frac{1}{4})X^*][1 - \frac{\delta}{4}Y^*]u^{j+1} = [1 + (\frac{\delta}{4} + \frac{1}{4})X^*][1 + \frac{\delta}{4}Y^*]u^j \quad (25)$$

We can obtain the following approximations using the Galerkin's method as described in Ref.4

$$2(\Delta z)^2 \frac{\partial}{\partial z} \left(\frac{1}{\rho} \frac{\partial}{\partial z} \mu u \right) \cong$$

$$(\frac{1}{\rho_{m-1,l}} + \frac{1}{\rho_{m,l}})\mu_{m-1,l}u_{m-1,l} - (\frac{1}{\rho_{m-1,l}} + 2\frac{1}{\rho_{m,l}} + \frac{1}{\rho_{m+1,l}})\mu_{m,l}u_{m,l} + (\frac{1}{\rho_{m+1,l}} + \frac{1}{\rho_{m,l}})\mu_{m+1,l}u_{m+1,l} \quad (26)$$

$$2(\Delta\theta)^2 \frac{\partial}{\partial\theta}(\frac{1}{\rho}\frac{\partial u}{\partial\theta}\mu)$$

$$\cong (\frac{1}{\rho_{m,l-1}} + \frac{1}{\rho_{m,l}})\mu_{m,l-1}u_{m,l-1} - (\frac{1}{\rho_{m,l-1}} + 2\frac{1}{\rho_{m,l}} + \frac{1}{\rho_{m,l+1}})\mu_{m,l}u_{m,l} + (\frac{1}{\rho_{m,l+1}} + \frac{1}{\rho_{m,l}})\mu_{m,l+1}u_{m,l+1} \quad (27)$$

Equation (25) may be solved in two steps as follows:
Write Equation (25) in the matrix form

$$A^* B^* u^{j+1} = ABu^j \quad (28)$$

consider $v^{j+1} = B^* u^{j+1}$, $v^j = Bu^j$ we have

$$A^* v^{j+1} = Av^j \quad (29)$$

Now, we can solve Equation (25) as follows:
Step 1

Solve the system $A^* v^{j+1} = Av^j$ for the vector v^{j+1} using a tridiagonal solver

Where the lower diagonal、 main diagonal and upper diagonal can be written as
In the left of equation (29):

$$AL = \frac{1}{2h^2 k_0^2} \frac{\rho}{\mu} (\frac{1}{\rho_{m-1,l}} + \frac{1}{\rho_{m,l}})\mu_{m-1,l}(-\frac{\delta}{4} + \frac{1}{4})$$

$$AU = \frac{1}{2h^2 k_0^2} \frac{\rho}{\mu} (\frac{1}{\rho_{m+1,l}} + \frac{1}{\rho_{m,l}})\mu_{m+1,l}(-\frac{\delta}{4} + \frac{1}{4})$$

$$AM = 1 + (-\frac{\delta}{4} + \frac{1}{4}) \left\{ n^2(r,z) - 1 - \frac{1}{2h^2 k_0^2} \frac{\rho}{\mu} (\frac{1}{\rho_{m-1,l}} + 2\frac{1}{\rho_{m,l}} + \frac{1}{\rho_{m+1,l}})\mu_{m,l} \right\}$$

In the right of equation (29)

$$AL = \frac{1}{2h^2 k_0^2} \frac{\rho}{\mu} (\frac{1}{\rho_{m-1,l}} + \frac{1}{\rho_{m,l}}) \mu_{m-1,l} (+\frac{\delta}{4} + \frac{1}{4})$$

$$AU = \frac{1}{2h^2 k_0^2} \frac{\rho}{\mu} (\frac{1}{\rho_{m+1,l}} + \frac{1}{\rho_{m,l}}) \mu_{m+1,l} (+\frac{\delta}{4} + \frac{1}{4})$$

$$AM = 1 + (+\frac{\delta}{4} + \frac{1}{4}) \left\{ n^2(r,z) - 1 - \frac{1}{2h^2 k_0^2} \frac{\rho}{\mu} (\frac{1}{\rho_{m-1,l}} + 2\frac{1}{\rho_{m,l}} + \frac{1}{\rho_{m+1,l}}) \mu_{m,l} \right\}$$

EXAMPLE

We Test the validity of the interface treatment with a example of the Acoustical Society of America (ASA) Benchmark Wedge as shown by figure 5. The base solution is computed by the image method

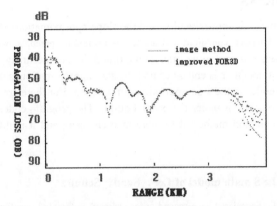

Figure 6   Three-dimensional sound propagation in a wedgelike ocean

Azimuth angle 0^0, Depth of receiver 30m

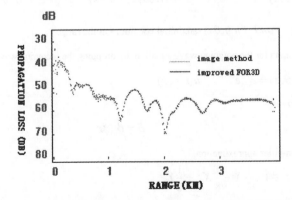

Figure 7 Three-dimensional sound propagation in a wedgelike ocean

Azimuth angle 30^0 Depth of receiver 30m

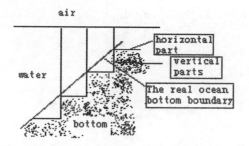

Figure 8 the bottom interface in the stair approximation method

IV. CONCLUSIONS

Two methods (the stair approximation and the slope approximation) have been developed to handle the irregular interface conditions for the three-dimensional parabolic wave - equation. The energy-conserving method is used in the stair approximation. The accuracy of the PE method improved by the two methods is enhanced and the stair approximation is better. It is easier to make the program codes in the stair approximation method. But there are still oscillations of the field when the receiver is under the ocean bottom. The program codes must be rewritten in the slope approximation method if the interface changed, and it is difficult to make the codes.

APPENDIX A: the LSS math model of Lee、 Saad、 Schultz :

Eq. (21) can be considered as a second-order ordinary differential equation with respect to the variant r. The field at $r + \Delta r$ has the form

$$u(r + \Delta r, \theta, z) = e^{-ik_0\Delta r}e^{ik_0\Delta r\sqrt{1+X+Y}}u^+(r, \theta, z) + e^{ik_0\Delta r}e^{-ik_0\Delta r\sqrt{1+X+Y}}u^-(r, \theta, z) \qquad (A1)$$

Where $u^+(r,\theta,z)$ *and* $u^-(r,\theta,z)$ are initial conditions at the range r

The one-way solution for the outgoing wave equation ,by dropping the incoming wave term, has the form:

$$u(r + \Delta r, \theta, z) = e^{-\delta}e^{\delta\sqrt{1+X+Y}}u(r, \theta, z) \qquad (A2)$$

The parameter δ is

$$\delta = ik_0\Delta r$$

Applying the high-order approximation:

$$\sqrt{1+X+Y} \cong 1 + \frac{1}{2}X - \frac{1}{8}X^2 + \frac{1}{2}Y$$

The formula (A2)becomes

$$u(r + \Delta r, \theta, z) = e^{-\delta}e^{\delta(1+\frac{1}{2}X-\frac{1}{8}X^2+\frac{1}{2}Y)}u(r, \theta, z)$$

We can obtain the LSS 3-dimensional wide-angle wave equation in operator form by Eq. (3)

$$u_r = ik_0[-1 + (1 + \frac{1}{2}X - \frac{1}{8}X^2 + \frac{1}{2}Y)]u \qquad (A3)$$

Assuming that the index of refraction $n(r,z,\theta)$ varies slowly with respect to θ, the operators X and Y are nearly commutative, the formula (A2) becomes

$$u(r+\Delta r,\theta,z) = e^{-\delta} e^{\delta(1+\frac{1}{2}X-\frac{1}{8}X^2)} e^{\delta\frac{1}{2}Y} u(r,\theta,z) \tag{A4}$$

For convenience we write

$$G(\delta,X) = e^{\delta(1+\frac{1}{2}X-\frac{1}{8}X^2)} \tag{A5}$$

Applying a Taylor series expansion the next formula is obtained

$$G(\delta,X) = e^{\delta}[1+\frac{1}{2}\delta X + \frac{1}{2!}(\frac{\delta^2}{4}-\frac{\delta}{4})X^2] + O(X^3) \tag{A6}$$

We also can make a rational function approximation to $G(\delta,X)$

$$G(\delta,X) = e^{\delta}\frac{1+PX}{1+\overline{P}X} \tag{A7}$$

Where P is a complex number to be determined and $\overline{P}$ is the complex conjugate of P

We impose the condition that the right side of (A6) is equal to the right side of (A7)(neglect the high order item of the Eq. (A6))

$$[1+\frac{1}{2}\delta X + \frac{1}{2!}(\frac{\delta^2}{4}-\frac{\delta}{4})X^2] = \frac{1+PX}{1+\overline{P}X} \tag{A8}$$

It can be determined that P is

$$P = \frac{\delta}{4} + \frac{1}{4}$$

We obtain

$$G(\delta,X) = e^{\delta}\frac{1+(\frac{\delta}{4}+\frac{1}{4})X}{1+(-\frac{\delta}{4}+\frac{1}{4})X} \tag{A9}$$

A similar development for the term in Eq.(A4)

$$H(\delta,Y) = e^{\frac{\delta}{2}Y}$$

We obtain

$$H(\delta,Y) = \frac{1+QY}{1+\overline{Q}Y} \tag{A10}$$

Where $Q = \frac{\delta}{4}$. We obtain

$$H(\delta,Y) = \frac{1+\frac{\delta}{4}Y}{1-\frac{\delta}{4}Y} \tag{A11}$$

Therefore, The final form of the local solution (A4) to the LSS equation (A3) has the form

$$u(r+\Delta r,\theta,z)=(\frac{1+(\frac{\delta}{4}+\frac{1}{4})X}{1+(-\frac{\delta}{4}+\frac{1}{4})X})(\frac{1+\frac{\delta}{4}Y}{1-\frac{\delta}{4}Y})u(r,\theta,z)$$ **(A12)**

Equation (A12)can be rewritten as

$$u(r+\Delta r,\theta,z)=(\frac{1+PX}{1+\bar{P}X})(\frac{1+QY}{1+\bar{\bar{Q}}Y})u(r,z)$$ **(A13)**

Where

$$X=n^2(r,z)-1+\frac{1}{k_0^2}\rho\frac{\partial}{\partial z}\left(\frac{1}{\rho}\frac{\partial}{\partial z}\right) \quad Y=\frac{1}{k_0^2 r^2}\frac{\partial}{\partial\theta}(\frac{1}{\rho}\frac{\partial u}{\partial\theta})$$

REFERENCES

[1] Tappert,F.D.,"The parabolic equation approximation methord,"in 《Wave Propagation and Underwater Acoustics,Lecture Notes in Physics》, Springer-Verlag,Heidelberg,1977 Vol.70, eds.J.B.Keller and .Spapadakis

[2]M.D.Collins and E.K.Weatwood,"A high-order energy-conserving parabolic equation for range-dependent ocean depth,soundspeed,and density," J.Acoust.Soc. 1991，Am.89.P1068-1075

[3] Lee.D.and M.H.Schulyz, 《Numerical Ocean Acoustic Propagation in Three Dimensions》 World Scientific,Singapore,1995

[4] M.D.Collins,"A high-order parabolic equation for wave propagation in an ocean overlying an elastic bottom," J.Acoust.Soc. 1989，Am.86.P1459-1464

[5] G.B.Deane and M.J.Buckingham,"An analysis of the three-dimensional sound field in a penetrable wedge with a stratified fluid or elastic basement," J.Acoust.Soc. 1993，Am93

[6] Lee,D.and S.T.McDaniel, 《Ocean Acoustics Propagation by Finite Difference Methods》 Pergamon Oxford, 1988

Theoretical and Computational Acoustics 2001
E.-C. Shang, Qihu Li and T. F. Gao (Editors)
© 2002 World Scientific Publishing Co.

Sound Wave Propagation in Non-Uniform Flow Using an Eulerian-Lagrangian Description

Mabrouk Ben Tahar, Fabien Treyssede, Gwénaël Gabard

Université de Technologie de Compiègn, Laboratoire Roberval UMR 6066,

Secteur Acoustique BP 20529, 60205 Compiègn Cedex, France

{fabien.treyssede, gwenael.gabard, mabrouk.bentahar}@utc.fr

The aim of this paper is to present a non-standard equation for studying the acoustic propagation in multidirectional non-uniform mean flows. This equation is obtained using a mixed Eulerian-Lagrangian description. Some elementary concepts are recalled, as well as their physical significance and the specificity of the Eulerian-Lagrangian description is outlined. Then, equation that governs sound propagation is derived from Euler equations. When the associate variational formulation is directly used, solutions are corrupted by spurious rotational modes. In this paper, a mixed variational formulation is proposed to avoid this problem. Finally, results obtained by finite element discretization are compared with semi-analytical solutions obtained from the Pridmore-Brown equation. The ability of the method to take into account convection and refraction phenomena is shown. Furthermore, wave propagation in a complex geometry is investigated.

1 Introduction

Flow non-uniformities may play a significant role on acoustic fields. Actually, sound propagation in non-uniform flows has been studied for almost half a century. Three major kinds of equations are most often used: the direct linearized Euler equations, the potential equation and the Pridmore-Brown equation.

The direct linearized Euler equations seemingly represent the only formulation that supports rotational, multidirectional flow. Some authors have developed a simplified numerical method for solving acoustic propagation in slowly varying ducts[1]. Only few authors have chosen to directly solve the linearized Euler equations and have successfully implemented the direct formulation using a FEM method[2,3]. Unfortunately, their numerical results have been limited to relatively simple mean flows and deliver no real analysis of refraction effects.

The linearized Euler equations constitute a system of at least 4 unknowns. Such a system is very difficult to solve and, in fact, few authors use those equations to study propagation in non-uniform flows. However, equations can be greatly simplified when flow irrotationality is assumed. The formulation obtained from this assumption is called the full-potential equation and is much more widespread in the literature.

In the full-potential formulation, the total and mean flow velocities are supposed to be irrotational and hence, derived from a scalar potential. It can be shown that the linearized Euler equations can be simplified to a propagation equation written only in term of the acoustic velocity potential. Some authors have studied the effect of flow variation and multidirectionality upon the acoustic characteristics in ducts with variable cross-section[4-6]. Other authors have restricted their study to slowly varying cross-section ducts in order to obtain some analytical or semi-analytical results[7-9]. Besides, several authors have extended the analysis to radiation[10-14].

It can be noted that this equation remains difficult to solve and lots of authors simplify it by making the plane wave assumption, by neglecting the radial component of the flow velocity (theories with slowly varying cross-sections) or more radically, by considering a uniform flow (theories with exhaust systems involving transfer matrices). However, the main drawback of the full-potential approach is that it cannot take into account the effect of boundary layer refraction because of the irrotationality assumption. As shown by the Pridmore-Brown equation, those effects may be important.

The Pridmore-Brown[15] equation with no doubt constitutes the simplest formulation to study the effect of flow non-uniformity. In this equation, the mean flow is supposed to be unidirectional and axially uniform but no restriction is made on the velocity profile. The formulation represents a one-dimensional eigenvalue problem in term of the axial wave number. Thus, it appears to be suited for the analysis of shear and boundary layer effects in infinite straight ducts. It has been studied for more than forty years and results point out the importance of refraction on acoustic attenuation in lined ducts[15-21].

In conclusion, the direct linearized Euler equations, the full-potential flow equation and the Pridmore-Brown equation are the most widely used formulations in the literature. Each of them has its own advantages and drawbacks. When geometries and/or flows become complex, it appears that sound propagation is hardly studied in a rigorous way. The only propagation equation, which is valid in this case, is the linearized Euler equations. However, another wave equation can be obtained from a mixed Eulerian-Lagrangian description[22,23]. This original concept consists in considering lagrangian perturbations of quantities (associated to the same particle) but still written in eulerian coordinates. In the classical linearized Euler equation, perturbations are implicitly eulerian, i.e. associated to the same geometric point but not the same particle.

In this paper, the "mixed" formulation is chosen. The theory inherent to the mixed Eulerian-Lagrangian description is first briefly recalled. The equation is solved by a FEM method. The numerical method presented is quite general but implemented for two-dimensional problems. Comparison with known results for straight ducts obtained from Pridmore-Brown theory shows that the method is able to handle refraction phenomena in a very accurate way. The case of a more complex duct is then presented.

2 Theory

First of all, the mixed Eulerian-Larangian description is presented in a general way. Some fundamental relations are recalled, and then applied to Euler equations in order to derive a propagation equation, for which the only unknown is the lagrangian perturbation of the displacement.

2.1 *The Eulerian-Lagrangian description*

In a continuous medium, two kinds of variables can be used to describe physical quantities: Lagrangian and Eulerian variables. The Lagrangian point of view, usually used for media at rest, consists in following the particle path from a reference time. Physical quantities are thus expressed in term of (a,t), where a is the position occupied by the particle at the reference time. The Eulerian variables, usually used in fluid mechanics, correspond to the geometrical position x at time t of the particle a (x is time dependent). For a perturbed field, one can chose either non-perturbed variables (x_0,t) or perturbed Eulerian variables (x,t). Physical quantities can be expressed in term of any variables and the transformation from one to another is simply made with a change of variable:

$$(a,t) \leftrightarrow (x_0,t) \leftrightarrow (x,t).$$

However, one must be careful of which variables are used, in particular for time derivatives. To be more precise, x_0 is the geometrical position of the particle a in the mean flow (or non-perturbed) configuration while x is the position of this same particle a in the perturbed configuration. Then, if u^L is the perturbation of the particle displacement vector, x_0 and x are related by:

$$x = x_0 + u^L.$$

As a consequence, the total perturbed quantities $\Psi(a,t)$ and $\Psi(x,t)$ are equal because both of them deals with the same particle (in the perturbed configuration, the particle a is located at x). In the mean flow configuration, the particle a is located at x_0 and then, mean flow quantities $\Psi_0(a,t)$ and $\Psi_0(x_0,t)$ are equal. These equalities, which are true for any physical quantity Ψ (tensor of arbitrary order), are summarized below:

$$\Psi(a,t) = \Psi(x,t),$$
$$\Psi_o(a,t) = \Psi_o(x_0,t). \tag{1}$$

In the remainder of this article, mean flow (or non-perturbed) quantities are distinguished from their total (or perturbed) counterparts by the subscript o. From now, two kinds of perturbation can be defined:

$$\Psi^E = \Psi(\mathbf{x},t) - \Psi_o(\mathbf{x},t),$$
$$\Psi^L = \Psi(\mathbf{a},t) - \Psi_o(\mathbf{a},t). \tag{2}$$

Superscripts E and L denote respectively Eulerian and Lagrangian perturbations. From these definitions, Eulerian perturbations are clearly associated to the same geometrical point but not the same particle, whereas Lagrangian perturbations are clearly associated to the same particle. It can be noticed that these notations are coherent with that of $\mathbf{u}^L$ chosen above, for the perturbation of displacement is associated to the same particle.

Then, relation's (1) and (2) yield:

$$\Psi^L = \Psi(\mathbf{x},t) - \Psi_o(\mathbf{x_0},t). \tag{3}$$

Thus, using (3), a Taylor expansion to the first order gives the fundamental relation between Eulerian and Lagrangian perturbation:

$$\Psi^L = \Psi^E + \mathbf{u}^L.\nabla\Psi_o. \tag{4}$$

This relation is true when any Eulerian variables $(\mathbf{x_0},t)$ or $(\mathbf{x},t)$ are used. In the remainder, the type of variables used to express quantities will be omitted for clarity when no confusion can be made.

In conclusion, Ψ^L represents the Lagrangian perturbation of the physical quantity Ψ expressed in term of Eulerian variables. This description is thus mixed and is often called the mixed Eulerian-Lagrangian description.

For the mixed Eulerian-Lagrangian description, i.e. when Lagrangian perturbations are written in Eulerian variables, the perturbation of derivatives is not direct. To the first order, it can be shown that:

$$\begin{cases} \left(\dfrac{d\Psi}{dt}\right)^L = \dfrac{d_o\Psi^L}{dt} & \text{where} \quad \dfrac{d_o\Psi}{dt} = \left(\dfrac{\partial}{\partial t} + \mathbf{v_0} \cdot \nabla\right)\Psi \\[4mm] \left(\dfrac{\partial\Psi}{\partial x_j}\right)^L = \dfrac{\partial\Psi^L}{\partial x_j} - \dfrac{\partial\mathbf{u}^L}{\partial x_j} \cdot \nabla\Psi_o \quad j = 1,2,3 \end{cases} \tag{5}$$

Besides, whatever the kind of variables, the perturbation to the first order of a product is:

$$(\Psi\Phi)^L = \Psi_o\Phi^L + \Psi^L\Phi_o. \tag{6}$$

2.2 Derivation of the propagation equation

We now derive the propagation equation in the case of a perfect fluid (viscosity and thermal conductivity are ignored). Thus, conservation equations reduce to Euler equations (assuming Einstein notations):

$$\begin{cases} \dfrac{d\rho}{dt} + \rho\dfrac{\partial v_j}{\partial x_j} = 0 \\[4mm] \rho\dfrac{dv_i}{dt} + \dfrac{\partial p}{\partial x_i} = 0 \end{cases} \tag{7}$$

Mean flow quantities also satisfy those equations. The Lagrangian perturbation of Euler equations is made directly using the rules of perturbation (5), (6) and yields after a few rearrangements:

$$\begin{cases} \dfrac{d_o}{dt}\left(\rho_o\left(\rho^L + \rho_o \dfrac{\partial u_j^L}{\partial x_j} \right) \right) = 0 \\[3mm] \rho^L \dfrac{d_o v_{oi}}{dt} + \rho_o \dfrac{d_o v_i^L}{dt} + \dfrac{\partial p^L}{\partial x_i} - \dfrac{\partial u_j^L}{\partial x_i}\dfrac{\partial p_o}{\partial x_j} = 0 \end{cases} \tag{8}$$

Integration of the first equation gives an explicit relation for the density perturbation, assuming a zero mean value of this quantity:

$$\rho^L = -\rho_o \frac{\partial u_j^L}{\partial x_j}, \tag{9}$$

Which simply expresses the fact that dilatation fluctuations directly balance density fluctuations. If transformations are supposed to be isentropic, the equation of state provides the well-known relation:

$$p^L = \rho^L c_o^2. \tag{10}$$

This equation is the same as the one obtained from the classical Eulerian description. However, unlike the latter, the mixed description has the great advantage of keeping this equation valid for inhomogeneous media and/or non-homentropic flows. For a lagrangian perturbation, the equation of state applies for a given particle so that the thermodynamic system is closed. This is not the case of Eulerian perturbations.

Then, using (9) and (10), the density and pressure perturbations can be eliminated from the momentum equation (8b). The velocity perturbation is expressed in term of the displacement perturbation. This yields, with the use of (7b) in the mean flow configuration:

$$\rho_o \frac{d_o^2 u_i^L}{dt^2} - \frac{\partial}{\partial x_i}\left(\rho_o c_o^2 \frac{\partial u_j^L}{\partial x_j} \right) + \frac{\partial u_j^L}{\partial x_j}\frac{\partial p_o}{\partial x_i} - \frac{\partial u_j^L}{\partial x_i}\frac{\partial p_o}{\partial x_j} = 0. \tag{11}$$

This propagation equation has the particularity to be written only in term of the Lagrangian perturbation of the displacement. Though it's relative simplicity, it remains quite general and valid for arbitrary perfect fluid flows. Note that flows are allowed to be rotational, sheared and even unsteady.

Further simplifications can be achieved with the homentropic and incompressibility assumptions, and if only homogeneous fluid (absence of admixture diffusion) are considered:

$$\rho_o \frac{d_o^2 u_i^L}{dt^2} - \rho_o c_o^2 \frac{\partial^2 u_j^L}{\partial x_i \partial x_j} = 0. \tag{12}$$

3   Variational Formulation

The propagation equation (12) is solved using a finite element method. The ability of such a method to deal with complex geometries and to be used as a black box makes this choice attractive. In this part, the variational formulation associated to the problem is presented.

It is assumed that fluctuating quantities have a time dependence $e^{-i\omega t}$ and that the mean flow is steady. Equation (12) becomes:

$$- \rho_o \omega^2 \mathbf{u}^L - 2i\omega \rho_o \mathbf{v_0} \cdot \nabla \mathbf{u}^L + \rho_o \mathbf{v_0} \cdot \nabla \left(\mathbf{v_0} \cdot \nabla \mathbf{u}^L \right) - \rho_o c_o^2 \nabla \left(\nabla \cdot \mathbf{u}^L \right) = 0. \tag{13}$$

Unfortunately, the direct implementation of this equation in a finite element code does not give proper results, particularly in the no-flow or uniform flow case. This problem is well known in electromagnetics[24] when a vector formulation is used. This is due to a total corruption by spurious rotational fields. In this paper, a mixed formulation is proposed to overcome this difficulty. It consists of replacing the last term of the above equation by the pressure gradient and adding the conservation of mass equation (9), which yields the following system of two equations:

$$\begin{cases} - \rho_o \omega^2 \mathbf{u}^L - 2i\omega \rho_o \mathbf{v_0} \cdot \nabla \mathbf{u}^L + \rho_o \mathbf{v_0} \cdot \nabla \left(\mathbf{v_0} \cdot \nabla \mathbf{u}^L \right) + \nabla p^L = 0 \\ p^L + \rho_o c_o^2 \nabla \cdot \mathbf{u}^L = 0 \end{cases} \tag{14}$$

Let $\mathbf{w}^*, \lambda$ be two trial fields. Then, these equations are respectively multiplied by these fields and integrated over the domain. After integrating by part, the following expression can be obtained:

$$\begin{cases} \displaystyle\int_\Omega \mathbf{w}^* \cdot \nabla p^L d\Omega - \omega^2 \int_\Omega \rho_o \mathbf{w}^* \cdot \mathbf{u}^L d\Omega - 2i\omega \int_\Omega \rho_o \mathbf{w}^* \cdot \left(\mathbf{v_0} \cdot \nabla \mathbf{u}^L \right) d\Omega \\ \quad - \displaystyle\int_\Omega \rho_o \left(\mathbf{v_0} \cdot \nabla \mathbf{w}^* \right) \cdot \left(\mathbf{v_0} \cdot \nabla \mathbf{u}^L \right) d\Omega + \int_S \rho_o (\mathbf{v_0} \cdot \mathbf{n_0}) \, \mathbf{w}^* \cdot \left(\mathbf{v_0} \cdot \nabla \mathbf{u}^L \right) dS = 0 \quad \forall \mathbf{w}^* \\ \displaystyle\int_\Omega \lambda p^L d\Omega - \int_\Omega \rho_o c_o^2 \nabla \lambda \cdot \mathbf{u}^L d\Omega + \int_S \rho_o c_o^2 \lambda \mathbf{u}^L \cdot \mathbf{n_0} dS = 0 \qquad\qquad \forall \lambda \end{cases} \tag{15}$$

Where $\mathbf{n_0}$ is the outward normal. Two boundary integrals appear respectively noted I_1 and I_2 in order of appearance. Each of them may vanish according to the type of boundary condition imposed on S. In a general case, three types of boundaries can be distinguished: fixed normal displacement and fixed pressure boundaries, absorbing wall boundary.

For a rigid wall boundary, the mean flow normal velocity as well as the normal lagrangian displacement are zero. This condition is implicitly verified by eliminating I_1 and I_2. A fixed normal displacement can also be imposed by specifying the normal displacement in I_2. The fixed pressure boundary is directly imposed at nodes in the final dicretized system.

For an absorbing wall, the admittance condition is taken into account by I_2. With the assumption of continuity of normal lagrangian displacements, this condition becomes:

$$p^L \Big|_{\Sigma_2} = -i\omega Z \mathbf{u}^L \cdot \mathbf{n_0} . \tag{16}$$

Note that this relation is much simpler to implement than Myers condition[25] in the Eulerian description.

The variational formulation is discretized using piecewise linear triangular elements. After assembling and applications of boundary conditions, the global discretized variational formulation yields the following algebraic system:

$$[A_r(\omega)]\{\xi_r\} = \{F_r\}. \tag{17}$$

The unknown nodal displacement vector is eventually obtained from inverting the frequency-dependent complex band matrix $[A_r]$. It is important to point out that $[A_r]$ is unsymmetrical when flow is present.

4   Results

Various simulations were carried out with the Eulerian-Lagrangian formulation. A first group of results is intended to assess the validity of the method for describing acoustic wave propagation in straight ducts with uniform or sheared flows. The reference solutions are provided by the Pridmore-Brown equation.

A more complex case is then presented in order to illustrate the capability of the lagrangian eulerian formulation to handle arbitrary geometries and flows.

4.1 *Validation*

The FEM method is compared with semi-analytical solutions obtained from the Pridmore-Brown equation. This equation gives modes occuring in a straight duct and in presence of flow. It is solved with an iterative Runge-Kutta method. This equation is an interesting way to validate the method because flows may be sheared and walls can be lined.

The geometry considered here is a simple straight duct of width $H=1$m and length $L=2m$. Modal pressure obtained from the semi-analytical solution is enforced at both the inlet ($y=0m$) and outlet ($y=2m$) of the duct (it is equivalent to impose a non-reflecting condition at the outlet for the mode being considered). On plots, y-axis is vertical, x-axis is horizontal, and pressure is given in Pa and displacement in m.

First results concern a mode 2 at $f=350Hz$ and is shown on Figure 1. Walls are lined and have a specific admittance of $0.1+0.1i$. The averaged Mach number is -0.4, which means that wave propagation is upward. The first range of results is obtained with a uniform flow profile, the second one with a parabolic profile. Though a parabolic profile is unrealistic (the Reynolds number is too high), it clearly illustrates refraction phenomena due to flow shear, as will be shown.

Results show a perfect agreement between semi-analytical and numerical results in both the uniform and shear cases. Besides, it can be observed that the mode is less attenuated when the flow is shear. As recalled in introduction, shear effects tend to increase pressure at the center of the duct for upward propagation, which yields a smaller attenuation when a lining is present. In our specific case, this attenuation can be quantified by attenuation coefficients: $\alpha_u=7.8dB/m$ and $\alpha=5.2dB/m$ respectively for uniform and parabolic profile. Thus, an error of about $5dB$ is made at the duct outlet when a uniform profile is assumed. Other refraction effects can be noted when compared to the uniform case: pressure nodal lines are shifted toward the center and the y-lagrangian displacement is increased at the walls (x-displacement is not shown here because it is not affected much).

Those results prove the method capacity to propagate sound waves when arbitrary flows are present. In particular, they show that the method is able to handle both convection and refraction phenomena caused by the mean flow boundary layer.

Furthermore, another flow effect is pointed out by a second group of results shown on Figure 2. Here, mode 1 is enforced in a rigid wall duct at $f=140Hz$. In the no-flow case, this mode is clearly evanescent (the first cut-off frequency is $170Hz$). Agreement between numerical and Pridmore-Brown solutions is very good.

Then, the same mode is enforced with an opposite uniform flow $M=-0.6$ (upward propagation). Note that only the numerical solution is shown for conciseness (but agreement with reference solution is still excellent). It can be observed that the mode is not attenuated any more. This is explained by the fact

that flow tends to decrease cut-off frequencies by a $\sqrt{1-M^2}$ factor. Thus, the new cut-off frequency is about $136Hz$, which is less than that chosen for calculations: the mode is now propagated. Consequently, this effect is also accurately taken into account by the method.

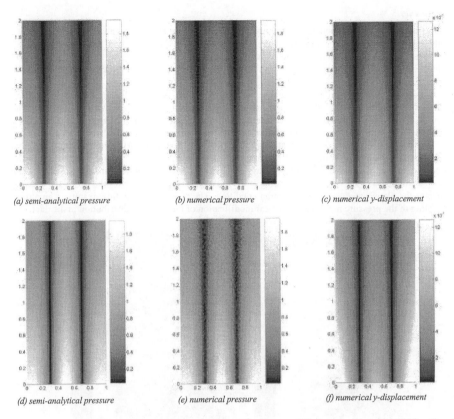

(a) semi-analytical pressure   (b) numerical pressure   (c) numerical y-displacement

(d) semi-analytical pressure   (e) numerical pressure   (f) numerical y-displacement

Fig. 1. Mode 2 at f=350Hz with wall specific admittance β=0.1+0.1i and averaged Mach number M=-0.4. (a)-(b)-(c): uniform flow profile. (d)-(e)-(f): parabolic flow profile

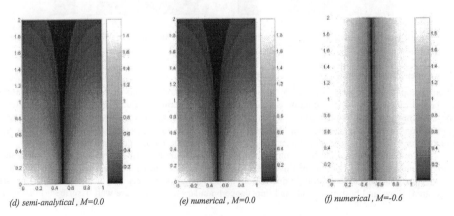

(d) semi-analytical , M=0.0   (e) numerical , M=0.0   (f) numerical , M=-0.6

Fig. 2. Pressure of mode 1 at f=140Hz with rigid walls and uniform flow.

4.2 *Varying duct*

Here, one considers the two dimensional propagation of acoustic waves in a potential flow confined by a duct with varying section. The geometry of the duct is inspired by the turbofan inlet profile used by Rienstra[9], see Figure 3. Though we are not solving an axisymmetric problem, this geometry is a good opportunity to show the influence of the flow on sound propagation and to illustrate the capability of the method.

A uniform pressure oscillation is imposed at the inlet (y=0). A non-reflecting boundary condition is required at the outlet (y=2), so the specific impedance corresponding to plane wave is imposed. The frequency of 200Hz is sufficiently low to ensure that only the plane wave propagates. The surface x=0 constitutes the equivalent of a symmetry condition. The other boundaries are rigid walls.

The present problem is first simulated without flow (see Figure 3 left) and then with a potential flow with a uniform velocity profile of 0.3 Mach at the inlet (Figure 3 right). The fluid is flowing in the increasing y direction and waves are propagating in the same direction (downward propagation). The pressure distribution have almost constant profile at the outlet section, this validates the use of the specific impedance of the plane wave as a non-reflection condition. The presence of the flow results in an almost exact doubling of the pressure modulus and a clear modification of the pressure distribution in the duct.

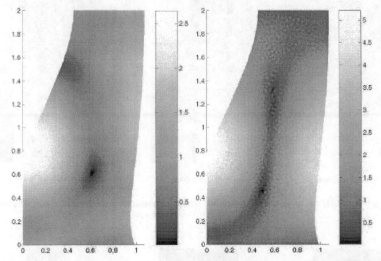

Fig. 3. Planar pressure modes at 200Hz in a varying duct: without flow (left) and with a potential flow of Mach number 0.3 (right).

5   Conclusion

A wave propagation equation derived by means of lagrangian perturbations of the conservation equations has been presented. This wave equation is valid for any flow governed by Euler equations. Compared to the standard linearized Euler equations this formulation represents a simpler set of equations for the lagrangian perturbation of the displacement. In the case of fluids with thermodynamical in-homogeneities this equation remains valid whereas the linearized Euler equations requires an additional equation.

To avoid spurious modes, a mixed variational formulation and its finite element discretization were introduced. The validations of this model have been carried out by comparison with the Pridmore-Brown equation in the case of straight ducts. Both models match accurately without flow or with (uniform or sheared) flows and both rigid wall and impedance boundary conditions are well described. Finally, the case of a varying section duct with a mean potential flow is considered to illustrate the influence of the flow on sound propagation and to show the capability of the method to handle arbitrary geometry.

As the flow velocity increases the finite element model requires rather fine meshes; therefore some further examinations may help to choose a better numerical model especially for the advective terms. It will also be worth investigating the capability of the present formulation with axisymmetric geometries as well as fluid-structures interactions and free-field radiation.

References

1. A. H. NAYFEH, B. S. SHAKER AND J. E. KAISER 1980 *American Institute of Aeronautics and Astronautics Journal* **18**(5), 515-525. Transmission of sound through nonuniform circular ducts with compressible mean flows.
2. A. L. ABRAHAMSON 1977 *AIAA 4th Aeroacoustics Conference, Atlanta*. A finite element algorithm for sound propagation in axisymmetric ducts containing compressible mean flow.
3. R. J. ASTLEY AND W. EVERSMAN 1981 *Journal of Sound and Vibration* **74**(1), 103-121. Acoustic transmission in non-uniform ducts with mean flow, part II: the finite element method.
4. R. K. SIGMAN, R. K. MAJJIGI AND B. T. ZINN 1978 *American Institute of Aeronautics and Astronautics Journal* **16**(11), 1139-1145. Determination of turbofan inlet acoustics using finite elements.
5. A. CABELLI 1982 *Journal of Sound and Vibration* **82**(1), 131-149. The influence of flow on the acoustic characteristics of a duct bend for higher order modes - a numerical study.
6. Z. L. JI, Q. MA AND Z. H. ZHANG 1995 *Journal of Sound and Vibration* **185**(1), 107-117. A boundary element scheme for evaluation of four-pole parameters of ducts and mufflers with low mach number non-uniform flow.
7. Y. C. CHO AND K. U. INGARD 1983 *American Institute of Aeronautics and Astronautics Journal* **21**(7), 970-977. Mode propagation in nonuniform circular ducts with potential flow.
8. A. F. GLADENKO AND A. F. SOBOLEV 1993 *Sov. Phys. Acoust.* **39**(6), 548-550. Green's function for a smoothly nonuniform flow duct.
9. S. W. RIENSTRA 1999 *Journal of Fluid Mechanics* **380**, 279-296. Sound transmission in slowly varying circular and annular lined ducts with flow.
10. S. J. HOROWITZ, R. K. SIGMAN AND B. T. ZINN 1986 *American Institute of Aeronautics and Astronautics Journal* **24**(8), 1256-1262. An iterative finite element-integral technique for predicting sound radiation from turbofan inlets in steady flight.
11. R. J. ASTLEY 1985 *Journal of Sound and Vibration* **103**(4), 471-485. A finite element wave envelope formulation for acoustic radiation in moving flows.
12. M. BEN TAHAR 1991 *Thèse de docteur d'état, Université de Technologie de Compiègne*. Formulation variationnelle par équations intégrales pour le rayonnement acoustique en présence d'écoulement non-uniforme.
13. P. ZHANG, T. WU AND L. LEE 1996 *Journal of Sound and Vibration* **192**(1), 333-347. A coupled FEM/BEM formulation for acoustic radiation in a subsonic non-uniform flow.
14. W. EVERSMAN 1999 *Journal of Sound and Vibration* **224**(4), 665-687. Mapped infinite wave envelope elements for acoustic radiation in a uniformly moving medium.
15. D. C. PRIDMORE-BROWN 1958 *Journal of Fluid Mechanics* **4**, 393-406. Sound Propagation in a fluid flowing through an attenuating duct.
16. P. MUNGUR AND G. M. L. GLADWELL 1969 *Journal of Sound and Vibration* **9**(1), 28-48. Acoustic wave propagation in a sheared fluid contained in a duct.
17. P. MUNGUR AND H. E. PLUMBEE 1969 *NASA* SP-207, 305-327. Propagation and attenuation of sound in a soft-walled annular duct containing a sheared flow.
18. A. H. NAYFEH, J. E. KAISER AND D. P. TELIONIS 1975 *American Institute of Aeronautics and Astronautics Journal* **13**, 130-153. Acoustics of aircraft engine-duct systems.
19. G. R. GOGATE AND M. L. MUNJAL 1993 *Journal of Sound and Vibration* **160**(3), 465-484. Analytical solution of sound propagation in lined or unlined circular duct with laminar mean flow.
20. A. BIHHADI AND Y. GERVAIS 1997 *Acustica* **83**, 1-12. Analysis of the distribution and attenuation of acoustic energy flux in lined duct containing inhomogeneous medium by the finite difference method.
21. N. K. AGARWALL AND M. K. BULL 1989 *Journal of Sound and Vibration* **132**(2), 275-298. Acoustic wave propagation in a pipe with fully developed turbulent flow.
22. H. GALBRUN 1931 *Gauthier-Villars, Paris*. Propagation d'une onde sonore dans l'atmosphère et théorie des zones de silence.
23. B. POIREE 1985 *Acustica* **57**, 5-25. Les équations de l'acoustique linéaire et non linéaire dans un écoulement de fluide parfait.
24. J. JIN 1993 *John Wiley & Sons, Inc.* The finite element method in electromagnetics.
25. M. K. MYERS 1980 *Journal of Sound and Vibration* **71**(3), 429-434. On the acoustic boundary condition in the presence of flow.

Theoretical and Computational Acoustics 2001
E.-C. Shang, Qihu Li and T. F. Gao (Editors)
© 2002 World Scientific Publishing Co.

Experimental Verifying of Acoustic Impedance Inversion

Guo Yonggang, Wang Ning, Lin Junxuan
Ocean University of Qingdao, China

Abstract:

Experiments were performed to verify the validity of our strategy to improve numerical stability on impedance inversion and investigate the influence of different extracting impulse response on the inversion. The acoustic impedance profiles are reconstructed from impulse responses, the results identify to sample data or handbook data with high precision. It is shown that our algorithm is effective in practice and the inverted impedance is less dependent on the methods of extracting the impulse responses.

Introduction

Over the last decades, the inverse scattering problem for plane wave through inhomogenous one-dimensional lossless media has been considered in many practical fields and a number of works have been contributed to it. Among these methods, there are mainly three basic algorithms: the exact inverse scattering method [1-4], the inverse sequential method based on characteristics[5-9] and approximated method[10-13]. The inverse sequential method of characteristics, undoubtedly the most straightforward approach for wave propagation analysis, is specially suitable in the layered medium model, and it features without assuming any empirical knowledge about medium. But it has two disadvantages: numerical instability[14][15]and strong restriction on the input wave[16].The instability comes from error propagation in the recursive procedure. Indeed, error propagation is an intrinsic feature of sequential procedures and cannot be completely prevented. For this case, the classical sequential methods often fails. To improve the numerical stability, we have present a strategy, which extract impulse response from a practical signal at first and then invert impedance profile from

276

impulse response[15][17].

The purpose is to verify experimentally the improved inverse algorithm and investigate the influence of different extracting impulse response on the inversion in this paper. This article is organized as follows. The Goupillaud inverse problem and deconvolution methods which extract impulse sequence from incident wave and reflective wave are discussed in Sec. I . The experiments and inverse results are described in Sec. II . Conclusions are presented in Sec.III.

I. Goupillaud inverse problem and Deconvolution of impulse response

As discussed in this paper, the one dimensional inverse scattering problem is illustrated in Fig.1.The medium is assumed to be a Goupillaud layered medium, which means a homogeneous half-space of impedance $\rho_0 c_0$ in contact with a sequence of n homogenous layers of impedances $\rho_1 c_1, \rho_2 c_2 \cdots \rho_n c_n$ and terminates with half-space of impedance $\rho_{n+1} c_{n+1}$. The sequence of n homogenous parallel layers have equal travel time.

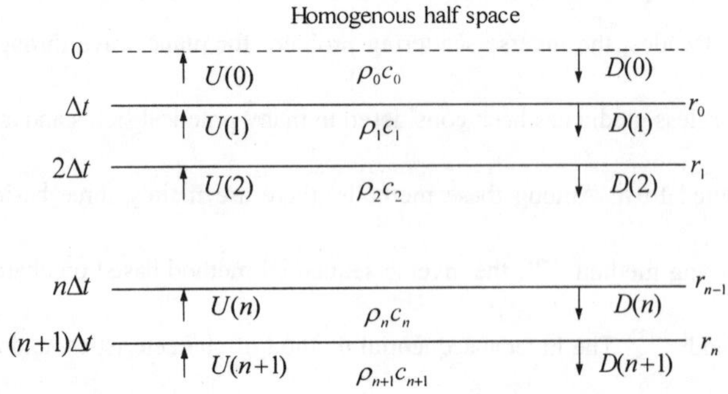

Fig.1 The Goupillaud layered medium (layers have equal travel time)

Ware and Aki[18] have carried out an exact formula to invert impedance profile from an impulse response. The formula is given by

$$r_n = \frac{\sum_{i=0}^{N} f(n-1,i)h(n-i)}{\prod_{i=1}(1-r_i^2)}$$

(1)

where $h(i)$ is the impulse response sequence, the function $f(n-1,i)$ satisfies the following

recursive relations

$$f(n+1,i) = f(n,i) + r_{n+1}g(n,i-1);$$
$$g(n+1,i) = g(n,i-1) + r_n f(n,i)$$

(2)

with the initial values

$$f(0,0) = r_0 r_1, \qquad g(0,0) = r_0$$

(3)

For an arbitrary incident wave, the classical sequential methods often fail for error propagation.

According to earlier studies we at first extract the impulse response to decrease the numerical

instability. The incident wave $x(t)$ and the reflective wave $y(t)$ are considered as input and output

in the view of a linear time-invariant causal system. $y(t)$ is the convolution of $x(t)$ and the

impulse response $h(t)$

$$y(t) = \int_0^\infty x(t-\tau)h(\tau)d\tau$$

(4)

Eq. (4) can be rewritten as the operator form

$$y(t) = \hat{A}h(t)$$

(5)

where $\hat{A}$ denotes the convolution operator.

It's a deconvolution problem to solve impulse response $h(t)$ when $x(t)$ and $y(t)$ are given.

For the noise, the signals will be polluted, so the direct deconvolution often fails. There are several

approaches to perform the deconvolution. The approached may be grouped primarily into

time-domain optimization and domain transform methods. In this paper, we adopt the former

method.

1.1 Conjugate gradient method [19]

Conjugate gradient method estimates an impulse response by directly minimizing the cost

functional $F(h)$

$$F(h) = < R, R > = < \hat{A}h - y, \hat{A}h - y > = \|R\|^2 \tag{6}$$

where $< \cdot, \cdot >$ denotes the standard inner product.

Assuming the initial value of h is h_0, the corresponding initial modified vector P_0 is given by

$$P_0 = -b_0 \hat{A} * R_0 = -b_0 \hat{A} * (\hat{A}h_0 - y) \tag{7}$$

where $\hat{A}*$ represents the correlation operator.

The iterative approximate procedure is given as follows

$$h_{i+1} = h_i + \alpha_i P_i \tag{8}$$

where

$$P_{i+1} = P_i - b_{i+1} \hat{A} * R_{i+1} \tag{9}$$

$$\alpha_i = \frac{1}{\|\hat{A}P_i\|^2} \tag{10}$$

$$R_{i+1} = R_i + \alpha_i \hat{A}P_i \tag{11}$$

$$b_i = \frac{1}{\|\hat{A}*R_i\|^2} \tag{12}$$

1.2   LMS iteration method[20]

Supposing $h(n)$ is an impulse response, and regarding the estimation $\hat{h}(n)$ as the weight coefficients of a FIR filter. The input $x(n)$ is assumed as the input sequence of the filter, the corresponding output sequence $y'(n)$ is defined by,

$$y'(n) = \sum_{k=1}^{N} \hat{h}(k)x(n+1-k) \quad (1 \le k \le N) \tag{13}$$

and the error is defined as $e(n) = y(n) - y'(n)$. The impulse response $h(n)$ is then estimated by Widrow's algorithm

$$\hat{h}_{i+1}(k) = \hat{h}_i + 2\mu e_i(k)x(n-k+1) \tag{14}$$

$$i = 1 \cdots N, \quad k = 1 \cdots N \tag{15}$$

where μ is the iteration stride factor.

For practical signal, we at first perform average filtering to reduce the noise before deconvolution.

$$\overline{m}(n) = \frac{1}{l}\sum_{i=0}^{l-1} m(n+i) \tag{16}$$

where $m(n)$ represents input $x(n)$ or output $y(n)$, l is the length of the average filter time window. But l should be chosen carefully, otherwise the real signals would be distorted. The averaged signal, then is considered as the data of deconvolution .

The last problem we address in this section is how to terminate the iteration, In practical calculation, we find that too many iterations would introduce spurious oscillations. On the contrary, too few iterations may give a solution that is not good enough. The criterion to stop the iterative process is that the magnitude of error power function $F(h)$ is less than the limitative threshold.

$$F(h) \le c \quad \text{(a constant)} \tag{17}$$

To the slowly varying medium, the impulse response wave should be smoothing, so we add a new criteria

$$P_c = \sum_{n=0}^{N-1} [s(n)]^2 \le \sigma . \tag{18}$$

where $s(n)$ represents the smoothness,

$$s(n) = d(n) * h(n) \tag{19}$$

where $d(n)$ is second order difference operator, the symbol * denotes convolution. In numerical computation, values c and σ are determined case by case.

II Experiments and inverse results

The first experiment is conducted in a sewage pond of a paint factory.(see **Fig.2**). The pond is approximately 9m wide, 60m long, and 2m deep, and the sound source is a spark source. There is a low speed sediment layer at the sub-surface. Sample measurement shows that the sound speed is approximately $c = 1.27 \times 10^3$ m/s, the density of the sediment $\rho = 1.08 \times 10^3$ kg/m, and the impedance is $\rho c = 1.37 \times 10^6$ Pa.s/m.

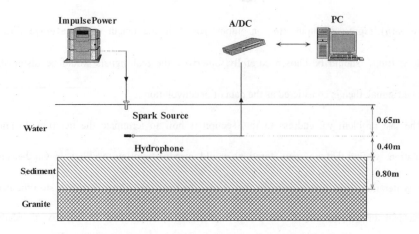

Figure 2. The experiment-1 set up for inverting sediment impedance

The setup for our second experiment is shown in **Fig.3**. A multi-sections sound tube is designed to admit only plane wave propagating in it. By filling different materials in the different sections, we can construct different layered models. The medium in this experiment is a three layers medium, the first layer is water, the second layer is a 20cm glycerin layer, and the third layer is the alloy base plate which is about 3cm thick. The sound source is also a spark source placed just on the top of the water surface, and the hydrophone is located about 56cm away from the source.

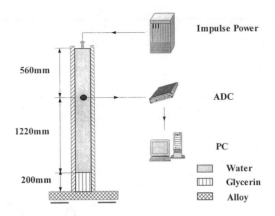

Figure 3. The experiment-2 set up for inverting glycerin impedance

Fig.4 (a-b) show the examples of recorded sound signals. The first impulse waves in both figures are the source signals; the corresponding power spectrums are shown in **Fig.5 (a-b)**. Both the sources have broadband frequency. The reconstructed impedance profiles using the different methods are shown in **Fig.6**. A comparison of the impedance values at the sampling point for the experiment-1 and that with handbook value for the experiment-2 are given in **Table-1** and **Table-2**, respectively.

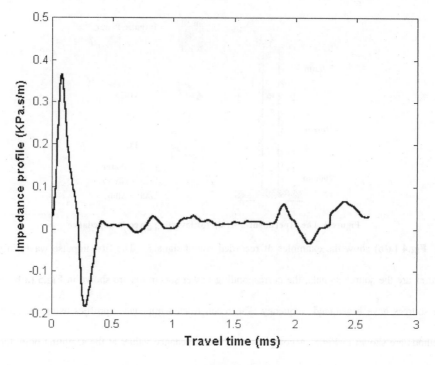

a: record of exp-1

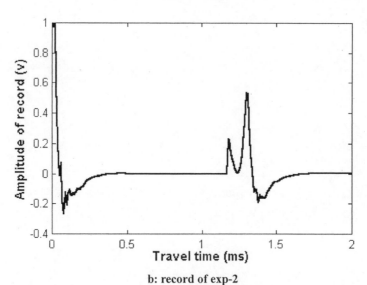

b: record of exp-2

Fig.4 Recorded reflective signals

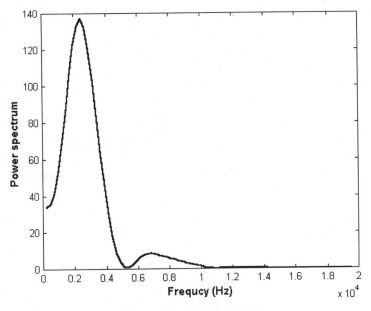

a: The spectrum of exp-1 input wave

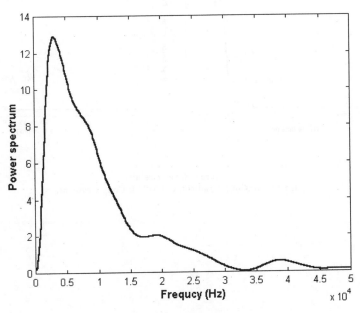

b: The spectrum of exp-2 input wave
Fig.5 Power spectrum of input signals

284

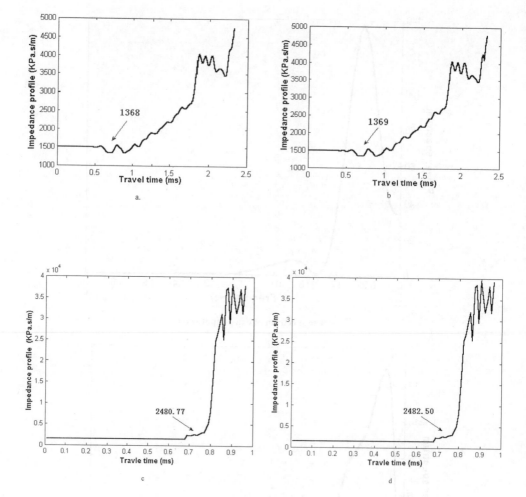

Figure 6. Inverse results
a: LMS; b: Conj. Gradient; c: LMS; d: Conj. Gradient;

Table-1: Comparison of the different deconvolution methods about Experiment-1.

	LMS	Con-gradient
Computing cost	78.98 s	2.97 s
error	1.5%	0.8%
The results of the sample $(MKg/m^2 s)$	1.368	1.369

The impedance of sample $1.37 \times 10^6 Kg/m^2 s$

Table-2: Comparison of the different deconvolution methods about Experiment-2.

	The result of glycerin $(MKg/m^2 s)$	Error(%)	Computing cost(s)
LMS	2.48077	2.55	32.68
Conj-Gradient	2.48250	2.62	0.55

The impedance of glycerin is $2.419 \times 10^6 Kg/m^2 s$ in the manual.

From the above results, acoustic impedance profile from impulse response is well reconstructed and identify well with the measured or the handbook data.

III. Conclusions

In this paper, experimental studies on acoustic impedance inversion were done. The conclusion follows with our experiment.

(1) Impedance inversion is less dependent on the methods to extract impulsive response.

(2)The average filtering joint conjugate gradient (CG) method is an effective deconvelution method.

(3) The stable solution criterion are useful for slowly varied medium to extract impulse response.

References

1. R.John and W.Kohn, "Construction of a Potential from a Phase Shift", Physical Review, 87,pp.977 - 922, 1952.
2. L.D.Faddeev, "The Inverse Problem in the Quantum Theory of Scattering", J.Math.Phys. Vol.4,pp.71 - 104, 1963.
3. I.M.Gelfand and B.M.Levitan, "On the Determination of a Differential Equation by its Spectral Function" ,Doklady Akademii Nauk SSR 77, pp.557 - 560,1951.

4. Roger G. Newton, "Inverse Scattering in One Dimension", J. Math. Phys.Vol.21,No.3,pp.493 - 506,1980.

5. G.Kunetz and I. D'Erceville, "Sur Certain Property D'uneonde Acoustic Plane de Compression Dan s un Milieu Stratify",Ann. Geophys, Vol.18,pp.351 - 359, 1962.

6. G.Kunetz,"Quelques   Exemples   d'analyse   d'enregistrements   sismeques", Geophys.Prospect,Vol.11,No.4,pp.409-422,1963.

7. E.A.Robinson, *Multichanel Time Series Analysis with Digital Computer Programs*,Hodden-Day,San Francisco,CA.1967

8. J.Claerbout . *Fundamentals of Geophysical Data Processing*, McGraw Hill,Inc.1976.

9. Roberto A.Tenenbaum and Moyes.Zindeluk, "A fast algorithm to solve the inverse scattering problem in layered media with arbitrary input", J.A.S.A.Vol.92, No.6, pp.3371 - 3378,1992.

10. H.Levine, *Unidirectional Wavemotion*,NorthHolland,Amsterdam,1978 .

11. J.Mendel. *Optimal seismic deconvolution: An Estimation Based Approach*. New York :Academic ,1983.

12. J.M.Candel, "Bremmer Series deconvolution of solution to the lossless wave equation in layered media", *IEEE Trans*. Geosci.Electron.GE-16,pp.103-112, 1978.

13. S.M.Candel,F.Defillipi and A.Launay, "Determination of the inhomogeneous structure of a medium from its plane wave reflection response Part 2:A numerical application", Journal of Sound and vibration Vol.68,No.4,pp.583-595, 1980.

14. A.Bamberger,G. Chavent, and P.Lailly, "About the stability of the inverse problem in 1-D wave equation-Application to the interpretation of seismic profiles", Appl.Math.Optim.5,pp.1-45,1979.

15. Wang N, etc, "Goupillaud inverse problem with arbitrary input", J.A.S.A. Vol.101, No.6, pp.3255 -3260, 1997.

16. N.E.Nahi.J.M.Mendel,and L.M.Silverman, "Recursive derivation of reflection coefficients from noisy seismic data", *IEEE Int.Conf.Acoust. Speech Sig,Process*,Tulsa,1978

17. Lin junxuan, "Adaptive iteration methods for enhancing the stability of inverse algorithm", Chinese journal of acoustics, Vol.22, No.3 pp.193-197.1997.

18. Jerry A. Ware and Keiiti Aki, "Continuous and Discrete Inverse-Scattering Problems in a stratified Elastic Medium 1.Plane Waves at Normal Incidence" J.A.S.A.Vol.11,pp.911 – 921,1969.

19. T.K.Sarkar, "Deconvolution of Impulse Response from Time-Limited Input and Output: Theory and Experiment", *IEEE .Trans .IM* ,Vol.IM-34, No.4, pp.541-546, 1985.

20. B.widrow, "A Comparison of Adaptive Algorithms based on the Methods of Steepest Descent and Random Search", *IEEE Trans*, Vol.AP-24, pp.615-637, Sept.1976.

Theoretical and Computational Acoustics 2001
E.-C. Shang, Qihu Li and T. F. Gao (Editors)
© 2002 World Scientific Publishing Co.

A Rhythm Estimator of Ship Noise

Chen Xixin, Jiang Guojian, Chang Daoqing, Lin Jianheng, Li Xuejun

Qingdao Acoustic Laboratory, Institute of Acoustics, Academia Sinica,

No. 8, Shangqing Road, Qingdao, 266023, China

E-mail: *jianglin@qd.col.com.cn*

Abstract

The 1st class $\mu(D)$ and 2nd class $\mu(D^{-1})$ rhythm estimators of time sequence signals are given in this paper, on the basis of transforming signal properly. It is pointed out that the product $\mu(D) \mu(D^{-1})=1$ for the periodic signals with rhythms, and $\mu(D) \mu(D^{-1})>1$ for no-periodic signals. Accordingly the ship noise rhythms are analyzed, and the results are obtained with this method to processing noise of a certain type ship .

Rhythms estimator

It's well known that the rhythm sense of ship noise is a kind of period or quasi-period magnitude modulation component, and it is one of the unique properties of ship noise. In some specified speed, the propeller noise and blades cavitation noise, as the main component of ship noise, are modulated by axis frequency[1]-[5] . Extraction and analysis of the rhythms is helpful for distinguishing between ship objects. A method of time sequence signal processing, given in this paper, can be used to extract and analyze ship noise rhythms.

Sliding window function $\varphi[a, b](x)$ is introduced for time sequence signals:

$$\varphi[a(n), a(n)+w](x) = \begin{cases} 0 & a \leq x \leq b \\ 1 \end{cases} \quad (1)$$

where

$$a(n) = \alpha n + \beta \quad (2)$$

is a integer liner function defined in integer set Z; and w window width; $n \in Z$. Assuming w and $a(n)$ not variable, then φ_n is expressed with $\varphi[a(n), a(n)+w]$, and

$$\varphi_n = \varphi[a(n), a(n)+w]$$
$$\varphi_{n+1} = \varphi[a(n+1), a(n+1)+w] \quad (3)$$

are called adjacent window.

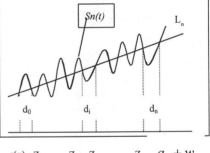

Fig. 1 Sliding window function

Ship noise signal $s(t)$, as a time sequence function, can be formed a new function by transformation with window $\varphi[a, b]$,

$$s_n^* = s_n(t) - \varphi_n L_n \quad (4)$$

where

$$s_n(t) = \varphi_n s(t) \quad (5)$$

L_n is a liner function , best close to s_n in least square criterion. Namely it satisfies that

$$\sum_{k \in z} [s_n(t) - \varphi(k) L_n(t)]^2 \quad (6)$$

is smallest . Signal $s_n{}^*$ can be seen in figure 1, it is the deviation of $s(t)$ from L_n.
The zero points of $s_n{}^*$ in window $\varphi_n=\varphi[a(n), a(n)+w]$ are

$$Z_1 < Z_2 < \cdots\cdots < Z_M \tag{7}$$

Defining d_i as the distance of the zero points z_i and z_{i+1}, d_0 as the distance between $a(n)$ and z_1, d_M as the distance between z_M and $a(n)+w$, then we can define

$$\mu(D) = \frac{1}{M+1}\sum_{i=0}^{M} d_i = d \tag{8}$$

as the first class estimator of time sequence signal rhythms. It is the average rhythm of signal $s(t)$ in the n of the window, with the dimension of time. Also we define

$$\mu(D^{-1}) = \frac{1}{M+1}\sum_{i=0}^{M} \frac{1}{d_i} \tag{9}$$

as the second class estimator, with the dimension of frequency.

Now , assume that

$$\varepsilon_i = \frac{d - d_i}{d}, \quad \text{namely} \quad d_i = d(1 - \varepsilon_i), \tag{10}$$

to every i, $|\varepsilon_i| < 1$, then

$$\mu(D^{-1}) = \frac{1}{M+1}\sum_{i=0}^{M} \frac{1}{d_i} = \frac{1}{(M+1)d}\sum_{i=0}^{M} \frac{1}{1-\varepsilon_i}$$

$$= \frac{1}{(M+1)d}\sum_{k=0}^{\infty}\left\{\sum_{i=0}^{M}(\varepsilon_i)^k\right\} \tag{11}$$

For $\sum\limits_{i=0}^{M}\varepsilon_i = 0$,consequently it is approximate to quadratic components,

$$\mu(D^{-1}) \approx \frac{M+1}{W} + \frac{1}{W}\sum_{i=0}^{M}\varepsilon_i^2$$

$$= \frac{1}{\mu(D)} + \frac{1}{W}\sum_{i=0}^{M}\varepsilon_i^2 \tag{12}$$

Obviously, if time signal $s(t)$ has equal interval zero points in the n of window, that means it has periodic characteristics, then $\varepsilon_i=0$ can be obtained, so

$$\mu(D^{-1}) = \frac{1}{\mu(D)},$$

namely $\qquad\qquad \mu(D)\mu(D^{-1}) = 1 \tag{13}$

Otherwise, if time signal $s(t)$ is not periodic in the n of window, intervals of zero points are not equal, then $\varepsilon_i \neq 0$, so

$$\mu(D^{-1}) > \frac{1}{\mu(D)}$$

namely

$$\mu(D)\mu(D^{-1}) > 1 \qquad (14)$$

Thus it can be seen that non-dimension value $\mu(D)$ $\mu(D^{-1})$ could be used to detect periodic and quasi-periodic properties in time sequence signals . The products $\mu(D)$ $\mu(D^{-1})$ is equal close to 1 for ship noise with periodic or quasi-period, and it is larger than 1 for the non periodic or non rhythms.

Data processing method

The block diagram is used to extract and analyze rhythms of ship noise, according to the principles of rhythms estimator above (see Figure 2).

The radiated noise of ship is amplified first, then filtered and shaped, and the envelope is extracted last. Shaping and envelope extracting are illustrated in figure 3.

After the signal is filtered with band-pass and low-pass, envelope, namely the slow changing magnitude modulated part of ship noise is extracted. Generally, the envelope includes the axis frequency of ship propeller ,and the blade frequency also.

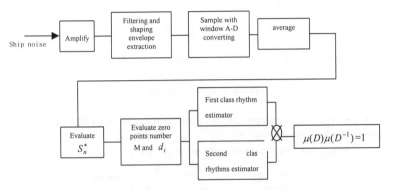

Fig. 2   Block diagram for analyzing ship noise rhythm

In Figure 2, steps from amplifying to A-D converting can be carried out by hardware. Hardware as well as software both can be used in the following steps. When the ship noise is processed by means of this method, programming (software method) is used after A-D converting step and the effect is good. When the product $\mu(D)\cdot\mu(D^{-1})$ is equal or very close to 1, the output of first class estimator is the average period of ship noise rhythms, and it's reciprocal is axis frequency.

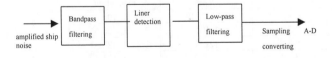

Fig. 3   Filtering, shaping and envelope extraction

Figure 4 illustrate waveform change of the noise signal with rhythms, in the processing of rhythms estimation with this method. After the waveform of the noise signal is changed and sampled with window mentioned above, it can be used to extract rhythms.

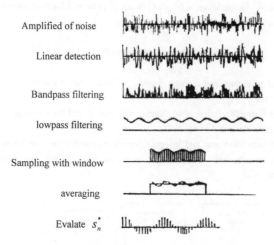

Amplified of noise

Linear detection

Bandpass filtering

lowpass filtering

Sampling with window

averaging

Evalate s_n^*

Fig. 4   Waveform change of the noise signal with rhythms, in the processing of rhythms estimation

Conclusions and discussion

The measured radiated noise of a certain type ship are analyzed with above method(table 1), and the results are compared with ocean ambient noise(table 2).

Table 1   Axis frequency estimation of a certain type ship

Estimated axis frequency with rhythms estimator	Actual rotational speed(rpm)	Ship speed (knot)	$\mu(D)\,\mu(D^{-1})$ estimation
8. 20	495/495	15. 2	0. 998
5. 15	310/310	9. 6	0. 998
4. 30	260/260	8. 1	1. 002
6. 30	380/0	9. 6	0. 998
4. 30	310/0	7. 5	1. 001
6. 00	300/300	8. 8	0. 997
4. 65	280/0	6. 8	0. 995

Table 2   Analysis of ocean ambient noise with rhythms estimator

wind speed （m/s）	Sea state	Noise level(re 1 μ pa)	$\mu(D)\,\mu(D^{-1})$ estimation
1. 1	0—1	117. 7 dB	1. 105
2. 5	2	119. 3	1. 135
4. 0	3	122. 7	1. 128
5. 8	3—4	125. 3	1. 151
8. 4	4	127. 6	1. 195

The conclusions are as follows:

(1) The product $\mu(D)\,\mu(D^{-1})$ is 0.995~1.002 for ship noise in various work state, and all product is close to 1(see table 1); but for ocean ambient noise data, the product $\mu(D)\,\mu(D^{-1})>1$(see table 2).

(2) The estimated ship axis frequency with estimator in this paper is in good agreement with actual rotational speed, and the results are satisfied well (see table 1).

(3) The method in this paper can be used to estimate blade frequency, but the difference between blade frequency and axis frequency should be paid attention to.

(4) When signal to noise ratio is low, the autocorrelation processing can be carried out first, to increase the "rhythm" sense to some extent. Then the rhythm is estimated with above method.

(5) This method needs small storage capacity. The processing is simple, convenient and rapid.

(6) This method provides a way to extract and analyze the "rhythm" characteristic of ship noise, and it also can be widely applied to other time series signal processing.

References

[1]Urick,R.J.,Principles of Underwater Sound,3nd Ed. McGraw Hill Book Co.,New York 1983.

[2]Ross,D., Principles of Underwater noise, Ocean press,1983.

[3]JIANG Guojian, et.al, Theoretical model of Noise Caused by ship Propeller cavitation , Chinese Journal of Acoustics,18(1999),37-46.

[4]JIANG Guojian, et.al ,Low frequency linear spectrum of ship noise,(Research report)

[5]TAO Duochun,"A Study on Ship Radiation Noise Rhythms(I)——Mathematical Model and Power spectrum Density ,ACTA ACUSTICA,8(1983),No.2

Theoretical and Computational Acoustics 2001
E.-C. Shang, Qihu Li and T. F. Gao (Editors)
© 2002 World Scientific Publishing Co.

A High-Resolution Beamforming Method and its Experiment on a Non-Uniform Line Array*

Ling Xiao, Linghao Guo, Zaixiao Gong, Yu Chen
National Key Laboratory of Acoustic, Institute of Acoustics,
Chinese Academy of Sciences, P.O. Box 2712, Beijing, China

The paper presents a high-resolution beamforming method for non-uniform line array. This method first predicts the signal on the uniform position from the actual receive data, and then use Levinson algorithm to estimate the degree of arrival (DOA). Using multi-frequency to accumulate the output, we can get robust high-resolution beamforming. Experiment is also conducted to show that the method can play a robust role in real passive DOA detection, even in cases that the line array are not uniform.

1.Introduction

Get the bearing of targets is an important task in sonar signal processing. The resolution of traditional beamforming is reverse ratio to the wavelength of the signal. In ocean environment, people have to work hard to deal with low frequency signal in order to detect the target of far away. In the mean time, developing methods with high resolution becomes even more important.

When the target's bearing is estimating, the sensors correspond to the sampling in time domain. The detecting problem is equivalent to the spectrum estimation problem. Many methods for high-resolution spectrum analysis can be translated to solve the problem easily.

Still, two sorts of difficult need to be deal with bearing detection. One difficult is that, when the high-resolution method is used, the variability error should be deduced, especially when the snapshot's number is limited. Another difficult is due to fact that the receiver is not precisely a uniform line array (ULA). Since in time domain, spectrum analysis devotes mostly to uniform samplings. Simple translation from to the spectrum analysis can only be used for the uniform line array.

For the first difficult, the variability of the estimation can be deduced by using multi-frequency signal. Being properly transformed, wide-band signal can be considered as different samples. One of such transformations will be given in the paper.

For the second difficult, in case of non-uniform line array, maximal likelihood estimation can be used to get high resolution. In this paper, a method with low computational cost is given in the paper. The method first interpolates a "virtual" uniform line array's data, then uses a uniform line array method to estimate the bearing.

All the above method is tested and works well in an experiment, which was conducted in 2000. In the following part of the paper, section 2 discusses the mathematics modal of the bearing estimation. Section 3 describes the summarize processing of multi-frequency for uniform line array. Section 4 introduces the interpolation formula for non-uniform line array. The method, which was described in section 3, can then be applied to get high resolution bearing estimation for non-uniform line array. Section 5 briefly described the experiment 2000, which was conducted for testing the methods in the paper.

2.Mathematic Model

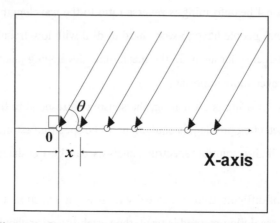

Figure 1. Sensors on line array in the DOA detection

Suppose the receive array is a line array. The signal comes from θ direction on the x-position 0 is $s(f,t)$. f is the frequency of the signal and t is the time. Then on the x-position, the received signal is

---

* This work is supported by National NSF of China, 10023004

$$s_{x,f,t} = s(f,t+\frac{x\cos\theta}{v}) = e^{i(2\pi ft+\phi_f)} \bullet e^{i(2\pi f\frac{\cos\theta}{v}x)} = s(f,t)\bullet e^{i(2\pi f\frac{\cos\theta}{v}x)} \quad (1)$$

It can be seen from (1) that when the line array is uniform, the signal can be considered as harmonic, due to the space sample x.(Figure 1)

3.Passive DOA Detection on Uniform Line Array

In passive DOA detection, wide-band signal need to be used because the source of signal is unknown and the SNR on a single frequency is not always be high. When the wide-band signal is used for high resolution bearing estimation, the variability need to be deduced by properly summing up each frequency's information. This can be done by: modifying the space spectrum for each frequency to a cost function, adding a small delta to the denominator, and then summing up all the cost functions. In the mean time, the high-resolution property can still be maintained. The following shows the whole process, aimed to get high resolution bearing estimation.

DOA Algorithm for ULA

Suppose a uniform line array is consisted by n sensors, the interval between each two adjacent sensors is d. The M frequencies of the wide-band signal are f_1, f_2, ..., f_M. Let the complex signal of frequency f_m on the n-th sensor be $s_m(n)$, $m=1,...,M$, $n=1,...,N$。

Step 1: Given out the follow three parameters:

1. The order of the AR model $p=2\times N/3$;

2. The parameter , $\delta_0>0$, which is used to control the maximal of each spectrum on a single frequency. In our experiment, we assign $\delta_0=0.01$, meaning that the maximal contribution of each single frequency is *100*;

3. The parameter α, which is used to control the linear weight. Generally we can assigned $\alpha=0.1$。

Step 2: For each frequency $f_m, m=1,...,M$, estimate the best parameters by Levinson algorithm:

$$a_1^{(m)}, a_2^{(m)}, ..., a_p^{(m)} \quad (2)$$

The best parameters are the parameters with

$$\min \left| s_m(n) + \sum_{j=1}^{p} a_j^{(m)} \times s_m(n-j) \right|^2 \qquad (3)$$

Step 3: For each frequency $f_m, m=1,...,M$, compute the cost function corresponding the direction from angle θ,

$$T_m(\theta) = \frac{1}{\left| 1 + \sum_{j=1}^{p} a_j^{(m)} \times e^{i2\pi f \frac{\cos\theta}{v} jd} \right|^2 + \delta_0} \qquad (4)$$

Step 4: Compute the linear weight coeficiences λ_m, $m=1,...,M$, for each frequencies:

$$\lambda_m = \frac{\sum_{n=1}^{N} |s_m(n)|^\alpha}{\sum_{m_1=1}^{M} \sum_{n=1}^{N} |s_{m_1}(n)|^\alpha} \qquad (5)$$

Step 5: Sum up the cost value of each frequency, using the linear weight which is given in Step 4, and give out the trust function for the direction of angle θ by:

$$T(\theta) = \sum_{m=1}^{M} \lambda_m T_m(\theta) \qquad (6)$$

4. Interpretation Formula for Non-uniform Line Array

It is difficult in real cases to get uniform line array. Even when we try to get a line array, to make it exactly uniform is not always easy. For non-uniform line array, the interpolation can be first used to estimate the signal on a "virtual uniform line array", from the real sensors' received data. The former method can then be used to get bearing estimation with high resolution.

The complex signal can be described by two parameters, the phase α and the energy R. Two interpolation formula, corresponding these parameters, are given by

$$R = R_1 + (R_2 - R_1) \times \sin(\pi \times (\frac{x - x_1}{x_2 - x_1} - \frac{1}{2})) \qquad (7)$$

$$\alpha = \alpha_1 + (\alpha_2 - \alpha_1) \times \frac{x - x_1}{x_2 - x_1} \qquad (8)$$

Here $-\pi < \alpha_2 - \alpha_1 < \pi$. R_1, R_2 are the energy of real sensors' signal and R is the energy of "virtual" sensor's signal. α_1, α_2 are the phases of real sensors' signal and α is the phase of "virtual" sensor's signal. x_1, x_2 are the coordinates of the real sensors' position and x is that of the "virtual" sensor's position (Figure 2). In the two formula, a linear interpolation is used for phase estimation and a non-linear interpolation is used for energy estimation.

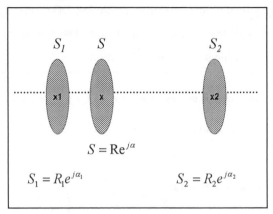

Figure 2. Estimate s from s_1 and s_2

5. Test Experiment

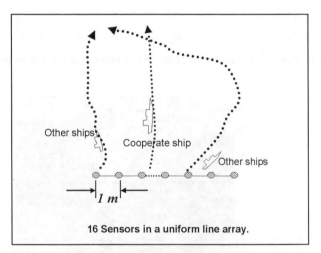

Figure 3. The experiment in 2000

During the experiment conducted in 2000, data from a HLA is used to test the interpolation formula and the high-resolution process. In the experiment, a cooperated ship is about $90°$ in the experiment area. There are some other ships around the line array. All the ships' noise is recorded in our experiment.

A 16-sensored uniform line array is used, and the interval between two adjacent sensors is 1 m. Among the 16 sensors, 8 sensors are chosen to consist a reference uniform line array. The interval between each two adjacent sensors is 2 m. Another combination of 8 sensors is used to test the interpolation formula (Figure 4). Two "virtual" sensors' data are interpolated to form a uniform line array. The "virtual" uniform line array s then used to get high resolution DOA. Finally, the "real" array result is compared with the "virtual" array result. The frequencies from 200Hz to 300Hz are processed for the passive DOA detection.

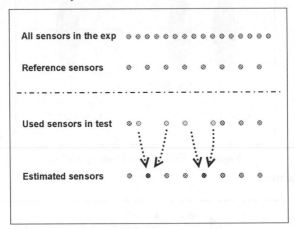

Figure 4. The combinations of sensors for reference ULA and non-ULA

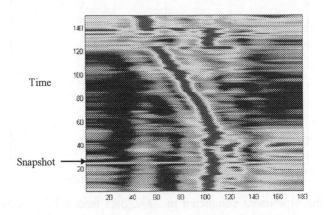

Figure 5. Tradition beamforming result for reference ULA

Figure 5 is the beamforming result of traditional method, from reference uniform line array. Since the adopted frequency is too low, the targets cannot be distinct well in the image.

Figure 6 shows the high-resolution beamforming from the reference uniform line array. The cooperate ship lies in about 90°, which can be easily distincted in the result.

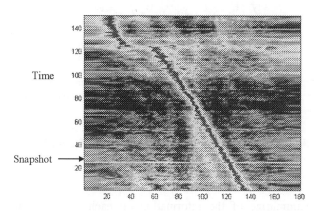

Figure 6. Result of high resolution process for reference ULA

Figure 7 is the result from the non-uniform line array. Compare Figure 6 with Figure 7, it can be seen that the interpolation works well in the experiment.

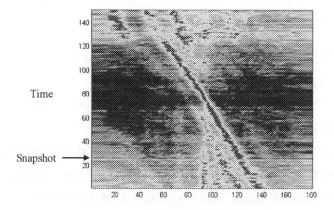

Figure 7. Result for the used Non-ULA

Three snapshots are compared in Figure 8, the traditional method, the reference uniform line array, and the non-uniform line array. It is shown that even in a snapshot figure, the interpolation and the summarize process which is given in the paper can get DOA estimation with low variability and high resolution.

300

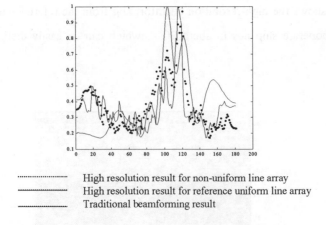

.... High resolution result for non-uniform line array
—————————— High resolution result for reference uniform line array
—————————— Traditional beamforming result

Figure 8. Compare snapshots on Figure 5,6 and 7.

6.Summary

Appropriate summarize of the information on each single frequency can get bearing estimation with high resolution for uniform line array. This summarize process can delimit the variability error on each single frequency. After the interpolation formula of this paper are used for non-uniform line array, the summarize process of each single frequency's information can also get DOA estimation with high resolution.

Reference

[1]Ianniello J. P., "Recent developments in sonar signal processing", IEEE signal processing, 15(1), 27-40 (1998)

[2]Schmidt R. O., "Multiple emitter location and signal parameter estimation", IEEE Trans. Antennas Propag. AP-34(3), 276-280(1986)

[3]A. H. Nuttall, and J. H. Wilson, "Adaptive beamforming at very low frequencies in spatially coherent, cluttered noise environments with low signal-to-noise ratio and finite-averaging times", J. Acoust. Soc. Am., 108(5), 2256-2265 (2000)

[4]Johnson, D. H., "The application of spectral estimation methods to bearing estimation problems", Proc. IEEE., 70(9), 1018-1028 (1982)

[5]Urick, R. J., "Principles of underwater sound", (3rd ed.) McGraw-Hill, New York, 1983

Theoretical and Computational Acoustics 2001
E.-C. Shang, Qihu Li and T. F. Gao (Editors)
© 2002 World Scientific Publishing Co.

Wave Propagation in Poroelasticity: Equations and Solutions

Andrzej Hanyga

Insitute of Solid Earth Physics, University of Bergen,

Allégaten 41, N5007 Bergen, Norway

A method previously developed for asymptotic Green's functions is applied to the determination of the time-domain asymptotic Green's function of Biot's poro-elasticity.

A method based on fractional calculus is applied to the generalized finite-difference implementation of equations describing pore gas motion in a rigid porous matrix.

1. Introduction

We demonstrate applications of some techniques developed for integro-differential equations with singular convolution kernels[1,2,3] for the calculation of time-domain solutions in Biot's poroelasticity and in poroacoustics. In the first case we shall derive a time-domain asymptotic expression for the Green's function. In the second case we shall construct a numerical method of solving the equations of motion.

Poroelasticity provides many examples of very detailed internal friction models, represented by frequency-dependent coefficients, which become memory kernels in the time domain. On the other hand the gap between the physical models considered in current literature and the available mathematical methods of solving the corresponding wave propagation is far from being closed. Commonly used asymptotic methods do not correctly account for the memory effects, in spite of the fact that the singularity of memory kernels appearing in poroelasticity (and frequently in viscoelasticity) allows asymptotic expansions based on real ray tracing[1,2,3]. This possibility does not exist for regular memory kernels[4]. Finite-difference schemes for porous media[5,6,7] assume a simplified representation of internal friction by a first-order derivative term in the equations. Tentative attempts at including more complicated memory effects via a simple manipulation of the frequency-domain solution and FFT or hidden variable methods (Zener kernel) have been made[8,9,10]. Finite-difference schemes for singular memory can be constructed by modern techniques based on Fractional Calculus and Lubich's discrete convolution methods. We shall present two examples illustrating various methods of accounting for singular memory effects in numerical methods. Both methods use a direct time-domain approach.

In Biot's poroelasticity memory effects are expected to be relevant for signals involving frequencies of order of the Biot critical frequency or exceeding it, hence for frequencies

$f \gtrsim 10^5$ Hz. There is however an increasing theoretical[11,12] and experimental[13] evidence that for rocks, which are heterogeneous porous materials on scales intermediate between pore dimensions and the wavelength, the associated attenuation reduces the threshold of memory effects to frequencies of order of 10 Hz. Consequently memory effects are relevant not only for ultrasound and acoustic well logs but also for seismic wave propagation.

It is shown in Refs[1,2] that the solutions of hyperbolic integro-differential equations with singular convolution kernels are C^∞-smooth at the wavefronts. This is due to the unboundedness of attenuation for frequency tending to infinity.

2. Asymptotic Green's functions for Biot's poroelasticity

2.1. *Introduction. Norris' formula for the time-domain Green's function*

Several papers have been devoted to the calculation of asymptotic Green's functions for Biot's poroelasticity in the time-domain. The solutions were either based on oversimplified equations[14,15,16] or the result was not presented in a form suitable for numerical implementation[17,18]. An elegant frequency-domain expression for the Green's function is presented in Norris' paper[19]. We shall transform Norris' solution to the time domain applying the techniques developed for solutions of integro-differential equations with singular memory[1,2,3].

Biot's equations are assumed in the same formulation as in Norris' paper:

$$
\begin{aligned}
\rho\, u_{i,tt} + \rho_\mathrm{f}\, w_{i,tt} &= \sigma_{ij,j} + \rho\, f_i & \rho_\mathrm{f}\, u_{i,tt} + N_{ij} * w_{j,tt} &= -p_{,i} \\
\zeta_t &= -w_{k,kt} + R & \sigma_{ij} = c_{ijkl}\, e_{kl} + M_{ij}\, \zeta \qquad p &= M_{ij}\, e_{ij} + m\, \zeta
\end{aligned}
\tag{2.1}
$$

where **u** denotes the matrix displacement vector and **w** denotes the relative fluid displacement vector, accounting for porosity. The variable ζ denotes fluid dilatation. We shall consider the initial-value problem in three-dimensional space with zero initial values and the medium at rest before $t < 0$, with a point source term

$$
L\mathbf{y} = \mathbf{F}\,\delta(\mathbf{x})\,H(t); \quad \mathbf{y}(t,\mathbf{x}) = 0 \qquad \text{for } t \le 0
\tag{2.2}
$$

where L denotes the integro-differential operator of Biot's equations (2.1) and $\mathbf{y} = [\mathbf{u}, \mathbf{w}]^\mathsf{T}$.

Norris' Green's function in its general form, applicable to poroelasticity, piezoelasticity and thermoelasticity, is given by the formula:

$$
\mathbf{G}(\mathbf{x}, \omega) =
$$
$$
\frac{1}{8\pi^2 r} \int_\mathcal{C} \sum_{r=1}^N \frac{\mathbf{q}_r \otimes \mathbf{q}_r}{\lambda_r}\, d\theta(\mathbf{n}) + \frac{1}{8\pi^2} \int_\mathcal{H} \sum_{r=1}^N \frac{\mathrm{i}k_r}{\lambda_r} \mathbf{q}_r \otimes \mathbf{q}_r\, \exp\left(\mathrm{i}\,k_r\, \mathbf{n} \cdot \mathbf{x}\right)\, d\Omega(\mathbf{n})
\tag{2.3}
$$

with the integrations extending over the unit circle and the half-sphere

$$
\mathcal{C} := \left\{ \mathbf{n} \mid \mathbf{n}^2 = 1, \mathbf{n} \cdot \mathbf{x} = 0 \right\}
\tag{2.4}
$$
$$
\mathcal{H} := \left\{ \mathbf{n} \mid \mathbf{n}^2 = 1, \mathbf{n} \cdot \mathbf{x} > 0 \right\}
\tag{2.5}
$$

and with

$$\lambda_r := \frac{1}{2} \mathbf{q}_r^\mathsf{T} \left(v_r^2 \mathcal{M} + \mathbf{K} \right) \mathbf{q}_r \tag{2.6}$$

The propagation speeds v_r and the polarization vectors $\mathbf{q}_r$ are defined below.
For the Biot poroelastic medium the matrix

$$\mathbf{K} = \begin{bmatrix} \mathbf{Q} & 0 \\ 0 & \kappa \end{bmatrix} \tag{2.7}$$

is defined in terms of the acoustic tensor

$$Q_{ik}(\mathbf{n}) := \left(c_{ijkl} - m^{-1} M_{ij} M_{kl} \right) n_j n_l \tag{2.8}$$

and the scalar

$$\kappa(\mathbf{n}) := \hat{N}_{ij}^{-1}(\omega) \, n_i \, n_j \tag{2.9}$$

while

$$\mathcal{M} = \begin{bmatrix} \rho \mathbf{I} - \rho_{\mathrm{f}} \hat{\mathbf{N}}^{-1} & 0 \\ 0 & m^{-1} \end{bmatrix}, \quad \mathbf{C} = \begin{bmatrix} 0 & \mathbf{b} \\ \mathbf{b}^\mathsf{T} & 0 \end{bmatrix} \tag{2.10}$$

where

$$b_i := \left[m^{-1} M_{ij} + \rho_{\mathrm{f}} \, \hat{N}_{ij}^{-1}(\omega) \right] n_j \tag{2.11}$$

The dispersion relation assumes the form

$$\det \mathbf{P}(k, \omega) = 0 \tag{2.12}$$

where

$$\mathbf{P}(k, \omega) := \omega^2 \mathcal{M} - \omega k \, \mathbf{C} - k^2 \, \mathbf{K} \tag{2.13}$$

Fixing the frequency ω and the wavenumber vector direction $\mathbf{n}$ eq. (2.12) can be solved for the wavenumber

$$k = k_r(\omega), \quad r = 1, 2, 3, 4 \tag{2.14}$$

The propagation speeds are now given by the formula

$$v_r := \omega / k_r \qquad r = 1, \ldots 4 \tag{2.15}$$

while the polarization vectors are the corresponding normalized null vectors of $\mathbf{P}_r(k, \omega)$

$$\mathbf{P}(k_r, \omega) \, \mathbf{q}_r = 0$$
$$\mathbf{q}_r^2 = 1 \quad \text{for } r = 1, \ldots 4$$

2.2. *Time-domain expressions for asymptotic Green's functions*

Singular memory plays an important role in the transformation from the frequency domain to te time domain. For simplicity we shall only assume memory effects implicit in the viscodynamic operator

$$N_{ij}(\tau) = N_{ij}^{(0)} \, \delta\left(\tau/\tau_{\mathrm{cr}}\right) + N_{ij}^{(1)} \, \left(\tau/\tau_{\mathrm{cr}}\right)_+^{-1/2}/\sqrt{\pi} + N_{ij}^{(2)} \, H(\tau) + N^{(3)} \, \left(\tau/\tau_{\mathrm{cr}}\right)_+^{1/2}/\Gamma(3/2) + \dots$$
$$(2.16)$$

for $\tau \to 0$. The general form of the viscodynamic operator is determined by the presence of viscous boundary layers in the pore fluid flow[15,20] while the shape of the pore cross-section determines the coefficients of the expansion. Matrix viscosity will be neglected because we lack an a priori information about the convolution kernels involved. It can however be taken into account if the associated memory kernels have a similar asymptotic expansion in terms of half-integer powers of τ.

In the frequency-domain the asymptotic expansion of the viscodynamic operator for $\omega \to \infty$ assumes the following form

$$\hat{N}_{ij}(\omega) = N_{ij}^{(0)} + N_{ij}^{(1)} \left(-i\omega/\omega_{\mathrm{cr}}\right)^{-1/2} + N_{ij}^{(2)} \left(-i\omega/\omega_{\mathrm{cr}}\right)^{-1} +$$
$$+ N_{ij}^{(3)} \left(-i\omega/\omega_{\mathrm{cr}}\right)^{-3/2} + \dots \quad (2.17)$$

with $\omega_{\mathrm{cr}} := 2\pi/\tau_{\mathrm{cr}}$. For example, in the case of cylindrical pores in an isotropic matrix

$$\hat{N}_{ij} = \hat{N}(\omega) \, \delta_{ij} \qquad (2.18)$$

$$\hat{N} = -\frac{\rho_{\mathrm{f}}}{\phi} \frac{J_0\left(e^{i\pi/4}(\omega/\omega_{\mathrm{cr}})^{1/2}\right)}{J_2\left(e^{i\pi/4}(\omega/\omega_{\mathrm{cr}})^{1/2}\right)}, \qquad \omega_{\mathrm{cr}} := \eta\,\phi/\left(4K\rho_{\mathrm{f}}\right) \qquad (2.19)$$

The other explicit cases (slits, triangular and rectangular cross-sections) can be found in Ref.[21].

Substituting (2.19) in (2.12) leads to an equation which has an asymptotic expansion in terms of inverse half-integer powers of ω. The dispersion relation can be solved asymptotically by the ansatz

$$k_r = \omega \left[p_r^{(0)} + s^{-1/2}\, p_r^{(1)} + s^{-1}\, p_r^{(2)} + \dots\right] = i \left[s\, p_r^{(0)} + s^{1/2}\, p_r^{(1)} + p_r^{(2)} + \dots\right] \quad (2.20)$$

where $s := -i\omega$. The exponential term in (2.3) can be expanded as follows

$$e^{ik_r \,\mathbf{n}\cdot\mathbf{x}} = e^{-s\, p_r^{(0)}\, \mathbf{n}\cdot\mathbf{x} - s^{1/2}\, p_r^{(1)}\, \mathbf{n}\cdot\mathbf{x}} \; e^{-p_r^{(2)}\, \mathbf{n}\cdot\mathbf{x}} \left[1 - s^{-1/2}\, p_r^{(3)} - s^{-1}\, p_r^{(4)} + \dots\right] \quad (2.21)$$

Note the complex-valued phase function $T_r + \sqrt{2}\, s^{-1/2}\, A_r$, parameterized by two real-valued functions, the plane-wave travel time $T_r = p_r^{(0)}\, \mathbf{n}\cdot\mathbf{x}$ and the plane-wave attenuation $A_r = p_r^{(1)}\, \mathbf{n}\cdot\mathbf{x}/\sqrt{2}$. The double role of A_r as a time delay factor and a high-frequency damping factor is more transparent in the following expression for the first factor in eq. (2.21)

$$\exp\left(i\,\omega\left(T_r + \omega^{1/2}\, A_r\right)\right) \exp\left(-\omega^{1/2}\, A_r\right)$$

The last factor in (2.21) has the form of a series of ray-asymptotic amplitudes of increasing order. The middle factor in eq. (2.21) is independent of frequency and represents an exponential plane-wave amplitude decay.

We note that in Norris' formula(2.3) $\mathbf{n} \cdot \mathbf{x} \geq 0$. The direction of wave propagation is implicitly chosen to coincide with $\mathbf{n}$, hence $p_r^{(0)} > 0$. Thermodynamic arguments imply that $p_r^{(1)}, p_r^{(2)} \geq 0$. The proof is rather awkward and complicated in this formulation of the dispersion relations, based on the displacement-pressure formulation of Biot's theory. Reverting to the equivalent formulation based directly on eqs. (2.1) and using the results of Ref.[22] one can prove that $p_r^{(1)} \geq 0$. Considering both high and low frequencies the inequality $p_r^{(2)} \geq 0$ can be established.

We now substitute (2.21) in (2.3) and transform the result to the time domain.

The time-domain Green's function can be expressed in terms of the functions

$$f_k(t,y) := \frac{1}{2\pi \mathrm{i}} \int_{-\mathrm{i}\infty}^{\mathrm{i}\infty} \mathrm{d}s \, s^{-k/2} \, e^{st - y \, s^{1/2}}, \quad k = 0, 1, 2, \ldots \tag{2.22}$$

defined for $y \geq 0$, cf Ref.[1,2]. The functions f_k are C^∞-smooth and $f(t,y) = 0$ for $y \leq 0$. The functions f_0, f_1, f_2 and f_3 can be expressed in terms of elementary functions and the error function

$$
\begin{aligned}
f_0(t,y) &= \tfrac{y}{2\sqrt{\pi}} \, t^{-3/2} \, e^{-y^2/(4t)}; \quad f_1(t,y) = \tfrac{1}{\sqrt{\pi}} \, t^{-1/2} \, e^{-y^2/(4t)} \\
f_2(t,y) &= \operatorname{erfc}\left(y/\left(2\sqrt{t}\right)\right); \quad f_3(t,y) = \tfrac{2}{\sqrt{\pi}} \, t^{1/2} \, e^{-y^2/(4t)} + y \operatorname{erfc}\left(y/\left(2\sqrt{t}\right)\right)
\end{aligned}
\tag{2.23}
$$

for $t > 0$, with $f_k(t,y) = 0$ for $t < 0$ (Fig. 1). For $\zeta \to 0+$ the functions f_0, f_1 and f_2 converge to the distributions $\delta(t)$, $t_+^{-1/2}/\sqrt{\pi}$ and $\theta(t)$, while $f_k \to t_+^{k/2-1}/\Gamma(n/2)$ for $n \geq 3$ in the sense of distributional convergence. Following Ref.[23] we denote by t_+^λ the function

$$t_+^\lambda = \begin{cases} t^\lambda & t > 0 \\ 0 & t < 0 \end{cases} \tag{2.24}$$

while $\theta(t)$ denotes the unit step function. We can thus view the functions f_k as the result of smoothing the corresponding distributions by the dissipative mechanism implicit in the model.

The time-domain Green's function has the form of an expansion

$$\frac{1}{2\pi \mathrm{i}} \int_{-\mathrm{i}\infty}^{\mathrm{i}\infty} e^{\mathrm{i}k_r \, \mathbf{n} \cdot \mathbf{x}} \, e^{st} \, \mathrm{d}s = g_0 - p_r^{(3)} \, \mathbf{n} \cdot \mathbf{x} \, g_1 - p_r^{(4)} \, \mathbf{n} \cdot \mathbf{x} \, g_2 + \ldots \tag{2.25}$$

in terms of the functions

$$g_k(t, \mathbf{x}) = G_k(t, \mathbf{n} \cdot \mathbf{x}) \tag{2.26}$$

where

$$G_k(t, \xi) := f_k(t - p_r^{(0)} \xi, p_r^{(1)} \xi) \, e^{-p_r^{(2)} \xi} \tag{2.27}$$

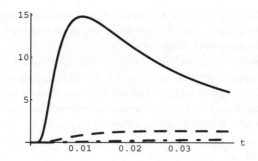

Fig. 1. Basis functions $f_k(t, y, k = 0, 1, 2,)$ for a fixed y.

The inequality $p_r^{(1)}\,\mathbf{n}\cdot\mathbf{x} \geq 0$ satisfied by the integrands of eq. (2.3) ensures that the arguments lie in the domain of definition of the functions G_k.

Eq. (2.3) involves gradient operators acting on g_k. In order to evaluate the result of applying the gradient operators to g_k we shall use the recurrence relations

$$\partial f_k/\partial t = f_{k-2} \tag{2.28}$$
$$\partial f_k/\partial y = f_{k-1} \tag{2.29}$$

Both recursion relations follow easily from the definition (2.22).

3. Determination of the dispersion/attenuation and amplitude decay

For many applications the terms involving G_k, $k > 0$, can be neglected. In this case the frequency-dependent coefficients in (2.3) can be replaced by their elastic parts, but the dispersion relation must be solved to second order in order to determine the parameters $p^{(1)}$ and $p^{(2)}$ appearing in the arguments of f_0. The first of them controls frequency-dependent dispersion and attenuation while the second one controls an exponential amplitude decay.

Using the compact notation

$$\mathcal{M} = \mathcal{M}_0 + s^{-1/2}\,\mathcal{M}_1 + s^{-1}\,\mathcal{M}_2 + \ldots\,; \quad \mathbf{C} = \mathbf{C}_0 + s^{-1/2}\,\mathbf{C}_1 + s^{-1}\,\mathbf{C}_2 + \ldots\,;$$
$$\mathbf{K} = \mathbf{K}_0 + s^{-1/2}\,\mathbf{K}_1 + s^{-1}\,\mathbf{K}_2 + \ldots\,; \quad \mathbf{P} = \mathbf{P}_0 + s^{-1/2}\,\mathbf{P}_1 + s^{-1}\,\mathbf{P}_2 + \ldots \tag{3.30}$$
$$\mathbf{R}_0 = \mathbf{C}_0 + 2\,p^{(0)}\,\mathbf{K}_0 \tag{3.31}$$

the equation for the simultaneous determination of the parameters $p_\alpha^{(1)}$ and $p_\alpha^{(2)}$ of the α-th mode and the associated polarization $\mathbf{w}_\alpha$ assume the following form:

$$\left\{ \mathbf{P}_0 + s^{-1/2}\, \mathbf{P}_1 + s^{-1}\, \mathbf{P}_2 - s^{-1/2}\, p^{(1)}\, \mathbf{R}_0 - s^{-1}\left[\left(p^{(1)2} + 2\, p^{(0)}\, p^{(2)}\right) \mathbf{K}_0 + p^{(2)}\, \mathbf{C}_0 + \right.\right.$$
$$\left.\left. p^{(1)}\, \mathbf{C}_1 + 2\, p^{(1)}\, p^{(0)}\, \mathbf{K}_1\right] + \dots \right\}\left[\mathbf{w}_\alpha^{(0)} + s^{-1/2}\, \mathbf{w}_\alpha^{(1)} + s^{-1}\, \mathbf{w}_\alpha^{(2)} + \dots\right] = 0 \quad (3.32)$$

Taking the scalar product of both sides with $\mathbf{w}_\alpha^{(0)}$ we have

$$p_\alpha^{(1)} = \frac{\mathbf{w}_\alpha^{(0)\mathsf{T}}\, \mathbf{P}_1\, \mathbf{w}_\alpha^{(0)}}{\mathbf{w}_\alpha^{(0)\mathsf{T}}\, \mathbf{R}_0\, \mathbf{w}_\alpha^{(0)}} \quad (3.33)$$

and

$$p_\alpha^{(2)} =$$
$$\frac{\mathbf{w}_\alpha^{(0)\mathsf{T}} \left(\mathbf{P}_1 - p_\alpha^{(1)}\, \mathbf{R}_0\right) \mathbf{w}_\alpha^{(1)} + \mathbf{w}_\alpha^{(0)\mathsf{T}}\, \mathbf{P}_2\, \mathbf{w}_\alpha^{(0)} - \mathbf{w}_\alpha^{(0)\mathsf{T}} \left(p_\alpha^{(1)2}\, \mathbf{K}_0 + p_\alpha^{(1)}\, \mathbf{C}_1 + 2\, p_\alpha^{(0)}\, p_\alpha^{(1)}\, \mathbf{K}_1\right) \mathbf{w}_\alpha^{(0)}}{\mathbf{w}_\alpha^{(0)\mathsf{T}}\, \mathbf{R}_0\, \mathbf{w}_\alpha^{(0)}}$$

$$(3.34)$$

provided the denominators do not vanish. The correction $\mathbf{w}_\alpha^{(1)}$ for the polarization vector can be found from the equation

$$\mathbf{P}_0\, \mathbf{w}_\alpha^{(1)} = -\left(\mathbf{P}_1 - p_\alpha^{(1)}\, \mathbf{R}_0\right) \mathbf{w}_\alpha^{(0)} \quad (3.35)$$

$$\mathbf{w}_\alpha^{(0)\mathsf{T}}\, \mathbf{w}_\alpha^{(1)} = 0 \quad (3.36)$$

provided the eigenvalue the associated is simple.

3.1. The isotropic case

For reasons of simplicity we now consider the Green's function of an isotropic poroelastic medium. The same method however carries over to the anisotropic case without any major modification. Let R denote the radial coordinate, i.e. the distance of the point from the point source.

In an isotropic medium

$$c_{ijkl} = \lambda\, \delta_{ij}\, \delta_{kl} + \mu\left(\delta_{ik}\, \delta_{jl} + \delta_{il}\, \delta_{jk}\right) \quad (3.37)$$
$$b_i = b\, n_i; \quad N_{ij} = N\delta_{ij}, \quad M_{ij} = M\, \delta_{ij} \quad (3.38)$$

Consequently

$$\mathbf{Q} = A\,\mathbf{n} \otimes \mathbf{n} + \mu\left(\mathbf{I} - \mathbf{n} \otimes \mathbf{n}\right) \quad (3.39)$$

with $A = \lambda + 2\mu - m^{-1} M^2$ and $\rho_{\text{eff}} = \rho - \rho_f^2/\hat{N}(\omega)$ and[19]

$$v_1^2 + v_2^2 = (A + b^2 M)/(\rho_{\text{eff}}) + M/\hat{N}(\omega); \quad v_1^2 v_2^2 = A M/\left(\hat{N}(\omega)\,\rho_{\text{eff}}\right);$$
$$v_3 = v_4 = (\mu/\rho_{\text{eff}})^{1/2} \tag{3.40}$$

where $v_\alpha = p_\alpha^{-1}$.

The information about the dispersion and exponential amplitude decay can be extracted from eqs (3.40).

In order to calculate the parameter $p_\alpha^{(1)}$, controling the pulse delay, frequency-dependent parameters must be expanded to order s^{-1}:

$$\hat{N}^{-1} = N^{(0)\,-1}\left[1 - s^{-1/2}\,N_1 + \ldots\right] \tag{3.41}$$

$$\rho_{\text{eff}}^{\;-1} = \rho_{\text{eff}}^{(0)\,-1}\left[1 - s^{-1/2}\,R_1 - \ldots\right]; \quad \rho_{\text{eff}}^{(0)} := \rho - \rho_f^2/N^{(0)}; \quad R_1 := \frac{\rho_f\,N_1}{\rho_{\text{eff}}^{(0)}\,N^{(0)}} \tag{3.42}$$

$$b^2 = b_0^2\left[1 - B_1\,s^{-1/2} - \ldots\right]; \quad b_0 := \frac{\hat{M}}{\hat{m}} + \frac{\rho_f}{N^{(0)}}; \quad B_1 := 2N_1\frac{\rho_f/N^{(0)}}{\hat{M}/\hat{m} + \rho_f/N^{(0)}} \tag{3.43}$$

where $N_1 := N^{(1)}/N^{(0)}$ is a pore shape factor, while $N^{(0)}$ is related to dynamic tortuosity. Expanding eq. (3.40) in an asymptotic series of half-integer powers of s we have

$$p_3^{(1)} = R_1\,p_3^{(0)}/2 \tag{3.44}$$

$$p_\alpha^{(1)} = \Bigg\{\left[B_1\,m\,b_0^2 + R_1\,(A + m\,b_0^2)\right]/\rho_{\text{eff}}^{(0)} + N_1\,m/N^{(0)} + (-1)^\alpha X^{-1/2} \times$$
$$\left[B_1\,m\frac{b_0^2\,(A + b_0^2\,m) + m\,\rho_{\text{eff}}^{(0)}/N^{(0)}}{\rho_{\text{eff}}^{(0)\,2}} + N_1\,m\left[1/N^{(0)} + (m\,b_0^2 - A)/\rho_{\text{eff}}^{(0)}\right]/N^{(0)} + \right.$$
$$\left. R_1\left[m(m\,b_0^2 - A)/N^{(0)} + (A + m\,b_0^2)^2/\rho_{\text{eff}}^{(0)}\right]/\rho_{\text{eff}}^{(0)}\right]\Bigg\}/\left(4v^{(0)\,2}\right), \quad \alpha = 1, 2 \tag{3.45}$$

where

$$X := \frac{A + m\,b_0^2}{\rho_{\text{eff}}^{(0)\,2}} + \frac{m^2}{N^{(0)\,2}} + 2m\frac{m\,b_0^2 - A}{\rho_{\text{eff}}^{(0)}\,N^{(0)}} \tag{3.46}$$

The propagation of the wavefronts is determined by the high-frequency limits $v_\alpha^{(0)}$ of the propagation speeds v_α, obtained by substituting $N^{(0)}$ for $\hat{N}$, $\rho_{\text{eff}}^{(0)}$ for ρ_{eff} and b_0 for b in (3.40). A comparison of the propagatiopn speeds shows that mode 1 corresponds to the fast longitudinal wave, while mode 2 corresponds to the slow wave. The delay of the pulse peak with respect to the wavefront given by the formula[2]:

$$\Delta_\alpha = p_\alpha^{(1)\,2} R^2/6 \tag{3.47}$$

where R denotes the distance of the wavefront from the source.

The frequency domain Green's function for an isotropic poroelastic medium is given in Norris' paper[19]. In the time domain

$$\mathbf{G}(\mathbf{x},t) = \rho_{\text{eff}}^{(0)-1}\left\{ \begin{bmatrix} v_3^{-2}\mathbf{E} + \mathrm{I}^2\,\nabla\otimes\nabla, & 0 \\ 0, & 0 \end{bmatrix} \Phi_3 + \right.$$

$$\left. \frac{m}{v_2^2 - v_1^2}\sum_{r=1}^2 (-1)^r \begin{bmatrix} \mathrm{I}^2\,(\kappa - v_r^2/m)\,\nabla\otimes\nabla, & -b\,\mathrm{I}\nabla \\ -b\,\mathrm{I}\nabla^{\mathsf{T}}, & \rho_{\text{eff}}^{(0)}\,\Omega^{-1} - A\,v_r^{-2} \end{bmatrix} \Phi_r \right\} \quad (3.48)$$

All the frequency dependent coefficients in this formula should now be interpreted as operators of the form $V = V_0\left[1 + W_1\,\mathrm{I}^{1/2} + W_2\,\mathrm{I} + \dots\right]$ where I denotes the indefinite integral:

$$\mathrm{I}f(t) := \int_0^t f(\tau)\,d\tau \quad (3.49)$$

and $\mathrm{I}^{1/2}$ is the fractional integral operator corresponding to multiplication by $s^{-1/2}$ in the frequency domain, $\mathrm{I}^{1/2}\,f_k = f_{k+1}$. In particular, $\rho_{\text{eff}}^{-1} = \rho_{\text{eff}}^{(0)-1}\,\Omega$ where

$$\Omega := 1 - \frac{\rho_f\,N^{(1)}}{\rho_{\text{eff}}^{(0)}\,N^{(0)2}}\,\mathrm{I}^{1/2} + \frac{\rho_f^2}{\rho_{\text{eff}}^{(0)}\,N^{(0)}}\left(\frac{N^{(1)2}}{N^{(0)2}} - \frac{N^{(2)}}{N^{(0)}}\right)\mathrm{I} + \dots \quad (3.50)$$

The functions Φ_r are defined by

$$\Phi_r := \frac{1}{4\pi R}\frac{1}{2\pi i}\int_{-i\infty}^{i\infty} e^{i\,k_r\,R}\,e^{st}\,ds = \frac{1}{4\pi R}\left[g_0 - p_r^{(3)}\,R\,g_1 - p_r^{(4)}\,R\,g_2 + \dots\right] \quad (3.51)$$

with $g_k := G_k(t, R)$. The operator $\rho_{\text{eff}}^{(0)-1}\,\Omega$ corresponds to the factor $1/\rho_{\text{eff}}(\omega)$ in the frequency domain.

Eq. (3.48) can be simplified by applying the identities:

$$\mathrm{I}\nabla g_k = Y := \left(-p_r^{(0)}\,g_k + p_r^{(1)}\,g_{k+1} - p_r^{(2)}\,g_{k+2}\right)\mathbf{x}/R \quad (3.52)$$

$$\mathrm{I}\nabla(g_k/R) = Y/R - g_{k+2}\,\mathbf{x}/R^2 \quad (3.53)$$

$$\mathrm{I}^2\,(\nabla\otimes\nabla)\,g_k = Z \quad (3.54)$$

and

$$\mathrm{I}^2\,(\nabla\otimes\nabla)\,(g_k/R) = Z/R - \left(g_{k+4}/R^2\right)\left(\mathbf{E} - 2\mathbf{x}\otimes\mathbf{x}/R^2\right)$$
$$+ 2\left[-\left(-p_r^{(0)}\,g_{k+2} + p_r^{(1)}\,g_{k+3} - p_r^{(2)}\,g_{k+4}\right) + g_{k+4}/R\right]R^{-2}\,(\mathbf{x}/R\otimes\mathbf{x}/R) \quad (3.55)$$

where

$$Z := \left[-p_r^{(0)}\,g_{k+2} + p_r^{(1)}\,g_{k+3} - p_r^{(2)}\,g_k\right]R^{-1}\left(\mathbf{E} - \mathbf{x}\otimes\mathbf{x}/R^2\right) +$$
$$[p_r^{(0)2}\,g_k - 2\,p_r^{(0)}\,p_r^{(1)}\,g_{k+1} + \left(p_r^{(1)2} + 2\,p_r^{(0)}\,p_r^{(2)}\right)g_{k+2} - 2\,p_r^{(1)}\,p_r^{(2)}\,g_{k+3} + p_r^{(2)2}\,g_{k+4}]\,(\mathbf{x}/R\otimes\mathbf{x}/R) \quad (3.56)$$

In order to calculate the coefficients of G_k, $k \leq k_{\max}$, the frequency-dependent parameters in Norris' formula have to be expanded to the order $k_{\max}$, but the dispersion relation has to be solved to order $k_{\max} + 2$. The final formula is too awkward to be presented here but the expressions given above can be readily programmed for numerical evaluation.

4. Poroacoustics

4.1. *Introduction*

In this section we shall focus on full-wave numerical implementations of problems involving singular memory. A method of constructing finite-difference time-stepping schemes for such problems will be demonstrated for a few poroacoustic models of a rigid gas-saturated matrix. In view of the scalar form of the underlying equations the ray asymptotic solutions derived in Refs[1,2,3,24] are also applicable for an arbitrary integrable singularity of the memory kernel.

4.2. *Equations*

The equations of motion of poroacoustics assume the form

$$\rho * u_{,tt} = \nabla \cdot (K * \nabla u) \tag{4.57}$$

but various physical models found in the literature lead to different memory kernels $\rho(\tau)$ and $K(\tau)$. The kernel $\rho(\tau)$ depends on the viscous effects in the pore flow averaged over the pore cross section, while $K(\tau)$ incorporates the effects of thermal dissipation in the gas[25]. In particular one of Wilson's models[26] is defined by the Laplace-transformed memory kernels

$$\tilde{\rho}(s) = \rho_\infty / \left[1 - C_\rho (\gamma_\rho + s)^{-\mu} \right]; \quad \tilde{K}(s) = K_\infty / \left[1 + C_K (\gamma_K + s)^{-\mu} \right] \tag{4.58}$$

with $0 < \mu < 1$. The thermodynamic condition of non-negative dissipation in the high-frequency limit applied to the two kernels separately implies that $C_\rho, C_K \geq 0$.

Well-posedness of the initial-value problem (IVP)

$$u(t, \mathbf{x}) = 0 \quad \text{for } t < 0 \tag{4.59}$$

$$u(0+, \mathbf{x}) = u_0(\mathbf{x}), \quad u_t(0+, \mathbf{x}) = v_0(\mathbf{x}) \tag{4.60}$$

is established in Ref[27]. We shall construct a generalized finite-difference scheme for numerical solution of the IVP. The construction is based on some elementary facts from Fractional Calculus, presented in the next subsection.

4.3. *Some tools from the Fractional Calculus*

The Laplace-transformed memory kernels of several poroacoustic models[26,28,29] involve the algebraic functions $(s + \gamma)^{\pm\mu}$ with $0 < \mu < 1$. The same function was introduced in the context of Biot's poroelasticity with arbitrary distribution of pore cross-sections by Johnson *et al.*[30]. We shall therefore construct the time-domain operators $(D + \gamma)^\mu$ and

$(I + \gamma)^\mu$ corresponding to these functions and apply them to reformulate the IVP in a numerically implementable form.

The fractional integral

$$I^\mu f(t) := \int_0^t \left[\tau^{\mu-1}/\Gamma(\mu) \right] f(t-\tau) \, d\tau \tag{4.61}$$

is a generalization of an n-fold indefinite integral I^n, where the superscript denotes the power. The Riemann-Liouville fractional derivative[31,32,33] is a left inverse operator for I^μ

$$D^\mu f(t) := D\, I^{1-\mu} f(t) \tag{4.62}$$

where D denotes an ordinary derivative. The above definition is valid for $0 < \mu \leq 1$ only. Since the fractional integrals have the semigroup property $I^\mu I^\nu = I^{\mu+\nu}$ it is easy to see that $D^\mu I^\mu f = D\, I\, f = f$ for sufficiently regular functions f. For the sake of convenience we shall use the notation $D^{-\mu} := I^\mu$. We are now in a position to define the operators

$$(D + \gamma)^{\pm\mu} f(t) := e^{-\gamma t} D^{\pm\mu} \left[e^{\gamma t} f(t) \right] \tag{4.63}$$

Eq. (4.63) reduces to an identity for integer values of μ.

The Laplace transform of $I^\mu f(t)$ is $s^{-\mu} \tilde{f}(s)$. The Laplace transform of $D^\mu f(t)$ is $s^\mu \tilde{f}(s) - \left(I^{1-\mu} f \right)(0+)$. The last term is the limit $\lim_{\epsilon \to 0} I^{1-\mu} f(\epsilon)$ of an integral with an integrable integrand. For a continuous function f this limit vanishes.

Applying the results of the preceding paragraph to (4.63) it is easy to see that the Laplace transformation reduces the operator $(D + \gamma)^{\pm\mu}$, $0 < \mu < 1$, to the multiplication by $(s + \gamma)^{\pm\mu}$, for continuous functions f in the plus case.

In order to examine the relaxation process associated with the operator $(D + \gamma)^\mu$ we consider the fractional differential equation with an initial condition

$$(D + \gamma)^\mu y(t) = f(t) \tag{4.64}$$
$$y(0+) = a \tag{4.65}$$

with given right-hand sides. Applying the Laplace transformation we get the equation

$$(s + \gamma)^\mu \, \tilde{y}(s) - a \, (s + \gamma)^{\mu-1} = \tilde{f}(s) \tag{4.66}$$

whence

$$\tilde{y}(s) = \frac{\tilde{f}(s)}{(s+\gamma)^\mu} + a \, \frac{1}{s+\gamma} \tag{4.67}$$

In the time domain

$$y(t) = a \, e^{-\gamma t} + \int_0^t g(t-\sigma) f(\sigma) \, d\sigma \tag{4.68}$$

where the function $g(t) \div (s + \gamma)^{-\mu}$ can be explicitly calculated by deforming the Bromwich contour $\mathcal{B} = \{s \mid \mathrm{Re}\, s = 0\}$ to the Hankel contour encircling the cut $\mathrm{Im}\, s = 0, \mathrm{Re}\, s < -1/\tau$ and applying the reflection formula for the Gamma function (Ref.[34], 6.1.17):

$$g(t) = \mathrm{e}^{-\gamma t}\, \frac{t^{\mu-1}}{\Gamma(1-\mu)} \quad \text{for } t > 0 \tag{4.69}$$

with $g(t) = 0$ for $t < 0$. Eq. (4.68) represents the relaxation associated with the operator $(\mathrm{D}+\gamma)^{\mu}$.

The generalized fractional operator (4.63) can be discretized by invoking an alternative concept of a fractional derivative, due to Grünwald and Letnikov. Grünwald and, independently, Letnikov defined the fractional derivative by generalizing the finite difference approximation of the ordinary derivatives

$$\mathrm{D}^{\mu}_{\mathrm{RL}} f = \lim_{h \to 0} h^{-\mu}\, (T_{-h} - \mathrm{Id})^{\mu}\, f \tag{4.70}$$

where $T_h\, f(t) := f(t-h)$. The limit usually exists in an averaged sense (in a Lebesgue space). For locally integrable functions the Riemann-Liouville and Grünwald-Letnikov derivatives coincide. It is also important to note that the fractional integral $\mathrm{I}^{\mu} f$ can be similarly approximated by replacing μ by $-\mu$ in eq. (4.70)[35]. A first candidate for a discretized fractional derivative is obtained by expanding the fractional power in a power series

$$\mathrm{D}^{\pm\mu} f(t) = \lim_{h \to 0} h^{\mp\mu} \sum_{n=0}^{\infty} (-1)^n \binom{\pm\mu}{n} f(t - nh) \tag{4.71}$$

with

$$\binom{\alpha}{\beta} := \frac{\Gamma(\alpha+1)}{\Gamma(\beta+1)\,\Gamma(\alpha-\beta+1)}$$

Stirling's formula shows that

$$\binom{\pm\mu}{n} \sim n^{-1\mp\mu}$$

for large n. For the IVP problem at hand the sum breaks off at $n = [t/h]$ (the integer part of t/h). The accuracy of the finite-step Grünwald-Letnikov approximation is $\mathrm{O}[h]$, but higher-order approximations are also available[35].

The discretized version of the generalized fractional operator follows directly from the definition (4.63):

$$(\mathrm{D}+\gamma)^{\pm\mu} f(t) = \lim_{h \to 0} h^{\mp\mu} \sum_{n=0}^{\infty} (-1)^n \binom{\pm\mu}{n} f(t - nh)\, \mathrm{e}^{-nh} \tag{4.72}$$

Note an improved, exponential convergence of the sum.

5. Fractional partial differential equations for the Wilson model

In one-dimensional space eqs (4.57) with the memory kernels (4.58) are equivalent to the following pair of fractional partial differential equations:

$$\left[1 - C_\rho \left(D + \gamma_\rho\right)^{-\mu}\right] v_x = \rho_\infty D^2 u$$
$$\left[1 + C_K \left(D + \gamma_K\right)^{-\mu}\right] v = K_\infty u_x \tag{5.73}$$

where v represents $K * u_x$. A formula in the previous section implies that these equations are consistent with their Laplace or Fourier domain counterparts. Since our task is to formulate a physical model in the time domain rather than to transform one IVP into another, we can safely ignore the initial values in the Laplace domain.

The finite-difference approximation of eqs (5.73) in one dimension is obtained by converting them to the form

$$\left[1 - C_\rho D^{-\mu}\right] \left(e^{\gamma_\rho t} v_x\right) = \rho_\infty e^{\gamma_\rho t} D^2 u$$
$$\left[1 + C_K D^{-\mu}\right] \left(e^{\gamma_K t} v\right) = K_\infty e^{\gamma_K t} u_x \tag{5.74}$$

using eq. (4.63). Substituting the backward Euler derivatives for D^2 and eq. (4.70) for $D^{-\mu}$, and solving for the values at $n h$ we have

$$-\rho_\infty u_n + h^2 \left(1 - h^\mu C_\rho\right) w_n =$$
$$\rho_\infty \left(u_{n-2} - 2 u_{n-1}\right) + h^{2+\nu} C_\rho \sum_{j=1}^{[t/h]} (-1)^j \binom{-\mu}{j} w_{n-j} e^{-j\gamma_\rho h} \tag{5.75}$$

$$-K_\infty \left(u_x\right)_n + \left(1 + h^\mu C_K\right) v_n = -h^\mu C_K \sum_{j=1}^{[t/h]} (-1)^j \binom{-\mu}{j} v_{n-j} e^{-j\gamma_K h} \tag{5.76}$$

where $w = v_x$.

Setting $\mathbf{v} = K * \operatorname{grad} u$, $w = \operatorname{div} \mathbf{v}$ and applying analogous transformations to the three-dimensional eq. (4.57) the following time-stepping scheme is obtained:

$$-\rho_\infty u_n + h^2 \left(1 - h^\mu C_\rho\right) w_n =$$
$$\rho_\infty \left(u_{n-2} - 2 u_{n-1}\right) + h^{2+\mu} C_\rho \sum_{j=1}^{[t/h]} (-1)^j \binom{-\mu}{j} w_{n-j} e^{-j\gamma_\rho h} \tag{5.77}$$

$$-K_\infty \left(\nabla u\right)_n + \left(1 + h^\mu C_K\right) \mathbf{v}_n = -h^\mu C_K \sum_{j=1}^{[t/h]} (-1)^j \binom{-\mu}{j} \mathbf{v}_{n-j} e^{-j\gamma_K h} \tag{5.78}$$

6. Fractional partial differential equations for the Allard-Champoux model

The Allard-Champoux model for an air-saturated rigid-frame porous medium is based on the following dynamic density and bulk modulus

$$\tilde{\rho} = \rho_\infty + a_1 \, s^{-1} \, (s + \gamma_\rho)^{1/2} \tag{6.79}$$

$$\tilde{K}(s)^{-1} = K_\infty^{-1} + \frac{1}{1 + b_2 \, s^{-1} \, (s + \gamma_K)^{1/2}} \tag{6.80}$$

(Ref.[29]). The coefficients a_j, b_k can be expressed in terms of physical parameters[29].

Setting $\mathbf{v} = K * \operatorname{grad} u$ and substituting eqs (6.79–6.80) in eq. (4.57) we obtain a formal pair of equations in the Laplace domain

$$\rho_\infty \, s^2 \, \tilde{u} + a_1 \, (s + \gamma_\rho)^{1/2} \, s \, \tilde{u} = \operatorname{div} \tilde{\mathbf{v}} \tag{6.81}$$

$$\left\{ K_\infty^{-1} \left[s + b_2 \, (s + \gamma_K)^{1/2} \right] - b_1 \, s \right\} \tilde{\mathbf{v}} = \left[s + b_2 \, (s + \gamma_K)^{1/2} \right] \operatorname{grad} \tilde{u} \tag{6.82}$$

Since the physical models are derived in the Fourier domain we can ignore the initial conditions in eqs (4.57) and formulate the problem in the time domain as follows

$$\rho_\infty \, \ddot{u} + a_1 \, (D + \gamma_\rho)^{1/2} \, \dot{u} = \operatorname{div} \mathbf{v} \tag{6.83}$$

$$\left\{ K_\infty^{-1} \left[D + b_2 \, (D + \gamma_K)^{1/2} \right] - b_1 \, D \right\} \mathbf{v} = \left[D + b_2 \, (D + \gamma_K)^{1/2} \right] \operatorname{grad} u \tag{6.84}$$

Substituting eqs (4.62,4.70) and $\dot{u}_k \cong (u_k - u_{k-1})/h$ and solving for the variables at $t = n h$, we have

$$\left(\rho_\infty + a_1 \, h^{1/2} \right) u_n - h^2 \, (\operatorname{div} \mathbf{v})_n =$$

$$\rho_\infty \, (2u_{n-1} - u_{n-2}) + a_1 \, h^{1/2} \, u_{n-1} - a_1 \, h^{1/2} \sum_{j=1}^{[t/h]} (-1)^j \binom{1/2}{j} e^{-j \gamma_\rho \, h} \times \tag{6.85}$$

$$(u_{n-j} - u_{n-j-1})$$

$$\left(K_\infty^{-1} - b_1 + h^{1/2} \, b_2/K_\infty \right) \mathbf{v}_n - \left(1 + b_2 \, h^{1/2} \right) (\operatorname{grad} u)_n =$$

$$\left(K_\infty^{-1} - b_1 \right) \mathbf{v}_{n-1} - (\operatorname{grad} u)_{n-1} - b_2 \, h^{1/2} \sum_{j=1}^{[t/h]} (-1)^j \binom{1/2}{j} e^{-j \gamma_K \, h} \times \tag{6.86}$$

$$\left[K_\infty^{-1} \, \mathbf{v}_{n-j} - (\operatorname{grad} u)_{n-j} \right]$$

The high frequency limit of the Biot-Allard model is somewhat more complicated than the models considered in Refs[2,3]. It involves a memory effect in the refraction index. In the Laplace domain, ignoring the terms involving the initial values we have

$$\left(\rho_\infty \, s^2 + a_1 \, s^{3/2} \right) \tilde{u} = \operatorname{div} \left[\frac{s^{1/2} + b_2}{\left(K_\infty^{-1} - b_1 \right) s^{1/2} + b_2/K_\infty} \operatorname{grad} \tilde{u} \right] \tag{6.87}$$

Assuming however that b_1, b_2 and K_∞ are constant we recover the equations of Refs[2,3], which implies that the equations (6.79–6.80) are hyperbolic with C^∞-smooth solutions.

7. A related model with an exact analytic solution

For testing the accuracy of the time-stepping algorithm it is convenient to compare the result of numerical integration which has an exact explicit solution. Exact solutions can be constructed for some modified versions of the equations discussed above.

Applying a method from Ref.[3] it is possible to construct a fractional partial differential equation with an exact solution of the initial-value problem $u(0, \mathbf{x}) = 0$, $u_t(0, \mathbf{x}) = \delta(\mathbf{x})$. We assume that all the coefficients are independent of $\mathbf{x}$.

Consider a wavefield of the form

$$u(t, \mathbf{x}) = \frac{1}{4\pi r} \frac{1}{2\pi i} \int_B w(s) \, e^{s\,(t - r\,\phi(s))} \, ds \tag{7.88}$$

In view of the identity [3]

$$\nabla^2 U(r) = -\delta(\mathbf{x}) + s^2 \, \phi(s)^2 \, U(r) \tag{7.89}$$

where

$$U(r) = \frac{1}{4\pi r} \, e^{-s\,r\,\phi(s)} \tag{7.90}$$

eq. (7.88) satisfies the equation (4.57) with the initial data $u(0, \mathbf{x}) = 0$, $u_t(0, \mathbf{x}) = \delta(\mathbf{x})$ provided

$$\tilde{\rho} = \tilde{K}\, w \tag{7.91}$$
$$w = \phi^2 \tag{7.92}$$

Choosing $\phi = a + b\,s^{-1}\,(\gamma + s)^{1/2}$ and

$$\tilde{\rho} = \rho_\infty \left[a + b\,s^{-1}\,(\gamma + s)^{1/2} \right] \tag{7.93}$$

$$\tilde{K} = K_\infty \left[a + b\,s^{-1}\,(\gamma + s)^{1/2} \right]^{-1} \tag{7.94}$$

we have an analytic solution of the problem

$$u(t, \mathbf{x}) = \left[a^2 \, f_0(t - ar, br, \gamma) - 2a\,b\, f_{-1}(t - ar, br, \gamma) + b^2 \, f_{-2}(t - ar, br, \gamma) \right] / (4\pi r) \tag{7.95}$$

where

$$f_0(t, y, \gamma) = \frac{y}{2\sqrt{\pi}} \, \theta(t) \, t^{-3/2} \, e^{-\gamma t} \, e^{-y^2/4t} \tag{7.96}$$

$$f_{-1}(t, y, \gamma) = D^{-1} \frac{\partial}{\partial y} f_0(t, y, \gamma) \tag{7.97}$$

$$f_{-2}(t, y, \gamma) = D^{-2} \frac{\partial^2}{\partial y^2} f_0(t, y, \gamma) \tag{7.98}$$

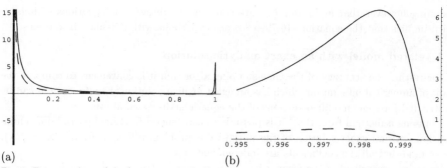

Fig. 2. Two snapshots of the fundamental solution at $t = 1.0$ (solid line) and $t = 2.0$ (dashed line) in terms of the scaled distance $x = r/t$; for $\gamma = 100\,\mathrm{s}^{-1}$. (a): a full snapshot (b): its wavefront part.

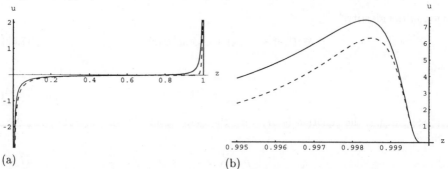

Fig. 3. Comparison of the snapshots at $t = 1.0$ for $\gamma = 100$ (solid line) and $\gamma = 0$ (dashed line). (a) a full snapshot, (b) its wavefront part.

where D^{-1} is the indefinite integral (from 0 to t) and $\theta(t)$ denotes the Heaviside function.

In the model considered here the wavefield exhibits exactly the same behavior at the wavefront as the models considered in Refs[2,3], but it has an additional exponential damping of the signal tail, which effectively shortens the signal duration. The damping rate depends on the characteristic frequency γ and is a low-frequency correction. In contrast, the parameter b controls the smoothing of the signal at the wavefront. The model has two characteristic time scales a^2/b^2 and $1/\gamma$ and the behavior of the solutions is sensitive to the relation between the two. In Figs 3 two snapshots of the fundamental solution (7.95) expressed in terms of a parameter ζ, defined by the relation $\gamma = \zeta\, b^2/a^2$, are shown. The values of the slowness $a = 1\,\mathrm{s/km}$ and the attenuation parameter $b = 0.1\,\mathrm{s/km}^2$ are fixed, while $\zeta = 0$ and 10^4. The value $\zeta = 0$ ($\gamma = 0$) corresponds to the model considered in Ref[2].

In Fig. 2 two snapshots of the fundamental solution (7.95) for $a = 1\,\mathrm{s/km}$, $b = 0.1\,\mathrm{s/km}^2$ taken at $t = 1\,\mathrm{s}$ and $t = 2\,\mathrm{s}$ are compared. The fundamental solution is expressed in terms of the scaled distance $x = r/t$.

7.1. *Conclusions*

An alternative Biot dynamic viscosity factor is suggested in Ref.[36]

$$F(s) = 1 + 0.25 (\tau s)^{1/2} \tag{7.99}$$

The above expression for $F(s)$ allegedly matches the low- and high-frequency asymptotics as well as the high porosity limit of the exact expressions. The corresponding fractional operator is $(1 + \tau^{1/2} D^{1/2}) u - (\tau/\pi t)^{1/2} u(0+)$, where u denotes the filtration velocity. The initial value term can be discarded and the resulting fractional differential equations fall within the scope of Refs[2,3].

The method of developing finite-difference schemes presented above can be adapted to the stress relaxation functions for porous media with fractal pore surfaces suggested in Refs[30,37,26]. The heuristic fractal Wilson model[26] requires replacing the operator $(D + 1/\tau_j)^{-\nu}$ by a product $(D + \gamma_j^{(1)})^{-\mu_1} (D + \gamma_j^{(2)})^{-\mu_2}$ for $j = \rho, K$, with $\mu_1 + \mu_2 = 1/2$, $\mu_1 = (3 - d)/2$ (d denotes the fractal dimension of the pore surface). The model of Johnson *et al.*[30] leads a dynamic tortuosity operator $\alpha = \alpha_\infty \left(1 + C_j D^{-1}(D + 1/\tau)^{-(\delta - 1/2)}\right)$, with a dynamic fractal dimension δ satisfying the inequality $2 \leq \delta \leq d < 3$. Finally, the paper[37] assumes the Cole-Cole transfer function[38], which requires the operator $\tau^\nu D^\nu + 1$, $0 \leq \nu \leq 1$ and its inverse, with $\nu = 1/d$.

The discretization method applied to poroacoustics illustrates the efficacy of fractional calculus methods for developing numerical schemes for a class of poroelastic and viscoelastic models involving an algebraic dependence on frequency.

Higher accuracy schemes can be based on Lubich's discretized convolution calculus[39,40], often however at the expense of a slower rate of decay of the weights appearing in the sums representing convolutions with memory kernels. Lubich's convolutions are more convenient in the context of viscoelastic problems, where they do not affect directly the construction of the time-stepping scheme.

Acknowledgments

The research was supported by Norsk Hydro under the contract NHT-B44-5003496-00 A travel grant from Meltzer fond is also gratefully acknowledged.

References

1. A. Hanyga and M. Seredyńska. *Wave Motion* **30**, pp. 175–195 (1999).
2. A. Hanyga and M. Seredyńska. *Geophys. J. Int.* **137**, pp. 319–335 (1999).
3. A. Hanyga and M. Seredyńska, "Asymptotic wavefront expansions in hereditary media with singular memory kernels." To appear in *Quart. Appl. Math.* (2001).
4. A. Hanyga and M. Seredyńska, *Pure appl. Geophys.* **157**, pp. 679–717 (2000).
5. S. Hassanzadeh, *Geophysics* **56**, pp. 424–435 (1991).
6. X. Zhu and G. E. McMechan, *Geophysics* **56**, pp. 328–339 (1991).
7. J. M. Carcione and G. Quiroga-Goode, *J. Computational Acoustics* **3**, pp. 261–280 (1995).
8. J. M. Carcione and G. Quiroga-Goode, *Geophysical Prospecting* **44**, pp. 99–129 (1996).
9. J. M. Carcione, *J. Acoust. Soc. Am.* **99**, pp. 2655–2666 (1996).

318

10. J. M. Carcione, *Eur. J. Mech., A/Solids* **12**, pp. 53–71 (1993).
11. B. Gurevich and S. L. Lopatnikov, *Geophys. J. Int.* **121**, pp. 933–947 (1997).
12. S. Gelinsky, S. A. Shapiro, T. Müller and B. Gurevich, *Int. J. Solids and Structures* **35**, pp. 4739–4751 (1998).
13. M. Batzle, R. Hofmann, D.-H. Han and J. Castagna, *The Leading Edge* **20**, pp. 168–171 (2001).
14. R. Burridge and C. A. Vargas, *Geophys. J. R. astr. Soc.* **58**, pp. 61–90 (1979).
15. A. N. Norris, *J. Acoust. Soc. Am.* **121**, pp. 359–370 (1993).
16. G. D. Manolis and D. E. Beskos, *Acta Mechanica* **76**, pp. 89–104 (1989).
17. G. Bonnet, *J. Acoust. Soc. Am.* **82**, pp. 1758–1762 (1987).
18. C. Boutin, G. Bonnet and P. Y. Bard, *Geophys. J. R. astr. Soc.* **90**, pp. 521 550 (1987).
19. A. N. Norris, *Proc. Roy. Soc. London Ser. A* **447**, pp. 175–188 (1994).
20. A. Hanyga, "Asymptotic theory of wave propagation in viscoporoelastic media," in *Theoretical and Computational Acoustics '97*, eds. Y.-C. Teng, E.-C. Shang, Y.-H. Pao, M. H. Schultz and A. D. Pierce (World-Scientific, Singapore) (1999).
21. M. Stinson, *J. Acoust. Soc. Am.* **89**, pp. 550 558 (1991).
22. A. Hanyga and M. Seredyńska, "Thermodynamics and Asymptotic Theory of Wave Propagation in Viscoporous Media," in *Theoretical and Computational Acoustics '97*, eds. Y.-C. Teng, E.-C. Shang, Y.-H. Pao, M. H. Schultz and A. D. Pierce (World-Scientific, Singapore) (1999). Proc. 3rd Int. Conf. on Computational and Theoretical Acoustics, Newark, NJ, July 14–18, 1997.
23. I. M. Gel'fand and G. E. Shilov, *Generalized Functions*, vol. I (Academic Press, New York) (1964).
24. A. Hanyga, *J. Comput. Acoustics* **9**, pp. 495–513 (2001).
25. K. Attenborough, *J. Acoust. Soc. Am.* **73**, pp. 785–799 (1983).
26. D. K. Wilson, *J. Acoust. Soc. Am.* **94**, pp. 1136–1145 (1992).
27. A. Hanyga. "Wave propagation in media with singular memory." To appear in *J. Comp. Mech. and Math.* (2001).
28. D. K. Wilson, *Applied Acoustics* **50**, pp. 171 188 (1997).
29. J.-F. Allard and Y. Champoux, *J. Acoust. Soc. Am.* **91**, pp. 3346–3353 (1992).
30. D. L. Johnson, J. Koplik and R. Dashen, *J. Fluid Mech.* **176**, pp. 379–402 (1987).
31. K. S. Miller and B. Ross, *An Introduction to the Fractional Calculus and Fractional Differential Equations* (J. Wiley & Sons, New York, 1993).
32. S. G. Samko, A. A. Kilbas and O. I. Marichev, *Fractional Integrals and Derivatives, Theory and Applications* (Gordon and Breach, Amsterdam, 1993).
33. I. Podlubny, *Fractional Differential Equations*, (Academic Press, San Diego, 1999).
34. M. Abramowitz and I. Stegun, *Mathematical Tables* (Dover, New York, 1970).
35. R. Gorenflo. "Fractional Calculus: Some Numerical methods", in *Fractals and Fractional Calculus in Continuum Mechanics*, eds. A. Carpinteri and F. Mainardi (Springer, Wien, 1997).
36. J. G. Berryman, L. Thigpen and R. C. Chin, *J. Acoust. Soc. Am.* **84**, pp. 360–373 (1988).
37. C. Ruffet, Y. Gueguen and M. Darot, *Geophysics* **56**, pp. 758–768 (1991).
38. K. S. Cole and R. H. Cole, *J. Chem. Phys.* **9**, pp. 341 351 (1941).
39. C. Lubich, *Numer. Math.* **52**, pp. 129–146 (1988).
40. C. Lubich, *Numer. Math.* **52**, pp. 413–426 (1988).

Theoretical and Computational Acoustics 2001
E.-C. Shang, Qihu Li and T. F. Gao (Editors)
© 2002 World Scientific Publishing Co.

An Anisotropic Cole-Cole Model of Seismic Attenuation

Andrzej Hanyga

Institute of Solid Earth Physics, University of Bergen,

Allégaten 41, N5007 Bergen, Norway

A simple model of seismic wave attenuation combining anisotropy with anelastic effects is constructed. The anelastic response is based on the Cole-Cole relaxation function. Time-stepping finite-difference and ray-asymptotic methods of numerical solution are discussed.

1. Introduction

Current technologies of seismic aquisition allow detection of anisotropic and anelastic effects in seismic wave propagation. Anisotropy and anelasticity play an important role in the AVO. In this paper mechanical behavior of rocks is represented by a highly inhomogeneous fluid- or gas-saturated porous medium. For the purposes of seismic modeling details of relative fluid motion are often irrelevant and the rock response is expected to be adequately represented by a linear hereditary viscoelastic medium.

The importance of memory effects in fluid-saturated rocks in the seismic frequency range has been recently demonstrated[1]. On the other hand, our knowledge of the microstructure and fluid motion in rocks provides some physical models of the relaxation and creep of rocks. Unfortunately these models are far too complicated for rigorous mathematical treatment and numerical implementation. Furthermore, many current physical models of rocks (such as Biot's poroelasticity) do not account for the attenuation observed in rocks, while attempts at incorporating additional attenuation mechanisms (squirt, fast-to-slow wave conversion) lead to difficult or incomplete mathematical models that can hardly be formulated in terms of initial/boundary value problems for a system of differential or pseudo-differential equations. For example, squirt[2] is often formulated in terms of flow orthogonal to the wave vector, which is an asymptotic concept and cannot be expressed in terms of the differential or pseudo-differential operator. The final result of the Gurevich-Lopatnikov model[3] is a dispersion relation, which can be expressed as a integro-differential equation in the spirit of hereditary viscoelasticity (that is, in terms of differential and convolution operators) only after appropriate simplifications[4]. It is therefore reasonable to choose a middle way between elasticity and sophisticated physical models and develop phenomenological anelastic anisotropic models, whose solutions have the important properties of the physical models but are amenable to rigorous analytic and numerical treatment. A relation between the phenomenological coefficients and the physical parameters (permeability, porosity, saturation, subwavelength inhomogeneities) should be the objective of a separate investigation.

Physical models provide some guidance for the choice of the model. The problem of estimating the order of magnitude of the critical frequency, at which the behavior of the medium changes, is more difficult. The expression for the critical frequency in Biot's theory involves readily measurable physical parameters, such as permeability, porosity and viscosity. Substituting the known values one comes up with a value of critical frequency well above the seismic range but relevant for the ultrasonic ranges. All the models of additional non-Biot attenuation involve some additional parameter that is hard to estimate. The models accounting for sub-wavelength inhomogeneities[3,5] involve a correlation length of the inhomogeneity distribution. The squirt model and related models involve the characteristic squirt-flow length[2]. The first of these characteristic lengths can perhaps be estimated from well-log data but the second one has to be found by matching the theoretical response of the rock with the observed one. Additional uncertainties are introduced by ad hoc assumptions about the pore shape. It is therefore necessary to rely on laboratory experiments. It is however expected that the characteristic frequencies in both cases are rather low and lie in the seismic frequency range, especially for heavy oil saturated rocks in the case of squirt flow[3,6]. Relevance of memory effects for seismic frequencies is also indicated by recent experimental evidence[1].

We can thus conclude that a correct seismic modeling of porous rocks should account for memory effects. The physical models are however not suitable for solving wave propagation problems. The best we can do is a phenomenological model inspired by the physical models.

Since our information about seismic anisotropy and attenuation comes from different data or from different processing of these data it is reasonable to treat them separately. The final set of equations for seismic wave propagation must however include both kinds of effects. We shall therefore construct the model by combining elastic anisotropy with anelastic effects expressed in terms of scalar relations between special types of strains and stresses[7]. Incidentally, this approach also simplifies the problem of ensuring causality of the anelastic response.

Seismic attenuation can be modeled by a hereditary viscoelastic medium. The stress-strain law in such a medium is represented by a relaxation law. Among various phenomenological relaxation laws in use some have adequate flexibility and economic parameterization. The first of these properties is absent in the Standard Linear Solid model of relaxation. For the second property, an important requirement to be imposed on phenomenological laws is relevance of all the parameters for wave propagation and the possibility of reliable parameter estimation. This requirement is not met by the methods of Day and Minster[8] and Emmerich and Korn[9], adopted in seismology merely on account of their convenience in numerical modeling. The last method is based on an oversimplified Debye model of relaxation. Debye relaxation is very uncommon in condensed matter, as shown in many papers and books dedicated to this subject[10,11,12,13].

Simple speculations based on general physical principles[14,15,12] and on statistics[16,17,18,19,20] suggest some non-Debye universal relaxation laws. In contrast to the approach of Day and Minster[8] and Emmerich and Korn[9] such theories have a predictive power since they imply certain relations that can be confirmed or refuted experimentally. In dielectric relaxation such models imply some universal asymptotic relations between the imaginary and real part of complex susceptibility and their derivatives[12,21,22], which have been confirmed by experiments.

In particular, one has to realize that the actual relaxation involved in delayed response is seldom a molecular relaxation, expressible in terms of pointwise exponential relaxation. More likely, we are dealing with cumulative effects of many relaxation effects, leading either to the Kohlrausch-Williams-Watts (KWW) law[12,17,23], introduced on experimental basis in the context of creep tests already in mid-19th century[24,25] and in the context of dielectric relaxation of polymers[26], or to the Cole-Cole law and its generalizations. Polymers, metals and semiconductors mostly conform to a general law with the asymptotics $(-i\omega)^{-\alpha}$ for $\omega \to \infty$ and $(-i\omega)^{-\beta}$ for $\omega \to 0$ with $0 < \alpha, \beta \leq 1$ (Refs[27,28]). All these relaxation laws can be represented by the Havriliak-Negemi law (Ref.[12])

$$\hat{\chi}(\omega) = \left[1 + (-i\omega\tau)^{\beta} \right]^{-\alpha/\beta} \tag{1.1}$$

The Cole-Cole law is a special case of the Havriliak-Negemi law with $\alpha = \beta$. For rocks, nearly flat loss otherwise known as constant Q is most often reported[29,30]. This might be due to fractal microstructure of rocks. It will be shown in Sec. that the constant Q property can be approximated by a Cole-Cole law.

The most controversial aspect of the Emmerich-Korn model is the parallel arrangement of relaxing mechanisms. Hierarchic combinations of relaxing mechanisms, in which a number of faster relaxations may trigger a slower one, were considered by Palmer et al.[31] (cf. also Klafter and Shlesinger[32]). Hierarchic structures arise quite naturally in spatially distributed relaxation due to cluster structure and a hierarchy of relaxations, which implies, for example, that larger subsystems relax before their components. There are several theoretical explanations of the Kohlrausch-Williams-Watts relaxation law, commonly observed in mechanical and dielectrical relaxation. These models are based either on tentative dynamical models[31,33,34] or, because of their universality encompassing a large variety of phenomena, on purely statistical considerations[21,22,35,36,37,38]. Statistical arguments are attractive because they lead to simple abstract probabilistic models of the relaxation spectrum consistent with experimental evidence. A generalized central limit theorem reduces the number of possible relaxation spectra to a stable probability law with one or two parameters. The most satisfactory and rigorous statistical analysis of the models which lead to the KKW law is given in Weron et al.[39].

Relaxation phenomena in complex and disordered media are most often distributed in space and involve some sort of local diffusion or heat propagation[14,33,15]. If anomalous diffusion is allowed for, these considerations lead to a general version of the Cole-Cole law. Normal diffusion leads to a special case of the Cole-Cole law. A spatially non-local Cole-Cole law, parametrically depending on the wavelength, was derived directly from the Continuous Random Walk Model of anomalous diffusion by K. Weron and Kotulski[19] (cf also Gomi and Yonezawa[40], Fujiwara and Yonezawa[41], K. Weron and Jurlewicz[38]). Non-locality resulted from the fact that the random walk problem was solved in an unbounded space. Replacing this problem by a physically more plausible boundary value problem for in subsets of finite size one obtains a superposition of Mittag-Leffler functions with characteristic time scales determined by the eigenvalues of the spatial part of the diffusion operator. Relaxation is then approximately governed by the slowest relaxation. This relaxation law should then be averaged over an ensemble of random dimensions of the subsets. One can thus expect a weighted sum of Mittag-Leffler relaxation functions with the slowest one dominating.

The Cole-Cole law has been very successful in modeling dispersive properties of dielectrics[10] and anelastic mechanical response in polymers[42]. It allows a very accurate modeling of electromagnetic or mechanical response over a very large range of frequencies provided the material exhibits only a single transition region and a single minimum of the Q parameter. It is therefore increasingly popular with seismologists applying phenomenological models of the dispersive stress-strain relation in rocks[43,1]. In a related context of dielectric properties of porous media it appears in Ref.[44]

The Cole-Cole law has the advantage of a very economical parameterization: it involves four parameters, the glassy and rubbery modulus, a coefficient related to the characteristic frequency at the transition zone and the exponent, controling the width of the transition zone and, *eo ipso*, the depth of the maximum of the specific dissipation Q^{-1}.

The Cole-Cole law developed below can be expressed in terms of fractional derivatives. Consequently, it can be discretized by the use of Grünwald-Letnikow derivatives or by a discrete implementation of convolutions with generalized Mittag-Leffler exponentials. Using this property we shall present an algorithm for solving the anisotropic Cole-Cole equations by generalized finite differences in time and an appropriate scheme in spatial variables. For some values of the exponent the anisotropic Cole-Cole model also allows solution by ray asymptotic methods. For the exponent 1/2 this follows directly from the results of Hanyga and Seredyńska[45,46]. This value of the exponent is fairly common as it results from attenuation associated with viscous boundary layers, e.g. in Biot's poroelasticity. For the exponents 2/3 and 1/3 one can solve a scalar equation for each mode separately[47] and couple all the modes at interfaces. The exponent 2/3 is very close to the values observed in creep tests and has been found[4] as the best power law fit to the Gurevich-Lopatnikov dispersion law[3]. In contrast to the exponent 1/2 the reasons for the frequent occurrence of the exponents $\cong 2/3$ is not known.

We shall therefore use the method of Carcione *et al.*[7] for separating the anisotropy from the anelastic response and apply the Cole-Cole law to model the latter. In Sec. 2 the relevant properties of the Cole-Cole transfer function and the Cole-Cole relaxation function are discussed. In Sec. 3 the specific dissipation is discussed. It is showed that the constant Q model is fairly well approximated by the Cole-Cole model. In Sec. 4 the Cole-Cole law is incorporated in a general anisotropic stress-strain constitutive relation. Thermodynamic restrictions are discussed in Sec. 5. Two generalized FD time-stepping schemes are constructed in Sec. 6.

2. The Cole-Cole law

The Cole-Cole transfer function can be expressed in the following form

$$M(\omega) = M_0 \frac{1 + a\,(-i\omega\tau)^{-\alpha}}{1 + (-i\omega\tau)^{-\alpha}} \tag{2.2}$$

where $M(\omega)$ stands for some frequency-dependent modulus (e.g. Young's modulus in Bagley and Torvik's polymer model), M_0 is the high-frequency limit of $M(\omega)$, τ represents a characteristic relaxation time and $0 < \alpha < 1$. Thermodynamic arguments[42,48] imply that the exponent α in the numerator and in the denominator are equal. Furthermore, the relaxed modulus $M_\infty = a\,M_0$ must be smaller than the instantaneous modulus M_0. This is known both from thermodynamics[49,42] and from familiar experimental evidence of stress relaxation. Incidentaly, we have found a physical interpretation of the parameter $a = M_\infty/M_0$. The exponent α controls the width of the transition zone between M_0 and M_∞.

The asymptotic behavior of the Cole-Cole transfer function for $\omega \to \infty$

$$M(\omega) \sim M_0 \left[1 - (1 - a)\,(-i\omega)^{-\alpha}\right] \tag{2.3}$$

The first term corresponds to an immediate elastic response. The delayed response is dominated by a power law, which is a well-known universal law for relaxations, including mechanical relaxation[12]. Universality of this feature has prompted many scientists to develop a statistical rather model of relaxation based on anomalous diffusion or, more directly, on continuous time random walks[41,40,19].

The Cole-Cole relaxation law can also be expressed in terms of a time convolution kernel. Let $E_{\gamma,\beta}(z)$ denote the generalized Mittag-Leffler function[50,51]

$$E_{\gamma,\beta}(z) := \sum_{n=0}^{\infty} \frac{z^n}{\Gamma(\gamma n + \beta)} \tag{2.4}$$

for arbitrary complex z. The Mittag-Leffler function proper $E_\gamma(z) := E_{\gamma,1}(z)$ is a special case of the generalized Mittag-Leffler function.

The identity

$$\int_0^\infty e^{-st}\, t^{\gamma-1}\, E_\gamma\left(-b\,t^\gamma\right)\,\mathrm{d}t = \frac{s^\gamma}{s^\gamma + b} \tag{2.5}$$

(op. cit.) implies that

$$\int_0^\infty e^{-st}\, t^{\gamma-1}\, E_\gamma'\left(-b\,t^\gamma\right)\,\mathrm{d}t = \frac{1}{s^\gamma + b}$$

In view of the relation

$$M = M_0 \left[1 + (a - 1)\frac{1}{1 + (\tau s)^\alpha}\right] \tag{2.6}$$

with $s = -i\omega$, the Cole-Cole relaxation law can be expressed as a time convolution with the kernel

$$\breve{M}(t) = M_0 \left[\delta(t) + \alpha\frac{1 - a}{\tau}\,(t/\tau)^{\alpha-1}\, E_\alpha'\left(-(t/\tau)^\alpha\right)\right]$$

$$\equiv M_0 \left[\delta(t) - \frac{1 - a}{\tau}(t/\tau)^{\alpha-1}\, E_{\alpha,\alpha}\left(-(t/\tau)^\alpha\right)\right] \tag{2.7}$$

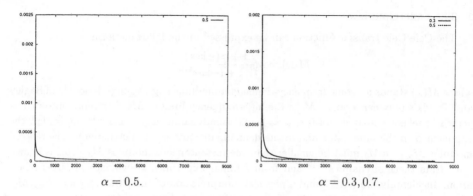

$$\alpha = 0.5. \qquad\qquad\qquad \alpha = 0.3, 0.7.$$

Fig. 1. The Cole-Cole relaxation function for $\alpha = 0.3, 0.5, 0.7$.

and the relaxation function is $R(t) = M_0 [1 + (1 - a) E_\alpha (-(t/\tau)^\alpha)]$. For $f(t)$ such that $t^\alpha f(t) \to 0$ for $t \to \infty$ and f is right-continuous at 0, the expression $M(\omega) \hat{f}(\omega)$ translates to

$$a\, M_0\, f(t) + (1 - a)\, M_0 \int_0^\infty E_\alpha \left(-(\theta/\tau)^\alpha\right) f'(t - \theta)\, d\theta \qquad (2.8)$$

in the time domain (use Theorem 1.4 of Ref.[51]).

In Appendix B it is shown that $E_{\alpha,\alpha}(-z^\alpha) \geq 0$ for $z > 0$. In view of the inequality $a < 1$ this implies that the relaxation function $M(t)$ represents an *actual* stress relaxation: $M(t) \leq M(0)$ and $M(t)$ is non-increasing.

The function $E_\alpha \left(-(t/\tau)^\alpha\right)$ is a natural fractional generalization of the exponential function $\exp(-t/\tau)$. The Mittag-Leffler relaxation function has a small-time asymptotic behavior similar, but not identical, to the stretched exponential of the Kohlrausch-Williams-Watts law

$$E_\alpha \left(-(t/\tau)^\alpha)\right) \sim 1 - (t/\tau)^\alpha/\Gamma(\alpha+1) \quad \text{vs} \quad \exp\left(-(t/\tau)^\alpha\right) \sim 1 - (t/\tau)^\alpha \qquad (2.9)$$

and a power-type large-time asymptotic behavior $E_\alpha \left(-(t/\tau)^\alpha\right) \sim (t/\tau)^{-\alpha}/\Gamma(1 - \alpha)$ (the asymptotics can be found in Ref.[51]. In Fig. 1 we have plotted a few examples of this fuction calculated by an algorithm developed for the Mittag-Leffler function in Ref.[52], (an alternative but more cumbersome method is presented in Ref.[54]). For numerical applications it is however more important to express the Cole-Cole relaxation function in terms of fractional differential operators, which can then be discretized.

3. Specific dissipation function $1/Q$ for the Cole-Cole law

For small attenuation the specific dissipation function Q can be approximated by the loss tangent

$$Q^{-1} \cong \tan \delta = \operatorname{Im} M(\omega)/\operatorname{Re} M(\omega) \qquad (3.10)$$

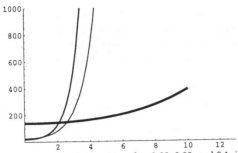

Fig. 2. Q^{-1} as a function of frequency for three values of α, 0.66, 0.33, and 0.1, in the logarithmic frequency scale (n represents 10^n Hz).

cf Ref[55]. In Fig. 2 the values of Q^{-1} for various values of the exponent α are shown. For small values of the exponent the transition zone becomes very wide and the specific attenuation Q becomes effectively constant in the seismic frequency range.

4. The anisotropic Cole-Cole model

The stiffness tensor c_{klmn} can be considered as a 6×6 positive semidefinite symmetric matrix C_{rs}, for example by rewriting it in the Voigt representation. The matrix C_{rs} can be decomposed in terms of dyadics $v_r^{(p)} v_s^{(p)}$, where $v^{(p)}$ denote the unit eigenvectors of C_{rs}. Reverting to tensor notation for the three-dimensional space

$$c_{klmn}^0 = \sum_{p=1}^{6} \Lambda_p^0 e_{kl}^{(p)} e_{mn}^{(p)} \tag{4.11}$$

where Λ_p^0, $p = 1, \ldots 6$, are the eigenvalues of C_{rs} and the eigenstresses $e_{kl}^{(p)}$ are the eigenvectors $v_r^{(p)}$ expressed in 3×3 tensor notation. They are assumed unitary in the sense of

$$e_{kl}^{(p)} e_{kl}^{(q)} = \delta_{pq} \tag{4.12}$$

Ref[56]. An eigenstrain is defined by the property that the corresponding stress differs from it by a scalar factor Λ.

We shall now use the shorthand notation $E = \{e_{kl}\}$, $E^{(p)} = \{e_{kl}^{(p)}\}$, $\langle E, E' \rangle = e_{kl} e_{kl}'$ and $C[E] = \{c_{klmn} e_{mn}\}$. For the elastic part of C we shall reserve the symbol C^0. Let $\mathcal{G}$ denote the symmetry group of the elastic medium defined by the stiffness tensor C^0. For a solid in a non-deformed configuration[57] $\mathcal{G} \subset SO(3)$ and the property $G \in \mathcal{G}$ is equivalent to $C\left[\hat{G}[E]\right] = \hat{G}\left[C[E]\right]$ for every symmetric E, where $\hat{G}[E] := G E G^{\mathsf{T}}$ is a rotation in the six dimensional space of symmetric tensors E, Refs[58,57,59]. Obviously

$$C^0[E] = \sum_{p=1}^{6} \Lambda_p^0 \langle E, E^{(p)} \rangle E^{(p)} \tag{4.13}$$

or, in a more compact notation, $C^0 = \sum_p \Lambda_p^0 E^{(p)} \otimes E^{(p)}$.

The eigenvalues Λ_p^0 can be divided into groups in such a way that Λ_p^0, Λ_r^0 are equal if they belong to the same group and have different values otherwise. We now associate the parameters a_p, α_p, τ_p with each group and define

$$\Lambda_p := \Lambda_p^0 \frac{1 + a_p\,(-\mathrm{i}\omega\tau_p)^{-\alpha_p}}{1 + (-\mathrm{i}\omega\tau_p)^{-\alpha_p}} \tag{4.14}$$

The anelastic stiffness coefficients are now given by the formula

$$C = C(\omega) = \sum_p \Lambda_p E^{(p)} \otimes E^{(p)} \tag{4.15}$$

In view of Theorem A.1 every $Q \in \mathcal{G}$ is a material symmetry of C' and vice versa, i.e. $C(\omega)$ and C^0 have the same material symmetry for every ω.

This method of introducing rheological effects in an anisotropic medium was applied in Carcione et al.[7], albeit without a rigorous justification given here.

5. Stability and dissipation

Multiplying both sides of the momentum balance

$$\rho \ddot{u}_k = \sigma_{kl,l} \tag{5.16}$$

by $\dot{u}_i$ and summing over i we have

$$\frac{1}{2}\rho \frac{\mathrm{d}\dot{\mathbf{u}}^2}{\mathrm{d}t} + \dot{u}_{i,j}\,c_{ijkl} * u_{k,l} + F_{j,j} = 0$$

where $F_j := -\dot{u}_i\,c_{ijkl} * u_{k,l}$ is the energy flux. Substituting $c_{ijkl} = c_{ijkl}^0\,\delta(t) + c_{ijkl}^{\mathrm{D}}(t)$ in the above identity, we get

$$\frac{\mathrm{d}E}{\mathrm{d}t} + \operatorname{div}\mathbf{F} + D = 0 \tag{5.17}$$

where

$$E = \frac{1}{2}\left[\dot{\mathbf{u}}^2 + c_{ijkl}^0\,u_{i,j}\,u_{k,l}\right] \tag{5.18}$$

is the energy density and

$$D = \dot{u}_{i,j}\,c_{ijkl}^{\mathrm{D}} * u_{k,l} \tag{5.19}$$

is the energy dissipation rate. The requirement that in a periodic quasi-plane wave the time-averaged dissipation rate is non-negative leads to Graffi's inequality

$$\operatorname{Im}\hat{c}_{ijkl}^{\mathrm{D}} \leq 0 \tag{5.20}$$

where the inequality states that the Fourier transform of the matrix-valued convolution kernel is negative semi-definite[60]. The matrix has to be interpreted as indexed by two indices ij and kl (or in the Voigt 6×6 form). The same condition can be derived from the consideration of hypothetical quasi-static cycles[49].

Applying inequality (5.20) to the stiffness coefficients (4.15) we have

$$\operatorname{Im} \hat{c}^{D}_{ijkl} = \operatorname{Im} \hat{c}_{ijkl} = \sum_p \Lambda_p \, e^{(p)}_{ij} \, e^{(p)}_{kl} \leq 0 \tag{5.21}$$

whence

$$\operatorname{Im} \Lambda_p = \Lambda^0_p \operatorname{Im} \frac{1 + a_p \, (-i\omega\tau_p)^{-\alpha_p}}{1 + (-i\omega\tau_p)^{-\alpha_p}} = \frac{\operatorname{Im} \left[a_p \, (-i\omega\tau_p)^{-\alpha_p} + (i\omega\tau_p)^{-\alpha_p} \right]}{|1 + (-i\omega\tau_p)^{-\alpha_p}|^2} < 0$$

or

$$\operatorname{Im} \left[a_p \, e^{i\pi\alpha_p/2} + e^{-i\pi\alpha_p/2} \right] = -(1 - a_p) \, \sin(\pi\alpha_p/2) < 0$$

Since $0 < \alpha_p \leq 1$, Graffi's inequality is equivalent to $a_p \leq 1$.

6. Time-stepping in the generalized finite-difference implementation

We shall discuss here only the time-stepping aspect of numerical schemes for the anisotropic Cole-Cole model.

The equations of motion are obtained by substituting in the energy balance (5.16) the constitutive relation

$$\sigma_{kl} = \sum_p \Lambda^0_p \, V_p * e^{(p)}_{kl} \tag{6.22}$$

where

$$\hat{V}_p = \frac{1 + a_p \, (-i\omega\tau_p)^{-\alpha_p}}{1 + (-i\omega\tau_p)^{-\alpha_p}} \, \hat{W}_p; \quad W_p = e^{(p)}_{kl} \, u_{k,l} \tag{6.23}$$

The circumflex denotes the Fourier transformation from the time domain to the frequency domain.

The operator (6.23) will now be expressed in the time domain. Convolution with a generalized exponential is not convenient in a numerical implementation. The time-domain counterpart of $(-i\omega)^{-\alpha}$ is the fractional integral operator I^α:

$$(I^\alpha f)(t) := \frac{1}{\Gamma(\alpha)} \int_0^\infty \theta^{\alpha-1} f(t - \theta) \, d\theta \tag{6.24}$$

Unfortunately, the dicretization of $I^\alpha f$ involves a series of terms with the k-th term of order $k^{-1+\alpha}$.

A better formulation can be obtained by expressing (6.23) in terms of Riemann-Liouville fractional derivatives[61,51]:

$$D^\alpha f = D \, I^{1-\alpha} f \tag{6.25}$$

where $D = \partial/\partial t$ and $0 < \alpha < 1$. The Laplace transform of the Riemann-Liouville derivative (6.25) is $s^\alpha \tilde{f}(s) - \lim_{\epsilon \to 0} I^{1-\alpha} f(\epsilon) = s^\alpha \tilde{f}(s)$ if f is right-continuous at 0.

Assume that the wavefield is at rest for $t < 0$. Identifying s with $-i\omega$ and multiplying both sides of eq. (6.23) by $(-i\omega\tau)^{\alpha_p} + 1$ we have after passing to the time domain

$$(\tau^{\alpha_p} D^{\alpha_p} + 1) V_p = (\tau^{\alpha_p} D^{\alpha_p} + a_p) W_p \tag{6.26}$$

For $V_p, W_p \in \mathcal{L}^q(\mathbb{R})$, $q \geq 1$, at fixed $\mathbf{x}$, the Riemann-Liouville derivative coincides with the Grünwald-Letnikov fractional derivative

$$D^\alpha f(t) = \lim_{h \to 0} h^{-\alpha} \sum_{k=0}^{\infty} (-1)^k \binom{\alpha}{k} f(t - kh) \tag{6.27}$$

where the limit is taken in the $\mathcal{L}^q$ sense for some $q \geq 1$. For a finite step h the error is $O[h]$.

Standard procedures lead to the following time-stepping algorithm:

$$W_p^n = e_{kl}^{(p)} u_{k,l}^n \tag{6.28}$$

$$[1 + (\tau/h)^{\alpha_p}] V_p^n = [a_p + (\tau/h)^{\alpha_p}] W_p^n + (\tau/h)^{\alpha_p} \sum_{k=1}^{[t/h]} (-1)^k \binom{\alpha_p}{k} W_p^{n-k} - $$
$$-(\tau_p/h)^{\alpha_p} \sum_{k=1}^{[t/h]} (-1)^k \binom{\alpha_p}{k} V_p^{n-k} \tag{6.29}$$

$$u_k^{n+1} = 2u_k^n - u_k^{n-1} + (h^2/\rho) \sum_p \Lambda_p^0 V_{p,l}^n e_{kl}^{(p)} \tag{6.30}$$

where $[t/h]$ denotes the largest integer $\leq t/h$. The computation of W_p^n from u_k^n and the computation of $V_{p,l}^n$ can be implemented by a pseudospectral method[62,63] or by FD.

By the reflection formula for the Gamma function and Stirling's asymptotic formula[64]

$$\binom{\alpha}{k} = \frac{\Gamma(\alpha + 1)}{k! \, \Gamma(\alpha + 1 - k)} \sim \pi^{-1} \sin\left(\pi(k - \alpha)\right) \Gamma(1 + \alpha) \, e^{\alpha - 1} \, k^{-1-\alpha}$$

Consequently, the best performance is expected at $\alpha \cong 1$.

A more efficient and accurate scheme can be obtained applying the method of operational convolutions[66,67]. For this purpose we express the operator $1/[1 + (\tau s)^\alpha]$ as a discrete convolution in the time domain

$$V_p^n = M_0 \left[W_p^n - \frac{1 - a}{\tau} \sum_{k=0}^{n} \Omega_k(h) W_p^{n-k} \right] \tag{6.31}$$

for $0 \leq n \leq N$, where N is a fixed number corresponding to the upper limit $T = Nh$ of the time range considered. The expressions $\Omega_k(h)$ are defined by the formula

$$\frac{1}{1 + [\tau g(z)/h]^\alpha} = \sum_{k=0}^{\infty} \Omega_k(h) \, z^k \tag{6.32}$$

where $g(z)$ is the generating function of a multistep method which is q-consistent, with $1 \le q \le 6$, and has appropriate stability properties[66]. In particular

$$g(z) = \sum_{l=1}^{q} \frac{1}{l}(1-z)^l \tag{6.33}$$

has the required properties. For small $k \le 10$ the expressions $\Omega_k(h)$ can be calculated symbolically from the defining formula, but a large number of them are needed. Consequently, they are calculated approximately by the FFT using the formula

$$\Omega_k(h) = \rho^{-k} \frac{1}{N} \sum_{k=0}^{N-1} F_n e^{-2\pi i k n / N} \tag{6.34}$$

where

$$\rho = \epsilon^{1/(2N)} \tag{6.35}$$

$$F_n = \frac{1}{1 + \left[\tau g(\rho e^{2\pi i n/N})/h\right]^\alpha} \tag{6.36}$$

and ϵ is the relative accuracy of F_n, $n = 0, \ldots N-1$. The entire set of weights $\Omega_k(h)$ can be generated at the beginning and stored for repeated use in FFT convolutions. The convolution has marginally better convergence properties since the time-domain counterpart $\theta^{\alpha-1} E_{\alpha,\alpha}(-\theta^\alpha)$ of $1/(1+s^\alpha)$ is asymptotically $-\theta^{-1} - O\left[\theta^{-1-\alpha}\right]$, cf[51]. Eq. (6.29) can therefore be replaced by

$$V_p^n = W_p^n - (1-a) \sum_{k=0}^{n} \Omega_k(h) W_p^{n-k} \tag{6.37}$$

A possible choice of parameters is $N = 512$ or 1024, $h = T/N$, where $[0, T]$ is the range of time of interest.

The discrete convolution implicit in the above approach can be eliminated by applying FFT to $\{V_p^n\}$, multiplication by the Cole-Cole transfer function and an inverse FFT to obtain $\{W_p^n\}$.

7. Conclusions

The Cole-Cole relaxation function has been incorporated in an anisotropic viscoelastic equation of motion. The memory effects can be handled by several alternative time-stepping algorithms. The spatial discretization can be based on pseudospectral or finite-difference methods. For some values of the exponent ray-asymptotic methods are also available.

Acknowledgment

Financial support by the Norsk Hydro AS under the contract NHT-B44-5003496-00 is gratefully acknowledged.

Appendix A

The method of incorporating rheological effects in the anisotropic stress-strain relations in Sec. depends on the following theorem:

Theorem A.1 *Let $\mathcal{G}$ be the material symmetry group of the stiffness tensor*

$$C = \sum_{r=1}^{6} \lambda_r\, E^{(r)} \otimes E^{(r)}$$

and suppose that the coefficients μ_r satisfy the condition
$\lambda_r = \lambda_s$ implies that $\mu_r = \mu_s$.

$\mathcal{G}$ is contained in the material symmetry group of the stiffness tensor

$$C' := \sum_{r=1}^{6} \mu_r\, E^{(r)} \otimes E^{(r)}$$

Proof. The strains $E^{(r)}$, $r = 1, \ldots, 6$, are eigenstrains of C. For every $Q \in \mathcal{G} \subset SO(3)$

$$C\left[\hat{Q}\left[E^{(r)}\right]\right] = \hat{Q}\left[C[E]\right] = \lambda_r\, \hat{Q}\left[E^{(r)}\right]$$

Consequently $\hat{Q}\left[E^{(r)}\right]$ is an eigenstrain of C with the same eigenvalue λ_r and

$$\hat{Q}\left[E^{(r)}\right] = \sum_{\{s|\lambda_s=\lambda_r\}} \alpha_{rs}(Q)\, E^{(s)} \tag{A.1}$$

From the identity

$$C\left[\hat{Q}[E]\right] = \sum_r \lambda_r\, \langle E^{(r)}, \hat{Q}[E]\rangle E^{(r)} = \sum_r \lambda_r \langle \widehat{Q^{-1}}\left[E^{(r)}\right], E\rangle =$$

$$\sum_r \sum_s {}' \lambda_r \alpha_{rs}\left(Q^{-1}\right) \langle E^{(s)}, E\rangle E^{(r)} =$$

$$\hat{Q}\left[C[E]\right] = \sum_r \langle E^{(r)}, E\rangle\, \hat{Q}\left[E^{(r)}\right] = \sum_r \sum_s {}' \lambda_r\, \alpha_{rs}(Q)\, \langle E^{(r)}, E\rangle E^{(s)}$$

where $\sum_s'$ is a shorthand for $\sum_{\{s|\lambda_s=\lambda_r\}}$, we conclude that

$$\alpha_{rs}\left(Q^{-1}\right) = \alpha_{sr}(Q)$$

We are now ready to prove that every $Q \in \mathcal{G}$ is a material symmetry for C':

$$C'\left[\hat{Q}[E]\right] = \sum_r \mu_r \langle E^{(r)}, \hat{Q}[E]\rangle E^{(r)} = \sum_r \sum_s {}' \mu_r\, \alpha_{rs}\left(Q^{-1}\right) \langle E^{(s)}, E\rangle E^{(r)} =$$

$$\sum_r \sum_s {}' \mu_r\, \alpha_{sr}(Q) \langle E^{(s)}, E\rangle E^{(r)}$$

Renaming s and r to r and s and using the assumption on μ_r, the right-hand side becomes

$$\sum_s \sum_r{}' \mu_s \langle E^{(s)}, E \rangle \hat{Q} \left[E^{(s)} \right] = \hat{Q} \left[C[E] \right]$$

which proves q.e.d. $\square$.

Appendix B

Theorem B.1 *Let* $0 < \alpha < 1$.
The function $f(z) := z^{\alpha-1} E_{\alpha,\alpha}(-z^\alpha)$ *is non-negative.*

Proof. The proof is based on the Bernstein theorem[68]. According to the Bernstein theorem it is sufficient to prove that the Laplace transform $h(y) := 1/(1+y^\alpha)$ of $f(z)$ is completely monotone, i.e. $(-1)^n \, d^n h(y)/dy^n \geq 0$ for $n = 0, 1, 2 \ldots$ and $y \geq 0$. The function $h(z) = f(g(z))$, where $f(x) := 1/(1+x)$, and $g(z) := z^\alpha$. Now f is completely monotone and g is a Bernstein function: $g \geq 0$ and g' is completely monotone. Indeed, $f \geq 0$, $(-1)^n f^{(n)}(x) = n! \, (1+x)^{-n-1} \geq 0$ for $n = 1, 2 \ldots$. On the other hand, $g \geq 0$, $g' \geq 0$ and then the derivatives of successive orders have opposite signs (provided $0 < \alpha < 1$.

It is a well-known fact[69], easy to verify by elementary calculi, that a superposition $f \circ g$ of a completely monotone function f and a Bernstein function g is competely monotone. $\square$.

References

1. M. Batzle, R. Hofmann, D.-H. Han and J. Castagna, *The Leading Edge* **20**, pp. 168–171 (2001).
2. G. Mavko, T. Mukerji and J. Dvorkin, *The Rock Physics Handbook* (Cambridge University Press, Cambridge) (1998).
3. B. Gurevich and S. L. Lopatnikov, *Geophys. J. Int.* **121**, pp. 933–947 (1997).
4. A. Hanyga and V. E. Rok, *J. Acoust. Soc. Amer.* **107**, pp. 2965–2972 (2000).
5. S. Gelinsky and S. A. Shapiro, *Geophys. J. Int.* **128**, pp. F1–F4 (1997).
6. Z. Wang and A. Nur, *J. Acoust. Soc. Am.* **87**, pp. 2384–2395 (1990).
7. J. M. Carcione, F. Cavallini and K. Helbig, *Acustica - acta acustica* **84**, pp. 495–502 (1996).
8. S. M. Day and J. B. Minster, *Geophys. J. R. astr. Soc.* **78**, pp. 105–118 (1984).
9. M. Emmerich and M. Korn, *Geophysics* **52**, pp. 1252–1264 (1987).
10. K. S. Cole and R. H. Cole, *J. Chem. Phys.* **9**, pp. 341–351 (1941).
11. A. K. Jonscher, *Dielectric Relaxation in Solids* (Chelsea Dielectrics Press, London, 1983).
12. A. K. Jonscher, *Universal Relaxation Law* (Chelsea Dielectrics Press, London, 1996).
13. T. V. Ramakrishnan, *Non-Debye Relaxation of Condensed Matter* (World-Scientific, Singapore, 1984).
14. M. Y. Kelbert and I. Y. Chaban, *Izv. Ak. Nauk, ser. Mechanics of fluids and gases* **5**, pp. 153–160 (1986).
15. M. Kelbert and I. Sazonov, *Pulses and Other Wave Processes in Fluids: An Asymptotical Approach to Initial Problems* (Kluwer Academic Publishers, Dordrecht, 1996).
16. A. Weron and K. Weron, "A statistical approach to relaxation in glassy materials," in *Mathematical Statistics and Probability Theory*, eds. P. Bauer. *et al* (Springer Verlag, Berlin, 1987), pp. 245–254.
17. K. Weron, *Acta Physica Polonica* **A70**, pp. 529–539 (1986).

332

18. V. U. Uchaikin and V. M. Zolotarev, *Chance and Stability: Stable Distributions an their Applications* (VSP, Utrecht, 1999).
19. K. Weron and M. Kotulski, *Phys. A* **232**, pp. 180–188 (1996).
20. W. Glöckle and T. Nonnnenmacher, *J. Stat. Phys.* **71**, pp. 741–757 (1993).
21. E. W. Montroll and J. T. Bendler, *J. Stat. Physics* **34**, pp. 129–162 (1984).
22. A. Jurlewicz and K. Weron, *J. Stat. Phys.* **73**, pp. 69–81 (1993).
23. K. L. Ngai, "Evidences for universal behaviour of condensed matter at low frequencies/long times," in *Non-Debye Relaxation in Condensed Matter*, eds. T. V. Ramakrishnan and M. R. Lakshmi (Bangalore 1987), pp. 23–191.
24. R. Kohlrausch, *Annalen der Physik* **12**, pp. 393– (1847).
25. F. Kohlrausch, *Poggendorfer Annalen (Annalen der Physik und Chemie)* **119**, pp. 337– (1863).
26. G. Williams and D. C. Watts. *Trans. Faraday Soc.* **66**, pp. 80– (1970).
27. J. D. Ferry. *Viscoelastic Properties of Polymers* (J. Wiley & Sons, New York, 1980).
28. R. M. Hill, *J. Materials Science* **17**, pp. 3630– (1982).
29. J. H. Schoen, *Physical Properties of Rocks: Fundamentals and Principles of Petrophysics* (Pergamon, London, 1996).
30. R. S. Carmichael, *Practical Handbook of Physical Properties of Rocks and Minerals* (CRC Press, Boca Raton, 1996).
31. R. G. Palmer, D. Stein, E. S. Abrahams and P. W. Anderson, *Phys. Rev. Lett.* **53**, pp. 958–961 (1984).
32. M. F. Shlesinger and J. Klafter, "The nature of hierarchies underlying relaxation in disordered systems," in *Fractals in Physics*, eds. L. Pietronero and E. Tosatti (North-Holland, Amsterdam, 1986), pp. 393–398.
33. J. Klafter and M. F. Shlesinger, *Proc. Natl Acad. Sci USA* **83**, pp. 848–851 (1986).
34. R. Metzler and J. Klafter, *Physics Reports* **339**, pp. 1–77 (2000).
35. K. Weron, *J. Phys.: Condens. Matter* **3**, pp. 9151–9162 (1991).
36. K. Weron, *J. Phys.: Condens. Matter* **3**, pp. 221–223 (1991).
37. K. Weron, *J. Phys.: Condensed Matter* **4**, pp. 10507–10512 (1993).
38. K. Weron and A. Jurlewicz, *J. Phys. A: Math. Gen.* **26**, pp. 395–410 (1993).
39. A. Weron, K. Weron and W. A. Woyczynski, *J. Stat. Phys.* **78**, pp. 1027–1038 (1995).
40. S. Gomi and F. Yonezawa, *Phys. Rev. Lett.* **74**, pp. 4125–4128 (1995).
41. S. Fujiwara and F. Yonezawa, *Phys. Rev. Lett.* **74**, pp. 4229–4232 (1995).
42. R. L. Bagley and P. J. Torvik, *J. of Rheology* **27**, pp. 201–210 (1983).
43. T. Toverud and B. Ursin, "Estimation of viscoelastic parameters from zero-offset VSP data," in *Norwegian Petroleum Research Seminar, Bergen 15-17 Nov. 2000* (2000), pp. 79–82.
44. C. Ruffet, Y. Gueguen and M. Darot, *Geophysics* **56**, pp. 758–768 (1991).
45. A. Hanyga and M. Seredyńska, *Wave Motion* **30**, pp. 175–195 (1999).
46. A. Hanyga and M. Seredyńska, *Geophys. J. Int.* **137**, pp. 319–335 (1999).
47. A. Hanyga and M. Seredyńska, "Asymptotic wavefront expansions in hereditary media with singular memory kernels", To appear in *Quart. Appl. Math.* (2001).
48. R. L. Bagley and P. J. Torvik, *J. of Rheology* **30**, pp. 133–155 (1986).
49. M. Fabrizio and A. Morro, *Mathematical Problems in Linear Viscoelasticity* (SIAM, Philadelphia, 1992).
50. K. S. Miller and B. Ross, *An Introduction to the Fractional Calculus and Fractional Differential Equations* (J. Wiley & Sons, New York, 1993).
51. I. Podlubny, *Fractional Differential Equations* (Academic Press. San Diego, 1998).
52. K. Diethelm and Y. Luchko, "Numerical solution of linear multi-term differential equations of fractional order." Preprint (2001).
53. C. Lubich, *SIAM J. Math. Anal.* **17**, pp. 704–719 (1986).

54. R. Gorenflo, I. Louchko and Y. Luchko, "Numerische Berechnung der Mittag-Leffler Funktion $E_{\alpha,\beta}(z)$ und ihrer Ableitung," Preprint http://www.math-fu.berlin.de/publ/preprints/1998 (1998).

55. C. M. Zener, *Elasticity and Anelasticity of Metals* (Chicago University Press, Chicago, 1948).

56. K. Helbig, *Foundations of Anisotropy for Exploration Seismics* (Pergamon, London, 1994).

57. C. Truesdell, *A First Course in Rational Mechanics* (John Hopkins University, Baltimore, MA, 1972).

58. A. J. M. Spencer, *Continuum Mechanics* (Longmans, London, 1980).

59. A. Hanyga, *Mathematical Theory of Nonlinear Elasticity* (Ellis Horwood, Chichester, 1984).

60. A. Hanyga and M. Seredyńska, "Thermodynamics and Asymptotic Theory of Wave Propagation in Viscoporous Media," in *Theoretical and Computational Acoustics '97*, eds. Y.-C. Teng, E.-C. Shang, Y.-H. Pao, M. H. Schultz and A. D. Pierce (World-Scientific, Singapore, 1999).

61. S. G. Samko, A. A. Kilbas and O. I. Marichev, *Fractional Integrals and Derivatives, Theory and Applications* (Gordon and Breach, Amsterdam, 1993).

62. D. Funaro, *Polynomial Approximation of Differential Equations* (Springer-Verlag, 1992).

63. B. Fornberg, *A Practical Guide to Pseudospectral Methods* (Cambridge University Press, Cambridge, 1996).

64. M. Abramowitz and I. Stegun, *Mathematical Tables* (Dover, New York,1970).

65. R. Gorenflo. *Fractional Calculus: Some Numerical methods*, in *Fractals and Fractional Calculus in Continuum Mechanics*, eds. A. Carpinteri and F. Mainardi (Springer, Wien, 1997).

66. C. Lubich, *Numer. Math.* **52**, pp. 129–145 (1988).

67. C. Lubich, *Numer. Math.* **52**, pp. 413–425 (1988).

68. D. V. Widder, *The Laplace Transform* (Princeton University Press, Princeton, 1946).

69. J. Prüss, *Evolutionary Integral Equations* (Birkhäuser Verlag, Basel, 1993).

Theoretical and Computational Acoustics 2001
E.-C. Shang, Qihu Li and T. F. Gao (Editors)
© 2002 World Scientific Publishing Co.

Quantifying the Effects of Pore Structure and Fluid Saturation on Acoustic Wave Velocity in Carbonates

Y. F. Sun[1,*], J. L. Massaferro[2], G. Eberli[3], and Y. C. Teng[4]

[1]Lamont-Doherty Earth Observatory, Columbia University, Palisades, NY 10964, USA

[2]Shell International Exploration and Production B.V., Volmerlaan 8, 2280 AB Rijswijk, The Netherlands

[3]Rosenstiel School of Marine and Atmospheric Science, University of Miami, Miami, FL 33149, USA

[4]Aldridge Laboratory of Applied Geophysics, Columbia University, NY, NY 10025, USA

*Author for correspondence: *sunyf@ldeo.columbia.edu*

Abstract

Recent studies reveal that the effect of pore structure on acoustic wave velocity at low effective pressure can be much greater than that of porosity in fractured and/or porous rocks. For two limestones of a given porosity, the difference in compressional velocity can be as large as 2.5 km/s or even larger. In such cases, seismic interpretation for lithological analysis and fluid detection are much complicated. In this paper, we propose a theoretical model for the velocity-porosity relationship including the effect of pore structure on acoustic wave propagation. A formation geometrical factor is introduced to characterize the influence of pore structure on wave velocity, which is effectively quantified as a polynomial function of porosity. We show that pore structure plays an important role in controlling sonic and seismic velocities, as observed in sonic and acoustic measurements. Using numerical modeling, we conclude that the pore structure effects on wave velocity profoundly affect acoustic signals in both amplitude and phase. These results can help to explain many pitfalls of amplitude versus offset (AVO) analysis.

Introduction

Changes in ocean chemistry, sea level, and climate are best recorded in carbonates. The structural and petrophysical properties of carbonates provide clues and footprints of past changes in physiochemical events, fauna, depositional facies, diagenesis, sea level fluctuations, and climatic events. About half of the world's largest hydrocarbon-producing reservoirs are also carbonates. Because of such importance of carbonate platforms, advances in carbonate geology have been made in last several decades (e.g., Choquette and Pray, 1970; Simo et al., 1993). The international Ocean Drilling Program (ODP) has devoted several carbonate legs to investigating the cause and effects of sea level changes and flow processes in carbonate platforms (e.g., Davies et al., ODP Leg 133, 1991; Sager, et al., ODP Leg 143, 1993; Premoli Silva et al., ODP Leg 144, 1993; Eberli et al., ODP Leg 166, 1997; Feary et al., ODP Leg 182, 1999; Isern and Anselmetti, ODP Leg 194, 2001).

Despite the scientific and economic significance of carbonate formations, "predicting porosity in subsurface carbonates in frontier basins remains difficult because current diagenetic models are largely qualitative, rather than quantitative" (Budd et al., 1995). It has been known for a long time that this unpredictability of porosity in carbonates is caused by the diversity of rock structures existing in carbonate formations. The quantitative relationship between seismic velocity and in-situ petrophysical properties in carbonates, however, is far less understood than in siliciclastic rocks. For two limestones of a given porosity, the differences in seismic compressional velocity can be as large as 2.5 km/s or even larger. Many such observations in carbonates sharply contradict intuition. And many theoretical and empirical models of velocity-porosity relationship developed in the last century are proven to be less successful for carbonate formation than for siliciclastic rocks and other very-low-porosity rocks (Wang, et al., 1991; Anselmetti and Eberli, 1993).

A handful of experimental studies on carbonates have been published recently (Rafavich et al., 1984; Wang et al., 1991; Anselmetti and Eberli, 1993, 1997). Anselmetti and Eberli (1993, 1997) measured acoustic properties of over 300 carbonate samples and conclude the important effect of pore types on velocity. Wang et al. (1998) combine core velocity measurements with the Gassmann model (Gassmann, 1951) for application of seismic monitoring of a CO_2 flood in a carbonate reservoir. Ramamoorthy and Murphy (1998) use log data and the Gassmann model to study the feasibility of fluid detection using 4D or time-lapse seismic studies in carbonate reservoirs. They all emphasize the importance of pore structure effects on wave velocities.

However, few attempts have been made to quantify the pore-structure effects on acoustic measurements. Recently, Sun and Goldberg (1997a) reported some theoretical results compared with classical experiments on low-porosity limestones and provide quantitative explanations of the dynamic effect of the changes of aspect ratio and pore size on seismic velocity. Their work is essentially based upon a dynamical theory of fractured porous media developed by Sun (1994), using a topological characterization of pore structures (Sun and Goldberg, 1997b). In this report, we briefly summarize the theoretical model and provide the basic working formula. We apply this theoretical model to sedimentary carbonates, using log and seismic data obtained in a production carbonate reservoir field.

Theoretical Model

Carbonates possess much more complex pore structures than sandstones. Whereas this structural complexity provides richer geological and petrophysical information than otherwise plain sands, it presents immense difficulty for systematic quantitative analysis. It remains a challenging task to analyze and interpret laboratory and field measurements on carbonates. In contrast to sandstones for which three factors (porosity, fluid content, and rock mineralogy) are practically enough to predict seismic velocities (V_p and V_s), recent studies show that two more factors describing the pore structures (e.g., aspect ratio and pore size) are indispensable to evaluate successfully the seismic responses in carbonates and many other fractured hard rocks (e.g., Sun and Goldberg, 1997a, 1997b).

It is commonly known that besides rock mineralogy and fluid type, the major factor controlling velocity variations is porosity. This is true for both suspensions and unconsolidated sediments. Wood's compressibility-average equation has thus been successfully used for marine sediments (Wood, 1941). For consolidated sediments and high-porosity sandstones, Gassmann theory, further developed by Hamilton, provides good approximation, by considering the consolidation effects of mechanical couplings between grains on wave velocities (Gassmann, 1951; Hamilton, 1971; Stoll, 1989; Hyndman et al., 1993; Sun et al., 1994). For intermediate-porosity (<40%) homogeneous sandstones saturated with liquid phase, Wyllie's time-average equation can be a good approximation (Wyllie et al., 1956). It is the simplest and the most intuitive velocity-porosity relationship that can exist. Since it considers no physical mechanisms of wave propagation other than a high-frequency approximation of ray theory, it cannot explain velocity measurements in structural carbonates of even similar porosity ranges. Many more sophisticated theoretical models of seismic properties of porous or fractured media have been applied with limited success in both experimental and field studies. Most of the effective medium theories either consider non-interacting cracks or are valid only for porosities of less than a few percent (e.g., Kuster and Toksoz, 1974; O'Connell and Budiansky, 1974; Bruner, 1976; Hudson, 1980). Because of its relative simplicity among the field theories (e.g., Kosten and Zwikker, 1941; Frenkel, 1944; Biot, 1956; Dvorkin and Nur, 1993), the Gassmann model as a limiting case of the Biot theory has been recently used to model acoustic responses of hydrocarbon reservoirs (e.g., Ramamoorthy and Murphy, 1998).

Starting from first principles, Biot's elegant approach (1956) to the mechanical properties of porous media has many theoretical advantages over other models. However, it does not consider the presence of cracks and/or fractures in a porous medium. In fact, unlike bulk modulus, the effective shear modulus of a porous medium is left untouched in his treatment. This is quite reasonable considering the historical context that Biot developed this new field from the point-view of soil mechanics and unconsolidated sediments (Biot, 1941). Therefore, this limits its applicability to rocks at low differential pressures, especially to carbonates (Gregory, 1976; Murphy, 1984). In order to include these geometrical factors in the fundamental governing equations, the Biot theory has to be extended and generalized (Sun, 1994; Sun and Goldberg, 1997b).

The extended Biot theory is based on a more general treatment of mechanics and thermodynamics of fractured porous media (Sun, 1994). The general theory starts from a topological characterization of structural media that provides a representation of the internal structure of a fractured porous medium at a finer scale. Physical entities such as velocity, mass, stress tensor, and energy are defined using the fundamental structural tensors. It is found that the stress tensor in a general structural medium is no longer symmetric and there exists intrinsic angular momentum in such medium. The law of balance of angular momentum is therefore independent from the law of balance of linear momentum, unlike in classical mechanics.

This theory gives a detailed account of possible internal motions existing in structural media. Many new phenomena associated with the existence of internal structures are predicted and ready for experimental verification.

In general, when a fractured porous medium is set in motion, the medium is immediately subject to four kinds of internal forces and acquires internal circulatory motions. These four internal forces originate from the existence of internal structures, the space-time variation of these structures, and inertial coupling between different phases. They are sources of instability of fractured porous media. As carbonates have complex internal structures, many "scattered" plots of physical property measurements are manifestations of these internal forces and motions. This general theory offers quantitative understanding of the causes and effects of structural complexity. It provides considerable insight and sufficient mechanism to define and analytically determine key model parameters for specific applications.

A more detailed exposition of this general theory and the extended Biot theory has been given elsewhere (Sun et al., 2001a). In the following, we present only a few very simplified working formulas that are believed to be sufficient for present-day practical applications in sedimentary sediments and rocks including carbonates. They can be effectively used in principle for rocks of full porosity range, to obtain high-resolution petrophysical and stratigraphic information from well log and seismic data.

Consider a fractured porous medium of two phases. Assume that one phase is solid and the other is fluid (air, gas, or liquid). For the case of both phases being solids, the following formulas are not valid and need to be modified. Let ρ_s, K_s, μ_s be the density, bulk modulus, and shear modulus of the solid phase respectively. Let ρ_f and K_f be the density and bulk modulus of the pore fluid, respectively. Let ϕ be porosity and ρ bulk density. As usual, let K and μ be the bulk and shear moduli, respectively. The compressional and shear wave velocity V_p and V_s are

$$V_p = \sqrt{\frac{K + \frac{4}{3}\mu}{\rho}},$$ (1)

$$V_s = \sqrt{\frac{\mu}{\rho}}$$ (2)

where

$$\rho = (1-\phi)\rho_s + \phi\rho_f$$ (3)

$$K = (1-\phi_k)K_s + \phi_k K_f$$ (4)

$$\phi_k = F_k \phi$$ (5)

$$F_k = \frac{1-(1-\phi)f_k}{[1-(1-\phi)f_k]\frac{K_f}{K_s} + (1-\frac{K_f}{K_s})\phi}$$ (6)

$$\mu = \mu_s(1-\phi)f_\mu.$$ (7)

There are eight parameters in Equations (1-7). Five basic parameters are related to the intrinsic solid and fluid properties, namely, ρ_s, K_s, μ_s, ρ_f, and K_f. One parameter is the porosity of the pore space, ϕ. We introduce two additional parameters, f_k and f_μ, that describe the pore geometrical effect on elastic wave velocities. We call these two parameters f_k and f_μ the bulk and shear effective formation geometrical factor, respectively. They are generally functions of porosity ϕ and physical states of the medium. When measurements on shear wave velocity are not available, we assume, as a first-order approximation, that f_μ is proportional to f_k, e.g.,

$$f_\mu = f_k = f .$$ (8)

It means that the Poisson's ratio of the skeletal frame under jacketed test is the same as or proportional to the Poisson's ratio of the solid grain. Even though this agrees with common practice in practical applications, this assumption should always be scrutinized by experiments and field data validation.

Physically, the formation geometrical factor f is a complicated function of porosity and pore shape, in addition to other factors such as pore size, saturation, mineralogy or lithology, pressure, and temperature. Its exact mathematical expression exists only for very simplified rock models. For natural rocks like carbonates exhibiting a variety of internal structures, it is rather impossible to obtain an exact expression of the function f that is valid for practical applications. However, previous studies tend to show that in many circumstances, the pore-shape distribution is statistically related to the porosity. To a large extent, this assumption may hold macroscopically for rocks without significant amount of fractures. Therefore, as a first step to understand the complexity of the geometrical effects on acoustic responses in carbonates, we adopt the following polynomial form of the formation geometrical factor f in terms of porosity:

$$f = A + B\phi + C\phi^2 . \tag{9}$$

The solid matrix shear and bulk moduli can be obtain in practice by solving $\mu_s = \rho_s V_{ss}^2$ and $K_s = \rho_s V_{ps}^2 - (3/4)\mu_s$ given the compressional wave velocity V_{ps}, shear wave velocity V_{ss}, and the density ρ_s of the solid matrix. Similarly, the fluid modulus K_f can be calculated using $K_f = \rho_f V_f^2$ from the fluid velocity V_f and density ρ_f.

Equations (1-9) define a unique porosity-velocity relationship for compressional wave velocity

$$V_p = V_p(\phi) ,$$

given the density ρ_s, compressional wave velocity V_{ps}, and shear wave velocity V_{ss} of the solid matrix and the density ρ_f, wave velocity V_f of the fluid, and estimates of the three polynomial coefficients in Eq. (9). As listed below, the range of these model parameters have been tested for many rock types, including, e.g., clay, sands, carbonate sediments, siliciclastic rocks, limestone, chalk, dolomite, granite, oceanic diabase gabbro at high pressure. Batzle and Wang (1992) document in detail some of the seismic properties of pore fluids. The polynomial coefficients in Eq. (9) can be determined using core physical property and sonic measurements and analysis of reservoir rock type.

ρ_s = solid matrix density (2.75: 2.65-3.1 g/cc)
V_{ps} = solid matrix compressional wave velocity (6.5: 6.2-7.3 km/s)
μ_{ss} = solid matrix shear wave velocity (3.3: 3.0-4.0 km/s)
ρ_f = fluid density (1.024: 0.001-0.75-1.024 g/cc)
V_f = fluid wave velocity (1.521: 0.34-1.521 km/s)
ϕ = porosity (given: 0 – 1.0)

Numerical Analysis

Over the past several years, this extended Biot theory has been successfully used for real-world applications. It has been essential for validating the feasibility of 4D or time-lapse seismic monitoring of reservoirs in siliciclastic sediments and rocks (Anderson et al., 1996). Using sonic, density, and porosity logs and reservoir production history of oil and gas saturation, pressure and temperature, the extended Biot theory was successfully used to generate time-lapse acoustic impedance models for a reservoir in Gulf of Mexico. These impedance and density models were used as input for 2-dimensional finite-element modeling. The seismic difference between two synthetic seismic sections simulating two seismic surveys of three years apart clearly shows the oil/gas/water changes. The extended Biot model has been crucial for simulating accurately the impedance changes under in-situ reservoir conditions. In this paper, we report its application to carbonate rocks and study the complex phenomena introduced by effects of pore structure on seismic velocities.

Log and seismic data

To further test the assumptions and simplifications in the theoretical model, Equations (1-9), we recently applied this model to analyzing the log and seismic data obtained from a production carbonate reservoir field. The location of the reservoir field is not relevant to this study. We use some of the results from three studied wells to demonstrate the important effect of pore structure on wave velocity measurements.

The overall quality of the log data is high for the entire logged interval in these three wells. The data points in Figures 1, 2 and 3 show the scattered velocity-porosity cross-plots obtained from sonic logs and density-derived porosity logs for the studied depth interval of the reservoir. Detailed log integration analysis shows that these scatters are not resulted from measurement errors and hole environment conditions.

The five model parameters related to the intrinsic solid and fluid properties are estimated from basic log analysis of the density, porosity, and sonic logs. An average value of 2.75 g/cc for the matrix (grain) density was estimated from a grain density log. The elastic properties of the rock matrix were assumed to be constant for the three wells with solid compressional and shear wave velocities being 6.5 km/s and 3.3 km/s, respectively. A density of 1.024 g/cc and velocity of 1.521 km/s were used for saturating fluid, which is assumed to be water.

Wyllie time-average equation is used, as a reference, to fit the velocity-porosity data for each well. Figures 1, 2, and 3 indicate that Wyllie equation does not result in a good fit. This also indicates the fact that these log data cannot be best presented by a cross-plot using any single one-parameter velocity-porosity model without considering pore structure effects.

Figure 1 shows a typical example that demonstrates the pore structure effects on acoustic velocity measurements in carbonate rocks. It is evident that at least another dimension describing the pore structure factor is required to develop a comprehensible velocity model. It is found that the polynomial function of the formation geometrical factor introduced in the proposed theoretical model can be used to characterize the large variation of velocity for the studied reservoir. The best-fit polynomial coefficients value for the formation geometrical factor f in Well A are found to be $f[A,B,C] = f[0.26,0.52,-2.23]$. The upper limit of the porosity-velocity relationship is characterized by the polynomial f-model $f_+[A,B,C] = f[0.39,0.78,-3.35]$. The lower limit of the porosity-velocity relationship is characterized by the polynomial f-model $f_-[A,B,C] = f[0.18,0.36,-1.56]$. For the same reservoir, f-model exhibits a different characteristic in Well B (Figure 2). The best-fit polynomial coefficients value for the formation geometrical factor f in Well B are found to be $f[A,B,C] = f[0.15,2.48,-7.14]$. The upper limit of the porosity-velocity relationship is characterized by the polynomial f-model $f_+[A,B,C] = f[0.18,2.97,-8.57]$. The lower limit of the porosity-velocity relationship is characterized by the polynomial f-model $f_-[A,B,C] = f[0.11,1.86,-5.35]$. In Well C, the best-fit f-model is $f[A,B,C] = f[0.31,-0.57,0.50]$ (Figure 3). The upper limit of the porosity-velocity relationship is characterized by the polynomial f-model $f_+[A,B,C] = f[0.37,-0.68,0.61]$. The lower limit of the porosity-velocity relationship is characterized by the polynomial f-model $f_-[A,B,C] = f[0.26,-0.48,0.43]$.

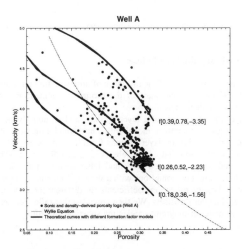

Figure 1. The velocity-porosity relationships with different formation geometrical factor models in comparison with the Wyllie equation fit for Well A.

With the five basic parameters for the intrinsic matrix and fluid property being estimated or assumed, the proposed model is then reduced to a unique velocity-porosity relationship for a given polynomial f-model. Using this relationship and density or porosity logs, we can make further comparison of calculated velocity with in-situ logs to justify the validity of the relationship and its model parameters. In this paper, we instead compare the synthetic seismogram generated using the pseudo-velocity log with the field seismic data at Well C. As shown in Figure 4, the synthetic seismogram agrees very well with the field seismic trace. The synthetic seismograms are generated using the convolution model with the density log and the pseudo-velocity log derived from the proposed porosity-velocity relationship with the best-fit f-model. A Ricker wavelet with a central frequency of 35 Hz is assumed for the source. The overall agreement between the synthetic seismograms and field data further indicates the important effects of pore structure on sonic and seismic velocity.

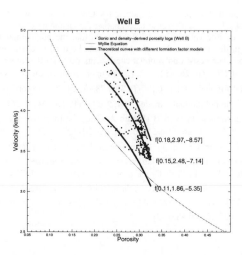

Figure 2. The velocity-porosity relationships with different formation geometrical factor models in comparison with the Wyllie equation fit for Well B.

Effect of pore structure on seismic amplitude and phase

To gain a better understanding of the influence of pore structure on acoustic wave propagation, we use a two-layer model to demonstrate the importance of the formation geometrical factor on seismic reflection amplitude variation versus offset or incident angles.

Assume that a two-layer model consists of a layer of shale and an underlying layer of porous carbonate rock. For illustration purposes, we use a geometrical model with a horizontal dimension of 500 m and a thickness of 200 m for the shale layer. The porous carbonate layer extends from the shale/carbonate-layer interface to infinity vertically. The maximum angle of reflection that can be observed is therefore 51.1^0. This porous carbonate rock layer can be either water-saturated or gas-saturated. We model the acoustic reflection from the layer interface between the tow layers for both the saturation conditions with a varying formation geometrical factor. In this paper, we use Zoeppritz equation to calculate the reflection coefficients for different incident angles to obtain a reflection series for a set of varying reflected angles or offsets. We then convolve this reflection series with a Ricker wavelet of 25 Hz to obtain a section of synthetic seismograms, neglecting spherical spreading.

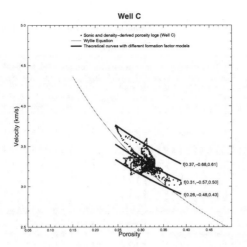

Figure 3. The velocity-porosity relationships with different formation geometrical factor models in comparison with the Wyllie equation fit for Well C.

The petrophysical model of the shale/carbonate-layer system is defined as follows. The density, compressional or acoustic velocity, and the shear velocity of the shale layer are assumed to be 2.06 g/cc, 2.743 km/s, and 1.394 km/s, respectively. We assume that the porous carbonate rock layer has a constant porosity of 30%. The intrinsic density, compressional velocity, and shear velocity of the solid matrix are assumed to be 2.60 g/cm^3, 6.5 km/s, and 3.3 km/s, respectively. For the case of water saturation, a density of 1.024 g/cc, and an acoustic velocity of 1.521 km/s are used for the pore fluid. For the case of gas saturation, the pore fluid is replaced by air, i.e., a density of 0.001 g/cc, and sound speed in air, 0.34 km/s are assumed for simplicity. For any given value of the formation porosity, the bulk properties of the porous carbonate layer (density, compressional and shear velocities) can then be calculated using the proposed porosity-velocity relationship with an estimated polynomial f-model. To illustrate the important effects of pore structure on AVO in this paper, we simply choose three f values, 0.55, 0.30, and 0.14 to cover the range of f variations estimated from the log data for the studied carbonate reservoir. In practical applications, however, the geometrical formation factor f should be calculated from porosity log using the best-fit polynomial f-model. Figures 5, 6, and 7 show the synthetic seismograms for both the water- and gas-saturated conditions for the given geometrical factor, f=0.55, 0.30, and 0.14, respectively.

Figure 5 shows the synthetic seismogram for a given formation geometrical factor f=0.55. When it is water-saturated, the porous carbonate layer has a bulk density of 2.127 g/cc, a bulk compressional velocity of 4.61 km/s and a bulk shear velocity of 2.26 km/s. The critical angle for the reflected P-wave from an incident P-wave in this case is 36.5^0 as shown in Figure 6 (a). When it is gas-saturated, the carbonate layer has a bulk density of 1.820 g/cc, a bulk compressional velocity of 4.82 km/s and a bulk shear velocity of 2.45 km/s. The critical angle in this case is 34.7^0. These parameters and similarly those for the synthetic seismograms shown in Figures 6 and 7 for f=0.30 and 0.14 respectively are given in Table 1.

Table 1. Predicted acoustic properties affected by the formation geometrical factor f

f	Water-saturated				Gas-saturated			
	ρ (g/cc)	V_p (km/s)	V_s (km/s)	θc	ρ (g/cc)	V_p (km/s)	V_s (km/s)	θc
0.55	2.127	4.61	2.26	36.5	1.820	4.82	2.45	34.7
0.30	2.127	3.61	1.67	49.4	1.820	3.56	1.81	50.4
0.14	2.127	2.81	1.14	77.5	1.820	2.43	1.23	none

342

Field data   Synthetic seismogram   Field data

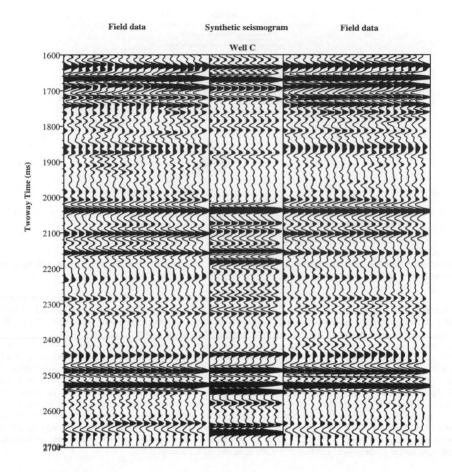

Figure 4. Comparison of synthetic seismograms generated using the method in Figure 4 with a portion of the seismic inline section across Well C.

(a) Water-saturated, f=0.55

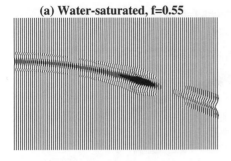

(a) Water-saturated, f=0.30

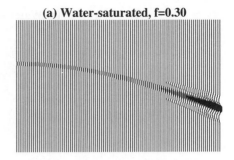

(b) Gas-saturated, f=0.55

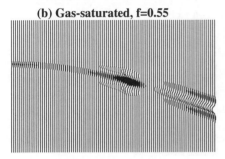

(b) Gas-saturated, f=0.30

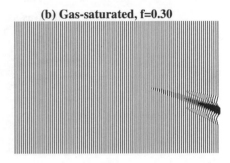

Figure 5. P-wave synthetic seismograms generated using the Zoeppritz equation for a two-layer model consisting of an encasing shale and an underlying carbonate reservoir of a given porosity of 30% and a given formation geometrical factor f=0.55.

Figure 6. P-wave synthetic seismograms generated using the Zoeppritz equation for a two-layer model consisting of an encasing shale and an underlying carbonate reservoir of a given porosity of 30% and a given formation geometrical factor f=0.30.

344

(a) Water-saturated, f=0.14

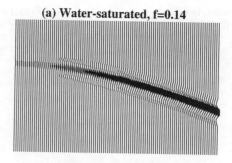

(b) Gas-saturated, f=0.14

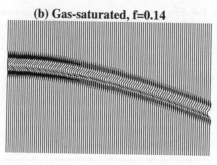

Figure 7. P-wave synthetic seismograms generated using the Zoeppritz equation for a two-layer model consisting of an encasing shale and an underlying carbonate reservoir of a given porosity of 30% and a given formation geometrical factor f=0.14.

The predicted acoustic properties in Table 1 indicate one important fact that pore structure can affect both compressional and shear velocities dramatically. Shear wave velocity is less sensitive to pore structure factor than the compressional wave. And compressional wave is more affected by pore structures when the propagating medium is gas-saturated than it is water-saturated. Density, on the other hand, is not affected by the pore structure for a given porosity and given intrinsic solid and fluid properties. Figure 5 shows that it is very difficult to distinguish gas saturation from water saturation when formation geometrical factor is large. In this case, the pore structure factor plays more dominant role than saturation on controlling the acoustic responses and the difference between the variations of amplitudes with offset under the two saturation conditions is not apparent on the seismograms. When the formation geometrical factor is intermediate (e.g., f=0.30), the saturation effect on seismogram becomes evident (Figure 6). In fact, the reservoir appears to be a "dim out" in this case as shown in Figure 6 (b). However it may thus be indistinguishable from the overlying or encasing shale. When the formation geometrical factor is relatively small (e.g., f=0.14), the water-saturated carbonate rock gives a false AVO anomaly as the seismic amplitudes strongly increase with offset. And it seems rather perplexing that the acoustic waveform becomes a "phase-reversal" and the amplitudes decrease as offset increases.

Concluding Remarks

We have shown that the effect of pore structures on acoustic wave velocity can be much greater than that of porosity in carbonate rock. The proposed porosity-velocity relationship with a polynomial f-model considers such effect. Using this theoretical model, we can indeed explain the observed large deviations of sonic velocity measurements from the theoretical fit using Wyllie time-averaging equation. This model can also be used to predict acoustic wave velocity if the density and porosity logs are available and the general lithology and saturation conditions are known. For the studied reservoir, synthetic seismograms generated using the predicted velocity-depth profile or pseudo-velocity log are in good agreement with field seismic data around the wells.

Numerical modeling results indicate that the pore structure factor can affect acoustic wave phenomena in a very profound way. In fact, it is the formation geometrical factor that controls how the seismic amplitudes and phases can be used for lithological interpretation and fluid detection. The variation of amplitude versus offset or incident angle cannot be used to distinguish hydrocarbon-bearing or dry rocks from water-saturated or wet formations when the geometrical factor is quite large. The "dim-out" phenomena occur when the geometrical factor is intermediate. Complete phase-reversal of acoustic signal occurs when the geometrical factor is small. These conclusions are drawn from this case study of a high-porosity carbonate reservoir. They help to explain and understand the pitfalls of AVO analysis and the corresponding mechanisms.

The theoretical model may also be valid for applications in other carbonate environments. For future investigation, three-dimensional full-waveform modeling of the pore structure effects on elastic wave propagation should be considered (Sun et al., 2001b).

Acknowledgements

The work in this report was made possible by partial support from Shell International Exploration and Production, B.V. Lamont Contribution No. 6218.

References

Anderson, R. N., Boulanger, A., He, W., Sun, Y. F., Xu, L., and Hart, B., 1996. 4-D seismic monitoring of reservoir production in the Eugene Island 330 Field, Gulf of Mexico, in: Weimer, P., and Davis, T. L., (eds.), Applications of 3-D Seismic Data to Exploration and Production, AAPG Studies in Geology No. 42 and SEG Geophysical Development Series No. 5, AAPG/SEG, Tulsa, 9-20.

Anselmetti, F. S., and G. P. Eberli, 1993. Controls on sonic velocity in carbonates, Pure and Applied Geophysics, 141, 287-323.

Anselmetti, F.S. and Eberli, G.P., 1997. Sonic velocity in carbonate sediments and rocks, in: Palaz, I. and Marfurt

K.J. (eds.), Carbonate Seismology, SEG Geophysical Developments Series, 6, 53-74.

Batzle, M. L., and Wang, Z., 1992. Seismic properties of pore fluids, Geophysics, 57, 1396-1408.

Biot, M. A., 1941. General theory of three dimensional consolidation, J. Appl. Phys., 12, 426-430.

Biot, M.A., 1956. Theory of propagation of elastic waves in a fluid-saturated porous solid, I. Low frequency range; II. High frequency range, J. Acoust. Soc. Am., 28: 168-191.

Bruner, W.M., 1976. Comment on 'Seismic velocities in dry and saturated cracked solids' by Richard J. O'Connell and Bernard Budiansky, J. Geophys. Res., 81: 2573-2576.

Budd, D. A., Saller, A. H., and Harris, P. M., 1995, Uncomformities and Porosity in Carbonate Strata, AAPG Memoir 63.

Choquette, P. W., and Pray, L. C., 1970. Geologic nomenclature and classification of porosity in sedimentary carbonates, AAPG Bulletin, 54, 207-250.

Davies, P. J., McKenzie, J. A., Palmer-Julson, A., et al., 1991. Northeast Australian Margin, Proceedings of the Ocean Drilling Program, Initial Reports, 133, College Station, TX (Ocean Drilling Program).

Dvorkin, J., and Nur, A., 1993. Dynamic poroelasticity: A unified model with the squirt mechanism and the Biot Mechanisms, Geophysics, 58: 524-533.

Eberli, G. P., Swart, P. K., Malone, M. J., et al., 1997. Bahamas Transect, Proceedings of the Ocean Drilling Program, Initial Reports, 166, College Station, TX (Ocean Drilling Program).

Feary, D., Hine, A. C., Malone, M. J., et al., 1999. Great Australian Bight: Cenozoic Cool-Water Carbonates, Proceedings of the Ocean Drilling Program, Initial Reports, 182, College Station, TX (Ocean Drilling Program).

Frenkel, J., 1944. On the theory of seismic and seismoelectric phenomena in a moist soil, Journal of Physics (USSR), 8: 230-241.

Gassmann, F., 1951. Elastic waves through a packing of spheres, Geophysics, 15: 673-685.

Gregory, A.R. 1976. Fluid saturation effects on dynamic elastic properties of sedimentary rocks, Geophysics, 41, 895-921.

Hamilton, E.L., 1971. Elastic properties of marine sediments, J. Geophys. Res., 76: 579-604.

Hudson, J.A., 1980. Overall properties of a cracked solid, Mathematical Proceedings of Cambridge Philosophical Society, 88: 371-384.

Hyndman, R.D., Moore, G.F., and Moran, K., 1993. Velocity, porosity, and pore-fluid loss from the Nankai subduction zone accretionary prism, in: Hill, I.A., Taira, A., Firth, J.V., et al. (Eds.), Proc. ODP. Sci Results, 131: College Station, TX (Ocean Drilling Program), 211-220.

Isern, A., and F. Anselmetti, 2000. Ocean Drilling Program Leg 194 Scientific Propspectus: Marion Plateau, Scientific Prospectus No. 94, College Station, TX (Ocean Drilling Program).

Kosten, C.W., and Zwikker, C., 1941. Extended theory of the absorption of sound by compressible wall-coverings, Physica, 8: 968-978.

Kuster, G.T., and Toksoz, N.M., 1974. Velocity and attenuation of seismic waves in two-phase media, I. Theoretical Formulations, Geophysics, 39: 607-618.

Murphy, W.F. 1984. Acoustic measures of partial gas saturation in tight sandstones, Journal of Geophysical Research, 89, 11549-11559.

O'Connell, R.J., and Budiansky, B., 1974. Seismic velocities in dry and saturated cracked solids, J. Geophys. Res., 79: 5412-5426.

Premoli Silva, I., Haggerty, J. A., Rack, F. R., et al., 1993. Northwest Pacific Atolls and Guyots, Proceedings of the Ocean Drilling Program, Initial Reports, 144, College Station, TX (Ocean Drilling Program).

Rafavich, F., Kendall, C. H. St. C., and Todd, T. P., 1984. The relationship between acoustic properties and the petrographic character of carbonate rocks, Geophysics, 49, 1622-1636.

Ramamoorthy, R., and Murphy, W. F. III, 1998. Fluid identification through dynamic modulus decomposition in carbonate reservoirs, in: Transactions of the SPWLA Annual Logging Symposium 39; Pages Q. 1998.

Sager, W. W., Winterer, E. L., Firth, J. V., et al., 1993. Northwest Pacific Atolls and Guyots, Proceedings of the Ocean Drilling Program, Initial Reports, 143, College Station, TX (Ocean Drilling Program).

Simo, J. A., Scott, R. W., and J. P. Masse, 1993. Cretaceous Carbonate Platforms, AAPG Memoir 56.

Stoll, R.D., 1989. Sediment Acoustics: Springer-Verlag, Berlin, Heidelberg.

Sun, Y.F. 1994. On the Foundations of the Dynamical Theory of Fractured Porous Media and the Gravity Variations Caused by Dilatancies, Ph.D. dissertation, Columbia University.

Sun, Y.F. and D. Goldberg, 1997a. Effects of aspect ratio changes on wave velocities in fractured rocks, SEG Expanded Abstract, 67, 925-928.

Sun, Y.F. and D. Goldberg 1997b. Estimation of aspect-ratio changes with pressure from seismic velocities, in:

Lovell, M. A. and Harvey, P. K. (eds.), Developments in Petrophysics, Geological Society Special Publication No. 122, 131-139.

Sun, Y.F., Kuo, J.T., and Teng, Y.C., 1994. Effects of porosity on seismic attenuation, J. Comput. Acoust., 2: 53-69.

Sun, Y.F., J.L. Massaferro, G. Eberli, and Y.C. Teng, 2001a. Theoretical modeling of pore structure effect on elastic wave propagation, Journal of Computational Acoustics, in preparation.

Sun, Y.F., Y.C. Teng, J.L. Massaferro, and G. Eberli, 2001b. 3D numerical modeling of pore structure effect on elastic wave propagation, Journal of Computational Acoustics, in preparation.

Wang, Z., Hirsche, W. K., and Sedgwick, G., 1991. Seismic velocities in carbonate rocks, J. Can. Petro. Tech., 30, 112-122.

Wang, Z., Cates, M. E., and Langan, R. T., 1998. Seismic monitoring of a CO_2 flood in a carbonate reservoir: A rock physics study, Geophysics, 63, 1604-1617.

Wood, A. B., 1941. A Textbook of Sound, G. Bell and Sons.

Wyllie, M.R., Gregory, A. R. and Gardner, L. W., 1956. Elastic wave velocities in heterogeneous and porous media, Geophysics, 21: 41-70.

Theoretical and Computational Acoustics 2001
E.-C. Shang, Qihu Li and T. F. Gao (Editors)
© 2002 World Scientific Publishing Co.

The Normalization of Time-Lapse Seismic Data and Its Application

Chen Xiaohong, Yi Weiqi and Zhang Guocai
University of Petroleum, Beijing, 102200, P. R. China

Time-lapse seismic monitoring is a technique to monitor reservoir with repeat seismic prospecting in different time during the period of reservoir production. In this paper, the method of normalization of time-lapse data is presented. And normalization for amplitude, frequency and phase of time-lapse seismic data in non-reservoir is done to obtain the equalized data. The normalization processing was applied in time-lapse seismic data of TPT block in Daqing Oilfield. The results indicate that the differences of time-lapse seismic data in non-reservoir can be eliminated with the normalization processing. While, the differences of time-lapse seismic data occur, which are related with the distribution of production wells. These reasonable identity and difference of these seismic data can analyze and interpret the dynamic variation in reservoir.

1. Introduction

Time-lapse seismic monitoring is a technique to monitor reservoir with repeat seismic prospecting in different time during the period of reservoir development. The changes of seismic responses with time can character the changes of fluid properties in reservoir, which can describe the variation of lithologic parameters (such as porosity, permeability, saturation, pressure and temperature) in reservoir.

In theoretical, the subtraction of time-lapse seismic data can directly image the dynamic fluid properties (fluid saturation, pressure, temperature and so on) of reservoir. In practice, however, the seismic data were acquired and processed in different time, which results in the variations in seismic profiles. These variations in seismic amplitude, velocity, frequency and phase are unexpected and undeserved in time-lapse seismic monitoring. Thus the equalization of time-lapse seismic data should be done to obtain the two 3-D equalized data. The reasonable identity and difference of these data can analyze and interpret the dynamic variation in reservoir..

In this paper, the time-lapse seismic data of TPT block in Daqing Oilfield are processed and analyzed. The results indicate that the differences of time-lapse seismic data in non-reservoir can be eliminated with the equalization processing. While, the differences of time-lapse seismic data occur in faith in interbed reservoir. And the differences are related with the distribution of production wells.

2. The method of normalization of time-lapse data

Due to the acquisition and processing of data in different time, the travel-time, amplitude, frequency and phase of time-lapse data in non-reservoir area have big differences in seismic profiles. The key for success to applying time-lapse data in reservoir monitoring is to equalize the two seismic data sets. Aimed at the differences in travel-time, amplitude, frequency and phase of time-lapse data, the normalization operators are applied in two profiles. The normalization operator is to find a best match filter, which can make effective seismic signals of each survey line match that of the reference survey line in non-reservoir part.

Let $G_1(t)$, $G_2(t)$ be time-lapse seismic data, the normalization operator P is to be found to minimizing following objective functional:

$$E(t) = \left\| G_1(t) - PG_2(t) \right\|$$ (1)

Here $\left\| \bullet \right\|$ is a norm, such as L^2-norm. We define match filter $P(t)$ as following

$$E(t) = \int_0^T \left[G_1(t) - P(t) * G_2(t) \right]^2 dt = \min$$ (2)

Here $*$ is convolution operator, T is maximum survey time of seismic prospecting. The minimum solution of (2) will be arrived from zero of Frechet derivative of functional (2) with respect $P(t)$. Thus we can obtain the match filter $P(t)$ though

$$\partial E / \partial P = 0$$

If we get the match filter $P(t)$, we can equalize the data $G_2(t)$ with the reference data $G_1(t)$. Then the normalized profile $G_2^*(t) = P(t) * G_2(t)$ will be obtained.

3. Field time-lapse seismic data

In TPT block of Daqing Oilfield, the seismic reconnaissance was carried out during 1960 to 1973. From 1988 to 1990, the digital detail survey was done in this area. Again, the high-resolution development seismic survey was done from the winter of 1998 to spring of 1999 in same area. The 2-D seismic profiles of 1988 and

1999 can form the exact time-lapse seismic data. In TPT block, there are many production wells. Thus the block is suitable for time-lapse seismic monitoring. The main target reservoirs are Putaohua and Fuyu oil-bearing layers.

In both profiles of 1988 and 1999, the static properties of strata are generally consistent. The faults, structure trend and form interpreted in both profiles are almost same, which reflects the consistency of old and new seismic data. These show that the data of old and new profiles have well repeatability.

However, time-lapse seismic data were acquired and processed in different time. The position and combination of the faults are not exactly same in two profiles. Due to progresses in the seismic acquisition technique in 1999's survey, the resolution is improved, the positions of faults are clear and accurate, the combination of faults changes a little.

The main reasons of above differences are: (a) The instruments and methods of acquisition are not same, which results in the different quality of acquisition. (b) The processing flow and main parameters are more reasonable in second survey. The geological phenomena are reflected more clear and reasonable in the new seismic profile.

In order to eliminate the differences in time-lapse seismic data due to the different processing flow and main parameters as can as possible, the seismic data of 1988 ware reprocessed with same flow, parameters as that of the data of 1999, and same person as well. The repeatability of two profiles is largely improved.

4. Seismic- geological settings and reservoir conditions

In TPT block, exploration and production wells were drilled in 1960 and 1970's. The sonic logging were not done in all wells. Thus the synthetic data with different frequency (40-75Hz) wavelet were done using resistivity and sonic logging curves in only six wells. These synthetic data are to calibrate the target layers and Putaohua and Fuyu oil-bearing layers.

In this area, the marker T_1 corresponding to the reflection of the top of Yaojia Formation is very stable in whole region. The energy of reflection is strong, the continuity is good and features of waveform are obvious. Marker T_1 is above the Putaohua reservoir. Thus it should not change with time when the reservoir was developed. It can be taken as a standard layer in time-lapse seismic data processing and analyses.

Reflector corresponding to the reflection of the top of Putaohua reservoir T_{11} is a main target we consider. The apparent frequency of T_{11} reflector is about 60-70 Hz. The amplitude is middle and strong. The continuity is good. There are four phases between Putaohua reservoir and marker T_1. The time space between them is about 80ms. The features of waveform of Putaohua are obvious. The reflection time in seismic profile is from 920 to 1020ms. Putaohua reservoir is below the T_{11} reflector. It changes with time in time-lapse seismic monitoring. It should be processed carefully.

Marker corresponding to the reflection of the top of Quantou Formation is affected by reservoir. It should be changed with time in time-lapse seismic monitoring.

In this paper, Putaohua reservoir is main target in time-lapse seismic processing. It is main oil-bearing layer in TPT block. There are 53 wells drilled in this area. Putaohua reservoir was penetrated in all wells. The capacity of some wells is high. Some wells are shuted due to difficult production. In this area, Putaohua oil-bearing layer is thin interbed reservoir. The cumulative thickness of oil layer is about 55-66m. The thickness of sandstone of Putaohua reservoir is 1.6-20.3m. The effective thickness is 0.9-12.3m. The quality of seismic profile is good. The apparent frequency of Putaohua reservoir is about 65-70 Hz. Three events can be seen in profile clearly.

5.   Processing and analyses for the time-lapse seismic data

Using the method we developed as previous, we process and analysis the time-lapse seismic data in TPT block. The superposition surveys done in 1988 and 1999 were used as time-lapse seismic data.

The acquisition technique of 2-D seismic data of 1988 is as following. Shot point is in the end of line. Coverage times is 30. Offset distance is 100m. The group interval is 25m. The sample rate is 1ms. The combination of geophones is linear. The number of geophones group is 24. The geophone distance is 2m.

The acquisition technique of 2-D high-resolution seismic data of 1999 is as following. Shot point is in the middle of line. Coverage times is 30. Offset distance is 40m. The group interval is 20m. The sample rate is 1ms. The number of geophones group is 3. The geophone distance is 2m.

The acquisition parameters of two surveys are different. Especially the trace space is not same. The spaces of CDP of surveys of 1988 and 1999 are 12.5m and 10m respectively. Thus the bin should be designed, which makes two surveys to have same CDP. In this paper, same traces are selected from two profiles. One trace is selected from every four traces in 1988's profile and every five traces in 1999's profile. This may form a set of matched time-lapse seismic data.

Aiming at the differences of time, amplitude, frequency and phase in time-lapse seismic data, the equalization operator was designed to match two profiles. The designing principle is to take marker T_1 as non-reservoir standard. The marker T_1 is above Putaohua reservoir and very stable in the whole area. It is processed as non-reservoir standard and is not changed with time in time-lapse seismic monitoring. Figure 1 shows the windowed seismic profiles of marker T_1 and Putaohua reservoir for two surveys in TPT block.

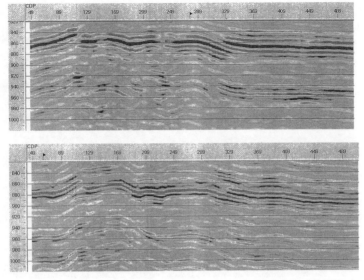

Figure 1   Time-lapse seismic data of line 108 surveyed in 1988(up) and 1999 (window 840-1040ms)

The equalization method is to find a best match filter. This filter reforms effective source signal in every line, which makes it same to the source signal of reference line. When the filter is obtained from non-reservoir window in profile, it will be applied to reservoir to correction two profiles. The differences of amplitude, frequency and phase in non-reservoir will be eliminated as can as possible, while the remained differences in reservoir should be interpreted as the changes caused by variation of oil, water and gas in reservoir. Figure 2 shows the comparison of the windowed seismic profiles of marker T_1 and Putaohua reservoir for two surveys after equalization. The identity of the marker T_1 in two profiles is largely improved.

354

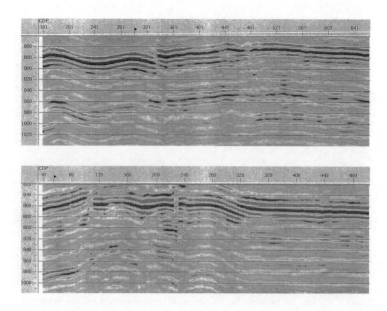

Figure 2　Time-lapse seismic data of line 108 surveyed in 1988(up) and 1999 after equalization（window 840-1040ms）

Figure 3 shows the difference profile of time-lapse seismic after equalization with taking marker T_1 as non-reservoir standard. This indicates that the difference in marker T_1 as non-reservoir standard is eliminated. However, the differences in Putaohua reservoir still remain. The differences are related to distribution of wells. This accords with the actual situation, which demonstrates that time-lapse seismic monitoring can be carried out in thin interbed reservoir. And it may succeed in TPT block of Daqing Oilfield.

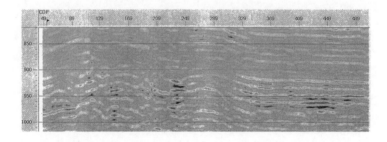

Figure 3　Difference profile of time-lapse seismic after equalization with taking marker T_1 as non-reservoir standard.

6. Conclusion

Normalization of time-lapse seismic data is the key step for time-lapse seismic reservoir monitoring. In the paper, the normalization of time-lapse data is presented by finding a best match filter. The filter reforms effective source signal in every seismic survey, which makes it same to the source signal of reference seismic survey. The normalization for amplitude, frequency and phase of time-lapse seismic data in TPT block in Daqing Oilfield was carried out with the match filter. The results indicate that the differences of time-lapse seismic data in non-reservoir can be eliminated with the normalization processing. While, the remaining differences of time-lapse seismic data are related with the distribution of production wells. This shows that the match filter operator is very effective in normalization processing. With these reasonable identity and difference of seismic data, we can analyze and interpret the dynamic variation in reservoir.

Acknowledgment

This work was supported by National Natural Science Foundation of China grant 49974030.

References

1. Pai Kristiansen, Philip Christie, et al, Foinaven 4-D; Processing and analysis of two designer 4-Ds, : 70th Ann. Internat. Mtg., Expanded Abstracts, Soc. Expl. Geophys., 2000.
2. Madhumita Sengupta, Gary Mavko, et al, Integrating time-lapse seismic and flow simulation to map saturation changes: A reservoir monitoring case study: 70th Ann. Internat. Mtg., Expanded Abstracts, Soc. Expl. Geophys., 2000.

Theoretical and Computational Acoustics 2001
E.-C. Shang, Qihu Li and T. F. Gao (Editors)
© 2002 World Scientific Publishing Co.

Modelling of Elastic-Wave Multiple Scattering in Metal-Matrix Composites

Arnaud Lange[1], Anthony Harker[2] and Nader Saffari[1]

[1]Department of Mechanical Engineering, University College London,
London WC1E 7JE, UK

[2]Department of Physics & Astronomy, University College London, London
WC1E 6BT, UK

Abstract. Deterministic multiple scattering formulations based on Twersky's approach and the polymerisation technique are presented. These lead to full-scale simulations of elastic-wave multiple scattering in fibre-reinforced composites. It is shown that the convergence and accuracy of the composite models depend on the geometry of the composite structure and the strategy used for the polymerisation. The paper explains why the multiple scattering formulation may not converge if precautions are not taken during polymerisation. A strategy is proposed to improve convergence. Results are presented for the case of an incident harmonic SH wave impinging on a bounded Ti/SiC composite region, the plane of propagation being orthogonal to the fibres' axis.

INTRODUCTION

Recent developments in Metal-Matrix Composite (MMC) technology has lead to challenging problems in the Non-Destructive Evaluation (NDE) of such materials. Considerable effort is being directed towards development of numerical models that can be used as analysis tools for optimising NDE techniques for these composites. Considering an incident wave that impinges on a bounded MMC region, multiple scattering occurs between a large number of fibres, and also between the internal layers of a single fibre. To investigate effects of various composites configurations on the total scattered wave field, deterministic models are needed, where the positions and the nature of the scatterers are known.

In the past few decades, several deterministic multiple scattering formulations have been developed. Twersky [1] introduced the concept of "ordered scattering" for acoustic wave problems, where the overall multiple scattering problem is decomposed into several single scattering problems in the frequency domain. Using this approach, the solution is computed recursively. Cheng [2] extended this concept to the case of elastic waves. Varadan, Varadan & Pao [3] derived an implicit form of the multiple scattering solution. Waterman [4] formulated the scattering problem using an integral equation approach, and Peterson & Ström [5] applied this approach to some multiple scattering problems. Bose & Mal [6] expressed all the wave fields in every scatterer's local coordinate system, and obtained a system of equations to solve for the final solution by applying the boundary conditions.

Although all these formulations lead to the multiple scattering solution, none of them can efficiently compute the solution for a large number of scatterers. The convergence of the recursive forms of the solution is too slow, and the other forms are limited by the computer memory. To cope with this problem, Cai & Williams [7] developed a scatterer polymerisation technique that allows a single transition matrix to be defined that corresponds to the response of an assemblage of scatterers. The computer memory can then be used more efficiently and large-scale multiple scattering problems can be solved [8, 9, 13]. However, the convergence of the multiple scattering formulations and the polymerisation technique strongly depends on the geometry of the problem.

In this study, Twersky's approach and the polymerisation technique are exposed. Geometrically dependent conditions for convergence of the formulation are analysed, and an approach is proposed to improve the convergence. Numerical results are given for the case of an incident SH wave field propagating in the cross-section plane and scattered by a bounded Titanium MMC region reinforced with Silicon Carbide fibres. The Titanium matrix and the various materials involved in the scatterers are taken to be homogeneous and isotropic.

THEORY

Wave Field Representations

Considering an incident harmonic SH wave, no mode conversion occurs during scattering, and the out-of-plane displacement w can be expressed in the form :

$$w(\vec{r},t) = \phi(\vec{r})e^{-i\omega t}, \tag{1}$$

where $\vec{r}$ is the position vector, t is time, ω is the angular frequency, and the displacement amplitude ϕ satisfies the scalar Helmholtz equation :

$$\nabla^2\phi + k^2\phi = 0, \tag{2}$$

with $k=\omega/\beta$ the shear wavenumber and β the shear velocity.

For cylindrical scatterers, the displacement amplitudes of the various fields can be expanded in terms of the cylindrical solutions of the Helmholtz equation i.e. the Bessel functions. For a single-scatterer problem, the incident wave amplitude ϕ^{inc} is expanded in terms of the regular Bessel functions of the first kind :

$$\phi^{inc} = \sum_{n=-\infty}^{+\infty} A_n J_n(kr)e^{in\theta} = \{A\}^T\{J(r,\theta)\}, \tag{3}$$

and because the temporal term has been chosen with a negative sign i.e. $e^{-i\omega t}$, the scattered wave amplitude ϕ^s is expanded in terms of the singular Hankel functions of the first kind :

$$\phi^s = \sum_{n=-\infty}^{+\infty} B_n H_n^{(1)}(kr)e^{in\theta} = \{B\}^T\{H(r,\theta)\}, \tag{4}$$

where (r,θ) are the polar coordinates of the observation point in the scatterer's local coordinate system, whose origin must lie inside the scatterer.

The transition matrix $[T]$ of the scatterer relates the wave expansion coefficients of the incident and scattered waves as :

$$\{B\} = [T]\{A\}. \tag{5}$$

Ordered Scattering Approach

Using Twersky's approach [1], different orders of scattering are defined according to the number of times the incident wave has been scattered. More clearly, the first order of scattering for scatterer i is the scattered wave field $\phi_i^{(1)}$ corresponding to the scattering of the incident wave ϕ^{inc} by scatterer i. Thus, $\phi_i^{(1)}$ has the following representation :

$$\phi_i^{(1)} = \{C_i^{(1)}\}^T \{H(r_i, \theta_i)\}, \tag{6}$$

where :
$$\{C_i^{(1)}\} = [T_i]\{A_i\}, \tag{7}$$

(r_i, θ_i) are the local polar coordinates related to scatterer i, $[T_i]$ is the transition matrix of scatterer i, $\{A_i\}$ is the column matrix that contains the expansion coefficient for the incident wave expressed in scatterer i's local coordinate system.

The second order of scattering for scatterer j is the scattered wave field $\phi_j^{(2)}$ corresponding to the scattering of all the first order scattered field of the remaining scatterers by scatterer j. Thus, $\phi_j^{(2)}$ has the following representation :

$$\phi_j^{(2)} = \sum_{\substack{i=1 \\ i \neq j}}^{Ns} \phi_{ji}^{(2)} = \sum_{\substack{i=1 \\ i \neq j}}^{Ns} \{C_{ji}^{(2)}\}^T \{H(r_j, \theta_j)\} = \{C_j^{(2)}\}^T \{H(r_j, \theta_j)\}, \tag{8}$$

where :
$$\{C_{ji}^{(2)}\} = [T_j][R_{ij}]^T \{C_i^{(1)}\} = [L_{ij}]\{C_i^{(1)}\}, \tag{9}$$

$$[R_{ij}]_{nm} = e^{i(n-m)\theta_{ij}} H_{n-m}^{(1)}(kd_{ij}), \tag{10}$$

$$\{C_j^{(2)}\} = \sum_{\substack{i=1 \\ i \neq j}}^{Ns} \{C_{ji}^{(2)}\} = \sum_{\substack{i=1 \\ i \neq j}}^{Ns} [L_{ij}]\{C_i^{(1)}\}, \tag{11}$$

with Ns the number of scatterers, $[R_{ij}]$ the local coordinate translation matrix from scatterer i to scatterer j, $[L_{ij}]$ is the inductance matrix of scatterer j on scatterer i, (d_{ij}, θ_{ij}) is the location of the origin of the local coordinate system of scatterer j with respect to the local coordinate system of scatterer i. Note that Eq. 10 is based on Graf's addition theorem [7] and is subjected to the validity condition $d_{ij} > r_j$.

Similarly, for a general order of scattering p, for scatterer j :

$$\phi_j^{(p)} = \{C_j^{(p)}\}^T \{H(r_j, \theta_j)\}, \tag{12}$$

with :
$$\{C_j^{(p)}\} = \sum_{\substack{i=1 \\ i \neq j}}^{Ns} [L_{ij}]\{C_i^{(p-1)}\}, \quad p \geq 2, \tag{13}$$

the total wave in the host medium is then:

$$\phi^{tot} = \phi^{inc} + \sum_{j=1}^{Ns} \sum_{p=1}^{+\infty} \{C_j^{(p)}\}^T \{H(r_j, \theta_j)\} = \phi^{inc} + \sum_{j=1}^{Ns} \{C_j\}^T \{H(r_j, \theta_j)\}, \quad (14)$$

with :

$$\{C_j\} = \sum_{p=1}^{+\infty} \{C_j^{(p)}\}. \quad (15)$$

Eq. 7, 13 and 14 give the recursive form of the multiple scattering solution. An implicit form of the solution can also be deduced [3, 7]. Note that Twersky's approach is recovered from the implicit formulation by using the Born series algorithm [14, 15].

Polymerisation Technique

This technique is based on the following observation: expressing the total wave field in a global coordinate system i.e.

$$\phi^{tot} = \phi^{inc} + \sum_{j=1}^{Ns} \{C_j\}^T \{H(r_j, \theta_j)\} = \phi^{inc} + \sum_{j=1}^{Ns} \{C_j\}^T [Q_j] \{H(r, \theta)\}, \quad (16)$$

with :

$$[Q_j]_{nm} = e^{i(n-m)\theta_{j0}} J_{n-m}(kd_j), \quad (17)$$

(r, θ) being the polar coordinates associated with the global coordinate system, $[Q_j]$ is the coordinate transformation matrix from scatterer j to the global coordinate system, (d_j, θ_{j0}) is the location of the origin of the global coordinate system with respect to the local coordinate system of scatterer j. Note that Eq. 17 is based on Graf's addition theorem and is subjected to the validity condition $r > d_j$. Defining $\{B^{total}\} = \sum_{j=1}^{Ns} [Q_j]^T \{C_j\}$, the multiple scattering solution takes the form of a single scattering solution :

$$\phi^{total} = \phi^{inc} + \{B^{total}\}^T \{H(r, \theta)\}, \quad (18)$$

and the total transition matrix for the assemblage of scatterers is then defined by :

$$\{B^{total}\} = [T^{total}]\{A\}, \quad (19)$$

where $\{A\}$ is the column matrix that contains the expansion coefficients for the incident wave expressed in the global coordinate system.

The strategy of the polymerisation technique is as follows: first, solve the multiple scattering problem for a small number of scatterers. This assemblage of scatterers defines a cell, and the transition matrix for such a cell is then calculated from the multiple scattering solution. Then, one can solve the multiple scattering problem for a small number of cells, and deduce the transition matrix that corresponds to this assemblage of cells, and this process is repeated until the transition matrix for the whole composite structure has been obtained. Note that the process that builds the final composite structure is not unique, different types of assemblage can be used. Cai & Williams [8] studied the transition matrix conformity errors, with regards to the reciprocity and the energy conservation properties, at the final stage of various scatterer polymerisation procedures to deduce the optimal one.

This polymerisation technique, together with Twersky's approach, allow the full-scale simulations of elastic-wave multiple scattering in fibre-reinforced composites to be computed. The transition matrix of every individual scatterer has been assumed to be known and expressed in the Bessel functions basis. The transition matrix for cylindrical scatterers of various cross-sections and nature can be conveniently determined in the Bessel functions basis using the Multiple Multipole Expansion method [10-13].

Validity Conditions

Two stages of the present formulation are subjected to validity conditions :

i) In Twersky's approach, Eq. 10 is a consequence of Graf's addition theorem, where a source field emanating from the origin of scatterer i is approximated by standing waves via regular Bessel functions in scatterer j's coordinate system :

$$H_n^{(1)}(kr_i)e^{in\theta_i} = \sum_{m=-\infty}^{+\infty} e^{i(n-m)\theta_{ij}} H_{n-m}^{(1)}(kd_{ij})J_m(kr_j)e^{im\theta_j}. \tag{20}$$

This series only converges for $d_{ij} > r_j$ [16]. Since Eq. 20 is used for applying the boundary conditions along scatterer j, this validity condition is satisfied if scatterer i's origin is not in the disk circumscribing scatterer j [7].

ii) In the polymerisation process, Eq. 17 also results from the Graf's addition theorem, where a source field emanating from scatterer j's origin is approximated by a source field emanating from the origin O of the global coordinate system :

$$H_n^{(1)}(kr_j)e^{in\theta_j} = \sum_{m=-\infty}^{+\infty} e^{i(n-m)\theta_{jo}} J_{n-m}(kd_j)H_m^{(1)}(kr)e^{im\theta}. \tag{21}$$

This series only converges for $r > d_j$ i.e. the response of the cell given by this representation only converges outside the disk centered at O and of radius $R = \max d_j$.

In practice, these conditions must be overly satisfied in order to obtain satisfactory results at a reasonable expansion order. Typically, the circumscribing disks of two different scatterers should not overlap in i) [3], and the response of a cell is efficiently represented via Eq. 21 outside its circumscribing disk for low and mid-frequency problems. This latter requisite affects the strategy to use for polymerisation. As shown on Fig. 1, a direct polymerisation from each scatterer to the center of the cell is applied for the case of MMCs via Eq. 16.

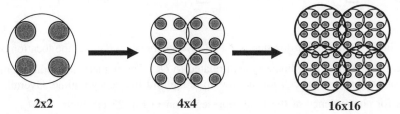

2x2   4x4   16x16

Figure 1. Direct polymerisation strategy leads to overlapping of the numerical validity domains for MMCs.

362

The numerical validity domains overlap more and more as the polymerisation is applied, and even overlap with other fibres in the neighbourhood. Thus the field emanating from the cell will not be satisfactorily represented in the neighbourhood of those fibres. And as the boundary conditions are applied via the T-matrix, numerical errors will occur which will be amplified in the next polymerisation procedures. To avoid such overlapping and hence to improve convergence, the strategy for polymerisation must be adapted to the geometry.

Polymerisation Strategy for MMCs

Extensive studies have been conducted on how to obtain satisfactory convergent field representations in the form of cylindrical Bessel expansions, especially in solid-state physics [17].

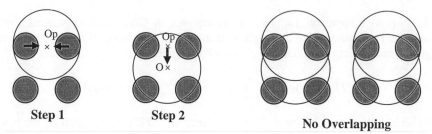

Figure 2. Numerical validity domains of a two-steps polymerisation strategy that improves convergence on left and right sides of the cell in MMCs models.

Considering an intermediate point O_p between two adjacent fibres as depicted in Fig. 2, polymerisation is performed in two steps : first the responses from the fibres are approximated from O_p i.e.

$$\phi^s = \sum_{n=-\infty}^{+\infty} B_n H_n^{(1)}(kr_i)e^{in\theta_i} = \sum_{n=-\infty}^{+\infty} B_n \left\{ \sum_{m=-\infty}^{+\infty} e^{i(n-m)\theta_{iO_p}} J_{n-m}(kd_{iO_p})H_m^{(1)}(kr_p)e^{im\theta_p} \right\},$$

(22)

thus the validity condition in this case is $r_p > d_{iO_p}$, where (d_{iOp}, θ_{iOp}) is the location of O_p with respect to the local coordinate system of scatterer i. The numerical validity domain is a disk circumscribing both fibres such that it does not overlap with other cells on the left and right sides. A similar procedure avoids overlapping at the top and the bottom of the cell. The second step is then to approximate this response from the centre of the cell :

$$\phi^s = \sum_{n=-\infty}^{+\infty} B_n \left\{ \sum_{m=-\infty}^{+\infty} e^{i(n-m)\theta_{iO_p}} J_{n-m}(kd_{iO_p}) \left[\sum_{l=-\infty}^{+\infty} e^{i(m-l)\theta_{O_pO}} J_{m-l}(kd_{O_pO})H_l^{(1)}(kr)e^{il\theta} \right] \right\},$$

(23)

thus the validity condition for this step is $r > d_{O_pO}$, where (d_{OpO}, θ_{OpO}) is the location of O with respect to the local coordinate system related to O_p. The numerical validity domain is a disk confined within the cell for the low up to the mid-frequency regime. Overall, the conditions for convergence of the field representation in Eq. 23 are both $r > d_{O_pO}$ and $r_p > d_{iO_p}$. The series in Eq. 23 do not commute.

NUMERICAL RESULTS

In the following, a representative factor for the rate of convergence of the scattered field is plotted. This factor C_r is defined as :

$$C_r(N) = \sqrt{\frac{\tilde{\mathcal{E}}(B_N H_N^{(1)})}{\tilde{\mathcal{E}}(\sum_{k=0}^{N} B_k H_k^{(1)})}}, \tag{24}$$

where N is the truncation order, and $\tilde{\mathcal{E}}$ is the normalized energy in the SH wave case :

$$\tilde{\mathcal{E}}(\phi) = \frac{2}{\mu |\phi^{inc}|^2} \mathcal{E}(\phi) = \frac{1}{|\phi^{inc}|^2} \int_A |k|^2 |\phi(r,\theta)|^2 + |\partial_r \phi(r,\theta)|^2 + \left|\frac{1}{r}\partial_\theta \phi(r,\theta)\right|^2 dA, \tag{25}$$

on a domain A around the region of interest.

The following figures show the convergence rate versus the expansion order for different cells at frequencies 1 MHz and 22.7 MHz. These frequencies correspond to wavelengths of 22 and 1 times the fibre diameter, respectively. The fibre diameter is 140µm. The distance of separation between the fibres is 132µm.

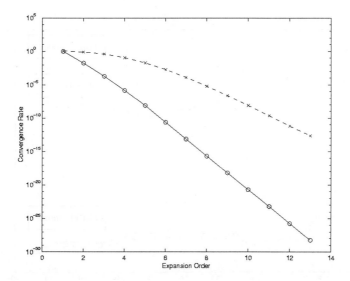

Figure 3. Convergence rate versus expansion order for the representation of the scattered field from a single SiC fibre in Titanium at 1 MHz (solid line) and 22.7 MHz (dashed line).

Fig. 3 shows the evolution of the convergence rate for a single fibre. It is a well-known fact that the convergence of the field representation in terms of Bessel functions expansions is slower as the frequency increases [17].

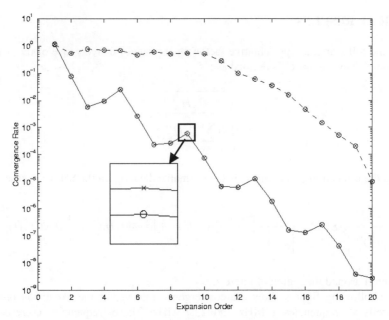

Figure 4. Convergence rate versus expansion order for the representation of the scattered field from a polymerised cell of four fibres at 1 MHz (solid line) and 22.7 MHz (dashed line), via direct polymerisation (o) and intermediate polymerisation (x).

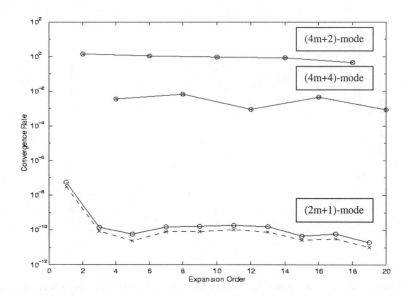

Figure 5. Convergence rate versus expansion order for the (4m+2), (4m+4) and (2m+1)-modes of the scattered field from a polymerised cell of 16 fibres at 22.7 MHz, via direct polymerisation (solid line,o) and intermediate polymerisation (dashed line,x).

Fig. 4 shows the convergence rate of the scattered field from a cell of four fibres. Results are presented for two different representations of the scattered field obtained using direct polymerisation as in Eq. 16, and an intermediate one as in Eq. 23. No noticeable differences occur at this stage since both representations are derived from the exact T-matrix for a fibre. Hence no validity domains overlap yet. However, a more accurate plot shows, as expected, that direct polymerisation performs slightly better at this stage. It is due to the double-series form of Eq. 23, which is more approximate than the single-serie form of Eq. 16 for representing the same field.

At the next stage, a cell of 16 fibres is obtained from the polymerisation of four cells containing 4 fibres each. Noticeable differences are observed at high frequencies for the odd modes of the scattered field representation only (Fig. 5). The overlapping of the validity domains is not important enough in order to generate significant errors.

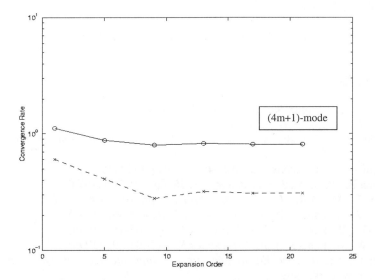

Figure 6. Convergence rate versus expansion order for the (4m+1)-modes of the scattered field from a polymerised cell of 64 fibres at 1 MHz , via direct polymerisation (solid line,o) and intermediate polymerisation (dashed line,x).

Significant differences between both polymerisation procedures can be seen at low frequencies after three polymerisations in Fig. 6, for the (4m+1)-modes of the scattered field. With direct polymerisation, the validity domains overlap with fibres in the neighbourhood. This generates errors in the representation of the fields and these errors amplify as the boundary conditions are applied. It has also been observed that, in general, the convergence of the recursive form of the solution becomes poorer as the polymerisation procedure is applied to larger and larger cells, even at low frequencies. The implicit form of the solution provides better results [8,13], since the solver minimizes the overall error. The authors are presently working on a method to control better the convergence properties and thus the accuracy of the models obtained from this formulation.

CONCLUSION

Deterministic multiple scattering formulations based on Twersky's approach and the polymerisation technique lead to full-scale simulations of elastic-wave multiple scattering in fibre-reinforced composites [7,13]. Because the convergence properties of these formulations depend on the geometry of the composite models and on the strategy chosen for the polymerisation, precautions must be taken in order to obtain convergent results. A procedure has been proposed to improve the convergence properties of MMCs models. Although the methods and the results have been presented in the frequency domain for harmonic SH waves only, time-domain solutions are sought and are obtained using an inverse Fourier-Laplace transform. Moreover, the extension to P/SV waves is straightforward [7]. The results will be validated experimentally.

ACKNOWLEDGEMENTS

This work is supported by the Engineering and Physical Sciences Research Council, grant no. GR/M83049.

REFERENCES

1. Twersky V., *J. Acoust. Soc. Am.* **24**, pp. 42-46 (1952).
2. Cheng S. L., *J. Appl. Mech.* **36**, pp. 523-527 (1969).
3. Varadan V. K., Varadan V. V. & Pao Y.-H., *J. Acoust. Soc. Am.* **63**, pp. 1310-1319 (1978).
4. Waterman P. C., *J. Acoust. Soc. Am.* **45**, pp. 1417-1429 (1969).
5. Peterson B., Ström S., *J. Acoust. Soc. Am.* **56**, pp. 771-780 (1974).
6. Bose S. K., Mal A. K., *Int. J. Solids & Struct.* **9**, pp. 1075-1085 (1973).
7. Cai L.-W., Williams J. H. Jr., *Ultrasonics* **37**, pp. 453-462 (1999).
8. Cai L.-W., Williams J. H. Jr., *Ultrasonics* **37**, pp. 463-482 (1999).
9. Cai L.-W., Williams J. H. Jr., *Ultrasonics* **37**, pp. 483-492 (1999).
10. Hafner C., *Post-modern Electromagnetics : Using Intelligent Maxwell Solvers*, John Wiley, New York, 1999.
11. Imhof M. J., *J. Acoust. Soc. Am.* **100**, pp. 2969-2979 (1996).
12. Imhof M. J., Toksöz M. N., *J. Acoust. Soc. Am.* **101**, pp. 1836-1846 (1997).
13. Lange A., Harker A. and Saffari N., *Rev. Progress in Quantit. NDE* **20**, eds. D.O. Thompson and D.E. Chimenti, AIP Conference Proceedings, New York, pp. 43-50 (2001).
14. Kitahara M., Nakagawa K., *Rev. Progress in Quantit. NDE* **17**, eds. D.O. Thompson and D.E. Chimenti, Plenum Press, New York, pp.43-50 (1998).
15. Schuster G.T., *J. Acoust. Soc. Am.* **77**, pp. 865-879 (1985).
16. Watson G.N., *A Treatise on the Theory of Bessel Functions*, 2nd ed., Cambridge University Press (1944).
17. Gonis A., Butler W.H., *Multiple Scattering in Solids*, Grad. Texts in Cont. Phys., Springer (2000).

Theoretical and Computational Acoustics 2001
E.-C. Shang, Qihu Li and T. F. Gao (Editors)
© 2002 World Scientific Publishing Co.

Elastic Wave Diffraction Fields Generated by Surface Excitation of a Crystal

Huang Xin, Wang Chenghao
Institute of Acoustics, The Chinese Academy of Sciences, Beijing 100080

The angular spectrum theory is currently used to analyse the diffraction fields generated by a finite aperture source on surface of the anisotropic medium. Only is the anisotropy of the propagation velocity regarded in such a sclar theory. So this theory is insufficient and exists many shortcomings. Since the elastic wave in solid is the vector field, the diffraction patterns for various wave modes, various components of each mode and various excited sources are different generally. The angular spectrum theory neglected actual physical field can only give a relative distribution of phenomenological field.

According to the generalized excitation theory in our previous works, the diffraction elastic wave fields excited by the surface finite-aperture source are obtained in this paper. As an example, the calculation results of the diffraction fields for YZ quartz are given and compared with that of the angular spectrum theory.

Introduction

The angular spectrum theory is usually used to analyse the diffraction fields generated by a finite aperture source on surface of the anisotropic medium[1][2]. However, such theoretical analysis is insufficient and exists many shortcomings. Firstly, it adopts a scalar field model as one does with the split diffraction, while the acoustic wave field in crystal is a vector field. The diffraction patterns for various components of vector are different generally. Secondly, for various wave models, the diffraction patterns may be great different, but the angular spectrum theory only consider the anisotropic difference of acoustic velocity. Thirdly, the nature of source are totally ignored in the angular spectrum theory. In fact, the "excitation efficiencies" of different force sources on surface are considerable distinction. Lastly, the angular spectrum theory neglected actual physical field can only give a relative value of phenomenological field.

In 1985[3][4], Wang et.al. developed a theory for generalized Green's function of surface excitation of a crystal. This theory concludes that the generalized wave field can be expressed as the convolution of the generalized sources to the generalized Green's function, and the expression of the generalized Green's function that only depends on the orientation and constants of materials is obtained. So the diffraction elastic wave fields generated by a finite aperture source of the surface of anisotropic medium can be given adequately and precisely.

In this paper, the generalized Green's function theory is reviewed above all. Then, the directivity patterns of longitudinal bulk wave Green's function and the diffraction patterns for YZ-Quartz are numerical simulated by the generalized Green's function theory. As a comparison, the results of the angular spectrum theory are also given.

The elastics field in anisotropic media generated by surface source

The following is a brief review of a rigorous theory of surface excitation. The related details can be found in references [3] and [4].

It is supposed that a one-dimension source distribution vector locates on the surface of the anisotropic medium,

$$T(x_1) = \{T_p\} = \{T_1, T_1, T_1\} = \{T_{31}, T_{32}, T_{33}\} .$$ (1)

$$(|x_1| < a)$$

It is also assumed that the source distribution is along x_1 direction with a finite aperture of $2a$, the coordinate system is shown as Fig.1.

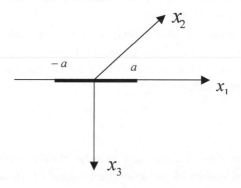

Fig.1 The source on the surface of half-space medium

The elastic wave field in wavenumber β-space generated by this source distribution can be written as

$$[\tilde{U}_m(\beta, x_3)] = [\tilde{U}_1, \tilde{U}_2, \tilde{U}_3] .$$ (2)

The boundary condition is $T_{3j} \neq 0$ (3)

The wave field solution can be expressed as following

$$\tilde{U}_j(\beta, x_3) = \sum_{n=1}^{3} C_n B_{jn} \exp(j\alpha_n \beta x_3) ,$$ (4)

here, the coefficient C_n will be determined by the boundary condition, B_{jn} are normalized eigenvector which are specified by:

$$[\Gamma] \cdot [B] = 0 .$$ (5)

Γ is the Christoffel matrix, α_n are the roots of the eigenequations:

$$\det[\Gamma] = 0 .$$ (6)

Substitution of Eq.(4) into Eq.(3) leads to

$$\left[\tilde{T}_m(\beta)\right] = \left[\tilde{\Pi}_{mn}\right] \cdot \left[C_n\right], \tag{7}$$

thus

$$\left[C_n\right] = \left[\tilde{\Pi}_{mn}\right]^{-1}\left[\tilde{T}_m(\beta)\right], \tag{8}$$

where $\left[\tilde{\Pi}_{mn}\right]^{-1} = \dfrac{\left[\tilde{W}_{nm}\right]}{\det[\Pi]}$, W_{nm} are the algebraic cofactors of $\left[\tilde{\Pi}_{mn}\right]$. The matrix of boundary

conditions is

$$\Pi_{in} = C_{3jk1}B_{kn} + C_{3jk3}B_{km}\alpha_{mn}.$$

Let

$$\left[\tilde{U}_m(\beta, x_3)\right] = \begin{bmatrix} \tilde{U}_1 \\ \tilde{U}_2 \\ \tilde{U}_3 \end{bmatrix} = \begin{bmatrix} \tilde{U}_1(\beta, x_3) \\ \tilde{U}_2(\beta, x_3) \\ \tilde{U}_3(\beta, x_3) \end{bmatrix}, \tag{9}$$

we get

$$\left[\tilde{U}_m\right] = \left[\tilde{G}_{mp}\right] \cdot \left[\tilde{T}_p\right]. \tag{10}$$

here $\tilde{G}_{mp}$ is the Green function in β-space

$$\tilde{G}_{mp} = \sum_{n=1}^{3} \frac{B_{mn} \cdot W_{nm}}{\det[\tilde{\Pi}_{mn}]} e^{j\alpha_n \beta x_3}. \tag{11}$$

By the inverse Fourier transform of Eq.(10), the elastic wave filed solution in real-space can be obtained

$$U_m(x_1, x_3) = \sum_{p=1}^{3} \int_{-\infty}^{+\infty} G_{mp}(x_1 - x_1', x_3)T_p(x_1', 0)dx_1', \tag{12}$$

where force source is

$$\tilde{T}_p(x_1, 0) = \frac{1}{2\pi} \int_{-\infty}^{+\infty} \tilde{T}_m(\beta)e^{j\beta x_1} d\beta, \tag{13}$$

the Green function G_{mp} are

$$G_{mp}(x_1, x_3) = \frac{1}{2\pi} \int_{-\infty}^{+\infty} G_{mp}(\beta, x_3)e^{j\beta x_1} d\beta$$

$$= \frac{1}{2\pi} \int_{-\infty}^{+\infty} (\sum_{n=1}^{3} \frac{B_{mn} \cdot W_{np}}{\det[\tilde{\Pi}_{mn}]})e^{j(\alpha_n \beta x_3 + \beta x_1)} d\beta$$

$$= \frac{1}{2\pi} \int_{-\infty}^{+\infty} (\sum_{n=1}^{3} \frac{B_{mn} \cdot W_{np}}{\det[\tilde{\Pi}_{mn}]})e^{j(\alpha_n \beta x_3 + \beta x_1)} d\beta. \tag{14}$$

The Eq.(12)-Eq.(14) are the results given by our exact vector theory, i.e., the elastic wave field in half-space media generated by surface source. It is the convolution of surface source vector to the Green function G_{mp}.

For angular spectrum theory, the scalar U can be represented as

$$U^{(p)} = \int_{-\infty}^{+\infty} G(x_1 - x_1', x_3) T_p(x_1') dx_1', \tag{15}$$

or

$$U^{(p)} = \frac{1}{2\pi} \int_{-\infty}^{+\infty} \tilde{G}(\alpha, x_3) \tilde{T}_p(\alpha) e^{j\alpha x_1} d\alpha$$

$$= \frac{1}{2\pi} \cdot \sum_{n=1}^{3} C^{(n)} \int_{-\infty}^{+\infty} \tilde{T}_p(\alpha) e^{j(\alpha x_1 + \gamma_n x_3)} d\alpha, \tag{16}$$

in which

$$G(x_1, x_3) = \int_{-\infty}^{+\infty} \tilde{G}(\alpha, x_3) e^{j\alpha x_1} d\alpha \tag{17}$$

$$\tilde{G}(\alpha, x_3) = \sum_{n=1}^{3} C^{(n)} e^{j\gamma_n(\alpha) x_3}, \tag{18}$$

where $C^{(n)}$ are indeterminable constants.

Thus, the differences between our exact theory from the angular theory can be concluded as following: 1)The elastic wave is a vector field, but the angular spectrum theory only include a scalar U; 2)For angular spectrum theory, the actual physical value of wave fields can't be carry out because the coefficient $C^{(n)}$ are indeterminable; 3)The angular spectrum theory only consider the difference of the acoustics velocity of various wave mode in anisotropic media. For our theory, there exist the various excitation factors $g_{mp}^{(n)}(\theta)$, which respond to various wave modes, $k_n (n = 1,2,3)$, field components, $U_m (m = 1,2,3)$, and force sources, $T_p(p = 1,2,3)$, respectively, The $g_{mp}^{(n)}(\theta)$ that depended on the material characteristics are anisotropic also.

It is assumed that the force source with the aperture $2a$ vibrates uniformly. So

$$T_p(x_1) = C \cdot \Gamma ect(\frac{x_1}{2a}), \tag{19}$$

$$\tilde{T}_p(\alpha) = C' \cdot \frac{\sin \alpha a}{\alpha a}. \tag{20}$$

In general, the force source distribution can be divided into enough small element sources with uniform amplitude and phase. Every element sources generate the elastic wave field $\{\delta U_m\}$. The contribution of whole diffraction wave field can be given by the vector superposition of each element field. Furthermore, at the far field, the $\{\delta U_m\}$ can be obtained by the stationary phase method when in the polar coordinate system:

For our exact theory,

$$\delta U_m^{(p)}(r,\theta) \approx$$

$$j\sqrt{\frac{1}{2\pi r}} \sum_j^3 \frac{1}{\sqrt{a_j}} \frac{B_{mj}(\beta_j)W_{jp}(\beta_j)}{\det[\Pi(\beta_j)]} \cdot (\frac{2\sin(\alpha_n\Delta)}{\alpha_n} \delta T_p) \cdot \exp[jr(\alpha_j(\beta_j)\beta_j\sin\theta + \beta_j\cos\theta) - j\frac{\pi}{4}] \qquad (21)$$

For angular spectrum theory,

$$\delta U_m^{(p)}(r,\theta) \approx$$

$$j\sqrt{\frac{1}{2\pi r}} \sum_j^3 \frac{C^{(n)}}{\sqrt{a_j}} \cdot (\frac{2\sin(\alpha_n\Delta)}{\alpha_n} \delta T_p) \cdot \exp[jr(\alpha_j(\beta_j)\beta_j\sin\theta + \beta_j\cos\theta) - j\frac{\pi}{4}] \qquad (22)$$

where the α_n satisfies the saddle point condition:

$$[\frac{d(\alpha_j\beta)}{d\beta}]\bigg|_{\beta=\beta_j} = -ctg\theta,$$

and $a_j = |d^2(\alpha_j(\beta)\beta)/d\beta^2|_{\beta=\beta_j} \cdot \sin\theta$.

Numerical results

As an example, the numerical results of the diffraction generated by the surface source on YZ-Quartz are given. Fig.2 and Fig.3 show the slowness and group velocity curves of YZ-Quartz. It can been seen that the slowness curves are not circle, so the energy flow angle is not the same as that of the phase velocity unless in the pure mode direction. In some direction range, the wave that have several difference wave vector directions, i.e., there are several wave fronts for one wave motion.

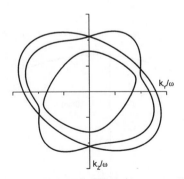

Fig.2 Slowness curves of YZ-Quartz

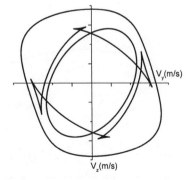

Fig.3. Group velocity curves of YZ-Quartz

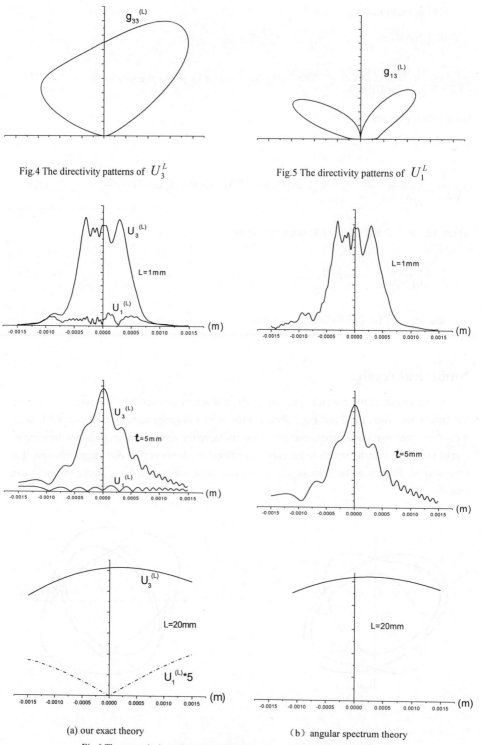

Fig.4 The directivity patterns of U_3^L

Fig.5 The directivity patterns of U_1^L

(a) our exact theory

（b）angular spectrum theory

Fig.6 The numerical results of the longitudinal wave diffraction fields
in various distance on the YZ-Quartz

The directivity patterns of U_3^L, U_1^L, the two components of the longitudinal wave field, are shown in Fig.4 and Fig.5. The numerical results of the longitudinal wave diffraction fields in various distance obtained by our exact theory and the angular spectrum theory are shown in Fig.6(a) and (b), respectively. The frequency of force source is 200MHz and the aperture is 1mm. Compared the results of our theory to those of angular spectrum theory, it can be seen that the results of U_3^L are obviously different in the near field, but they are similar in the far field. The results of the U_1^L component can't be carried out by the angular spectrum theory. However, the actual vector wave field can be given by our theory.

References

[1] Papadakis, E.P., "Ultrasonic diffraction loss and phase change in anisotropic materials", J. Acoust. Soc. Am., 40(1966), 863-876

[2] Kharusi,M.S. and Farnell,G.M., "Plane ultrasonic transducer diffraction fields in highly anisotropic srystals", J. Acoust. Soc. Am., 48(1970), 665-670

[3] Wang, Chenghao and Chen, Dongpei, "Analysis of surface excitation of elastic wave field in a half space of piezoelectric crystal", Part I&II, Chinese J. Of Acoustics, 4(1985), No.3, 232-243 & 4(1985), No.4, 297-313

[4] Wang Chenghao etal, "SAW diffraction field generated by source with finite aperture on a piezoelectric crystal surface", Chinese J. Of Acoustics, 11(1992), No.1, 1-10

Theoretical and Computational Acoustics 2001
E.-C. Shang, Qihu Li and T. F. Gao (Editors)
© 2002 World Scientific Publishing Co.

Study of Zigzag Dispersion Curves in Rayleigh Wave Exploration

Laiyu Lu, Bixing Zhang and Honglang Li

Institute of Acoustics, The Chinese Academy of Sciences, Beijing, 100080,

People's Republic of China

SUMMARY

The zigzag dispersion curves of Rayleigh wave are usually obtained in practical exploration especially in multi-layered media contained the low-velocity zones. The mechanism of zigzag dispersion curves of Rayleigh wave in multi-layered media is studied in this paper, and the characteristics of zigzag shapes in different models are also discussed. It is found that the number of zigzag shapes decreases with the decrease of the number of the low-velocity zones in multi-layered media. The shapes of the zigzag dispersion curves can reflect the position of the low-velocity zones. In addition, the effects of parameters of medium models on the zigzag dispersion curves are investigated in detail.

INTRODUCTION

Rayleigh wave is also named as the surface wave which energy decreases rapidly with the depth from the free surface increases. Rayleigh wave exploration is a shallow seismic exploration method that can be used to interpret the geological information below the free surface. The method has been focused by many researchers because of its' high resolution, simple operation, nondestruction and other characteristics[1-7]. Rayleigh wave with enriched frequencies can be excited by sources in practical exploration. The different wavelength (λ) corresponds the different exploration depth. The velocity-wavelength profile can be obtained by cross correlation method in data processing. The effective exploration depth of Rayleigh wave is related to its wavelength, so $V - \lambda$ curve is usually known as the velocity-depth ($V - z$) profile which indicates the information of different depth. Generally speaking, the $V - z$ profiles of Rayleigh wave obtained in actual exploration are non-continual zigzag curves for multi-layered medium contained low-velocity zones[3-5]. The zigzag shapes can give the possible positions of the low-velocity structures (fractures, oil, gas, etc). However, for a given mode of the guided wave, the dispersion curve calculated by the elastic theory is a smooth and continual curve without zigzag shapes. Mechanism of zigzag dispersion curves in actual exploration has been studied theoretically by few authors. In this paper, the mechanism of zigzag dispersion curves in Rayleigh wave exploration will be investigated thoroughly.

It is proved theoretically that there usually exist infinite modes which are corresponding to the guided waves propagating along the stratified direction in multi-layered media[4, 8-11]. They are named as the fundamental (first), second, third, etc. modes from the low to high frequency range. It is normally assumed in previous applications that the fundamental mode dominates the recorded waveforms and higher modes can be ignored. In fact, however, the higher modes are always generated and can sometimes possess significant amounts of the energy. The signals received by the detectors would be the results of the superposition of all modes. The actual contribution of each individual mode can be represented by a complicated function of layered model, frequency and receiver locations[7]. The characteristics of zigzag shapes of $V - \lambda$ will be affected by these factors.

In this paper, dispersion curves in Rayleigh wave exploration are studied. At first, the formulations are stated. Then, numerical results are obtained. The properties of zigzag dispersion curves of multilayered media are analyzed, and the effects of some media parameters on the dispersion curves are investigated. The significant conclusions are obtained.

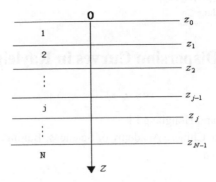

Fig. 1 Configuration of multi-layer media model.

FORMULATION

A semi-infinite medium made up of N parallel, homogeneous, isotropic layers are considered. All layers will be assumed to be elastic solid. The cylindrical coordinate system (r,θ,z) is adopted [Fig. 1]. The positive z axis is taken as the directed into the medium. The various layers and interfaces are numbered away from the free surface. For the jth layer, its' properties are denoted by V_{pj} (P-wave velocity), V_{sj} (S-wave velocity), ρ_j (density) and h_j (thickness). The argument (j) will be omitted whenever it is possible without causing confusion.

It is convenient to introduce B,P,C coordinate system[8-12]

$$\begin{cases} B = e_r \dfrac{\partial}{\partial(kr)} + e_\theta \dfrac{\partial}{kr\partial\theta}, \\ C = e_r \dfrac{\partial}{kr\partial\theta} - e_\theta \dfrac{\partial}{\partial(kr)}, \\ P = e_z. \end{cases} \qquad (1)$$

Only is P-SV wave considered in this paper. The displacement components in B,P,C coordinate system and cylindrical coordinate system (r,θ,z) satisfy

$$\begin{cases} u_r = \dfrac{1}{k}\dfrac{\partial u_B}{\partial r} + \dfrac{1}{kr}\dfrac{\partial u_c}{\partial\theta} \\ u_\theta = \dfrac{1}{kr}\dfrac{\partial u_B}{\partial\theta} - \dfrac{1}{k}\dfrac{\partial u_C}{\partial r} \\ u_z = u_P \end{cases} \qquad (2)$$

Define the following vectors of S (motion stress) and ϕ

$$\begin{cases} S = (U_B/k, U_P/k, \tau_P/\omega^2, \tau_B/\omega^2)^T, \\ \phi(z) = (Ae^{iaz}, Be^{-iaz}, Ce^{ibz}, De^{-ibz})^T = (\varphi^+, \varphi^-, \psi^+, \psi^-)^T. \end{cases} \qquad (3)$$

Then, the vectors S and ϕ satisfy

$$S = M\phi, \ \varphi(z_j) = \lambda\phi(z_{j-1}), \lambda = diag(P, 1/P, Q, 1/Q)$$

$$M = \begin{bmatrix} 1 & 1 & \gamma_s & -\gamma_s \\ \gamma_p & -\gamma_p & 1 & 1 \\ \rho(\gamma-1) & \rho(\gamma-1) & \rho\gamma\gamma_s & -\rho\gamma\gamma_s \\ \rho\gamma\gamma_p & -\rho\gamma\gamma_p & \rho(\gamma-1) & \rho(\gamma-1) \end{bmatrix} \qquad (4)$$

The superscript T in Eq.(3) represents the transposed-matrix, $c = \omega/k$, $P = e^{\gamma_p kh}$, $Q = e^{\gamma_s kh}$, $\gamma = 2(V_s^2/c^2)$, $\gamma_p = -(1 - c^2/V_p^2)^{1/2}$, $\gamma_s = -(1 - c^2/V_s^2)^{1/2}$, $ia = k\gamma_p$ and $ib = k\gamma_s$. An implicit time dependence $e^{-i\omega t}$ of the field is assumed. $\varphi = \varphi^+ + \varphi^- = Ae^{iaz} + Be^{-iaz}$ and $\psi = \psi^+ + \psi^- = Ae^{ibz} + Be^{-ibz}$ are the displacement potentials of P and SV waves in frequency wave number domain, respectively. φ^+ and ψ^+ represent waves propagating along the positive z axis, while φ^- and ψ^- represent waves propagating along the negative z axis. So the real parts of γ_p and γ_s should be smaller (or equal to) than zero.

By the boundary conditions of every interface, the vanishing of two stress components at the free surface and the potential at infinite, the dispersion equation can be obtained

$$E_6^{(1)} = 0 . \tag{5}$$

It means that the sixth component of vector E at the free surface is vanishing, and it can be can be obtained by propagator matrix using the relation

$$E^{(j-1)} = F^{(j)} E^{(j)} , \quad (j = 2,3,\cdots, N) . \tag{6}$$

Where matrix F is propagator matrix which can be decomposed into the following form

$$F = \frac{1}{4\rho^2 \gamma_p \gamma_s} U \lambda^* V . \tag{7}$$

The representations of matrices λ^*, U and V can be referenced the related literatures[8-11].

It is assumed that the source locates at $z = z_s, r = 0$ in the first layer medium. S_1 and S_2 components of vector S at the free surface can be obtained

$$\begin{cases} S_1 = \dfrac{\Delta_1}{E_6^{(1)}} = \dfrac{(-T_{11}E_6^{(1)}+T_{31}E_3^{(1)}-T_{41}E_2^{(1)})A_{sn}+(-T_{12}E_6^{(1)}-T_{32}E_3^{(1)}-T_{42}E_2^{(1)})B_{sn}}{E_6^{(1)}} , \\[2mm] S_2 = \dfrac{\Delta_2}{E_6^{(1)}} = \dfrac{(-T_{21}E_6^{(1)}+T_{31}E_5^{(1)}+T_{41}E_3^{(1)})A_{sn}+(-T_{22}E_6^{(1)}-T_{32}E_5^{(1)}+T_{42}E_3^{(1)})B_{sn}}{E_6^{(1)}} . \end{cases} \tag{8}$$

Where A_{sn}, B_{sn} and matrix T are the quantities related to the source[8]. Combined with the Eq.(2), it can be obtained the displacement components at the free surface in the cylindrical system

The modes of guided waves propagating in multi-layered media are all given by the dispersion equation (5). It can be proved that the dispersion function is real for the real horizontal wave number. Only are the modes of real horizontal wave number considered in this paper. These modes can be obtained easily by the bisection method in numerical simulation, and their excitation amplitudes can be obtained by residues of the poles which is determined by Eq.(5)

$$\begin{cases} u_r (r,\theta,z;\omega) = i\pi\Delta_1 [\dfrac{nH_n^{(1)}(kr)}{kr} - H_{n+1}^{(1)}(kr)]k^2 \cos(n\theta) \Big/ \dfrac{\partial E_6^{(1)}}{\partial k} , \\[2mm] u_z (r,\theta,z;\omega) = i\pi\Delta_2 H_n^{(1)}(kr)k^2 \cos n\theta \Big/ \dfrac{\partial E_6^{(1)}}{\partial k} . \end{cases} \tag{9}$$

The waveforms of time domain can be obtained by the Fourier transform of Eq.(9).

ZIGZAG DISPERSION CURVES ARE CAUSED BY MULTI-MODES IN LAYERED MEDIA

In practical application, zigzag dispersion curves of the guided waves can be used to locate the position of the low-velocity zones. It means that the zigzag shapes maybe exist only in layered media. How about the dispersion curves in the homogeneous half-space media? Why is the zigzag shapes exist in layered media? Two models are considered to answer these problems.

Tab.1 Parameters of homogeneous half-space and three-layer models

Model	Densities (kg/m³)	P-wave velocity (m/s)	S-wave velocity (m/s)	Thickness (m)
1	2500	4000	2500	∞
2	3000	6000	3500	5
	2370	3500	2000	2
	3000	6000	3500	∞
3	3000	6000	3000	5
	2370	3500	2000	2
	3000	6000	3500	∞
4	3000	6000	3000	5
	2370	3500	2000	2
	3000	5200	3500	∞

Firstly, a homogeneous half-space model (model 1) is considered, of which the parameters are shown in Tab.1. In this case, there exist only one kind of guided wave whose phase velocity is 2269m/s without dispersion phenomenon. Fig.2 (a) gives the displacement intensity curves. The result is obtained by the Eq. (9) and the offset (the distance from source to the detector) is $20m$. Fig. 2 (b) is the waveforms in time domain. The offsets are $8m, 10m, \cdots, 26m$ from down to up, respectively. The amplitudes have been normalized. By the signal show in Fig. 2 (b), the velocity-depth profile can be obtained by the cross correlation method, It is shown in Fig.2(c). There are no zigzag shapes in this case.

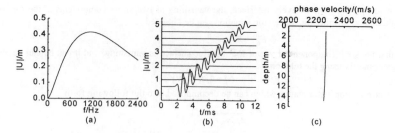

Fig. 2   The displacement intensity curve (a), waveforms (b) and the
velocity-depth profile (c) in model 1.

Then, another three-layer medium contained a low-velocity zone (model 2) is considered. The dispersion curves are depicted in Fig.3 (a). It can be seen that there exist infinite guided waves. The first mode has no cut-frequency and the phase velocity is $V_{3\infty}$ (the Rayleigh wave velocity in the case where there is only the third layer) when the frequency f=0. The other modes have the cut-frequency at which the phase velocities are all equal to that of the S-wave velocity of the last layer. The phase velocity of the model 4 is almost no related to frequency in high frequency range, and it is equal to $V_{1\infty}$ (the Rayleigh wave velocity in the case where there is only the first layer). But the phase velocities of the other modes tend towards the S-wave velocity of the low-velocity zone when the frequency tends towards infinite.

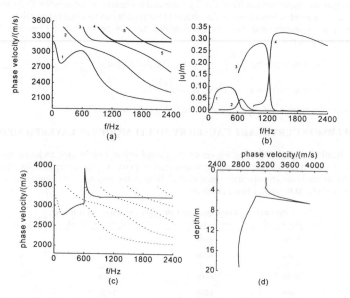

Fig.3   The dispersion curves (a,c,d) and displacement intensity curves (b) of modes in model2.

Fig.3 (b) gives the excitation intensities of the guided waves. It can be seen that excitation intensities of different modes have different distribution ranges. In low-frequency range, the dominant intensity of guided waves is the first mode, while in high frequency range the intensity of mode 4 is the dominion. So in different frequency ranges, the intensities of the modes excited by the source are different. If the source frequency increases from zero to infinite, the received guided waves maybe skipped among the modes, and the dispersion curves will be non-continual. Solid line in Fig.3 (c) is the dispersion curves obtained by the cross correlation method. It can be seen that the curve is skipped from the first mode to the third mode when the frequency is grater than 600Hz, and to the fourth mode when the frequency is grater than 1200Hz. The intensities of second mode and the modes whose orders are higher than 4 are relatively small [Fig. 2(b)]. So these modes has no influence to the dispersion curves. If this $V - f$ curve is transformed further into the $V - \lambda$ curve, the zigzag dispersion curve could be obtained [Fig. 3(d)]. This structure of zigzag dispersion curve of Rayleigh wave is very similar to that obtained in Rayleigh wave exploration.

From the above discussion, it is easy to see that the dispersion curve in homogeneous half-space has no zigzag shapes, while it is otherwise for layered medium contained a low-velocity zone. Why is it? It was known that there exist many modes for layered medium and whose excitation intensity and phase velocity for given frequency are different each other. The signals that the detector received are the superposition of all modes, so the dispersion curve obtained by the cross correlation method will not be the curve of some single mode, but the curve formed by paragraphly skip among all modes according to the contribution to the detector of each mode. It is apparent that the dispersion curve has no zigzag shapes for homogeneous half-space because in which there is only one mode without dispersion phenomenon. It can be concluded that the reason for the zigzag dispersion curves is related to the multi-modes in layered medium.

THE EFFECTS OF MEDIA PARAMETERS ON THE ZIGZAG DISPERSION CURVES

The dispersion equation is the function of four parameters for each layer: S-wave velocity, P-wave velocity, density and thickness. In Rayleigh wave exploration, the information of the media model can be deduced through the properties of zigzag dispersion curve. In our past works on this topic[4], the effects of layer thickness and Possion's ratio on zigzag dispersion curves have been studied. In this section, the effects of S-wave velocities, P-wave velocities and densities of different layers on zigzag dispersion curves will be investigated in detail.

The effects of S-wave velocities on dispersion curves

The Model 2 is still considered, the effects of S-wave velocities on zigzag dispersion curve are investigated by changing the S-wave velocities of each layer while other parameters are invariable. As comparison, the dispersion curves of the model 2 is given by the dash line in Fig. 4.

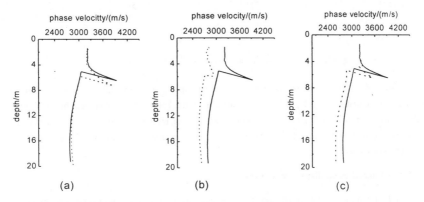

Fig. 4.The effects of S-wave velocities on the dispersion curves.

The solid line in Fig.4 (a) gives the zigzag dispersion curve obtained by changing the S-wave velocity of the third layer from 3500m/s to 4000m/s. It can be seen that the phase velocity near $20m$ increases due to $V_{3\infty}$ increases. Although the zigzag shape has a small shift towards to the depth direction, it can still reflect the location of the low-velocity zones. Near the free surface, the phase velocity equal to $V_{1\infty}$.

Fig.4 (b) gives the zigzag dispersion curves by changing the S-wave velocity of first layer from 3500m/s to

3000m/s. $V_{l\infty}$ becomes smaller in this case. It can be seen that the difference between the dispersion curves after and before the V_{s1} is changed is more significant than that of Fig.4 (a). The phase velocity at any depth is become smaller than that of before V_{s1} is changed. The location of the zigzag shape can also reflect the location of the low-velocity zone.

The dispersion curves by changing the S-wave velocity of low-velocity from 2000m/s to 1500m/s are shown in Fig.4 (c). The properties of zigzag dispersion curve near the free surface are mostly dominated by the parameters of the first layer, and the properties of dispersion curve at other depth are mostly dominated by the parameters of the second and the third layer.

The effects of P-wave velocities on dispersion curves

In order to investigate the effects of P-wave velocities on the dispersion curves, The model 3 is considered [Tab. 1]. The solid line in Fig. 5 gives the dispersion curves in the model 3. The dash line in Fig. 5 (a) is the dispersion curve by changing V_{p3} from 6000m/s to 5200m/s. It can be seen that the dispersion curves is almost coincidence with that of before V_{p3} is changed. This means that the effects of the P-wave velocity of the third layer is smaller than that of S-wave velocity of the third layer compared to Fig. 4 (a).

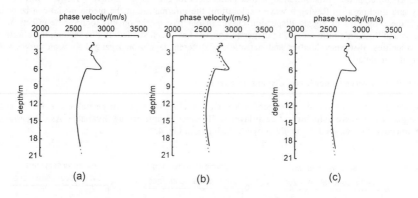

Fig. 5 The effects of P-wave velocities on the dispersion curves.

The dash line in Fig. 5(b) is the dispersion curve by changing V_{p1} from 6000m/s to 5200m/s. Similar to Fig. 4(b), the phase velocity at any depth become smaller than that before V_{p1} is changed, but it is apparent less than that of Fig. 4(b). Fig. 5(c) gives the dispersion curve by changing V_{p2} from 3500m/s to 3000m/s, It can be seen that the dispersion curve has no changes after V_{p2} decreases except that phase velocity at deeper location becomes small.

The effects of densities on dispersion curves

The model 4 shown in Tab. 1 is considered to investigate the effects of densities on dispersion curves which are given by solid lines in Fig. 6. The dash lines in Fig. 6(a), 6(b) and 6(c) are the dispersion curves by changing ρ_3 from 3000kg/m³ to 2500 kg/m³, ρ_1 from 3000kg/m³ to 2500 kg/m³ and ρ_2 from 2370kg/m³ to 1500 kg/m³, respectively. It can be seen that the location of the zigzag shape has no changes regardless of any changes of densities of any layers, but phase velocity of deep position has some corresponding changes.

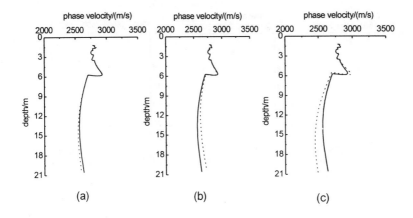

Fig. 6 The effects of densities on the dispersion curves.

On the base of the analyses in above paragraphs, it can be concluded that the S-wave velocity is the dominant parameter influencing the zigzag dispersion curves of Rayleigh wave. The changes of the S-wave velocities of each layer affect not only the amount of phase velocity but also the location of the zigzag shapes. The latter is similar to the changes in the thickness of each layer.

Compared to the effects of S-wave velocities on dispersion curves, the effects of P-wave velocities and the densities are small relatively. The changes of the P-wave velocities and the densities of each layer have no effects on the location of the zigzag shapes. Based on this reason, only are S-wave velocity and thickness of each layer considered in the inversion of Rayleigh wave exploration. On the other hand, the changes of the parameter of first layer will cause more apparent changes in zigzag dispersion curves compared to the same parameter of the second and the third layer.

PROPERTIES OF DISPERSION CURVES FOR MULTI-LAYERED MODELSS

Only is three-layer mode contained one low-velocity zone considered in the above discussion. In this case, the dispersion curves has only one zigzag shape whose location can indicate the position of the low-velocity zone. In this section, multi-layered models contained more than one low-velocity zones are considered.

Tab.2 Parameters of the multi-layered models

Model	Densities (kg/m³)	P-wave velocity (m/s)	S-wave velocity (m/s)	Thickness(m)
	2500	4000	2500	5
	1060	2350	1120	0.8
5	3000	6000	3500	3
	1370	1500	800	2
6	2800	5000	3000	∞ (3 or 4 for seven-layer model)
	1000	1200	650	1 or 0.8
	2800	5000	3000	∞

Firstly, a five-layer model (model 5) contained two low-velocity zones are considered. The parameters of each layer are shown in Tab.2. The solid line in Fig. 7 (a) gives the dispersion curves for this case. It can be seen apparently that there are two zigzag shapes which corresponding to two low-velocity zones, respectively. The first zigzag shape near the free surface located in $5m$ which is very the location of the first low-velocity layer. However, the second zigzag shape locates in $12m$ and it is deeper than the location (8.8m) of the low-velocity layer. It means that bigger errors will be got in the location of the second low-velocity zone than that of the first low-velocity zone.

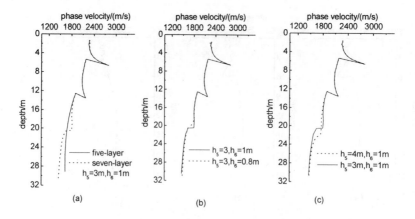

Fig. 7 The dispersion curves in multi-layered model.

The dash line in Fig.7 (a) is the dispersion curve for a seven-layer model (model 6) in which $h_5 = 3m$ and $h_6 = 1m$ as shown in Tab.2. It is found that there exist three zigzag shapes in dispersion curves. The first and the second zigzag shapes near the free surface are coincidence with that of the five-layer model in above, while the third zigzag shape is near 20m. Similar to the second zigzag shape, the location of the third zigzag shape is deeper than the actual position.

The dispersion curves of the seven-layer models that the third low-velocity zone has different thickness are given in Fig. 7 (b). As we expect that the locations of the first and second zigzag shapes of $h_6 = 0.8m$ are coincidence with that of $h_6 = 1m$. The third zigzag shape for two cases locate in the same depth, and the zigzag shape of $h_6 = 0.8m$ is smoother and vaguer than that of $h_6 = 1m$. In practical, the third zigzag shape will be vanished when the thickness of the third low-velocity zone decreases to a small value. It is the reason that the thin low-velocity zone is difficult to find in Rayleigh wave exploration. This phenomenon was investigated in detail in our previous work.

Fig.7 (c) gives the dispersion curves of the seven-layer models that the third low-velocity zone has different location (i.e. h_5 is different). It can be found that the third zigzag shape of $h_5 = 4m$ locates in deeper position compared to that of $h_5 = 3m$. In other words, for multi-layered model, the location of zigzag shape can also reflect the position of the low-velocity zone, and this is the same as the three-layer model.

It is found that there exist some errors in interpretation of the position of the low-velocity layer based on the location of the zigzag shapes in $V - \lambda$ curves. The deeper the low-velocity zone is, the bigger the error is. A solution to avoid or decrease the errors is to take the location of the zigzag shapes multiplied by a weight coefficient as the actual position of the low-velocity zone. The weight coefficient can be given by combining the personal experiences with the elastic properties of the work areas. According to above analyses of the multi-layered model, it can be found that better results can be obtained if the weight coefficient of deeper location bigger than that of shallow location for a same dispersion curves is adopted.

CONCLUSION

In Rayleigh wave exploration, longer wavelength penetrate deeper than shorter wavelength for a given mode. The longer and shorter wavelength are sensitive to the elastic properties of the deeper and shallower zones, respectively. For this reason, the location of the zigzag shapes can indicate the position of the low-velocity zone in $V - \lambda$ profile. Sometimes big errors may be obtained by this corresponding relation. So the ratio of the penetrating depth to the wavelength of Rayleigh wave is introduced. But the value of ratio is given only by the empirical formulation. The analyses of this paper show that S-wave velocity is a dominant factor to the properties of the zigzag dispersion curves. Theoretical basis is offered to determine the value of this ratio. In addition, the study of mechanism of zigzag dispersion curves will lay a fundamental theory to the inversion of Rayleigh wave exploration considering high modes.

ACKNOELEDGEMENT

This work was supported by the National Natural Science Foundation of China (19804019).

REFERENCES

[1] Abo-Zena A. Dispersion function computation for unlimited frequency values. Geophys. J. R. Astron. Soc., 1979,**58**:91-105.

[2] Menke W. Comment on 'Dispersion function computations for unlimited frequency values' by A Abo-Zena. Geophys. J. R. Astr. Soc., 1979,**59**:315-323

[3] Xiaoping Guan, Jiazheng Huang and Hongqiu Zhou. Probe the interpretation theory of steady-state Rayleigh wave method in engineering exploration., Chinese J. Geophys., 1993,**36**,(1):96-105.

[4] Bixing Zhang, Boxun Xiao and Wenjie Yang *et.al.*, Mechanism of zigzag dispersion curves in Rayleigh wave exploration and its inversion study., Chinese Journal of Geophysics., 2000,**43**(4):557-567.

[5] Jiazheng Huang, Hongqiu Zhou and Xiaoping Guan. Theoretical study of the Rayleigh-wave technique in engineering geology. Geophysical & Geochemical Exploration., 1991,**15**(4):268-277.

[6] Yinguan Wang, Chong Tian and Guomin Yao *et.al.* Reasearch of Rayleigh surface wave nondestructive measurement for plane residual stresses within a heat-patching disk. Acta Acoustica., 1999,**24**(1):53-58.

[7] Jianghai Xia, Richard D. Miller and Choon B. Park. Estimation of near-surface shear-wave velocity by inversion of Rayleigh waves. Geophysics.,1999,**64**(3):691-700.

[8] Bixing Zhang, M. Yu. and C. Q. Lan *et.al.*, Elastic wave and excitation mechanism of surface waves in multilayered media,. J. Acoust. Soc. Am., 1996,**100**(6):3527-3538.

[9] Bixing Zhang, Wei Xiong and M. Yu. *et.al.*, Study of energy distribution of guided waves in multilayered media.,J. Acoust. Soc. Am., 1998, **103**(1):125-135.

[10] Bixing Zhang, M. Yu. and X. Wei. *et.al.* Study of acoustic wave and surface waves in stratified media. Acta. Acoustica.,**22**(3):230-241.

[11] Bixing Zhang, M. Yu. and X. Wei. *et.al.*Trapped modes in multilayered media included a low-velocity formation. Chinese Journal of Nonferrous Metals., 1998, **8**(2):340-346.

[12] Youming Li, Peiyi Shu. On surface wave dispersion and body wave generalized reflection coefficient computations for layered media. Acta Geophysica Sinica. 1982, **25**(2):131-139.

Theoretical and Computational Acoustics 2001
E.-C. Shang, Qihu Li and T. F. Gao (Editors)
© 2002 World Scientific Publishing Co.

Time Reversal Self-Adaptive Focusing in Anistropic Elastic Solid Medium

Bixing Zhang and Chenghao Wang
Institute of Acoustics, The Chinese Academy of Sciences, 100080, Beijing,
People's Republic of China

Abstract

In this paper, the time reversal method in anisotropic elastic solid is theoretically studied for the first time. The transversely-isotropic anisotropic medium (6mm) is modeled as the anisotropic elastic solid. And unidirecional glass-reinforced epoxy-fiber is chosen as the material of the 6mm anisotropic medium. Time reversal acoustic field is numerically investigated by ray approximation method. The focused acoustic field has different characteristics in different direction. The focused field is also symmetric about the principal axes. It is found that the width of the principal lobe of the focused acoustic field reaches the minimum in the maximum group velocity direction and reaches the maximum in the minimum group velocity direction. The relation of time reversal acoustic field to the parameters of anisotropic medium is also studied in detail.

Introduction

Time Reversal (TR) is a novel method of self-adaptive focusing which doesn't require the priori knowledge about the properties and structures of the media and transducer. The sound beams can be bent and the focusing points can be defocused in the inhomogeneous media. This phenomena can create the phase aberrance and image distortion. TR method can overcome this difficult and realize the re-focusing. This is proved in theory and experiment[1-9].

However, the focusing beams can be also defocused by the anisotrpy. The focusing beam that is determined to be focused to some point will be not focused to the predetermined focusing point because the propagation velocity is related to the propagation direction. The different focusing results will be obtained in different propagation direction. The self adaptive focusing in anisotropy is more difficult and more complicated than that in isotropy. Many people studied the TR self adaptive focusing in isotropy, but the TR works in anisotropy have not been seen.

In this paper, the TR self adaptive focusing in anisotropy is investigated. At first, the theoretical formulation about the anisotropy and TR method is reviewed. Then, the numerical simulation results are given and analyzed.

Theoretical formulation

It is considered the half infinite anisotropic medium which is the non-piezoelectric 6mm crystal structure. It is also assumed that the angle between the crystallographic axis of the 6mm material and the free surface is equal to φ. Two Cartesian coordinate systems (x, y, z) whose x and y axes are on the free surface and (x', y', z') oriented along the crystallographic axis are adopted, respectively. The y and y' axes are superposed in the free surface. The anisotropic 6mm material is in the range $z \geq 0$ and the region $z < 0$ is the vacuum. It is only considered two dimensional wave propagation in this paper (i.e. the acoustic field is no related to the y coordinate). The transmission transducer array is on the free surface and extends to infinite along the y direction. The center of the transmission transducer array is on the origins of the Cartesian coordinate systems (x, y, z) and (x', y', z'). The configuration of the medium is shown in Fig. 1.

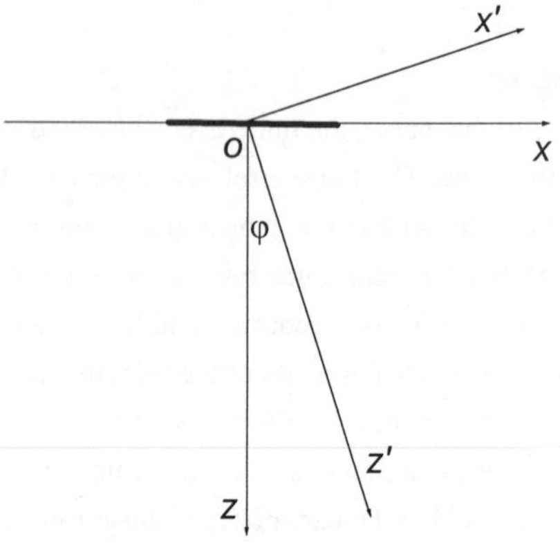

Fig. 1. The geometrical configuration of the medium

A. Acoustic wave in anisotropy

The modulus matrix of the 6mm material in $z \geq 0$ is constant in Cartesian coordinate system (x', y', z')

$$
C = \begin{pmatrix}
c_{11} & c_{12} & c_{13} & 0 & 0 & 0 \\
c_{12} & c_{11} & c_{13} & 0 & 0 & 0 \\
c_{13} & c_{13} & c_{33} & 0 & 0 & 0 \\
0 & 0 & 0 & c_{44} & 0 & 0 \\
0 & 0 & 0 & 0 & c_{44} & 0 \\
0 & 0 & 0 & 0 & 0 & c_{66}
\end{pmatrix}, \tag{1}
$$

where $c_{66} = (c_{11} - c_{12})/2$. There is only five elastic moduli in the medium.

It is convenient to use Cartesian coordinate system (x, y, z). The modulus matrix of Eq. (1) should be transformed to the system (x, y, z). It can be written as the following form

$$
C' = \begin{pmatrix}
c'_{11} & c'_{12} & c'_{13} & 0 & c'_{15} & 0 \\
c'_{12} & c'_{22} & c'_{23} & 0 & c'_{25} & 0 \\
c'_{13} & c'_{23} & c'_{33} & 0 & c'_{35} & 0 \\
0 & 0 & 0 & c'_{44} & 0 & c'_{46} \\
c'_{15} & c'_{25} & c'_{35} & 0 & c'_{55} & 0 \\
0 & 0 & 0 & c'_{46} & 0 & c'_{66}
\end{pmatrix}. \tag{2}
$$

Where

$$c'_{11} = c_{11}\cos^4\varphi + c_{33}\sin^4\varphi + 2(c_{13}+2c_{44})\sin^2\varphi\cos^2\varphi, \quad c'_{12} = c_{12}\cos^2\varphi + c_{13}\sin^2\varphi,$$

$$c'_{13} = c_{13}\cos^4\varphi + c_{13}\sin^4\varphi + (c_{13}+c_{33}-4c_{44})\sin^2\varphi\cos^2\varphi,$$

$$c'_{15} = [(c_{13}+2c_{44}-c_{11})\cos^2\varphi + (c_{33}-c_{13}-2c_{44})\sin^2\varphi]\sin\varphi\cos\varphi, \quad c'_{22} = c_{11},$$

$$c'_{23} = c_{12}\sin^2\varphi + c_{13}\cos^2\varphi, \quad c'_{25} = (c_{13}-c_{12})\sin\varphi\cos\varphi,$$

$$c'_{33} = c_{11}\sin^4\varphi + c_{33}\cos^4\varphi + 2(c_{13}+2c_{44})\sin^2\varphi\cos^2\varphi,$$

$$c'_{35} = [(c_{13}+2c_{44}-c_{11})\sin^2\varphi + (c_{33}-c_{13}-2c_{44})\cos^2\varphi]\sin\varphi\cos\varphi,$$

$$c'_{44} = c_{44}\cos^2\varphi + c_{66}\sin^2\varphi, \quad c'_{46} = (c_{44}-c_{66})\sin\varphi\cos\varphi,$$

$$c'_{55} = c_{44}\cos^2 2\varphi + (c_{11}+c_{33}-2c_{13})\sin^2\varphi\cos^2\varphi, \quad c'_{66} = c_{44}\sin^2\varphi + c_{66}\cos^2\varphi.$$

Then, the christoffel equation can be obtained

$$\Omega = \left|\Gamma_{ij} - \rho V^2 \delta_{ij}\right| = \Omega_1\Omega_2 = 0 , \tag{3}$$

where

$$\Omega_1 = c'_{44}\cos^2\theta + c'_{66}\sin^2\theta + c'_{46}\sin 2\theta - \rho V^2 , \tag{4}$$

$$\Omega_2 = \begin{vmatrix} c'_{55}\cos^2\theta + c'_{11}\sin^2\theta + c'_{15}\sin 2\theta - \rho V^2 & c'_{15}\sin^2\theta + c'_{35}\cos^2\theta + (c'_{13}+c'_{55})\sin\theta\cos\theta \\ c'_{15}\sin^2\theta + c'_{35}\cos^2\theta + (c'_{13}+c'_{55})\sin\theta\cos\theta & c'_{33}\cos^2\theta + c'_{55}\sin^2\theta + c'_{35}\sin 2\theta - \rho V^2 \end{vmatrix}, \tag{5}$$

here θ is the angle between the radius vector of the field point and the z axis. Let $\xi = \theta - \varphi$, the propagation phase velocities of the quasi-P and quasi-S waves are

$$\begin{cases} V_p^2 = \dfrac{1}{2\rho}\left[c_{11}\sin^2\xi + c_{33}\cos^2\xi + c_{44} + \sqrt{[(c_{11}-c_{44})\sin^2\xi - (c_{33}-c_{44})\cos^2\xi]^2 + (c_{13}+c_{44})^2\sin^2 2\xi}\right], \\[2mm] V_{s1}^2 = \dfrac{1}{2\rho}\left[c_{11}\sin^2\xi + c_{33}\cos^2\xi + c_{44} - \sqrt{[(c_{11}-c_{44})\sin^2\xi - (c_{33}-c_{44})\cos^2\xi]^2 + (c_{13}+c_{44})^2\sin^2 2\xi}\right], \end{cases} \tag{6}$$

and the phase velocity of the pure S wave is

$$V_{s2}^2 = (c_{44}\cos^2\xi + c_{66}\sin^2\xi)/\rho . \tag{7}$$

It is easy to obtain the slowness surfaces, ray surfaces and normal surfaces of the quasi-P, quasi-S and pure S waves by Eqs. (4)-(7). It can be seen that the propagation velocities of the quasi-P, quasi-S, and pure S waves are related the orientation of the radius vector relative to the crystallographic axis.

The displacement field in the medium can be obtained by the christoffel equation. It can be written as the following form in the k_x wavenumber domain

$$\begin{pmatrix} U_x \\ U_y \\ U_z \end{pmatrix} = \begin{pmatrix} e^{ik_{z1}z} & e^{ik_{z2}z} & 0 \\ 0 & 0 & e^{ik_{z3}z} \\ a_1 e^{ik_{z1}z} & a_2 e^{ik_{z2}z} & 0 \end{pmatrix} \begin{pmatrix} K_1 \\ K_2 \\ K_3 \end{pmatrix}, \tag{8}$$

where

$$a_i = -\frac{c'_{11}k_x^2 + c'_{55}k_{zi}^2 + 2c'_{15}k_x k_{zi} - \rho\omega^2}{c'_{15}k_x^2 + c'_{35}k_{zi}^2 + (c'_{13} + c'_{55})k_x k_{zi}} \ , \quad i = 1, 2 \ , \tag{9}$$

and k_{zi} $(i = 1,2,3)$ represent the wavenumbers in z direction of quasi-P, quasi-S, and pure S

waves, respectively.

The stress components can be obtained

$$\begin{pmatrix} \tau_{xz} \\ \tau_{yz} \\ \tau_{zz} \end{pmatrix} = \begin{pmatrix} \Delta_1 & \Delta_2 & 0 \\ 0 & 0 & \Delta_5 \\ \Delta_3 & \Delta_4 & 0 \end{pmatrix} \begin{pmatrix} iK_1 e^{ik_{z1}z} \\ iK_2 e^{ik_{z3}z} \\ iK_3 e^{ik_{z3}z} \end{pmatrix} . \tag{10}$$

Where $\Delta_1 = c'_{15}k_x + c'_{35}k_{z1}a_1 + c'_{55}(k_{z1} + k_x a_1)$, $\Delta_2 = c'_{15}k_x + c'_{35}k_{z2}a_2 + c'_{55}(k_{z2} + k_x a_2)$,
$\Delta_3 = c'_{13}k_x + c'_{33}k_{z1}a_1 + c'_{35}(k_{z1} + k_x a_1)$, $\Delta_4 = c'_{13}k_x + c'_{33}k_{z1}a_2 + c'_{35}(k_{z1} + k_x a_2)$,
$\Delta_5 = c'_{44}k_{z3} + c'_{46}k_{z1}k_x$. So it is obtained by Eqs. (8) and (10) that

$$\begin{pmatrix} U_x \\ U_y \\ U_z \end{pmatrix} = -i \begin{pmatrix} e^{ik_{z1}z} & e^{ik_{z3}z} & 0 \\ 0 & 0 & e^{ik_{z3}z} \\ a_1 e^{ik_{z1}z} & a_2 e^{ik_{z2}z} & 0 \end{pmatrix} \begin{pmatrix} \Delta_4/\Delta & 0 & -\Delta_2/\Delta \\ -\Delta_3/\Delta & 0 & \Delta_1/\Delta \\ 0_3 & 1/\Delta_5 & 0 \end{pmatrix} \begin{pmatrix} \tau_{xz} \\ \tau_{yz} \\ \tau_{zz} \end{pmatrix}_{z=0} , \tag{11}$$

where $\Delta = \Delta_1\Delta_4 - \Delta_2\Delta_3$. If only the stress component τ_{zz} is excited in the range

$[-l/2, l/2]$ at the free surface $z = 0$, i.e.

$$\tau_{zz}\big|_{z=0} = \int_{-l/2}^{l/2} e^{-ik_x x} dx = \frac{2}{k_x} \sin\frac{k_x l}{2} \ . \tag{12}$$

Therefore

$$\begin{cases} U_x(x,z) = \dfrac{1}{2\pi} \int_{-\infty}^{\infty} U_x(k_x,z)e^{ik_x x} dk_x = \int_{-\infty}^{\infty} \dfrac{i}{\pi k_x \Delta} \sin\dfrac{k_x l}{2} \left(\Delta_2 e^{ik_{z1}z} - \Delta_1 e^{ik_{z2}z}\right) e^{ik_x x} dk_x \ , \\[4mm] U_z(x,z) = \dfrac{1}{2\pi} \int_{-\infty}^{\infty} U_z(k_x,z)e^{ik_x x} dk_x = \int_{-\infty}^{\infty} \dfrac{i}{\pi k_x \Delta} \sin\dfrac{k_x l}{2} \left(a_1\Delta_2 e^{ik_{z1}z} - a_2\Delta_1 e^{ik_{z2}z}\right) e^{ik_x x} dk_x \ , \end{cases} \tag{13}$$

The corresponding displacement components of the compressional wave are

$$\begin{cases} U_x^P(x,z) = \displaystyle\int_{-\infty}^{\infty} \dfrac{i\Delta_2}{\pi k_x \Delta} \sin\dfrac{k_x l}{2} e^{i(k_x x + k_{z1}z)} dk_x \ , \\[4mm] U_z^P(x,z) = \displaystyle\int_{-\infty}^{\infty} \dfrac{ia_1\Delta_2}{\pi k_x \Delta} \sin\dfrac{k_x l}{2} e^{i(k_x x + k_{z1}z)} dk_x \ , \end{cases} \tag{14}$$

The integrals in Eq. (14) can be given by the ray approach method for the far field.

$$\begin{cases} U_x^P(x,z) = g_x(r,\theta)e^{i(\omega r/V_g+\pi/4)} & , \\ U_z^P(x,z) = g_z(r,\theta)e^{i(\omega r/V_g+\pi/4)} & , \end{cases} \tag{15}$$

where $g_x(r,\theta) = \sqrt{\dfrac{2}{\pi r f \cos\theta}} \dfrac{\Delta_2}{k_x\Delta} \sin\dfrac{k_x l}{2}$, $g_z(r,\theta) = a_1 g_x(r,\theta)$, $f = \left|\dfrac{d^2 k_{z1}}{dk_x^2}\right|$, $g_x(r,\theta)$ and

$g_z(r,\theta)$ are named as the directional factors in x and z axes. and the saddle point condition is determined by

$$\frac{dk_{z1}}{dk_x} = -tg\theta \quad . \tag{16}$$

It is indicated in Eqs. (15) and (16) that the compressional wave propagates as its group velocity V_g instead of the phase velocity V_p. The signal received by the receiver is the information about the group velocity[10].

B. Time reversal field

It is assumed that the transducer has M array elements. The acoustic field radiated by the nth array element ($n=1,2,...,M$) can be written as the following form in the time domain

$$U_n(\mathbf{r}_n,t) = g(\mathbf{r}_n)f(t-\tau_n), \tag{17}$$

where $\tau_n = r_n/V_n$ represents the propagation time from the nth element to the receiver point. $g(\mathbf{r}_n)$ is the directional factor in x or z axis. The total field excited by all the element of the transducer at the receiver point is

$$U(\mathbf{r},t) = \sum_{n=1}^{M} U_n(\mathbf{r}_n,t) = \sum_{n=1}^{M} g(\mathbf{r}_n)f(t-\tau_n), \tag{18}$$

The acoustic waves excited by different array element arrive at the receiver point in different time. They can't be superposed coherently and formed a widened pulse. If the time signal of the each element is reversed

$$U_n(\mathbf{r}_n,T-t) = g(\mathbf{r}_n)f(T-t-\tau_n), \tag{19}$$

where T is a long time. This reversal signal is normalized with the maximum of 1, and is then re-transmitted by the same array element

$$f_n^{TR}(t) = \frac{g(\mathbf{r}_n)}{|g(\mathbf{r}_n)|} f(T-t-\tau_n), \tag{20}$$

So the total field is

$$U^{TR}(\mathbf{r}',t) = \sum_{n=1}^{M} \frac{g(\mathbf{r}_n)}{|g(\mathbf{r}_n)|} g(\mathbf{r}_n')f(T-t+\tau_n'-\tau_n), \tag{21}$$

If $r' = r$, it is easy to obtain that

$$U^{TR}(\mathbf{r},t) = \left[\sum_{n=1}^{M} |g(\mathbf{r}_n)| \right] f(T-t). \tag{22}$$

This shows that the waves radiated by the different element arrive at the receiver synchronously and superpose coherently. The acoustic field is refocused to the original point.

Numerical simulation

In this section, the numerical results are given. The unidirecional glass-reinforced epoxy-fiber is chosen as the material of the anisotropic medium[11]. This material is 6mm anisotropy. The elastic constants and the density of this material are given in Table 1. The corresponding values of a kind of isotropic material are also given as the comparison analyses. In numerical simulation, the received signals are the displacement components of quasi-P wave in z axis. The transmission pulse is a sine packet with 10 periodic[7,8]. The frequency is $1MHz$.

Table 1. The values of elastic constants and the densities of the materials.(The units of elastic constants and densities are $10^{10} N/m^2$ and kg/m^3, respectively.)

materials	c_{11}	c_{12}	c_{13}	c_{33}	c_{44}	ρ
epoxy-fiber	2.58	2.42	0.70	6.01	0.49	1900
isotropy	6.01	5.03	5.03	6.01	0.49	1900

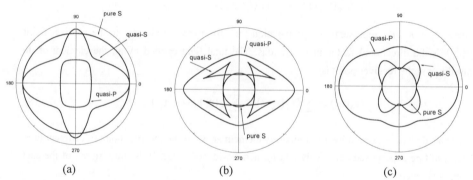

(a) (b) (c)

Fig. 2. The slowness surfaces (a), ray surfaces (b) and normal surfaces (c) of the quasi-P, quasi-S and pure S waves of unidirecional glass-reinforced epoxy-fiber.

The slowness surfaces, ray surfaces and normal surfaces of the quasi-P, quasi-S and pure S waves of this material are shown in Fig. 2. Fig. 2(b) is important for this paper because the received information is about the group velocity. It can be seen that the group velocity reaches the maximum of $\sqrt{c_{33}/\rho}$ in the direction $\xi = 0$ (i.e. $\theta = \varphi$) and reaches the minimum of $\sqrt{c_{11}/\rho}$ in the direction $\xi = 90°$ (i.e. $\theta = 90^0 + \varphi$).

It is assumed that the length of the transducer array is in the range [-0.5, 0.5] in numerical simulation, i.e. the length of the transducer is $L=0.1m$. And the transducer is divided into 64 array elements. In the first step of TR, 64 array elements of the transducer transmit a sine packet wave, respectively. Fig. 3(a) displays the displacement components U_z^p of the quasi-P wave received at point R ($x = 0, z = 0.4m$) in the case that φ and θ are both equal to zeroes. It shows that the quasi-P waves radiated by different elements arrive to R in different time. In the second step of TR, the above signals are reversed and retransmitted by each element. In this time, the received displacement z components at the same point R are shown in Fig. 3(b). It is clearly shows that the signals radiated by the different elements arrive synchronously in phase. They are enhanced each other and refocused. After TR, the signals from the different elements superpose coherently. Fig. 3(c) gives the waveforms in the case that all the elements transmit the signals synchronously before and after TR. It can be concluded that the coherent peak of the signal after TR is more enhanced. The good effect of self-adaptive focusing is obtained.

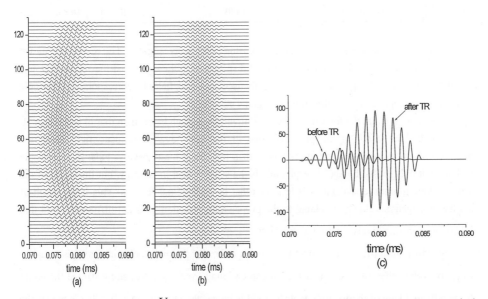

Fig. 3. Displacement components U_z received at point (0, 0.4). (a)Signals received when 64 elements are excited respectively in the first. (b)signals received when 64 reversed signals are retransmitted respectively after TR. (c)Signals received when 64 elements are excited simultaneously before and after TR.

It is convenient to adopt the focusing gain

$$G = 20 \log \frac{Max[P^{TR}(t)]}{Max[P(t)]} \, , \tag{23}$$

where $Max[P(t)]$ and $Max[P^{TR}(t)]$ are the maximums of the signals received before and after TR. Let g_M be the maximum of the time series of Eq. (18), the focusing gain can be written as the following form by Eq. (22)

$$G = 20 \log \left[\frac{1}{g_M} \sum_{n=1}^{M} |g(r_n)| \right] . \tag{24}$$

So for the Fig. 3, it is easy to obtain that the focusing gain is equal to 14.4dB.

The focusing gain is related to the distance from the focusing point to the transducer. In general, the more the distance from the focusing point to transducer is, the less the focusing gain is. The focusing gain is also related to the orientation of the focusing point. Table 2 gives the values of the focusing gains in some points in anisotropic medium.

Table 2. The values of focusing gain in some focusing points.

(x_0, y_0)	$\varphi = 0°$			$\varphi = 90°$		
	(0, 0.4)	(0.02, 0.4)	(0.04,0.4)	(0, 0.4)	(0.02, 0.4)	(0.04,0.4)
Focusing gain	14.4dB	13.4dB	12.4dB	11.0dB	10.4dB	10.8dB

The time reversal acoustic field near the focusing point is investigated. The results are shown in Fig. 4. The curves 1 and 2 represent the field distribution in the cases that the focusing points are (0,0.4) and (0.04,0.4), respectively. The focused acoustic field in corresponding isotropic medium is also depicted in Fig. 4(a) (curves 3 and 4).

It is easy to see that the focusing effect in x direction is much better than that in z direction. It is known that the width of the principal lobe of the time reversal acoustic field along the longitudinal axis in isotropic medium is longer than that along the transverse axis. In anisotropic medium, not only exists the above results that is the same as that in isotropy, but the width of the principal lobe in transverse (x) axis is less than that in longitudinal (z) axis. This is the result of that the propagation velocity is related to the propagation direction. It is not difficult to find that field to be more effectively focused along x axis.

Figs. 4(c) and 4(d) are for $\varphi = 90°$. In this case, the group velocity in z axis is equal to $\sqrt{c_{11}/\rho}$ which is less than that in x axis. It can be found from Figs. 4(c) and 4(d) that the width of the principal lobe of the focused acoustic field is bigger than that in $\varphi = 0$. That is to say, the width of the principal lobe of the focused acoustic field is different in different direction in the anisotropic medium, but the width of the principal lobe reaches the minimum in the maximum group velocity direction and reaches the maximum in the minimum group velocity direction.

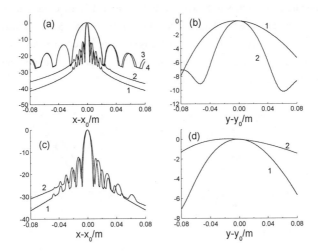

Fig. 4. Time reversal acoustic field along the x [(a) and (c)] and z [(b) and (d)] axes. The curves 1 and 2 represent the field distribution in the cases that the focusing points are (0,0.4) and (0.04,0.4), respectively. The curves 3 and 4 are similar to that of 1 and 3 in isotropy. φ is equal to zero for (a) and (b) and is $90°$ for (c) and (d).

The curves 1 and 2 in Fig. 4 are for the focusing points (0,0.4) and (0.04,0.4), respectively. When the focusing point is not in z axis, the width of the principal lobe of the focused field in x axis is bigger than that when the focusing point is in z axis [Figs. 4(a) and 4(c)]. However, the width of the principal lobe in z axis is bigger than that when the focusing point is in z axis for

$\varphi = 90°$ [Fig. 4(d)] and less than that when the focusing point is in z axis for $\varphi = 0$ [Fig. 4(b)].

To understand the distribution of the focused acoustic field near the focusing point, the isolines of the focused acoustic field are given in Fig. 5. Figs. 5(a) and 5(b) are for $\varphi = 0$ and 5(c) and 5(d) for $\varphi = 90°$, respectively. The focusing points are on z axis in Figs. 5(a) and 5(c) and are not on z axis in Figs. 5(b) and 5(d). The closed curves in Fig. 5 from the inner to external circles represent that the focusing gains are equal to –1dB, -2dB, -3dB, -4dB, -5dB, -6dB, respectively. It is easy to see that the isolines are concentrated in $\varphi = 0$, i.e. the focusing width is the mimmum in $\varphi = 0$ (i.e. the maximum group velocity direction). This is consistent to the results in Fig. 4. It is noticed that the difference of the distribution range of the isolines between x and z axes is more significant. The range of isolines in z axis is about more than 20 times as that in x direction. This characteristics can be also seen in Fig.4. It can be seen from Figs. 5(b) and 5(d) that the isolines are not symmetric about x or z axis when the focusing point is not on z axis. One of the reason is that the radius vector from transducer to focusing point is not perpendicular to the transducer array. The similar result can be obtained in isotropic medium.

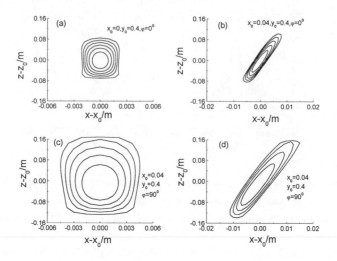

Fig. 5.   Isolines of the focused acoustic field.

Fig. 6 displays the relation of the focusing gain to the number of the array elements. It is indicated that the focusing gain increases as the number of the array elements increase when the number is about less than 40. If the number is greater than 40, the focusing gain tends towards to a given value as the number increases. So the saturated phenomena could appear in the focusing gain when the number of the transducer array elements increases because the transmission pulse has a definite width. This phenomena should appear in isotropic medium[8].

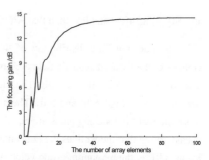

Fig.6. The relation of the focusing gain to the number of the array elements.

Conclusions

Unidirecional glass-reinforced epoxy-fiber is chosen as the material of the 6mm anisotropic medium. The time reversal method in this material is theoretically studied for the first time in this paper. The relation of time reversal acoustic field to the parameters of anisotropic medium is analyzed in detail. It is found that the width of the principal lobe of the focused acoustic field reaches the minimum in the maximum group velocity direction and reaches the maximum in the minimum group velocity direction.

The works in this paper about the time reversal method are very elementary. There are many problems to be done in our future works. Time reversal self-adaptive focusing in anisotropic medium is more complicated and more interesting. This paper lays a foundation for further study of the time reversal in anisotropic medium.

Acknowledgment

This work was supported by the President Foundation of Institute of Acoustics, The Chinese Academy of Sciences.

References

[1] Mathias Fink, Time reversal of ultrasonic fields- part I: basic principles, IEEE Trans. UFFC, 39(5), 555-566, 1992

[2] Francois Wu, Jean-Louis Thomas, and Mathias Fink, Time reversal of ultrasonic fields- part II: experimental results, IEEE Trans. UFFC, 39(5), 567-578, 1992

[3] Ros K. Ing and Mathias Fink, Time reversal Lamb waves, IEEE Trans. UFFC, 45(4), 1032-1043, 1998

[4] Najet Chakroun, Mathias Fink, and Francois Wu, Time reversal processing in ultrasonic nondestructive testing, IEEE Trans. UFFC, 1087-1098, 1995

[5] W. A. Kuperman, W. S. Hodgkiss, and Hee Chen Song, Phase conjugation in the ocean: experimental demonstration of an acoustic time reversal mirror, J. Acoust. Soc. Am., 103(1), 25-40, 1998

[6] Hee Chen Song, W. A. Kuperman, and W. S. Hodgkiss, A time reversal mirror with variable range focusing, J. Acoust. Soc. Am., 103(6), 3234-3240, 1998

[7] Wei Wei and Wang Chenghao, Self adaptive focusing by time reversal through interface between different media, Chinese J. of Acoustics, 19(1), 2000, 83-88

[8] Wei Wei, Liu Chen and Wang Chenghao, Self-focusing of acoustical beam in solid by time reversal processing, Chinese J. of Acoustics, 19(1), 2000, 89-95

[9] Bixing Zhang, Chenghao Wang, and Minghui Lu, Acoustic wave time reversal self-focusing in underwater waveguide, IEEE 2000 Ultrasonics Symp. Proc.

[10] Wang Chenghao and Chen Dongpei, Generalized Green's functions of surface excitation of elastic wave field in a piezoelectric half-space, Chinese J. Of Acoustics, 4(4), 298-313, 1985

[11] J. E. Zimmer and J. R. Cost, Determination of the constants of a unidirectional fiber composite using ultrasonic velocity measurements, J. Acoust. Soc. Am., 47(3), 795-803, 1970

Theoretical and Computational Acoustics 2001
E.-C. Shang, Qihu Li and T. F. Gao (Editors)
© 2002 World Scientific Publishing Co.

Shear Wave Velocity Estimation in HTI Media

Qizhen Du, Yuan Dong, Huizhu Yang
Department of Engineering Mechanics, Tsinghua University, 100084, P. R. China

Yong Wang
Jiangsu Oil Exploration Bureau, Nanjing, 210000, P. R. China

The importance of anisotropic phenomena in wave propagation and processing of seismic data is now widely recognized by the exploration community. Transverse isotropy with a horizontal symmetry axis (HTI media) is the simplest azimuthally anisotropic model used to describe fractured reservoirs that contain parallel vertical cracks. This paper puts forward the double profile for the horizontal reflectors in HTI media with any strength of anisotropy. The double profile is obtained through doubling the offset and traveltime of seismic data. Moreover, the equations of P-S wave normal-moveout (NMO) velocity with arbitrary symmetry and any strength of anisotropy are obtained. The converted-wave NMO velocity is controlled by the azimuthal angle, the vertical velocity of incident wave, the vertical velocity of reflected wave and anisotropic parameters (Thomsen parameters). Using azimuthal NMO velocities of P-P wave, P-SV wave and P-SH wave in multicomponent seismic data, we can estimate the velocities of SV-, SH-wave and anisotropy parameters, including Thomsen parameter γ, which has close relationship with crack density.

For multilayered media, the recursive equations of NMO velocity of each layer are developed, which can be applied to HTI media with arbitrary symmetry and any strength of anisotropy. The formula developed here is the same as Dix equation. The numerical results indicate that the S-wave velocity can be exactly computed by using P-P wave and P-S wave NMO velocities in various HTI media, which provides reliable basis for the inversion of Thomsen parameters in 2D multicomponent seismic data.

Key Words HTI media, P-S wave, NMO velocity, S-wave velocity

1. Introduction

The importance of anisotropic phenomena in wave propagation and processing of seismic data is now widely recognized by the exploration community. In recent years, researchers pay more attention to fractured reservoirs. The prominent mark of fractured strata is S-wave splitting (Crampin, 1985). S-wave used to be found application in detecting fracture. But it is expensive and the signal/noise (S/N) ratio is low. Since 1994, 3D P-wave has been widely applied to fracture detection (Lefeurve, 1994). Some convincing results have been obtained in P-wave Amplitude (AVO, AVA) (Lynn, 1996), traveltime (Li, 1997) and velocity (Mallick et. al., 1998). However, the application of P-wave reflects some characteristics of fracture indirectly. Thomsen parameter γ can be estimated in the special case of the vanishing parameter ε, corresponding to thin cracks and negligible equant porosity (Tsvankin, 1997). Along with the development of multicomponent acquisition, processing and interpretation (Bertagne and Sparkman, 1999), it becomes feasible to make use of P-S wave to detect fracture. That is to say, by usingmulticomponent seismic data, the velocities of P-wave, SV-wave and SH-wave and anisotropy parameters can be estimated.

Transverse isotropy with a horizontal symmetry axis (HTI media) is the simplest azimuthally anisotropic model used to describe fractured reservoirs that contain parallel vertical cracks. Thomsen (1988) developed the equations of P-wave and S-wave NMO velocities of horizontal reflectors in symmetry plane. Sena (1991), Li and Crampin (1993) studied the more general case. They gave the non-hyperbolic equations in non-symmetry plane in the case of weak anisotropy. Using spherical harmonic expansions, Sayers and Ebrom developed P-wave non-hyperbolic NMO equations in HTI media. Tsvankin (1997) developed pure modes reflected waves NMO velocity equations for horizontal reflectors with any strength of anisotropy, and pointed out that azimuthally dependent P-wave NMO velocity is controlled by the direction of principal axis, the vertical velocity and an anisotropy parameter δ. Furthermore, Al-Dajani and Tsvankin (1998) studied non-hyperbolic expressions and the case of mutilayered media. Grechka et. al. (199b) pointed out that the formula, which is the same as Dix equation, could be applied to common-midpoint-point (CMP) gathers of pure modes in HTI media with arbitrary symmetry and any strength of anisotropy.

For converted-wave processing, Tessmer and Behle (1998) developed common-converted-point (CCP) stack method for P-SV wave that can be applied to isotropic media. Moreover, they obtained the relation among P-wave, SV wave and P-SV wave. Levin (1989), Serrif and Sriram (1991) studied the velocity of converted-wave in VTI media and developed NMO equation of P-SV wave and obtained SV wave velocity by usingvelocities of P-P wave and P-SV wave. According to above-mentioned works, Tsvankin and Thomsen (1994) developed non-hyperbolic equation of P-SV wave in VTI media and studied the relation among P-P wave, P-SV wave and SV wave.

The above-mentioned works have founded solid basis for fracture detecting. However, some critical issues still remain unsolved. According to the researches done by Grechka et. al., for the orthorhombic media, the azimuthal variation of NMO velocities of converted-wave and pure modes generally have an elliptical form. In the direction of the principal axis, all the NMO velocities of P-SV wave, P-SH wave, P-wave, SV-wave and SH-wave followed the same rule developed by Seriff and Sriram (1991), (Grechka et. al., 1999). In the direction of non-principal axis, the relations among these waves are the critical issues of this paper. Besides, the relations are essential in using 2D seismic data to detect fracture.

In this paper the authors present the double profile, which is obtained through doubling the offset and traveltime of seismic data. On the basis the double profile, the azimuthally dependent NMO velocity in HTI media with arbitrary symmetry and any strength of anisotropy is estimated. The azimuthally dependent converted-wave NMO velocity is controlled by the azimuthal angle, the vertical velocity of incident wave, the vertical velocity of reflected wave and anisotropic parameters (Thomsen parameters). Moreover, this paper develops the relationship among P-SH wave, P-SV wave, SH-wave, and SV-wave. For the multilayered media, the recursive equation of NMO velocity is developed.

2. Principle

2.1. *HTI media*

Transversely isotropic medium has a rotational symmetry. The stiffness matrix of TI media includes 5 independent constants. The propagation is independent of azimuthal angle in TI media with vertical symmetry; however, it is dependent on azimuthal angle in TI media with horizontal symmetry (Fig. 1). Thus, this anisotropy is usually called "azimuthally anisotropic media" or "EDA (Extensive Dilatancy Anisotropy) media" so that the different properties can be emphasized.

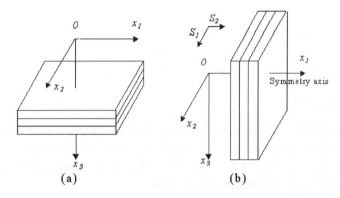

(a)　　　　　　　(b)

Fig.1. Symmetry planes and isotropic planes in TI media: (a) VTI model; (b) HTI model. Symmetry plane [x_1, x_3] of a transversely isotropic medium with the symmetry axis points either in the x_3 (VTI) or in the x_1 direction (HTI). Plane [x_2, x_3] in HTI media is an isotropic plane. Plane [x_1, x_2] in VTI media is an isotropic plane.

Thomsen (1986) gave the definitions of the anisotropy parameters of transverse isotropy: ε, γ and δ. Making use of these parameters in HTI media, Tsvankin (1997) and Rú ger(1997) gave the same expression of parameters as Thomsen(1986). Parameters in the HTI media denote $\varepsilon^{(R)}$, $\gamma^{(R)}$ and $\delta^{(R)}$, while parameters in the VTI media denote $\varepsilon^{(V)}$, $\gamma^{(V)}$ and $\delta^{(V)}$. These parameters can be denoted as:

$$\varepsilon = \frac{C_{11} - C_{33}}{2C_{33}}, \tag{1}$$

$$\delta = \frac{(C_{11} + C_{55})^2 - (C_{33} - C_{55})^2}{2C_{33}(C_{33} - C_{55})}, \tag{2}$$

$$\gamma = \frac{C_{66} - C_{44}}{2C_{44}}, \tag{3}$$

where C_{ij} is the component of the stiffness matrix. If the media is VTI media, (1), (2) and (3) are the parameters defined by Thomsen (1986); If the media is HTI media, equations (1), (2) and (3) are the parameters defined by Tsvankin (1997) and Rú ger(1997).

There are three body waves in HTI media: P-wave, SH-wave and SV-wave. In the isotropy plane, their velocities are:

$$V_{Pvert} = \sqrt{\frac{c_{33}}{\rho}} = \alpha_0 \quad , \tag{4}$$

$$V_{SVvert} = \sqrt{\frac{c_{44}}{\rho}} = \beta_0 \quad , \tag{5}$$

$$V_{SHvert} = \sqrt{\frac{c_{66}}{\rho}} \quad , \tag{6}$$

where C_{ij} are the 5 independent components of the stiffness matrix.

2.2 *The double profile*

In the propagation of P-S wave, incident wave is different from reflected wave, and each of them has different polarization and velocity. Thus the travelpath is unsymmetric. Even if the reflectors are horizontal, the CMP gather of converted-wave is not common-reflected-point (CRP) gather. To solve this problem, we introduce the double profile.

The double profile is obtained through doubling the offset and traveltime of seismic data. The depth of the equivalent media is two times that of original media, and the original reflector is regarded as imaginary interface. Above the imaginary interface, we make use of initial incident wave as incident wave, and converted-point is in the interface. Under the imaginary interface, we make use of the original converted-wave as incident wave, and converted-point is in the imaginary interface (Fig. 2). Grechka studied NMO velocity in orthorhombic media by using this idea (Grechka et. al., 1999). In the double profile, traveltime-offset curve and NMO velocity followed the same rule as in the initial profile. To the imaginary interface, incidence wave, reflected wave and refracted wave followed Snell's law. Because the reflected angle is equal to the incident angle, the existence of imaginary interface has no effect on the travelpath to the pure modes. Each, however, can be treated as pure mode wave propagates above or under the imaginary interface, if the wave is converted-wave.

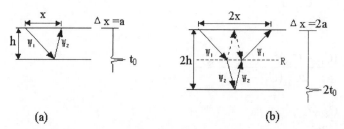

(a)   (b)

Fig.2. The sketch of the double profile. (a) Original profile: incident wave is W1, and reflected wave is W2. (b) The double profile: incident wave and reflected wave are W1 in upper media, and incident wave and reflected wave are W2 in lower media.

2.3. *NMO velocity of converted-wave in single-layer HTI media*

Because traveltime-offset curve and NMO velocity followed the same rule as in the initial profile, we can make use of the double profile to study NMO velocity of converted-wave. In the original profile, CMP gathers of converted-wave are not CMP ones. But in the double profile, the equivalent travelpath is symmetric. Accordingly, CMP gathers of converted-wave are CMP ones. Unless stated in advance, our discussion is restricted to the double profile case, under the assumption that the vector of group velocity is situated in the incident plane. Al-Dajani and Tsvankin (1998) use similar assumption in developing the recursive NMO velocity in mutilayered HTI media. If the wave propagates in the non-principal axis plane, the vector of phase velocity is usually off the incident plane. In this case, we obtain the equation of NMO velocity of converted-wave in HTI media:

$$V_{nmo}^2 = \frac{1}{t_0}(t_{01}V_{1vert}^2 \frac{1+A_1}{1+A_1 \sin^2\alpha} + t_{02}V_{2vert}^2 \frac{1+A_2}{1+A_2 \sin^2\alpha}) \ , \tag{7}$$

where α is the azimuthal angle between the seismic line and the symmetry axis in HTI media; A_1 and A_2 are anisotropy terms of upper and lower media in the double profile. The type of wave can be either one kind of P-, SH- or SV-wave. The expression is:

$$A_i = \frac{1}{V_i}\frac{d^2V_i}{d\theta^2}\Big|_{\theta=90^\circ}, \ (i = P, SV, SH) \ , \tag{8}$$

where $A_P = 2\delta^{(V)}, A_{SV} = 2\sigma^{(V)}, A_{SH} = 2\gamma^{(V)}, \sigma^{(V)} = (\frac{V_{Pvert}}{V_{SVvert}})^2(\varepsilon^{(V)} - \delta^{(V)})$.

V_i is phase velocity, and it is as a function of phase angle θ. V_{1vert} and V_{2vert} are vertical velocity of W1 and W2 in upper and lower media respectively. t_{01} and t_{02} are two-way vertical traveltime of W1 and W2 in upper and lower media respectively. $t_0 = t_{01} + t_{02}$, It is twice of original traveltime of converted-wave.

Equation (7) can be applied to reflected wave in HTI media with any strength of anisotropy. If the waves in the upper and lower media are of the same type, equation (7) is the one given by Tsvankin (1997). In addition, according to the equation of Tsvankin (1997), equation (7) becomes:

$$V_{nmo}^2 = \frac{1}{t_0}(t_{01}V_{1nmo}^2 + t_{02}V_{2nmo}^2) \ , \tag{9}$$

V_{1nmo} and V_{2nmo} are NMO velocities of W1 and W2 in upper and lower media respectively.

This formula can be applied to VTI media or isotropic media. For VTI media, equation (9) is the formula given by Seriff and Sriram (1991). For isotropic media or isotropic plane in HTI media ($\alpha = 90^\circ$), equation (9) becomes:

$$V_{nmo}^2 = \frac{1}{t_0}(t_P V_P^2 + t_{SV} V_{SV}^2) = \sqrt{V_P V_{SV}} \quad , \tag{9a}$$

This equation coincides with that developed by Tessmer and Behle (1998).

Formula (9) makes it clear that this relation can be applied to non-principal axis in HTI media. According to this formula, we can use converted-wave and one kind of wave of upper and lower media to obtain the NMO velocity of another kind of wave. It is feasible to get the NMO velocities of SH-wave and SV-wave, via converted-wave velocity and NMO velocity of P-wave. Accordingly, γ can be estimated by using NMO velocities of P-P wave and P-SH wave in detecting fracture:

$$V_{SHnmo}^2 = \frac{1}{t_0 - t_P}(t_0 V_{nmo}^2 - t_P V_{Pnmo}^2) = V_{SHvert}^2 \frac{1 + 2\gamma^{(V)}}{1 + 2\gamma^{(V)} \sin^2 \alpha} \quad , \tag{10}$$

where t_0 is two-way vertical traveltime of converted-wave in the double profile and t_P is two-way vertical traveltime of P-wave in the double profile.

2.4. *Multilayered NMO velocity in HTI media*

In multilayered media, an NMO velocity of one interface is integration of velocity of each layer above the interface. In isotropic media, this integration is described by Dix formula (Dix, 1955). In VTI media, there is similar equation (Hake et. al., 1984). It is valid in the symmetry plane in HTI media (Tsvankin, 1995). Based on the assumption that the vector of group velocity is situated in the incident plane, pure modes NMO velocity of non-symmetry plane in HTI media has the similar expression with Dix formula. (Al-Dajani and Tsvankin, 1998; Grechka et. al., 1999). Similarly, based on the same assumption, converted-wave NMO velocity of non-symmetry plane in HTI media is developed in this paper, by making use of the double profile. The expression is similar to Dix formula:

$$V_{nmo}^2 = \frac{1}{t_0} \sum_{i=1}^{n} \Delta t_{0_i} V_{nmo_i}^2 \quad , \tag{11}$$

where t_0 is the two-way vertical traveltime of nth interface in the original profile, V_{nmo_i} the NMO velocity of ith layer in the original profile, and Δt_{0_i} two-way vertical traveltime of ith layer. By usingconverted-wave NMO velocity equation (9) of single-layer, equation (11) becomes:

$$V_{nmo}^2 = \frac{1}{t_0}(\sum_{i=1}^{n} \Delta t_{01_i} V_{1nmo_i}^2 + \sum_{i=1}^{n} \Delta t_{02_i} V_{2nmo_i}^2) \quad , \tag{12}$$

where the first term is summation of NMO velocity of incident wave of each layer; the second term is summation of NMO velocity of reflected wave of each layer. Formula (11) can be written in the expression similar to equation (9) as:

$$V_{nmo}^2 = \frac{1}{t_0}(t_{01} V_{1nmo}^2 + t_{02} V_{2nmo}^2) \quad , \tag{13}$$

This formula can be applied tc multilayeded HTI media. In the equation, two terms correspond to the incident wave and reflected wave respectively. This formula shows that converted-wave NMO velocity can be obtained by summation of NMO velocities of incident wave and reflected wave in the double profile. Accordingly, in multilayered HTI media, via formula (13), making use of converted wave and either the incident wave or the reflected wave, we can obtain NMO velocity of the other.

3. Numerical Results

The numerical results of converted-wave in HTI media are given in this paper. These illustrations indicate that the converted-wave NMO velocity is correct even for moderate anisotropy in HTI media, in the case of the offset equaling to the depth. Moreover, SH-wave NMO velocity can be computed from P-SH wave and P-P wave NMO velocities by using formula (13), no matter whether it is the single-layer or multilayered media. This provides convenience to obtain anisotropy parameter γ in detecting fracture by usingmulticomponent data.

Traveltime is computed by using 3D anisotropy ray-tracing method. The NMO velocity of synthetic data is obtained via fitting traveltime-offset curve by employing least-square method.

3.1. *Single HTI layer*

The propagation in HTI media is related to azimuthal angle of seismic line. We discuss converted-wave NMO velocity in single-layer at first. Besides the model parameters from Al-Dajani and Tsvankin (1998), we add parameter γ (Table 1) to depict the effect of S wave. Model 1 corresponds to weak anisotropy, and the degree of anisotropy of Model 2 is stronger than that of Model 1.That is, if they correspond to fracture strata, fracture density of Model 2 is bigger than Model 1. To get azimuthal NMO velocity, seismic lines are laid along with different azimuth (Fig. 3).

Table 1The parameters of two models of HTI media, where D=1500m is the reflector depth.

Parameters	Model1	Model2
$\varepsilon^{(V)}(\varepsilon^{(R)})$	-0.143(0.2)	-0.143(0.2)
$\delta^{(V)}(\delta^{(R)})$	-0.184(0.1)	-0.318(-0.2)
$\gamma^{(V)}(\gamma^{(R)})$	-0.143(0.2)	-0.167(0.25)
$V_{Pvert}(V_{P0})$ m/s	2662(2250)	2958(2500)
$V_{SVvert}(V_{S0})$ m/s	1500(1500)	1500(1500)

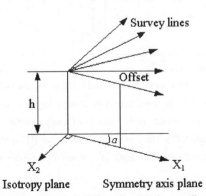

Fig.3. Seismic lines layout sketch of single-layer in HTI media. (Maximum offset is equal to the depth, namely,

1500m)

The numerical calculations were performed to the models, and the results were shown in Fig. 4 and Fig. 5. For P-SH wave (Fig 3c, Fig. 5c), in the case of conventional array (offset/depth<1), the NMO velocity determined by fitting a hyperbola to the exact traveltime-offset curve with least-square method is quite close to the calculations from formula (7). The result of P-P wave is completely the same as that in Tsvankin (1997), Al-Dajani and Tsvankin (1998). These state that the formula (7) and the ray-tracing method are correct. When the azimuth angle is equal to zero, i.e. in the symmetry plane, the difference between the analysis and the fitting is maximal. The reason is that NMO curve severely deviates the fitting curve. However, when the azimuth angle is equal to $90°$, i.e. in the isotropy plane, they completely coincide each other. The difference between the analysis of SH-SH wave and the fitting result is very small. The curves are basically consistent. This indicates that the NMO curve of SH-SH wave is ideally hyperbolic regardless of what azimuthal angle is.

The NMO velocity of SH-SH wave obtained from formula (10) is shown in Fig. 4d and Fig. 5d. Moreover, the result from formula (10) also is shown in these figures. They are very close to each other. Even for the strong anisotropy of Model 2, the maximal error is less than 1.5%   which case occurs at the $0°$ azimuthal angle. Thus, using azimuthal NMO velocities of P-P wave, P-SV wave and P-SH wave in multicomponent seismic data, we can estimate the velocities of SV, SH and anisotropy parameters, including Thomsen parameter γ , which has close relationship with crack density.

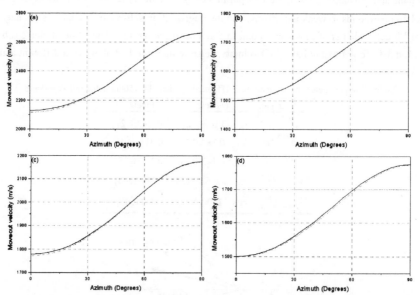

Fig.4. Get SH-wave normal-moveout velocity from P- and P-SH wave. The solid curve in (a) (b) and (c) is the moveout velocity as a function of azimuth determined by fitting a hyperbola to the exact t^2-x^2 curves; the solid curve in (d) is the SH-wave NMO velocity from P and P-SH using equation (10). The dot curve is the NMO (zero-spread) velocity from equation (7). The curves are calculated for HTI model 1 (Table 1) and on the spread length X/D=1, where D=1500m is the reflector depth: (a) P-P wave; (b) SH-SH wave; (c) P-SH wave; (d) SH-SH wave.

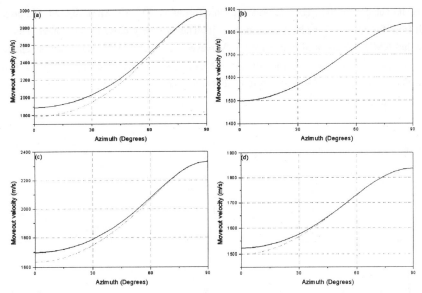

Fig.5. Get SV-wave normal-moveout velocity from P- and P-SV wave. The solid curve in (a) (b) and (c) is the moveout velocity as a function of azimuth determined by fitting a hyperbola to the exact t^2-x^2 curves; the solid curve in (d) is the SV-wave NMO velocity from P and P-SV using equation (10). The dot curve is the NMO (zero-spread) velocity from equation (7). The curves are calculated for HTI model 2 (Table 1) and on the spread length X/D=1, where D=1500m is the reflector depth: (a) P-P wave; (b) SV-SV wave; (c) P-SV wave; (d) SV-SV wave.

3.2. *Multilayered HTI media*

The propagation in multilayered HTI media is complicated, which is related to the azimuthal angle between the symmetry and the seismic line in each layer. Besides the same parameters from Al-Dajani and Tsvankin (1998), we add parameter γ (table 2). Seismic lines are laid along with different azimuthal angle. The maximal offset is equal to the depth, where D=1500m is the reflector depth.

Table 2 Parameters of 3 layered model

Parameters	layer1	layer2	layer3
$\varepsilon^{(V)}(\varepsilon^{(R)})$	-0.143(0.2)	-0.045(0.05)	-0.143(0.2)
$\delta^{(V)}(\delta^{(R)})$	-0.184(0.1)	-0.203(-0.15)	-0.318(-0.2)
$\gamma^{(V)}(\gamma^{(R)})$	-0.083(0.1)	-0.0455(0.05)	-0.167(0.25)
$V_{Pvert}(V_{P0})$ (m/s)	2000(1690)	2500(2348)	3000(2535)
$V_{SVvert}(V_{S0})$ (m/s)	1150(1150)	1400(1400)	1525(1525)
Depth (m)	500	1000	1500

Fig. 6 is the numerical results of multilayered HTI media. For P-SH wave (Fig 6c), in the case of conventional array (offset/depth<1), the NMO velocity determined by fitting a hyperbola to the exact traveltime-offset curve with least-square method is quite close to the result from formula (11). The result of P-P wave is completely the same as that in Tsvankin (1997), Al-Dajani and Tsvankin (1998). When the azimuth angle is equal to $0°$, that is, in the symmetry plane, the difference between the analysis and the fitting is maximal.

The NMO velocity of SH-SH wave obtained from formula (13) is shown in Fig. 5d. Moreover, the result from formula (12) is also shown in Fig. 5d. They are very close to each other. Thus, using azimuthal NMO velocities of P-P wave, P-SV wave and P-SH wave in multicomponent seismic data, we can estimate the velocities of SV, SH and anisotropy parameters, including Thomsen parameter γ, which has close relationship with crack density.

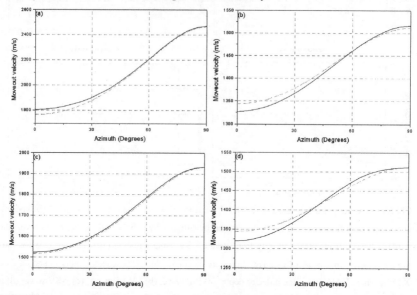

Fig. 6. Get SH-wave normal-moveout velocity from P- and P-SH wave. The solid curve in (a) (b) and (c) is the moveout velocity as a function of azimuth determined by fitting a hyperbola to the exact t^2-x^2 curves; the solid curve in (d) is the SH-wave NMO velocity from P and P-SH using equation (13). The dot curve is the NMO (zero-spread) velocity from equation (12). The curves are calculated for HTI 3 layers model (Table 2) and on the spread length X/D=1, where D=1500m is the max reflector depth: (a) P-P wave; (b) SH-SH wave; (c) P-SH wave; (d) SH-SH wave.

4. Discussion and Conclusions

Based on the assumption that the vector of group velocity is situated in the incident plane, by usingthe double profile, the converted-wave NMO velocity equation of the non-symmetry plane in single-layer HTI media is developed. The converted-wave NMO velocity is controlled by the azimuthal angle, the vertical velocity of incident wave, the vertical velocity of reflect wave and anisotropic parameters (Thomsen parameters). Moreover, the converted-wave NMO velocity in the non-symmetry plane is developed, which has similar expression as Dix formula in multilayered HTI media. Via this formula, using azimuthal NMO velocities of P-P wave, P-SV wave and P-SH wave in multicomponent seismic data, we may estimate the velocities of SV, SH and anisotropy parameters may be estimated, including Thomsen parameter γ, which has close relationship with crack density. This provides theoretical basis for inversion of anisotropy parameters from multicomponent seismic data.

The numerical calculations show that the converted-wave NMO velocity is correct even for moderate anisotropy in HTI media. For the multilayered model, the equations developed here are employed for accurate computation S-wave velocity from P-S wave velocity and P-P velocity.

In single-layer HTI media, pure modes NMO velocities are non-hyperbolic. To the multilayered media, the character of non-hyperbolic is obvious. Thus, by using converted-wave to establish S-wave velocity, fracture may be detected. But in the case of moderate or long array (offset/depth>1), the effect of non-hyperbolic of traveltime-offset curve should be considered in developing the NMO velocity.

Recently, 3D multicomponent technique develops rapidly. People pay more attention to the method that use converted-wave to detect fracture. Furthermore, it has the potential to obtain the anisotropy of P-wave and S-wave (Grimm et. al., 1998). Thus, multicomponent technique will become one of the principal domains in seismic exploration (Bertagne and Sparkman, 1999).

Acknowledgements

We would like to thank Dr. Guo Jiuying and Dr. Ni Yi of CNPC-BGP for the valuable suggestions to our work. We also thank Dr. Lin Xiaozhu of the Petroleum Institute of Jianghan, and the discussion with him gives us some advice.

References

1 A. Al-Dajani, and I. Tsvankin, 1998, "Nonhyperbolic reflection moveout for horizontal transverse isotropy", *Geophysics*, 63(5)(1998), 1738~1753.

2 A. Bertagne and G. Sparkman, "An introduction to this special section: Multicomponent offshore", *The Leading Edge*, 18(11)(1999), 1272.

3 S. Crampin, "Evidence for aligned cracks in the earth's crust", *First Break*, 3(3)(1985), 12-15.

4 C. H. Dix, "Seismic velocities from surface measurements", *Geophysics*, 10(1955), 68~86.

5 V. Grechka, P. Contreras and I. Tsvankin, , "Inversion of normal moveout for monoclinic media", *69th Ann. Internat. Mtg., Soc. Expl. Geophys., Expanded Abstracts*, (1999), 1883~1887.

6 V. Grechka, V. Theophanis, and I. Tsvankin, "Joint inversion of P- and PS-waves in orthorhombic media: Theory and a physical modeling study", *Geophysics*, 64, (1999), 146~161.

7 V. Grechka, I. Tsvankin, and J. K. Cohen, "Generalized Dix equation and analytic treatment of normal-moveout velocity for anisotropic media", *Geophysical Prospecting*, 47(2)(1999), 117~148.

8 R. E. Grimm, H. B. Lynn, C. R. Bates , G. M. Mayko, and V. Kuuskraa, "Detection and Analysis of Naturally Fractured Gas Reservoirs: Summary and Synthesis", *68th Ann. Internat. Mtg., Soc. Exp. Geophys.*, (1998) 1555~1558.

9 F. Lefeurve, "Fractured related anisotropy detection and analysis:" and if the P-wave were enough?" *64th Ann. Internat. Mtg., Soc. Exp. Geophys., Expanded Abstracts*, (1994), 942~945.

10 X. Li, and S. Crampin, "Approximations to shear-wave velocity and moveout equations in anisotropic media", *Geophysical Prospecting*, 41(1993), 833~857.

11 XiangYang Li, "Viability of azimuthal variation in P-wave moveout for fracture detection", *67th Ann. Internat. Mtg., Soc. Exp. Geophys.*, (1997) 1555~1558.

12 H. B. Lynn, et al., 1996, "Correlation between P-wave AVOA and S-wave anisotropy in a naturally fractured gas reservoir", *The Leading Edge*, 15(8)(1996), 931~935.

13 S. Mallick, K. L. Craft, L. J. Meister, R. E. Chambers, "Determination of the principal directions of azimuthal anisotropy from P-wave seismic data", *Geophysics*, 63(1998), 692-706.

14 C. M. Sayers, D. A. Ebrom, 1997, "Seismic travel time analysis for azimuthally anisotropic media: Thory and experiment", *Geophysics*, 62(5)(1997), 1570~1582.

15 A. G. Sena, "Seismic traveltime equations for azimuthally anisotropic and isotropic media: Estimation of interval elastic properties", *Geophysics*, 56(12)(1991), 2090~2101.

16 A. J. Seriff, K.P. Sriram, 1991, "P to SV reflection moveouts for Transverse isotropy media with a vertical symmetry axis", *Geophysics* , 56(1991), 1271-1274.

17 G. Tessmer, A. Behle, "Common reflection point data-stacking technique for converted waves", *Geophysical Prospecting*, 36(1998), 671-688.

18 L. Thomsen, "Weak elastic anisotropy", *Geophysics*, 51(1986), 1954-1966.

19 L.Thomsen, "Reflection seismology over azimuthally anisotropic media", *Geophysics*, 53(1988), 304-313

20 I. Tsvankin, "Anisotropic parameters and P-wave velocity for orthorhombic media", *Geophysics*, 62(4)(1997), 1292~1309.

21 I. Tsvankin, L. Thomsen, 1994, "Nonhyporbolic reflection moveout in anisotropic media", *Geophysics*, 59(6)(1994), 1290~1304.

22 Ke Yin, Huizhu Yang, Yuan Dong, "Reflection coefficient of P-wave in anisotropic media", *Qinghua Daxue Xuebao/Journal of Tsinghua University*, 38(2)(1998), 12-16(In Chinese).

23 A. Rüger, "P-wave reflection coefficients for Transverse isotropy models with vertical and horizontal axis of symmetry", Geophysics, 62(1997), 713~ 722.

Theoretical and Computational Acoustics 2001
E.-C. Shang, Qihu Li and T. F. Gao (Editors)
© 2002 World Scientific Publishing Co.

Physical Aspects in Active Noise and Vibration Control

G. Rosenhouse
Faculty of Civil Engineering, Technion, Haifa, 32000, Israel

Abstract

Active Noise and Vibration Control (ANVC) means generation of intentional sound and/or
vibration fields in order to change those caused by existing sources. It has applications in the
industry, including the car and aircraft industries, structural design and audio engineering among
others. Active noise and vibration control involves several disciplines that link mechanics with Hi-
Tech. It necessitates knowledge of physical and theoretical acoustics, electro acoustics, control
engineering and electrical engineering.

Basic concept:

A new principle in favor of ANVC [7] emerges by combination of Huygens' principle, the dynamic
Saint-Venant's principle and wave superposition.
It states that if there is any acoustic source, that generates a propagating linear wave front, the source
can be replaced by a suitable equivalent discrete system of sources over a selected wave front. This
front is defined as the sources' surface (e.g., a sphere within the source domain). Each source in
this array of acoustic sources radiates a wavelet. After a short time and at a distance from the source
surface, which is large enough, a new wave front which approximates that radiated by the original
source will be superimposed by the wavelets of the discrete equivalent system.
For the purpose of cancellation of the original field, a destructive interference, at the prescribed
control points, can be obtained by inversion of polarity of the collocated replacement (or secondary)
sources that belong to the equivalent system. An electronically supported optimization procedures is
added if necessary for adaptive modification of the secondary sources. It is supposed that the far
field magnitude as a whole will at least reduce, since under the conditions that Saint Venant's
principle is satisfied, the local effect of the discretization tends to disappear as the distance from the
equivalent system and the original source increases.
A partial "coverage" of the wave front by the equivalent system leads to a partial equivalence in
the sound field. For example, more attenuation in a certain direction from the source is obtained if
the system is optimized and the equivalent system is denser in this direction. This leads to a local
"quiet zone", while in other places acoustic intensity might be increased.

Active devices:

Sensors of ANVC sense the disturbance field emitted by the primary sources and transmit the
information to the controllers of the system, as data to be processed. The processed information is
then delivered to the system's actuators that radiate in turn an anti-disturbance. In order to achieve
good results, this process has to be confined within the margin of an allowed error. Since the control
system functions electrically, the analysis is based on an electrical signal processing.

The signals of the unwanted steady or time varying (acoustic or vibrational) noise are collected by microphones or by vibration sensors and transmitted to the computing unit by a linking interface. They are then digitally processed by an adaptive electronic system that defines the canceling signals to be radiated by secondary or auxiliary sources, such as loudspeakers for sound and shakers or ceramic actuators for vibration, into the control region. This step is made by an interface that changes the digital results into analog output. Again, the combined field is detected and a feedback of the error is used for correction of the error created by the control system.

ANVC systems have deficiencies that can prevent their being the suitable solution in certain cases and especially in satisfying the allowed margin of error in phase and amplitude of the intentional field. This situation happens mainly in time varying sound fields, very high amplitude waves or nonlinear fields and when the domain has more than one dimension. It is also difficult or even impossible to apply ANC against high audible frequency waves and a very dense array of cancellers is needed under such circumstances. The same situation applies also to vibrations that are to be reduced. In this context, causality becomes an important factor in design of ANVC algorithms. It means the ability to sample the unwanted noise by a time interval earlier than the radiation of the anti-noise so as to enable data processing. Theories of feedback and feedforward systems, including the use of devices which consist of electrical filters, such as Finite Impulse Response (FIR) and Infinite Impulse Response (IIR) and additional adaptive means in complicated control interfaces had to be developed. Interest has been moving now towards more complicated algorithms, including MIMO feedforward control with a multi-rate controller.

In cases of significant non-linear effects generic non-linear filter structures, such as polynomial or neural network filters can be used.

Simulation:

The aim of the ANVC is to obtain an optimized secondary acoustic field that cancels an unwanted noise in a certain domain. For that purpose modeling of real system has been performed. Efforts were done in order optimize by mathematical formulation the solution. However, the analytical approach becomes extremely difficult (or even practically impossible) in complicated domains. [7] suggests using simulation based on Virtual Reality (VR) methods that enable approaching the optimized field by changing the simulated properties and the locations of the secondary sources on the computer screen.

The control part in active noise and vibration control:

ANVC definitely belongs to the fields of control systems engineering and signal processing, and it has to remove unwanted acoustic disturbances and to improve quality of heard sound and eliminate damaging vibrations. Good results can be obtained only by minimizing the errors by using a chosen cost function. To reach this goal it is necessary to construct proper control units (using combinations of feedback, feedforward and hybrid components, among others) and develop learning capabilities. The resulting control algorithms are designed to bring existing inputs as close as necessary to the wanted ones, if a good procedure is provided. A design failure for example, due to lack of a comprehensive physical understanding of the problem, may lead to negative effects, such as losing the stability of the process that causes the collapse of the acoustic outputs. Hence, a physical insight to the control system is of major importance.

Passive/active and semi-active devices:

A most effective way of ANVC is its coupling with passive isolation means. Under such schemes the active control plays a corrective role, while the passive isolator is in charge of the heavy duty of main amplitude reduction. This hybrid isolator which is based on active/passive coupling, is used typically against strong excitations under conditions of uncertainty. Examples are building isolation against the non-deterministic vibration during earthquakes suspension systems and machinery soft mounts against resonant structural dynamics.

The semi-active control is based on changing the acoustic response of a system by electrically controlling its shape and properties. An example is design of moving cores for the cells' silencer. The motion of the core adjusts the lengths of the cells in accordance with the spectrum to be attenuated. An a-priori adaptive process attains an optimized silencer first. Next, adjustments can be made continuously by the electrical system with the engine working. The control system can be combined with a conventional active noise control unit.

Patents and technologies:

The existing solutions and patents that apply active noise and vibration control, include active ear muffs, elastic mountings, ANC in ducts, control of panels vibrations, and acoustics in three - dimensional spaces (in airplanes, cars and buildings) and hybrid control of machines.

The philosophy of an extended definition of active noise and vibration control

The books about ANVC up to now (all published during the last decade) emphasized noise and vibration cancellation [1-6, 8] by generating of an anti-phased sound field, using secondary sources. Usually these sources are a part of an electro-acoustic system that includes sensors, a control unit and sound or vibration radiators. Audio systems (surround systems and so on) are exceptional since they are supposed to improve the sound field by using ANC.

A much broader definition is given in [7]:

Active Sound (or noise as a sub group) and Vibration Control (ASVC) includes any type of generation of secondary acoustic fields in order to change intentionally the primary acoustic field in favor of acoustic or other needs.

This definition expands the frame of ANVC in a way that allows for an overwhelming number of application, and goes far beyond the basic definition of Lueg in 1933. This way ANVC couples with other areas of physics, such sonoluminescence, thermodynamics (also in reducing combustion noise), magnetism, and also in biology, such as animal sonar systems (of bats, dolphins), speech and hearing and tinnitus. In fact in certain cases tinnitus might be partially relieved by using artificial sounds (shower and shaving electrical razor noise among others).

Summary:

The general progress in electronics and control has advanced the construction of ANVC control units, actuators and sensors, and modern DSP improves the ability of ANVC to produce an almost real time control of unwanted signals of very complicated forms. However, this does not suffice and physical argumentation is not only necessary but can also contribute to development of new ideas and ANVC devices.

References:

Comte-Bellot, C., Olivari, D., Aeroacoustics and Active Noise Control, Rohde Saint Genese, Belgium, 1997

Elliott, S., Signal Processing for Active Control, Academic Press, San Diego, 2001

Fuller, C.R., Elliott, S.J., Nelson, P.A., Active Control of Vibration, Academic Press, London, 1996

Hansen, C.H., Snyder, S.D., Active Control of Noise and Vibration, Chapman & Hall, 1996
Kuo, S.M., Morgan, D.R., Active Noise Control Systems-Algorithms and DSP Implementations, Wiley 1996
Nelson, P.A., Elliott, S.J., Active Control of Sound, Academic Press, London, 1992
Rosenhouse, G., Active Noise Control. Vol 1: Active Noise Control - Fundamentals for Acoustic Design, WIT Press, UK and Computational Mechanics, USA, 2001
Tokhi, M.O., Leitch, R.R., Active Noise Control, Oxford Science Publications, Clarendon Press, Oxford, 1992

Theoretical and Computational Acoustics 2001
E.-C. Shang, Qihu Li and T. F. Gao (Editors)
© 2002 World Scientific Publishing Co.

Study on Method of Crosshole Seismic Tomography by Wavelet Transform

Zhenglin Pei[1], Qinfan Yu[2]
[1] Key Lab of Geophysical Exploration, University of Petroleum, Beijing, 100083, China
[2] China University of Geosciences, Beijing, 100083, China

Summary

Combining the multiresolution analysis (MRA) with tomography, this paper proposes a new method of crosshole seismic tomography named the wavelet multiscale seismic traveltime tomography. The method overcomes the drawbacks of linearized inversion depending on initial model and easily trapped by local minimum, and greatly improves the performance of linearized inversion. The results of numerical modeling and field application show that the main advantages of the new algorithm are: reaching the global minimum; inversion results less depending on the initial model; good stability; high resolution and high quality of tomography, and is suitable for non-uniform media with high contrast velocity, and also provides different resolution inverse results benefiting geological interpretation of tomographic images.

Introduction

Crosshole seismic (or acoustic) tomography(CT) is a powerful geophysical exploration tool. It has been used in China for more than twenty years and has found many successful applications in mineral exploration and engineering and environmental geology investigation. The CT is a very complex and nonlinear inverse problem, and its non-linearity always associates with the inhomogeneity in media. The conventional linearized inverse methods are only suitable for smooth inhomogeneous media, and are not suitable for inhomogeneous media with high contrast velocity because of the presence of numerous local minima in the objective function. The multi-grid method (Bunks, et al, 1995) is used to successfully deal with multi-modal objective functions in seismic inversion. The wavelet transform is also a multiresolution transform (Mallat, 1989). Thus, the resolution spaces of the wavelet transform provide a natural framework for multigrid analysis. This property of wavelet transforms has been used to solve the tomography problem in MRA model space (Meng and Scales, 1996). To reduce the number of local minimum in objective function in tomographic inversion and overcome the shortcomings of linearized inversion for complex models, a new tomography methods in MRA model space and MRA data space. is proposed in this paper. As will be shown, it is very suitable for inhomogeneous media with high contrast velocity and has obtained excellent results.

Wavelet Multiscale Tomography

The linear traveltime tomography problem can be represented as

$$Ax = b \tag{1}$$

Where $x \in M$, the model space, $b \in D$, the data space, A is the known Jacobian matrix.

The traditional method is to solve for the model x directly from the linear system (1). In this paper, we apply wavelet transform to the linear system (1).

$$WAx = Wb \tag{2}$$

Where W is a compactly supported, orthonormal second order spin 1-D wavelet transform with symmetry.

For 1-D orthonormal wavelet transform, we have $W^T W = I$, and hence

$$WAW^T Wx = Wb \tag{3}$$

Let $\tilde{b} = Wb, \tilde{x} = Wx, \tilde{A} = WAW^T$.

Where $\tilde{b}, \tilde{x}$ denote 1-D wavelet transform to data vector and model vector respectively, $\tilde{A}$ denotes 1-D wavelet transform to the row and column of matrix A respectively. Thus

$$\tilde{A}\tilde{x} = \tilde{b} \tag{4}$$

The linear system (4) is a representation of the wavelet coefficients at different scale to the linear system (1), i.e. the original CT inverse problem can be decomposed into a set of inverse problems at different scales by MRA. We apply LSQR algorithm (Paige and Saunders, 1982) to solve Eq. (4) in wavelet domain, and then obtain the tomographic image at different scale by using formula

$$x = W^T \tilde{x} \tag{5}$$

For the coarsest scale there are a few local minima which are apart from each other. Thus, at the coarsest scale linearized methods can get closer to the neighborhood of the global minimum, and its inverse result is implemented as an initial model of the inverse problem of the second coarsest scale. This process is repeated until the inverse problem of the finest scale is solved. This algorithm is named MLSQR algorithm.

Numerical Modeling and Field Application

The velocity model consists two low velocity bodies in layer media. The intervals of both the receiver and the source are 1m in a crosshole manner. The results of the numerical modeling prove that the tomography method proposed above is correct and effective (shown in Fig. 1).

One of the real cases is as follows. The purpose of the survey is to investigate the distribution of fracture zone and karst in baserock of hydraulic project in Shi Chuan, Sorth West China. The intervals of the receiver and the exploding source are 3m and 9m in a crosshole manner, respectively (Fig. 2a). The better distributions of the ray density (Fig. 2b) and the ray orthogonality (Fig. 2c) in the surveying system besides up and bottom two small areas ensure that the reliability of tomogrphic image is high. The ray density represents the number of the rays passing thought a velocity cell. The ray orthogonality is measured by the maximum sine of angles between the rays passing thought a velocity cell. The results of the LSQR method (initial velocity is 5 km/s) (Fig.3a) and the MLSQR method (Fig.3b) are basically consistent, but the resolution of the MLSQR method is very high. The distribution of fracture zone and karst (low velocity area with velocity less than 4.5km/s) in tomographic images are in good agreement with the placewhere the index of rock quality in borehole is less than 20%(Fig. 4).

Conclusions

The crosshole seismic (or acoustic) traveltime tomography method proposed above is effective and suitable for inhomogeneous media with high contrast velocity. The method is reaching the global minimum. Its inverse result is less depending on the initial model; good stability; high resolution and high quality of tomography. And it also provides different resolution inversion results benefiting geological interpretation of tomographic images.

Acknowledgments

This work was supported by the ninth Five-Year-Plan Project of National Science and Technology under grant No. 95913060203.

References

Bunks, C., Saleck, F.M., Zaleski, S., and Chavent, G. , 1995, Multiscale seismic waveform inversion: Geophysics, 60 (5), 1457-1473.

Mallat S. A, 1989, multiresolution signal decomposition: The wavelet representation, IEEE Trans. Patt. Anl. Machine Intell, 11 (7), 674-693.

Moser T. J., 1991, Shortest path calculation of seismic rays, Geophysics, 56 (1), 59-67.

Paige, C.C., and Saunders, M.A., 1982, LSQR: Sparse linear equations and least square problems. ACM Trans. Math. Softw., 8, 43-71

Zhaobo Meng and Scales J. A., 1996, 2-D tomography in multi-resolution analysis model space, SEG 66th annual meeting, Expanded Abstracts, 1126-1129.

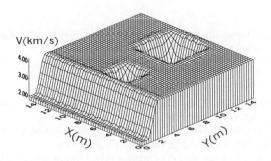

(a) Velocity model

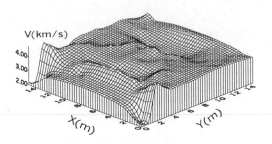

(b) Tomography by LSQR method

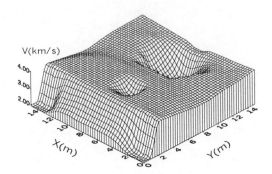

(c) Tomography by MLSQR method

Fig. 1. Tomography by   (b) LSQR method, (c)MLSQR method
for (a) numerical model

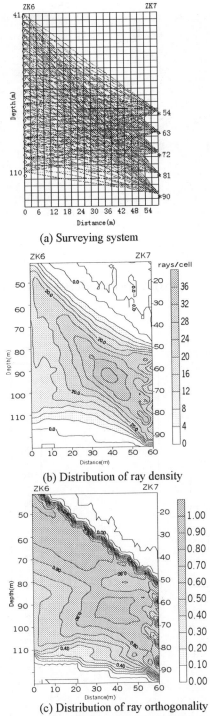

(a) Surveying system

(b) Distribution of ray density

(c) Distribution of ray orthogonality

Fig.2. The surveying system and the estimated reliability of tomogrphic image

418

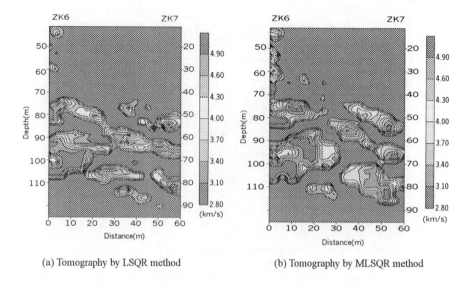

(a) Tomography by LSQR method   (b) Tomography by MLSQR method

Fig.3.Tomography by (a) LSQR method, (b) MLSQR method for the real example

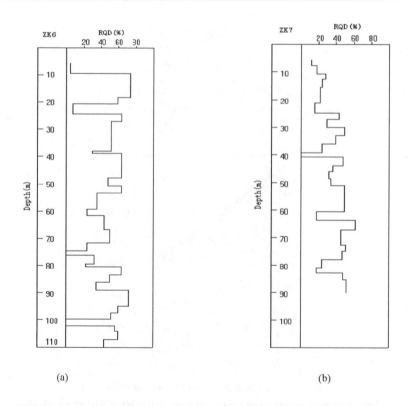

(a)   (b)

Fig. 4 Curves of the index of rock quality in borehole ZK6 and ZK7 via depth

Theoretical and Computational Acoustics 2001
E.-C. Shang, Qihu Li and T. F. Gao (Editors)
© 2002 World Scientific Publishing Co.

Directivity and Frequency Property of Scattering Field by a Penny-Shaped Crack

Hui Deng

Institute of Acoustics, Chinese Academy of Sciences,

17 Zhongguancun Street, Beijing, 100080, P. R. China

Jian-Zhong Shen

Institute of Acoustics, Chinese Academy of Sciences,

17 Zhongguancun Street, Beijing, 100080, P. R. China

Scattering field by penny-shaped crack gains much concern in NDT. In this paper, the surface of the penny-shaped crack is regarded as pressure-released. Because it coincides with a degenerate coordinate surface of oblate spheroidal coordinate system, the separation of variable method is applied to solve the scalar wave equation to calculate scattering field. One kind of oblate angle functions, which can be expanded as the series of associated Legendre functions, is chosen as the solution of one separated differential equation. It is a solution around the ordinary point and convergent within the neighborhood area of origin. The oblate radial functions are asymptotic expansions to the solutions around the irregular singularity. They can be expressed as the series of spherical functions. The coefficients of these series are calculated by continued fraction method and normalized by properly chosen terms respectively. For the desired accuracy in numerical computation, the series are truncated according to the characteristics of asymptotic series. For normal and oblique incidence, under various kinds of conditions, the scattering field is calculated and analyzed. The directivity and low-frequency characteristics of the scattering far field on different conditions are discussed and compared with each other.

1. Introduction

In ultrasonic NDT, how to model the whole testing procedure, from testing pulse injection into material, wave propagation then interaction with the defect, till acceptance of response wave by transducer, is crucial to this technology. Based on this, quantities characterizing different kinds of defects can be extracted for further classification.

The key point of the procedure is characteristic of the scattering field forming after sound wave encounters with the defects. Cracks, as a special kind of defects, usually have acute edges and corners. The resulting mathematical models on scattering fields usually possess singularities of higher order. This on one hand leads to difficulty in finding solutions; on the other hand, the singularities are usually closely related to the peculiarities of the field. The penny-shaped crack, as a kind of typical crack, has gained much more attention. Series expansion method has been used to solve the scattering field [1]. But it is limited by difficulty in disposing the boundary conditions and usually poor convergence feature of the series. Solutions have also been deduced from the integral equations directly by special transformations[2,3,4], but on many cases the singularities lead to difficulties in integral. In this paper, separation of variable method in oblate spheroidal system is adopted to gain analytical solution of the scattering field. The details about singularities of equations are given. Based on this, numerical results of the scattering field are obtained. Some conclusions about directivity and frequency characteristic of scattering field by penny-shaped crack are drawn.

2. Mathematical Description on Scattering Field by Penny-shaped Crack

In unbounded homogeneous medium, a penny-shaped crack is located at x-y plane; symmetric about z-axis with its center coincided with the origin. The diameter of the crack is $2a$. The incident plane wave is inclined to z-axis at the angle θ_i (shown in Fig. 2.1.).

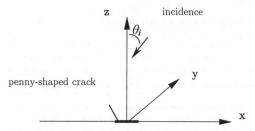

Fig. 2.1. Incident wave and penny-shaped crack in unbounded medium

The incident wave is in the form

$$u^i(\vec{r}) = e^{ik\hat{k}\cdot\vec{r}} = e^{ik(x\sin\theta_i + z\cos\theta_i)} . \tag{2.1}$$

Governing equation of the field is scalar Helmholz equation

$$\left(\nabla^2 + k^2\right)u(\vec{r}) = 0 . \tag{2.2}$$

Assuming the thickness of the crack is zero, the surface of the crack S is pressure-released

$$2\mu\hat{m}\cdot\nabla u(\vec{r}) + \lambda\hat{n}\left(\nabla\cdot u(\vec{r})\right) + \mu\hat{m}\times\left(\nabla\times u(\vec{r})\right) = 0 . \tag{2.3}$$

In the infinite far field, the following Sommerfeld radiation conditions are satisfied

$$\lim_{r\to\infty} u(\vec{r}) = 0 . \tag{2.4}$$

$$\lim_{r\to\infty} r\left(\frac{\partial u(\vec{r})}{\partial r} - iku(\vec{r})\right) = 0 . \tag{2.5}$$

Denoting the incident field and scattering field by u^i and u^s respectively, the total field in the medium is

$$u(\vec{r}) = u^i(\vec{r}) + u^s(\vec{r}). \tag{2.6}$$

3. Analytical Solution

In the oblate spheroidal coordinate system, three coordinate variables are described by η 、 ξ 、 φ respectively. The Helmhotz equation becomes (where $c = ka$)

$$\left[\frac{\partial}{\partial \eta}(1-\eta^2)\frac{\partial}{\partial \eta} + \frac{\partial}{\partial \xi}(1+\xi^2)\frac{\partial}{\partial \xi} + \frac{(\xi^2+\eta^2)}{(1+\xi^2)(1-\eta^2)}\frac{\partial^2}{\partial \varphi^2} + c^2(\xi^2+\eta^2) \right] u = 0. \tag{3.1}$$

Let:
$$u = S_{mn}(c,\eta)R_{mn}(c,\xi)\Phi(\varphi). \tag{3.2}$$

The solution of ordinary differential equation of Φ is

$$\Phi(\varphi) = A\cos m\varphi + B\sin m\varphi. \tag{3.3}$$

The ordinary differential equation of η has two singularities at $\eta = \pm 1$ respectively. When the value of $c = ka$ is small, there is a solution of Gegenbauer expansion around the ordinary point $\eta = 0$. In order that the values of u are finite at $\eta = \pm 1$, the following expansion is chosen as the solution

$$S_{mn}(-ic,\eta) = \sum_{r=0,1}^{\infty} a_r^{mn}(-ic)P_{m+r}^m(\eta). \tag{3.4}$$

In this paper, summation operator $\sum\limits_{r=0,1}^{\infty}$ means, when $n-m$ is odd, summation is over odd terms of r. when $n-m$ is even, then over even terms of r and the term $r = 0$. P_{m+r}^m is $m+r$ order m degree associate Legendre functions.

This solution series is convergent within the circle centered at $\eta = 0$.

The equation of ξ has three singularities. At the irregular singularity $\xi = \infty$, a solution in the form of series of Bessel expansion exists. Considering the radiation condition, the following expression is selected as the solution

$$R_{mn}(-ic,i\xi) = \frac{1}{\sum\limits_{r=0,1}^{\infty}\frac{(2m+r)!}{r!}a_r^{mn}(-ic)} \left(\frac{\xi^2+1}{\xi^2}\right)^{\frac{1}{2}m} \sum_{r=0,1}^{\infty} i^{r+m-n}\frac{(2m+r)!}{r!}a_r^{mn}(-ic)j_{m+r}(c\xi). \tag{3.5}$$

In (3.5), if j_{m+r} is $m+r$ order spherical Bessel functions, R_{mn} is denoted by $R_{mn}^{(1)}$. If j_{m+r} is $m+r$ order Neumann functions, R_{mn} is denoted by $R_{mn}^{(2)}$.

In essence, it is the asymptotic expansion of the real solution. In numerical computation, the number of terms is truncated according to the optimization principle for truncation of asymptotic series.

In two expansions of R_{mn} and S_{mn} above, the coefficients are determined by continued fraction method. The eigenvalues should be determined firstly as follows.

Substituting S_{mn} series into the ordinary differential equation of η, applying the properties of associated Legendre function, the difference equation about a_r^{mn} can be deduced

$$A_r^m a_{r+2}^{mn}(-ic) + B_r^m a_r^{mn}(-ic) + C_r^m a_{r-2}^{mn}(-ic) = 0 , \tag{3.6}$$

Where

$$A_r^m = \frac{(2m+r+2)(2m+r+1)}{(2m+2r+3)(2m+2r+5)}c^2 , \tag{3.7}$$

$$B_r^m = \left[(m+r+1)(m+r) - \lambda_{mn}(c) + \frac{2(m+r)(m+r+1) - 2m^2 - 1}{(2m+2r-1)(2m+2r+3)}c^2 \right] , \tag{3.8}$$

$$C_r^m = \frac{r(r-1)c^2}{(2m+2r-3)(2m+2r-1)} . \tag{3.9}$$

Dividing two sides by a_{r-2}^{mn}, then the recursion formula of a_r^{mn} is

$$A_r^m \frac{a_{r+2}^{mn}(-ic)}{a_r^{mn}(-ic)} \frac{a_r^{mn}(-ic)}{a_{r-2}^{mn}(-ic)} + B_r^m \frac{a_r^{mn}(-ic)}{a_{r-2}^{mn}(-ic)} = -C_{r-2}^m , \tag{3.10}$$

$$\frac{a_r^{mn}(-ic)}{a_{r-2}^{mn}(-ic)} = -\frac{C_{r-2}^{mn}}{B_r^{mn} + A_{r+2}^{mn} \dfrac{a_{r+2}^{mn}(-ic)}{a_r^{mn}(-ic)}} . \tag{3.11}$$

Multiplying two sides by $\dfrac{(2m+r)(2m+r-1)}{(2m+2r-1)(2m+2r+1)}c^2$,

Let

$$\alpha_r^m = \frac{(2m+r)(2m+r-1)}{(2m+2r-1)(2m+2r+1)}c^2 \frac{a_r^{mn}(-ic)}{a_{r-2}^{mn}(-ic)} , \tag{3.12}$$

$$\beta_r^m = \frac{(2m+r)(2m+r-1)}{(2m+2r-1)(2m+2r+1)}c^2 C_{r-2}^{mn} , \tag{3.13}$$

$$\gamma_r^m = \lambda_{mn} - B_r^{mn} . \tag{3.14}$$

Then the forward recursion containing λ_{mn} is

$$\alpha_{r+2}^m = \cfrac{\beta_{r+2}^m}{\gamma_{r+2}^m - \lambda_{mn} - \cfrac{\beta_{r+4}^m}{\gamma_{r+4}^m - \lambda_{mn} - \cdots}} . \tag{3.15}$$

In similar method, the backward recursion containing λ_{mn} can also be found

$$\alpha_{r+2}^m = \gamma_r^m - \lambda_{mn} - \cfrac{\beta_r^m}{\gamma_{r-2}^m - \lambda_{mn} - \cfrac{\beta_{r-2}^m}{\gamma_{r-4}^m - \lambda_{mn} - \cdots}} . \tag{3.16}$$

When c is small, expanding λ_{mn} as power series of c in two recursion formulas (3.15) and (3.16), the values of λ_{mn} can be determined by comparing the coefficients of the same power of c.

Then the coefficients a_r^{mn} can be calculated by their recursion formulas. Although the recursion formulas are both in the form of infinite continuous fraction, they are convergent. For the small values of c, the convergence is fast enough for available numerical computation.

By applying the boundary condition (which can be simplified to Neumann condition), the scattering far field C_s (i.e., when $\xi \to \infty$) is

$$C_s = 2i \sum_{m=0}^{\infty} \sum_{n=0}^{\infty} \frac{\varepsilon_m}{\tilde{N}_{mn}} \frac{R_{mn}^{(1)}(-ic,i0)}{R_{mn}^{(3)}(-ic,i0)} S_{mn}(-ic,\cos\theta_i) S_{mn}(-ic,\eta)\cos m\varphi. \tag{3.17}$$

Where $R_{mn}^{(3)} = R_{mn}^{(1)} + iR_{mn}^{(2)}$.

4. Numerical Results

Based on the analytical solution, computed results are given as following four groups

(i) On the condition of normal incidence, for certain value of ka, the variance curve of the scattering far field against the position coordinate η is calculated. Some selected results are shown in Figure 4.1. Different subplots are for different values of ka. In oblate spheroidal system, since η reflects the azimuth angle of certain orientation, these results in fact represent the directivity under corresponding conditions.

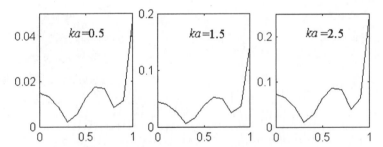

Fig. 4.1. Scattering field variation with η for three different ka $x: \eta \sim y: |C_s|$

(ii) On certain direction, the variance of the scattering far field against ka in case of normal incidence is also analyzed. Figure 4.2 gives some typical results. Different subplots are for several different values of η, i.e., different direction. This group of results expresses the frequency spectrum on different orientation.

(iii) On the condition of oblique incidence, for different incident angles and different values of ka, the scattering far field distributions (field variance along with both η and φ) are calculated. Figure 4.3 shows some typical results.

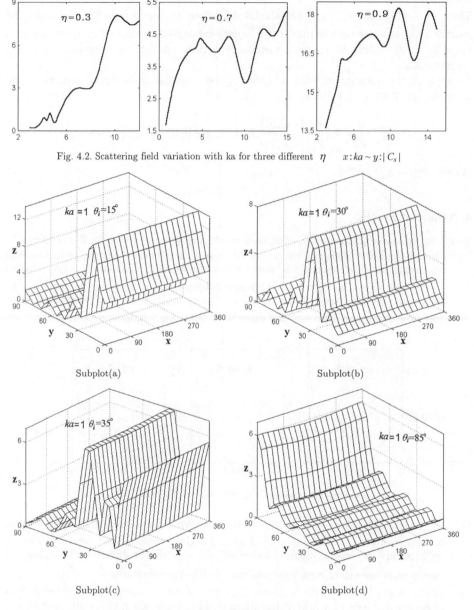

Fig. 4.2. Scattering field variation with ka for three different η $x:ka \sim y:|C_s|$

Subplot(a)

Subplot(b)

Subplot(c)

Subplot(d)

Fig. 4.3. Distribution of scattering far field for different incident angles and different values of ka
subplots (a) incident angle=15° ka=1 (b) incident angle=30° ka=1 (c) incident angle=35° ka=1

(d) incident angle=85° ka=1 $x: \varphi(o) \sim y: \cos^{-1}\eta\,(o) \sim z: |C_s|$

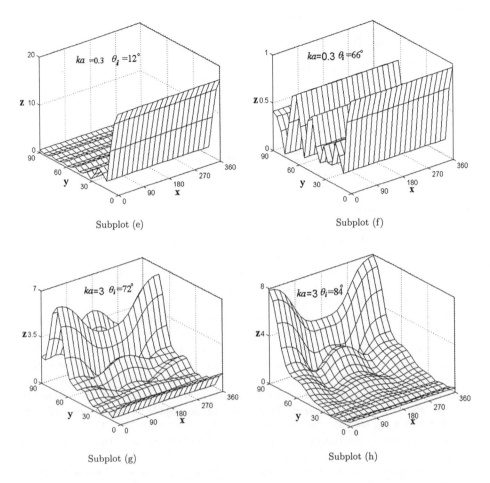

Subplot (e)

Subplot (f)

Subplot (g)

Subplot (h)

Fig. 4.3. Distribution of scattering far field for different incident angles and different values of ka
subplots (e) incident angle=12° ka=1 (f) incident angle=66° ka=1 (g) incident angle=72° ka=3

(h) incident angle=84° ka=3 $x: \varphi(o) \sim y: \cos^{-1}\eta\ (o) \sim z: |C_s|$

(iv) For different incident angles, supreme amplitude of scattering field and the position of the supreme point are studied. Figure 4.4 shows the relation between the supreme amplitude of scattering field to incident angle for two different values of ka.

5. Discussion

Computation results indicate that

426

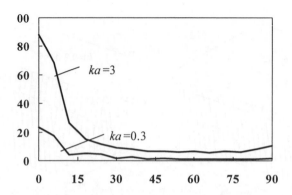

Fig. 4.4. Supreme amplitude of scattering field against incident angles x:incident angle (°)~y: max $|C_s|$

(i)   On condition of normal incidence, for different values of ka, the directivity of scattering far field by penny-shaped crack carries small changes in shape (see Fig. 4.1.). But there are obvious differences in amplitude. Strong scattering concentrates around the z-axis. When the value of ka increases, the directivity becomes keener and the scattering amplitude larger.

(ii)   On condition of normal incidence, the variation of the amplitude of scattering field against ka expresses frequency property of scattering field. The numerical computation results show that, when ka increases, the amplitude of scattering field increases as a whole. Thus it can be concluded that the scattering field of higher frequency components is stronger than that of lower frequency components. But at some certain frequency values, the amplitude fluctuates irregularly and violently. (See Fig. 4.2.) The frequency dependence of amplitude is various on different directions (i.e., for different values of η) in a complicated manner. Based on this point, in NDT, the central frequency of ultrasonic transducer should be chosen carefully to ensure the test sensitivity.

(iii) For oblique incidence, when the incident angle begins to increase from zero degree, within a certain variance range, there is no obvious change in directivity of scattering field, i.e., strong field peak keeps around the z-axis (as subplot (e) in Fig. 4.3. shows). Until the incident angle reaches a certain value (this angle value is referred to as critical angle), the strong scattering peak moves to the reflection direction of the incidence (the same azimuth angle to z-axis). For instance, by computation results, when $ka = 1$, this critical angle is between 12° and 15°. In Figure 4.3, subplot (a) and (b) show the case that when incident angles are 15° and 30° respectively, the scattering peak are at $\cos^{-1}\eta = 15^o$ and $\cos^{-1}\eta = 30^o$ correspondingly. When incident angle approaches to grazing incidence angle (i.e., approaches to 90°), there exists another critical angle (the second). Once the incident angle exceeds this angle, the scattering peak begins to point to x-axis and keep to it. (Subplot (d) and (h) show such case). The value of ka has influence on the two critical incident angles. The larger the value of ka is, the smaller the first critical incident angle is and the larger the second one. But no intensive influence is found.

Further, when ka is small, the variance of scattering field against the direction of φ is little. With the increment of ka, the amplitude variance becomes larger. Comparing subplots for small ka to those for large ka in Figure 4.3, this point is known. This variance is also affected by incident angle. But it is not so much more affected by incident angle than by ka.

(iv)   For quite a few incident angles, strong scattering peak not only occurs at the reflection direction of the incidence as above gives, but also around the corresponding axis. The peak around the axis is called subordinate peak. Subplot (c) and (g) in Figure 4.3 show two typical results when the subordinate peak is at z-axis and x-axis respectively. Under several conditions, the strength of scattering around the axis even exceeds that at the reflection direction. (As subplot (f) in Figure 4.3 gives) The subordinate peak depends on both the incident angles and ka. When ka is small, it occurs more frequently and its amplitude is larger (compared to the scattering at the anti-symmetrical direction).

(v)   For arbitrary incidence, when the incident angle begins to increase, the maximal intensity of scattering field decreases rapidly and then approaches to a constant (as Fig. 4.4.). The larger the ka is, the steeper the descent is.

Acknowledgments

The research is supported by National natural scientific fundation of P.R.China No.19974062 and by fundation of Institute of Acoustics, Chinese Academy of Sciences.

References

1. S. K. Datta, "Diffraction of plane elastic waves by ellipsoidal inclusions," *J. Acoust. Soc. Am.* **61**(6) (1977), 1432-1437
2. P. A. Martin, *et al.*, "Diffraction of elastic waves by a penny-shaped crack analytical and numerical results," *Proc. R. Soc. Lond.* **A390** (1983), 91-129
3. P. S. Keogh, "High-frequency scattering of a normally incident plane compressional wave by a penny-shaped crack," *J. Mech. Appl. Math.* **39**(4) (1986), 535-566
4. A. S. Eriksson, "Natural frequencies of penny-shaped crack with spring boundary condition," *J. of Appl. Mech.* **62** (1995), 59-63
5. S. T. Tsinopoulos, "An advanced boundary element axisymmetric formulation for acoustic radiation and wave scattering problems," *J. Acoustic Soc. Am.* **105**(3) (1999), 1517-1525
6. G. Dassios *et al.*, *Low Frequency Scattering* (Oxford, Clarendon Press, 2000)
7. M. M. Philip, *Methods of Theoretical Physics* (New York, McGraw-Hill Book Co., 1953)
8. C. Flammer, *Spheroidal Wave Functions* (Stanford U Press, 1957)

Theoretical and Computational Acoustics 2001
E.-C. Shang, Qihu Li and T. F. Gao (Editors)
© 2002 World Scientific Publishing Co.

An Adaptive Wavelet Method for Numerical Solution of Wave Propagation in Multi-Layered Media with General Boundary Condition

Jian-Wei Ma, Hui-Zhu Yang

Department of Engineering Mechanics, Tsinghua University, Beijing, 100084,
P. R. China

A fast adaptive wavelet-based method, named Multiresolution Finite Difference (MRFD) is first proposed to simulate the wave propagation in multi-layered media with general boundary. It is a promising method for complex media where large gradient and dramatic variety occur because of little computational burden and robustness. Numerical results derived from the geophysics exploration, proves the efficacy and potential of new scheme.

1. Introduction

Conventional numerical methods such as finite difference (FD), finite element and spectral methods have been successfully applied for wave propagation. But it is difficult to obtain an efficiency solution with large gradient, boundary layer and dramatic variety.

Theory of wavelet transform (WT) has been developed rapidly during the last few years and has been applied successfully in many different fields [1-5]. Recently, Numerous papers about partial differential equations (PDEs) basis on wavelet multiresolution analysis (MRA) have been published [6-9]. The solutions of many PDEs are very smooth in a large, but is not as smooth globally. The wavelet-based method allows to obtain an efficient sparse but accurate approximate representation of some functions and operators due to the vanishing moments, localization, and MRA of the wavelet. High resolution   computations are performed only in regions where sharp transitions occur. Thus, the computational effort and memory requirements can be optimized. Moreover, using wavelet decomposition it is possible to detect singularities and irregular structure. However, at present, most of wavelet algorithms can handle periodic boundary condition (BC) and uniformity media only. The effective treatment of general BC is still an open question, especially for the bounded region problems, even though different possibilities of dealing with this problem have been studied [9].

This paper is devoted to the simulation of wave propagation in complex media, using FD scheme and Daubechies' compactly supported orthogonal WT to discretise the time and space dimension of wave equation, respectively. It is thus that the problem is solved in the wavelet domain rather than the traditional Euclidian space. It is fast adaptive algorithm due to the difference operator and solution vector is sparseness in wavelet domain. The scheme handles general BC also.

2. Solving wave propagation using MRFD scheme

(1) MRFD scheme for wave propagation

We consider the following problem of one dimensional wave propagation problem in inhomogeneous dissipation media with general B.C. and initial condition in geophysics exploration.

$$u_{xx} - \mu_0 \sigma(x) u_t - \mu_0 \frac{1}{v^2(x)} u_{tt} = s(x,t) \tag{1}$$

$$u_t(0,t) = 0 \quad , \quad \frac{1}{v(x)} \frac{\partial u}{\partial t} + \frac{\partial u}{\partial x} = 0 \quad x \in \Omega \tag{2}$$

$$u(x,0) = 0 \quad , \quad u_t(x,0) = 0 \tag{3}$$

Here, $u(x,t)$ is displacement, $s(x,t)$ is seismic wavelet, $v(x)$ is velocity, and Ω_b denote absorbing boundary region in bottom. Assuming FD scheme, Daub4 WT is used to discretise the time and space variable, respectively. Let $\Delta t v^2(x) = k(x), \Delta t v(x) = p(x)$, and $u_J(x,t)$ represent the wavelet approximation of the solution from scale zero to scale J. We then obtain:

$$u_J^{l+1}\left(x_m^{(J)}\right) = \frac{2u_J^l\left(x_m^{(J)}\right)}{1+k\left(x_m^{(J)}\right)\sigma\left(x_m^{(J)}\right)} - \frac{u_J^{l-1}\left(x_m^{(J)}\right)}{1+k\left(x_m^{(J)}\right)\sigma\left(x_m^{(J)}\right)} + \frac{\Delta t k(x)u_{Jxx}^l\left(x_m^{(J)}\right)}{\mu_0\left(1+k\left(x_m^{(J)}\right)\sigma\left(x_m^{(J)}\right)\right)} - \frac{k\left(x_m^{(J)}\right)s^l\left(x_m^{(J)}\right)}{\mu_0\left(1+k\left(x_m^{(J)}\right)\sigma\left(x_m^{(J)}\right)\right)} \tag{4}$$

$$u_J^{l+1}(0) - u_J^l(0) = 0 \quad , \quad u_J^{l+1}\left(x_m^{(J)}\right) = u_J^l\left(x_m^{(J)}\right) - p\left(x_m^{(J)}\right)u_{Jx}^l\left(x_m^{(J)}\right) \tag{5}$$

$$u_J^0\left(x_m^{(J)}\right) = 0 \quad , \quad u_J^1\left(x_m^{(J)}\right) = 0 \tag{6}$$

where, $x_m^{(J)} = A2^{J-1}m, \left(m = 0,1,\cdots,N/2^{J-1}, j \in (0,J]\right), A, N$ denote sampling interval and length under zero scale (i.e., former signal), respectively. $u_{Jx}\left(x_m^{(J)}\right) = D_{Jx} \cdot u_J\left(x_m^{(J)}\right), u_{Jxx}\left(x_m^{(J)}\right) = D_{Jxx} \cdot u_J\left(x_m^{(J)}\right), D_{Jx}$ and D_{Jxx} denote the representation of d/dx and d^2/dx^2 in orthogonal wavelet bases.

(2) The computation of coefficient matrix

We consider representations of operators in the nonstandard form, following [3]. Figure 1 shows the sparse matrix with narrowband D_{Jxx}, i.e., the nonstandard form of the operator d^2/dx^2 in wavelet bases for $J=3$. In practice, the coefficient matrix in eqn.4 can be become more sparse by applying the threshold. It is one of keys to keep the high-speed of algorithms

3. Numerical results and discussion

For a simple case in geophysics, the three-layered nondissipation media with 500m depth and Richer seismic wavelet $s(t) = \left[1 - 2(\pi f t)^2\right]e^{-(\pi f t)^2}$ are considered. Here, the main frequency $f = 3892Hz$ and $\Delta t = 5.303 \times 10^{-4}, N = 200$. The 1~4 linetype in Fig.2 show the instantaneous slice of seismic wave propagation at $t = j * 0.05(j = 1,2,3,4)$, respectively. It easy can be see from Fig.2 that the effects of absorbing BC and interface between different layer.

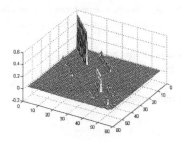

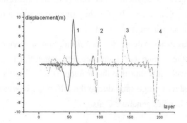

Fig.1 Sparse matrix D_{Jxx} Fig.2 Instantaneous wave

For the second example, Fig.3 shows a multi-layered velocity model that contains some thin layers and lower-velocity layers. In this example, the seismic received profile that often used in exploration is obtained under multiscale.

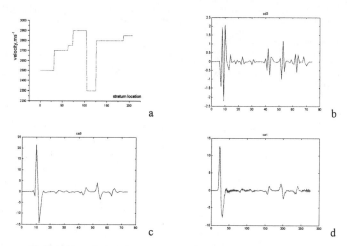

Fig.3 Multi-layered velocity model and the results of MRFD under different scale.

a Velocity model, b Detail, scale=3, c Approximation, scale=2, d Approximation, scale=1

From the Fig.3, we see, the evolving process of response from coarser scale to finer scale is clearly displayed. It also shows the potential of MRFD in picking up the thin layer. Since MRFD pays more attention to the coherent relation between scales, it is more efficient and freedom compared to FD under signal scale. The phenomena of pseudo-oscillation in Fig.2 and Fig.3 be mainly caused by non-2 power sampling and finite length of former signal, general physical BC, low order of FD and so on.

4. Conclusions

A hybrid MRFD method for numerical solution of seismic wave propagation in layered media with general BC has been first proposed in this paper. The computational burden and efficiency were improved since the sparse representation of operator and function in wavelet bases. Hence, the MRFD is an efficient and competitive method. Finally, several important directions for future research must include a introduction of interpolation wavelet and multiwavelet in order to handle arbitrary BC and a study of how reduce the effect of non-2 power sample using boundary wavelet design.

Acknowledgment: The authors would like to thank the China National Natural Science Foundation for supporting this work under Grant 19872037.

References
1 DAUBECHIES, I. "Orthonormal bases of compactly supported wavelets", *Comm. Pure Appl. Math.*.41(1998), 909-996.
2 BEYLKIN,G. "On the representation of operators in bases of compactly supported wavelets", *SIAM J. Numer. Anal.* 6(1992),1716-1740.
3 YU-PING WANG, "Image representations using multiscale differential operators", *IEEE Trans. on Image Processing* 18(12)(1998),1757-1771.

432

4   J W MA, H Z YANG, Y P ZHU, "A discussion about mutliscale seismic waveform inversion", *Progress in Geophysics , Chinese J.*, **15**(4)(2000),54-61.
5   J W MA, Y P ZHU, H Z YANG, "Multiscale-combined seismic waveform inversion using orthogonal wavelet transform", *Electron. Lett.*, **37**(4)(2001),161-261.
6   Y Z WANG, W B WANG, "Analysis of the propagation and reflection of a monopulse in layered and lossy midia using a multiscale wavelet collocation method", *Electron. Lett.*, **33**(6)(1997),497-498.
7   LIANG, J.,ELANGOVAN,S. "Application of wavelet transform in travelling wave protection", *Elctrical Power and Energy Systems*, **22**(2000),537-542.
8   KOSMANIS, I. T, "A hybrid FDTD-wavelet galerkin technique for the numerical analysis of field singularities inside waveguides", *IEEE Trans. magnetics,* **36**(4)(2000), 902-906.
9   OLEG, V.,SAMUEL P., "A multilevel wavelet collocation method for solving PDEs in finite domain", *J Comput.Phys*,**120**(1995),33-47.

Theoretical and Computational Acoustics 2001
E.-C. Shang, Qihu Li and T. F. Gao (Editors)
© 2002 World Scientific Publishing Co.

Multiscale-Combined Inversion of Convolutional Model Based on Orthogonal Wavelet Transform

Jianwei Ma, Huizhu Yang

Department of Engineering Mechanics, Tsinghua University, Beijing, 100084, P. R. China

Yaping Zhu

Center of Wave Phenomenon, Department of Geophysics, Colorado School of Mines, Colorado, 80401, USA

Abstract —Traditional techniques of multiscale inversion pay a lot attention to the truth that the local minima at coarser scale are far apart from each other, thus provide opportunities for finding the global minimum efficiently at the coarser scale and then switch to the finer scales. But they have not taken full advantage of the coherent relation between scales. In this paper, a method of multiscale-combined inversion (MCI) is first proposed to exploit the coherent relation between scales, compared to the traditional techniques, which we prefer to call multiscale-independent inversion (MII). Since the model (or signal) at coarser scale is a "smoother" version of the model (or signal) at finer scale, inversion result derived from the coarser scale may be considered as certain kind of constraints in the following scales, not merely as a good initial guess for the next scale. MCI is a promising method because of some advantages such as little dependence on initial model, efficiency of convergence and robustness. As one realization of MCI, a new formulation of scale-constrained Least Mean Square (LMS) method has been constructed, which we point out is a certain kind of generalizations of the traditional finite power constrained LMS. Several numerical results show the effectiveness and potential of the method.

Index Terms—Convolutional model, Wavelet Transform, scale-constrained LMS, MCI.

I. INTRODUCTION

Theory of Wavelet transform (WT) has been developed rapidly during the last few years and has been applied successfully in many different fields, such as image compression and processing, optimum signal smoothing, and speech analysis. The review of the overall theory can be found in [1~3]. Also, wavelet multiscale analysis has been largely applied in signal and image processing [4~6] and in PDE field [7~9]. Recent applications in geophysics include seismic signal analysis [10], data compression [11]. Numerous papers about multiscale optimal inversion basis on wave equation modeling have been published [11~14]. These methods mentioned all belong to MII.

The concept of multiscale inversion in wavelet domain is still under development. Though it can be performed with different kind of forward strategies such as convolutional model and wave equation simulation, its central idea is based on decomposing a problem into scale sequence and solving the problem from one scale to another. At long scales there are fewer local minima and those that remain are further apart from each other. Thus, at long scales iterative methods can get closer to the neighborhood of the global minimum. In other words, beginning at large scales and using this information to guide the solution to finer detail. All of it can be performed better by MCI.

434

Deconvolution is one of the most important aspects of seismic signal processing [15~17]. The objective of the deconvolution procedure is to remove the obscuring effect of the wavelet's replica making up the seismic trace and therefore obtain an estimate of the reflection coefficient sequence. This paper is devoted to MCI with an emphasis on convolutional model using orthogonal wavelet transform.

The paper is organized as follows. First, a briefly review of convolution model and the wavelet transform multiresolution theory is given. Second, we transform the parameter of convolution model into wavelet domain by scale decompose. Thus, the inversion can be re-directed into the optimization of the significant coefficients. A new formulation of scale-constrained LMS has been constructed, in which we firstly present the idea of multiscale-combined inversion (MCI). Third, several examples are discussed. Finally, we present our conclusions, and give our perspectives for future work.

II. CONSTRAINED LMS OF CONVOLUTION MODEL

A. The Convolution Model

In its simplest form, the 1-D convolution problem can be written as

$$p(t) = s(t) * r(t).$$ (1)

Where * denotes convolution, t is integer-valued time index, $p(t)$ is received valid seismic trace, $s(t)$ is seismic wavelet in which source effects, absorption and multiple reflection effects may be included, $r(t)$ is reflectivity function, i.e., the sequence of reflection coefficients representing the desired lithology. Additionally, there is a direct relationship between reflectivity and impedance for the 1-D problem as $r = (Z_2 - Z_1)/(Z_2 + Z_1)$. Here, Z is defined as ρv. In this paper, the constant density is assumed. Thus, for convenience, the velocity is shown sometimes in the following instead of reflectivity function r due to the above expression.

Of course, the convolution model (1) can be rewritten into the form of matrix

$$\mathbf{p} = \mathbf{Sr},$$

i.e.,

$$\begin{Bmatrix} p_0 \\ p_1 \\ \vdots \\ p_{n-1} \end{Bmatrix} = \begin{bmatrix} s_0 \\ s_1 & s_0 \\ \vdots & \vdots & \ddots \\ s_{l-1} & s_{l-2} & & s_0 \\ & s_{l-1} & & s_1 \\ & & \ddots & \vdots \\ & & & s_{l-1} \end{bmatrix} \begin{Bmatrix} r_0 \\ r_1 \\ \vdots \\ r_{m-1} \end{Bmatrix}.$$ (2)

Expression (2) is an overdetermined linear equatoin group. In theory, the amplitude of reflectance can be obtained by using the amplitude of valid wave divided by that of seismic wavelet. Unfortunately, there are some values equal or close to zeros in wavelet amplitude spectrum. It will cause instability in calculation process. Thus, we often need add white noise to seismic wavelet in process of deconvolution. So the amplitude level of wavelet can be augmented, thus keep away from these zeros. But, in this way, the quite large error will be resulted from it. Thus, we should be very carefully in estimating the inversion results we get from the noise-added model, because there may be some parameters lying in the null-space of the model, which may largely pollute the result.

If solving expression (2) in time domain, we may find that the conditional number of wavelet matrix is so large that the result is very susceptible to the small error in the model or data. In summary, all of the above reasons have limited the application of deconvolution.

B. Constrained LMS

The inversion of above problem can be treated with the technique of linear LMS firstly.

$$J = \|\mathbf{p} - \mathbf{Sr}\|_p = (\sum_{i=0}^{n-1} \left| p_i - \sum_{k=0}^{m-1} s_{ik} r_k \right|^p)^{\frac{1}{p}}, \ p \geq 1 , \tag{3}$$

where $p = 2$, P and $\tilde{P} = Sr$ are the observed and synthetic data, respectively. Given the velocity v, the synthetic data are calculated from equation (1). The velocity field which minimize the object function in equation (3) is the desired solution. In theory, the LMS solution $\mathbf{r} = (\mathbf{S}^T \mathbf{S})^{-1} \mathbf{S}^T \mathbf{p}$ can be obtained by minimizing the object function J. However, this method is unpractical because the matrix $\mathbf{S}^T \mathbf{S}$ is likely to be singular. Even if not, it's too sensitive of the small error of computation to gain enough desirable solution. Especially, the problem is discovered in answering large equations.

For real seismic signal, the vector of reflection coefficients $\mathbf{r}$ is always provided with finite power, i.e.,

$$\|\mathbf{r}\|^2 = \mathbf{r}^T \mathbf{r} = k' \leq k, \ k > 0, \quad k', \ k = const . \tag{4}$$

Using Lagrange multiplier, the object function constrained finite power can be written as

$$J = (\mathbf{p} - \mathbf{Sr})^T (\mathbf{p} - \mathbf{Sr}) + \lambda(\mathbf{r}^T \mathbf{r} - k') . \tag{5}$$

Then we have

$$J \Rightarrow \mathbf{S}^T \mathbf{Sr} - \mathbf{S}^T \mathbf{p} + \lambda \mathbf{r} = 0$$

$$\Rightarrow r = (\mathbf{S}^t \mathbf{S} + \lambda \mathbf{I})^{-1} \mathbf{S}^T \mathbf{p}. \tag{6}$$

Here, Lagrange multiplier λ is added to the diagonal elements of $\mathbf{S}^T \mathbf{S}$. That is to say, the stability of inverse process can be improved by adding direct current (DC) to the entire systems by λ.

III. MULTISCALE ANALYSIS THEORY

Classical notations and definitions related to orthogonal multiresolution are used [3]: $V_j, j \in Z$ stands for the sequence of embedded approximation spaces of $L^2(R)$ and $W_j, j \in Z$ defined as $V_{j-1} = V_j \oplus W_j$. $\phi_j(x)$ stands for the generating scaling function, i.e., $\phi(x - k), k \in Z$ is an orthonormal basis of V_0, and $\psi(x)$ stands for the wavelet function. $2^{j/2} \psi(2^j x - k), k \in Z, j \in Z$ is then an orthonormal basis of

$L^2(R)$. Here, the index $j = 0$ for the coarsest scale is arbitrary and chosen for later convenience. A briefly review of the wavelet transform is given in following.

A wavelet is a function $\psi \in L^2(R)$ with a zero mean

$$\int_{-\infty}^{\infty} \psi(t)dt = 0 . \tag{7}$$

It can be normorlized, i.e., $\|\psi\| = 1$. A family of time-frequency atoms is obtained by scaling ψ with a and translating it with b :

$$\psi_{a,b}(t) = \frac{1}{\sqrt{a}} \psi\left(\frac{t-b}{a}\right). \tag{8}$$

The wavelet transform of $f \in L^2(R)$ at the scale a and time b is

$$Wf(a,b) = \langle f, \psi_{a,b} \rangle = \int_{-\infty}^{+\infty} f(t) \frac{1}{\sqrt{a}} \psi^*\left(\frac{t-b}{a}\right) dt \tag{9}$$

Any wavelet, orthogonal or not, generates a direct sum decomposition of $L^2(R)$. For each $j \in Z$, let us consider the closed subspaces of $L^2(R)$.

$$V_j = \cdots \oplus W_{j+2} \oplus W_{j+1}, \qquad j \in Z \tag{10}$$

These subspaces clearly have the following properties, i.e., multiresolution analysis (MRA):

$$
\begin{aligned}
&(1) \cdots \subset V_1 \subset V_0 \subset V_{-1} \subset \cdots; \\
&(2) clos_{L^2}\left(\bigcup_{j \in Z} V_j\right) = L^2(R); \\
&(3) \bigcap_{j \in Z} V_j = \{0\}; \\
&(4) V_{j-1} = V_j \oplus W_j, j \in Z; \\
&(5) f(x) \in V_j \Leftrightarrow f(2x) \in V_{j-1}, j \in Z
\end{aligned}
\tag{11}
$$

(6) There exists a scaling function $\varphi \in V_0$ such that $\{\varphi(x-k)\}_{k \in Z}$ is a Riesz basis of V_0. In our work, we only use orthonormal bases and will require the basis of condition (6) to be an orthonormal rather than just a Riesz basis.

An immediate consequence of conditions 1~6 is that the function φ may be expressed as a linear combination of the basis functions of V_{-1},

$$\varphi(x) = \sqrt{2} \sum_{k=0}^{L_f-1} h_k \varphi(2x-k). \tag{12}$$

Similarly, we have

$$\psi(x) = \sqrt{2} \sum_{k=0}^{L_f - 1} g_k \varphi(2x - k) \tag{13}$$

The coefficients $H = \{h_k\}_{k=1}^{L_f}$ and $G = \{g_k\}_{k=1}^{L_f}$ are the quadrature mirror filters (QMFs) of length L_f. In general, the sums (13) and (14) do not have to be finite and, by choosing $L_f < \infty$, we are selecting compactly supported wavelets, see e.g.[1][3]

Due to the MRA structure, any signal $f_J \in V_J$ can be written as

$$f_J(x) = \sum_k c_{J,k} \phi_{J,k}(x) = \sum_k c_{0,k}(x) + \sum_{j=0}^{J-1} \sum_k d_{j,k} \psi_{j,k}(x) \tag{14}$$

For dyadic wavelet transform, we often adopt Mallat's pyramid algorithm. The decomposition of coefficients is defined by the following recursive,

$$c_{m,n} = \sum_k \overline{h}_{k-2n} c_{m-1,k}$$
$$d_{m,n} = \sum_k \overline{g}_{k-2n} d_{m-1,k} \tag{15}$$

Reconstruct relation is

$$c_{m-1,n} = c_{m,n} + d_{m,n} = \sum_k c_{m,k} h_{n-2k} + \sum_k d_{m,k} g_{n-2k} \tag{16}$$

In fact, above algorithm only is a filter operation. But, different general filter, it constructed by wavelet is a special filter with adaptive bandwidth. The decomposition algorithm is depicted by space division and double channel filter banks in Fig.1.

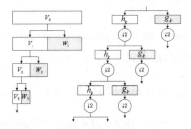

Fig.1. Space division and double channel filter banks

Where, $\downarrow 2$ stands for sub-sample, V_i, W_i stand for the space of approximate(A_i) and detail(D_i) of decomposition coefficient respectively.

IV. THE MCI METHOD BASED ON CONVOLUTIONAL MODEL

Transform convolution model $\mathbf{p} = \mathbf{Sr}$ into wavelet domain: $\mathbf{Wp} = \mathbf{WSW}^T \mathbf{Wr}$, i.e.,

$$\tilde{\mathbf{p}} = \tilde{\mathbf{S}}\tilde{\mathbf{r}}. \tag{17}$$

Here, $\widetilde{\mathbf{p}} = \mathbf{Wp}$ and $\widetilde{\mathbf{r}} = \mathbf{Wr}$ individually correspond to the WT of valid wave and the sequence of reflection coefficients. $\widetilde{\mathbf{S}} = \mathbf{WSW}^T$ is the WT of seismic wavelet matrix. The expression (17) depicted the 1-D convolution model in wavelet domain.

Commonly, the seismic inverse problem has several components that may be decomposed by scale. The observed seismic data have both temporal and spatial features, the source has temporal feature, and the velocity model has spatial features. Thus, the decomposition by scale can be applied to the source, to the observed data, to the velocity model, or to any combination of the three. In this paper, the three components are all decomposed. Finally, it is important to note that the Daub4 wavelet [3] is adopted in this paper.

MII answer $\widetilde{\mathbf{p}}_j = \widetilde{\mathbf{S}}_j \widetilde{\mathbf{r}}_j$ under scale j first, and then fine it by stepwise scale. It is obvious that the process of solution didn't take full advantage of inherent relation cross scales. In this paper, the idea of MCI is first introduced.

Assume the range of scale sequence from 0 to J. The inversion start at rather coarse scale j_0 ($0 < j_0 \le J$), advance to fine scale little by little. The inverse solution of rather coarse scale is noted $\widetilde{\mathbf{r}}_{j^c}$, and that of fine scale is noted $\widetilde{\mathbf{r}}_{j^f}$, usually $j^f = j^c - 1$.

One key of MCI is in search of inherent relation cross scales. Here, $\widetilde{\mathbf{r}}_{j^c}$ is taken into account restriction of currently optimized scale. Furthermore, we can adopt the united restriction of significant coefficient in several scales. Extended $\widetilde{\mathbf{r}}_{j^c}$ to the same dimension of $\widetilde{\mathbf{r}}$ (i.e., optimized parameters in currently scale),

$$\widetilde{\mathbf{r}}'_{j^c} = \begin{bmatrix} \widetilde{\mathbf{r}}_{j^c} \\ \mathbf{0} \end{bmatrix}. \tag{18}$$

Combine the restriction of finite power, the object function J can be rewritten,

$$J = (\widetilde{\mathbf{p}} - \widetilde{\mathbf{S}}\widetilde{\mathbf{r}})^T (\widetilde{\mathbf{p}} - \widetilde{\mathbf{S}}\widetilde{\mathbf{r}}) + \lambda(\widetilde{\mathbf{r}}^T \widetilde{\mathbf{r}} - k') + \gamma(\widetilde{\mathbf{r}} - \widetilde{\mathbf{r}}'_{j^c})^T (\widetilde{\mathbf{r}} - \widetilde{\mathbf{r}}'_{j^c})$$

$$J_{\min} \Rightarrow \mathbf{S}^T \mathbf{S} \mathbf{r} - \mathbf{S}^T \mathbf{p} + \lambda \mathbf{r} + \gamma(\mathbf{r} - \widetilde{\mathbf{r}}'_{j^c}) = 0$$

$$\Rightarrow r = (\mathbf{S}^t \mathbf{S} + \lambda \mathbf{I} + \gamma \widetilde{\mathbf{I}}_{j^c})^{-1} (\mathbf{S}^T \mathbf{p} + \gamma \widetilde{\mathbf{r}}'_{j^c}), \tag{19}$$

where $\widetilde{\mathbf{I}}_{j^c}$ is part restriction operator of unit matrix. It is composed by 0 or 1. Then, the number 1 is corresponding restrictive location of rather coarse scale j^c. λ、γ stand for Lagrange multiplier.

It is obvious, λ throw equivalence restriction to full diagonal of matrix $\mathbf{S}^T \mathbf{S}$. But, γ only throw restriction of part diagonal. We also understand that it throw different restriction to diagonal element. Thus, the inversion of multiscale constrained LMS has more convenient and flexible than the case of single scale. Aim at the characteristic of singularity value of matrix $\mathbf{S}^T \mathbf{S}$, we enhanced remedy measure in different location. Accordingly, the stability and nicety of inversion were balanced efficiently.

Different LMS under Single scale independence processed in usually physical space, the MUIM carried through in wavelet domain. Here, the object constrained by λ and γ is $\tilde{\mathbf{r}}$ obviously.

For example, Fig.2 shows the constrained matrix when the inversion under scale j^f is restricted by the solution under scale j^c ($j^f = j^c - 1$). Here, a_{ij} are the elements of $\mathbf{S}^T\mathbf{S}$. When we enhanced γ_j restriction to several different coarse scale, the general form of scale-constrained is shown in Fig.3.

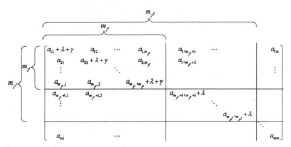

Fig.2. Matrix constrained by a coarser scale

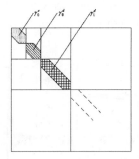

Fig.3. The general form constrained by several scales

V. NUMERICAL EXAMPLES

A. *Example 1*

A simple three-layered velocity model was generated to check the formalism developed in the previous section and to compare the technique of MCI, MII, and single-scale inversion (SSI). Fig. 4 a~c) show the model, wavelet and response, respectively.

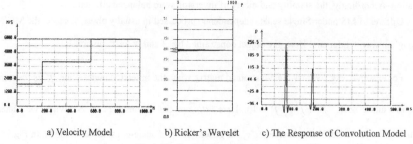

a) Velocity Model   b) Ricker's Wavelet   c) The Response of Convolution Model

Fig.4. Test model of example 1

The Fig.5 shows that the results of SSI and MII by Householder transform. For convenience, the latter is integrated display from scale 7 to 0.

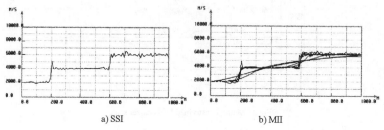

a) SSI   b) MII

Fig.5. The results of SSI and MII

In the same way, the results of MCI shown in Fig.6.

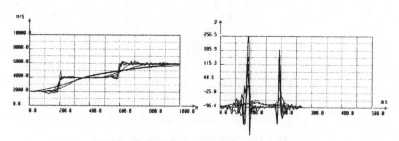

Fig.6. a) the results of inverse velocity model: scale 7~0. b) the results of inverse seismic response: scale 7~0.

According this example, several conclusions can be drawn.

1）The evolving process of inversion from coarser scale to finer scale is clearly displayed in the results of multiscale inversion.

2) Under the same conditions, the result of MCI in the finest scale is equal to that of single scale.

3) In fact, MCI is a generalized form of LMS solution constrained by finite power under single scale.

4) Since MCI pays more attention to the coherent relation between scale, and since it applies constraints across the scales, it is more efficient compared to the method of MII.

5）Real seismic data is band-limited and is inevitably polluted by the noise, hence the inversion result at the finest scale may not always represent the best estimation of the true model. We often observe the truth that an inversion solution of an ill-posed problem becomes so instable that it is totally unacceptable. In some cases, the results under certain middle scale may be more desirable compared to the results at the finest scale. It's because that the singular values of the model got from the middle scale have fairly smaller condition number and are less affected by the noise, while those singular values at the finer scales may not be so lucky.

B. Example 2

Fig.7 shows a multi-layered velocity model which contains some thin layers and lower-velocity layers. In this example, spacial grid step size is 10m, the period of seismic wavelet is 15ms, the minimal model velocity is 1800m/s, the restriction of $\frac{1}{4}$ wavelength is about 7m.

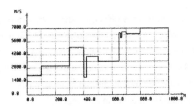

Fig.7. Test model of example 2

Fig.8 a~h) show the result of inverse velocity model under scale 7~0, respectively.

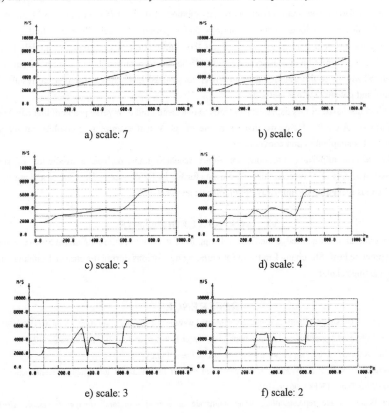

a) scale: 7 b) scale: 6

c) scale: 5 d) scale: 4

e) scale: 3 f) scale: 2

442

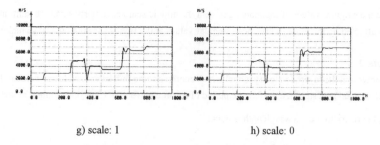

g) scale: 1   h) scale: 0

Fig.8. The results of inverse velocity under scale 7~0 respectively

As one can see, the inversion result clearly shows the potential of MCI in picking up the thin layer. It also can be seen from Fig.8 that the thin layer of low velocity in depth 400~420 m is shown step by step from scale 3 to scale 0. And the thin layer in depth 650~660m is checked under scale 1~0.

Additionally, The MCI based on convolution model can be used in post-stack seismic profile yet. But it is another issue for which we will not address in detail in this paper.

VI. CONCLUSIONS

A method of multiscale-combined inversion based on the theory of wavelet transform has been first proposed in this paper. The main advantage of this method is that it links information of different scale coherently. Guided by some estimation criterions and experiences, researchers can find out the most satisfactory inversion model at some scales, or get the combination of the inversion models in the scale sequence. . A new formulation of scale-constrained LMS has been constructed, which we point out is a certain kind of generalizations of the traditional finite power constrained LMS. Several numerical results show the effectiveness and potential of our methods.

The method discussed in this paper easy extends to one large kind of inverse problem that can be deduced into the form of $\mathbf{A}\mathbf{x} = \mathbf{b}$. Here, we discussed one of its typical examples: convolutional model-based inversion only for simpleness and convenience.

Finally, several important directions for future research must include a prediction of significant coefficients and a study of how reduce the inverse data further by EZW encode for velocity and more relativity of seismic data is picked up by multiwavelet multiscale analysis.

VII. ACKNOWLEDGMENTS

We are grateful to acknowledge the financial support of the China National Natural Science Foundation. We also appreciate Prof. Shi-chang Lin for the stimulating discussions during his stay in Tsinghua University as a senior visiting scholar.

REFERENCES

[1] I. Daubechies, "Orthonormal bases of compactly supported wavelets", *Commun. Pure Appl. Math.* **41**(1988),.909-996.

[2] Y. Meyer, "Wavelets Algorithms and Applications", Society for Industrial and Applied Mathematics, 1993

[3] S. Mallat, "A wavelet tour of signal processing", Academic Press, 1998.

[4] M. Bhatia, W.C. Karl, A.S. Willsky, "A wavelet-based method for multiscale tomographic reconstruction", *Medical Imaging, IEEE Trans.* **15**(1)(1996),92-101.

[5] Yu-Ping Wang, "Image representations using multiscale differential operators", *Image Processing, IEEE Trans.*

8(12),(1999),1757-1771.

[6] Jianwei Ma, Huizhu Yang,Yaping Zhu, "Simulation of acoustic wave propagation in complex media using MRFD method", *Acta physic sinica(Chinese J.)*, **8**(2001).

[7] J.Frohlich, K.Schneider, "An adaptive wavelet-Vaguelette algorithm for the solution of PDEs", *J. Comput. Phys.* **130**(1997), 174-190.

[8] S.Lazaar, P.Ponenti, P.J.Liandrat, "wavelet algorithms for numerical resolution of PDEs", *Comput.Mechods Appl. Mech.Eng*, **116**(1994). 309-314.

[9] M. Holmstrom, "Solving hyperbolic PDEs using interpolating wavelets", *SIAM J. Sci. Comput.* **21**(2)(1999),405-420.

[10] K. Yomogida, "Detection of anomalous seismic phases by the wavelet transform", *Geophys. J. Int.* **116**(1994), 119-130.

[11] C.Bosman, E.Reiter, "Seismic data compression using wavelet transform", *63th Ann. Internat. Mtg.,Soc.Expl.Geophys*, 1261-1264.

[12] C. Bunks, F. M. Saleck, S. Zaleski, &G. Chevent, "Multiscale seismic waveform inversion", *Geophysics,* **60**(5)(1995), 1457-1473.

[13] X. Li, M. D. Sacchi, T. J.Ulrych. "Wavelet transform inversion with prior scale information", *Geophysics,* **61**(5)(1996), 1379-1385.

[14] Jianwei Ma, Huizhu Yang, Yaping Zhu, "A discussion about mutliscale seismic waveform inversion", *Geophysics research, Chinese J.*, **15**(4)(2000),55-61.

[15] Jinghuai Gao, Xiaolong Dong, WenBing Wang, "Instantaeous parameters extraction via wavelet transform", *Geoscience and Remote Sensing, IEEE Trans.* **37**(2)(1999), 867-870.

[16] K.N.Plataniotis, S.K.Katsikas,D.G.Lainiotis, "Optimal seismic deconvolution: distributed algorithms", *Geoscience and Remote Sensing ,IEEE Trans.***36**(3)(1998),779-792.

[17] QianSheng Cheng, Rong Chen, Ta-Hsin Li, "Simultaneous wavelet estimation and deconvolution of reflection seismic signals", *Geoscience and Remote Sensing, IEEE Trans.* **34**(2)(1996), 377-384.

Contact Address

Jianwei Ma was born in Zhejiang, China, in 1976. He is currently pursuing the Ph.D. degree in engineering mechanics at Tsinghua University, Beijing, P.R.China. His research interests include wavelet analysis, geophysics, inversion, Hamilton system, Symplectic geometry.

Address: Department of Engineering Mechanics, Tsinghua University, Beijing, P.R.China.100084

E-mail: mjw_wavelet@263.net

Theoretical and Computational Acoustics 2001
E.-C. Shang, Qihu Li and T. F. Gao (Editors)
© 2002 World Scientific Publishing Co.

Elastic Wave Simulation in Heterogeneous Viscoelastic Media with a Curved Traction-Free Surface

Xiuming Wang

CSIRO Petroleum, P.O. Box 1130, Technology Park, Bentley, WA 6102, Australia

E-mail: *Xiuming.wang@per.dpr.csiro.au*

Numerical modelling of elastic wave propagation in heterogeneous viscoelastic media with free-surfaces has been studied by using various schemes in the past. The principal methods are finite-element, finite-difference, pseudo-spectral and spectral-element methods. Finite-element methods are very flexible in handling perfectly elastic models with free surfaces. Spectral-element methods are special higher-order finite-element formulations. The latter is one of the best methods to tackle free surface problems as they balance the requirements of complexity, accuracy, and computation time. However the implementation is complicated compared with finite difference methods. Pseudo-spectral methods are accurate but time-consuming, and have difficulty in handling strongly curved or rugged stress-release surfaces although they handle smooth free surfaces reasonably well. Finite-difference methods use fine grids to treat irregular stress-release surfaces and are easily implemented. This work focuses attention on using a staggered finite-difference method. The finite-difference method was chosen because it is simple and the codes are portable. Moreover, this method provides a convenient environment to implement complicated boundary conditions. Based on the variable-order algorithms in finite-difference methods, we extended the existing imaging and variable-order finite-difference algorithms to heterogeneous viscoelastic media with rugged free-surfaces. The implementation of finite-difference algorithms incorporated with curved stress-release boundaries are studied. Under stable conditions with constraints, the proposed method works effectively for modelling wave propagation for 2-D viscoelastic models. With the use of gradually variable orders in space, the unstable problem associated with spatial derivative calculations has not been seen. Results obtained by our algorithms are compared with those using vacuum schemes for representing free surfaces. We also test the viscoelastic model against a perfectly elastic model for high values of quality factor. Our numerical investigations demonstrate that the algorithm is efficient and effective. It can also be extended to 3-D viscoelastic heterogeneous media without problems.

1. Introduction

Computational acoustics and wave propagation simulation are used in the fields of non-destructive ultrasonic testing, oil exploration, and earthquake seismology. Despite an ever-growing power of conventional computers, the challenge for accurately simulating the elastodynamic problem still exists. An example in oil exploration occurs with the simulation of seismic wave propagation in a complex large-scale three-dimensional structure incorporating a curved stress-release boundary. In seismic exploration, seismology, and non-destructive ultrasonic detection, experimental observation points are located at or near the stress-release surface, and this surface may have a strong effect on the received signals. Modelling and understanding this effect has been one of the major issues in seismic exploration (Tessmer et al, 1992; Hestholm and Ruud, 2000; and Robertsson, 1996), seismology (Komatitsch and Vilotte, 1998), and non-destructive ultrasonic detection (Kishore et al, 2000).

Several approaches have been proposed for simulating wave propagation in heterogeneous media with a curved stress-release boundary. These include finite-element methods (FEM), boundary element methods (BEM), finite-difference methods (FDM), pseudo-spectral methods (PSM), and spectral element methods (SEM).

Finite-element methods handle complex models with curved stress-release boundaries occurring at the air-solid boundary, with a high degree of accuracy. Although more suited to heterogeneous elastic and poroelastic media with complex geometries (Mu, 1984; Teng, 1988; Atalla et al, 1996; Panneton and Atalla, 1997; Shao and Lan, 2000), these methods, based on a weak variational formulation of the wave equation, allow a natural treatment of free-boundary conditions. However this comes with the cost of excessive computational costs in comparison with the explicit finite-difference methods (Graves, 1996). This is the main reason they have attracted little attention from geophysical modellers in oil exploration. Low-order finite-element methods also exhibit poor dispersion properties (Marfurt, 1984). Higher-order conventional finite-element methods unfortunately generate spurious waves (Komatitsch and Vilotte, 1998). Recently, a spectral-element method has been proposed (Dauksher and Emery, 1997; Seriani, 1998; Komatitsch and Vilotte 1998). This method combines a finite-element scheme with spectral expansion. The basic idea of spectral element method is the choice of shape functions. It is similar to the sine and cosine terms of a Fourier series, which leads to a high rate of convergence of a series to represent the solution. The procedure of the method is, first, decompose the computational domain into many sub-domains, second, express the sought solution as a truncated expansion of a product of Chebyshev polynomials (Dauksher and Emery, 1997; Seriani, 1998) or of Lagrange polynomials (Komatitsch and Vilotte, 1998) on each sub-domain. Then the solution is computed by solving the variational formulation of the orthogonal problem via the Galerkin approach. So the spectral element method is a higher-order variational method for the spatial approximation of the elastic wave equation. This method reduces the stiffness matrix to a diagonal, which reduces the computational cost drastically. Compared to the explicit higher-order finite-difference and rapid expansion methods (Tal-Ezer, 1986; Kosloff et al, 1989; and Carcione et al, 1992), one is still not sure which method is more time efficient.

A pseudo-spectral method based on the FFT technique, introduced from fluid dynamics around two decades ago, has been proposed for application in elastodynamics (Gazdag, 1981; Kosloff and Baysal, 1982). This pseudo-spectral method has been one of the most important numerical techniques introduced due to its accuracy and the minimum number of grid points needed to represent the Nyquist wavelength for non-dispersive propagation. A FFT procedure is used to calculate the spatial derivatives, so that the number of points per minimum wavelength can be two. A Chebyshev expansion is used in the time domain (Tal-ezer et al, 1986), which leads to the rapid expansion method (REM). Unfortunately, this method cannot directly handle the model with curved stress-release surfaces. In order to overcome this difficulty, a set of algebraic polynomials such as Chebyschev polynomials in space is used to replace the original Fourier series (Kosloff et al, 1990; Tessmer et al, 1992). The spatial differencing in the horizontal direction is calculated by the FFT technique; while the vertical derivatives are performed by the Chebychev transform to incorporate boundary conditions into the numerical scheme. The implementation of a curved free surface is done by mapping a rectangular grid onto a curved one. Because of this, it needs a lot of computational time. On the other hand, instead of using spectral expansion techniques, the calculations of the spatial derivatives can also be carried out by using higher-order finite-difference methods to improve computation efficiency. However, we note that this procedure is limited to free-surfaces with only smoothly varying topography and cannot be used for a rugged topographic case. As is remarked in the paper by Hestholm and Ruud (2000), the total model depth should be large enough so that the topographic undulations do not exceed a certain percentage of its size. Also, it depends on the ruggedness of the topographic free-surfaces used.

Boundary integral equations (BIE) and boundary element methods (BEM) are the alternative schemes to simulate wave propagation for a curved stress-release surface. These methods are based on an integral equation representation of the problem relating quantities on physical boundaries. Integral formulations employ fundamental solutions and Green's theorem to represent the wave field. Also, boundary-integral equation techniques with the discrete wave-number Green's function representation have been used to study wave propagation in multi-layered media having irregular interfaces (Bounch et al., 1989; Durand et al, 1999). Although efficient, methods of this kind are most often limited to linear and homogeneous problems. It is well known that this method has non-uniqueness of the solutions of the continuous boundary integral equations at resonance frequency of the corresponding interior problems, which leads to ill-conditioned discrete equations if left uncorrected.

A finite-difference scheme is one of the most powerful methods in seismic wave simulations. Compared with the finite-element method, it is simpler and more flexible. Although there are stability and dispersion problems, more and more finite-difference schemes have been proposed for various scenarios to improve its efficiency. Similar to the development of finite-element methods, the finite-difference methods have been developed mostly using conventional grids and then progressed to the use of staggered grids, from lower-order to higher-order in both time and space domains. The staggered-grid scheme with various orders in space both for second order and first order systems of the equation of wave motion are popular (Madariaga, 1976; Virieux, 1986; and Levender, 1988) and are much better than the conventional ones in reducing spatial dispersion. In order to simulate the wave propagation in the model having a rugged stress-release topographic surface, various FD approaches have been employed, such as imaging methods (Levander, 1988; Robertsson, 1994), vacuum methods (Graves, 1996), and transformed methods by using Chebyshev spectral expansion (Tessmer et al, 1992) and finite-difference procedures (Hestholm and Ruud, 2000). In the case of a fluid pressure-release surface, a simple conformal procedure was proposed (Hastings et al, 1997; Schneider and Wangner, 1998). Compared with other methods, the advantages of the staggered grid finite-difference formulation are, (a) Source is straightforward to introduce and can be expressed in terms of displacement, particle velocity or stress. (b) A stable and accurate representation for a planar free surface can be implemented easily by using second order in the space domain . (c) It is flexible to use lower or higher order differential operators both in time and in space domains. (d) The FD operators are local, and the entire model does not have to reside in the core memory all at once so the boundary conditions can be implemented easily. (e) The method can be easily implemented with other numerical techniques such as finite element or boundary element, since it can calculate the wave field at any position.(f)

The method is easily implemented on scalar, vector or parallel computers, so the computation efficiency is improved greatly. As mentioned above, although there are some FD algorithms for elastic wave modelling with curved stress-release surfaces, they have some limitations, especially for transformed method based finite-difference procedures having a serious stability problem for the rugged topographic stress-release surface. The imaging methods and vacuum methods are not limited to strongly curved stress-release surfaces. However, both methods use very fine grids in order to obtain accurate results. Also, the vacuum method is unstable if the velocities of P- and S- waves are taken to be too small. The variable grid method (Hayashi and Burns, 1999) seems to be efficient, but it needs to apply an averaging or weighting technique to reduce the unstable problems.

Based on existing methods, we have developed a gradually variable-order finite-difference method to simulate the wave field in viscoelastic heterogeneous media with curved stress-release surfaces. The model attenuation properties are described by quality factors of compressional and shear waves respectively, i.e., Q_p and Q_s.

Since earth materials have been shown to have a nearly constant quality factor over the seismic frequency range, A viscoelastic model consisting of a series of standard linear solids in parallel can approximate a constant quality factor over a specific frequency range effectively. First, we give basic summary of the theory for viscoelastic models, then we give the wave equation of motion for viscoelastic media, after that we discuss the implementation of variable-order finite-difference algorithms and the treatment of curved stress-release boundary conditions. Finally we give numerical examples and discuss the effects of the stress-release surface on wave propagation.

2. Constitutive Relation of the Viscoelastic Medium

For elastic media, stress is directly proportional to instantaneous strain but independent of the rate of strain, and the mechanical energy is stored without dissipation. It is well know that, for a perfectly viscous fluid the stress is proportional to the rate of strain and independent of the strain itself. However, the energy is completely dissipated. So for the real earth, we may consider the combination of the mechanical properties of elastic solids and viscous liquids (Carcione, 1988). For equivalent viscoelastic media, we may think that the stress depends on both the strain and the rate of strain, as well as the higher time derivatives of the strain. The basic hypothesis of a theoretical anelastic model in a phenomenological viscoelastic theory is that the current stress tensor is related to the history of the strain tensor. This can be written as (Carcione, 1988)

$$T_{ij} = C_{ijkl} * \dot{e}_{kl} = \dot{C}_{ijkl} * e_{kl}, \quad i,j,k,l \in \{1,2,3\}, \tag{2.1}$$

where $*$ denotes the time convolution, T_{ij} and e_{kl} are the stress and strain tensors, respectively, and C_{ijkl} is fourth-order tensor-valued function of time called the relaxation function. The relaxation function is analogous to the elastic constants in perfectly elastic media. This function determines the behaviour of a material that can be written as

$$C = M_R \left(1 - \sum_{n=1}^{N} \left(1 - \frac{\tau_{en}}{\tau_{Tn}} \right) e^{-t/\tau_{Tn}} \right) H(t), \tag{2.2}$$

where M_R is relaxation modulus of the medium, $H(t)$ is the Heaviside function. The relaxation function in Eq. (2.2) is equivalent to a series of N standard linear materials linked in a parallel way, and τ_{Tn}, τ_{en} are stress and strain relaxation times of the nth mechanism of the medium. According to the theory of viscoelasticity, Eq.(2.1) can also be reformed into

$$T_{ij} = \dot{\lambda}_v * e_{ll}\delta_{ij} + 2\dot{\mu}_v * e_{ij}, \tag{2.3}$$

where λ_v and μ_v are two independent functions obtained from the relaxation function C_{ijkl} for a homogeneous medium, and

$$A_v = \lambda_v + 2\mu_v. \tag{2.4}$$

If the standard linear model described by Eq.(2.2) is used, the following holds

$$A_v = (\lambda + 2\mu) \left[1 - \sum_{n=1}^{N} \left(1 - \frac{\tau_{en}^P}{\tau_{Tn}} \right) e^{-t/\tau_{Tn}} \right] H(t), \tag{2.5}$$

$$\mu_v = \mu \left[1 - \sum_{n=1}^{N} \left(1 - \frac{\tau_{en}^S}{\tau_{Tn}} \right) e^{-t/\tau_{Tn}} \right] H(t), \tag{2.6}$$

where λ and 2μ are the elastic Lame constants. The values of A_v and μ_v are the functions of strain relaxation time of compressional waves, τ_{en}^P and that of shear waves, τ_{en}^S. Using geometric equations and the constitutive relations in Eqs. (2.3), we have

$$\dot{e}_{ij} = (\partial_j V_i + \partial_i V_j)/2 , \tag{2.7}$$

$$\dot{T}_{ii} = (\dot{A}_v - 2\dot{\mu}_v) * \partial_i V_i + 2\dot{\mu}_v * \partial_i V_i , \tag{2.8a}$$

$$\dot{T}_{ij} = \dot{\mu}_v * (\partial_i V_j + \partial_j V_i) , \quad i \neq j . \tag{2.8b}$$

Substitution of Eqs.(2.5) and (2.6) into Eqs. (2.8a) and (2.8b) yields

$$\frac{\partial T_{ii}}{\partial t} = \left\{ (\lambda + 2\mu) \left[1 - \sum_{n=1}^{N} \left(1 - \frac{\tau_{en}^P}{\tau_{Tn}} \right) \right] - 2\mu \left[1 - \sum_{n=1}^{N} \left(1 - \frac{\tau_{en}^S}{\tau_{Tn}} \right) \right] \right\} \nabla \cdot \vec{V} + 2\mu \left[1 - \sum_{n=1}^{N} \left(1 - \frac{\tau_{en}^S}{\tau_{Tn}} \right) \right] \nabla \cdot \vec{V} + \sum_{n=1}^{N} R_{iin} , \tag{2.9a}$$

$$\frac{\partial T_{ij}}{\partial t} = \mu \left[1 - \sum_{n=1}^{N} \left(1 - \frac{\tau_{en}^P}{\tau_{Tn}} \right) \right] (\partial_j V_i + \partial_i V_j) + \sum_{n=1}^{N} R_{ijn}, \quad i \neq j , \tag{2.9b}$$

$$\frac{\partial R_{iin}}{\partial t} = \frac{1}{\tau_{Tn}} \left[(\lambda + 2\mu) \left(1 - \frac{\tau_{en}^P}{\tau_{Tn}} \right) - R_{iin} + 2\mu \left(\frac{\tau_{en}^S}{\tau_{Tn}} - 1 \right) \right] \nabla \cdot \vec{V} + \frac{2\mu}{\tau_{Tn}} \left(1 - \frac{\tau_{en}^S}{\tau_{Tn}} \right) \nabla \cdot \vec{V}, \quad n \in \{1,2,3,...,N\} , \tag{2.10a}$$

$$\frac{\partial R_{ijn}}{\partial t} = \frac{1}{\tau_{Tn}} \left[\mu \left(1 - \frac{\tau_{en}^S}{\tau_{Tn}} \right) - R_{ijn} \right] (\partial_j V_i + \partial_i V_j) , \quad n \in \{1,2,3,...,N\}, i \neq j, \tag{2.10b}$$

and

$$R_{iin} = \left[\frac{(\lambda + 2\mu)}{\tau_{Tn}} \left(1 - \frac{\tau_{en}^P}{\tau_{Tn}} \right) - \frac{2\mu}{\tau_{Tn}} \left(1 - \frac{\tau_{en}^S}{\tau_{Tn}} \right) \right] e^{-t/\tau_{Tn}} H(t) * \nabla \cdot \vec{V} + \left[\frac{2\mu}{\tau_{Tn}} \left(1 - \frac{\tau_{en}^S}{\tau_{Tn}} \right) \right] e^{-t/\tau_{Tn}} H(t) * \nabla \cdot \vec{V} , \quad n \in \{1,2,3,...,N\},$$

$$R_{ijn} = \frac{\mu}{\tau_{Tn}} \left(1 - \frac{\tau_{en}^S}{\tau_{Tn}} \right) e^{-t/\tau_{Tn}} H(t) * (\partial_j V_i + \partial_i V_j), \quad n \in \{1,2,3,...N\}, i \neq j ,$$

where R_{ijn} is a memory variable. The momentum conservation equation in a viscoelastic medium is the same as that in an elastic medium, i.e.,

$$\rho \frac{\partial V_i}{\partial t} = \partial_j T_{ij} + f_i, \tag{2.11}$$

where ρ is a density, and f_i is a volume force. For a 2-dimensional medium, if we take N to be one, i.e., one standard linear solid is taken into account, Eqs. (2.9) and (2.10) can be simplified into

$$\frac{\partial T_{xx}}{\partial t} = (\lambda + 2\mu) \frac{\tau_e^P}{\tau_T} \left(\frac{\partial V_x}{\partial x} + \frac{\partial V_z}{\partial z} \right) - 2\mu \frac{\tau_e^S}{\tau_T} \frac{\partial V_z}{\partial z} + R_{xx} , \tag{2.12a}$$

$$\frac{\partial T_{zz}}{\partial t} = (\lambda + 2\mu) \frac{\tau_e^P}{\tau_T} \left(\frac{\partial V_x}{\partial x} + \frac{\partial V_z}{\partial z} \right) - 2\mu \frac{\tau_e^S}{\tau_T} \frac{\partial V_x}{\partial x} + R_{zz} , \tag{2.12b}$$

$$\frac{\partial T_{zx}}{\partial t} = \mu \frac{\tau_e^S}{\tau_T} \left(\frac{\partial V_x}{\partial z} + \frac{\partial V_z}{\partial x} \right) + R_{zx} , \tag{2.12c}$$

and

$$\frac{\partial R_{xx}}{\partial t} = -\frac{1}{\tau_T} \left[R_{xx} + (\lambda + 2\mu) \left(\frac{\tau_e^P}{\tau_T} - 1 \right) \left(\frac{\partial V_x}{\partial x} + \frac{\partial V_z}{\partial z} \right) + \frac{2\mu}{\tau_T} \left(\frac{\tau_e^S}{\tau_T} - 1 \right) \frac{\partial V_z}{\partial z} \right], \tag{2.12d}$$

$$\frac{\partial R_{zz}}{\partial t} = -\frac{1}{\tau_T} \left[R_{zz} + (\lambda + 2\mu) \left(\frac{\tau_e^P}{\tau_T} - 1 \right) \left(\frac{\partial V_x}{\partial x} + \frac{\partial V_z}{\partial z} \right) + \frac{2\mu}{\tau_T} \left(\frac{\tau_e^S}{\tau_T} - 1 \right) \frac{\partial V_x}{\partial x} \right], \tag{2.12e}$$

$$\frac{\partial R_{zx}}{\partial t} = -\frac{1}{\tau_T} \left[R_{zx} + \mu \left(\frac{\tau_e^S}{\tau_T} - 1 \right) \left(\frac{\partial V_x}{\partial z} + \frac{\partial V_z}{\partial x} \right) \right]. \tag{2.12f}$$

In the above equations, τ_e^P and τ_e^S are viscoelastic strain relaxation times for compressional and shear waves, respectively, and τ_T, viscoleastic stress relaxation time for both compressional and shear waves. The parameters can be determined by quality factors of Q_p and Q_s. A procedure by Blanch et al (1995) for approximating constant Q_p and Q_s over a predetermined frequency range for an arbitrary number of N standard linear solids are used in this work, which is very effective for time domain seismic wave modelling. The general relation between these parameters can be found in (Blanch et al, 1995). In order to simplify our discussion, we take into account one standard linear solid as was done in Robertsson (1994) and Hestholm and Ruud (2000). The approximate relations between quality factors and relaxation times are

$$\tau_T = \left(\sqrt{1+\frac{1}{Q_p^2}} - \frac{1}{Q_p}\right)\frac{1}{2\pi f_0}, \tag{2.13a}$$

$$\tau_e^P = \frac{1}{(2\pi f_0)^2 \tau_T}, \tag{2.13b}$$

$$\tau_e^S = \frac{1+2\pi f_0 Q_s \tau_T}{2\pi f_0 Q_s - (2\pi f_0)^2 \tau_T}, \tag{2.13c}$$

where f_0 is the centre frequency of the wavelet function spectrum. By using Eqs.(2.13a)-(2.13c), We can calculate the strain and stress relaxation times provided the wavelet centre-frequency and the quality factors of compressional and shear waves are given.

3. Implementation of Variable-order Finite-difference Algorithms

According to the function convolution or Taylor expansion, the derivatives of a function $f(x)$ with respect to x can be written as (Holberg, 1987; Igel et al, 1995; Wang, 2001)

$$D_x f(x+\Delta x/2) = \sum_{n=1}^{L/2} a_n \left[f(x+n\Delta x) - f(x-(n-1)\Delta x) \right], \tag{3.1a}$$

$$D_x f(x-\Delta x/2) = \sum_{n=1}^{L/2} a_n \left[f(x+(n-1)\Delta x) - f(x-n\Delta x) \right], \tag{3.1b}$$

where a_n is a weighting coefficient that can be optimised when frequency bands and the relative errors are known (Holberg, 1978), and its leading term is the same as that in conventional staggered grids; $L/2$ are the length of the derivative operator. For the second-, fourth-, and sixth-orders, L is taken to be 2, 4, and 6. Based on the staggered algorithms (Madariaga, 1972; Levander, 1988), the formulations of finite-difference of velocity-stress equations can be read

$$V_{xi+1/2,j}^{n+1/2} = V_{xi+1/2,j}^{n-1/2} + \left[D_x T_{xx}^n + D_z T_{zx}^n \right] \Delta t/\rho, \tag{3.2a}$$

$$V_{zi,j+1/2}^{n+1/2} = V_{zi,j+1/2}^{n-1/2} + \left[D_z T_{zz}^n + D_x T_{zx}^n \right] \Delta t/\rho, \tag{3.2b}$$

where V_x and V_z are particle velocity components in x- and z-axis, respectively, at time $(n+1/2)$. The stress derivatives with respect to x and z can be written as

$$D_x T_{xx}^n = \sum_{l=1}^{L} a_l^z \left[T_{xx}^n(i+l-1,j) - T_{xx}^n(i-l,j) \right], \tag{3.3a}$$

$$D_z T_{zz}^n = \sum_{l=1}^{L} a_l^z \left[T_{zz}^n(i,j+l-1) - T_{zz}^n(i,j-l) \right], \tag{3.3b}$$

$$D_x T_{zx}^n = \sum_{l=1}^{L} a_l^x \left[T_{zx}^n(i+l,j) - T_{zx}^n(i+1-l,j) \right], \tag{3.3c}$$

$$D_z T_{zx}^n = \sum_{l=1}^{L} a_l^x \left[T_{zx}^n(i,j+1) - T_{zx}^n(i,j+1-l) \right]. \tag{3.3d}$$

For the time-derivatives of stress and memory variables, they can be written as

$$T_{xxi,j}^{n+1} = \alpha_1 \left[D_x V_x^{n+1/2} + D_z V_z^{n+1/2} \right] \Delta t - \left[\alpha_2 D_z V_z^{n+1/2} - R_{xxi,j}^n \right] \Delta t + T_{xxi,j}^n, \tag{3.4a}$$

$$T_{zzi,j}^{n+1} = \alpha_1 \left[D_x V_x^{n+1/2} + D_z V_z^{n+1/2} \right] \Delta t - \left[\alpha_2 D_x V_x^{n+1/2} - R_{zzi,j}^n \right] \Delta t + T_{zzi,j}^n, \tag{3.4b}$$

$$T_{zxi+1/2,j+1/2}^{n+1} = \mu \frac{\tau_e^S}{\tau_T} \left[D_x V_z^{n+1/2} + D_z V_x^{n+1/2} \right] \Delta t + R_{zxi+1/2,j+1/2}^n \Delta t + T_{zxi+1/2,j+1/2}^n, \tag{3.4c}$$

$$R_{xxi,j}^{n+1} = \alpha_3 \left(D_x V_x^{n+1/2} + D_z V_z^{n+1/2} \right) \Delta t + \left(\alpha_4 D_z V_z^{n+1/2} - R_{xxi,j}^n / \tau_T \right) \Delta t + R_{xxi,j}^n, \tag{3.4d}$$

$$R_{zzi,j}^{n+1} = \alpha_3 \left(D_x V_x^{n+1/2} + D_z V_z^{n+1/2} \right) \Delta t + \left(\alpha_4 D_x V_x^{n+1/2} - R_{zzi,j}^n / \tau_T \right) \Delta t + R_{zzi,j}^n, \tag{3.4e}$$

$$R_{zxi+1/2,j+1/2}^{n+1} = -\alpha_4/2 \left(D_z V_x^{n+1/2} + D_x V_z^{n+1/2} \right) \Delta t - R_{zxi+1/2,j+1/2}^n \Delta t/\tau_T + R_{zxi+1/2,j+1/2}^n, \tag{3.4f}$$

where $\alpha_1 = (\lambda+2\mu)\frac{\tau_e^P}{\tau_T}$, $\alpha_2 = 2\mu\frac{\tau_e^S}{\tau_T}$, $\alpha_3 = \frac{(\lambda+2\mu)}{\tau_T}\left(1-\frac{\tau_e^P}{\tau_T}\right)$, and $\alpha_4 = \frac{2\mu}{\tau_T}\left(\frac{\tau_e^S}{\tau_T}-1\right)$.

A staggered cell and definitions of each stress, memory variable and particle-velocity component are shown in Fig.1. Note that the definitions of these physical arguments are not located at the same positions. For example, the particle-velocity components are defined between the locations of stress components, so that the derivatives of the stress components with respect to x or z can be used for updating the particle-velocity components. Therefore, each physical argument is linked with the neighbours of the other arguments. The staggered grid algorithms can reduce the spatial dispersion more than the conventionally centred-finite difference algorithm because of the definition of the physical arguments are derived from their neighbours. The locations of stress, particle-velocity, and memory variable components in a staggered grid are located at Position A, B, C, and D for ¼ cell shown in Fig.1.

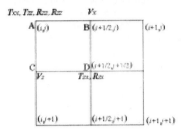

Fig.1. Location of various physical arguments. A: Location of T_{xx}, T_{zz}, R_{xx}, and R_{zz} ; B: Location of V_x ; C: Location of V_z ; D: Location of T_{zx} and R_{zx} .

4. The Free-surface Condition and its Approximation Implementations

In a 2-dimensional model, the viscoelastic stress-release surface causes appropriate boundary conditions for stress-components. These are the same as those in a perfect elastic model, while there are additional conditions for the variable memory arguments, i.e., they should also be zero. So the whole boundary conditions for the viscoelastic case can be written as

$$
\begin{cases}
T_{xx} = 0 \\
T_{zz} = 0 \\
T_{zx} = 0 \\
R_{xx} = 0 \\
R_{zx} = 0
\end{cases}
\tag{4.1}
$$

Substitution of Eq. (4.1) into Eq. (2.11) yields

$$
\begin{cases}
\dfrac{\partial V_x}{\partial z} = -\dfrac{\partial V_x}{\partial z} \\[2mm]
\dfrac{\partial V_z}{\partial z} = \left(\dfrac{\tau_e^S}{\tau_e^P} \dfrac{2\mu}{\lambda+2\mu} - 1 \right) \dfrac{\partial V_x}{\partial x}
\end{cases}
\tag{4.2}
$$

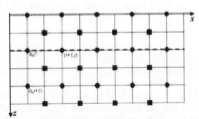

Fig.2. Location of the stresses and memory variables in the vicinity of a horizontal free surface. The solid circle stands for the location of normal stress and memory variable components, while the solid square stands for the shear stress and memory variable components

During the numerical modelling, implementation of the free-surface boundary condition is the key issue for stability of the FD algorithms. Improper implementation of this boundary condition may lead to serious stability problems. Taking into account the work done in Levander (1988), Robertsson (1996), and Hayashi and Burns (1999), we use both imaging and direct methods and variable-order FD algorithms to tackle this issue. During the calculation, the free surface is meshed into either horizontal or vertical directions, and we force each grid point, related to the stress and memory variable components on the surface, to satisfy the free boundary conditions. For the cross point between horizontal and vertical free-surfaces with which the location of the normal stress and memory variable components coincide, these components are set to zero and specific imaging techniques for the others should be taken into account. If we use the 4th-order staggered FD, the location of stress and memory variable components are not at the same point, which is shown in Fig.1. For example, the location of the normal stress and memory variable components in the vicinity of the free surface is shown in Fig.2, where the medium above the line is a vacuum. In this case, it is impossible to take both normal and shear stress and corresponding memory variables to be zero at the same time. In order to tackle this problem, we use an imaging method to make the shear components zero indirectly. The imaging algorithm to guarantee the free-boundary conditions for the horizontal free-surface along $z= j\Delta z$ can be written as

$$\begin{cases} T_{zz}(i,j) = 0, \\ T_{zz}(i,j-1) = -T_{zz}(i,j+1), \\ R_{zz}(i,j) = 0, \\ R_{zz}(i,j-1) = -R_{zz}(i,j+1), \end{cases} \tag{4.3a}$$

and

$$\begin{cases} T_{zx}(i,j-1/2) = -T_{zx}(i,j+1/2), \\ T_{zx}(i,j-3/2) = -T_{zx}(i,j+3/2), \\ R_{zx}(i,j-1/2) = -R_{zx}(i,j+1/2), \\ R_{zx}(i,j-3/2) = -R_{zx}(i,j+3/2). \end{cases} \tag{4.3b}$$

If we digitise the curved free surface into a combination of horizontal and vertical free surfaces, there are seven possibilities encountered during implementation of the stress-release boundary conditions into the FD algorithm, which is shown in Fig.3.

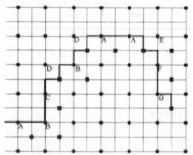

Fig.3. Staggered finite-difference grids in the vicinity of the curved stress-release surface. There are 7 cases that maybe be encountered. The solid circle stands for the location of the normal stresses and memory variables of T_{xx}, T_{zz}, R_{xx}, and R_{zz}; while the solid square stands for the shear stresses and memory variables of T_{zx}, and R_{zx}

The horizontal boundary conditions for both stresses and memory variables can be described in Eqs. (4.3a) and (4.3b), respectively, while the vertical free surface can be treated in a similar way as for the horizontal case by interchanging the x and z, also, the mirroring direction should be taken into account in accordance with the solid areas being in left or right side of the free-surface. Special care should be taken into account for the corner points. According to Fig.3, any kinds of rugged stress-release surface, if the dip angle is less than $\pi/2$, it can be constructed with a combination of the grid points shown in Fig.1. There are seven special cases. Detailed implementation into the staggered FD algorithms for these points are described as the following:

In case A, it is a typical horizontal free surface. Theoretically, stress-release boundary conditions are to set T_{zz}, T_{zx}, R_{zz} and R_{zx} to be zero and then imaged in vertical direction, while the other stress and memory variable components are updated by using the fourth-order finite difference approximation along the boundary surface. The particle-velocity components are calculated using Eqs.(3.2a) and (3.2b). Note that only the stress and memory variable components in vertical directions are imaged.

In case B, since the location of the normal stress and corresponding memory components are defined at the corner and they are on the free surface, these arguments should be set to zero, while the others are updated in a normal way. In this case, no imaging is taken into account because all of the stress and memory variable components are defined properly at their own locations.

In case C, the situation is similar to that in case A, where the normal components of T_{xx} and R_{xx} in horizontal directions are set to zero and the imaging for T_{xx}, R_{xx}, T_{zx}, and R_{zx} takes place respectively along horizontal direction. The normal stress and corresponding memory variable components of T_{zz} and R_{zz} are updated in the same way as the T_{xx} and R_{xx} in case A.

In case D, although the location of the normal stress and corresponding memory variable components are above the stress-release surface, they can not be set to zero but imaged according to the path way of the free surface, i.e., all of the stress and memory variable components are imaged from the symmetric location with respect to the vertically and horizontally free surface paths, respectively. Since the shear stress and corresponding variable memory components pass through the free surface, they are set to zero and imaged accordingly.

In case E, the situation is somehow similar to case D. However, the stress-release surface only passes location V_z. The T_{zx} and R_{zx} to the left of the free surface boundary point are set to zero since the free surface passes their locations, while stress and memory variable components of T_{xx}, T_{zx}, R_{xx}, and R_{zx} are imaged horizontally with respect to the free surface. The stress and memory variable components of T_{zz} and R_{zz} are imaged vertically with respect to the free surface.

In case F, this is a similar one as in case C. The imaging and updating procedures for the stress and particle velocity are the same as in case C. The only difference is that the imaging locations are in opposite directions.

In case G, since the free surface passes the location of the normal stress and corresponding memory variable components, they are set to zero. The imaging of the shear stress and corresponding memory variable components are carried out along the vertical direction.

For the areas including stress-release surface, we use very fine grids and 4th-order FD, and followed by using 6th-order, and then 8th-order gradually. In each computational zone, the grids are regular, and for high-order, the coarse girds can be used. In this case, the gradually variable-order FD and staggered grids are used so that the stability problem in Hayashi and Burns (1999) can be solved.

5. Stability and Dispersion Analyses

To extrapolate particle-velocities and stresses in time, the second-order "leap-frog" procedure is used. The equations for the memory variables become stiff for small stress relaxation times compared with the time step Δt. Our discussion is limited to one linear standard medium that corresponds to one memory variable. We used the "leap-frog" and Crank-Nicolson algorithms to update the system of equations. Note that small changes in Eq.(3.4d)-(3.4f) can lead to either conditionally stable or conditionally unstable results, therefore, the whole finite-difference formulations are conditionally stable. More discussion on this issue can be found in Robertsson (1994) and the corresponding references. In fact, the stability criteria for the conditionally stable schemes are similar to that for elastic schemes. The Courant number is also $c\Delta t / \Delta x$, where c is the velocity for viscoelastic schemes. However, c should be taken to be the highest phase velocity. This can be found at infinite frequency (Blanch, et al, 1995). Therefore,

$$c_{\max} = \sqrt{\tau_e^P (\lambda + 2\mu)/(\tau_T \rho)} \ . \tag{5.1}$$

For the time-step selection, the following equation should be satisfied in 2d case for second-order in time:

$$\frac{\Delta t c_p}{\Delta h} \leq \left[\sqrt{2} \sum_{l=1}^{L} |a_l| \right]^{-1} , \tag{5.2}$$

where c_p is the highest phase velocity of compressional waves, a_l is the weighting coefficients, and L is the order of FD in space shown in Eqs.(3.1a) and (3.1b). Since the variable-order and variable grids are used. The minimized time step satisfying the highest-order FD schemes is employed so that the time stability can guaranteed

454

6. Numerical Modelling Analyses

First, we use the vacuum method and the proposed method to simulate a model with a flat free surface. The wavelet function is the first derivative of the Gaussian function. The highest frequency is four times the fundamental or centre frequency. It is perfectly symmetric about the middle, which crosses zero value. For the artificial boundary, we use a very small quality factor and damping methods to reduce these boundary effects in an absorption areas near the artificial boundaries. The snapshot at time of 1.44 second by using the vacuum method (4a) and our proposed method (4b) are shown in Fig.4. During the calculations, the source is at the free surface with a centre frequency of 10Hz. The velocities of compressional and shear waves are 5.00km/s and 2.89km/s, respectively. The density is 2780 kg/m^3. The size of the model is 20.48km in length by 9.64 km in width. We use 10 points per minimised wavelength. The results are in agreement with each other. However, the amplitude of the Rayleigh waves in Fig.4a is weaker than that in Fig.4b, which is because of the leakage of energy in the vacuum method in which P-wave velocity cannot be set zero directly for stability issue.

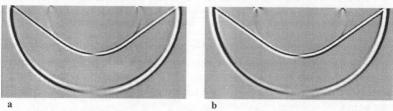

a   b

Fig.4. Snapshot calculated at the time of 1.44s using vacuum method (a) and the proposed method (b)

We also simulated perfectly elastic and viscoelastic models with a horizontal interface, respectively. A source is above the interface with a distance of 0.71km; the source centre frequency is 15Hz; the grid length for Δx and Δz is 10m, respectively; the time sampling of Δt is $8.5\,e^{-4}$ second; the velocities of compressional and shear waves in up layer are 3.5km/s, 2.08km/s, respectively; while those in down layer are 4.0km/s, 2.21km/s, respectively, the densities for the up and down layers are 2300 kg/m^3, and 2780 kg/m^3, respectively. The modelling results are shown in Fig.5. The blue lines stand for the modelling results for perfectly elastic media; while the red lines, for the viscoelastic media with quality factors of 1000 (4a) and 20 (4b), respectively. The trace number from top to bottom corresponds to the receiver location from the source to right in the horizontal direction. The distance for the nearest receiver to the source is 0.48km, and the offset of the receivers is 0.16km. The horizontal receiver line passes through the source location, and is above the interface with a distance of 0.71 km. It is shown in Fig.5 that when the quality factors are large such as 1000, the two lines coincide with each other. The larger the quality factors, the less the attenuation of the models. These figures show for large Q, the viscoelastic algorithm agrees with the perfectly elastic FD method independently implemented using perfectly elastic theory (Wang. 2001).

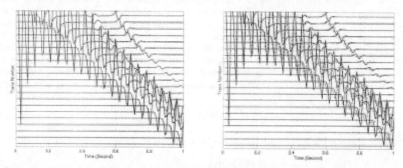

Fig.5.Common shot gather of the particle-velocity components in the horizontal direction for elastic and viscoelastic models with an interface. The blue lines stand for a perfectly elastic model, while the red lines for viscoelastic models with the quality factor Q of 1000 (5a) and 20 (5b), respectively. For a large Q, the two lines in Fig.5a coincide with each other since larger Q means less attenuation.

Common shot gathers of the particle-velocity components in vertical directions are shown in Fig.6 for various quality factors of Q of 1000 (6a), 20 (6b) and 10 (6c), respectively. The model used is the same as in Fig.5. In addition to strong attenuation for low Q, these figures show that, the time arrivals for lower Q are smaller than those for higher Q, which is also seen in Fig.5.

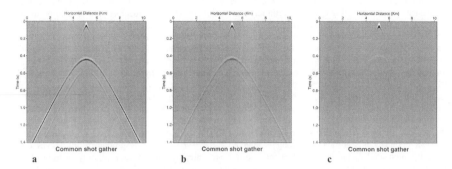

a   b   c

Fig.6. Common shot gathers of the particle-velocity components in the vertical direction for various quality factors Q of 1000 (a), 20 (b) and 10 (c), respectively. For a small Q in Fig.6c, the attenuation is so strong that it is hard to see the reflected wave signals.

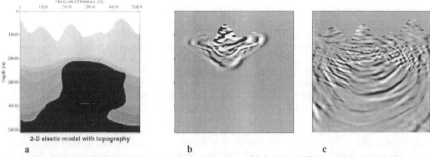

a   b   c

Fig.7. Physical model (7a) and the snapshot at the time of 0.4 second (7b) and 1.2 second (7c), respectively. There are 7 layers in the model, from the top to bottom the compressional velocities being 3.2km/s, 3.5km/s, 3.8km/s, 4.1km/s, 4.3km/s, 4.5 km/s, 5.5km/s, respectively. The density is 2500 kg/m^3. The Poisson's ratio is 0.25 for all of the media. The quality factors of compressional and shear waves are 800.

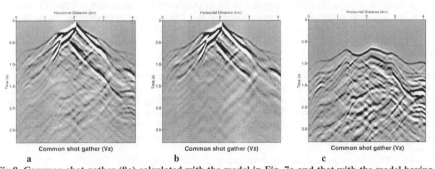

a   b   c

Fig.8. Common shot gather (8a) calculated with the model in Fig. 7a and that with the model having only the traction-free surface (8b). Fig.8c is the results of Fig.8a subtracted by Fig. 8b where the amplitude is enhanced so that the reflected arrivals from the formations are clearly seen.

456

Fig.7a is a 2-D viscoelastic model with a curved free surface. There are 7 layers in the model, from the top to bottom the compressional velocities being $3.2km/s$, $3.5km/s$, $3.8km/s$, $4.1km/s$, $4.3km/s$, $4.5\ km/s$, $5.5km/s$, respectively. The density is $2500\ kg/m^3$. The Poisson's ratio is 0.25 for all of the media. The quality factors of compressional and shear waves are constant with the value of 800. The highest compressional velocity is for the salt dome. We used our codes to calculate the wave field snapshots for this model. They are shown in Fig.7a (0.4 second) and Fig.7b (1.2 second), respectively. From these snapshots we see that the effects of the curved stress-release surface is significant. Fig.8 are the common shot gathers in which the receivers are located on the free surface for the modelling results (8a) with the model shown in Fig.7a, the modelling ones only with the topographic effects (8b), and those for the subtraction of Fig.8a with Fig.8b. By comparison Fig.8a with Fig.8b we know that because of the existence of the curved stress- release surfaces, it is hard to see the reflected wave signals from the layered media and the salt dome. In order to see more clearly the reflected arrival, we subtracted the common shot gather in Fig.8a with the one in Fig.8b, which is proposed by Fu et al (1999). In this case, the scattering wave signals from the curved stress-release surface can be removed if the distance between the free surface and the real layered media are big enough that the scattering waves from the free-surface and the reflected arrivals are not overlapped. Fig.8c is the subtracted result from which we can clearly see the first arrivals from the layered media and the salt dome. Note the waveforms in Fig.8c are amplified. In fact, the reflected arrivals are weaker than the wave signals scattered from the stress-release surface. The reason why the scattered wave signals is significantly strong is that the surface waves such as ground-rolling waves is not geometrically attenuated along the surface if they do not encounter fractures or discontinuity in front of their propagation directions and there is no energy leakage along the free-surface; while for reflected arrivals from layered media or salt dome, velocity contrasts are not so big and there is energy leaking into the deep layers unless the incident angle is larger than the critical shear angle.

7. Conclusions and Discussions

Based on the analysis for variable algorithms in finite-difference methods, we extended the existing imaging methods and the second-order in time, variable-order in space finite-difference algorithms with staggered grids to heterogeneous viscoelastic media with irregular free-surfaces. Under the conditionally stable conditions, the proposed method works effectively and efficiently for modelling wave propagation for 2-D viscoelastic models with a curved stress-release surface. By using the gradually variable orders (4th-, 6th-, and 8th orders) in space, we have not observed the instability due to spatial derivative calculations. For 4th-order, the grid points can be the 20-40 points per minimized wavelength in the vicinity of curved free surface to preserve the accuracy of the calculations. For the normal areas, the 6th-order FD needs 4 points per minimised wavelength; while for 8th-order, 3 points per minimized wavelength. However, the gradually variable grids should be used in order to reduce spatial dispersion problems in this special implementation for curved free-surface model. According to our numerical modelling, free-surface effects on wave propagation are very strong especially for those combined with low velocity channels in the vicinity of the surface. It is difficult to use directly the common shot gather to abstract the underground formation structures without detailed treatment of the rugged topographic effects on seismic data.

Acknowledgements

The Author wishes to express his sincere gratitude to Kevin Dodds and Bension Singer of CSIRO Petroleum, Boris Gurevich and John McDonald of Curtin University of Technology for reviewing the paper. He also appreciates the helpful discussions with Yi Zeng and Li-Yun Fu of CSIRO Petroleum.

References

1. Atalla, N., Panneton, R., and Debergue, P., 1996, A mixed Displacement-pressure formulation for Biot's poroelastic equation: JASA, 104, 1444-1452.
2. Blanch, J. O., Robertsson, J. O. A., and Symes, W. W., 1995, Modelling of a constant Q: Methodology and algorithm for an efficient and optimally inexpensive viscoelastic technique: Geophysics, 60, 176-184.
3. Bouchon, M., Campillo, M., and Gaffet, S., 1989, A boundary integral equation-discrete wave number representation method to study wave propagation in multilayered media having irregular interfaces: Geophysics, 54, 1134-1140.
4. Carcione, J. M., 1988, Seismic modelling in viscoelastic media: Geophysics, 58, 110-120.
5. Carcione, J. M, Kosloff, D., Behle A., and Seriani G., 1992, A spectral Scheme for wave propagation simulation in 3-D elastic-anisotropic media: Geophysics, 57,1593-1607.

6. Dauksher W., and Emery A. F., 1997, Accuracy in modelling the acoustic wave equation with Chebyshev spectral finite elements: Finite Elements in Analysis and Design, 26, 115-128

7. Durand, S., Gaffet, S., and Virieux, J., 1999, Seismic diffracted waves from topography using 3-D discrete wavenumber-boundary integral equation simulation: Geophysics, 64, 572-578.

8. Fu, L. Y, Wu, R. S., and Guan, H, M., 1999, Removing rugged-topography scattering effects in surface seismic data: 69th Ann. Internat. Mtg., Soc. Expl. Grophys., Expanded Abstracts, 528-531.

9. Gazdag, J. 1981, Modelling of the acoustic wave equation with transform methods: Geophysics, 46, 854-859.

10. Graves, R., 1996, Simulation seismic wave propagation in 3D elastic media using staggered-grid finite differences: Bull. Seism. Soc. Am., 86,1091-1106.

11. Hastings, F. D., Schneider, J. B., and Broschat, S. L., 1997, A finite-difference time-domain solution to scattering from a rough pressure-relase surface: J. Acoust. Soc. Am., 102, 3394-3400.

12. Hayashi, K, and Burns, D. R., 1999, Variable grid finite-difference modelling including surface topography: 69th Ann. Internat. Mtg., Soc. Expl. Grophys., Expanded Abstracts, 528-531.

13. Hestholm, S. O., and Ruud, B. O., 2000, 2D finite-difference viscoelastic wave modelling including surface topography: Geophys. Prosp., 48, 341-373.

14. Holberg, O., 1978, Computational aspects of the choice of operator and sampling interval for numerical differentiation in large-scale simulation of wave equation: Geophysical prospecting, 35,629-655.

15. Igel, H., Mora, P., and Riollet, B., 1995, Anisotropic wave propagation through finite-difference grids: Geophysics, 60, 1203-1216.

16. Kishore, N. N., Sridhar, I., and Iyengar, N. G. R., 2000, Finite element modelling of the scattering of ultrasonic waves by isolated flaws: NDT & E International, 33, 297-305.

17. Komatitsch, D., and Vilotte J. P., 1998, The spectral element method: an efficient tool to simulate the seismic response of 2D and 3D geological structures: Bull. Seism, Soc. Am., 88, 368-392.

18. Kosloff D., Filho, A. Q., Tessmer E., and Behle, A., 1989, Numerical solution of the acoustic and elastic wave equations by a new rapid expansion method: Geophys. Prosp., 37, 383-394.

19. Kosloff, D., Kessler, D., Filho, A. Q., Tessmer, E, Behle, A., and Strahilevitz, R., 1990, Solution of the equations of dynamic elasticity by a Chebychev spectral method: Geophysics, 55, 734-748.

20. Levander, A. R., 1988, Fourth-order finite-difference P-SV seismograms: Geophysics, 53, 1425-1436.

21. Madariaga, R., 1976, Dynamics of an expanding circular fault: Bull. Seism. Soc. Am., 65, 163-182.

22. Mu, Y. G., 1984, Elastic wave migration with finite element method: Acta Geophysica Sinica, 27, 268-278.

23. Marfurt, K., J., 1984, Accuracy of finite-difference and finite-element modelling of the scalar and elastic wave equations: Geophysics, 49, 533-549.

24. Orszag, S. A., 1980, Spectral methods for problems in complex geometries: J. Comput. Phys., 37, 70-92.

25. Panneton, R., and Atalla, N., 1997, An efficient finite element scheme for solving the three dimensional poroelasticity problem in acoustics: J. Acoust. Soc. AM., 101, 328-3298.

26. Robertsson, J. O. A., Blanch, J. O, and Symes, W.W., 1994, Vsicoelastic finite-difference modelling: Geophysics, 59, 1444-1456

27. Robertsson, J. O. A, 1996, A numerical free-surface condition for elastic/viscoelastic finite-difference modelling in the presence of topography: Geophysics, 61, 1921-1934.

28. Seriani, G., 1998, 3-D large-scale wave propagation modelling by spectral element method on Cray T3E: Comp. Meth. Appl. Mech. and Eng., 164, 235-247.

29. Schneider J. B., and Wagner C. L., 1998, Simple conformal methods for finite-difference time-domain modelling of pressure-release surfaces: J. Acoust. Soc. Am., 104, 3219-3226.

30. Shao, X. M., Lan, Z. L, 2000, Finite-element methods for the equations of waves in fluid-saturated porous media: Chinese J. of Geophysics, Vol.43, 264-277.

31. Tal-Ezer, H., 1986, Spectral methods in time for hyperbolic problems: SIAM J. Numer. Anal., 23, 12-26.

32. Teng, Y. C., 1988, Three-dimensional finite element analysis of waves in an acoustic media with inclusion: J. Acoust. Soc. Am., 86, 414-422.

33. Tessmer, E., Kosloff, D., and Behle, A., 1992, Elastic wave propagation simulation in the presence of surface topography: Geophys. J. Internat., 108, 621-632.

34. Virieux, J., 1986, P-SV wave propagation in heterogeneous media: Velocity-stress finite-difference method: Geophysics, 51 889-901

35. Wang, X. M., 2001, Seismic wave simulation in anisotropic media with heterogeneity using a high-order finite-difference method: Proceedings of 5th SEGJ International Symposium, 113-120, Tokyo, Japan.

Theoretical and Computational Acoustics 2001
E.-C. Shang, Qihu Li and T. F. Gao (Editors)
© 2002 World Scientific Publishing Co.

Theoretical Simulation of the Behavior of Micromachined Electrostatic Ultrasonic Transducers

Li-Feng Ge

School of Electronic Engineering and Information Science, Anhui University,
Hefei 230039, People's Republic of China

A general three-dimensional (3D) theoretical model for simulating the behaviour of micromachined electrostatic ultrasonic transducers has been developed. The model takes into account the transducer's geometry, the properties of material used, tension in diaphragm, and bias voltage, and then gives a complete mathematical formulation. Some special cases of it are briefly discussed also. The paper further applies the general model and its simplified varieties to predict natural frequencies of a surface micromachined ultrasonic transducer. The comparison between these theoretical values and measured results indicates that the 3D model gives accurate prediction and is a generalized model for describing the behaviour of various electrostatic audio and ultrasonic transducers.

PACS numbers: 43.38.Bs

INTRODUCTION

At the beginning of the twentieth century, Wente first reported a condenser microphone based on electrostatic principle [1]. The significant innovation, as commended by Hunt, "ushered in a new and thrilling era for the quantitative measurement of acoustical phenomena". [2] Such a microphone comprises a tightly stretched metallic diaphragm exposed to the sound field and a closely spaced backplate electrode with annular or radial grooves. Sound pressure deflects the diaphragm, resulting in a detectable change of capacitance and output voltage of the microphone; conversely, an alternating voltage applied between the two electrodes causes the diaphragm to vibrate, generating a sound wave. In 1937[3] Sell made a solid dielectric film coated with metal electrode touch directly with the roughened or grooved metal backplate at a number of high points (or ridges), leading to applications in the ultrasonic frequency range.

In the middle of the last century, Kuhl et al. [4] performed a systematic experimental study on such electrostatic (or capacitive) ultrasonic transducers, and observed that such transducers have a multi-peaked response, and indicated qualitatively that its frequency response depends on the flexural compliance and the mechanical stress of the film, and the compliance formed by the air cushion within the grooves, excepting the mass of the film. After that Matsuzawa [5] also observed that grooved transducers have the multi-peaked response, so that he concluded that such transducers are "practically useless". Nevertheless, he found that transducers using metal backplates with finely, uniformly roughened surfaces show a single resonance system, and presented a Helmholtz resonator model for roughened transducers, which indicates that its fundamental frequency can be determined by the mass of the diaphragm, and the mean thickness of the layer of air. In early 1990s, however, an enhanced investigation was focused on transducers with parallel V-shape grooves, and the Helmholtz resonator model was usually used for V-grooved transducers [6-8]. But it is clear that such a simple single-degree-of-freedom model cannot explain experimentally observed complex phenomena, such as a multi-peaked response and unusual wide bandwidth.

Since 1970's micromachining techniques have brought revolutionary improvements in the field of transducer technology, naturally also for acoustic transducers. At the end of 1980s, a micromachined silicon-based condenser microphone with a silicon nitride diaphragm, and a silicon-based electrostatic ultrasonic transducer with a polymer diaphragm were reported, respectively [9,10]. Micro-machining techniques enable the fabrication of transducers with high precision and reproducibility, and motivate further the interest in establishing a reliable model and realizing optimal design.

During the last decade, the theoretical research on the operation mechanism of micromachined ultrasonic transducers has been enhanced and several quantitative models presented also. Anderson et al., using a 2D treatment, presented a plate-under-tension model for their U-grooved transducers; and concluded that for such small air-gaps the Helmholtz model is unsuitable, and that the bending stiffness of diaphragm is the most substantial factor, deciding the fundamental, but they also indicated there are still discrepancies between their predictions and measured results [11]. Ge indicated that from the mechanical viewpoint, a sub-diaphragm of such an ultrasonic transducer behaves as a thin plate founded on an air cushion, in other words, a plate supported by an air-spring; or from acoustical view, an individual air-gap structure can be regarded as a short tube enclosing air, which is terminated by the flexible motive diaphragm and the backplate as a rigid wall, and then presented a plate-on-air-spring model for V-grooved transducers and a tensile-plate-on-air-spring model for U-grooved transducers respectively [12-15]. In addition, a leading group [16] applied the classical membrane-under-tension model, which has been used normally for condenser microphones, to a surface micromachined ultrasonic transducer; and later, further applied the plate-under-tension model to such a device with vacuum-sealed cavities [17].

Even though the electrostatic ultrasonic transducers have complex and diversified configurations, we may classify them to three basic categories according to the geometry of backplate textures, which are roughened, grooved, or pit-array textured. Such a classification is beneficial for theoretical modelling of the transducer behaviors. Clearly, the tensile-plate-on-air-spring model for grooved transducers is a 2D model, and can be extended, without any substantial difficult, to a 3D model for micromachined ultrasonic transducers with pit-array structures [18]. This paper focuses on developing further such a general 3D model.

I. THEORY OF ELECTROSTATIC ULTRASONIC TRANSDUCERS

A typical micromachined ultrasonic transducer consists of an array of hundreds or thousands capacitive sensing cells connected in parallel. Each cell is of a micro-air-gap structure, sketched as Fig.1 (a), and its air-gap is cylindrical, pyramidal or hexagonal shape. The air-gap and the metal-coated sub-diaphragm on top of it constitute a dynamic system. Therefore, it is clear that all of sub-diaphragms will vibrate with the same phase and same magnitude when either an electrical signal, in the transmitting mode, or an external acoustic pressure, in the receiving mode, excites the transducer.

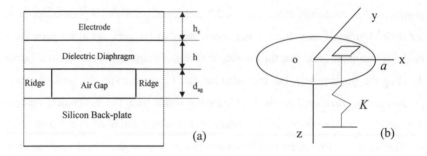

Fig. 1 (a) The typical micro air-gap structure of a micromachined ultrasonic transducer; (b) Physical model of the sub-air-gap structure, regarded as a sub-dynamic system

Further, the dynamic sub-system can be regarded as a thin tensile plate supported by an elastic medium, its physical model is shown as Fig. 1(b). Thus, the motion of the sub-diaphragm can be described by the following four-order partial differential equation,

$$DV^4 w - TV^2 w + \rho_s \frac{\partial^2 w}{\partial t^2} + R_M \frac{\partial w}{\partial t} + K_{ag} w = P_A + P_E \qquad (1)$$

where $\nabla^4 = \nabla^2 \nabla^2$, ∇^2 is the Laplacian operator, $\nabla^2 = \frac{\partial^2}{\partial x^2} + \frac{\partial^2}{\partial y^2}$; $w(x, y; t)$ is the lateral displacement of the diaphragm, t is time; D is the bending stiffness of the diaphragm, T the in-plane tension in it, which is normally provided by the residual stress in the diaphragm; ρ_s

is the area specific density of the diaphragm; R_M is the mechanical damping of the sub-dynamic system; K_{ag} the compressibility of air in the cavity, $K_{ag} = \dfrac{\rho_0 c_0^2}{d_{ag}}$ for one-order approximation, where d_{ag} the average air-gap thickness, ρ_0 the density of air, c_0 the sound velocity in air. P_A is the pressure loading to the sub-diaphragm by acoustic radiation to the surrounding medium. For an air medium, P_A can be estimated by using a baffled rigid piston model. Thus, the acoustic reactance of the load can be neglected, so we have $P_A = -\rho_0 c_0 \dfrac{\partial w}{\partial t}$.

Another external pressure P_E is produced by the electrostatic attraction. It can be represented by its first order approximation under small-signal and small-deflection assumptions, i.e.

$$\mathbf{P_E} \approx -\left(\frac{\varepsilon_0 V_0^2}{2 x_0^2} + \frac{\varepsilon_0 V_0}{x_0^2} \mathbf{V_t} - \frac{\varepsilon_0 V_0^2}{x_0^3} \mathbf{w_t} \right)$$ [13], here complex variables are bold-faced, and $\mathbf{V_t} = V e^{j\omega t}$ is the time-varying signal voltage, $\mathbf{w_t} = w e^{j\omega t}$ the time-varying displacement; V_0 the dc bias voltage, ε_0 the permittivety of air; $x_0 = d_{ag} + w_0 + h/\varepsilon_r$, is the effective separation between two plate, where w_0 is the static deflection of diaphragm caused by bias voltage, ε_r the relative dielectric constant of the dielectric diaphragm. It can be seen that the electrostatic force includes three parts, the first stems from bias, producing the static deflection of the diaphragm; the second comes from the time-varying voltage, resulting the diaphragm oscillation around the static deflection position; the last one is caused by a electrostatic negative stiffness, which is defined as $K_{neg} = -\dfrac{\varepsilon_0 V_0^2}{x_0^3}$ [15], which reduces the affection of air-spring stiffness. Thus, the governing equation (1) becomes

$$D\nabla^4 w - T\nabla^2 w + \rho_s \frac{\partial^2 w}{\partial t^2} + \left(R_M + \rho_0 c_0 \right)\frac{\partial w}{\partial t} + Kw = -\frac{\varepsilon_0 V_0^2}{2 x_0^2} - \frac{\varepsilon_0 V_0}{x_0^2} V_t \qquad (2)$$

where $K = K_{ag} + K_{neg}$, which is used as a characterized mathematical expression of the elastic medium. It may be assumed that the mechanical damping of the sub-dynamic system does not change the natural frequencies, and only reduces the resonant peaks. So, we may neglect the damping term while determining natural frequencies of the sub-system. Now, considering a sub-diaphragm on an individual cylinder-shaped cavity, which corresponds to a circular plate, the general solution of the equation in polar co-ordinates is [19]

$$W(r,\theta) = \sum_{n=0}^{\infty} \left[A_n J_n(\alpha r) + C_n I_n(\beta r) \right] \cos n\theta \qquad (3)$$

where J_n is the Bessel functions of the first kind, and I_n is modified Bessel functions of the first kind; the coefficients A_n and C_n are undetermined constants; the characteristic values

$$\alpha, \beta = \sqrt{\frac{T}{2D}\left[\sqrt{1+\frac{4D(\rho_s\omega^2-K)}{T^2}}\mp 1\right]}, \text{ where } \omega \text{ is angular frequency.}$$

Applying the clamped boundary conditions at the circumference of the sub-diaphragm, i.e. $W(a)=0$, $\partial W(a)/\partial r = 0$, to equation (3) yields the characteristic equation

$$\frac{I_n(\beta a)}{J_n(\alpha a)} + \frac{\beta I_{n+1}(\beta a)}{\alpha J_{n+1}(\alpha a)} = 0 \tag{4}$$

where a is the radius of the sub-diaphragm, and n indicates the number of nodal diameter, $n = 0,1,2,\ldots$.

Consequently, lowest natural frequencies, mainly concerned by us, can be determined by solving equation (4) while $n = 0$ (i.e. no nodal diameter case), which is

$$\frac{I_0(\beta a)}{J_0(\alpha a)} + \frac{\beta I_1(\beta a)}{\alpha J_1(\alpha a)} = 0 \tag{5}$$

Then, the corresponding mode shapes are determined by the following eigenfunctions

$$\phi_{0s}(r) = J_0(\alpha_{0s}r) - \frac{J_0(\alpha_{0s}a)}{I_0(\beta_{0s}a)}I_0(\beta_{0s}r) \tag{6}$$

These solved natural frequencies are donated as ω_{0s}, and s represents the number of nodal circle, $s = 0,1,2,\ldots$.

Following the theory and method given by [14], we may develop an electrical equation of the transducer, and then, based on both dynamic and electrical equations, then determine all impedance parameters to characterize the behavior of the transducer, and finally determine the acoustic receiving sensitivity and transmitting response.

II. SOME SPECIAL CASES

The theoretical simulation for pit-array-textured transducers developed in the preceding section fully takes into account three mechanical stiffness effects, T, D, and K_{ag}, and an electrostatic negative stiffness effect, K_{neg}, thereby possesses general significance. In the model, $T \neq 0$, $D \neq 0$, and $K \neq 0$, so the model may be called the TDK model for short. Some special cases of it are further briefly discussed in the following.

A. 3D special cases

A.1 3D Plate-under-tension or TD model ($T \neq 0, D \neq 0, K = 0$)

In this case, the K-item in the governing dynamic equation (1) can be removed, so that the equation reduces to

$$DV^4w - TV^2w + \rho_s \frac{\partial^2 w}{\partial t^2} + R_M \frac{\partial w}{\partial t} = P_A + P_E \tag{1.A1}$$

The characteristic values will be $\alpha, \beta = \sqrt{\frac{T}{2D}\left[\sqrt{1+\frac{4D\rho_s\omega^2}{T^2}}\mp 1\right]}$.

A.2 3D Plate-on-air-spring or DK model ($T = 0, D \neq 0, K \neq 0$)

In this case, the T-item in the governing dynamic equation (1) can be removed, so the corresponding equation will be

$$DV^4w + \rho_s \frac{\partial^2 w}{\partial t^2} + R_M \frac{\partial w}{\partial t} + K_{ag}w = P_A + P_E \tag{1.A2}$$

The corresponding characteristic value will be $\lambda = \sqrt{\sqrt{\frac{\rho_s\omega^2 - K}{D}}}$.

A.3 3D Membrane-under-tension or T model ($T \neq 0, D = 0, K = 0$)

In this case, the K- and D-item in the governing dynamic equation (1) can be removed, so that the equation degenerates to a two-order partial differentiate equation:

$$-TV^2w + \rho_s \frac{\partial^2 w}{\partial t^2} + R_M \frac{\partial w}{\partial t} = P_A + P_E \tag{1.A3}$$

This is just classical membrane model, for a circular plate the fundamental $f_{01} = \frac{2.405}{2\pi a}\sqrt{\frac{T}{\rho_s}}$.

B. 2D special cases

Furthermore, for V- or U-grooved transducers, since the length of sub-diaphragm is much larger than its width, i.e. $y \gg x$, we can apply 2D approximation. Thus, for its general case of $T \neq 0, D \neq 0, K \neq 0$, the equation (1) reduces to a 2D representation:

$$D\frac{\partial^4 w}{\partial x^4} - T\frac{\partial^2 w}{\partial x^2} + \rho_s \frac{\partial^2 w}{\partial t^2} + R_M \frac{\partial w}{\partial t} + K_{ag}w = P_A + P_E \tag{1.B}$$

This is the case developed by [15] for U-grooved transducers.

B.1 2D Plate-under-tension or TD model ($T \neq 0, D \neq 0, K = 0$)

In this case, the K-item in the governing dynamic equation (1.B) can be removed, so the equation will be simplified as

$$D\frac{\partial^4 w}{\partial x^4} - T\frac{\partial^2 w}{\partial x^2} + \rho_s\frac{\partial^2 w}{\partial t^2} + R_M\frac{\partial w}{\partial t} = P_A + P_E \qquad (1.B1)$$

B.2 2D Plate-on-air-spring or DK model ($T = 0, D \neq 0, K \neq 0$)

In this case, the T-item in the governing dynamic equation (1.B) can be removed so the equation will be simplified as

$$D\frac{\partial^4 w}{\partial x^4} + \rho_s\frac{\partial^2 w}{\partial t^2} + R_M\frac{\partial w}{\partial t} + K_{ag}w = P_A + P_E \qquad (1.B2)$$

This is the case developed by [13]- [14] for V-grooved transducers.

B.3 2D Membrane-under-tension or T model ($T \neq 0, D = 0, K = 0$)

In this case, the K- and D-item in the dynamic equation (1.B) can be removed so that the equation will be further simplified to a two-order partial differential equation:

$$-T\frac{\partial^2 w}{\partial x^2} + \rho_s\frac{\partial^2 w}{\partial t^2} + R_M\frac{\partial w}{\partial t} = P_A + P_E \qquad (1.B3)$$

C. 1D special case:

Helmholtz (or Mass-spring resonator) or K model ($T = 0, D = 0, K \neq 0$)

Finally, if both T and D can be neglected, the simplest 1D case will occur. Thus, the T- and D-item can be removed from the governing dynamic equation (1), so that the equation degenerates finally to a two-order differential equation,

$$\rho_s\frac{\partial^2 w}{\partial t^2} + R_M\frac{\partial w}{\partial t} + K_{ag}w = P_A + P_E \qquad (1.C)$$

This is a well-known single-degree-of-freedom resonator model; the diaphragm is regarded as a rigid body, and the elastic medium K as stiffness.

III. A NUMERICAL EXAMPLE

The above-mentioned surface micromachined ultrasonic transducer, developed by [16], serves a good numerical example for the understanding and verification of the theory developed by this paper. The literature [16] listed also a set of typical data in its Table 1, and reported that experimentally determined fundamental frequency is 4.6MHz. The micro air-gap structure of such a transducer consists of a 0.75μm thick nitride film, coated with 50nm gold electrode, and a cylindrical air-gap with 52 mm radius and 1μm depth. The residual stress of 280 Mpas provides the tension in the film. The static capacitance of the device is about 124pF and the bias applied is 40V. Now, let us apply the general theory and its varities to the set of data for predicting the natural frequencies.

First try the classical membrane-under-tension or T model described by (1.A3). The literature [16] applied the model to such a transducer, and claimed that there is excellent agreement between the measured results and the theoretical predictions. But, the problem is that they neglected the mass of gold electrode. If not including the gold mass, the fundamental frequency calculated is 4.6MHz. Nevertheless, the mass of 50nm thick gold electrode is already 40% of the mass of 750nm thick nitride film, so that cannot be neglected. Including the mass of gold electrode then, the fundamental frequency calculated by the classical model will be 3.63874MHz. The literature [17] developed further a mathematical description of the mechanical impedance of a membrane of such a device with vacuum-sealed cavities, which includes only the tension and bending stiffness effects, and did not give a theoretical formula for the fundamental frequency. Let us apply the special TD model given by this paper; the fundamental resolved from the dynamic equation (1.A2) is 4.49532MHZ, which agrees well with the measured value. Finally, apply further the general TDK model to the set of data. The lowest natural frequencies can be determined by the characteristic equation (5). The first two roots are 0.971963MHz and 4.59919MHz respectively. Their corresponding mode shapes can be found from the eigenfunction (6), shown in Fig. 2.

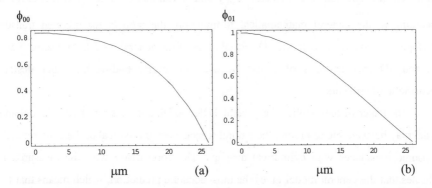

Fig.1. First two mode shapes of a typical micromachined ultrasonic transducer with cylindrical air-gaps (a=26μm): (a) ϕ_{00}, the fundamental mode, corresponding to the piston-like motion of its diaphragm, at f_{00}=0.971963MHz; (b) ϕ_{01}, the second mode, corresponding to the first bending vibration of the diaphragm, at f_{01}=4.59919MHz.

It is clear that the second resonant frequency agrees excellently with the reported measured result. In addition, the theoretical simulation found a lower resonant frequency, which depends on K, corresponding to the piston-like motion of the diaphragm, and so should be regarded as the fundamental frequency. The theoretical result also consists with ones obtained by this author's preceding works [12-15] for V- and U-grooved transducers. Evidently, if the first and second resonant frequencies are well separated, then two single peaks clearly appear in the frequency response curve, so we may say that the transducer has a so-called "*single-peak response*" around each of the two frequencies. And if the two frequencies are relatively close to each other, or the two peaks overlap, a so-called "*multi-peak response*" occurs. This analysis suggests that such a transducer has two single operation modes that can be utilized, i.e. a *fundamental mode* and a *first-bending mode*, and a *combination mode*, if the two frequencies are close enough to each other [20]. Therefore, the theoretical model provides a reasonable explanation why such micromachined ultrasonic transducers usually have a multi-peak response and unusually wide bandwidth.

IV. CONCLUSIONS

A general three-dimensional (3D) theoretical model for simulating the behaviour of micromachined electrostatic ultrasonic transducers has been developed. The model takes into account the tension in the diaphragm, T, the bending stiffness of the diaphragm, D, and the stiffness of elastic medium, K (including both the compressibility of air in the cavity, K_{ag}, and the electrostatic negative stiffness, K_{neg}), and so can be called the TDK model for short. According to the general mathematical formulation, the models for grooved transducers developed previously are only the 2D special cases of the general model. And some other 3D, 2D, and 1D special cases of it can be used for corresponding particular designs of electrostatic transducers.

The paper further applies the general TDK model, and its simplified T model and TD model may be available to predict the natural frequencies of a typical surface micromachined ultrasonic transducer with cylindrical air-gaps. The comparison with the measured result indicates that the general model gives the most accurate prediction, which means that for the particular transducer all of the four stiffness effects must be taken into account. It also suggests that the theoretical simulation gives a general model for describing the behaviour of various electrostatic audio and ultrasonic transducers.

ACKNOWLEDGEMENTS

The work was supported by the National Natural Science Foundation of China (69974001), National Climbing Plan (1999-045), Anhui Natural Science Foundation (99043522), and Anhui University 211 Foundation (99022021).

References

1. E. C. Wente, A condenser transmitter as a uniformly sensitive instrument for the absolute measurement of sound intensity, *Phys. Rew.*, **10**:39-63, 1917.

2. F. V. Hunt, *Electroacoustics, Acoust. Soc. Am.*, New York, 1954, 1982.

3. H. Sell, Eine neue kapazitive methode umwandlung mechanischer schwingungeg in elektrische und umgekehrt, *Z. Tech. Phys.* **18**:3-10, 1937.

4. W. Kuhl, G.R. Schodder and F. -K. Schroder, Condenser transmitters and microphones with solid dielectric for airborne ultrasonics. *Acoustica*, **4**(5): 519-532, 1954.

5. K. Matsuzawa, Condenser microphones with plastic diaphragms for airborne ultrasonics, *J. Phys. Soc. Japan*, **13**(12): 1533-1543, 1958; **15**(1): 167-174, 1960.

6. M. Rafic and C. Wykes, The performance of capacitive ultrasonic transducers using V-grooved backplates, *Meas. Sci. Technol.* **2**:168-174, 1991

7. J. Hietanen, J. Stor-Pellinen, and M. Luukala, A model for an electrostatic ultrasonic transducer with a grooved backplate, *Meas. Sci. Technol.* **3**:1095-1097, 1992.

8. P. Mattila, F. Tsuzuki, H. Vaataja, K. Sasaki, Electroacoustic model for electrostatic ultrasonic transducers with V-grooved backplates , *IEEE Trans. Ultrason. Ferroelect. Freq. Contr.*, **42**(1): 1-7, 1995.

9. D. Hohm and G. Hess, A subminiature condenser microphone with silicon nitride membrane and silicon back plate, *J. Acoust. Soc. Am.*, **85** (1): 476-480, 1989.

10. K. Suzuki, K. Higuchi, and H. Tanigawa, A silicon electrostatic ultrasonic transducer, *IEEE Trans. Ultrason. Ferroelect. Freq. Contr.*, **36**(6): 620-627, 1989.

11. M. J. Anderson, J. A. Hill, C. M. Fortunko, N. S. Dogan, R. D. Moore, Broadband electrostatic transducers: Modelling and experiments, *J. Acoust. Soc. Am.*, **97**:262-272, 1995.

12. L.-F. Ge, DK model for electrostatic ultrasonic transducers with V-grooved backplates, *J. Acoust. Soc. Am.*, **100**(4): 2809, 1996.

13. L. -F. Ge, A theoretical model of electrostatic ultrasonic transducers with micro air-gap structures, *Chinese Scientific Bulletin*, **43**(9): 728-731, 1998; in Chinese **42**(22): 2387-2390, 1997.

14. L.-F. Ge, Electrostatic airborne ultrasonic transducer: modeling and characterization, *IEEE Trans. Ultrason. Ferroelect. Freq. Contr.*, **46**(5): 1120-1127, 1999.

15. L.-F. Ge, Dynamic mechanism and its modelling of micromachined electrostatic ultrasonic transducers, *Sciences in China (Series A)*, **42**(12): 1308-1315, 1999; In Chinese, **29**(11): 1013-1019, 1999.

16. M. I. Haller and B. T. Khuri-Yakub, A surface micromachined electrostatic ultrasonic air transducer, *IEEE Trans. Ultrason. Ferroelect. Freq. Contr.*, **43**(1): 1-6, 1996.

17. I. Ladabaum, X. Jin, H. Soh, and B. T. Khuri-Yakub, Surface micromachined capacitive ultrasonic transducers, *IEEE Trans. Ultrason. Ferroelect. Freq. Contr.*, **45** (3): 678-690, 1998.L.-F. Ge, Progress in electrostatic transducers for sound and ultrasound (in English), *J. Anhui University*, **24**(3): 74-84, 2000.

18. L.-F. Ge, Progress in electrostatic transducers for sound and ultrasound (in English), *J. Anhui University*, **24**(3): 74-84, 2000.

19. A. Leissa, *Vibration of Plates*, (Acoust. Soc. Am. NY, 1993).

20. L.-F. Ge and J. Shao, Behaviour characteristics of silicon micromachined electrostatic ultrasonic transducers, The 139th Meeting of ASA, Atlanta, 2000; *J. Acoust. Soc. Am.*, **107**: 2911, 2000.

Theoretical and Computational Acoustics 2001
E.-C. Shang, Qihu Li and T. F. Gao (Editors)
© 2002 World Scientific Publishing Co.

The Group Velocities of Anisotropic Media as a Function of Their Propagation Direction

J. Z. Shen

Institute of Acoustics, Chinese Academy of Sciences, 17 Zhongguancun Street, Beijing 100080, China

J. Z. Niu

Department of Physics, Central University for Nationalities, 27 Baishiqiao Rd., Haidian, Beijing, 100081, China

W. Arnold

Fraunhofer-Institute for Non-destructive Testing, Bldg. 37 University, D-66123 Saarbrücken, Germany

For anisotropic media the knowledge of group velocities and phase velocities of ultrasonic waves is very desirable. Knowing one set of these velocities, the elastic constants can be obtained. In this paper, the magnitude of the group velocity as a function of its direction is established. The magnitude of the group velocity and as well as the phase velocity for any given direction can be numerically calculated by this set of equations.

1. Introduction

The measurements of elastic constants is increasingly important because of the widespread use of composite materials. The knowledge of the elastic constants are important for the design of composites materials for manufacturing and after manufacturing for in-service inspection of these materials. The most promising way to non-destructively determine the elastic constants is to measure ultrasonic velocities and then using an inversion technique to extract the elastic data.

When ultrasonic waves propagate in anisotropic media, it is well known that there exist two velocities, the phase velocity and the group velocity. In general, these two velocities differ both in direction and in magnitude, except in some special high-symmetry directions. Using phase velocity data to recover elastic constants is easier than using group velocity data. In the so-called direct problem, i.e., when the elastic constants are given, the theory to calculate all phase velocities as well as the corresponding group velocities for any given direction of k-vector is well established [1,2] providing a set of equations to calculate the phase velocities of the ultrasonic wave for any k-vector direction as well as the corresponding group velocities. In contrast, even for the case of known directions of the group velocity, there exist no corresponding equations and the situation becomes very complicated.

In the so-called the inverse problem, the velocities are given and the elastic constants are wanted. For the case of known phase velocity data, the inverse procedure can be carried out without difficulty and some optimizing techniques can be used to recover the elastic constants. For the case of known group velocity data, the difficulties rise considerably in recovering the elastic data. As there exists no closed-form solution of group velocities related to its direction, one needs to find firstly the direction of phase velocity by an optimizing procedure in a set of assumed elastic constants, and then secondly an optimizing procedure in order to examine the deviation between the calculated and the measured group velocities rendering the recovery of the elastic constants complicated. This situation has been pointed out previously by several authors[3,4].

In most situations, time-of-flight (TOF) measurement are used to measure ultrasonic velocities and mostly the group velocity is determined. The difference between group and phase velocity can only be neglected if the anisotropy is very weak. To measure in some special high-symmetry directions, where the two velocities are identical, is a convenient way out. But this usually requires to measure in several perpendicular directions or to use several samples with special shapes to get all phase velocity data required. In general, however, there are not enough high-symmetry directions for recover all elastic constants.

There is another way to find the phase velocity direction from the known group velocities by exploiting the ray surface. The k-vector must always be normal to the ray surface. However, one must know a section of the ray surface in order to find its normal. In other words, additional group velocity data are needed, which does not help further.

Finding the corresponding phase velocity direction from a known group velocity, some numerical optimizing techniques have been presented[4]. This could work if the elastic constants were known. However, in the inverse problem, the elastic constants are the wanted quantities, one has to recover the phase velocity direction by assuming the elastic constants. Then by using these recovered phase velocity one eventually obtains the correct elastic constants by numerical optimizing techniques.

It is more convenient to find an equation allowing to discern the magnitude of the group velocity from its own direction. Many efforts were made, but they were not very successful[4]. In this paper, we try to overcome this problem. In the next section, a set of equations is deduced allowing to find the group velocity in any direction. The equations are valid for all kind of crystal structures. As the set of equations is both nonlinear and complicated, no explicit solution could be obtained so far, and it may only be solved numerically.

2. Theory

The equations of elastic waves for an elastic anisotropic solid are well established [1]. The value of the phase velocity $V_p = \omega/k$, where $\bar{k}$ is wave vector, $k = (k_1^2 + k_2^2 + k_3^2)^{1/2}$ and ω is the angular frequency, is determined by the characteristic equation

$$\det\left|\Gamma_{ij}(c_{ij}, \vec{l}) - \rho V_p^2 \delta_{ij}\right| = 0 \tag{1}$$

where ρ is the density of the medium. The direction of the phase velocities is the same as the wave vector. The elements Γ_{ij} are the Christoffel coefficients which are functions of the elastic constants c_{ij}, and the direction of the wave vector $\vec{l}$, $\vec{l} = l_i \hat{e}_i$.

By introducing a vector P and a matrix M, where the matrix M depends only on the elastic constants c_{ij} and P only on the wave direction and by writing the Christoffel coefficients Γ_{ij} into a vector G, we get

$$G = M \cdot P \tag{2}$$

where

$$G = (G_1, G_2, G_3, G_4, G_5, G_6)^T = (\Gamma_{11}, \Gamma_{22}, \Gamma_{33}, \Gamma_{23}, \Gamma_{13}, \Gamma_{12})^T \tag{3}$$

$$P = (l_1^2, l_2^2, l_3^2, l_2 l_3, l_1 l_3, l_1 l_2)^T \tag{4}$$

The superscript T denotes transposition. In the most general case

$$M = \begin{pmatrix} c_{11} & c_{66} & c_{55} & 2c_{56} & 2c_{15} & 2c_{16} \\ c_{66} & c_{22} & c_{44} & 2c_{24} & 2c_{46} & 2c_{26} \\ c_{55} & c_{44} & c_{33} & 2c_{34} & 2c_{35} & 2c_{45} \\ c_{56} & c_{24} & c_{34} & c_{44} + c_{23} & c_{36} + c_{45} & c_{25} + c_{46} \\ c_{15} & c_{46} & c_{35} & e_{45} + c_{36} & c_{13} + c_{55} & c_{14} + c_{56} \\ c_{16} & c_{26} & c_{45} & c_{46} + c_{25} & c_{14} + c_{56} & c_{12} + c_{66} \end{pmatrix} = (M_1, M_2, M_3, M_4, M_5, M_6)^T \tag{5}$$

In order to calculate the group velocity, we rewrite G in the following form

$$G = M \cdot P = \frac{1}{V_p^2} M \cdot P_V \tag{6}$$

and

$$G_i = \frac{1}{V_p^2} M_i \cdot P_V \tag{7}$$

where P_V has a similar form as P

$$P_V = ((V_1^P)^2, (V_2^P)^2, (V_3^P)^2, V_2^P V_3^P, V_1^P V_3^P, V_1^P V_2^P)^T \tag{8}$$

The V_i^P's are the components of the phase velocity, $V_i^P = V_p l_i$. and $V_p^2 = V_i^P V_i^P$.

In practice, the form of the matrix M could be simplified for the crystal classes possessing higher symmetries, For example, for orthorhombic symmetry, there are 9 independent elastic constants and the matrix M can be simplified as

$$M = \begin{pmatrix} c_{11} & c_{66} & c_{55} & 0 & 0 & 0 \\ c_{66} & c_{22} & c_{44} & 0 & 0 & 0 \\ c_{55} & c_{44} & c_{33} & 0 & 0 & 0 \\ 0 & 0 & 0 & c_{44} + c_{23} & 0 & 0 \\ 0 & 0 & 0 & 0 & c_{13} + c_{55} & 0 \\ 0 & 0 & 0 & 0 & 0 & c_{12} + c_{66} \end{pmatrix} \tag{9}$$

For tetragonal, hexagonal, cubic symmetries, so-called transverse isotropic materials, and for isotropic materials, all M matrices have this form, only the number of independent elastic constants are different. For example, $c_{22} = c_{11}$, $c_{55} = c_{44}$, $c_{23} = c_{13}$ for tetragonal symmetry. Furthermore, $c_{22} = c_{11}$, $c_{33} = c_{11}$, $c_{55} = c_{44}$, $c_{66} = c_{44}$, $c_{12} = c_{13}$, $c_{23} = c_{13}$ for cubic symmetry.

The group velocity is defined as

$$\vec{V}^G = \nabla_k \omega \tag{10}$$

This expression is convenient to use if the dispersion relation is given explicitly. In most cases, the dispersion relation is only given in an implicit form. Then the group velocity can be obtained by implicit differentiation leading to [1]:

$$\vec{V}^G = -\frac{\nabla_k \Omega}{\partial \Omega / \partial \omega}$$

(11)

with

$$\Omega = \det \begin{vmatrix} k^2\Gamma_{11} - \rho\omega^2 & k^2\Gamma_{12} & k^2\Gamma_{13} \\ k^2\Gamma_{12} & k^2\Gamma_{22} - \rho\omega^2 & k^2\Gamma_{23} \\ k^2\Gamma_{13} & k^2\Gamma_{23} & k^2\Gamma_{33} - \rho\omega^2 \end{vmatrix}$$

(12)

or

$$\Omega = (k^2G_1 - \rho\omega^2)(k^2G_2 - \rho\omega^2)(k^2G_3 - \rho\omega^2) + 2k^6G_4G_5G_6 -$$
$$- (k^2G_1 - \rho\omega^2)k^4G_4^2 - (k^2G_2 - \rho\omega^2)k^4G_5^2 - (k^2G_3 - \rho\omega^2)k^4G_6^2$$

(13)

In most cases, the group velocity differs from the phase velocity both in amplitude and in direction. It should be noted that Eq. (11) is as a function of phase velocity direction, the wave vector, instead of the direction of the group velocity.

Using Eqs (6) and (7) and some additional relations, such as $V_i^P = V_p l_i$ etc., the group velocity $\vec{V}^G$ can be calculated from Eq. (11) and expressed as

$$\vec{V}^G = V_g \vec{e}^G = \frac{1}{2\rho V_p^2 F} E \cdot M \cdot D$$

(14)

where the group velocity and its components are

$$\vec{V}^G = (V_1^G, V_2^G, V_3^G) = V_g \vec{e}^G = V_g(e_1^G, e_2^G, e_3^G)$$

(15)

Here, E and F are functions of phase velocity and stiffness, respectively.

$$F = (G_2 - \rho V_p^2)(G_3 - \rho V_p^2) + (G_1 - \rho V_p^2)(G_3 - \rho V_p^2)$$
$$+ (G_1 - \rho V_p^2)(G_2 - \rho V_p^2) - G_4^2 - G_5^2 - G_6^2 \tag{16}$$

$$E = (E_1, E_2, E_3, E_4, E_5, E_6) \tag{17}$$
$$E_1 = (G_2 - \rho V_p^2)(G_3 - \rho V_p^2) - G_4^2,$$
$$E_2 = (G_3 - \rho V_p^2)(G_1 - \rho V_p^2) - G_5^2,$$
$$E_3 = (G_1 - \rho V_p^2)(G_2 - \rho V_p^2) - G_6^2,$$
$$E_4 = 2G_5 G_6 - 2(G_1 - \rho V_p^2)G_4,$$
$$E_5 = 2G_6 G_4 - 2(G_2 - \rho V_p^2)G_5,$$
$$E_6 = 2G_4 G_5 - 2(G_3 - \rho V_p^2)G_6 \tag{18}$$

where all G_i's are expressed as in Eq. (7). The matrix D is a function of the components of the phase velocity.

$$D = \begin{pmatrix} 2V_1^P & 0 & 0 & 0 & V_3^P & V_2^P \\ 0 & 2V_2^P & 0 & V_3^P & 0 & V_1^P \\ 0 & 0 & 2V_3^P & V_2^P & V_1^P & 0 \end{pmatrix}^T \tag{19}$$

There is an additional relation between the group and phase velocities

$$V_P = V_g \vec{e}^G \cdot \vec{l} \tag{20}$$

Eq. (20) can be rewritten as

$$V_P^2 = V_g e_i^G V_i^P \tag{21}$$

In the Eqs. (14), (16), (18), V_P^2 is replaced by Eq. (21). Then, combining Eqs. (14) and (21), a set of four equations can be formed with four unknown variables, namely the magnitude of the group velocity V_g and the three components of the phase velocity V_i^P provided the elastic constants and the direction of group velocity are known. Solving these equations one can calculate the group velocity along any given direction, for example, by means of a modified numerical Newton method.

3. Discussion

The classical Christoffel matrix can be separated into two parts: a matrix M containing the material properties, Eq. (5), and a vector P determined by the wave propagation direction, Eq. (4). Using this fact and expressing some equations in a different way, we obtain Eq. (14) in order to calculate the group velocities from the known phase velocities and wave vectors. By combining Eqs. (14) and (21) which includes a set of four equations, a general form is established. In this set of equations, the magnitude of the group velocity is given in its direction. Usually, in such algorithms, the phase velocity as a function of group velocity direction is determined. The set of four equations can be solved by numerical methods, for example by the Newton method allowing one to determine the magnitude of the group velocity and the phase velocity for any given group velocity direction. However, the set of equations is both nonlinear and multi-valued. How to obtain the proper roots of these equations is a key problem. One should also pay attention to the situations that symmetries of the crystal might cause some kind of simplification of the implicit dispersion relation (13). In other words, the general formulas deduced are more suitable for arbitrary wave direction and low-symmetry crystal structures.

From Eqs. (14) and (21), it is also suggested that the wave velocity depends on a larger number of elastic constants in non-symmetric directions than in symmetric directions. Hence the velocity data in different arbitrary direction contain more information about elastic constants. Hence, it is suggested that it would be more effective using velocity data along the non-symmetry direction for the recovery of elastic constants than in high-symmetry directions. This, however, requires to obtain an efficient way to solve the set of four equations derived in this paper.

4. Acknowledgement

This work was partly supported by the Foundation of Natural Sciences of China, contract No. 19874070. Part of this work grew out within a joint Sino-German collaboration project supported by the German Ministry of Science and Technology which we gratefully acknowledge.

5. References

[1]   B.A. Auld, "Acoustic Fields and Waves in Solids" (John Wiley, 1973).
[2]   A.G. Every, "General Closed-form Expressions for Acoustic Waves in Elastically Anisotropic Solids", *Phys. Rev. B*, **22** (1980) 1746-1760.
[3]   F. Chai and T.T. Wu, "Determinations of the Anisotropic Elastic Constants Using Laser-generated Surface Waves", *J. Acoust. Soc. Am.*, **95** (1994) 3332-3241.
[4]   A.G. Every and W. Sachse, "Determination of the Elastic Constants of Anisotropic Solids from Acoustic-wave Group-velocity measurements", *Physical Review B*, **42** (1990), 8196-8205.

Theoretical and Computational Acoustics 2001
E.-C. Shang, Qihu Li and T. F. Gao (Editors)
© 2002 World Scientific Publishing Co.

Enhancing Resolution in Ultrasonic Imaging Using Transducer Harmonic Components

Tadeusz Stepinski, Tomas Olofsson and Ping Wu

Signals and Systems, P.O. Box 528, Uppsala University, 751 20 Uppsala, Sweden

Abstract

Ultrasonic transmitting transducers produce harmonic components that are normally lost due to the limited bandwidth of the receiving transducer. In pulse-echo mode it is natural since the same transducer is used both as transmitter and receiver. However, when using a separate receiver the higher harmonic components emitted by the transmitter can be used for improving the temporal resolution in ultrasonic imaging. This paper presents an application of this technique to the non-destructive evaluation (NDE) for improving quality of ultrasound B-scans acquired in the immersion inspection of solids.

Firstly, the presence of higher components in the signal received from copper specimen is proven in a set-up consisting of two ultrasonic transducers with different center frequencies. A narrowband transmitter has lower center frequency while the receiver has a broader band and a higher center frequency. The received harmonic components, although much weaker than the echo in the fundamental frequency band, can be combined to enhance the signal to noise ratio and the resolution in B-scan images. Secondly, the broadband signals acquired in this set up are processed using algorithms enhancing the temporal resolution. It is shown that the presence of higher harmonics results in an improved temporal resolution. Practical application to NDE of electron beam weld in copper is used to illustrate the proposed technique.

Introduction

Ultrasonic transducers used in NDE can produce harmonic components that are normally not utilized due to the filtering effect of the receiver. In pulse-echo mode, where the same transducer acts both as a transmitter and receiver, the higher harmonic components of the received signal are suppressed by the transducer frequency characteristics. In pitch-catch or transmission mode a pair of transducers with the same center frequency is normally used.

In this paper, we propose a technique applying higher transducer harmonic components for improving the quality of ultrasound B-scans acquired in the NDE. We propose using a setup consisting of two transducers with different center frequencies and bandwidths. To test this idea we used a transmitter with lower center frequency and a narrowband that was excited with a broadband pulse. A broadband transducer with a higher center frequency was used as a receiver. In such a pitch-catch setup, the higher resonances of the transmitter were excited due to the broadband pulse and resulted in the higher harmonics. Because of the broadband frequency property the receiver captured echoes containing both the fundamental and harmonic components. The setup was used for the inspection of copper specimens and the fundamental and harmonic components were extracted and then used separately or in combination.

Our detailed ultrasonic investigation was made on copper specimens welded by means of electron beam (EB) technique. The interest for EB welded copper originates from our project concerned with ultrasonic inspection of copper canisters for spent nuclear fuel [1]. One of our main goals during ultrasonic assessment of EB welds is detecting small voids and porosity present in the weld zone. In this situation we are investigating all possible physical phenomena that could be used for improving the imaging of defects in the EB weld. Making use of higher harmonic components in the ultrasonic signals opened a new direction in our research. Although harmonic components may be created by the nonlinearity of wave propagation, they are small compared to those created by the high resonances of transducers in the present case [2, 3]. Thus, here we limit ourselves to the higher harmonic components generated by the transducer and to the investigation of their usefulness for improving temporal resolution. Similar analysis of the spatial resolution was presented elsewhere [2, 4].

Experimental method

A transducer pair (transmitter and receiver) aiming at the same volume in the metal specimen immersed in water was used. B-scans were acquired by scanning the transducer pair in water using a mechanical scanner. Ultrasonic array system ALLIN manufactured by R/D Tech, France was used in the experimental setup (Fig. 1).

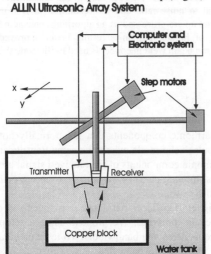

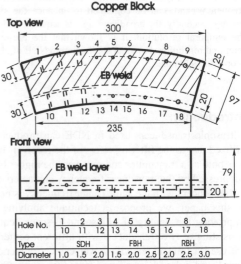

Figure 1. Experimental setup with two transducers, lower frequency focused transmitter (left) and higher frequency receiver (right).

Figure 2. Geometry of the inspected copper canister section with drilled holes.

A spherically focused transducer (PANAMETRICS V392) with a 38-mm diameter and a 190-mm focal length was used as a transmitter. It had a 0.85-MHz center frequency (measured), an 87%, –6-dB bandwidth (although the relative bandwidth is broad, the absolute bandwidth is not because of the low center frequency). A broadband planar transducer (PANAMETRICS V327) was used as the receiver. It had a nominal center frequency 10 MHz and the measured frequency 7 MHz. A 3-D RF data set was acquired for this transducer pair and used to create the A-, B- and C- scans of specimen. The inspected copper specimen was a section of a copper canister with an electron beam weld; it had a set of side-drilled holes (SDH) and bottom-drilled holes (BH) (see Fig. 2). Two identical sets of holes were drilled in the specimen, the holes #1 to #9 and #10 to #18 (upper and lower holes in Fig. 2, respectively). The only difference was that the upper holes were drilled in the welded zone while the lower holes were in solid copper. A substantial scattering from the weld structure was expected and the lower holes were made as a reference. The specimen was placed in the tank and inspected from the top in pitch-catch mode. The reflecting surfaces of the holes were approximately at a 60-mm depth beneath the top surface.

Experimental results - harmonic components

The experiments with the above described transducer pairs had two particular aims:
- To detect and evaluate the presence of higher harmonics in an ultrasonic signal scattered from a copper specimen immersed in water.
- To evaluate the information carried by the higher harmonics and its usefulness for improving the resolution during imaging the internal specimen structure.

To detect the presence of the higher harmonic components in the reflected ultrasonic signal a number of B-scan were acquired and spectra of the individual scans were estimated. The higher harmonic components were detected in the signal.

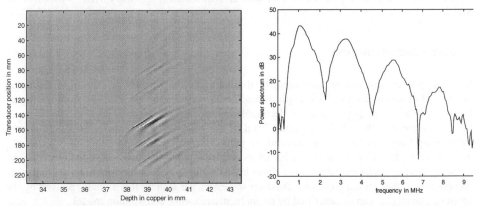

Figure 3. B-scan of the lower holes (left) and its amplitude spectrum (right). Signal from hole #10 corresponds to the lowest echo in the B-scan. Power spectrum was calculated for the A-scan at 150-mm (response of hole #12).

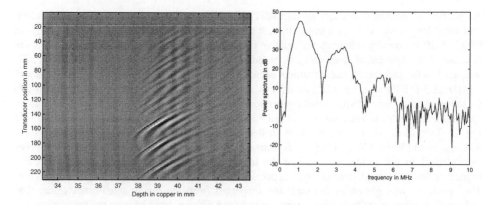

Figure 4. B-scan of the upper holes (left) and its amplitude spectrum (right). Signal from hole #1 corresponds to the lowest echo in the B-scan. Power spectrum was calculated for the A-scan at 150 mm (response of hole #3).

An example of this result is shown in Figures 3 and 4, where two B-scans of the specimen are shown together with power spectra of the selected A-scans. The B-scans were acquired along the specimen, so that they contained the responses of all drilled holes. The SDH's are the strongest reflectors (in the bottom part of Figures 3 and 4). The B-scan in Figure 3 corresponds to the area without the weld (lower holes) while the scan in Figure 4 was measured in the welded part of the specimen. The backscattering from the weld structure is clearly visible. In both cases the power spectra of A-scan corresponding to 150 mm were calculated using Welch method with Hanning window, and clear pronounced maxima are seen at frequencies of approx. 1 MHz, 3MHz and 5.5 MHz.

Theory - deconvolution

Now we will try to evaluate usefulness of the harmonic components for the high resolution imaging of discontinuities in NDE applications. Let us consider the contribution from the higher frequency bands to the performance of deconvolution of ultrasonic signals.
Let us assume that the whole frequency band of the received signal can be split into J non-overlapping frequency bands corresponding to individual harmonic components. In other words, we will use J artificial transducers, each with its center frequency at certain harmonic component. Creating these artificial transducers is helpful in analyzing the effects of including the harmonic components in the estimation of the reflection sequence. We assume that signals received by the artificial transducers can be described by a simple, discrete-time convolution model

$$x_1(k) = h_1(k) * r(k) + e_1(k)$$

$$\dots \qquad\qquad (1)$$

$$x_J(k) = h_J(k) * r(k) + e_J(k)$$

where:

$h_i(k)$ - transducer impulse response including its electro-acoustical impulse response and the electrical excitation

$r(k)$ - material reflection sequence that is to be estimated

$e_i(k)$ - measurement error associated with the transducer T_i

The equation set (1) can be written in a compact matrix-vector form

$$\boldsymbol{x_1} = \boldsymbol{H_1 r} + \boldsymbol{e_1}$$

$$\dots \qquad\qquad (2)$$

$$\boldsymbol{x_J} = \boldsymbol{H_J r} + \boldsymbol{e_J}$$

where time sequences have been replaced with the respective vectors and the matrices $\boldsymbol{H_i}$ are finite impulse response Toeplitz matrices associated with ith transducer T_i. Furthermore, assume that the reflection sequence $\boldsymbol{r}$ = col $[r(1)...r(N)]$ and the measurement errors $\boldsymbol{e_j}$ = col$[e_j(1)$ ... $e_j(N)]$ are independent random vectors with zero means and Gaussian distributions. Further, assume that their covariance matrices are known

$$\boldsymbol{C_r} = E[\boldsymbol{rr}^T] = \sigma_r^2 \boldsymbol{I} \qquad \boldsymbol{C_j} = E[\boldsymbol{e_j e_j^T}] = \sigma_j^2 \boldsymbol{I}$$

Let us consider the problem of estimating the reflection sequence $\boldsymbol{r}$, given the measurements $\boldsymbol{x_1, x_2,...x_J}$ and the transducer impulse responses $\{\boldsymbol{H_i}\}$. In other words, we will perform deconvolution of the sequence $\boldsymbol{r}$ based on the signals measured by the artificial transducers T_1 to T_J. It can be proven that this problem has a linear minimum mean square solution that takes the following form in the frequency domain [5, 6]

$$\hat{R}(\omega) = \frac{\displaystyle\sum_j \frac{1}{\sigma_j^2} H_j^*(\omega) X_j(\omega)}{\displaystyle\sum_j \frac{1}{\sigma_j^2} H_j^*(\omega) H_j(\omega) + \frac{1}{\sigma_r^2}} \qquad (3)$$

where capital letters denote Fourier transforms of the respective sequences:

$$\hat{R}(\omega) = \Im(\hat{r}); \quad X_j(\omega) = \Im(x_j); \quad and \quad H_j(\omega) = \Im(h_j(\cdot))$$

$\hat{r}$ is the estimate of $\boldsymbol{r}$, and $H_j^*(\omega)$ is a complex conjugate of $H_j(\omega)$. It is easy to see that the solution takes the form of a multi-variable Wiener filter. It is apparent that for $H_j(\omega)$ non-overlapping in frequency domain this solution should be equivalent to the classical Wiener filter developed for the whole frequency range.

However, the solution given by eq. (3), makes possible analyzing the effect of adding subsequent harmonic components to the transducer's fundamental frequency band. Let us look at the power spectral density of the estimation error $\tilde{r}$, $R_{\tilde{r}}(\omega)$. It can be shown that this density (which is the Fourier transform of the autocorrelation function of $\tilde{r}(k)$) for J transducers is given by

$$R_{\tilde{r},J}(\omega) = \frac{1}{\sum_J \frac{1}{\sigma_j^2} H_j^*(\omega)H_j(\omega) + \frac{1}{\sigma_r^2}} = \frac{1}{\sum_J \frac{1}{\sigma_j^2}\left|H_j(\omega)\right|^2 + \frac{1}{\sigma_r^2}} \qquad (4)$$

It is apparent that adding a subsequent transducer with nonzero frequency response $H_j(\omega)$ in certain frequency band results in decreasing $R_{\tilde{r}}(\omega)$ in this frequency band. This effect depends upon the level of measurement noise (σ_j) in this band. It can be easily proven that by adding a subsequent transducer $J+1$ we always obtain $R_{\tilde{r},J+1}(\omega) < R_{\tilde{r},J}(\omega)$, which means that we reduce variance of the estimation error $\Delta\sigma_{\tilde{r}}^2 = \frac{1}{2\pi}\int_{-\pi}^{\pi}(R_{\tilde{r},J+1}(\omega) - R_{\tilde{r},J}(\omega))d\omega < 0$ and thus we increase the estimation accuracy.

Experimental results – deconvolution

Signals (B-scans) acquired from the copper block shown in Fig. 2 in the setup shown in Fig. 1, using the PANAMETRICS V327 transducer as a receiver, were used to illustrate the above presented theory. The B-scans were deconvolved using the Wiener filter, eq. (3), using different bandwidths of the received signal. To achieve that, the transducer signal was filtered by low pass filters eliminating higher harmonic components. The LP filters were realized in frequency domain by windowing the A-scans FFT with rectangular windows. The cut-off frequency of these windows was set at the dips of the power spectrum, e.g., approx. 2.2 MHz for the first filter. The Wiener filter was designed using section of the A-scan # 176 corresponding to the SDH 11 as a prototype of the impulse response $h_i(k)$.

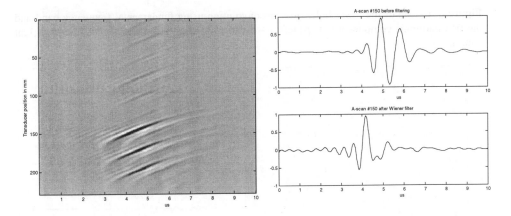

Figure 5. B-scan of the lower holes after deconvolution using transducer fundamental band (left panel). A-scan # 150 before and after Wiener filtering (right panel).

Some deconvolution results obtained for various frequency bands are shown in Figures 5 to 8. The deconvolved B-scans are shown in the left panels of Figures 5 to 8. The right panels are split in two parts, the upper presenting the A-scan #150 for the respective frequency band, and the lower the same A-scan after the deconvolution. The deconvolution, obtained using the information contained in the fundamental frequency band (0 to 2.2 MHz) is shown in Fig. 5, while the respective results obtained for the fundamental and the first harmonic (0 to 4.5 MHz), for the full useful frequency band (0 to 8.5 MHz), and for higher harmonics only (2.2 to 8.5 MHz) are presented in Figures 6, 7 and 8, respectively.

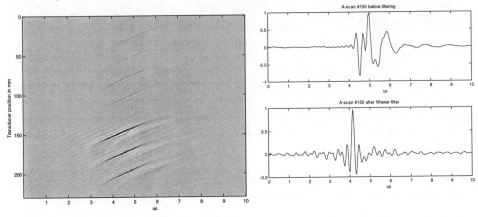

Figure 6. B-scan of the lower holes after deconvolution using fundamental frequency band and the first harmonic component (left panel). A-scan # 150 before and after Wiener filtering (right panel).

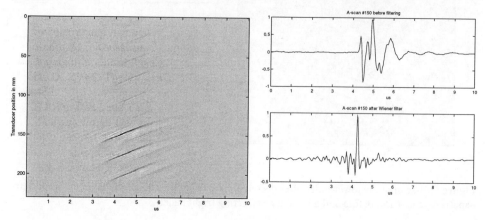

Figure 7. B-scan of the lower holes after deconvolution using whole transducer's frequency band (left panel). A-scan # 150 before and after Wiener filtering (right panel).

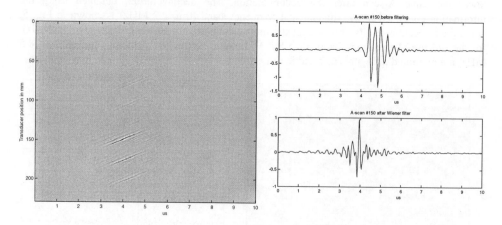

Figure 8. B-scan of the lower holes after deconvolution using higher transducer's frequency bands only (left panel). A-scan # 150 before and after Wiener filtering (right panel).

The results presented in the Figures 5 to 8 confirm the above presented theory – by adding subsequent frequency bands we obtain better deconvolution results. The deconvolved signals become more "spiky" coming closer to an ideal pulse. At the same time no apparent negative effects are pronounced, e.g., the increase of noise level after the deconvolution, that is a typical problem when using the Wiener filter.

Based on the above results we can draw a conclusion that all frequency bands, referred above as harmonic components, contain useful information about discontinuities in the inspected specimen. Temporal resolution can be considerably improved the using higher harmonic components. However, the higher harmonics starting from the third component have a very low energy and are corrupted by discretization noise. It should be also noted that these components are also affected by the strongest material attenuation.

Conclusions

The presence of high harmonic components in ultrasonic signals obtained in pitch-catch inspection of copper specimens in immersion was experimentally confirmed. These components mainly originated from the higher resonant modes of the transmitter excited by a broadband pulse. Although energy contained in those frequency bands was considerably lower than that in the fundamental one, a useful information about the specimen discontinuities was detected in higher harmonic bands. Moreover, it was proven theoretically and illustrated by experiment that using this information for the deconvolution considerably improves the obtained result.

The study reveals that the signals sent out by a transducer as transmitter (usually with limited bandwidth) contain much more information than that which is received in pulse-echo mode by the same transducer. In other words, much information (in higher harmonic components) contained in the signals sent out by a transmitter is lost in pulse-echo mode but can be captured in pitch-catch mode using another transducer that has broader frequency bandwidth. Therefore, the present study demonstrates an effective method of exploiting high frequency information contained in ultrasonic signals.

Acknowledgments

This research was supported by the Swedish Nuclear Fuel and Waste Management Co. (SKB). The authors wish to thank Eider Martinez who assisted in the measurements of the copper specimens.

References

1. T. Stepinski, and P Wu, "Ultrasonic Inspection of Nuclear Copper Canisters", SKB Projektrapport 97-06, August 1997.
2. P. Wu, F. Lingvall and T. Stepinski, "Inspection of copper canisters for spent nuclear fuel by means of ultrasound, Electron beam weld evaluation, harmonic imaging, materials characterization and ultrasonic modeling", *Technical Report TR-00-23*, Swedish Nuclear Fuel and Waste Management Co, Stockholm, December 2000.

488

3. T. Stepinski and P. Wu, "Ultrasonic Harmonic Imaging in Nondestructive Evaluation: Preliminary Experimental Study ", *IEEE Ultrasonic Symposium*, Puerto Rico, 22-26 October, 2000.
4. T. Stepinski, P. Wu and L. Ericsson, "Ultrasonic Imaging of Copper Material Using Harmonic Components", *The 2nd International Conference on NDE in Relation to Structural Integrity for Nuclear and Pressurized Components*, New Orleans, USA, May 24-26, 2000, pp. C-301-313.
5. T. Olofsson and T. Stepinski, "Maximum a posteriori deconvolution of ultrasonic signals using multiple transducers", *J. Acoust. Soc. Am.* 107 (6), June 2000, pp. 3276-3288.
6. T. Olofsson, "Maximum a posteriori deconvolution of ultrasonic data with applications in Nondestructive Testing", *Ph.D. Thesis,* Uppsala University, 2000.

Theoretical and Computational Acoustics 2001
E.-C. Shang, Qihu Li and T. F. Gao (Editors)
© 2002 World Scientific Publishing Co.

Angular Space Algorithm — A Novel Algorithm for the Angular Spectrum Approach for Axisymmetric Transducers

Ping Wu and Tadeusz Stepinski

Signals and Systems, P.O. Box 528, Uppsala University, 751 20 Uppsala, Sweden

In this paper a novel algorithm is proposed for efficiently and accurately implementing the angular spectrum approach (ASA) to calculating curved as well as planar transducers with axisymmetry. In contrast to the existing algorithms, called **k**-space algorithms, in which the angular spectrum approach (ASA) is implemented in the **k**-space (the wave vector space), the proposed algorithm is called angular space algorithm because the ASA is implemented in the angular space in which the polar angle is the only variable for an axisymmetric transducer. In the angular space algorithm the problem of undersampling the Green function can be overcome so that the aliasing error in the algorithm can be eliminated. The angular space algorithm is formulated from a general case where the transducer is arbitrarily curved, and a double integral needs to be calculated. In the case of axisymmetric transducers, the double integral reduces to the single integral with a single variable of polar angle. The **k**- and angular space algorithms have been applied to calculating the field from a spherically curved transducer, and the results from both algorithms have been compared with the analytical results. The comparison has shown that the efficiency and accuracy of the angular space algorithm are much better than those of the **k**-space algorithm.

1. Introduction

The angular spectrum approach (ASA) is one of the useful tools in modeling radiation and propagation of waves in acoustics and ultrasonics, and that has been investigated extensively and applied to both planar and curved transducers.[1-8] A rather comprehensive review of the ASA has been made by the authors.[8] For planar transducers, the ASA is usually implemented using the two dimensional (2-D) spatial Fourier-transform in the Cartesian coordinates. In the polar coordinates the 2-D spatial Fourier-transform reduces to the mth-order Hankel transform,[9] which is alternatively referred to as the Fourier-Bessel transform because the mth-order Bessel function is the kernel function in the integral transform.[10] When the planar transducers are axisymmetrical, the Hankel transform takes the zeroth-order.[7, 10, 11]

It is well known that the 2-D Fourier or Hankel transform only applies in a plane. For a curved transducer, therefore, the ASA that is based on the Fourier or Hankel transform is not applicable. To deal with the curved transducers, the authors[8] developed the extended ASA in the Cartesian coordinates, which is a double integral and applicable to arbitrarily shaped transducers (curved or planar), and applied it to a linear array with cylindrically concave surface.[12] For axisymmetrically curved transducers, e.g., spherical and conical transducers that are commonly used for ultrasonic focusing, the double integral in the Cartesian coordinates can be reduced to a single integral in the polar coordinates.

In the present work, we establish a general form of the extended ASA in the polar coordinates, and apply it to axisymmetrically curved transducers. We start with developing an algorithm, called k-space algorithm, that implements the extended ASA in the k-space (the wave-number space) in which the angular spectrum of plane waves is a function of spatial frequencies. Then we establish another algorithm, called angular space algorithm, that implements the ASA in the angular space in which the angular spectrum is a function of azimuth and polar angles. Finally we compare the efficiency and accuracy of the two algorithms.

2. Theory

2.1. General consideration

An acoustic field from a curved transducer with surface S represented by $z = f(x, y)$ in terms of pressure at frequency ω is expressed by the Rayleigh integral in the following manner,

$$\tilde{p}(\mathbf{r}) = -\frac{jk\rho_0 c}{2\pi} \iint_S \tilde{v}_n(\mathbf{r}') \frac{\exp(jkr_s)}{r_s} dS, \tag{1}$$

where $\tilde{v}_n(\mathbf{r}')$ is the normal velocity on the transducer surface, $r_s = |\mathbf{r} - \mathbf{r}'|$ $= \sqrt{(x - x')^2 + (y - y')^2 + [z - f(x', y')]^2}$ is the distance from source point $\mathbf{r}'$ on surface S to field point $\mathbf{r}$ in the medium, ρ_0 is the density, c is the sound velocity, and $k = \omega/c$ is the wave number. The Rayleigh integral for the curved transducer can be solved by the extended ASA[8]

$$\tilde{p}(x, y, z) = \frac{k\rho_0 c}{(2\pi)^2} \int_{-\infty}^{\infty} \int_{-\infty}^{\infty} V(k_x, k_y; z = z_1) \frac{\exp\left[j\left(xk_x + yk_y + (z - z_1)k_z\right)\right]}{k_z} dk_x dk_y, \ (z \geq z_1),$$
$$\tag{2}$$

$$V(k_x, k_y; z = z_1) = \iint_{S_{xy}} \frac{\tilde{v}_n(x', y', f(x', y'))}{\cos\theta_z} \exp\left[-j(x'k_x + y'k_y)\right] \exp\left[j(z_1 - f(x', y'))k_z\right] dx' dy',$$
$$\tag{3}$$

where $k_x = kn_x$, $k_y = kn_y$, $k_z = \sqrt{k^2 - k_x^2 - k_y^2} = k\sqrt{1 - n_x^2 - n_y^2}$ are the spatial frequencies in the x-, y- and z-directions, respectively, S_{xy} is the area of the projection of the surface S onto the x-y plane,

$$\cos\theta_z = \frac{1}{\sqrt{(f_x(x, y))^2 + (f_y(x, y))^2 + 1}}, \tag{4}$$

where $f_x(x, y)$ and $f_y(x, y)$ are the partial derivatives with respect to x and y, respectively. In Eq. (3), the condition $z_1 \geq \max[f(x', y')]$ must be met in order to ensure that $V(k_x, k_y; z = z_1)$ is always finite for all k_x and k_y. Eqs. (2) and (3) constitute the k-space algorithm of implementing the ASA and the extended ASA.

In the polar coordinates, using the following relations of the coordinate transform,

$$x = \rho \cos \varphi, \quad y = \rho \sin \varphi, \tag{5}$$

$$k_x = k_\rho \cos \psi, \quad k_y = k_\rho \sin \psi, \tag{6}$$

the transducer surface equation is of the form of $z = f(\rho, \varphi)$, and Eqs. (2)–(4) become,

$$\tilde{p}(\rho, \varphi, z) = \frac{k\rho_0 c}{(2\pi)^2} \int_0^{2\pi} \int_0^\infty V(k_\rho, \psi, z = z_1) \exp[j\rho k_\rho \cos(\varphi - \psi)] \frac{\exp[j(z - z_1)k_z]}{k_z} k_\rho \, dk_\rho \, d\psi, \tag{7}$$

$$V(k_\rho, \psi, z = z_1) = \int_0^{2\pi} \int_0^\infty \frac{\tilde{v}_n(\rho, \varphi, f(\rho, \varphi))}{\cos \theta_z} \exp[-jk_\rho \rho \cos(\varphi - \psi)] \exp[j(z_1 - f(\rho, \varphi))k_z] \rho \, d\rho \, d\varphi, \tag{8}$$

$$\cos \theta_z = \frac{1}{\sqrt{(f_\rho(\rho, \varphi))^2 + (f_\varphi(\rho, \varphi))^2 + 1}}, \tag{9}$$

where $k_\rho = kn_\rho$, $k_z = \sqrt{k^2 - k_\rho^2} = k\sqrt{1 - n_\rho^2}$, $f_\rho(\rho, \varphi)$ and $f_\varphi(\rho, \varphi)$ are the partial derivatives with respect to ρ and φ, respectively.

2.2. k-space and angular space algorithms for implementing the ASA for axisymmetric transducers

For a transducer that is axisymmetric, the surface equation $z = f(x, y)$ becomes $z = f(\sqrt{x^2 + y^2}) = f(\rho)$, and Eqs. (7)–(9) become,

$$\tilde{p}(\rho, z) = \frac{k\rho_0 c}{2\pi} \int_0^\infty V(k_\rho, z = z_1) J_0(\rho k_\rho) \frac{\exp[j(z - z_1)k_z]}{k_z} k_\rho \, dk_\rho, \tag{10}$$

$$V(k_\rho, z = z_1) = 2\pi \int_0^\infty \frac{\tilde{v}_n(\rho, f(\rho))}{\cos \theta_z} J_0(k_\rho \rho) \exp[j(z_1 - f(\rho))k_z] \rho \, d\rho, \tag{11}$$

$$\cos \theta_z = \frac{1}{\sqrt{(f_\rho(\rho))^2 + 1}}, \tag{12}$$

where $J_0(x) = (1/2\pi) \int_0^{2\pi} e^{jx\cos\theta} d\theta$ is the zeroth-order Bessel function. For a planar transducer located in the plane of $z=0$, $\cos\theta_z = 1$, $f(\rho) = 0$, and Eqs. (11) and (10) become the zeroth-order Hankel transform. Since this algorithm conducts the calculation of angular spectra and fields in the k-space, it is called *k-space* algorithm.

Our early research has shown that aliasing error appears because $\exp[j(z - z_1)k_z]$ in Eq. (10) is always undersampled, and this error can be reduced using some optimal selection of numerical parameters.[3,4] Here we propose another method to reduce the aliasing error so as to improve the calculation accuracy.

By making a variable transformation of

$$k_\rho = k\sin\theta,$$ (13)

Eq. (10) turns out to be

$$
\begin{aligned}
\tilde{p}(\rho, z) &= \frac{k\rho_0 c}{2\pi} \lim_{n_\rho \to \infty} \int_0^{\arcsin(n_\rho)} V(\theta, z_1) J_0(\rho k\sin\theta) \frac{\exp[j(z - z_1)k\cos\theta]}{k\cos\theta} k\sin\theta k\cos\theta d\theta \\
&= \frac{k^2 \rho_0 c}{2\pi} \left\{ \int_0^{\pi/2} V(\theta, z_1) J_0(\rho k\sin\theta)\sin\theta \exp[j(z - z_1)k\cos\theta] d\theta \right. \\
&\quad \left. + \int_0^\infty V(\beta, z_1) J_0(\rho k\cosh\beta)\cosh\beta \exp[-(z - z_1)k\sinh\theta] d\beta \right\}.
\end{aligned}
$$ (14)

In Eq. (14), the following things have been considered. Since the integration limits to k_ρ are from 0 to infinity, then $n_\rho = k_\rho / k = \sin\theta$ goes from 0 to infinity. Therefore, θ will be real valued for $n_\rho \leq 1$ and θ becomes imaginary for $n_\rho > 1$. When θ is real, it ranges from 0 to $\pi/2$, and in this case, θ is the angle of the propagation direction of a plane wave with respect to the z-axis, and $\theta = 0$ for $n_\rho = 0$, and $\theta = \pi/2$ for $n_\rho = 1$. When it is imaginary θ will be replaced with $j\beta$. Obviously, in Eq. (14) the first term represents the homogeneous waves and the second represents the inhomogeneous waves that decrease very rapidly as z increases because of $\exp[-zk\sinh\theta]$. It can be seen that in this case the singular point in Eq. (14) at the lower limit ($\theta = 0$) is eliminated.

The angular spectrum in Eq. (11) turns out to be

$$V(\theta, z_1) = 2\pi \int_0^\infty \frac{\tilde{v}_n(\rho, f(\rho))}{\cos\theta_z} J_0(k \sin\theta\rho) \exp[j(z_1 - f(\rho))k \cos\theta]\rho d\rho. \tag{15}$$

From the variable transformation in Eq. (13), it follows that, if θ is sampled in equal angular interval $\Delta\theta = constant$, then k_ρ (or n_ρ) is sampled in unequal interval. Since this algorithm conducts the calculation of angular spectra and fields in the θ-space, we propose the name θ-space algorithm. This will yield a good reduction of aliasing error, which will be shown in the following section.

It should be noted that the inhomogeneous parts, $\int_k^\infty (\cdot)dk_\rho$, in Eq. (10) and, $\int_0^\infty (\cdot)d\beta$, in Eq. (14) decreases very rapidly as z increases, and thus are neglected almost without loss of accuracy when z is not extremely close to the transducer surface. In the present study, therefore, only the homogeneous part in Eq. (14) will be used to construct the field.

2.3. Application to a special case: spherically focused, uniform transducers

Consider a spherically focused transducer that has a uniform normal velocity on its surface, a radius a and a focal length F, and is positioned as shown in Fig. 1.

Mathematically, the surface of the transducer can be represented by the equation,

$$z = f(\rho) = F - \sqrt{F^2 - \rho^2}, \quad (\rho \le a), \tag{16}$$

the normal velocity on the surface is expressed by

$$\tilde{v}_n(\rho, f(\rho)) = 1, \tag{17}$$

and inserting Eqs. (16) and (17) into Eq. (12), we have

$$\cos\theta_z = \sqrt{F^2 - \rho^2}/F. \tag{18}$$

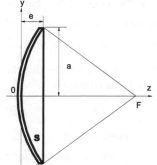

Fig. 1. Geometry of the spherically focused transducer

In this case Eqs. (10) and (11) (for the *k-space* algorithm) become

$$\tilde{p}(\rho, z) = \frac{k\rho_0 c}{2\pi} \int_0^k V(k_\rho, z_1) J_0(\rho k_\rho) \frac{\exp[j(z - z_1)k_z]}{k_z} k_\rho dk_\rho, \tag{19}$$

$$V(k_\rho, z_1) = 2\pi F \int_0^a \frac{J_0(k_\rho \rho)}{\sqrt{F^2 - \rho^2}} \exp\left[j\left(z_1 - F + \sqrt{F^2 - \rho^2}\right)k_z\right]\rho d\rho, \tag{20}$$

in which the inhomogeneous waves are neglected.

Eqs. (14) and (15) (for the θ-space algorithm) become

$$\tilde{p}(\rho, z) = \frac{k^2 \rho_0 c}{2\pi} \int_0^{\pi/2} V(\theta, z_1) J_0(\rho k \sin\theta) \sin\theta \exp\left[j(z - z_1)k\cos\theta\right]d\theta , \tag{21}$$

$$V(\theta, z_1) = 2\pi F \int_0^a \frac{J_0(\rho k \sin\theta)}{\sqrt{F^2 - \rho^2}} \exp\left[j\left(z_1 - F + \sqrt{F^2 - \rho^2}\right)k\cos\theta\right]\rho d\rho , \tag{22}$$

in which the inhomogeneous waves are neglected.

For a spherically annular transducer with inner and outer radii, a_1 and a_2, the algorithm only needs a small modification for the angular spectrum calculation, that is, replacing the lower limit 0 and the upper limit a in Eqs. (20) and (22) with a_1 and a_2, respectively. Actually, a spherically focused circular transducer in Fig. 1 can be thought of as a special case of the annular transducer with $a_1 = 0$ and $a_2 = a$.

To minimize the aliasing error, sampling frequency larger than or equal to Nyquist frequency should be used. For the k-space algorithm, it is $\exp\left[j(z - z_1)\sqrt{1 - k_\rho^2}\right]$ that should be sampled at a sufficiently fine sampling interval (i.e., a sufficiently high rate) in terms of k_ρ, but the undersampling can not be avoided, because the "instantaneous frequency" of $\exp\left[j(z - z_1)\sqrt{1 - k_\rho^2}\right]$ can be infinite.[3,4] For the θ-space algorithm, it is $\exp\left[j(z - z_1)k\cos\theta\right]$ that needs being sufficiently sampled in terms of θ. From Refs. 3 and 4, it follows that the "instantaneous frequency" of $\exp\left[j(z - z_1)k\cos\theta\right]$ in terms of θ can be determined by

$$f_I(\theta) = \frac{1}{2\pi}\frac{d}{d\theta}\left[(z - z_1)k\cos\theta\right] = \frac{(z - z_1)}{\lambda}\sin\theta . \tag{23}$$

which varies with z and θ. To eliminate the aliasing error the sampling frequency, F_{sample}, used needs to meet the Nyquist frequency requirement, that is, F_{sample} must always be larger than twice of the instantaneous frequency,

$$F_{sample} \geq 2f_I(\theta) = \frac{2(z - z_1)}{\lambda}\sin\theta , \tag{24}$$

where λ is the wavelength in the medium. For given F_{sample} we may assume the sampling interval to be $\Delta\theta = 1/F_{sample}$, and the number of sampling points for θ ranging from 0 to $\pi/2$ is

$$N_s = \frac{\pi/2}{\Delta\theta} = \frac{\pi}{2}F_{sample} \tag{25}$$

To ensure that Eq. (24) holds for all z and θ, and from Eq. (25), we take the number of sampling points that satisfies

$$N_s = \frac{\pi(z_{max} - z_1)}{\lambda},$$ (26)

where z_{max} is the maximum dimension of the field to be calculated in the z-direction. The above equation will be used in the following calculations.

To evaluate the accuracy and efficiency of the two algorithms, the on-axis acoustical fields of the spherically focused transducer are calculated from the algorithms and compared with the analytical solution,[13]

$$\tilde{p}(\rho, z) = \frac{\rho_0 c v_n e^{-j\alpha t}}{1 - z/F} \left\{ \exp[jkz] - \exp\left[jk\sqrt{z^2 + (1 - z/F)4F^2 \sin\left(\frac{\alpha}{2}\right)} \right] \right\},$$ (27)

where $\alpha = \arcsin(a/F)$.

3. Comparison and Discussions

The efficiency of the new algorithm (the θ-space algorithm), Eqs. (21) and (22), will be demonstrated by comparing with the k-space algorithm, Eqs. (19) and (20), in the case of the spherically focused transducer. The transducer is assumed to have a uniform normal velocity (v_n =constant) on the surface, a 9.86-mm radius and a 223.5-mm focal length.

The angular spectra calculated by the k- and θ-space algorithms are shown in Figs. 2(a) and (b), respectively. The difference is the horizontal axis, the direction cosine n_ρ in Fig. 2(a), whereas the angle θ in Fig. 2(b). The sampling points used are 4500 for the k-space algorithm and only 1500 for the θ-space algorithm. But the field constructed from the k-space algorithm shows large aliasing error (Fig. 3(a)), whereas the field from the θ-space algorithm has very small aliasing error (Fig. 3(b)). This demonstrates that the θ-space algorithm is much better than the k-space algorithm, both in efficiency and accuracy.

4. Conclusions

Starting from the extended ASA proposed by the authors[8] and transforming the Cartesian coordinates into the polar coordinates, two algorithms have been developed for implementing the ASA to calculating axisymmetrically curved transducers. The first is the k-space algorithm in which the ASA is implemented in the k-space, and this is simply the polar coordinate counterpart of the extended ASA. The other one is the angular space algorithm in which the ASA is implemented in the angular space where the polar angle is the variable for an axisymmetric transducer. Both algorithms have been applied to calculating the field from a spherically focused transducer, and the results from both algorithms have been compared with the analytical one. The comparison has shown that the angular space algorithm is much better in efficiency and accuracy than the k-space algorithm. This is because the Green function can be sufficiently sampled in the angular space and thus the aliasing error in the angular space algorithm can be eliminated.

496

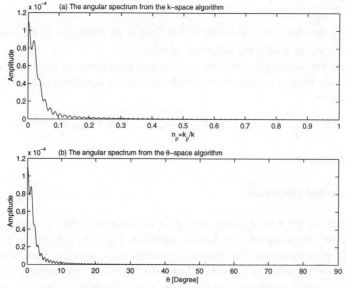

Fig. 2. Angular spectra of the spherically focused transducer calculated from (a) the k-space algorithm and (b) the θ-space algorithm.

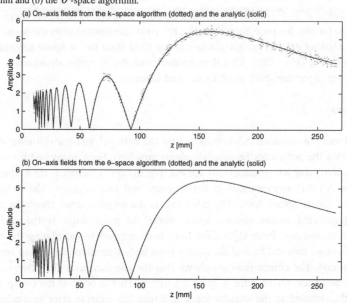

Fig. 3. Comparison of the on-axis fields of the spherically focused transducer calculated from the analytic solution (solid) and from (a) the k-space algorithm (dotted), and (b) the θ-space algorithm (dotted).

Acknowledgments

This work was sponsored by the Swedish Nuclear Fuel and Waste Management Co. (SKB).

References

1. E. G. Williams and J. D. Maynard, "Numerical evaluation of the Rayleigh integral for planar radiators using the FFT," J. Acoust. Soc. Am. **72**, 2020-2030 (1982).
2. D. P. Orofino and P. C. Pedersen, "Efficient angular spectrum decomposition of acoustic sources - Part I. Theory, and Part II. Results," IEEE Trans. Ultrason. Ferroelec. Freq. Contr. **40**, 238-257 (1993).
3. P. Wu, R. Kazys and T. Stepinski, "Analysis of the numerically implemented angular spectrum approach based on the evaluation of two-dimensional acoustic fields--Part I: Errors due to the discrete Fourier transform and discretization, and Part II: Characteristics as a function of angular range," J. Acoust. Soc. Am. **99**, 1339-1359, (1996).
4. P. Wu, R. Kazys and T. Stepinski, "Optimal selection of parameters for the angular spectrum approach to numerically evaluate acoustic fields," J. Acoust. Soc. Am. **101**, 125-134, (1997).
5. D.-L. Liu, and R.C. Waag, "Propagation and backpropagation for ultrasonic wavefront design," IEEE Trans. Ultrason. Ferroelec. Freq. Contr. **44**, 1-13, (1997).
6. P. Wu, and T. Stepinski, "Elastic fields in immersed isotropic solids from phased arrays: the time harmonic case," Res. Nondectr. Eval. **10**, 185-204, (1998).
7. P. T. Christopher and K. J. Parker, "New approaches to the linear propagation of acoustic fields," J. Acoust. Soc. Am. **90**, 507-521 (1991).
8. P. Wu and T. Stepinski, "Extension of the angular spectrum approach to curved radiators," J. Acoust. Soc. Am. **105**, 2618-2627, (1999).
9. A. V. Oppenheim, G. V. Frisk, and D. R. Martinez, "An algorithm for the numerical evaluation of the Hankel transform," *Proc. IEEE.* vol. 66, no. 2, pp. 264-265, (1978).
10. G. Arfken, *Mathematical Methods for Physicists*, 3rd Ed. Academic Press, San Diego, 1985, pp. 794-797.
11. J. Brunol, and P. Chavel, "Fourier transformation of rotationally invariant two-variable functions: computer implementation of Hankel transform," *Proc. IEEE.* vol. 65, pp. 1089-1090, (1977).
12. P. Wu and T. Stepinski, "Spatial impulse response method for predicting pulse-echo fields from a linear array with cylindrically concave surface," IEEE Trans. Ultrason. Ferroelec. Freq. Contr. **46**, 1283-1297, (1999).
13. H.T. O'Neil, "Theory of focusing radiators," J Acoust. Soc. Am. **21**, 516-526, (1949).

Theoretical and Computational Acoustics 2001
E.-C. Shang, Qihu Li and T. F. Gao (Editors)
© 2002 World Scientific Publishing Co.

Directionality Features for Acoustic Radiation from a Nonseparable Source on a Rigid Cylinder

Genshan Jiang[1], Jing Tian[2], Liansuo An[1], Yongbo Zhu[2], Kun Yang[1]

[1]North China Electric Power University, Baoding 071003

[2]Institute of Acoustics, Academia Sinica, Beijing 100080

In this paper, the sound directionality patterns produced by a nonseparable source located on a rigid cylinder of infinite length have been investigated for the case in which the source strength may be represented as a sum of separable sources by use of addition theorems. The result shows that the patterns of amplitude and phase distribution over the whole farfield space for the source are dependent of the frequencies of sound waves, circumference of the cylinder, and the strength of the source piston. It reveals that the observation points and the frequencies are important to the techniques of the boiler tube leak trace. This result can be applied to the technique of acoustic leak detection.
Key words: Acoustic radiation, Directionality, Nonseparable sources, Tube leak detection

1. Introduction

Circular cylinders occur in some components of industrial boilers, e.g. steam generators, evaporators, superheater, reheater and other types of heat exchangers. It is important to be able to monitor them so that internal damage, typically a leak from one of the cylinders, can be detected. Of the numerous leak detection schemes applied in the past, three types are currently being used today: Acoustic, Conductivity, and Mass Balance. Each approach has strengths and weaknesses. The acoustic detection method (ADM) uses sensors and software to detect tube leaks much the same way as operations personnel. While this can be a reliable method, the software must learn which sounds are considered leaks and which are considered normal operating sounds. Until the software is properly trained, this method can be susceptible to false tube leak indications and alarms. A leak often gives rise to noise. This makes it possible to use acoustic techniques for leak detection. Acoustic monitoring method gives results quickly and also offers the advantage of being non-invasive. Its techniques may be divided into two catgories: passive techniques which listen to the noise produced by the leak; and active techniques which rely on the modification of the sonic transmission path by hydrogen bubbles, hot or cold spots caused by the leak. Both techniques require a good understanding of the acoustic characteristics of the heat exchangers. The aim of this paper is to provide this understanding.

This paper presents the directionality patterns of three-dimension produced by a nonseparable source located on a rigid cylinder of infinite length. For the case where the source is a circular piston which can be expressed as a sum of separable sources represented as a separable function of azimuth angle and axial dimension. The two-dimensional patterns for the separable rectangular source had been given by Laird and Cohen[2]. The present work extends the theory to the case where the source is circular or any nonseparable piston. It is observed that the patterns of amplitude and phase over the whole farfield space of the leak source are dependent of the frequencies of sound waves, circumference of the cylinder, and the strength of the leak source. The result reveals that the observation points and the measured frequencies are important to the techniques of the boiler tube leak trace.

2. Theory for the separable sources

2.1 General considerations

Fig.1 shows a cylindrical coordinate system and a spherical coordinate system. For purposes of describing the far zone field, spherical coordinates (R, φ, θ) will eventually be used. The source piston is assumed to vibrate radially in such a way that its velocity distribution may be represented as a separable function of φ and z. The boundary conditions at $r = a$, where a is the radius of the cylinder, is then given by the expression,

$$u_r\big|_{r=a} = U_0 e^{-i\omega t}(\sum_{m=0}^{\infty} a_m \cos m\varphi)(\int_{-\infty}^{\infty} F(k_z)e^{ik_z z}dk_z) \quad (1)$$

The Fourier series, which for simplicity has been taken to be a cosine series, gives the dependence on φ. The Fourier integral represents the dependence on z, with $F(k_z)$, the transform of the velocity distribution in the z-direction, being independent of φ.

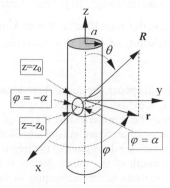

Fig.1 Rigid cylinder with a piston source

Having prescribed the boundary condition, it remains only to write down the general solution of the wave equation in cylindrical coordinates and to match this solution to the boundary condition. The general expression for a combination of outgoing cylindrical waves of even dependance on φ is given by [1]

$$p(\mathbf{R}) = p(r, \varphi, z) = e^{-i\omega t}(\sum_{m=0}^{\infty} \cos m\varphi) \int_{-\infty}^{\infty} A_m(k_z)H_m^{(1)}(k_r r)e^{ik_z z}dk_z \quad (2)$$

where $k_z^2 + k_r^2 = k^2 = \omega^2/c^2$, and where $H_m^{(1)}$ is the first Hankel function of order m.

Matching the solution (2) to the boundary condition (1) at $r = a$, introducing the wave zone approximation, and replacing the cylindrical coordinates by spherical coordinates, we obtain the solution,

$$p(\mathbf{R}) = p(R, \theta, \varphi) = i\omega\rho U_0 \left(\frac{2}{\pi R \sin\theta}\right)^{1/2} e^{-i\omega t} \times \sum_{m=0}^{\infty} a_m \exp[-i(m+\frac{1}{2})\pi/2]\cos m\varphi$$

$$\times \int_{-\infty}^{\infty} \frac{F(k_z)\exp\{iR[(k^2-k_z^2)^{1/2}\sin\theta + k_z\cos\theta]\}}{(k^2-k_z^2)^{3/4} H_m^{(1)'}[(k^2-k_z^2)^{1/2}a]}dk_z \quad (3)$$

Here we have adopted the usual asymptotic expression for the first Hankel function.

Acorrding to the reference [1], the infinite integral appearing in Eq. (3) can be asymptotically evaluated as

$$\int_{-\infty}^{\infty} \frac{F(k_z)\exp\{iR[(k^2-k_z^2)^{1/2}\sin\theta + k_z\cos\theta]\}}{(k^2-k_z^2)^{3/4} H_m^{(1)'}[(k^2-k_z^2)^{1/2}a]}dk_z =$$

$$= \left(\frac{2\pi}{R\sin\theta}\right)^{1/2} \frac{\exp\{i[kR-(\pi/4)]\}F(k\cos\theta)}{kH_m^{(1)'}(ka\sin\theta)} \quad (4)$$

Using this result, we have as a final description of the radiation for the separable sources:

$$p(R,\theta,\varphi) = 2\rho c U_0 \frac{e^{i(kR-\omega t)}}{R} \frac{F(k\cos\theta)}{\sin\theta} \sum_{m=0}^{\infty} \frac{a_m e^{-im\pi/2}}{H_m^{(1)'}(ka\sin\theta)} \cos m\varphi \tag{5}$$

where the ρc is the specific acoustic resistance of the surroundings around the cylinder.

It might seem pertinent to ask to what extent the validity of this result is affected by the approximation involved in the method of stationary phase. An examination reveals that the errors introduced approach zero as R is increased without limit. But, from general consideration, it can be argued that the patterns will not change with R once the wave zone is reached. It is therefore concluded that the approximations involved in evaluating the integrals are of the same order as those involved in the usual wave zone approximation and that the result obtained is a correct description of the wave zone field.

2.2 The case of a uniform rectangular separable source

Let us first consider the particular problem in which the source is a rectangular area over which the radial velocity is uniform and of amplitude U_0. The Fourier coefficients, a_m, describing the φ-dependence of such a sourse which subtends an angle 2α are given by

$$\left. \begin{array}{l} \alpha_0 = \alpha/\pi \\ \alpha_m = 2\sin m\alpha/m\pi, \quad m = 1,2,3,\cdots,\infty \end{array} \right\} \tag{6}$$

The z- dependence is described by the Fourier transform

$$F(k_z) = \sin k_z z_0 / \pi k_z \tag{7}$$

where $2z_0$ is the height of the rectangular source. The pattern for the whole far-field space may now be determined by inserting the expressions (6) and (7) into Eq. (5). This yields

$$p(R,\theta,\varphi)\big|_{a,z_0} = \frac{4\rho c U_0}{\pi^2} \frac{e^{i(kR-\omega t)}}{R} \frac{\sin(kz_0\cos\theta)}{k\cos\theta\sin\theta} \sum_{m=0}^{\infty} \frac{\sin m\alpha}{m\in_m} \frac{e^{-im\pi/2}}{H_m^{(1)'}(ka\sin\theta)} \cos m\varphi \tag{8}$$

Where $\in_0 = 2$, and $\in_m = 1$ ($m \neq 0$). This result can be applied to the radiation from a point source on a rigid cylindrical baffle and the radiation from a vibrating circular band on a rigid cylinder, but not suitable for the case of nonseparable source.

3. The acoustic radiation from a nonseparable source

For the case of nonseparable sources, in general insteading of the product of z and φ functions in Eq. (1), the Fourier coefficients, a_m, would be functions of z and would be themselves each expressible as Fourier integrals. So in this way it will be felt that the complication of this case is greatly increased. But as we know, nonseparable piston can be divided into many small rectangular sources, and each small source can be approximately regard as a uniform separable source. So we can use the addition theorems to calculate the acoustic radiation from a nonseparable source by use of the result of separable sources.

3.1 The sound field of a uniform circular nonseparable source

Since the cases of practical interest, we can consider that the source piston is a circular area over which the radial velocity is uniform and of amplitude U_0, as shown in Fig.2. According to the addition theorems, the radiation field can be written as

$$P(R,\theta,\varphi)\big|_D = \lim_{N\to\infty} \sum_{n=1}^{N} \left[p(\alpha_n, z_{0n}) - p(\alpha_n, z_{0n-1}) \right] \tag{9}$$

502

where

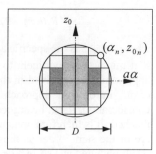

$$\left.\begin{array}{l} \alpha_n = \dfrac{D}{2a}\cos(\dfrac{\pi}{2N}n) \\[2mm] z_{0_n} = \dfrac{D}{2}\sin(\dfrac{\pi}{2N}n) \end{array}\right\} \quad n = 0,\ 1,\ 2,\ \cdots,\ N \qquad (10)$$

and N is the positive integer, D is the diameter of the circular leak area, a is the radius of the cylinder. $p(\alpha_n, z_{0_n})$ indicates the field function excited by the uniform rectangular source at $\alpha = \alpha_n$ and $z_0 = z_{0_n}$, and has the form

Fig.2 A uniform circular piston on a rigid cylinder

$$p(\alpha_n, z_{0_n}) = \frac{4\rho c U_0}{\pi^2}\frac{e^{i(kR-\omega t)}}{R}\frac{\sin(kz_{0_n}\cos\theta)}{k\cos\theta\sin\theta}\sum_{m=0}^{\infty}\frac{\sin m\alpha_n}{m\in_m}\frac{e^{-im\pi/2}}{H_m^{(1)\prime}(ka\sin\theta)}\cos m\varphi \qquad (11)$$

By setting θ equal to $\pi/2$, Eq. (9) can be written as

$$P(R,\pi/2,\varphi)\big|_D = \frac{4\rho c U_0}{\pi^2}\frac{e^{i(kR-\omega t)}}{R}\underset{N\to\infty}{Lim}\sum_{n=1}^{N}\left[(z_{0_n}-z_{0_{n-1}})\sum_{m=0}^{\infty}\frac{\sin m\alpha_n}{m\in_m}\frac{e^{-im\pi/2}}{H_m^{(1)\prime}(ka)}\cos m\varphi\right] \qquad (12)$$

3.2 The case of a circular symmetrical nonseparable source

If the source piston is a circular area over which the radial velocity is not uniform, but of amplitude $U_0 = U_0(r')$, where r' is the polar coordinate whose origin coincides with the centre of the piston, and $0 \le r' \le D/2$, as shown in Fig.3. According to the result of section 3.1, the radiation field for the nonuniform source can be written as

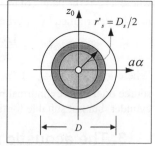

$$P(R,\theta,\varphi)\big|_D = \underset{S\to\infty}{Lim}\sum_{s=1}^{S}\left[p(U_{0_s})_{D_s} - p(U_{0_s})_{D_{s-1}}\right] \qquad (13)$$

$$\left.\begin{array}{l} U_{0s} = U_0(r_s') \\[2mm] r_s' = \dfrac{1}{2}D_s = \dfrac{D}{2S}s \end{array}\right\} \quad s = 0,\ 1,\ 2,\ \cdots,\ S \qquad (14)$$

Fig.3 A circular symmetrical piston on a rigid cylinder

where $p(U_{0_s})_{D_s}$ has the form of Eq. (9) and indicates the field function produced by the uniform circular piston which diameter is D_s and the radial velocity is uniform amplitude $U_{0s}=U_0(r_s')$.

Eq. (9) and (13) extend the separable source theory to the case where the source is any nonseparable function of the azimuth and axial dimensions, such as a circular piston.

4. Numerical examples and discussion

Applying the theory presented in above sections, we computed the farfield radiating patterns from a uniform circular piston on a rigid cylinder. In order to compare with the result of Laird [1], we set $\alpha_0 = 3.7°$ (degree), i.e. $D/a = 0.129$. Fig.4(a) shows the variation of the normalized amplitude with the frequency (or namely acoustic size ka) and azimuth (φ) in the horizontal plane, that is, the plane perpendicular to the axis of the cylinder, is obtained by setting θ equal

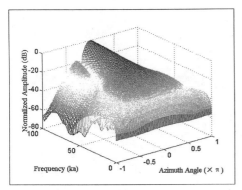

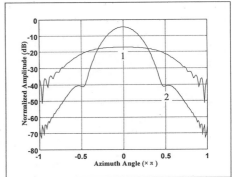

Fig.4(a) Horizontal amplitude pattern for a uniform circular source with $D/a = 0.129$.

Fig.4(b) Horizontal amplitude pattern for a circular source with 1. ka=14, 2. ka=60.

to $\pi/2$, i.e. Eq.(12). Fig.4(b) shows that the amplitude displayed in Fig.4(a) is respectively plotted against azimuth (φ) at ka=14 and ka=60. From the result we can realize that the directionality features in the horizontal plane will get strengthen while the wave frequency (ka) excited by the source piston is increased. For the comparision, this result is very similar to the result obtained by Morse [3] for the case of an infinite strip source, and by Laird [1] and Jiang [2] for the case of a uniform rectangular source. So we can conclude that for the case of a point, rectangular, circular, and an infinite strip source on a rigid cylindrical baffle the directionality features in the horizontal plane all have the same rules.

The amplitude patterns computed for the whole farfield space are shown in Fig.5(a) and Fig.6(a), which was normalized to $\theta = 0°$ at $\varphi = 0°$. The phase patterns obtained are shown in Fig.5(b) and Fig.6(b), and the phase was also arbitrarily normalized to the point of $\theta = 0°$ and $\varphi = 0°$. In these two examples, the cylinder is respectively of 90 and 14 wavelength circumference (ka), and the half angle subtended by the diameter of the circular source is $3.7°$.

From the Fig.5 and 6, we can realized that the maximum of amplitude (0 dB) occurs at the point with $\theta = 0°$ and $\varphi = 0°$, and where is also the point of the minimum of phase (0 rad). The range of amplitude and phase at high frequency is larger than at the low frequency. For example, when ka=90, the variation range of amplitude is near by 80 (dB), and the phase one is about 50π (rad). But when ka=14, the amplitude range is less than 40 (dB), and the phase one is less than 14π (rad). Moreover we can also find that the maximum of the phase occurs at the point of $\theta = 90°$ and $\varphi = \pm180°$ while ka=14, but the maximum of the phase does not occur at this point while ka=90.

Occurrence of the secondary maxima which are observed over the back portion of this pattern, shown in Fig.4(b), may be explained qualitatively in terms of two waves which are diffracted in opposite directions around the cylinder. At $\varphi = \pm180°$, these two waves arrive in phase, and a maximum is produced. A diffracted wave traveling 360° around the cylinder would travel ka (such as 14 or 60) wavelengths, and would therefore travel one wavelength in about $360°/ka$ (such as 26° or 6°). Therefore, if the field angle is changed by $180°/ka$ (such as 13° or 3°), one of the diffracted waves would be advanced by a half wavelength while the other would be retarded by a half wavelength, and the two would again meet in phase, produced another maximum. This effect is less pronounced as the front of the cylinder is approached, since the two waves become more unequalin amplitude.

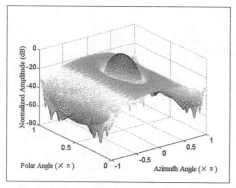

Fig.5(a) The amplitude pattern for a uniform circular source with $D/a = 0.129$, $ka = 90$.

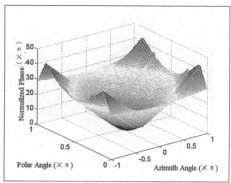

Fig.5(b) The phase pattern for a uniform circular source with $D/a = 0.129$, $ka = 90$.

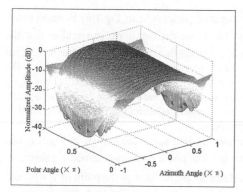

Fig.6(a) The amplitude pattern for a uniform circular source with $D/a = 0.129$, $ka = 14$.

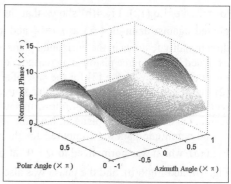

Fig.6(b) The phase pattern for a uniform circular source with $D/a = 0.129$, $ka = 14$.

5. Conclusions

The problem of the radiation of sound from a finite separable source and an infinite strip vibrating on the side of a rigid cylinder were respectively solved by Laird and Morse. This problem has now been extended to the case of a finite nonseparable source, such as a circular area of uniform or symmetric distribution amplitude. The horizontal pattern is found to be the same as that given by Laird and Morse. The whole farfield space patterns have been computed from the new expressions.

The analysis and numerical results show that the directionality features depend mainly on the wavelength circumference (ka), shape of the source, and distribution of the amplitude. Other cases of possible interest to which the general result, Eq.(9), can be applied are the radiation from a point and an ellipse source. The results have been found useful to the techniques of the boiler tube leak trace.

References

1. D.T. Laird, and H. Cohen, Directionality Patterns for Acoustic Radiation from a Source on a Rigid Cylinder [J]. *Journal of the Acoustical Society of America*, **24** (1), 1952, 46~49.
2. G.S. Jiang, Directionality Patterns for Acoustic Radiation from a Leak Source on a Heat-exchanger Cylinder in Boilers, *Proceedings of the Chinese Society for Electrical Engineering* [J], to be submitted.
3. P.M. Morse, Vibration and Sound [M]. McGraw-Hill, New York, 1948.

Theoretical and Computational Acoustics 2001
E.-C. Shang, Qihu Li and T. F. Gao (Editors)
© 2002 World Scientific Publishing Co.

A New Technique for the Design of Acoustic Matching Layers for Piezocomposite Transducers

Nicola Lamberti
Dip. di Ing. dell'Informazione ed Ing. Elettrica, Università di Salerno,
84084 Fisciano (SA), Italy

Francisco Montero de Espinosa
Instituto de Acustica, C.S.I.C., Serrano 144, 28006 Madrid, Spain

Giosuè Caliano and Riccardo Carotenuto
Dip. di Ing. Elettronica, Università di Roma Tre, Via della Vasca Navale 84,
00146 Roma, Italy

1. Abstract

At present a lot of piezoelectric broad band ultrasonic transducers for both, medical and NDE purposes, use as active material piezoelectric ceramic composites. A lot of scientific and technological research has been devoted to the optimisation of piezocomposite materials, in order to increase the transducer band and efficiency. In a typical transducer based on piezocomposites the active material is mounted on a soft lossy backing and one matching layer is placed on the front, radiating face of the transducer with the aim to match the acoustic impedance of the medium and to enlarge the bandwidth. In this paper an optimisation work is shown to demonstrate that a composite configuration can be used in the matching layer, in order to improve the efficiency and the band of the transducer. An approximated two–dimensional analytical model has been used to optimise the design of a composite–structured matching layer in the case of 2–2 composites, obtaining different results for the polymer and piezoceramic composite phases; a design technique is suggested in order to improve the transducer performance. With the aim to verify the proposed design criterion, a transducer prototype, based on a 2–2 piezocomposite, with a composite matching layer was realised. On this sample we measured the electrical input impedance and the insertion loss and we compared the obtained results with those of a transducer classically matched to the load. The obtained results confirm the computed improvements in the transducer performance and justify the proposed design approach.

506

2. INTRODUCTION

In the last years, most of the piezoelectric broad band ultrasonic transducers for both, medical and NDE purposes, use as active material piezoelectric ceramic composites. A lot of scientific and technological research has been devoted to the optimisation of piezocomposite materials, in order to increase the transducer band and efficiency [1]. In a typical transducer based on piezocomposites the active material is mounted on a soft backing and one matching layer is placed on the front, radiating face of the transducer, with the aim to match the acoustic impedance of the medium and, then, to enlarge the bandwidth. Standard transducer one dimensional models — KLM, Mason, etc. — are used to optimise the design of the matching layer, supposing that the piezocomposite is a homogeneous material. With these approaches the specific acoustic impedance of the matching material becomes too close, and generally higher than the one of the piezocomposite polymer phase, resulting, with some composite geometry, in a destruction of the composite concept.

In this paper we use an approximated two–dimensional analytical model, previously used to describe multielement array transducers [2] and multi frequency 2–2 composites [3], to optimise the design of a composite–structured matching layer in the case of 2–2 composites. The transducer structure is shown in Fig. 1: it is composed by piezoceramic strips separated by polymer strips with the same thickness and length, mounted on a lossy backing and with a matching layer on the radiating face.

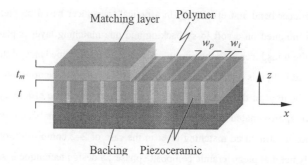

Fig. 1. Geometry of the analyzed 2–2 composite.

The approximated analytical model which we use, considers the piezoceramic element of the composite as a two–dimensional (in the x and z directions) resonator whose vibrations can be described, in the frequency domain, by means of a 5 x 5 matrix relating the forces on the external surfaces orthogonal to the x and z axes and the voltage to the velocities on the same surfaces and the current. In this way the piezoceramic element can be seen as a 5–bipole with four mechanical and one electrical ports. In the model the stress and electric boundary conditions are satisfied only in an integral form, but these approximations do not substantially affect the results [4], [5]. The polymer strips are also considered as two–dimensional structures and the 5x5 matrix describing their behaviour in the frequency domain is deduced from that of the piezoelectric element simply cancelling all the piezoelectric constants and by taking the strip capacitance into account [3]. The full composite can be seen as a cascade connection, in the x direction, of the piezoelectric 5–bipoles alternated to the polymer 5–bipoles. The connection between the ports of the multipoles represents the mechanical contact between the elements. In order to build up the matrix of the overall composite, we use an algorithm which computes the resulting matrix of the cascade of the two elemental matrices. Iterating this algorithm to all the composite elements, we obtain the total matrix [2]; with this matrix we are able to compute the composite electrical input impedance and all the transfer functions.

3. CLASSIC MATCHING LAYER DESIGN

The 2–2 piezocomposite that we have analysed, is constituted by nine piezoelectric (PZT–4 by Vernitron) strips separated by eight polymer (Araldite H by Ciba & Geigy) strips. The composite thickness t and length L are respectively: $t = 1.6$ mm and $L = 8.4$ mm, while the widths are $w_p = 1$ mm and $w_i = 0.7$ mm for the piezoceramic and the polymer strips respectively. These so large dimensions are not typical for a 2–2 piezocomposite, but we made this choice with the purpose to minimise measurement errors when we will realise the prototypes. In order to stress the influence of the matching layer on the transducer performance we have not inserted the backing in the transducer design. As a first step we designed the matching layer by considering the composite as a homogeneous material and by using the Souquet criterion [6]:

$$z_m = \sqrt{2 z_L^2 z_c} ,$$

(1)

where z_m is the specific acoustic impedance of the matching layer, while z_c and z_L are the specific acoustic impedance of the composite and the load respectively ($z_L = 1.5$ Mrayls — water). As a first step to compute the acoustic impedance of the composite, we computed the sound propagation velocity v_c from the knowledge of the antiresonance frequency $f_p = 1$ MHz:

$$v_c = 2 \cdot t \cdot f_p = 3200 \text{ m/s}.$$

(2)

508

The composite mass density is instead computed by making the mean, weighted on the volumes, of the PZT–4 and Araldite densities:

$$\rho_c = \frac{\rho_p \, w_p + \rho_i \, w_i}{w_p + w_i}. \tag{3}$$

With ρ_p = 7500 kg/m^3 (PZT–4 mass density) and ρ_i = 1170 kg/m^3 (Araldite mass density), we obtain for the composite ρ_c = 4894 kg/m^3. The composite specific acoustic impedance is therefore z_c = 15.7 Mrayls, and by applying the Souquet formula (1) we obtain z_m = 4.1 Mrayls. As far as the matching layer thickness is concerned, some of the authors showed that the maximum of the bandwidth is obtained if t_m = $\lambda/4$ at the frequency where the transfer function of the transducer without matching layer has the maximum [7]. In the present case we consider the Transmission Transfer Function (TTF), i.e. the ratio between the force exerted by the element on the load and the applied voltage; TTF is maximum at f_s = 828 kHz, and supposing a propagation velocity in the matching layer v_m = 2000 m/s, we obtain t_m = 0.6 mm.

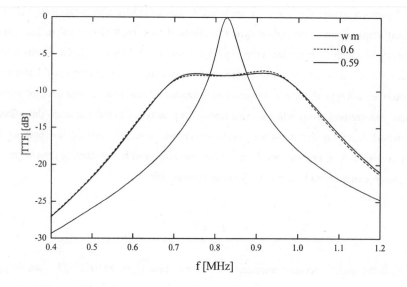

Fig. 2. Comparison between the normalized responses of the single piezoceramic element without matching layer (w m) and with layers with thickness t_m = 0.6 mm, and t_m = 0.59 mm.

In order to verify the transducer performance with the matching layer we compute the TTF of a single piezoelectric element and compared the result with that obtained without matching layer. Fig. 2 shows this comparison. As expected, the transfer function computed when the matching layer is present has a wider band and a lower efficiency. In the figure the result obtained with $t_m = 0.59$ mm is also shown; this thinner thickness let us to obtain a flatter response and therefore it is chosen as optimum result. The specific acoustic impedance of the matching layer is greater than that of the polymer one ($z_i = 3.15$ Mrayls) and therefore a mismatch is expected for this composite phase. In order to investigate this situation, we computed the TTF of a polymer element with the matching layer and compared it with that computed without matching layer. The results are shown in Fig. 3. As it can be seen, with the matching layer a larger bandwidth is obtained, but the result is poorer than that obtained for the piezoelectric element; the reason of this unsatisfactory behaviour is due to the mismatch between the polymer and the load, due to the fact that $z_m > z_i$.

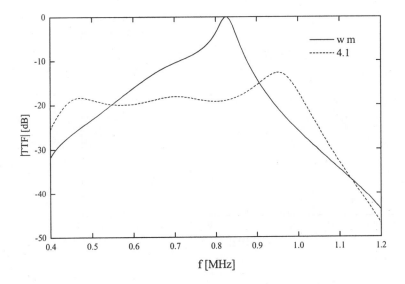

Fig. 3. Comparison between the normalized responses of the single polymer element without matching layer (w m) and with a matching layer of acoustic impedance $z_m = 4.1$ Mrayls.

4. COMPOSITE MATCHING LAYER DESIGN

Our idea in order to improve the transducer performance is to match the piezocomposite to the load by means of a "composite" matching layer, i.e. a matching layer composed by strips with the same thickness, but different specific acoustic impedances. The acoustic impedance of the material in front of the piezoceramic and polymer elements can be computed by separately applying the Souquet criterion to the two composite phases:

$$z_{mp} = \sqrt{2\, z_L^2\, z_p}\,,$$

(4)

$$z_{mi} = \sqrt{2\, z_L^2\, z_i}\,.$$

(5)

The obtained results with $z_p = 34$ Mrayls are: $z_{mp} = 5.4$ Mrayls and $z_{mi} = 2.4$ Mrayls. The emission of the piezoelectric element computed in this case is compared in Fig. 4 with the result obtained with $z_m = 4.1$ Mrayls.

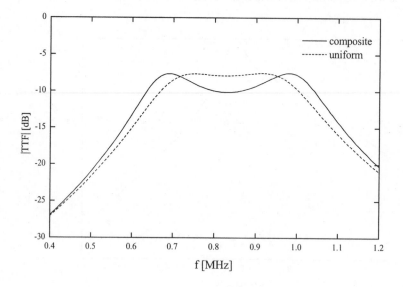

Fig. 4. Comparison between the normalised responses of the single piezoelectric element with a composite matching layer ($z_{mp} = 5.4$ Mrayls, $z_{mi} = 2.4$ Mrayls) and with a uniform layer with $z_m = 4.1$ Mrayls.

As it can be seen the best result is obtained when a uniform matching layer is used. Fig. 5 shows the same comparison for the emission of the polymer element. In this case the two results are comparable, even if the band obtained with the composite matching layer is greater.

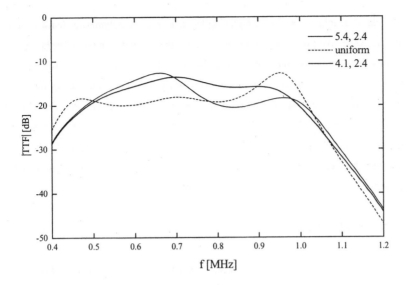

Fig. 5. Comparison between the normalized responses of the single polymer element with a composite matching layer with z_{mp} = 5.4 Mrayls and z_{mi} = 2.4 Mrayls, with a composite layer with z_{mp} = 4.1 Mrayls and z_{mi} = 2.4 Mrayls, and with a uniform layer with z_m = 4.1 Mrayls.

Finally, in order to improve the responses both of the piezoceramic and polymer elements we computed the responses with a composite layer with z_{mp} = 4.1 Mrayls and z_{mi} = 2.4 Mrayls. The result obtained for the piezoceramic is the same of that obtained with the uniform layer, while in Fig. 5 we report the one obtained for the polymer element. As far as the flatness of band is concerned, this last is the better result. The obtained results show that a composite matching layer can improve the transducer response and a criterion for the design of this layer can be proposed: the specific acoustic impedance of the phase in front of the piezoceramic elements can be computed by using the Souquet's expression (1), i.e. by considering the impedance of the composite, while the impedance of the phase in front of the polymer can be still computed by means of the Souquet's expression, but considering the impedance of the polymer itself (4). The results in Figs. 4 and 5 also show that the response of the piezoceramic element is not influenced by the polymer matching layer phase, while the response of the polymer depends on both the matching layer phases.

From a practical point of view a problem is to realise the composite matching layer with a uniform thickness ($\approx \lambda/4$) and two different materials, that generally means two different propagation velocities; in the computed results we supposed that the velocities in the two matching layer phases are the same. The classical solution to realise matching layers is to use a polymer loaded with some powder, varying the respective proportion in order to obtain the desired acoustic impedance. The propagation velocity of the obtained material accordingly varies. In literature there are some papers in which are reported the material mechanical properties in function of the components proportion: for example, in [8] it can be seen that by using a tungsten–vinyl composite with 92% in vinyl, an impedance of 4.1 Mrayls and a velocity 1650 m/s are obtained. By using a polyurethane polymer (Grace 70010) powdered with alumina, (10% of alumina), an impedance of 2.4 Mrayls and a velocity 1700 m/s are obtained. These data show the possibility to realise a matching layer with two phases of different acoustic impedance, but with the same propagation velocity.

5. EXPERIMENTAL RESULTS

In order to test the proposed matching layer design, we realised two transducers without backing, in order to stress the contribution of the matching layer to the bandwidth. Both transducers were realised with the same components and dimensions that we have used for the simulations; in one transducer the matching layer was realised following to the classical technique, while in the other we realised a composite matching layer, according to the proposed design criterion. According to (1), the uniform matching layer must have a specific acoustic impedance of 4.1 Mrayls; this is obtained by using a tungsten–vinyl composite with 92% in vinyl. The propagation velocity of this material is of 1650 m/s and, therefore, in order to obtain a thickness of the layer of about $\lambda/4$ @ f_s = 828 kHz, we lapped the layer to t_m = 0.5 mm. The composite matching layer was built by using the dice and fill technique: we coated on the radiating face of the transducer the tungsten–vinyl composite (z_{mp} = 4.1 Mrayls) in order to build the matching layer phase in front of the piezoceramic; the layer was then lapped to the thickness of 0.5 mm. In correspondence of the piezo–composite polymer phase the layer was diced by means of a high speed dicing machine, Berney T34; the width of the cut is w_l. The cuts were filled by using a polyurethane polymer (Grace 70010) powdered with the 10% of alumina (z_{mi} = 2.4 Mrayls). As already cited, the velocities of the two component of the matching layer are about equal: 1650 m/s for the tungsten–vinyl composite and 1700 m/s for the polyurethane with alumina; in this way the thickness of 0.5 mm is approximately $\lambda/4$ @ f_s = 828 kHz for both materials. As a first test, in order to compare the two transducer performances, we measured their electrical input impedance loading them with water. Fig. 6 shows the measured input resistance (i.e. the impedance real part). As it can be seen, the curve measured for the transducer with the composite matching layer has two peaks due to the presence of another resonator introduced in the structure by the second phase of the layer; this implies a larger bandwidth. In medical and NDT applications the transducer is generally used to transmit the ultrasonic pulse and to receive the echo reflected from the target; from this point of view the most useful transfer function to analyse the transducer performance is the insertion loss: it is defined as the ratio between the received and the exiting electrical voltages.

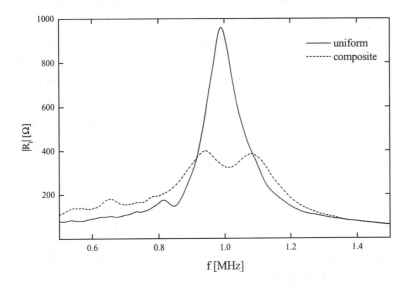

Fig. 6. Measured electrical input resistance of the transducer with the uniform matching layer compared with that of the transducer with the composite matching layer.

Fig. 7 shows the insertion losses measured for the two transducers; as it can be seen, the pulse received from the composite matched to the load with the composite matching layer is appreciably shorter than the pulse received from the classical transducer: the first one can be considered extinguished after about 8 μs, while the other go to regime after about 20 μs. Further also the sensitivity of the proposed transducer is better than the classical one: the peak to peak amplitude of the signal received by the new transducer is about the 40% greater.

6. CONCLUSIONS

In this paper an optimisation work is shown to demonstrate that composite configurations can be also applied to the matching layer in order to improve the efficiency and the band of transducers based on piezoceramic composites. An approximated two–dimensional analytical model has been

514

used to optimise the design of a composite–structured matching layer in the case of 2–2 composites, obtaining different results for the polymer and piezoceramic composite phases. A new design criterion is proposed: the acoustical impedance of the matching layer phase in front of the piezoceramic elements can be computed by applying the well–known Souquet's formula to the mean acoustic impedance of the composite; the impedance in front of the polymer can be computed by applying the same expression to the impedance of the polymer itself. The computational results also show that the response of the piezoceramic element is not influenced by the polymer matching layer phase, while the response of the polymer depends on the two matching materials. With the aim to verify the proposed design criterion, a transducer prototype, based on a 2–2 piezocomposite, with a composite matching layer was realised. On this sample we measured the electrical input impedance and the insertion loss and we compared the obtained results with those of a transducer classically matched to the load. The obtained results confirm the computed improvements in the transducer performance and justify the proposed design approach.

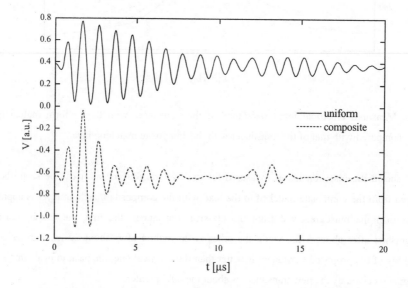

Fig. 7. Measured insertion loss of the transducer with the uniform matching layer compared with that of the transducer with the composite matching layer.

7. REFERENCES

[1] W. A. Smith, "The role of piezocomposites in ultrasonic transducers", *IEEE Ultrasonics Conf. Proc.*, pp. 755–766, 1989.

[2] N. Lamberti, V. Genovese, M. Pappalardo, "A two–dimensional model of the multielement piezoelectric transducer", *IEEE Ultrasonics Conf. Proc.*, pp. 785–789, 1990.

[3] N. Lamberti, F. R. Montero de Espinosa, A. Iula, R. Carotenuto: "Two–Dimensional Modelling of Multifrequency Piezocomposites"; *Ultrasonics,* Vol. 37, Is. 8, pp. 577–583, Jan. 2000.

[4] N. Lamberti and M. Pappalardo, "A General Approximated Two–Dimensional Model for Piezoelectric Array Elements," *IEEE Trans. on Ultrason., Ferroelec. Frequency Contr.* vol. 42, no. 2, pp. 243–252, Mar. 1995.

[5] N. Lamberti, F. R. Montero de Espinosa, A. Iula, R. Carotenuto: "Characterisation of Piezoelectric Ceramics by Means of a Two Dimensional Model"; *IEEE Trans. on Ultrason., Ferroelec. Frequency Contr.*, vol. 48, no. 1, pp. 113–120, Jan 2001.

[6] J. Souquet, P. Defranould and J. Desbois, "Design of Low–Loss, Wide–Band Ultrasonic Transducer for Noninvasive Medical Applications," *IEEE Trans. on Sonics and Ultrasonics,* vol. SU–26, no. 3, pp. 75–81, March 1979.

[7] N. Lamberti and M. Pappalardo, "Frequency Spectra Analysis of Vibrations in Piezoelectric+Matching Layer Plates for Optimum Transducer Design," *Ultrasonics International 89 Conference Proceedings*, pp. 460–465, 1989.

[8] S.Lees, R. Gilmore, and P.Kranz, "Acoustic properties of tungsten–vinyl composites" *IEEE Trans. on Sonics and Ultrasonics*, vol. SU–20, no. 1, pp. 1–2, 1973.

Theoretical and Computational Acoustics 2001
E.-C. Shang, Qihu Li and T. F. Gao (Editors)
© 2002 World Scientific Publishing Co.

Lateral Resolution Enhancement of Ultrasound Images Using Neural Networks

Riccardo Carotenuto*, Gabriele Sabbi, and Massimo Pappalardo
Dipartimento di Ingegneria Elettronica, Università degli Studi Roma Tre
Via della Vasca Navale, 84 - 00146 Roma, Italy

Spatial resolution in modern ultrasound imaging systems is limited by the high cost of large aperture transducer arrays, which means large number of transducer elements and electronic channels. A new technique to enhance the lateral resolution of pulse-echo imaging system is presented. The method attempts to build an image which could be obtained with a transducer array aperture larger than that physically available. We consider two images obtained imaging the same object with two different apertures, the full aperture and a sub aperture, of the same transducer. A suitable artificial neural network (ANN) is trained to reproduce the relationship between the image obtained with the transducer full aperture and the image obtained with a sub aperture. After a suitable training, the network is able to produce images with almost the same resolution of the full aperture transducer, but using a reduced number of real transducer elements. All the computations are carried out on envelope-detected decimated images: the overall computational cost is low and the method is suitable for real time applications. The proposed method was applied on experimental data obtained with the ultrasound synthetic aperture focusing technique (SAFT), giving quite promising results. Real-time implementation on a modern full digital echographic system is currently being developed.

I. INTRODUCTION

The finite aperture and the finite overall bandwidth of the ultrasonic transducer array used severely limit the resolution of the image generated by an echographic B-scan system. Assuming space invariance and linearity, the resolution capabilities of the system can be expressed in terms of the point spread function (PSF), i.e. the image of a point reflector, by the following relation:

$$g(x, z) = h(x, z) ** f(x, z) + n(x, z) \qquad (1)$$

where $f(x, z)$ is the spatial reflectance distribution of internal organs of the human body to be imaged, $g(x, z)$ is the degraded echographic image of the object $f(x, z)$, $h(x, z)$ is the blurring degradation function, i.e. the PSF, which accounts for the finite aperture and bandwidth of the transducer, and finally $n(x, z)$ describes the additive quantization and electronic noise. In the above equation the symbol ** denotes the convolution operator with respect to the variables x and z, representing the lateral and the axial coordinates. The z axis is related to the two way transit time $t = 2z/c$, where c is the average sound velocity in the human body [1-2].

Over the last years several attempts have been made to improve *a posteriori* the image resolution by means of the Fourier-based digitally implemented deconvolution approach [3-7]. This restoration technique applies the inverse process to the image $g(x, z)$ in order to retrieve the original information $f(x, z)$. Two main problems arise with this inverse technique; the first is to have a sufficiently good and reliable knowledge of the PSF, which depends not only on the geometrical characteristics of the transducer array, but also on the tissue under investigation. The second is related to the mathematical operation involved in the Fourier-based deconvolution, which requires

---

* Corresponding author: Tel: +390655177010; fax +39065579078; e-mail: r.carotenuto@uniroma3.it

to perform the ratio between the transformed image and the transformed PSF. Since this operation involves large amplification factors for frequencies where the transformed PSF approaches zero, small deviations from the assumed model can create very large errors and greatly amplify the noise.

In past years, neural networks have been extensively used in image processing. Traditionally, ANNs were applied in image compression or image restoration. Several reported application are direct replacements of conventional image processing techniques such as eigenvalue extraction, vector quantization, 2-D filtering [8-12]. Other approaches treat images as two dimensional surfaces instead of considering it as a series of individual pixels [13]. ANNs were also extensively used for quantitative analysis in medical imaging [14].

In this paper the resolution enhancement problem is approached by considering that there is some relationship between the low resolution image and the high resolution image of the same scene. The aim of this work is to propose an algorithm capable to identify and exploit this relationship in order to build images which could be obtained with a transducer array aperture larger than that physically available The proposed method works in the time domain, and it is nonlinear because it does not work on radio frequency data but directly, after decimation, on the envelope-detected signals. The effectiveness of the proposed nonlinear processing is demonstrated for a typical medical echographic phased array system by means of experimental data, obtaining a significant improvement of the spatial resolution.

II. DIRECT MAPPING FOR RESOLUTION ENHANCEMENT

The luminance of each pixel is a complicated function of many different factors: the reflectance of the corresponding region of the explored field, the attenuation suffered by the acoustic pulse in the round trip, the amplification of the system and the bandwidth of the transducer array, and finally the effective shape assumed by the PSF in the examined region. The PSF depends on many factors including the tissues under investigation, but it mainly depends on the array aperture D, i.e. on the number N of the active elements. On the other hand, the final resolution of the image depends on the PSF shape, i.e. on the number of elements: the higher is the number of active elements, the higher is the image resolution.

The basic idea of the proposed method to enhance the image resolution is to identify and reproduce the underlying non-linear mapping existing between a low resolution image obtained with a small array of N elements and the high resolution image obtained with a larger array of N' (with $N' > N$) elements. We can write

$$g_1(x, z) = h_1(x, z) ** f(x, z)$$

$$g_2(x, z) = h_2(x, z) ** f(x, z) \qquad (2)$$

where g_1 is the low resolution RF image obtained with a wide PSF h_1, and g_2 is the high resolution RF image obtained with a narrow PSF h_2; for simplicity, but without loss of generality, the additive noise is neglected. By using the Fourier Transform we obtain:

$$G_1 = H_1 F$$

$$G_2 = H_2 F \qquad (3)$$

$$G_2 = H_2 H_1^{-1} G_1 = K G_1$$

where K is the desired relationship between the low resolution image and the high resolution image. It is worth noting that K is image independent. If F and G are little sub-regions of the whole image, K can be seen as a local relationship. A direct computation of K starting from the two images is practically unsuitable due to the required inversion of matrix H_1, which is very sensitive to noise.

The proposed method is based on the use of a suitable ANN to reproduce the mapping: during the training phase, the low resolution image is scanned shifting by one pixel a rectangular window of $n_x \times n_z$ pixels, so that the image is completely covered.

The luminance value of the pixels belonging to the window is fed as inputs of the ANN. Each input window is related with the corresponding central pixel of the window in the high resolution image. The luminance value of this pixel is the desired output of the mapping relationship which the ANN is trained to reproduce (see Fig. 1).

After the training is completed, by providing as input a new low resolution image obtained with the same echographic system, the ANN is capable to produce an image which shows a resolution very similar to the image used during the training step.

The obtained mapping takes into account all the deterministic phenomena, such as diffraction, tissue absorption, the effective shape assumed by the PSF in the examined region, depending on the scatterer distribution, and the effects due to the acquisition and measuring chain. Shadowing and PSF shape aberration due to the scatterer pattern can also be represented if these phenomena are included in the training data set.

After the mapping is built, we can use an echographic probe with the same characteristics of that used to generate the high resolution images, but actually having only a subset of the active transmit-receive channels; a relevant hardware cost reduction both for the probe and the transmit-receive electronics can be obtained.

It is worth noting that the relationship between the two images is highly space variant in practice, i. e. the PSFs and the ratio K can significantly vary in different regions of the image. This problem can be overcome by training different ANNs on limited regions of the image, or by including in the input data some information about the current position of the input windows.

III. MAPPING REPRESENTATION WITH NEURAL NETWORKS

A static mapping is a function $f : \Re^n \to \Re$ from the input and the state space to the output space. In this paper, n is the number of the pixels of the input windows, whose luminance is related to the corresponding pixel luminance of the high resolution image.

We need a suitable and compact representation of a given mapping from experimental data gathered from measurements on a real echographic system.

Let us define an input and output range and a quantization level M for the luminance values. This can be reasonably assumed, if we consider that typical experimental values are the output of an Analog to Digital Converter (ADC). Such a quantization implies that the mapping f is considered only within a n-dimensional box consisting of M^n discrete points. A very troublesome problem, often called the curse of dimensionality, arises dealing with every nonparametric techniques for mapping representation: in general, we have a memory exponential complexity in the number of inputs. Even for small M and n, the exponential growth of memory requirements makes this approach impracticable in a direct way [15]. For example, if $M = 256$ and $n = 10$, then 2^{80} memory location should be allocated. Moreover, the acquisition effort of the input-output pairs also has an exponential growth.

At present, many different methods have been reported in literature, capable to compactly represent a given data set, from the classical interpolation functions to the n-dimensional splines [16-19].

Several theorems justify a great variety of computational structures, called artificial neural networks (ANN), which are currently used as mapping approximators (see for example [20-23]).

In this work, we use the Sum Decomposition (SD) net proposed by one of the authors et al. [24-25] and we compare the results obtained with a properly sized Multi-Layered Perceptron (MLP).

The learning procedure, i.e. the weight adjustment process in order to produce the correct input-output association, can be divided in two steps regardless the net used: the weight adjustment and the validation. For this purpose, the available data set, which is composed of input-output pairs,

520

LEARNING

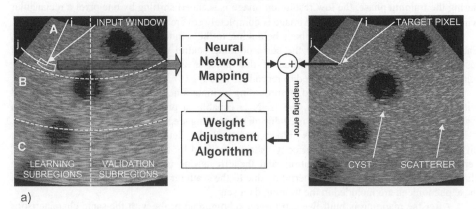

a)

OPERATION

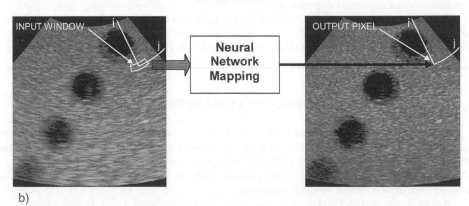

b)

Figure 1. Schematics of the proposed method: a) training on a suitable set of example data: here the same image provides both the learning and the validation data sets. The neural network output is compared to the target pixel of the high resolution image and the error is fed back to the weight adjustment algorithm; b) output generation during normal operation: the neural network generates the output pixel value associated with the input window.

is divided in two subsets. The first subset is fed to the learning algorithm during the weight adjustment. After the adjustment is completed, the second subset, never shown before to the net, is used to verify that the net effectively is capable to represent the mapping with the desired grade of approximation.

IV. EXPERIMENTAL RESULTS

The proposed algorithm has been applied to experimental data set acquired on a cyst phantom at the Biomedical Ultrasonics Laboratory, University of Michigan (http://www.bul.eecs.umich.edu) by using the synthetic aperture focusing technique (SAFT) [26]. The phantom is composed of an uniform parenchyma and some circular cylinders of absorbing material, plus one wire on the left side. Imaged by an echographic system, the absorbing cylinders appear as black circular regions, mimicking cysts in the human body, while the wire appears as a single scatterer white point.

The SAFT echographic data were acquired with a 3.5 MHz center frequency probe, 128 channels, pitch 0.22 mm, sampled at about 13 MHz, 4 bytes per sample, 2048 samples per channel. The data set is then composed of 16,384 sampled RF tracks, spanning a range of about 230 mm in depth. We obtained a phased array scanning image, composed of 200 lines of view, with 90 degree beam deflection, which, after envelope detection, decimation, log-compression, 8 bit quantization, and scan conversion, is a 512 x 512 pixels gray level image. In particular, dynamic focalization both in transmission and in reception was performed.

Figs. 2a and b show the large aperture image obtained with $N = 128$ elements, and the sub-array image obtained with $N = 64$ elements. The 128 elements image will be the target image and the 64 elements image will be the input image to be processed. Fig. 2d shows the normalized difference image between the 128 and the 64 elements images, where the gray levels are proportional to the differences and white pixels correspond to zero difference.

The differences between these images are mainly located in the lateral direction around the single high reflectance scatterer and cysts. Comparing the target image with the sub-array image, it can be seen that the target image shows a wider black region (absorbent cyst) and a narrower white region (reflecting cyst) than the sub-array image.

In order to apply the proposed method, we have to choose the size (i. e. the pixel number) of the input window. The window size is related to the width of the PSF of the system. As a matter of fact, the PSF size changes through the image: the PSF in regions very near or very far of the transducer, and in side zones, is larger than the PSF in central regions, due to diffraction.

The available image was first divided in three horizontal sub regions (see Fig. 1) A, B, and C, in each of them we suppose constant PSF, and for each sub region we compute the best size of the input window. A different mapping is then computed for each subregion.

We have only one image acquired with the same experimental set up, and each subregion was further halved in order to have two data sets: the learning set and the validation set.

The image is scanned with the input window and the values of the pixels belonging to the window are fed as current inputs to the neural network learning process. The net weights are adjusted by the learning rule until the desired error in reproducing the mapping is achieved. Each complete scanning process of the subregion and weight adjusting is called epoch.

The resolution enhancement due to the increased aperture of the transducer slightly affects the axial resolution, and we found that each pixel of the high resolution image is related mainly with pixels along the lateral direction, so that the input window was reduced to a single row of pixels along the lateral direction.

By using the proposed net, the learning phase took 5,000 epochs. After learning is completed, we carried out the validation: the neural network was tested on the validation set, which is composed by the pixels not used in the learning phase.

We applied different input windows to the three axial subregions A, B, and C, whose width was evaluated by a trial and error procedure: for each image region, we computed the validation error as a function of the width of the input windows, and we chose the knee value of the descending curve, which is a good trade-off between low error and computational effort.

With the used echographic system, we found the best compromise between mapping representation accuracy and computational efforts with the size of the input window $w = 13$ for region A, $w = 11$ for region B, and $w = 9$ for region C (see Fig. 1).

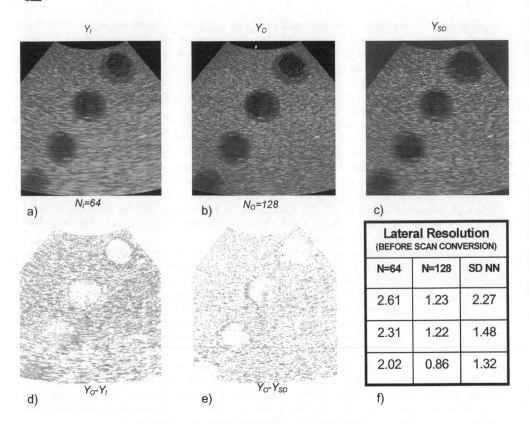

Figure 2. Input and target data sets: a) SAFT echographic image of the synthetic phantom with transducer aperture of 64 elements, dynamic focusing in transmission and reception; b) SAFT echographic image of the synthetic phantom with transducer aperture of 128 elements, dynamic focusing in transmission and reception; c) Sum Decomposition Neural Net computed image; d) difference image between the 64 and the 128 elements images; white pixels correspond to zero error; e) difference image between the 128 and the Sum Decomposition Neural Net computed image; white pixels correspond to zero error; f) lateral resolution Table.

By processing the whole image, the final image is computed (see Fig. 2c). Comparing the computed and the true 128 elements images, the general impression is very satisfying, i.e. the two images appear to be very similar as far as the global sharpness and the correspondence of each detail is concerned. In the computed image the absorbent and the reflecting cysts appear almost as large as in the true 128 elements image. The obtained improvement, in respect to the 64 elements image, is also evident. Fig. 2e shows the difference between the 128 elements image (Fig. 2b) and the neural network image (Fig. 2c). Comparing the difference images of Figs. 2d and 2e it is possible to see that with the computed image the errors around the single high reflectance scatterer and the absorbent cyst are greatly reduced; also for the reflecting cyst there is a significant reduction. The absorbent cysts in Figs. 2b and 2c are practically identical, and the same holds for the single scatterer: only few isolated point errors are visible. In particular, no artifacts in geometry and/or in intensity were generated, even if the processing is nonlinear.

In order to quantify the resolution increase we used a measure of the resolution based on the autocovariance function of sample horizontal lines of the image as defined in [7]. The autocovariance function of the image was normalized to 1.00. The lateral resolution R_l was defined as the width of the autocovariance function at l dB below the peak; we used $l = 6.00$. The computed resolution have been summarized in the Table of Fig. 2 for three different image horizontal lines (100, 323 and 420). Based on the numerical results, we conclude that the proposed algorithm improves the resolution.

The performances of the proposed neural network were compared with the performances of a properly sized standard MLP net. In particular, due to the heavy computational load of the MLP training procedure, we considered only the central region of the image; the input window size, which depends on the PSF width, is the same we used for the application of the proposed method.

For each trial, the learning phase took 100,000 epochs; in addition to the higher computational complexity of the output generation, the computational effort in the learning phase of MLP too is much more higher than the learning effort needed by the proposed net.

In Fig. 3 it can be seen the starting 64 elements aperture image (a), the target 128 elements image (b), the image provided by the proposed method (c), and the computed MLP image (d).

The results of the two neural networks are quite similar, showing that both the network acquired the capability to reproduce the underlying mapping from the low resolution to the high resolution image. However, the proposed network achieved the goal with a substantially lower computational effort than the MLP net, both in the learning phase and in the normal output generation.

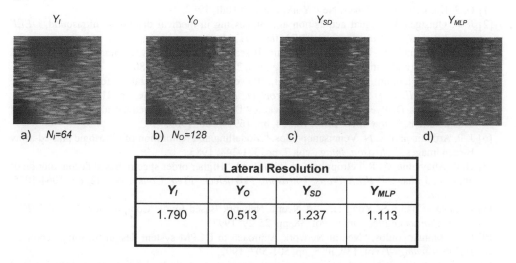

Lateral Resolution			
Y_I	Y_O	Y_{SD}	Y_{MLP}
1.790	0.513	1.237	1.113

Figure 3. Echographic images obtained with different methods: a) SAFT image with an aperture of 64 elements; b) SAFT image with an aperture of 128 elements; c) image computed by using the Sum Decomposition net; d) image computed by using a MLP net

V. CONCLUSIONS

A new method for echographic image resolution enhancement is presented. After a learning phase, in which a mapping between low and high resolution images is built, the method is able to provide images which could be obtained with a larger transducer aperture using information contained both in the image to be enhanced and in the mapping itself.

The ANN used to represent the underlying mapping between the low and the high resolution image shows good computational performances, both in training and in output generation, in comparison with a standard MLP net.

The proposed method show the capability to extract and represent from experimental data a mapping function which is image independent, i.e. it is mainly related to the system used to produce

524

the image, and not to the particular image under investigation, as demonstrated by the validation procedure.

The obtained results also show that the proposed algorithm provides a good resolution enhancement with no geometrical or intensity artifacts, which are indeed a well known problem of many nonlinear algorithms.

The proposed technique requires no hardware modifications of standard echographic systems, and it could be directly applied in modern digital systems; it is readily applicable to linear, convex and phased arrays and can easily be implemented in real-time. The computational cost is low, compared with blind, Fourier-based, deconvolution approaches: only multiplications and sums on real numbers are required. Moreover, contrary to Fourier-based methods, the entire processing is made on envelope-detected and decimated images, further lowering the total amount of data to be processed.

The neural network based resolution enhancement could represent a good trade-off between quality of the image and computational efforts.

Real-time echographic system implementation is currently being developed.

REFERENCES

[1] G. S. Kino, *Acoustic Waves*. New York: Prentice-Hall, 1987.

[2] J. U. Quistgaard, "Signal acquisition and processing in medical diagnostic ultrasound," *IEEE Signal Processing Magazine*, pp. 67-74, Jan. 1997.

[3] M. Fatemi and A. C. Kak, "Ultrasonic B-scan imaging: theory of image formation and technique for restoration," *Ultrason. Imag.*, vol. 2, n° 1, pp. 1-47, 1980.

[4] D. Iracà, L. Landini, and L. Verrazzani, "Power spectrum equalization for ultrasonic image restoration," *IEEE Trans. Ultrason., Ferroelect., Freq. Contr.*, vol. 36, pp. 216-222, Mar. 1989.

[5] T. Loupas, S. D. Pye, and W. N. McDicken, "Deconvolution in medical ultrasonics: practical considerations," *Phys. Med. Biol.*, vol. 34, pp. 1691-1700, 1989.

[6] J. P. Ardouin and A. N. Venetsanopoulos, "Modelling and restoration of ultrasonic phased-array B-scan images," *Ultrason. Imag.*, vol. 7, pp. 321-344, 1985.

[7] U. R. Abeyratne, A. P. Petropulu, and J. M. Reed, "Higher order spectra based deconvolution of ultrasound images," *IEEE Trans. Ultrason., Ferroelect., Freq. Contr.*, vol. 42, pp. 1064-1075, Nov. 1995.

[8] L. Cheng-Chang, S. Yong Ho, "A Neural Network Based Image Compression System," *IEEE Trans. Consumer Electronics*, vol. 38, pp. 25-29, 1992

[9] C. N. Manikopoulos, "Neural Network Approach to DCPM system design for image coding," *IEE Proceedings-I*, vol. 139, n° 5, pp. 501-507, 1992.

[10] G. Cottrell, "Principal components analysis of images via back propagation," *SPIE Visual Communications and Image Processing*, vol. 1001, pp. 1070-1076, 1988.

[11] D. T. Pham, E. J. Bayro-Corrochano, "Neural Computing for Noise Filtering, Edge Detection and Signature Extraction," *J. Systems Engineering*, vol. 2, pp. 111-122, 1992.

[12] J. G. Daugman, "Complete Discrete 2-D Gabor Transforms by Neural Networks for Image Analysis and Compression," *IEEE Trans. Acoust , Speech, Signal Processing*, vol. 36, pp. 1169-1179, 1988.

[13] E. S. Dunstone, "Image Processing using an Image Approximation Neural Network", *Proceedings IEEE ICIP-94*, vol. 3, pp. 912-916, 1994.

[14] C.S. Pattichis, A. G. Constantinides, "Medical Imaging with Neural Networks," *Proc. of the 1994 IEEE Workshop on Neural Networks for Signal Processing*, pp. 431-440, 1994.

[15] C.G. Atkeson and D.J. Reinkensmeyer, "Using Associative Content-Addressable Memories to control robots," in W.T. Miller III, R.S. Sutton & P.J. Werbos (Eds.), Neural Networks for Control. London, England: MIT Press, pp. 255-286, 1990.

[16] C. De Boor, "Bicubic spline interpolation, " *J. Math. Physics*, vol. 41, pp. 212-218, 1962

[17] J. Sard, and S. Weintraub, *A book of splines*. New York: Wiley, 1970.

[18]   K. Stokbro and K. Umberger, "Forecasting with weighted maps," *Proc. Conf. Nonlinear Modeling and Forecasting, SFI Studies in the Sciences of Complexity.* Eds. Casdagli M. & Eubank S., Addison-Wesley, vol. 12, pp. 73-93, 1992.

[19]   A. Bowyer, "Computing Dirichelet tessellations," *The Computer Journal*, vol. 2, pp. 162-166, 1981.

[20]   A. N. Kolmogorov, "On the representation of continuous functions of several variables by superimpositions of continuous of one variable and addition," *Dokl. Akad. Nauk SSSR*, vol. 108, pp. 179-182, 1957 (English transl.: *Amer. Math. Society Translations*, vol. 28, pp. 5-59, 1965).

[21]   D.A. Sprecher, "On the structure of continuous functions of several variables," *Transactions of the Amer. Math Soc.*, vol. 115, pp. 340-355, 1965.

[22]   K.I. Funahashi, "On the approximate realization of continuos mappings by neural networks," *Neural Networks*, vol. 2, pp. 183-192, 1989.

[23]   M. Nørgaard, "Neural network based system identification toolbox," *Technical Report 95-E-773, Inst. of Automation, Technical University of Denmark,* 1995.

[24]   R. Carotenuto, L. Franchina, and M. Coli, "Nonlinear System Process Prediction using Neural Networks," *Proc. of IEEE Int. Conf. on Neural Networks (ICNN'96)*, pp. 184-189, 1996.

[25]   R. Carotenuto, L. Franchina, and M. Coli, "Multidimensional Mapping Representation by Multiple 1-dimensional Decomposition for Complex System Modelling," *Proc. of Int. Conf. on Fractal and Chaos in Chemical Engineering (CFIC'96)*, Rome, Italy, September 2-5, World Scientific, vol. 1, pp. 518-529, 1996.

[26]   M. Karaman, P.-C. Li, and M. O'Donnel, "Synthetic aperture imaging from small scale systems," *IEEE. Trans. Ultrason., Ferroelectr., Freq. Contr.*, vol. 39, pp. 429-442, 1995.

Theoretical and Computational Acoustics 2001
E.-C. Shang, Qihu Li and T. F. Gao (Editors)
© 2002 World Scientific Publishing Co.

A Note on the Common Rule of Using Six Boundary Elements per Wavelength

Steffen Marburg

Institut für Festkörpermechanik, Technische Universität, 01062 Dresden, Germany

E-mail: *marburg@mfm.mw.tu-drescen.de*

1   Introduction

It is widely accepted that the element size in element–based acoustic computations should be related to the wavelength. Often, the element size is measured in a certain (fixed) number of elements per wavelength. In many cases, this number of elements per wavelength is given for constant or linear/bilinear elements. It varies between six and ten. Obviously, this number is closely related to a certain desired accuracy. Often the error is of an acceptable magnitude that depends on the user and that meets certain technical requirements. Examples can be given for finite element analysis [5, 9] and for the boundary element method [5, 6]. The idea of using a fixed number of elements per wavelength is most likely a consequence of SHANNON's sampling theorem. This theorem is of fundamental importance in vibration and acoustics for experimental measurements and frequency detection. It states that at least two points per wavelength (or period of oscillating function) are necessary to detect the corresponding frequency. However, a simple detection cannot be sufficient to approximate a function. SCHMIECHEN [8] investigated discretization of axisymmetric structures for modal analysis. He states that two points per wavelength are strictly sufficient, but would still not lead to accurate mode shapes. Another factor of 3 to 5 is advised. This is equivalent to the number of six to ten nodes per wavelength. We mention that the number of nodes was given for bilinear shell elements. In their paper on application of finite elements, WOJCIK et. al. [10] report computational results with five percent error using nine and two percent error using 18 linear elements per wavelength. It is mentioned that these results were obtained for a scattering problem and quantify the overall error including discretization and approximate radiation boundary conditions.

THOMPSON and PINSKY [9] discussed the topic for a one–dimensional bar investigating the behaviour of finite elements up to a polynomial degree of five. Their recommendation to ensure a certain acceptable accuracy especially when looking at the error in the phase angle was to use at least ten linear elements per wave. Furthermore, instead of using linear elements they found that three quadratic or two cubic elements provide similar error limits as the above mentioned ten linear elements do. However, to increase accuracy THOMPSON and PINSKY recommend the use of even higher polynomial orders while fixing the number of elements per wavelength to two. IHLENBURG's comprehensive study on finite element error analysis [4, Ch. 4] based on a series of joint papers with BABUŠKA starts with the above mentioned common rule in engineering practice. It is shown theoretically and in one–dimensional computational experiments that the rule of at least about six elements per wavelength, roughly equivalent to $kh < 1$, is reliable for finite element approximation at low wavenumbers, but it must be modified for higher ones. Although, this rule is insufficient for high wavenumbers it can also be shown that a rule $k^2 h < 1$ (which was a result from asymptotic mathematical estimates) is too strong. In one of his examples, IHLENBURG presents the number of one–dimensional finite elements to achieve a certain accuracy in the so–called H^1–seminorm. For a wavenumber of $kl = 30\pi$ in this one–dimensional case (15 waves over the whole length l), 14 linear, 3.2 quadratic or 1.7 cubic elements per wavelength are necessary to achieve less than 50% error. 33 linear, 5.1 quadratic or 2.3 cubic elements are required for less than 10% and 188 linear, 12 quadratic or 4.3 cubic elements for an error of about 1%. This H^1–seminorm is a stronger measure than the more commonly – at least in engineering – used L^2–norm (mean square of error or vector norm). For higher order polynomials, we refer to IHLENBURG [4, pp. 155–157]. More generally, he supplied the rule $kl\,(kh/p)^p < 1$ for polynomials of order $p > 1$. This implies a number of three quadratic or two cubic elements per wavelength to obtain reliable solutions. Comparison of the computational costs on this one–dimensional example show a minimum expenditure for quadratic and cubic elements, obviously depending on the desired accuracy.

Returning to boundary element methods, we find a number of articles incorporating rules that stem from finite element methods. So we notice the application of the six–elements– rule in [5] for linear elements and even for constant elements in [6]. MAKAROV and OCHMANN [6] indicate that they have checked the six–elements–rule for constant elements as well and they obtain an acceptable accuracy on their examples on wave scattering for wavenumbers $kr = 20\pi$ where r was the radius of a spherical obstacle. ZALESKI [11] investigated discretization requirements for sound radiation at complicated structures. He confirmed the rule of six linear boundary elements per wavelength to achieve an acceptable accuracy in the sound pressure for lower wavenumbers. Furthermore, a number of more than two and less than three quadratic elements is recommended to ensure a similar accuracy. All these papers did not explicitly mention, that the rule of a fixed number of elements per wavelength may be invalid for high wavenumbers. Error analysis on boundary element methods is published in several different papers, see for example HSIAO and KLEINMANN [3]. However, in most cases they do not provide an explicit rule for the mesh size. In his dissertation, GIEBERMANN [2] shows that even for high wavenumbers the BEM solution obtained by GALERKIN discretization has an error of order kh. This validates a rule of a fixed number of elements per wavelength independent of the frequency for GALERKIN–BEM. In this paper, we consider boundary element collocation method for constant, linear and quadratic elements. Investigations are mainly focused on computational examples.

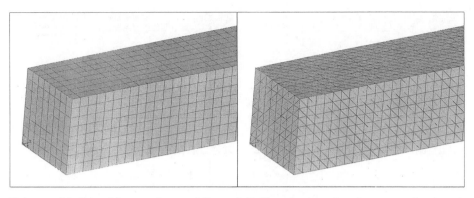

Figure 1: Meshes of duct–ends, quadrilateral (left) and triangular elements, edge–length of elements: 0.025 m

2 Regular mesh of long duct

We consider the propagation of sound in an air–filled duct of length $l = 3.4$ m and a 0.2m×0.2m square cross section. The material data of air being $\rho = 1.3 \, \mathrm{kg/m^3}$ and $c = 340 \, \mathrm{m/s}$, we can expect one wave over the length l at $f = 100 \, \mathrm{Hz}$. We assume $Y = 0$ on the entire surface with the exception of $Y(l) = (\rho c)^{-1}$. Furthermore, $v_s = 0$ except for $v_s(0) = 1 \, \mathrm{m/s}$. The exact solution of the corresponding one–dimensional problem is given by

$$p(x) = -v_s(0) \, \rho c \, e^{ikx} \ . \tag{1}$$

The sound pressure magnitude is constant in the tube and over the entire frequency range. The solution may be considered as waves traveling through the duct. The boundary condition at $x = l$ ensures that the wave is fully absorbed. We want to investigate triangular and quadrilateral elements. Starting with a mesh of 17 quadrilateral boundary elements over the length and one element over the width we obtain 70 square elements of a edge–length of 0.2 m. The successive refinement steps involve meshes of 280, 1120 or 4480 square elements, respectively. To create the corresponding triangular meshes, each square element is divided into two triangles. It seems useful to explain briefly the error measures that will be employed in the following. For a vector $\boldsymbol{f}$, $\{f_1, f_2, \ldots, f_n\}$, we use the Euclidean or the maximum norm

$$\|\boldsymbol{f}\|_2 = \sqrt{\sum_i f_i^2} \ , \qquad \|\boldsymbol{f}\|_\infty = \max_i |f_i| \tag{2}$$

respectively. The relative errors of complex valued results are computed for magnitude. We will denote the relative error for the sound pressure magnitude e_2^Γ and e_∞^Γ for surface points and e_2^Ω and e_∞^Ω for internal points. The error is normalized with the exact solution

530

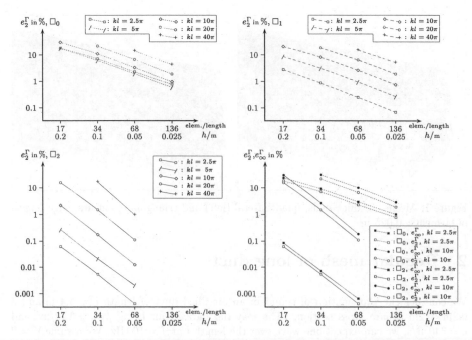

Figure 2: Convergence for selected wavenumbers, sound pressure magnitude error at the surface in Euclidean norm, for constant (upper left), bilinear (upper right) and biquadratic elements (lower left), comparison of Euclidean and maximum norm for constant and for biquadratic elements (lower right)

for example

$$e_2^\Gamma = \frac{\|p_{num}^\Gamma - p_{ex}^\Gamma\|_2}{\|p_{ex}^\Gamma\|_2}. \tag{3}$$

Comparison of results for the sound pressure at the surface using triangular and quadrilateral elements did not allow recognition of differences. Figure 2 gives an impression of the convergence behaviour with respect to the element size h while Figure 3 supplies information about the error dependence in terms of wavenumber k. With respect to both figures, we suggest the formulas like

$$e_2^\Gamma(k) \sim C_k(h,\ldots)\,k^{\alpha_2} \qquad \text{and} \qquad e_2^\Gamma(h) \sim C_h(k,\ldots)\,h^{\beta_2}. \tag{4}$$

The empirical values of α and β from our computational example can be found in Tables 1 and 2. While we can determine $\alpha = p+1$ (polynomial degree p) over the entire frequency range being $0 \leq kl \leq 80\pi$ we observe a slightly different behaviour for β. As can be found in the book by ATKINSON [1], we expect $\beta = p+1$. In particular, for regular meshes and even degree of polynomials we may expect $\beta = p+2$. Summarizing we can write the error

Table 1: Exponents α of equation (4) for different p and h

p	h=0.2m		h=0.1m		h=0.05m		h=0.025m	
	α_2	α_∞	α_2	α_∞	α_2	α_∞	α_2	α_∞
0	0.38	0.33	0.92	0.88	0.89	0.90	0.83	0.84
1	0.64	0.66	1.29	1.45	1.37	1.50	1.41	1.51
2	2.94	3.05	3.07	3.19	2.87	2.80	—	—

Table 2: Exponents β of equation (4) for different p and kl

p	$kl = 2.5\pi$		$kl = 5\pi$		$kl = 10\pi$		$kl = 20\pi$		$kl = 40\pi$	
	β_2	β_∞	β_2	β_∞	β_2	β_∞	β_2	β_∞	β_2	β_∞
0	1.56	1.56	1.70	1.71	1.78	1.78	1.82	1.82	1.73	1.49
1	1.87	1.86	1.83	1.90	1.80	1.90	1.76	1.88	1.58	1.67
2	3.70	3.47	3.36	3.07	3.83	3.99	3.76	3.85	4.14	4.87

dependence as

$$
\begin{array}{lll}
p = 0 : & \text{on arbitrary mesh:} & e^\Gamma(k,h) \sim C\,k\,h \\
 & \text{on regular mesh:} & e^\Gamma(k,h) \sim C\,k\,h^2 \\
p = 1 : & \text{on arbitrary mesh:} & e^\Gamma(k,h) \sim C\,k^2\,h^2 \\
p = 2 : & \text{on arbitrary mesh:} & e^\Gamma(k,h) \sim C\,k^3\,h^3 \\
 & \text{on regular mesh:} & e^\Gamma(k,h) \sim C\,k^3\,h^4
\end{array}
\tag{5}
$$

The above scheme may be extended for higher polynomial degrees. In general, we can suggest the use of elements of an even polynomial degree. Tables 3 and 4 give a survey of the number of elements per wavelength being required to remain below a certain error in the Euclidean norm and in the infinity norm, respectively. We further obtain the maximum frequency for certain numerical errors. Note, that the error functions appear as waves. For that reason, the number of constant or linear element for low error values is somewhat difficult to estimate. Concerning more detailed discussion of this example we refer to [7].

3 Conclusions and future investigations

It could be seen in the previous section as well as in [7] that the number of boundary elements being necessary for numeric calculations depends on the error measure and on the desired accuracy. A low error of less than five percent in each of the investigated error measures can be most efficiently achieved using biquadratic elements, three to four of them

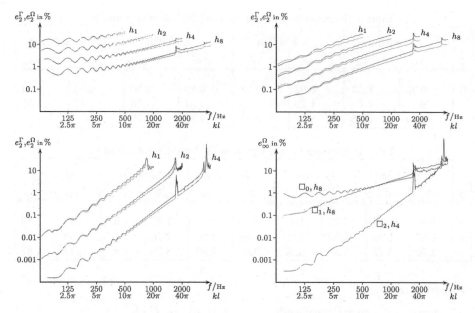

Figure 3: Convergence for different meshes ($h_1 = 0.2m$, $h_2 = 0.1m$, $h_4 = 0.05m$, $h_8 = 0.025m$), error of sound pressure magnitude at internal points (solid lines) and at the surface (dotted lines) in Euclidean norm, for constant (upper left), bilinear (upper right) and biquadratic elements (lower left), comparison of finest meshes for different interpolation degree (lower right)

per wavelength. If a greater error of about ten percent is allowed and higher frequencies are considered, constant elements seem to perform best. For ten percent error in the Euclidean norm of the sound pressure magnitude at the surface and at internal points, five constant elements are enough. In many engineering applications, it is important to remain below a certain maximum error. In that case, we refer to the error in the maximum norm. Accepting ten percent error in this norm, at least seven or eight constant elements should be used. Even more linear elements are necessary.

Returning to the common rule of six linear elements per wavelength, we resume that for the considered example this rule accepts an error of about 8–10% in the Euclidean norm and 15–20% in the maximum norm. The use of constant elements is recommended prior to bilinear elements. Orthogonal base functions and a well conditioned system matrix are arguments for the employment of constant elements. On a uniform mesh, the convergence rates for constant and bilinear elements are equal due to the superconvergence effect for even order of approximation. For irregular meshes, this effect will vanish. The convergence behaviour of constant and bilinear elements is different with respect to wavenumber k.

Table 3: Long duct, number of elements being required to remain below a certain error e_2^Γ of the sound pressure at the surface (* indicates estimates)

p	edge–length h	f_{max}/Hz and elements per wavelength for e_∞^Γ									
		< 1%		< 3%		< 5%		< 10%		< 20%	
		f_{max}	el/λ	f_{max}	el/λ	f_{max}	el/λ	f_{max}	el/λ	f_{max}	el/λ
0	0.20	—	—	7	(243)	10	(170)	15	(113)	180	9.4
	0.10	7	(486)	13	(262)	20	(170)	375	9.1	880	3.9
	0.05	13	(523)	320	21	680	10	1380	4.9	2600*	2.6
	0.025	375	36	1470	9.3	2325	5.8	4550*	3.0	9000*	1.5
min El./λ		> 30		> 8		> 5		3 … 5		1.5 … 3	
1	0.20	12	(142)	130	13	165	10.3	265	6.4	465	3.7
	0.10	135	25	250	14	350	9.7	590	5.8	1060	3.2
	0.05	255	27	550	12	810	8.4	1450	4.7	2410*	2.8
	0.025	630	22	1045	13	1850	7.4	3550*	3.8	6200*	2.2
min El./λ		> 20		≈ 13		7 … 10		4 … 6		2 … 3	
2	0.20	370	4.6	525	3.2	625	2.7	735	2.3	820	2.1
	0.10	865	3.9	1225	2.8	1420	2.4	1620	2.1	2320*	1.5
	0.05	2010	3.4	2510	2.7	3165	2.1	3550*	1.9	4500*	1.5
min El./λ		3 … 4		2.5 … 3		2 … 2.5		≈ 2		≈ 1.5	

Table 4: Long duct, number of elements being required to remain below a certain error e_∞^Γ of the sound pressure at the surface (* indicates estimates)

p	edge–length h	f_{max}/Hz and elements per wavelength for e_∞^Γ									
		< 1%		< 3%		< 5%		< 10%		< 20%	
		f_{max}	el/λ	f_{max}	el/λ	f_{max}	el/λ	f_{max}	el/λ	f_{max}	el/λ
0	0.20	—	—	6	(243)	8	(212)	12	(142)	25	(68)
	0.10	6	(567)	10	(262)	15	(227)	175	(19)	575	5.9
	0.05	10	(680)	120	(57)	430	16	970	7.0	1640	4.1
	0.025	175	78	975	14	1425	9.5	1975	6.9	3525*	3.9
min El./λ		> 50		> 10		> 8		≈ 7		≈ 4	
1	0.20	10	(170)	20	(85)	110	15	190	8.9	290	5.9
	0.10	20	(170)	150	23	205	17	335	10.1	515	6.6
	0.05	180	38	345	20	485	14	750	9.1	1210	5.6
	0.025	390	35	805	17	1135	12	1650	8.2	2625	5.2
min El./λ		> 30		> 15		> 10		≈ 8		5 … 6	
2	0.20	320	5.3	455	3.7	520	3.3	655	2.6	765	2.2
	0.10	725	4.7	1020	3.3	1215	2.8	1465	2.3	1630	2.1
	0.05	1625	4.2	2190	3.1	2525	2.7	3010	2.3	3325	2.0
min El./λ		> 4		3 … 3.5		2.5 … 3		≈ 2.3		≈ 2	

It is known that, under certain circumstances, discontinuous boundary elements show superconvergence effects for collocation methods. This effect will be studied more detailed in the future.

References

[1] K. E. Atkinson. *The Numerical Solution of Integral Equations of the Second Kind.* Cambridge University Press, 1st edition, 1997.

[2] K. Giebermann. *Schnelle Summationsverfahren zur numerischen Lösung von Integralgleichungen für Streuprobleme im R^3.* Dissertation, Universität Karlsruhe, 1997.

[3] G. C. Hsiao and R. E. Kleinmann. Error analysis in numerical solutions of acoustic integral equations. *International Journal for Numerical Methods in Engineering,* 37:2921–2933, 1994.

[4] F. Ihlenburg. *Finite Element Analysis of Acoustic Scattering,* volume 132 of *Applied Mathematical Sciences.* Springer Verlag, Berlin Heidelberg New York, 1998.

[5] LMS Numerical Technologies, Leuven. *SYSNOISE User's Manual, Rev. 5.4,* 1997.

[6] S. N. Makarov and M. Ochmann. An iterative solver for the Helmholtz integral equation for high frequency scattering. *Journal of the Acoustical Society of America,* 103(2):742–750, 1998.

[7] S. Marburg. Six elements per wavelength. Is that enough? *Journal of Computational Acoustics,* 2000. accepted for publication.

[8] P. Schmiechen. *Travelling Wave Speed Coincidence.* Phd–thesis, Imperial College of Science, Technology and Medicine, University of London, 1997.

[9] L. L. Thompson and P. M. Pinsky. Complex wavenumber Fourier analysis of the p–version finite element method. *Computational Mechanics,* 13:255–275, 1994.

[10] G. L. Wojcik, D. K. Vaughan, J. Mould Jr., F. Leon, Q.-D. Qian, and M. A. Lutz. Laser alignment modeling using rigorous numerical simulations. In *Optical/Laser Microlithography IV, Proceeding of the Symposium on Microlithography,* volume 1463, Bellingham, Washington, 1991. Society of Photo–Optical Instrumentation Engineers. (preprint available in the WWW at http://www.wai.com/AppliedScience/Software/Emflex/Papers/emflex-spie91.html).

[11] O. Zaleski. *Anforderungen an die Diskretisierung bei der Schallabstrahlungsberechnung.* Diploma thesis, Technische Universität Hamburg–Harburg, Arbeitsbereich Meerestechnik II – Mechanik, 8 1998.

Theoretical and Computational Acoustics 2001
E.-C. Shang, Qihu Li and T. F. Gao (Editors)
© 2002 World Scientific Publishing Co.

Adaptive Multigrid for Helmholtz Problems

Guido Bartsch*, Christian Wulf[†]
Institute of Technical Acoustics, Aachen University of Technology, Germany

1   Introduction

Acoustic simulations for interior problems traditionally have been done for the design of large rooms, e.g. concert halls, auditories and churches. Today, smaller rooms (cars, studios, etc.) and their acoustic behavior attract our attention. When scaling down the dimension of the enclosure to be computed, geometrical acoustics becomes less applicable since wave effects have to be taken into account. For a comprehensive description of the acoustic field within the room, a broadband sound field simulation is necessary. The problem here is the extreme requirement on computational resources (cpu, memory) and the weak convergence of standard iteration schemes for high wave numbers.

To solve the computational problem the application of problem-adapted data-structures and algorithms is highly recommended [1]. In our solver WaveSolve^{3D} we combine sparse-matrix operations employing the structure of finite element matrices, multigrid algorithms for fast convergence rates and adaptive grid refinements to minimize the global and local error. The need for a finite element calculation is finally limited by the frequency where the mode density is sufficient for the use of particle based models. But this limit hasn't been reached in room acoustic up to now, due to the lack of

---

*Guido.Bartsch@akustik.rwth-aachen.de, www.akustik.rwth-aachen.de/~bartsch
[†]Christian.Wulf@akustik.rwth-aachen.de

adequate algorithms. On the other hand, the limitation has one interesting aspect for the work: If it is sufficient to simulate the sound field below one certain frequency limit, the truncated modal basis approach is a very efficient way to solve the problem.

Once having a modal basis, one can easily make parametric studies for different boundary conditions and different source positions. This is due to the fact, that the projection of the original Helmholtz equation onto a truncated modal basis results in small systems of equations which can be solved quickly.

The dimension of these systems depends on the number of modes to be taken into account. But the effort to compute this modal basis is the most time consuming step in the whole calculation. For the final simulation result each mode has to be weighted and superimposed to the other ones. It should be noted that every mode contributes to the sound field for all possible frequencies. So, the simulation quality depends on the number of modes to be calculated. On the other hand, the influence of a mode decreases with increasing distance between its eigenfrequency and the considered frequency of interest. This discussion is not in the focus of this paper. Our task was to compute the m first eigenmodes numerically within given error bounds.

2   The modal basis approach

Interior acoustics simulation based on wave theory means to find solutions of the boundary value problem [8]:

$$\triangle \hat{p} + k^2 \hat{p} = 0 \quad \text{in} \quad \Omega \tag{1}$$

$$\frac{\partial \hat{p}}{\partial n} + i\,k\,\rho_0\,c\,\hat{v}_n = 0 \quad \text{on} \quad \Gamma_1 \tag{2}$$

$$\frac{\partial \hat{p}}{\partial n} + i\,k\,\rho_0\,c\,A_n\,\hat{p} = 0 \quad \text{on} \quad \Gamma_2 \tag{3}$$

$$\text{and} \quad \int_\Omega \hat{p}\,d\Omega = 0 \tag{4}$$

were k is the wave number as a parameter, Ω is the field-domain and $\Gamma = \Gamma_1 + \Gamma_2$ its boundary. A_n is the wall-admittance and $\hat{v}_n$ is the normal velocity as the driving force. The first step in the modal basis approach for the solution of this boundary value problem is to solve the eigenvalue problem

$$\triangle \hat{p} + \tilde{k}^2 \hat{p} = 0 \quad \text{in} \quad \Omega\,, \tag{5}$$

$$\frac{\partial \hat{p}}{\partial n} = 0 \quad \text{on} \quad \Gamma \tag{6}$$

which we can derive from (1) to (3) in the absence of any driving force $(\hat{v}_n = 0)$ and in the case of rigid walls $(A_n = 0)$. Denoting $H^1(\Omega)$ as the Sobolev-space of continuous and piecewise differentiable functions, then an equivalent weak formulation of this eigenvalue problem reads [5]:

Find $u_i \in H^1(\Omega) \setminus \{0\}$ and numbers $\lambda_i \in \mathbf{R}$, such that

$$a(u_i, v) = \lambda_i (u_i, v)_0 \quad \forall v \in H^1(\Omega). \tag{7}$$

Here,

$$(u, v)_0 := \int_{\Omega} u \, v \, d\Omega \tag{8}$$

denotes the $L_2(\Omega)$ inner product and the constant $\mu > 0$ forces

$$a(u, v) := \int_{\Omega} \nabla u \cdot \nabla v \, d\Omega + \mu (u, v)_0, \tag{9}$$

to be a $H^1(\Omega)$-elliptic bilinear form [3, 6]. On the orthogonality constraints $(u_i, u_j)_0 = \delta_{ij}$ the eigenfunctions form an orthonormal basis in $L_2(\Omega)$ [5]. Thus, we can construct a m-dimensional subspace $U_m(\Omega) \subset H^1(\Omega)$ of complex valued functions with vanishing mean average which is spanned by the eigenfunctions $\{u_i \mid i = 2, \ldots, m+1\}$. The eigenfunctions u_i coincide with the undamped modes defined by (5) and (6), therefore we call $U_m(\Omega)$ modal space and $\{u_i \mid i = 2, \ldots, m+1\}$ the corresponding modal basis.

To obtain an approximate solution of the Helmholtz problem (1) to (4), we transform it into its variational formulation:

$$\int_{\Omega} \nabla \hat{p}_m \cdot \nabla q \, d\Omega + ik\rho_0 c \int_{\Gamma_2} A_n \hat{p}_m \, q \, d\Gamma_2 - k^2 \int_{\Omega} \hat{p}_m \, q \, d\Omega$$

$$= -ik\rho_0 c \int_{\Gamma_1} \hat{v}_n q \, d\Gamma_1 \quad \forall q \in U_m(\Omega) \tag{10}$$

Making the ansatz $\hat{p}_m = \sum_{i=1}^{m} \beta_i u_{i+1}$ for the approximate solution and choosing $q = u_{j+1}$, $j = 1 \ldots m$, equation (10) reduces to a m-dimensional system of linear equations:

$$(S + i\,k\,C - k^2 M)\,x = -i\,k\,f \quad . \tag{11}$$

where S is the so called stiffness matrix, C the damping matrix, M the mass matrix, x the coefficient vector and f the acoustic force vector. Due to the orthogonality of the eigenfunctions, the approximation error $\epsilon_m = \hat{p} - \hat{p}_m$

takes its minimum in the mean square. Thus, the quality of our sound field calculation only depends on the number of eigenmodes building the modal space.

We treat the number of desired eigenmodes as a fixed number and try to compute these eigensolutions by a finite element method within given error bounds. Therefore we construct a sequence of finite element spaces

$$S^k \subset S^{k+1} \subset \ldots \subset H^1(\Omega) \tag{12}$$

until the approximate solutions of (7) reach the desired accuracy.

3 Hierarchical error estimation

We measure the discretization errors by estimating the distance between the true and approximated eigenvalues. Since we cannot compare our approximations with the unknown true solutions, we replace the latter ones by "better" approximations. In our case, we solve the eigenvalue problem (7) in finite element spaces $\mathcal{S}_L \subset H^1(\Omega)$ spanned by piecewise defined polynomials of linear order. To estimate the discretization error, we compare these solutions with those from an approximation of quadratic order. We construct these higher order finite element spaces $\mathcal{S}_Q \supset \mathcal{S}_L$ by a hierarchical extension with piecewise quadratic polynomials associated to the midnodes of edges in the underlying triangulation abbreviated as

$$\mathcal{S}_Q = \mathcal{S}_L \oplus \mathcal{Q}. \tag{13}$$

If we denote $\{\varphi_{L,i} \mid i = 1, \ldots, n_L\}$ as the FE-basis of $\mathcal{S}_L$ and $\{\varphi_{Q,i} \mid i = 1, \ldots, n_Q\}$ as the FE-basis of $\mathcal{Q}$, the corresponding discrete eigenvalue problem can be written in block form:

$$\begin{pmatrix} A_{LL} & A_{LQ} \\ A_{QL} & A_{QQ} \end{pmatrix} \begin{pmatrix} v_{L,i} \\ v_{Q,i} \end{pmatrix} = \lambda_{S_Q,i} \begin{pmatrix} M_{LL} & M_{LQ} \\ M_{QL} & M_{QQ} \end{pmatrix} \begin{pmatrix} v_{L,i} \\ v_{Q,i} \end{pmatrix}, \tag{14}$$

where the entries of the submatrices are defined by

$$(A_{LL})_{ij} = a(\varphi_{L,i}, \varphi_{L,j}) \quad (M_{LL})_{ij} = (\varphi_{L,i}, \varphi_{L,j})_0 \tag{15}$$

$$(A_{QQ})_{ij} = a(\varphi_{Q,i}, \varphi_{Q,j}) \quad (M_{QQ})_{ij} = (\varphi_{Q,i}, \varphi_{Q,j})_0 \tag{16}$$

$$(A_{LQ})_{ij} = a(\varphi_{L,i}, \varphi_{Q,j}) \quad (M_{LQ})_{ij} = (\varphi_{L,i}, \varphi_{Q,j})_0 \tag{17}$$

$$A_{QL} = (A_{LQ})^T \qquad M_{QL} = (M_{LQ})^T. \tag{18}$$

Now we assume that $\lambda_{S_Q,i}$ is closer to the real eigenvalues λ_i than $\lambda_{S_L,i}$ obtained from the approximation of linear order

$$A_{LL} v_{L,i} = \lambda_{S_L,i} M_{LL} v_{L,i} \tag{19}$$

That means we have a constant $\beta < 1$ such that

$$\lambda_{S_Q,i} - \lambda_i \quad \leq \quad \beta(\lambda_{S_L,i} - \lambda_i) \tag{20}$$

holds [9]. Let $\tilde{\lambda}_i$ be an approximation of $\lambda_{S_L,i}$ as a result of an iterative solution method for the discrete eigenvalue problem (19). With the relation

$$\lambda_i \leq \lambda_{S_Q,i} \leq \lambda_{S_L,i} \leq \tilde{\lambda}_i \tag{21}$$

as a consequence of the minimum principle of the Rayleigh-Quotient [5], the assumption (20) is equivalent to the following expression:

$$\tilde{\lambda}_i - \lambda_i \leq \frac{1}{1-\beta}(\tilde{\lambda}_i - \lambda_{S_Q,i}) \tag{22}$$

Therefore we can replace the estimation of the total error $\tilde{\lambda}_i - \lambda_i$ by the estimation of the iteration error $\tilde{\lambda}_i - \lambda_{S_Q,i}$. The derivation of an estimate for the latter one will now be described briefly.

If B_{LL} is a preconditioner for the matrix A_{LL} and D_{QQ} just the diagonal part of A_{QQ}, then

$$B := \begin{pmatrix} B_{LL} & 0 \\ 0 & D_{QQ} \end{pmatrix} \tag{23}$$

is a preconditioner for the matrix

$$A := \begin{pmatrix} A_{LL} & A_{LQ} \\ A_{QL} & A_{QQ} \end{pmatrix}, \tag{24}$$

which means that there are constants $\gamma_1 < 1$ and $\gamma_2 < 1$ such that

$$(1 - \gamma_1)(v, Bv) \leq (v, Av) \leq (1 + \gamma_2)(v, Bv) \tag{25}$$

holds for any vector v [2].

If $\tilde{v}_i$ is the eigenvector approximation corresponding to $\tilde{\lambda}_i$, we use $(\tilde{v}_i \quad 0)^T$ as initial guess for the higher order eigenvalue problem (14). Then the residual is defined by

$$r_i := \begin{pmatrix} r_{L,i} \\ r_{Q,i} \end{pmatrix} = \begin{pmatrix} A_{LL}\tilde{v}_i - \lambda_i k B_{LL}\tilde{v}_i \\ A_{LQ}\tilde{v}_i - \lambda_i k B_{LQ}\tilde{v}_i \end{pmatrix} \tag{26}$$

If $\tilde{\lambda}_i$ fulfills

$$\lambda_i \leq \tilde{\lambda}_i \leq \lambda_{i+1} \tag{27}$$

one can prove the following estimate [9]:

$$\tilde{\lambda}_i - \lambda_{\mathcal{S}_Q,i} \leq \frac{\lambda_{i+1}}{\lambda_{i+1} - \tilde{\lambda}_i} \frac{(r_i, B^{-1} r_i)}{(1 - \gamma_1)} \tag{28}$$

Now employing the block structure of B, we can write

$$(r_i, B^{-1} r_i) = (r_{L,i}, B_{LL}^{-1} r_{L,i}) + (r_{Q,i}, D_{QQ}^{-1} r_{Q,i}), \tag{29}$$

where the first part on the right side is the iteration error resulting from the solution procedure for (19) and the second part an indicator for the discretization error. Since the components of the latter one are completely decoupled, we can view them as the distribution of the error over the finite element mesh. According to this distribution we construct a new finite element space through adaptive mesh refinements.

4   Multigrid preconditioned subspace iteration

For the numerical solution of (19) we use a preconditioned subspace iteration as described in [4, 9]:

$$\tilde{v}_i^{k+1} = v_i^k - B_{LL}^{-1}(A_{LL} v_i^k - \lambda_i^k M_{LL} v_i^k) \tag{30}$$

Each defect correction step (30) is followed by a Rayleigh-Ritz procedure $\tilde{V}^{k+1} \rightarrow V^{k+1}$ to orthonormalize the columns of $\tilde{V}^{k+1} := \{\tilde{v}_1^{k+1} \ldots \tilde{v}_m^{k+1}\}$ with respect to the M_{LL}-inner product.

This iteration converges without any scaling, if the preconditioner fulfills

$$(1 - \gamma_1)(v, B_{LL} v) \leq (v, A_{LL} v) \leq (1 + \gamma_2)(v, B_{LL} v) \tag{31}$$

for some fixed number $\gamma = \gamma_1 = \gamma_2 < 1$. For the convergence rate one can derive estimations [9] of the form

$$\lambda_i^{k+1} \leq \lambda_i^k - (1 - \gamma)^2 \frac{(\lambda_i^k - \lambda_{L,i})(\lambda_{L,i+1} - \lambda_i^k)}{\lambda_{L,i+1}}. \tag{32}$$

To achieve a fast reduction of the iteration error it is desirable to have a preconditioner with a small number γ. Multigrid preconditioners guarantee this property independently of the dimension of A_{LL} and thus independently of the grid refinement depth [2, 7]. Due to our adaptive mesh refinements it is very easy to construct a multigrid preconditioner since we already have discretizations of our eigenvalue problem on different grid levels.

The iteration (30) is stopped, if the iteration errors

$$\epsilon_{L,i} := (r_{L,i}, B_{LL}^{-1} r_{L,i}) \qquad (33)$$

for all eigenvalues are negligibly small compared to the discretization errors

$$\epsilon_{Q,i} := (r_{Q,i}, D_{QQ}^{-1} r_{Q,i}) \qquad (34)$$

which occurred on the previous level of refinement. Thus, only a small number of iterations have to be performed on each grid level to keep a proper balance between iteration and discretization errors. This makes the resulting adaptive multigrid algorithm very effective.

5   Numerical results

As a simple test case, we have chosen two coupled rooms as shown in Fig. 1.

Figure 1: Coupled rooms: Geometry

To compare the quality of the our adaptive refined result, we first built a uniform mesh consisting of 56681 nodes. Afterwards we calculated the modal basis corresponding to the lowest 150 eigenvalues.

The visualize the error, we cut out a horizontal plane in 1m height over the floor. The error itself is shown in Fig. 2 as a topology plot. Here the uprisings correspond to large error components.

Now, starting with coarse mesh, we performed a second simulation with adaptive mesh refinements. The final mesh consisted of 54333 Nodes; the mesh density is shown in Fig. 3.

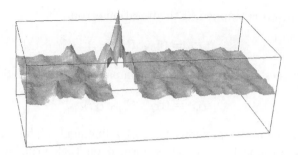

Figure 2: Coupled rooms: Error distribution / Uniform Mesh

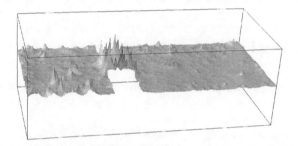

Figure 3: Coupled rooms: Mesh density / Adaptive Mesh

The reader may note, that the mesh density has been increased automatically in those regions with large error components of the former simulation as shown in Fig. 2. This mesh refinement reduces the heightened error components.

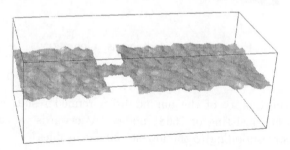

Figure 4: Coupled rooms: Error distribution / Adaptive Mesh

Compared to the uniform mesh result, our adaptive mesh refinements lead to a much more evenly distributed error distribution as shown in Fig. 4. A more quantitative result is shown in Tab. 1 for the largest computed eigenvalue λ_{150}. Although the total error ϵ_{150} in both cases is estimated

	Uniform Mesh	Adaptive Mesh
#Nodes	56681	54333
λ_{150}	15.372	15.341
ϵ_{150}	0.857	0.866
E_{max}	162.672	37.465

Table 1: Mesh-quality comparison

to be approximately the same, the largest error components E_{max} differ significantly[1].

6 Conclusions

When solving Helmholtz problems using the modal basis the number of modes and the accuracy of the modal basis directly influences the solution quality. It is well known that the representation of sound fields by modal basis functions u_i is optimal with respect to the L^2 error norm. So, it is necessary to have a FE basis which minimizes the discretization error when computing the modal basis.

We derived an error estimator for the eigenvalues of the modal basis functions which is based on the summation of error components distributed over the mesh. The error components itself can serve as an indicator where the mesh has to be adaptively refined. We have shown that this kind of adaptive mesh refinement leads to a much more evenly distributed error components in the final result compared to uniform meshes. Therefore static rules like 'six nodes per wavelength' are obsolete.

References

[1] G. Bartsch and A. Franck. Hybrid and parallel extension of the sound field simulation for interior problems. In *Proc. of the 7th International Congress on Sound and Vibration (ICSV)*, Garmisch-Partenkirchen, Germany, 2000.

[2] F. Bornemann, B. Erdmann, and R. Kornhuber. Adaptive multilevel methods in three space dimensions. *Intern. Journ. on Numerical Methods in Engineering*, 36:3187–3203, 1993.

---

[1]E_{max} is normalized, such that its value would be the total error, if all error components take this maximum.

[3] D. Braess. *Finite Elemente.* Springer-Verlag, Berlin, 1992.

[4] J. H. Bramble and A. V. Knyazev. A subspace preconditioning algorithm for eigenvector/eigenvalue computations. *Adv. Comput. Math,* 6:159–189, 1996.

[5] R. Courant and D. Hilbert. *Methoden der mathematischen Physik.* Springer-Verlag, Berlin, Heidelberg, New York, 1993.

[6] W. Hackbusch. *Theorie und Numerik elliptischer Differentialgleichungen.* Teubner, Stuttgart, 1996.

[7] Jung, U. Langer, A. Meyer, W. Queck, and M. Schneider. Multigrid preconditioners and their applications. Third Multigrid Seminar, pp. 11-52, Report R-MATH-03/89, Karl-Weierstrass-Institut für Mathematik, Berlin, 1989.

[8] H. Kuttruff. *Room Acoustics.* Spon Press, London, New York, 4. edition, 2000.

[9] K. Neymeyr. A posteriori error estimation for a preconditioned algorithm to solve elliptic eigenproblems. Technical report, Univ. Tübingen and Stuttgart, 1997.

Theoretical and Computational Acoustics 2001
E.-C. Shang, Qihu Li and T. F. Gao (Editors)
© 2002 World Scientific Publishing Co.

Application of Fast Methods for Acoustic Scattering Problems

Stefan Schneider

Technische Universität Dresden, Institut für Festkörpermechanik, D-01062 Dresden

Abstract

Our work is devoted to the solution of large scale ($kl = 10 \ldots 100\pi$) three dimensional radiation and scattering problems covered by the time harmonic Helmholtz equation. We present an application of the Regular Grid Method and Multilevel Fast Multipole Method to acoustic scattering problems. These methods lead to a memory requirement of $\mathcal{O}(N)$ what enables us to solve exterior scattering or radiation problems with several ten-thousands of unknowns. In two computational examples we show the efficiency of these methods.

1 Introduction

The classical boundary element method leads to a dense (non-symmetric) system matrix. Thus the cost to store it is $\mathcal{O}(N^2)$ if N is the number of unknowns of the problem. Further we solve the arising linear system with iterative solvers. For each matrix vector product in them the computational costs are $\mathcal{O}(N^2)$. For large N the classical boundary element method leads to unacceptable costs in terms of memory requirement and computation time. To overcome these problems a variety of so called fast methods was developed (see for example [1, 2, 3, 4]). All these methods construct a sparse approximation of the system matrix. This leads to $\mathcal{O}(N^p)$ with $p < 2$ memory requirement and number of operation to perform a matrix vector multiplication. The application of these methods to practical problems are very promising for engineers dealing with large problems.

In practical applications there often exists a finite element model of the surface of the scattering object. Because of these models were designed for static or dynamic analysis they consist of more elements that are needed for the discretization of the surface to solve radiation or scattering problems using the boundary element method. On the other hand even if one constructs boundary element models for large objects (i.e. machine housing or aeroplane) in a frequency range of $kl = 10 \ldots 100\pi$ (with the wavenumber k and the characteristic length l of the object) the number of elements needed increases rapidly. To insure an acceptable accuracy the common rule "six linear elements per wavelength" leads to a large number of unknowns ($N = 10^4 \ldots 10^6$). Thus the classical boundary element method is no longer applicable. As for our applications the memory cost is the most limiting quantity for the problem size in consideration we will investigate such fast methods leading to $\mathcal{O}(N)$ or $\mathcal{O}(N \log^p N)$ memory requirement.

Our paper is organized as follows. The second chapter will introduce the reader to the considered problem and the notation used. In the third chapter we give an overview of the Regular Grid Method (RGM) described in [5] and its modification for our problems. The fourth chapter gives a short introduction to the Fast Multipole and the Multilevel Fast Multipole Method (MLFMA). Some numerical results are presented in the fifth chapter.

2 Problem and Notation

We like to solve three dimensional wave propagation problems in fluids covered by the time harmonic Helmholtz equation. Thus the scattered sound pressure u^s has to fulfill

$$\Delta u^s + k^2 u^s = 0 \quad \text{in } \Omega \tag{1}$$

with boundary conditions

$$\frac{\partial u}{\partial n} - aYu \;\; = \;\; av_s \;\; \text{on}\, \Gamma \tag{2}$$

and the Sommerfield radiation condition

$$\lim_{r \to \infty} r\left(\frac{\partial u^s}{\partial r} - i\,ku^s\right) \;\; = \;\; 0 \tag{3}$$

with the wave number k and the total sound pressure u with $u = u^s + u^i$ where u^i represents an "incoming sound pressure". To reduce the dimension of the problem and to fulfill the Sommerfield radiation condition we will use the boundary integral representation of (1)

$$c(\mathbf{y})u^s(\mathbf{y}) + \int_\Gamma \left(\frac{\partial \phi}{\partial n}(\mathbf{x}, \mathbf{y})u^s - \phi(\mathbf{x}, \mathbf{y})\frac{\partial u^s(\mathbf{x})}{\partial n}\right) d\Gamma \;\; = \;\; 0 \tag{4}$$

with the fundamental solution (Green function)

$$\phi(\mathbf{x}, \mathbf{y}) \;\; = \;\; \frac{e^{ik|\mathbf{x}-\mathbf{y}|}}{4\pi|\mathbf{x}-\mathbf{y}|}\,. \tag{5}$$

With the boundary conditions (2) the equation (4) for the total sound pressure u reads as

$$c(\mathbf{y})u(\mathbf{y}) + \int_\Gamma \left(\frac{\partial \phi}{\partial n}(\mathbf{x}, \mathbf{y}) - aY\phi(\mathbf{x}, \mathbf{y})\right) u(\mathbf{x})\, d\Gamma =$$
$$c(\mathbf{y})u^i(\mathbf{y}) + \int_\Gamma \phi(\mathbf{x}, \mathbf{y})\left(av_s - \frac{\partial u^i}{\partial n}\right) d\Gamma + \int_\Gamma \frac{\partial \phi}{\partial n}(\mathbf{x}, \mathbf{y})u^i(\mathbf{x})\, d\Gamma\,. \tag{6}$$

To avoid the problem of irregular frequencies for exterior problems we are using the Burton/Miller formulation (see [6]) with the coupling parameter α chosen as $\alpha = i/k$

$$c(\mathbf{y})u(\mathbf{y})(1 + \alpha aY(y)) + \int_\Gamma \left(\frac{\partial \phi}{\partial n}(\mathbf{x}, \mathbf{y}) - aY\phi(\mathbf{x}, \mathbf{y})\right) u(\mathbf{x})\, d\Gamma +$$
$$\alpha\frac{\partial}{\partial n(y)}\int_\Gamma \left(\frac{\partial \phi}{\partial n}(\mathbf{x}, \mathbf{y}) - aY\phi(\mathbf{x}, \mathbf{y})\right) u(\mathbf{x})\, d\Gamma =$$
$$c(\mathbf{y})(u^i(\mathbf{y}) + \alpha(av_s - \frac{\partial u^i}{\partial n})) + \int_\Gamma \phi(\mathbf{x}, \mathbf{y})\left(av_s - \frac{\partial u^i}{\partial n}\right) d\Gamma +$$
$$\int_\Gamma \frac{\partial \phi}{\partial n}(\mathbf{x}, \mathbf{y})u^i(\mathbf{x})\, d\Gamma +$$
$$\alpha\frac{\partial}{\partial n(y)}\int_\Gamma \phi(\mathbf{x}, \mathbf{y})\left(av_s - \frac{\partial u^i}{\partial n}\right) + \frac{\partial \phi}{\partial n}(\mathbf{x}, \mathbf{y})u^i(\mathbf{x})\, d\Gamma\,. \tag{7}$$

To solve (6) or (7) we will use the collocation method with discontinuous ansatz functions for the sound pressure u thus $u = \sum_{i=1}^N \hat{u}_i \varphi_i$ leading to a linear system $A\hat{\mathbf{u}} = \mathbf{b}$ for the unknown sound pressure $\hat{u}_i$ at the collocation points. In a straight forward implementation one would first calculate and store all entries of the matrix A and second one solves the linear system. This implementation requires $\mathcal{O}(N^2)$ computer memory for the matrix storage and $\mathcal{O}(N^2)$ arithmetic operations to perform one matrix-by-vector multiplication. As we are interested in solving problems with $N > 10\,000$ the arising linear system will be solved using an iterative solver. Thus we only need the product $v = A\hat{\mathbf{u}}$ and not explicitly the entries of A. In the following we will describe two methods for an efficient calculation of $v = A\hat{\mathbf{u}}$.

3 Regular Grid Method

The Regular Grid Method for electromagnetic scattering problems is well described in [5] and as far as possible we will use the same notation as introduced there. For the sake of briefty we will only outline this method and modify it so that it is suitable to solve (6) and (7). The main idea of this method is that since the Green function (5) is defined on $\mathbb{R}^3$ with $\mathbf{x} \neq \mathbf{y}$ we can approximate its value at the boundary Γ by its values on an auxiliary uniform Cartesian grid. The interaction of the grid points will be calculated using the Fast Fourier transform. The method consists of the following 5 stages.

Stage 1: Splitting (5) in two parts

$$\phi(\mathbf{x}, \mathbf{y}) = \phi_1(\mathbf{x}, \mathbf{y}) + \phi_2(\mathbf{x}, \mathbf{y}) \tag{8}$$

where $\phi_1(\mathbf{x}, \mathbf{y})$ is singular and has local support[1] and $\phi_2(\mathbf{x}, \mathbf{y})$ is a smooth and bounded function. Now the matrix A is represented as $A = A_1 + A_2$.

Stage 2: All non-zero entries of A_1 are calculated in the conventional way.

Stage 3: Construction of an auxiliary Cartesian grid $\mathbb{R}_h^3$:

$$\mathbf{x}_{ijk} = (ih, jh, kh), \ i = -\infty, \infty, \ j = -\infty, \infty, \ k = -\infty, \infty \tag{9}$$

Stage 4: Two interpolation formulas are introduced to approximate the Green function and its normal derivative

$$\phi_2(\mathbf{x}, \mathbf{y}) \approx \sum_i c_{ip} \sum_j c_{jq} \phi_2(\mathbf{x}_i^p, \mathbf{y}_j^q) \tag{10}$$

$$\nabla\phi_2(\mathbf{x}, \mathbf{y}) \approx \sum_i c_{ip} \sum_j \mathbf{d}_{jq} \phi_2(\mathbf{x}_i^p, \mathbf{y}_j^q). \tag{11}$$

For the different choices of the interpolation formulas see [5]. We did use plane waves.

Stage 5: The entries of A_2 are calculated using the values of the Green function on the Cartesian grid and any known quadrature formula

$$
\begin{aligned}
a_{2,ij} &= \int_\Gamma \left(\nabla\phi_2(\mathbf{x}, \mathbf{y}_i)\nu(\mathbf{x}) - aY\phi_2(\mathbf{x}, \mathbf{y}_i)\right)\varphi_j(\mathbf{x}) \, d\Gamma \\
&\approx \sum_k c_{ki} \sum_n \gamma_n \left(\sum_l \phi_2(\mathbf{x}_i^k, \mathbf{y}_n^l)\mathbf{d}_{ln}\nu_n - aY\sum_l \phi_2(\mathbf{x}_i^k, \mathbf{y}_n^l)c_{ln}\right)\varphi_{jn} \\
&\approx \sum_{k=1}^Q \sum_{l=1}^P v_{ki} w_{lj} \phi_2(\mathbf{x}^k, \mathbf{y}^l)
\end{aligned}
\tag{12}
$$

with the outwards normal vector ν. To obtain (12) we have done the summation over n and introduced a global numeration of the Cartesian grid points $\mathbf{x}^p$.

The matrix A_2 can now be written in the following form

$$A_2 \approx \tilde{A}_2 = V^T B W, \quad V = (v_{ki}), \ W = (w_{lj}), \ B = (\phi_2(\mathbf{x}^k, \mathbf{y}^l)) \tag{13}$$

and finally the system matrix A takes the form

$$A \approx \tilde{A} = A_1 + V^T B W. \tag{14}$$

Within the iterative solution the product $\tilde{A}\mathbf{u}$ can be realized by a sequential multiplication of the matrices A_1, W, B and V. The appearing product of the matrix B and a vector $\mathbf{v} = W\mathbf{u}$ can be seen as a discrete convolution and can be realized by a Fourier transformation. For more details see [7, 8]. The Fourier transformation was realized using the fftw-package available at *www.fftw.org*.

---

[1] $\phi_1(\mathbf{x}, \mathbf{y}) \neq 0 \Leftrightarrow |\mathbf{x} - \mathbf{y}| \leq R$ where R is a constant depending on the wave number k

4 Fast Multipole Method

Instead of a local approximation of the kernel function as in the RGM the Fast Multipole Method uses two different expansion systems. One represents the near field pattern of a source point with a sufficient distance of the observation point while the other represents the far field pattern of a source point.

The application of the Fast Multipole method for scattering problems was presented by [3]. Some error analysis for this method can be found in [9].

In the following we will give a short introduction in this method for solving acoustic scattering problems in three dimensions. For a more detailed prescription we refer to [2]. The far field pattern of the radiating solution created by

$$u_\tau(y) = \int_\tau \left(\frac{\partial \phi}{\partial n}(\mathbf{x}, \mathbf{y}) - aY\phi(\mathbf{x}, \mathbf{y}) \right) \varphi(x) \, do(x) \tag{15}$$

is given by

$$\Psi^\tau(\hat{z}) = \frac{1}{4\pi} \int_\tau e^{ik(z_\tau - x)\hat{z}} \left(k^2 \hat{z}\nu(x) - aYik \right) \varphi(x) \, do(x) \tag{16}$$

with the unit vector $\hat{z} = z/|z|$. Further we can get a local solution $v(y)$ with $y \in \tau$ of the Helmholtz equation with it's near field pattern Υ^τ as follows

$$v(y) = \frac{1}{4\pi} \int_{S^2} \Upsilon^\tau(\hat{z}) e^{ik(y - z_\tau)\hat{z}} \, do(\hat{z}) . \tag{17}$$

The translation operator μ that translates the far field pattern into the near filed pattern was found by Rokhlin [3]. In this work it was shown that the near field pattern of a solution $v(y)$ with $y \in B(z_2, \rho_2)$ of the Helmholtz equation due to the radiating solution u in $\mathbb{R}^3 \setminus B(z_1, \rho_1)$ with $|z_1 - z_2| > \rho_1 + \rho_2$ can be written as

$$
\begin{aligned}
\Upsilon^{\tau_2}(\hat{z}) &= \left(\sum_{m=0}^n i^m (2m+1) h_m^{(1)}(k|z_2 - z_1|) P_m((\widehat{z_2 - z_1})\hat{z}) \right) \Psi^{\tau_1}(\hat{z}) \\
&= \mu_n(z_2 - z_1, \hat{z}) \Psi^{\tau_1}(\hat{z}) .
\end{aligned}
\tag{18}
$$

To apply the above formulas we have to introduce an admissibility condition. This means that two panels of the cluster-tree[2] τ_1, τ_2 are called η-admissible with $\eta \in (0, 1)$ if

$$(\rho_1 + \rho_2) < \eta |z_1 - z_2| \tag{19}$$

holds. The set of all η-admissible panels of a panel τ_i is called the far field $\mathcal{F}(\tau_i)$ of τ_i. The interaction of the far field $\mathcal{F}(\tau_i)$ and the panel τ_i is calculated using (16)-(18). The interaction of the panel τ_i with non-η-admissible panels is evaluated in the common way forming a sparse matrix A_{near}. Thus the i-th row of $v = A\hat{u}$ (which represents the discretization of (6,7) at the collocation point $y_i \in \tau_i$) becomes

$$
\begin{aligned}
v_i = (A\hat{u})_i &= (A_{near})_i \hat{u} + \\
&\sum_{\tau_j \in \mathcal{F}(\tau_i)} \sum_\alpha \underbrace{\frac{1}{4\pi} w_\alpha e^{ik(y_i - z_i)\hat{z}_\alpha}}_{=:S_{i,\alpha}} \underbrace{\mu(z_i - z_j, \hat{z}_\alpha)}_{=:D_{ij}} \Psi^{\tau_j}(\hat{z}_\alpha)
\end{aligned}
$$

with

$$
\Psi^{\tau_j}(\hat{z}_\alpha) = \sum_{q=1}^N \hat{u}_q \underbrace{\int_{supp\varphi_q \cap \tau_j} \frac{1}{4\pi} e^{ik(z_j - x)\hat{z}_\alpha} \left(k^2 \hat{z}_\alpha \nu(x) - aYik \right) \varphi_q(x) \, do(x)}_{=:T_{\alpha,q}}
$$

finally

$$v_i = (Au)_i = (A_{near})_i \hat{u} + (S \cdot D \cdot T \cdot \hat{u})_i$$

---

[2]For the definition of panel and cluster-tree and $B(z, \rho)$ see for example [2, 4]

with sparse matrices S, D, T as defined above.

As shown in [2] the amount of non-zero entries of the matrices S, D, T for the Fast Multipole method is $\mathcal{O}(N^{3/2})$ due to the large number of interactions. One idea to overcome this problem is to use the Fast Multipole method on $\mathcal{O}(\log N)$ levels of the cluster-tree. Doing so we end up with an $\mathcal{O}(N \log^2 N)$ method.

5 Numerical results

We tested the described methods with two computational examples. Our goal was to show that the amount of computer memory in use really behaves like $\mathcal{O}(N)$ and $\mathcal{O}(N \log^2 N)$ as predicted. To approximate the sound pressure we used linear discontinuous ansatz functions.

Our first example is an interior radiation problem. We are interested in the sound pressure distribution

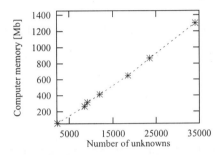

Figure 1: Memory requirement for the "car cabin" model using MLFMA

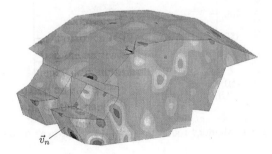

Figure 2: Real part of the solution for the "car cabin" model

on the interior surface of a car cabin. The excitation is a normal velocity of the left dash board (see Figure 2). The problem was solved using the Multilevel Fast Multipole Method described in chapter 4. Figure 1 shows that in the region of N investigated the memory requirement behaves like $\mathcal{O}(N)$. So far we did not try problems with larger N to see if the predicted behavior of $\mathcal{O}(N \log^2 N)$ holds.

The second example was solved using the RGM. It is the so called "cat-eye" (see Figure 3). We like to calculate the scattered field of an incoming plane wave $u^i(\mathbf{x}) = p_0 \exp(-i\, k\mathbf{d}\, \mathbf{x})$ with $\mathbf{d} = [1,1,1]$. All surfaces are assumed to be rigid. The geometry was modeled using four different meshes with 5880, 7680,

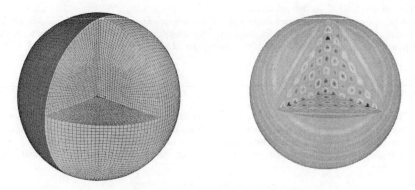

Figure 3: Geometry model of the "cat-eye" (left) and the real part of the solution (right)

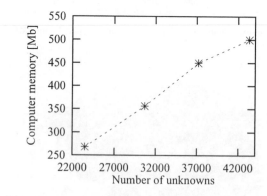

Figure 4: Memory requirement for the "cat-eye" model

9300 and 10830 rectangular elements. The grid size h was chosen as $h = \lambda/15$. The wave number k was chosen such that $\lambda = 2\pi/k$ was four times the element length. The required memory is shown in Figure 4. It is easy to see that the amount of computer memory we need is indeed $\mathcal{O}(N)$ as expected.

6 Conclusion

With the numerical examples we have analyzed so far we can prove the estimates of the computer memory needed.

As presented in [5] to solve a problem with N unknowns using the RGM we need $\mathcal{O}(N)$ of memory. The amount of the computational costs shows a strong dependence on the type of the object. The method seems to be well suited to solve problems with a 1D or 2D like geometry (for example the "radiating tube" or "aeroplane wing").

The Multilevel Fast Multipole Method behaves at least like $\mathcal{O}(N \log^2 N)$ in terms of memory consumption. But the total memory requirement is higher than for the RGM. This is caused by the memory cost for storing the interaction coefficients μ_n at the different levels. An alternative way is to calculate these coefficients as they are needed within an iteration step. In [10] an efficient method based on a one-dimensional multipole method is presented. Using this idea a significant reduction of the memory requirement can be achieved.

Especially for complicated shaped surfaces both methods works very efficient. For the RGM this is because of many surface nodes are sharing the same nodes of the auxiliary grid. In the MLFMA such a surface allows to assemble many elements in a panel which fits in a small ball $B(z, \rho)$ and only a few near field coefficients are needed. The same happens if the surface discretization is much finer than needed for an acoustic calculation what is often the case when the surface discretization is constructed out of FE-models built for static or dynamic problems.

Acknowledgements

The author is very grateful to A. Bespalov for its support by realizing the presented work on the RGM. Further I like to thank T. Nähring for the fruitful discussions.

References

[1] C. C. Lu E. Michelssen J. M. Song W. C. Chew, J. M. Jin. Fast solution methods in electromagnetics. *IEE Trans. Antennas Propag.*, 45(3):533–543, 1997.

[2] K. Giebermann. *Schnelle Summationsverfahren zur numerischen Lösung von Integralgleichungen für Streuprobleme im $\mathbb{R}^3$*. PhD thesis, Universität Karlsruhe, 1997.

[3] V. Rokhlin. Diagonal forms of the translation operators for the Helmholtz equation in three dimensions. *Applied and Computational Harmonic Analysis*, 1:82–93, 1993.

[4] S. Sauter. *Über die effiziente Verwendung des Galerkinverfahrens zur Lösung Fredholmscher Integralgleichungen*. PhD thesis, Universität Kiel, 1992.

[5] A. Bespalov. On the usage of a regular grid for implementation of boundary integral methods for wave problems. *Russ. J. Numer. Anal. Math. Modelling*, 15(6):469–488, 2000.

[6] A.J. Burton and G. F. Miller. *The application of integral equation methods to the numerical solution of some exterior boundary-value problems*. Number 323:201-220. Proceedings of the Royal Society of London, 1971.

[7] A. Brandt. Multilevel computation of integral transforms and particle interactions with oscillatory kernels. *Comp. Phys. Commun.*, 65:24–38, 1991.

[8] J. Cooley and J. Tukey. An algorithm for the machine calculation of complexes fouriers series. *Math. Comp.*, 19(90):297–301, 1965.

[9] J. Rahola. Diagonal forms of the translation operators in the fast multipole algorithm for scattering problems. *BIT*, 60:333–358, 1996.

[10] S. Velamparambil and W. C. Chew. A fast polynomial representation for the translation operators of an MLFMA. *Microwave and Optical Technology Letters*, 28(5):298–303, 2001.

Theoretical and Computational Acoustics 2001
E.-C. Shang, Qihu Li and T. F. Gao (Editors)
© 2002 World Scientific Publishing Co.

Applications of Fractional Differential Operators to the Damped Structure Borne Sound in Viscoelastic Solids

B. Nolte, S. Kempfle, I. Schäfer

Department of Mechanical Engineering, Institute of Mathematics,

University of the Federal Armed Forces Hamburg

E-mail: *bodo.nolte@unibw-hamburg.de*

ICTCA in Beijing, China

1 Abstract

Time dependent analysis of the dynamic damped behaviour of continua are mathematically modelled by partial differential equations. One gets uniqueness, existence and stability (well posed problems) by the implementation of the correct initial boundary conditions. However, by taking memory effects under consideration, any change in the past of the system changes the future dynamic behaviour. Classical damping descriptions failed when describing the behaviour of many materials, like teflon. This is because in classical theory the operators are local ones. The implementation of fractional time derivatives into the partial differential equations is an alternative technique to overcome these problems. Now the time derivative operator is a global one, memory effects in structure borne sound are calculable.

In this presented paper the theory of fractional time derivative operators and their application in continuum mechanics is briefly sketched. The main result when using this methods for damping behaviour is that a global operator is needed which takes the whole history into account. We call this theory the functional calculus method instead of the well known fractional calculus with the use of initial conditions.

To show the effectivity of this method the impulse response of a viscoelastic rod is compared with measurement. It is shown that the damping behaviour is described very much better than by other models with comparable few parameters. Moreover it is the only one that works in a wide frequency range as well as it can describe the dispersion of the resonance frequencies. The implementation of this damping description in a Boundary Element Code is an application in dynamics of 3D continua in frequency domain.

2 Introduction

In figure 1 several mechanical models for common damping description are shown. They all consist of string and damper elements. The first one is the Maxwell model, which contains one string and one damper element in a row, the second one is the often used Kelvin-Voigt model which is built of a string and a damper element parallel to each other. The latter one is the so-called three paramenter model (two strings and one damper).

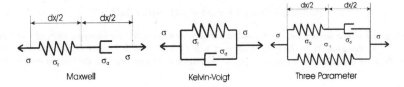

Figure 1: Damping models

Another damping model is the model which consists only of one string element with a constant and complex parameter E for each frequency. This model results in a constant damping work (not frequency depended). Therefore it is called the model of constant hysteresis (or the model of constant complex Young modulus). Let us call this model in this paper the classical model. In common theory the string force is proportional to the displacement or in a more common sense, it is proportional to the zero order time derivative of the displacement, while the damper force is proportional to the velocity that means to the first time derivative of the displacement. This type of damper is called a viscous one. To reach a more general material damping description, the damping force should now be no longer proportional to the first time derivative, it is turned to be proportional to a time derivative of a fractional order α.

In this paper the fractional derivative theory is first applied to the one mass oscillator (chap. 4), which is based on ordinary differential equations, and in the following chapter it is applied to the damped rod, which is mathematically based on partial differential equations.

3 The fractional derivative theory

In this chapter a very brief sketch about fractional derivative theory, used in this paper, is given. It is based on the Beyer–Kempfle definiton

$$\frac{d^\alpha f(t)}{dx^q} = f_{,x^\alpha}(t) = \mathfrak{F}^{-1}\left\{(i\omega)^\alpha \mathfrak{F}(f(t))\right\} \tag{1}$$

$$= \frac{1}{2\pi}\int\limits_{-\infty}^{\infty} e^{i\omega t}(i\omega)^\alpha \int\limits_{-\infty}^{\infty} e^{-i\omega t} f(t)\, dt\, d\omega \tag{2}$$

for the fractional derivative of order α, which is equivalent for a wide function class with the global operator

$$_{-\infty}\mathbf{D}_t^\alpha\left(f(t)\right) = \frac{1}{\Gamma(k-\alpha)}\frac{d^k}{dt^k}\int\limits_{-\infty}^{t}(t-\tau)^{k-\alpha-1}f(\tau)\, d\tau, \quad (k-1 \le \alpha < k),\ k \in \mathbb{N},\ \alpha \in \mathbb{R}^+. \tag{3}$$

As an example the fractional derivative of the sine-function ends this introduction. Using the above definitions the fractional derivative of the sine-function results in

$$_{-\infty}\mathbf{D}_t^\alpha\left(\sin(\omega x)\right) = \frac{1}{2}e^{-i\omega x - i\frac{\pi}{2}\alpha}\omega^\alpha i - \frac{1}{2}e^{i\omega x + i\frac{\pi}{2}\alpha}\omega^\alpha i = \omega^\alpha \sin(\omega x + \frac{\pi}{2}\alpha) \text{ for } \omega > 0. \tag{4}$$

However it should be mentioned that this fractional definition of derivative theory makes sure that the translation invariance and the halfgroup property hold, both properties are very important and necessary for physical applications. (some references should be mentioned [1], [3], [7], [8], [4])

4 The fractional damped one mass oscillator

4.1 Impulse response, stable case

The first application of fractional derivative theory to damping description is shown in fig. 2. A simple one mass oscillator with mass m, string c and damper d is excited by an ideal impulse. The corresponding ordinary differential equation to this physical problem is

$$m u_{,tt}(t) + d u_{,t^\alpha}(t) + c u(t) = f(t) \quad \text{with } m, d, c > 0. \tag{5}$$

The impulse response, see fig. 3, is mathematically described by the superposition of two functions k_1 and k_2. The function k_1 is the oscillating part while function k_1 is a relaxational one describing the memory effect; k_1 oscillates with respect to k_2. It should be mentioned that in the integer case k_2 vanishes and k_1 oscillates with respect to the t-axis. The entire solution is obtained from the residue theorem from complex analysis. The integral obtaining k_2 is solved numerically. The zeros of the corresponding 'fractional polynom' are $s = \rho \pm i\sigma$.

$$k(t) = H(t)\,[k_1(t) + k_2(t)]$$

$$k_1(t) = \frac{2e^{\rho_1 t}}{c_1^2 + b_1^2}(c_1 \cos(\sigma_1 t) + b_1 \sin(\sigma_1 t)) \text{ for } \sigma_1 \neq 0$$

$$c_1 = \mathrm{Re}(2ms_1 + \alpha d s_1^{\alpha-1}) \; ; \; b_1 = \mathrm{Im}(2ms_1 + \alpha d s_1^{\alpha-1}) \quad \text{with} \quad s_1 = \rho_1 + i\sigma_1$$

$$k_2(t) = \frac{1}{\pi} \int_0^\infty \frac{e^{-rt} \cdot d \cdot r^\alpha \sin(\alpha\pi)}{[c + mr^2 + d \cdot r^\alpha \cos(\alpha\pi)]^2 + [d \cdot r^\alpha \sin(\alpha\pi)]^2}\,dr \tag{6}$$

It should be emphasized that the entire solution, shown in fig. 3, is stable and causal. No answer occurs for t less than zero, of course the impact excites the mechanical system when time t equals zero.

4.2 Impulse response, unstable case

Moreover the unstable case is also calculable. The case of instability is obtained from a negative sign in front of the damping force

$$mu_{,tt}(t) - du_{,t^\alpha}(t) + cu(t) = f(t) \quad \text{with } m,d,c > 0. \tag{7}$$

In fig. 4 the result for an impuls excitation is shown on the left hand side. This is not the answer one would expected. The solution is stable but there is no causality. One obtains causality and an unstable solution by adding a special homogeneous solution to this 'impulse response' and the result is depicted in the right hand side of fig. 4. This is a total new application of homogeneous solutions and this theory will be described in detail in a following paper. Homogeneous solutions are normally used to fulfill the initial conditions. However in the fractional case there are no initial conditions. The entire history of a system has to be known.

$$k(t) = \begin{cases} k_-(t) & \text{for } t < 0 \\ k_2(t) & \text{for } t > 0 \end{cases} \quad , \quad k_{\text{causal}}(t) = \begin{cases} k_-(t) - u_{\text{hom}}(t) = 0 & \text{for } t < 0 \\ k_2(t) - u_{\text{hom}}(t) & \text{for } t > 0 \end{cases}$$

$$k_-(t) = u_{\text{hom}}(t) = -\frac{2e^{\rho_1 t}}{c_1^2 + b_1^2}(c_1 \cos(\sigma_1 t) + b_1 \sin(\sigma_1 t)) \text{ for } \sigma_1 \neq 0$$

$$c_1 = \mathrm{Re}(2ms_1 - \alpha d s_1^{\alpha-1}) \; ; \; b_1 = \mathrm{Im}(2ms_1 - \alpha d s_1^{\alpha-1})$$

$$k_2(t) = \frac{1}{\pi} \int_0^\infty \frac{-e^{-rt} \cdot d \cdot r^\alpha \sin(\alpha\pi)}{[c + mr^2 - d \cdot r^\alpha \cos(\alpha\pi)]^2 + [d \cdot r^\alpha \sin(\alpha\pi)]^2}\,dr$$

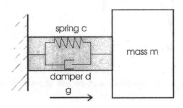

Figure 2:

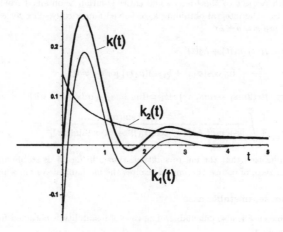

Figure 3: Impulse response

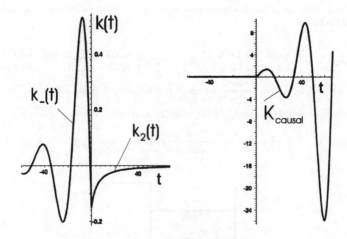

Figure 4: Uncausal and causal impulse response

4.3 Frequency response functions

The second example is a comparison between the frequency response functions of the classical excitation conditions of the simple mass oscillator. The excitation types are shown in fig. 5. Excitations A and B result in the frequency response function V_1, depicted in fig. 6, excitation C in V_3 and excitation D in V_2, with damping ratio D and frequency ratio η, the ratio of the current frequency and the undamped natural frequency. The phase function φ, also shown in fig. 6, is the same for all three cases. In the case of V_1 the peaks increase with respect to the frequency ratio when the damping ratio increases ($\alpha = 0.7$ or 0.3), instead of a decreasing behaviour in the integer case. In the case of V_2 there are parts where V_2 is greater than one. This does not appear in the integer case ($\alpha = 1.0$). The function V_3 behaves like V_1. The point where $\eta = 1$ and $\varphi = \pi/2$ ($\alpha = 1.0$) belonging to the phase function is no longer a fixed point. It is now a function of the fractional order α. These effects are amplified when α decreases.

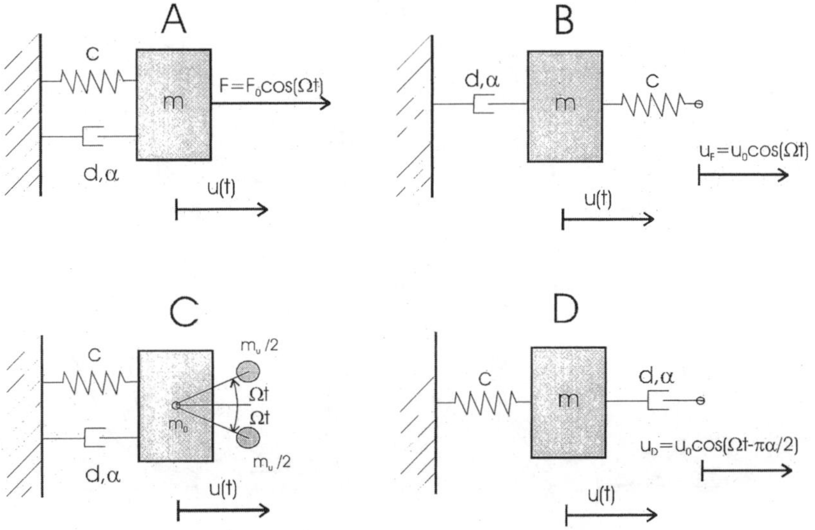

Figure 5: Excitation conditions

4.4 Comparison of measurement and calculations

The question is: Does a real material behaves fractional? The impulse response of a one mass oscillator is shown in fig. 7. The material under consideration (marked grey in fig. 2, g is the earth-acceleration) is 12 cm long and its mass is neglegtible compared to the big mass m, from an engineering point of view this is a one mass oscillator. In fig. 7 the measured impulse response, the fractional impulse response (fractional Kelvin-Voigt model) and the classical impulse response (model of constant complex Young modulus) are depicted. Only the fractional case gives a good correspondence to the measurement. The classical model fails totally.

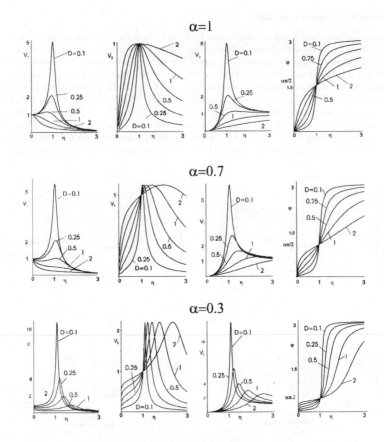

Figure 6: Fractional frequency response functions

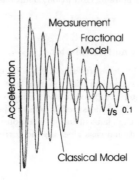

Figure 7: Impulse response

5 The fractional damped rod

Frequency response functions ([2], [9], [10]) of a 1D rod are shown in fig. 9 and 10 for serveral damping models in comparison to the measured frequency response function. The partial differential equation of the fractional three parameter model yields

$$\rho u_{,tt}(x,t) = E_1 u_{,xx}(x,t) + \eta \left(2 + \frac{E_1}{E_2}\right) u_{,xxt^\alpha}(x,t) - \frac{\eta}{E_2} \rho u_{,ttt^\alpha}(x,t).$$

Taking E_2 to infinity this equations represents the fractional Kelvin-Voigt model, while if $E_1 = 0$ this equation corresponds to the fractional Maxwell model. The frequency response function is obtaind by transforming this partial differential equation into the frequency domain. Two parameters have to be indentified, the viscous modulus η and the fractional derivative order α. They are obtained by the first natural eigenfrequency and by the upper envelope or hull function O_E. Again the fractional Kelvin-Voigt model gives the best result, especially for the phase function φ.

$$O_E = \frac{-\Omega^2}{\phi \lambda A \sinh(\mathrm{Re}\,(\lambda)\,l)} = \frac{\Omega^2}{A} \sqrt{\frac{2}{\left[\mathrm{Re}^2\,(\phi\lambda) + \mathrm{Im}^2\,(\phi\lambda)\right] \left[\cosh(2l\,\mathrm{Re}(\lambda)) - 1\right]}} \tag{8}$$

$$\text{with} \quad \phi(\omega) = \frac{E_1 + \eta \left(2 + \frac{E_1}{E_2}\right)(i\omega)^\alpha}{1 + \frac{\eta}{E_2}(i\omega)^\alpha} \quad , \quad \lambda = \omega \sqrt{\frac{-\rho \left(1 + \frac{\eta}{E_2}(i\omega)^\alpha\right)}{E_1 + \eta \left(2 + \frac{E_1}{E_2}\right)(i\omega)^\alpha}}$$

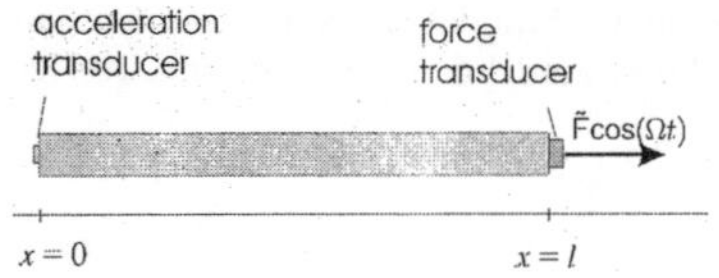

Figure 8: Measurement set up

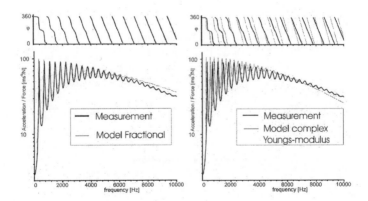

Figure 9: Frequency response function 1

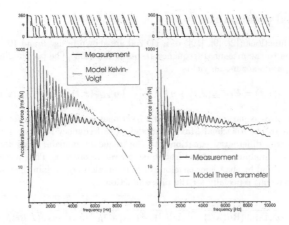

Figure 10: Frequency response function 2

5.1 Dispersion

The dispersion effect is shown in fig. 11. Again only the fractional Kelvin-Voigt model is able to fit the measured eigenfrequencies. The classical model (constant complex Young modulus) is not able to show dispersion. The classical Kelvin-Voigt model does, but not with the identified parameters obtained from the measurement.

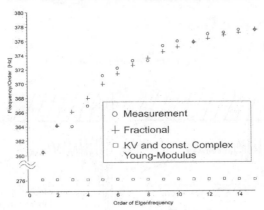

Figure 11: Dispersion

5.2 Comparison 1D-3D case

Naturally a real rod is a 3D structure and the influence of a Poisson's ratio ν is not neglegtible. In fig. 12 the influence of this parameter is shown. The rod under consideration has a length l of 1 m and a diameter of 0.01 m. The boundary element model has 300 linear elements per length and one element at both ends ([5], [6]). The heigth h is 0.0088622 m. Because of dispersion effects the six element per wavelength rule of thumb does not hold.

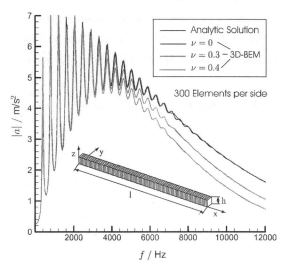

Figure 12: BEM calculation

6 Conclusions

The question was: Does a real material behaves fractional? The answer is: Yes it does. Especially modern and relatively strongly damped materials like PTFE (Teflon), Teflon with bronce, plastic materials, PVC, PMMA, PA, PE, PU 70, PU90 etc. (The material investigated in this paper is teflon). It should be emphasized that this fractional theory applied to describe damping behaviour is still a phenomenological one.

References

[1] BEYER, H.; KEMPFLE, S.: Physically consistent damping laws. *ZAMM* 75 (1995) 8, 623-635.

[2] GAUL, L.; KLEIN, P.; KEMPFLE, S.: Impulse response function of an oscillator with fractional derivative in damping description. *Mech. Res. Commun.* 16 (1989) 5, 297-305.

[3] KEMPFLE, S.; SCHÄFER, I.: Functional calculus method versus Riemann-Liouville approach, FCAA 2(4), 1999, pp. 416-427

[4] KEMPFLE, S.; SCHÄFER, I.: Fractional differential equations and initial conditions, FCAA 3(4), 2000, pp. 387-400

[5] NOLTE, B.: *Randelementberechnungen und Nahfeldmessungen zur akustischen Fluid-Struktur-Interaktion*, PhD-Thesis, Department of Mechanical Engineering, University of the Federal Armed Forces, Hamburg, 1998

[6] NOLTE, B.; SCHÄFER, I.: Modellierung fraktionaler Körperschalldämpfung: Vergleich 1D analytischer Lösung mit 3D Randelementverfahren, In: *Anwendungen der Akustik in der Wehrtechnik*, Meppen, Germany, Sept. 2000, pp. 97-118

[7] OLDHAM, K. B.; SPANIER, J.: *The fractional calculus.* Academic Press, New York 1975.

[8] PODLUBNY, I.: *Fractional differential equations.* Academic Press, New York 1999.

[9] SCHÄFER, I.: Beschreibung der Dämpfung in Stäben mittels fraktionaler Zeitableitungen. *ZAMM* 80 (2000) 5, 356-360.

[10] SCHÄFER, I; SEIFERT, H. J.: Description of the Impulse Response in Rods by Fractional Derivatives, accepted for publication: *Zeitschrift für Angewandte Mathematik und Mechanik*

Theoretical and Computational Acoustics 2001
E.-C. Shang, Qihu Li and T. F. Gao (Editors)
© 2002 World Scientific Publishing Co.

Indirect Bem Formulation for 2D Elastodynamic Problem

Mabrouk Ben Tahar, Slim Soua
Université de Technologie de Compiègne, Laboratoire Roberval UMR 6066,
Secteur Acoustique BP 20529, 60200 Compiègne Cedex France
E-mail: *Mabrouk.BenTahar@utc.fr*

This paper deals with a 2d scattering problem of elasto-acoustic plane waves in elastic isotropic and homogenous media. An indirect BEM formulation is applied to obtain the solution of a problem with mixed boundary condition in both bounded and unbounded domains. A system of integral equations with singular kernels is derived. The variational formulation allows direct integration of singular terms and Hypersingular integral, coming from the traction representation, is treated based on analytical transformation. An associated algorithm is implemented. Several examples and discussions, showing stability and accuracy of this method are presented.

1.Introduction

The improvement of wave scattering models in elastic media was the subject of many research works, in fact a very interesting application is the inverse problem in non-destructive testing nevertheless the direct problem need to be performed and discussed before investigating reflections in inverse problem. Among numerical approaches, the boundary element method BEM proved to be a specific and efficient method for the study of elastic wave scattering by obstacles. Since 1980 [NIW80] [ANT85], the resolution of the bidimensionnal elastodynamic problem in time domain and using BEM formulations was the subject of many works

Direct time-domain algorithms using transient kernels with a collocation approach was developed by many authors [ANT85], [NIW80], [ISR90], [BIR98,99], [CAR00], [COD96], [RIZ94], [PEI97], [FRA99], [MAN83,98], [WAN90] and [REZ86]. Linear and constant temporal interpolation functions were used to approximate displacements, and usually a non symmetric matrix was to inverse. Some improvements of those algorithms where performed on numerical aspects: quadratic interpolation was employed to describe variables [ISR90], [BIR99],[CAR00], more attention was given to regularization procedure, as discussed by Frangi in [FRA99]. He gave regular decomposition in time domain for dynamic and static cases. Traction discontinuity in time is coped by career [CAR00] and in [BIR98], analytical development proved to be efficient and avoided singularity problems. Others authors treated fundamental solution using improved integration of Greens functions [COD96], [RIZ94]. Many studies proved non-stability of time stepping results, caused essentially by dimension of such problems.
Many papers are interested in numerical irregular frequencies [REZ86] and some means to avoid those irregular eigenfrequencies are suggested. They consist in using interpolation technique [NIS88] or in adding the direct integral equation for points of the external domain to the final equations [SCH86]. Finally this problem leads to a unique solution.
The singular character of fundamental solution was the interest of many works and some numerical, analytical or combined procedures were performed [FRA99],[DAV99].
Nishimura and Kobayashi [NIS89] exploited the efficiency of the variational approach to reduce singularity. Later BECACHE [BEC93] presented a variational formulation with hypersingular kernel applied to Neuman's boundary condition problem. They performed analytical regularization. The bilinear variational form was splitted into two parts containing weak singularities with regard to order of integration and finally a symmetric system easier to solve is obtained.
In this paper, we introduce the last procedure in a mixed-problem solving formulation based on an indirect approach, in the harmonic time domain.

First, we set the equations of elastodynamic problem derived from application of FOURIER-LAPLACE transform and the associated indirect integral solution. Then a description of numerical procedure is developed and finally some examples showing the validity of the described formulation are presented.

2. Elastodynamic boundary element formulation

A 2D isotropic and linear elastic body Ω, Fig 1, with regular boundary $\Sigma = \Sigma_u \cup \Sigma_t$. The harmonic time dependence is of the form $e^{-i\omega t}$ and will be omitted in what follows.

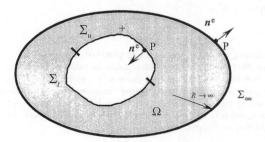

Fig. 1. General elastic problem.

In a fixed rectangular coordinate system x_i, the equation of motion is given by:

$$div \, \boldsymbol{\sigma}_M(u(M)) \; + \rho\omega^2 u(M) = \vec{0}, \forall M \in \Omega \tag{1}$$

Where ρ is the mass density $\boldsymbol{\sigma}_M$ and u denote the component of stress tensor and displacements vector respectively, $\boldsymbol{\omega}$ is the circular pulsation and n^e is the outward unit normal. We assume that no body force is applied.

Hook's law gives the stress displacement relation for homogeneous isotropic and linear elastic solid. Therefore (1) can be rewritten as

$$\Delta_N(u) + \rho\omega^2 u = 0 \tag{2}$$

Where

$$\Delta_N(u) = (\lambda + \mu).grad \, div(u) + \mu\Delta(u) \tag{3}$$

Mixed boundary conditions are also considered:

$$u(M) = u_0(M) \text{ on } \Sigma_u \text{ and } T^{n_M}(u(M)) = t_0(M) \text{ on } \Sigma_t \tag{4}$$

The boundary traction vector are denoted as follows:

$$T^{n_M}(u(M)) = \tau(M) \tag{5}$$

We assume plane strain deformation.

For the scattering problem we note u the scattered field and satisfies the radiation condition at infinity, furthermore the initial incident field is always present. The total displacement is:

$$u_T(M) = u(M) + u^{inc}(M) \tag{6}$$

The scattered displacement solution to this problem is derived from betti's reciprocity theorem [NISH86] and can be represented in terms of simple and double layer elastic potentials:

$$u(M) = \int_{\Sigma_u} \hat{G}(M,P).[\tau(P)]d\Sigma_p - \int_{\Sigma_t} \hat{T}(M,P).[u(P)]d\Sigma_p \tag{7}$$

And when M lies on the boundary:

$$\frac{1}{2}(u^+(M) + u^-(M)) = \int_{\Sigma_u} \hat{G}(M,P).[\tau(P)]d\Sigma_p - \int_{\Sigma_t} \hat{T}(M,P).[u(P)]d\Sigma_p \tag{8}$$

Where $[f] = f+ - f-$ denotes the jump of a function f over Σ.

$\bar{G}$ is the matrix of green fundamental solution, where $\bar{G}_{ij}$ is the displacement at P in the jth direction due to a unit concentrated transformed dynamic load at M in the ith direction, and $\hat{T}_{ij}$ is the resulting traction at this point on a boundary.

The expression of $\bar{G}_{ij}$ in 2d problem is given by:

$$\bar{G}_{ik} = \frac{i}{4\mu}\left[\delta_{ik}H_0^1(k_S r) + \frac{1}{k_S^2}\left(H_0^1(k_S r) - H_0^1(k_P r)\right)_{,ik}\right] \qquad (9)$$

Where $r = |MP|$, k_P and k_S represent the wave number of the pressure and shear wave respectively, λ and μ are the Lame elastic constant:

$$k_P^2 = \frac{\omega^2 \cdot \rho}{\lambda + 2\mu} \qquad k_S^2 = \frac{\omega^2 \cdot \rho}{\mu} \qquad (10)$$

The expression of $\hat{T}_{ik}$ is given by

$$\hat{T}_{ik} = \boldsymbol{T}^k(M, P, \boldsymbol{n}_P^e).\boldsymbol{e}_i = (\boldsymbol{\sigma}_P(\boldsymbol{G}^k(M, P)).\boldsymbol{n}_P^e).\boldsymbol{e}_i \qquad (11)$$

Where $\boldsymbol{n}_P^e$ is the unit normal vector at P.

Equation (8) is the integral representation for displacements on the boundary, where integrals are taken in Cauchy principal sense. We obtained a second integral equation by applying the stress operator defined in (5) to (8)

$$\frac{1}{2}\left(\boldsymbol{\tau}^+(M) + \boldsymbol{\tau}^-(M)\right) = \int_{\Sigma_u}\hat{Z}(M, P, \boldsymbol{n}_M).[\boldsymbol{\tau}(P)]d\Sigma_p - \int_{\Sigma_t}\hat{\Sigma}(M, P, \boldsymbol{n}_M).[\boldsymbol{u}(P)]d\Sigma_p \qquad (12)$$

Where $\hat{\Sigma}(M, P, \boldsymbol{n}_M) = \boldsymbol{\sigma}_M(\boldsymbol{T}^k(M, P), \boldsymbol{n}_M)$ and $\hat{Z}(M, P, \boldsymbol{n}_M) = \boldsymbol{\sigma}_M(\boldsymbol{G}^k(M, P), \boldsymbol{n}_M)$

In order to perform a BEM analysis, an algebraic system of equation is deduced by applying (8) and (12) to boundary points located on Σ_u and Σ_t respectively (Equation (4)).

$$\boldsymbol{u}_0(M) = \int_{\Sigma_u}\bar{G}(M, P).[\boldsymbol{\tau}(P)]d\Sigma_p - \int_{\Sigma_t}\hat{T}(M, P).[\boldsymbol{u}(P)]d\Sigma_p , \; \forall M \; on \; \Sigma_u \qquad (13)$$

$$\boldsymbol{t}_0(M) = \int_{\Sigma_u}\hat{Z}(M, P, \boldsymbol{n}_P).[\boldsymbol{\tau}(P)]d\Sigma_p - \int_{\Sigma_t}\hat{\Sigma}(M, P, \boldsymbol{n}_P).[\boldsymbol{u}(P)]d\Sigma_p , \; \forall M \; on \; \Sigma_t \qquad (14)$$

In the following expressions, $\boldsymbol{\psi}(P)$ and $\boldsymbol{\varphi}(P)$ will denote the traction and displacement jumps respectively. Let $\bar{\boldsymbol{\psi}}(P)$ and $\bar{\boldsymbol{\varphi}}(P)$ be a trial functions defined respectively on Σ_u and Σ_t , thus after integration over Σ the variational formulation associated to the system of equations (13) and (14) is

$$\int_{\Sigma_u}\bar{\boldsymbol{\psi}}(M)\boldsymbol{u}_0(M)d\Sigma_M - \int_{\Sigma_t}\bar{\boldsymbol{\varphi}}(M)\boldsymbol{t}_0(M)d\Sigma_M = \int_{\Sigma_u}\int_{\Sigma_u}\bar{\boldsymbol{\psi}}(M).\bar{G}(M, P).\boldsymbol{\psi}(P)\,d\Sigma_p d\Sigma_M$$

$$- \int_{\Sigma_u}\int_{\Sigma_t}\bar{\boldsymbol{\psi}}(M).\hat{T}(M, P).\boldsymbol{\varphi}(P)\,d\Sigma_p d\Sigma_M$$

$$- \int_{\Sigma_t}\int_{\Sigma_u}\bar{\boldsymbol{\varphi}}(M).\hat{Z}(M, P).\boldsymbol{\psi}(P)\,d\Sigma_p d\Sigma_M \qquad (15)$$

$$+ \int_{\Sigma_t}\int_{\Sigma_t}\bar{\boldsymbol{\varphi}}(M).\hat{\Sigma}(M, P).\boldsymbol{\varphi}(P)\,d\Sigma_p d\Sigma_M$$

It should be pointed out that the resulting integral terms on two apart boundaries are not singular, the self-effect coefficients on Σ_u contain weak singularity. The double layer interaction term is hyper-singular and will be developed according to on the transformation submitted in [NIS89] and summarized in [BEC94] .

The boundary Σ is divided into N linear elements. A piecewise linear variation along each boundary element is assumed for the jumps of displacement and traction. The boundary displacement and traction along an element is interpolated with linear shape functions.
A numerical schema using three Gauss points on each element is adopted to perform integration.
Once the transformation is achieved, for singular integrals, the numerical integration is performed subdividing the current boundary element into sub-elements and dealing with each of them independently.

Finally (15) we can be written in a symmetric matrix form as:

$$
\begin{bmatrix} AS & SD \\ \hline SD^t & -(AD_1 + AD_2) \end{bmatrix} \cdot \left\{ \begin{array}{c} \uparrow \\ \boldsymbol{\psi} \\ \downarrow \\ \uparrow \\ \boldsymbol{\varphi} \\ \downarrow \end{array} \right\} = \left\{ \begin{array}{c} \uparrow \\ u_0 \\ \downarrow \\ \uparrow \\ t_0 \\ \downarrow \end{array} \right\}
\tag{16}
$$

3. Numerical examples

3.1. Scattering of elastic waves with mixed boundary condition

The first example simulates the propagation of the transient pressure wave in a 2d tube made of an isotropic elastic material. Mixed boundary conditions are applied to the tube Fig 2.

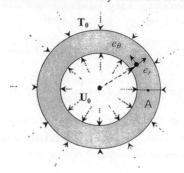

Fig. 2. Tube with mixed boundary condition

The general analytical solution of such a problem is a linear combination of HANKEL functions

$$
u(r,\boldsymbol{\omega}) = A(k_P).H_1^1(k_P r) + B(k_P).H_1^2(k_P.r)
\tag{17}
$$

Where $a < r < b$
The physical parameters of this example are $E = 2.068^E + 11 Pa$, a mass density $\rho = 7800$ (Kg/m^3), Poisson's ratio is taken as $v = 0.3$ and $a = 0.1m$, $b = 0.2m$ represent respectively the outer and inner disc radius.
All results are given with dimensionless values. Fig. 3. shows radial displacement in the frequency range [0,8] at the point A. we represent spatial mapping at a given frequency f=7.5 Table 1.
Our model seems to generate an error on the first eigenfrequency of observed range with a meshing criterion of minimum 14 boundary elements per transversal wavelength to obtain satisfactory results [BEN99].

The comparison of the analytical and numerical models shows that far from the eignfrequencies, good agreement is achieved. In fact he eigenfrequencies of the internal Dirichlet or Neumann problem, are irregular frequencies for the mixed problem. The computed results around these irregular frequencies do not represent physical values.

Table 1. Spatial field variation mapping	Numerical results	Analytical results
Displacement u_x		
Stress σ_{xx}		

3.2. Plane wave scattering by cylindrical cavity

A cylindrical cavity Ω of radius a is impacted by an homogenous plane wave of magnitude A^{inc}. d^{inc} and k^{inc} are respectively the polarization vector and the wave number in the same direction as the wave propagation. The geometry of this third example is given in Fig. 4.

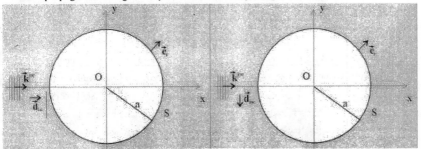

Fig. 4. a) A pressure plane wave impacting a cylindrical cavity for $\theta_{inc} = \pi/2$

b) A shear plane wave impacting a cylindrical cavity for $\theta_{inc} = \pi/2$

Thus the incident displacement at a fixed point $M(x, y)$ can be written as

$$\vec{u}^{inc}(x, y) = A^{inc} d^{inc} e^{\left(i\vec{k}^{inc}.\vec{OM}\right)} \qquad (20)$$

568

Two cases are considered: the incident wave is either a pressure wave or a shear wave. The propagation is in the $\vec{x}$ direction and $\theta_{inc} = \dfrac{\pi}{2}$ represents the angle of incidence between k^{inc} and $\vec{y}$ Fig. 4. The pressure and shear displacement wave field have respectively the following expressions in the Cartesian coordinate system ($\vec{x}, \vec{y}$)

$$\vec{u}^{inc}(x, y) = A^{inc} e^{(i.k_p.x).\vec{x}}, \quad \vec{u}^{inc}(x, y) = A^{inc} e^{(i.k_s.x).\vec{y}} \tag{21}$$

$\left| \vec{u}^{inc} \right|$ normalizes all results given through the mappings in the space domain.

Table 2. (respectively Table 3.) represents the radial and tangential components of the total displacement field for a pressure (respectively shear) incident plane wave.

Since the studied domain is defined in polar coordinates, our mappings have a circular aspect.

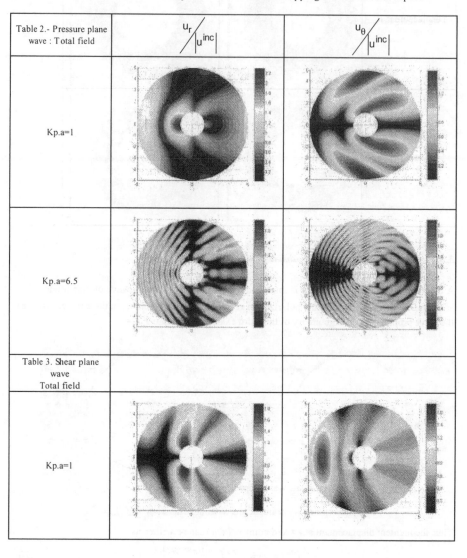

Table 2.- Pressure plane wave : Total field	$u_r / \left| u^{inc} \right|$	$u_\theta / \left| u^{inc} \right|$
Kp.a=1		
Kp.a=6.5		
Table 3. Shear plane wave Total field		
Kp.a=1		

Kp.a=6.5	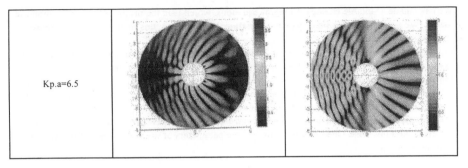

3.3. *Scattering of elastic wave by crack distribution*:

A cylindrical region is submitted to an harmonic pressure Po around its boundary. This geometry is made of an isotropic elastic material whose physical parameters are the same as in the first example. A distribution of traction free cracks is buried in this region, and we aim to show the interaction of elastic wave with this distribution.

The mappings in Table 4. represent at the dimensionless frequency kp.a=9.5, the real and imaginary parts of the displacement for various crack distributions at different location within the cavity. The more constrained areas appear on these mappings.

In this example, we show on one hand the capacity of our computational code and on the other hand this allows the constitution of acoustic signature database of this type of defect.

Table 4. Elastic wave scattered by a crack distribution	*Ux*	*Uy*
1 crack	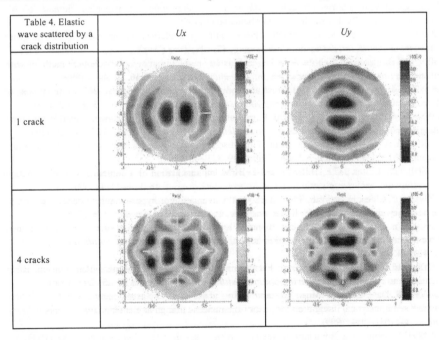	
4 cracks		

4. Conclusion

In this paper a BEM formulation applied to transient elastodynamic problem has been suggested. The given formulation takes into account of the radiation condition on displacement allowing the simulation of scattering problems in infinite domains.

In addition the choice of a jumps form for the variable description seems to be of good interest as two problems are simultaneously solved, and the computational effort on source term is reduced.

We have made a suitable choice of integral equations, which is in the origin of the final symmetric matrix. We associate to the final system a variational formulation; this mainly aims to increase the order of integration.

In fact, as a result we obtain only one singular term. Analytical regularization was achieved on the associate hypersingular kernel, and stable results are the outcome of this transformation.

For our study, time transient response represents an additional extension because our aim is to solve the harmonic problem. Furthermore, the application in non-destructive testing with harmonic perturbation frequency domain seems to be of a big interest.

The developed numerical model can give a contribution in non-destructive building up a database of defects at a low cost. One can easily modify the parameters of the defect (its geometry, its type or its location…).

References

[ANT85]. H. Antes, A boundary element procedure for transient wave propagation in two-dimensional isotropic elastic media, *Finite Element Anal. Des.*, **1**, 313-322 (1985).

[BEC93]. E. Becache, J. C. Nedelec, N. Nishimura, Regularization in 3D for anisotropic elastodynamic crack and obstacle problems, *Journal of elasticity*. **31**, 25-46 (1993).

[BEC94]. E. Becache, T. Ha. Douong, A space-time variational formulation for the boundary integral equation in a 2D elastic crack problem, *Mat. Mod. Numer. Anal.* **28**, 141-176 (1994).

[BEN99]. M. Ben Tahar, C. Granat, T. Ha-Duong. Variational integral formulation in the problem of elastic scattering by a buried obstacle, *Mathematical aspects of boundary element methods*, 53–65. Chap-man & Hall/CRC Research Notes in Mathematics.(1999).

[BIR98]. B. Birgisson, S. L. Crouch, elastodynamic boundary element method for piecewise homogenous media, *Int. J. Numer. Meth. Engng.* **42**, 1045-1069 (1998).

[BIR99]. B. Birgisson, E. Siebrits, A. P. Peirce, Elastodynamic direct boundary element methods with enhanced numerical stability properties, *Int. J. Numer. Meth. Engng.* **46**, 871-888 (1999).

[BON95]. M. Bonnet, Regularized direct and indirect symmetric variational BIE formulations for three-dimensional elasticity, *Engineering Analysis with boundary Elements*, **15**, 93-102 (1995).

[CAN95]. M. Cannarozzi, M. Mancuso, Formulation and analysis of variational methods for time integration of linear elastodynamics, *Comput. Methods. Appl. Mech. Engrg.* **127**, 241-257 (1995).

[CAR00]. J. A. M. Carrer, W. J. Mansur, Time discontinuous linear traction approximation in time-domain BEM: 2-D *Int. J. Numer. Meth. Engng.* **49**, 833-848 (2000).

[CAS00]. G.S. Castor, J.C.F. Telles, The 3-D BEM implementation of a numerical Green's function for fracture mechanics applications, *Int. J. Numer. Meth. Engng.* **48**, 1199-1214 (2000).

[CHI91]. C.C. Chien, H. Rajiyah, S.N. Atluri, On the evaluation of hyper-singular integrals arising in the boundary element method for linear elasticity, *Computational Mechanics*, **8**, 57-70 (1991).

[CHO00]. Y. Cho, J.L. Rose, An elastodynamic hybrid boundary element study for elastic guided wave interactions with a surface breakin defect, *International Journal of Solids and structures* . **37**, 4103-4124 (2000).

[COD96]. H.B. Coda, W.S. Venturini, Further improvements on three dimensional transient BEM elastodynamic analysis, *Engineering analysis with boundary elements* **17**, 231-243 (1996).

[DAV99]. K. Davey, M. T. Alonso Rasgado, I. Rosindale, The 3-D elastodynamic boundary element method: semi-analytical integration for linear isoparametric triangular elements, *Int. J. Numer. Meth. Engng.* **44**, 1031-1054 (1999).

[DOM93] J. Dominguez, Boundary Elements in Dynamics, *Comp Mec Pub*, Southampton, (1993).

[ENR00]. Enru Liu, Zhongjie Zhang, Elastodynamic BEM modelling of multiple scattering of elastic waves by spatial distributions of inclusions, *Journal of the Chinese institute of engineers*. Vol **23**, N° 3, pp. 357-368 (2000).

[FRA99]. A. Frangi, elastodynamics by BEM : A new direct formulation, *Int. J. Numer. Meth. Engng.* **45**, 721-740 (1999).

[HEY99]. E. Heymsfield, Influence matrix diagonal block elements for two-dimensional elastodynamic problems, *Int. J. Numer. Meth. Engng.* **44**, 1341-1358 (1999).

[ISR90]. A. S. M. Israil and P. K. Banerjee, Advanced time-domain formulation of BEM for two-dimensional transient elastodynamics, *Int. J. Numer. Meth. Engng.*, **29**, 1421-1440 (1990).

[KIT89]. M. Kitahara, K. Nakagawa, J.D.Achenbach, Boundary integral equation method for elastodtnamic scattering by a compact inhomogeneity, *Comp Mech*, **5**, 129-144 (1989).

[LIE00]. S. T. Lie, G. Yu, S. C. Fan, Further improvement to the stability of the coupling BEM/FEM scheme for 2-D elastodynamic problems, *Comp Mech*, **25**, 468-476 (2000).

[MAI91]. G. Maier, M. Diligenti, A. Carini, A variational approach to boundary element elastodynamic analysis and extension to multidomain problems, *Comp Meth Appl Mech Eng* **92**, 193-213 (1991).

[MAL99]. V. Mallardo, M.H. Aliabadi, A BEM sensitivity and shape identification analysis for acoustic scattering in fluid-solid problems, *Int. J. Numer. Meth. Engng.* **41**, 1527-1541 (1999).

[MAN83]. W. J. Mansur, A time-stepping technique to solve wave propagation problems using the boundary element method ,Ph.D. Thesis, Southampton University, U.K., 1983.

[MAN98]. W.J. Mansur, J.A.M. Carrer, E.F.N.Siqueira, Time discontinuous linear traction approximation in time-domain BEM scalar wave propagation analysis, *Int. J. Numer. Meth. Engng.* **42**, 667-683 (1998).

[NIS88]. N. Nishimura, Application of the boundary integral equation method to solid mechanics, PH.D Thesis, Kyoto University, (1988).

[NIS89]. N. Nishimura, S. Kobayashi, A regularized boundary integral equation methods in elastodynamics for crack problems, *Comp. Mech.*, **4**, (1989).

[NIW80].Y. Niwa, T.Fukuri, S. Kito and L. Fujiki, An application of the integral equation method to two-dimensional elastodynamics Theoret. *Appl. Mech.* Tokyo Univ., **28**, 281-290 (1980).

[PEI97]. A. Peirce, E. Siebrits, Stability analysis and design of time-stepping schemes for general elastodynamic boundary element models, *Int. J. Numer. Meth. Engng.* vol. **40**, 319-342 (1997).

[REZ86]. M. Rezayat, D.J. Shippy, F.J. Rizzo, On time-harmonic elastic-wave analysis by the boundary element method for moderate to high frequencies, *Comp Meth Appl Mech Eng.* vol. **55**, 349-367 (1986).

[RIZ94]. D C Rizo, D.L.Carabalis, An advanced direct time domain BEM formulation for general 3-D elastodynamic problems, *Comp Mech* **15**, 249-269 (1994).

[SAE99]. A. Saez, J. Dominguez, BEM analysis of wave scattering in transversally isotropic solids, *Int. J. Numer. Meth. Engng.* **44**, 1283-1300 (1999).

[SCH86]. H.A.Shenck, , Improved integral formulation for acoustic radiation problems, *J.A.S.A* 1986, **44**, 41-58.

[SCH93]. P.J.Schafbuch, R.B. Thomson, F.J. Rizzo, Elastic scatterer interaction via generalized born series and far-field approximations, *J.A.S.A* 1993, **1**, 295-307.

[WAN90]. H. Wang, Prasanta, K. Banerjee, axisymetric transient elastodynamic analysis by boundary element method, *International Journal of Solids and structures* . **4**, 401-415 (1990).

[WAN95]. C.-Y. Wang, J. D. Achenbach, 3-D time-harmonic elastodynamic Greens functions for anisotropic solids, *Proc. Roy. Soc. London*, **A449**, 441-458 (1995).

[WAN96a]. C.-Y. Wang, J. D. Achenbach, S. Hirose, Two dimensional time domain BEM for scattering of elastic waves in solids of general anisotropy, *Int. Jour. Sol. Struc*, **26**, 3843-3864 (1996).

[WAN96b]. C.-Y. Wang, J. D. Achenbach, Lamb's problem for solids of general anisotropy, *Wave motion*, **24**, 227-242 (1996).

Theoretical and Computational Acoustics 2001
E.-C. Shang, Qihu Li and T. F. Gao (Editors)
© 2002 World Scientific Publishing Co.

Improvement of Reconstruction Accuracy by Using the Nonsingular BEM in the Near-Field Acoustical Holography Based on the Inverse BEM

Jeong-Guon Ih[1] and Sung-Chon Kang[2]

[1]Center for Noise and Vibration Control, Department of Mechanical Engineering,
 Korea Advanced Institute of Science and Technology, Science Town, Taejon 305-701, Korea
 E-mail: *ihih@sorak.kaist.ac.kr*

[2]Engine R&D Center, Daewoo Heavy Industries and Machinery Ltd., 7-11 Hwasoo-dong,
 Dong-gu, Incheon, 401-702, Korea

Abstract

The nearfield acoustical holography (NAH) based on the boundary element method (BEM) is one of the indirect identification methods for vibro-acoustic source properties. In this method, the sound radiation and transmission between a vibrating source and a hologram plane is modeled by the vibro-acoustic transfer matrix using the boundary integral equation (BIE). Distribution of surface velocities of the source can be reconstructed by multiplying the inverse of the calculated vibro-acoustic transfer matrix and the measured field pressure vector. In principle, the pressure data can be measured at a surface with any shape of the nearfield plane including the conformal one. In this technique, the field pressure should be measured as close to the source as possible in order to acquire precise field information including the non-propagating evanescent waves for an accurate source reconstruction. In three-dimensional problems, a propagating wave in the free field can be a Green function satisfying the Helmholtz equation. In the acoustical BIE, i.e. Kirchhoff-Helmholtz integral equation, monopole and dipole Green functions are proportional to $1/R$ or $1/R^2$ when R denotes the distance between source and field point. This means that there is a serious singularity problem in the nearfield. In order to overcome this dilemma, the non-singular boundary integral formulation utilizing propagating plane waves is employed in the NAH. With this kind of BIE, all singularities included in the conventional acoustical BIE can be removed. Consequently, the field pressure can be determined precisely in the very closed field and the acoustical parameters can be continuously calculated on the surface. This can result in the improvement of the BEM-based NAH. In this study, the nonsingular BIE is reformulated to deal with the foregoing matter and the holography equation is derived by using this formulation. Simulations and experiments are performed for an exterior acoustic model in order to demonstrate the advantageous characters of the present method. These include the prediction accuracy of nonsingular BEM (NBEM), and the improvement of reconstruction result in NAH. When the NBEM is used for NAH, it is observed that the reconstruction error is improved more and more by making the hologram plane to nearing the source surface. In particular, the improvement of reconstructed result is distinguishable when the distance becomes less than about 20 % of the characteristic length.

INTRODUCTION

The near-field acoustical holography (NAH) has been recently applied to many areas as an indirect method for identifying the vibro-acoustic parameters of structure-borne sound sources. This technique is equivalent to solving an inverse or backward problem in order to recover the vibro-acoustic properties at the source plane from measured pressure data at the hologram plane, either with conformal or regular shape.

Two major techniques have been suggested in realizing the NAH. One is based on the spatial Fourier transform [1-5] in which the discretely and uniformly measured field pressures on a regular hologram plane are decomposed into the space-wave number domain, and the other is based on the BEM. In the BEM-based NAH, the sound radiation and propagation between the vibrating source surface and the measurement field are modeled as the vibro-acoustic transfer matrix by using the boundary element method (BEM). [6-14] The distribution of surface acoustic parameters of the source can be reconstructed by multiplying the inverse of transfer matrix and the measured field pressure vector. Conformal or any shape of the near field plane including the regular one can be adopted as the hologram plane. This conformal NAH method can deal with sound sources with arbitrarily shaped geometries that cannot be described by separable coordinates.

When a source field is being restored by the BEM-based NAH technique, one tends to position sensors as close to the source surface as possible in order to get profound information on the field. However, the conventional direct acoustic BEM based on the Kirchhoff-Helmholtz integral equation suffers from the strong singularity problem in the near field of the source surface. Computational error increases very steeply as approaching from the far field to the source surface. The problem is due to the singular kernel in the fundamental solution of BIE and it can influence the reconstruction accuracy. In the three-dimensional Kirchhoff-Helmholtz BIE, the monopole and dipole terms expressed by Green function and its derivative are proportional to $1/R$ or $1/R^2$, respectively, where R is the distance between source and field point. When the field points are located in the very near field, R is very small and thus calculation of Green functions diverges in the numerical integration. This invokes a dilemma that the field pressure should be measured in the near-field for the precise reconstruction, while the prediction accuracy can be deteriorated greatly in the near-field due to the bias error stemming from the singularity of Green function and its derivative. A non-singular boundary integral equation was formulated and suggested to solve this singularity problem of BEM in the near-field. [15,16] In this method, the level of singularity can be lowered by eliminating a propagating plane wave term from the integral identity. All singularities included in the conventional acoustic BIE can be changed to weak ones or can be entirely removed. As a result, the near-field pressure can be determined very precisely.

In order to introduce the reconstruction accuracy in the nearfield, the concept of non-singular BIE is introduced to reformulate the holography equation. Conventional and nonsingular methods are applied to a radiating box as a simulation example and the reconstructed results are compared with each other. The effectiveness of NAH based on the inverse, nonsingular BEM is investigated in the viewpoint of the reduction of reconstruction error. It is observed that the nonsingular holographic formulation can provide the accurate un-biased vibro-acoustic transfer matrix between source and field and thus certainly improve the resolution of the reconstructed source field.

I. REFORMULATION OF THE HOLOGRAPHY EQUATION

In order to remove the singularity existing in the conventional Kirchhoff-Helmholtz BIE, a propagating plane wave is deducted from the conventional BIE. Introducing an additional propagating plane wave further, which is directed orthogonal to the former propagating plane wave, and followed by several steps of calculations, the nonsingular BIE can be derived as follows: [15]

$$-j \int_{S_o} G(\mathbf{r},\mathbf{r}_o) [\rho_o \, \omega V_n(\mathbf{r}_o) + k \, \mathbf{n}(\mathbf{r}_o) \cdot \{\mathbf{h}(\mathbf{y}) \hat{p}(\mathbf{r}_o;\mathbf{y},\mathbf{h}(\mathbf{y}),p(\mathbf{y}))$$

$$- \mathbf{h}(\mathbf{y}) \hat{p}(\mathbf{r}_o;\mathbf{y},\mathbf{h}(\mathbf{y}),q) + \mathbf{n}(\mathbf{y}) \hat{p}(\mathbf{r}_o;\mathbf{y},\mathbf{n}(\mathbf{y}),q)\}] \, dS(\mathbf{r}_o)$$

$$- \int_{S_o} \frac{\partial G(\mathbf{r},\mathbf{r}_o)}{\partial \mathbf{n}(\mathbf{r}_o)} [p(\mathbf{r}_o) - \hat{p}(\mathbf{r}_o;\mathbf{y},\mathbf{h}(\mathbf{y}),p(\mathbf{y})) - \hat{p}(\mathbf{r}_o;\mathbf{y},\mathbf{h}(\mathbf{y}),q) + \hat{p}(\mathbf{r}_o;\mathbf{y},\mathbf{n}(\mathbf{y}),q)] \, dS(\mathbf{r}_o)$$

$$= \begin{cases} 0 & \text{for interior problem } (\mathbf{r} \in S_o, \mathbf{r} \notin V), \\ p(\mathbf{r}) - \hat{p}(\mathbf{r};\mathbf{y},\mathbf{h}(\mathbf{y}),p(\mathbf{y})) - \hat{p}(\mathbf{r};\mathbf{y},\mathbf{h}(\mathbf{y}),q) + \hat{p}(\mathbf{r};\mathbf{y},\mathbf{n}(\mathbf{y}),q) & \\ & \text{for interior problem } (\mathbf{r} \notin S_o, \mathbf{r} \in V), \\ p(\mathbf{r}) & \text{for exterior problem } (\mathbf{r} \in S_o, \mathbf{r} \notin V \text{ or } \mathbf{r} \notin S_o, \mathbf{r} \in V). \end{cases} \quad (1)$$

Here, $\mathbf{h}(\mathbf{y})$ is the unit tangential vector at $\mathbf{y}$, $\mathbf{n}(\mathbf{x})$ is the unit normal vector at $\mathbf{x}$, $p(\mathbf{y})$ is the sound pressure at $\mathbf{y}$, $q = -\rho_o c V_n(\mathbf{y})$, ρ_0 is the density of medium, c is the speed of sound, and $V_n(\mathbf{y})$ is the normal velocity at $\mathbf{y}$. When $\mathbf{r}$ represents a point on the surface, $\mathbf{y}$ coincides with $\mathbf{r}$, whereas $\mathbf{y}$ represents the closest surface point to $\mathbf{r}$ if $\mathbf{r}$ is a point in the domain. The propagating plane wave that satisfies both Helmholtz equation and Kirchhoff-Helmholtz integral equation can be expressed as $\hat{p}(\mathbf{x};\mathbf{y},\mathbf{a},\overline{p}) = \overline{p} \exp[jk(\mathbf{x}-\mathbf{y})\cdot\mathbf{a}]$, where $\overline{p}$ is the amplitude of plane wave, $\mathbf{y}$ is an arbitrary reference point, and $\mathbf{a}$ is the direction vector of the plane wave at $\mathbf{y}$. In spite of the fact that Eq. (1) can be used at any position of field point $\mathbf{r}$, it is applicable only to the smooth surface where the surface normal vector can be defined uniquely. In Eq. (1), all singularities are eliminated and thus the pressure can be estimated accurately even in the very close near-field to the source surface.

From the discrete form of Eq. (1), the field pressure, $\{p\}_f$, can be described only by the surface velocity,

$\{v\}_s$, provided $[D]_s^{-1}$ exists as follows:

$$\{p\}_f = ([M]_f + [D]_f [D]_s^{-1}[M]_s)\{v\}_s \equiv [G]\{v\}_s. \tag{2}$$

Here, $\{p\}_s$, is the surface pressure, $[D]_s$, $[M]_s$ mean the dipole and monopole matrices on the surface, $[D]_f$, $[M]_f$ are those corresponding to field data, respectively, and $[G]$ is the vibro-acoustic transfer matrix or complex frequency response functions correlating the surface normal velocity and the field pressure that contains the geometric information of the system as well. Once one knows the field pressure at m points, the surface velocity at n ($\leq m$) nodes can be uniquely determined. This can be accomplished by utilizing the over-determined least-squared solutions approach and the singular value decomposition (SVD) technique. [7,17,18] The SVD of $[G]$ provides the acoustic modal expansion between hologram plane and source field. The surface velocity can be inversely determined as

$$\{v\}_s = [G]^+\{p\}_f = ([G]^H [G])^{-1} [G]^H \{p\}_f = [W][\Lambda]^{-1} [U]^H \{p\}_f, \tag{3}$$

where $[\Lambda] = diag(\lambda_1, \lambda_2, \ldots, \lambda_n)$, $\lambda_1 \geq \lambda_2 \geq \ldots \geq \lambda_n \geq 0$, $\{u_i\}^H \{u_j\} = \delta_{ij}, \{w_i\}^H \{w_j\} = \delta_{ij}$. Here, δ_{ij} is the Kronecker delta, λ_i denotes a singular value of $[G]$, the superscript 'H' signifies the Hermitian operator, and the superscript '+' signifies the pseudo-inverse. If the transfer matrix $[G]$ is determined by the nonsingular BEM and the field pressure $\{p\}_f$ is measured, the surface velocity $\{v\}_s$ can be determined from Eq. (3). By using this surface velocity $\{v\}_s$, it is possible to determine other vibro-acoustic parameters on the surface or in the field.

The transfer matrix $[G]$ is ill-conditioned in general, the inversion of the measured field pressure contaminated with noise can amplify the high-order components greatly, which are comparable with the inverse of non-propagating wave components. Due to this reason, the reconstructed result will be distorted from the actual source field, but the problem can be partially sorted out by the regularization of the singular matrix. In this study, the optimal iterative regularization method [9,19] is employed in the final reconstruction although other regularization techniques [13,20,21] can be also used.

II. SIMULATION EXAMPLE

A parallelepiped radiator with the dimension of $500(w) \times 700(l) \times 320(h)$ mm is chosen as a demonstration example of foregoing nonsingular form of NAH formulation using the inverse BEM. The boundary element model is composed of 234 nodes and 464 linear, triangular, and isoparametric elements, in which the maximum characteristic length is $L_c = 93.8$ mm that limits the effective high frequency to 611 Hz under the $\lambda/6$-criterion. Only the top plane made of 1 mm thick steel plate vibrates with known velocity field and other faces are assumed rigid.

When the top plate is vibrating with (5,1) mode at 163 Hz, the error ratio of predicted field pressures calculated by the CBEM and the nonsingular BEM (NBEM) is shown in Fig. 1. Calculations are done for the evenly spaced 609 field points varying the distance z from the source plane. Fig. 2 shows the maximum, minimum, and mean errors (e_{mean}), and a 95% confidence interval of the prediction error, which corresponds to the range of $e_{mean} \pm 2\sigma_p$, where σ_p is the standard deviation. Here, one can find that the error in the predicted field pressure by CBEM increases when the field point becomes near to the surface for $z < 0.2L_c$. [15,16] Severe inaccuracy of the predicted field pressure implies that the modeled vibro-acoustic transfer matrix does not correctly represent the relation of field pressure and surface velocity in the near-field, and the resultant restored source strength image will be distorted very much from the original one.

For the holographic reconstruction, the field pressures are estimated for the evenly distributed 117

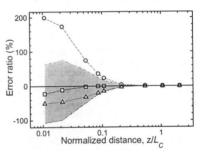

FIG. 1. Predicted error ratio of the field pressure calculated by CBEM and NBEM at 163 Hz, (5,1) mode. The shaded area illustrates 95 % confidence interval of the prediction error with varying z for 609 field points: —□—, mean; --○--, maximum; ····△····, minimum; — · —, (mean+2σ_p); — - - -, (mean-2σ_p).

576

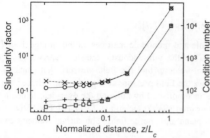

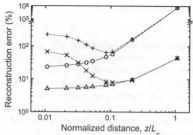

FIG. 2. Calculated singularity factor and condition number of the vibro-acoustic transfer matrix. Singularity factor: $-\times-$, CBEM; $-\bigcirc-$, NBEM. Condition number: $\cdots+\cdots$, CBEM; $-\square-$, NBEM.

FIG. 3. Reconstruction error in surface velocity. CBEM: $-+-$, before regularization; $-\times-$, after regularization. NBEM: $-\bigcirc-$, before regularization; $-\triangle-$, after regularization

points in 50X50 mm pattern above the vibrating plane, in which the field points from the surface are within the distance range of 1-200 mm or, equivalently, $0.0107L_c$-$2.13L_c$. The measured field pressure data are contaminated intentionally by the unbiased Gaussian random noise with a noise variance ($= \sum |n_i|^2$; $n_i =$ noise included in field pressure at i^{th} node) of 50.

The singularity of transfer matrix is compared in terms of the singularity factor ($= \sum \lambda_i^{-2}$) and the condition number in Fig. 2. The singularity factor is proportional to the square of velocity reconstruction error,[13,19] whereas the condition number represents the upper bound of excessive error that is expected in the inversion of transfer matrix. In the far-field, the transfer matrix generated by any type of BEM is nearly same and highly singular. In the very close near-field, the singularity of transfer matrix generated by CBEM is nearly constant for the distance variation, but the singularity related to NBEM decreases steadily with shortening the distance. This is because smaller high-order singular values are calculated by CBEM than those done by the nonsingular method

The errors in reconstructing the source velocity before and after the regularization are presented in Fig. 3. The optimal iterative regularization method[9,19] is employed in this study. Before applying the optimal wave-vector filter, the reconstruction error has a similar trend with that of the singularity of transfer matrix because small singular values affect the reconstructed results dominantly. The high-order singular values corresponding to the non-propagating wave components become smaller as the field point becomes away from the source surface. These small singular values excessively amplify the noise included in the measured field pressure in the inversion process. For the field plane located farther than about $0.2L_c$, the results of NBEM and CBEM calculations are nearly same. When the field plane is separated from the source more than about $0.6L_c$, the velocity reconstruction error is larger than $O(10^5)$% for both NBEM and CBEM calculations. It can be observed that the resultant error of NBEM is reduced steadily with becoming close to the source surface. This fact reveals that the precise information on the non-propagating wave components is very important to improve the accuracy of final reconstruction of the source. When using the CBEM in NAH, the reconstruction error is minimal at certain location near $z = 0.1L_c$ and it increases abruptly with becoming close to the source surface. Before applying the regularization technique, the minimum reconstruction error is as much as 63.2 % at $z \cong 0.107L_c$ when the CBEM is used although the error after applying the regularization is reduced to a fraction, i.e. 7.9 %, of that before the regularization. At $z \cong 0.05L_c$, the reconstruction error can be reduced from 90 % by using the CBEM to 32 % by using the NBEM without regularization, and after regularization it becomes from 14 % to 5.5 %. At an extremely near-field point such as $z = 0.01L_c$, the reconstruction error is 5.1 % in using NBEM and 67.4 % in using CBEM even after applying the regularization. At the nearest sensor point which is farther than about $z > L_c/30$, the regularization effect on the CBEM result is better than the effect of NBEM before applying the regularization, although the regularized NBEM result is far better than the regularized CBEM result. The actual and reconstructed velocity fields are compared in Fig. 4 for two boundary element methods, in which the aforementioned facts and statements can be clearly confirmed. Here, the position of measurement plane corresponds to $z = 0.043L_c$.

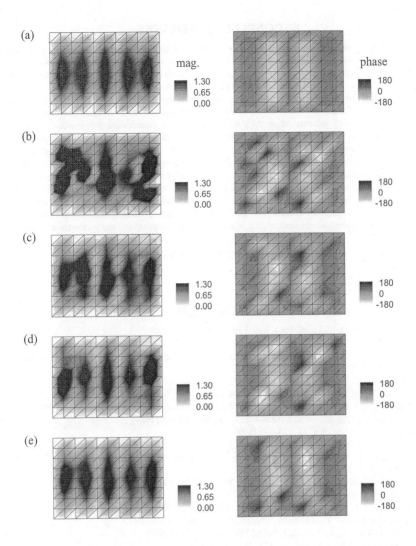

FIG. 4. Reconstructed surface velocity field from evenly distributed 117 measuring points at z=4 mm plane. (a) Measured, (b) CBEM without regularization (ε=110.8 %), (c) NBEM without regularization (ε=31.4 %), (d) CBEM with regularization (ε=18.2 %), (e) NBEM with regularization (ε=5.7 %).

III. CONCLUDING REMARKS

For the realization of source identification using the NAH based on the inverse BEM, which is powerful in dealing with complex shaped sources, it is utmost important to accurately model the vibro-acoustic transfer matrix that relates source and measurement field properties. However, in modeling this transfer matrix, the conventional BIE suffers from the singularity problem that is originated by the singularity of Green function near the integration points. The singularity problem induces inaccurate, highly biased evaluation of field properties in the close near-field of the source. Because one tends to position the sensors as close to the source surface as possible in order to get rich information on the evanescent-like wave components, this is a serious problem: a small modeling error can invoke a very large irrecoverable reconstruction error by the inverse calculation. In this study, using the fact that the near-field singularities can be avoided by employing the nonsingular BIE, the effectiveness of nonsingular BEM applied to the NAH has been investigated for the reduction of reconstruction error through a simulation example of the exterior problem. As the field points become close to the source surface, separated less than about $0.2L_c$, the resultant reconstruction error decreases monotonically with the use of NBEM. Because the BEM-based NAH inevitably requires the field pressure measured in the close proximity to the source surface, the present unbiased, viz. nonsingular, approach is strongly recommended for improving the accuracy of reconstructed source field in spite of the increase of computation time for the transfer matrix.

ACKNOWLEDGMENTS

This work was partially supported by NRL and BK21 Project.

REFERENCES

[1] E. G. Williams and J. D. Maynard, Phys. Rev. Lett. **45**, 554-557 (1980).

[2] J. D. Maynard, E. G. Williams, and Y. Lee, J. Acoust. Soc. Am. **78**, 1395-1413 (1985).

[3] J. Hald, "STSF – a unique technique for scan-based near-field acoustic holography without restrictions on coherence," B&K Tech. Rev., No.1 (1989).

[4] E. G. Williams, B. H. Houston and J. A. Bucaro, J. Acoust. Soc. Am. **86**, 674-679 (1989).

[5] S.-H. Park and Y.-H. Kim, J. Acoust. Soc. Am. **104**, 3179-3189 (1998).

[6] K. Gardner and R. J. Bernhard, Trans. ASME, J. Vib. Acoust. Stress Reliab. Des. **110**, 84-90 (1988).

[7] W. A. Veronesi and J. D. Maynard, J. Acoust. Soc. Am. **85**, 588-598 (1989).

[8] M. R. Bai, J. Acoust. Soc. Am. **92**, 533-549 (1992).

[9] S.-C. Kang and J.-G. Ih, J. Acoust. Soc. Am. **107**, 2472-2479 (2000).

[10] E. G. Williams, B. H. Houston, P. C. Herdic, S. T. Raveendra, and B. Gardner, J. Acoust. Soc. Am. **108**, 1451-1463 (2000).

[11] Z. Wang and S. F. Wu, J. Acoust. Soc. Am. **104**, 2020-2032 (1997).

[12] J. D. Maynard, J. Acoust. Soc. Am. **108**, 2504 (2000).

[13] B.-K. Kim and J.-G. Ih, J. Acoust. Soc. Am. **100**, 3003-3016 (1996).

[14] E. G. Williams, *Fourier Acoustics* (Academic, London, 1999), Chaps.3,5,7,8.

[15] B.-U. Koo, B.-C. Lee, and J.-G. Ih, J. Sound Vib. **192**, 263-279 (1996).

[16] S.-C. Kang and J.-G. Ih, J. Sound Vib. **233**, 353-358 (2000).

[17] D. M. Photiadis, J. Acoust. Soc. Am. **88**, 1152-1159 (1990).

[18] J.-G. Ih and S.-C. Kang, J. Acoust. Soc. Am. **108**, 2528 (2000).

[19] B.-K. Kim and J.-G. Ih, J. Acoust. Soc. Am. **107**, 3289-3297 (2000).

[20] S. H. Yoon and P. A. Nelson, J. Sound Vib. **233**, 669-705 (2000).

[21] E. G. Williams, J. Acoust. Soc. Am. **108**, 2503 (2000).

Theoretical and Computational Acoustics 2001
E.-C. Shang, Qihu Li and T. F. Gao (Editors)

Diffusion Effects of Acoustic Wave Reflection from a Totally Refracting Random Medium

Jeong-Hoon Kim
Department of Mathematics, Yonsei University, Seoul, Korea

Abstract Based upon a stochastic and asymptotic formulation of the acoustic model, the diffusion effects of a random multilayer structure on wave propagation are analyzed for a totally refracting random medium on which waves are incident obliquely. Using a two step decomposition method, the random phase processes of the reflection coefficients are asymptotically represented in terms of the infinitesimal generators of the Kolmogorov backward equations obtained in the two divided regions of propagating and evanescent waves. The marching limit process of the phase processes throughout the whole region is expressed by the product of two evolution operators and a transformation given at the turning point. Finally, the transition probability density of the random phase process solving the Komogorov-Fokker-Planck equation is explicitly approximated by an infinite dimensional functional construction.

1 Introduction

A large class of acoustic wave propagation problems are based upon the layered structure of a medium. When there is a point source above this inhomogeneous medium, one can think of it as launching rays in all directions, which leads to a consideration of an oblique wave incidence upon the medium. The following problem is considered. A plane wave is incident upon a layered slab from an upper homogeneous medium. The slab itself rests upon another homogeneous medium. In geometrical acoustics, an asymptotic representation of acoustic wave fields within the layered medium can be obtained explicitly. The approximation is referred to as the approximate solution of a WKB type. However, the medium considered in this article is a randomly-layered medium; the effects of fluctuations of plane waves in a random multilayer are to be studied. The average acoustiic parameters such as density and sound speed are assumed to increase with depth. This increase refracts the penetrating acoustic energy upward so that an obliquely incident wave, undergoing random multiple scattering, turns at some depth. The geophysical model exhibits this structure very well; over the ages the various physical deposition processes have created a heterogeneous material with a highy-laminated fine structure in the vertical or depth direction. The constitutive parameters vary with depth on two length scales; they posess a fine-scale layering structure superposed upon a gradual slow-scale increase induced by compactification effects.

The schematic picture of the problem is that an acoustic plane wave impinges upon a refracting rapidly-layered random slab occupying the region $-L \leq z \leq 0$ from above. In the transverse directions, the acoustic parameters are assumed to be homogeneous ones to a degree that makes one-dimensional modeling a reasonable approximation. The exterior regions $z > 0$ and $z < -L$ are homogeneous ones; in these regions the acoustic parameters are constant. The problem of interest is formulated as a stochastic boundary value problem for scattering variables and subsequently it is recast as a stochastic initial value problem for the reflection coefficient. The two step decomposition approach will be used to the problem. Problems above and below the turning point correspond to propagating and evanescent waves, respectively, and they are dealt with separately and then the two results are combined at the turning point by the continuity of wave fields. The scattering variables must be suitably defined in each region.

To see the variations of the large-scale structure of the medium, the wave needs to have short wavelength compared to the length scales of the average macrostructure but to acquire the statistical properties of the details of the random inhomogeneities, the wavelength needs to be long compared to the size of the random microstructure. From the point of analysis, this leads to combine both a geometrical acoustics-like accomodation of the large scale macroscopic variations and a central limit theorem or diffusion limit theorem for the fine scale microscopic fluctuations. The combined limits can reveal the very rich interactive structures of the problem for suitably defined quantities. An asymptotic limit theory, developed by Khasminskii[1], of stochastic differential equations for multiscale variables remains a main mathematical tool for the analysis. The finite dimensonal distribution of the solution process is given by the Kolmogorov backward equation with a certain infinitesimal generator. Two infinitesimal generators of the phase processes of the reflection coefficients will be obtained separately corresponding to the divided regions and a transformation of the two processes adjoining at the turning point will be given. The transition probability density of the random phase process throughout the whole region satisfies the Komogorov-Fokker-Planck equation and its explicit approximation is given by an infinite dimensional functional.

2   Asymptotic and stochastic formulation

The basic equations are linearized acoustic wave equations for a pressure and a particle velocity. Within the random slab, the acoustic parameters ρ(density) and c(sound speed) are assumed to vary with depth in a manner that superposes a randomly-fluctuating component upon a slowly-varying (nominally linear) increasing average value. A small parameter ϵ to quantify this two different scale dependence as follows:

$$\rho(z) = \rho(z, z/\epsilon^2), \qquad c(z) = c(z, z/\epsilon^2), \qquad -L \leq z \leq 0, \tag{1}$$

where ρ and c are unit correlation length random functions of the second arguments while the first arguments account for deterministic nonstationary modulation; the actual correlation length of the processes is, therefore, ϵ^2. Let the frequency of the incident wave also be scaled by $\omega = \tilde{\omega}/\epsilon$, where $\tilde{\omega}$ is $O(1)$ (the tilde will subsequently be omitted throughout). Consequently the orders $O(1)$, $O(\epsilon)$, $O(\epsilon^2)$ correspond to the large-scale structure of the medium, the wavelength of incident wave, and the correlation length of the random features of the medium, respectively. This implies that the wavelength is small enough to probe macroscopic variations of the medium and large enough to average over microscopic random effects. A type of limit is, therefore, possible which combines a geometric acoustics-like limit for large-scale deterministic variations and a diffusion limit for fine-scale random fluctuations.

The monochromatic, time-harmonic, plane wave incident upon the layer with an incidence angle θ measured from the normal. The assumed plane wave incidence and transverse homogeneity imply the following equations for the acoustic pressure P and the normal component of the particle velocity U_z in the random slab.

$$\partial P/\partial z = -i(\omega/\epsilon)\rho(z, z/\epsilon^2)U_z, \tag{2}$$

$$\partial U_z/\partial z = -i(\omega/\epsilon)K^{-1}(z, z/\epsilon^2)(1 - c_0^{-2}c^2(z, z/\epsilon^2)\sin^2\theta)P, \tag{3}$$

where c_0 is the sound speed of the upper homogeneous medium and K(bulk modulus)$= \rho c^2$.

There are two ways by which this two-point boundary value problem for the wave fields can be recast as an initial value problem for a suitably defined variable. The first method is to obtain the fundamental or propagator matrix and change the dependent variables by this matrix. This propagator matrix method replaces a linear two-point boundary value problem for a vecor-valued dependent variable with a linear initial value problem for a matrix-valued dependent variable. The second method involves appropriately chosen dependent variables from observation of the form of boundary condition as well as the structure of wave fields within the region of scattering medium. The new dependent variables in this method are scattering variables and define a reflection coefficient which solves a Riccati type nonlinear equation. The choice of the method depends on the nature of wave propagation. If the averaged acoustic parameters in equations (2)-(3) vary in such a way that there exists a turning point dividing the behavior of WKB solutions to the averaged equations, the first method will be useful for removing the rapid deterministic phase variations for propagating waves, while the second method may be more appropriate for evanescent waves. It is assumed in this article that the averaged density and sound speed increase in such a way that the expected value of $K^{-1}(1 - (c/c_0)^2 \sin^2 \theta)$ is monotone decreasing and changes sign from positive to negative as z decreases; this corresponds to the fact that the eigenvalues of the averaged coefficient matrix of (2)-(3) change from pure imaginary to real valued. Let z_T be the point where the eigenvalues vanish. The point z_T will remain the dividing point for the stochastic problem in the chosen scale.

If $\Phi \ (= (\phi_{ij}))$ is the propagator matrix of the effective problem (i.e., problem (2)-(3) in which mean zero terms are suppressed), the new dependent variables A and B defined by $[P, U_z]^T = \Phi[A, B]^T$ are scattering variables. The dependent variable A represents upward-propagating (or upgoing) wave amplitude while B while the dependent variable B represents downward-propagating (or downgoing) wave amplitude. These variables satisfy a system of coupled linear equations whose boundary condition can be produced by using the hyperbolicity of the wave equation. First, the natural choice of a boundary condition at the interface $z = 0$ is that $B = 1$ at $z = 0$. The hyperbolicity of (2)-(3) implies that the choice of a boundary condition at some depth $z = -L$ will not affect the reconstruction of the pressure and particle velocity fields at the interface $z = 0$ for any finite time t provided that L is large enough. Thus one can establish a boundary condition at $z = -L$ by terminating the medium at that point and assuming that the medium is uniform on $z < -L$. A total reflection occurs at $z = -L$; $|A/B| = 1$ at $z = -L$.

To characterize the reflected waves at $z = 0$ from the random medium, the scattering nature of the medium must be fully utilized. The invariant imbedding representation of the random reflection coefficient R defined by the ratio of the upgoing to downgoing wave amplitudes, i.e., $R = A/B$, can be used here. This reflection coefficient will satisfy a stochastic Riccati type nonlinear equation. In terms of $\lambda = i\omega/\epsilon$ and $\det \Phi = \phi$, this is indeed

$$\partial R/\partial z + \lambda \phi^{-1}[(\alpha_1 \phi_{22}^2 - \beta_1 \phi_{12}^2) + 2(\alpha_1 \phi_{21} \phi_{22} - \beta_1 \phi_{11} \phi_{12})R + (\alpha_1 \phi_{21}^2 - \beta_1 \phi_{11}^2)R^2] = 0, \qquad (4)$$

where α_1 and β_1 denote the random component of ρ and $K^{-1}(1 - (c/c_0)^2 \sin^2 \theta)$, respectively. The continuity of the wave fields at the interface $z = -L$ leads to the initial condition for this equation;

$$R(-L) = \frac{\phi_{12}(-L) + \kappa \phi_{22}(-L)}{\phi_{11}(-L) + \kappa \phi_{21}(-L)}, \qquad (5)$$

where κ is the acoustic impedance in the uniform region $z < -L$. Once this stochastic equation is solved and stored from $z = -L$ to $z = 0$, the total wave fields at any given depth z can be constructed by solving the following one-way equation for the downgoing wave from $z = 0$.

$$\partial B/\partial z + \lambda \phi^{-1}[(\alpha_1 \phi_{21} \phi_{22} - \beta_1 \phi_{11} \phi_{12}) + (\alpha_1 \phi_{21}^2 - \beta_1 \phi_{11}^2)R]B = 0. \qquad (6)$$

It is usually difficult to get the exact propagator matrix but the present high frequency regime allows one to have a WKB approximant for the required fundamental matrix. If the approximant belongs to a Lie group $SU(1,1)^2$ that leaves the Hermitian form invariant, it means that the approximant preserves the group structure of the propagator matrix of the exact problem. Then the underlying group structure of the resultant equation of the scattering variables can imply that the time average energy flux for a harmonic wave is invariant. This with a total reflection at one boundary point will lead to total reflection everywhere in the random medium. If this is the case, the problem for the random reflection coefficient will reduce to the problem of a random phase function.

One way of characterizing the statistical properties of the stochastic Riccati equation for either R or its phase is to replace the equation by an equation with a white noise type of random idealization (Brownian motion) and interpret the equation in the sense of either Ito or Stratonovich. This will lead to the Fokker-Planck equation for the transition probability density of a diffusion process. Then asymptotic analysis for the Fokker-Planck equation will reveal that the leading order equation becomes a variable coefficient Ornstein-Uhlenbeck equation. From a modeling point view, however, an asymptotic limit theory developed by Khasminskii covers more broad range of random fluctuations and moreover provides more appropriate approximation in such a way that the drift coefficient of the infinitesimal generator has a lower order modification contributed by the random fluctuations and the diffusion coefficient contains an information about the coupling of noise intensities with deterministic variations.

The problem of interest includes the existence of a propagating-evanescent transition point, that is, a turning point in the random medium. It is denoted by z_T. Randomization effect is expected to vary depending on the region above or below the turning point and also on how far it is from the turning point. A two-step decomposition approach should be useful for this type of problem. Two different scattering formulations are to be established corresponding to the regions adjoining at z_T. It is expected physically that the reflecion process above the turning point undergoes rapid phase wrapping[3], while the reflection process below the turning point to leading order decays exponentially. Significant diffusion will occur in the region above the turning point, while one expects little diffusion in the region below the turning point. The asymptotic behavior of the corresponding random phases can be characterized in terms of two limit theorems[4,5] that extend Khasminskii's result. For the frequency range of interest, the scale dependence of diffusion and drift terms of the process generator in each region will indicate the subtle interplay of refraction and random multiple scattering.

3 Total internal reflection

3.1 Propagating wave above the turning point

In order to obtain quantities exhibiting a limiting stochastic behavior as ϵ goes to zero, one first needs to obtain a propagator matrix of (2)-(3) with stochastic terms suppressed. Then the new equations for the scattering variables A and B defined by this matrix become centered. Since it is difficult to have such an exact propagator matrix, it is desirable to have a WKB approximant from which the rapid deterministic term of (2)-(3) is essentially negligible in the asymptotic limit of interest. Moreover, the present problem includes a turning point. Because stochastic effects in the propagating regime are expected to depend highly on the distance from the turning point, it is required to have an approximant that is uniformly valid in the both transition and outer regions. This means that the independent variable needs to be stretched with the transition region scale near the turning point.

Such an approximant can be actually obtained from the results of Lynn and Keller[6]. In terms of $\alpha = <\rho>$, $\beta = <K^{-1}(1-(c/c_0)^2\sin^2\theta)>$, and the mean travel time (measured upward from z_T) defined by

$$\tau(z) = \int_{z_T}^{z} <c^{-2} - c_0^{-2}\sin^2\theta>^{1/2} dz \qquad (7)$$

(under the assumption of independence of the relevant processes) the explicit representation of such a matrix $\Phi \ (= (\phi_{ij}))$ is given by

$$\phi_{1j} = \alpha^{1/2}\zeta\Theta_j(\lambda^{2/3}\delta), \qquad (8)$$

$$\phi_{2j} = \lambda^{-1}\alpha^{-1/2}(\alpha'\zeta/2\alpha + \zeta')\Theta_j(\lambda^{2/3}\delta) + \lambda^{-1/3}\alpha^{-1/2}\zeta^{-1}\Theta_j'(\lambda^{2/3}\delta) \qquad (9)$$

where $\delta^{3/2} = (3/2)\tau$, $\zeta^4 = \delta/\alpha\beta$, $\Theta_1 = -i\pi^{1/2}\lambda^{1/6}e^{-\lambda\tau(0)}\{Ai(\lambda^{2/3}\delta) + iBi(\lambda^{2/3}\delta)\}$, and $\Theta_2 = -\Theta_1^*$.

Now the above approximant is applied to the problem in the region $z_T < z \le 0$. Since the group structure of the above approximant implies the invariance of the energy flux, the problem reduces to a consideration of the random phase of the unimodular reflection coefficient. The random phase function ψ defined by $R = \exp(-i\psi)$ is then expressed by the new stretched independent variable $\xi = (\omega/\epsilon)^{2/3}\delta$ and solves the following stochastic Riccati equation on the interval $0 < \xi < \xi_0$.

$$d\psi/d\xi = (\omega/\epsilon)^{2/3}\pi\alpha\zeta^4 M^2(\xi)\beta_1\{1 - \cos(\psi + 2\Omega^\dagger(\xi))\}$$
$$+ \pi\alpha^{-1}\zeta^4 N^2(\xi)\alpha_1\{1 + \cos(\psi + 2\Psi^\dagger(\xi))\} \qquad (10)$$

where an initial value condition ψ_0 at the turning point satisfies $R(z_T) = \exp(-i\psi_0)$. Here M, $\Omega^\dagger$, and N, $\Psi^\dagger$ are the modulus and phase of $Ai(-\xi) + iBi(-\xi)$ and $Ai'(-\xi) + iBi'(-\xi)$, respectively, up to a constant. The deterministic functions α and ζ depend on a slow variable z and they behave like constants in the ξ-scale. Since asymptotically the leading order terms of M and N are $(1/\sqrt{\pi})\xi^{-1/4}$ and $(1/\sqrt{\pi})\xi^{1/4}$, respectively, one can observe that there is a competition between the two terms of the right hand side of (10). The first term has an order $\epsilon^{-2/3}$ and the second term has an order 1 when ξ is finite (i.e., in the transition region) but these terms become comparable to each other and have the same order as of $\epsilon^{-1/3}$ as ξ gets large (i.e., in the outer region). The second term can not be ignored anymore in the diffusion limit. Unless such a problem is dealt with separately in the two different regions, it is, therefore, required to have a diffusion limit theory[4] extended to an asymptotically unbounded interval.

One applies an extended limit theory to equation (10) and obtains an exact asymptotic form of the single frequency infinitesimal generator corresponding to the phase process (a diffusion type Markov pocess). This is an adjoint form of the generator of the Kolmogorov-Fokker-Planck equation for the transition probability density of the phase process. It is given by

$$L = [\gamma_{22}\omega^{4/3}\pi^2\alpha^2\zeta^8 M^4(1-\cos\Omega^*)^2$$
$$+ \epsilon^{2/3}(\gamma_{21} + \gamma_{12})\omega^{2/3}\pi^2\zeta^4 M^2 N^2(1-\cos\Omega^*)(1+\cos\Psi^*)$$
$$+ \epsilon^{4/3}\gamma_{11}\pi^2\alpha^{-2}N^4(1+\cos\Psi^*)^2]\partial^2_{\psi\psi} \qquad (11)$$
$$+ [\gamma_{22}\omega^{4/3}\pi^2\alpha^2\zeta^8 M^4(1-\cos\Omega^*)\sin\Omega^*$$
$$- \epsilon^{2/3}\omega^{2/3}\pi^2\zeta^4 M^2 N^2\{\gamma_{21}(1-\cos\Omega^*)\sin\Psi^* - \gamma_{12}\sin\Omega^*(1+\cos\Psi^*)\}$$
$$- \epsilon^{4/3}\gamma_{11}\pi^2\alpha^{-2}N^4(1+\cos\Psi^*)\sin\Psi^*]\partial_\psi,$$

where $\Omega^* = \psi + \Omega^\dagger$ and $\Psi^* = \psi + \Psi^\dagger$ and the (nonstationary) noise intensity matrix $(\gamma_{ij}(z))$ is the one generated by the ramdom processes α_1 and β_1.

This asymptotic generator is now uniformly valid in the both transition and outer regions above the turning point. Using the asymptotic behavior of the moduli M and N, one can notice that there is a competition among drift and diffusion coefficients. In terms of the z−scale, in which the order 1 corresponds to the order $\epsilon^{2/3}$ in the ξ−scale, the terms involving γ_{22}, γ_{21}, γ_{12}, and γ_{11} are of orders 1, $\epsilon^{2/3}$, $\epsilon^{2/3}$, and $\epsilon^{4/3}$, respectively, in the transition whereas all of these are of comparable order $\epsilon^{2/3}$ in the outer region. The singular scale dependence of the diffusion and drift terms in the transition region indicates that the refraction of the wave plays little role near the turning point. This implies physically that the randomization effect is enhanced as the rays become aligned with random layers. Also the multiple scattering is even more pronounced near the turning point than in the outer region as the frequency increases in the valid range of our analysis. At higher frequencies, the randomization effect due to the increased multiple scattering is expected to be much stronger even in the outer region so that the wave fields become relatively insensitive to the existence of the turning point.

3.2 Evanescent wave below the turning point

In the previous section, the reflection coefficient is assumed to be known as a deterministic value at the turning point. The assumed value should be a random one to be determined from the initial value problem in the region below the turning point. However, the scattering formulation established for the problem above the turning point is not appropriate for the problem below the turning point because the error term in the Lynn and Keller approximant is not small at all below the turning point. Below the turning point, therefore, different scattering variables are required for the same basic problem. Here the second method described in the second section will be used to obtain appropriate scattering variables.

Motivated by a particular case corresponding to the homogeneous medium, in which ρ_0 and c_0 denote the density and sound speed, respectively, let A (for plus) and B (for minus) be $(1/2)(P \pm \rho_0 c_0 U_z \sec \theta)$. Then A and B correspond to the upgoing and downgoing wave amplitudes, respectively. From the same argument as in the problem above the turning point, the coefficient matrix of the resultant equation will have the same group structure as in the case above the turning point. Then the problem again reduces to an evolution of the random phase of the unimodular reflection coefficient. Contrary to the case above the turning point, it is not required to have a stretched independent variable here due to the physical nature of the wave; one expects little diffusion in the region below the turning point because the equation will be a type of stochastic equation with a rapidly varying (not oscillating) component.

The corresponding initial value problem below the turning point is given by

$$d\psi/dz = (-k/\epsilon)[(\mu + \nu \cos \psi) + (\mu_1 + \nu_1 \cos \psi)], \tag{12}$$

where the average plus a mean-zero random part of $\pm \tilde{\rho} \cos \theta + (\tilde{K}^{-1} - \tilde{\rho}^{-1} \sin^2 \theta) \sec \theta$ is denoted by $\mu + \mu_1$ (for plus) or $\nu + \nu_1$ (for minus). Also, $k = \omega/c_0$, $\tilde{\rho} = \rho/c_0$, and $\tilde{c} = c/c_0$. The limiting behavior of solutions to the above equation is controlled by the solution of an effective problem.

An asymptotic limit theory[5] can show how the deterministic force of rapidly varying perturbation interplays with random noise intensity in terms of coupling of these two factors contained in the drift and diffusion coefficients of the generator. From this theory, the random noise of the evanescent wave below the turning point is now characterized by the following infinitesimal generator.

$$L = [(k/\epsilon)^2 \int_\sigma^{\sigma+\epsilon} ds(\gamma_{11} + \gamma_{12}^c + (\gamma_{21} + \gamma_{22}^c) \cos \psi) e^{-(k/\epsilon) \int_\sigma^s \nu \sin \psi^0 dt}] \partial_{\psi\psi}^2,$$

$$+[-(k/\epsilon)(\mu + \nu\cos\psi) + (k/\epsilon)^2 \int_\sigma^{\sigma+\epsilon} ds\{\gamma_{12}^s + \gamma_{22}^s\cos\psi \tag{13}$$

$$-(\gamma_{11} + \gamma_{12}^c + (\gamma_{21} + \gamma_{22}^c)\cos\psi)(k/\epsilon)\int_\sigma^s \nu\cos\psi^0 e^{-(k/\epsilon)\int_t^s \nu\sin\psi^0 du}dt\}]\partial_\psi$$

where (γ_{ij}) is a noise intesity matrix generated by μ and ν, and $\gamma_{ij}^s = \gamma_{ij}\sin\psi^0$ and $\gamma_{ij}^c = \gamma_{ij}\cos\psi^0$.

3.3 Transformation at the turning point

A wave propagation problem in the random slab has been dealt with in the two divided regions. Two different formulations were obtained and the infinitesimal generators of the random phase processes were represented corresponding to these two regions. Since the two problems adjoin at the turning point, it is, therefore, required to consider a transformation of these processes.

From the continuity of the wave fields at the turning point, one can obtain the algebraic relation between the phase $\overline{\psi}_T$ at the turning point, assumed to be known deterministically for the problem above the turning point, and the (random) phase $\underline{\psi}_T$ at the same point obtained by solving the problem below the turning point. It is expressed by $\overline{\psi}_T = h(\underline{\psi}_T)$ for some (multivalued) function h. Probability distributions of the limiting unified random phase, $\psi(0, -L, \psi_{-L})$, throughout the whole region are then represented in terms of this transformation and the limiting propagators, $L(-L, z_T)$ and $U(z_T, 0)$ corresponding to the regions below and above the turning point, respectively. For any initial phase ψ_{-L} and sufficiently smooth test function f,

$$< f(\psi(0, -L, \psi_{-L})) >= L(-L, z_T)H_T U(z_T, 0)f(\psi_{-L}), \tag{14}$$

where $H_T f = f \circ h$. The random phase process ψ can now march through the medium sequentially with L, H_T and U.

4 Probability distributions of the random phases

To obtain the probability distributions of the random phase processes, partial differential equations with the generators given in the previous section are to be solved in the two divided regions. These equations are the Kolmogorov backward equations. The corresponding forward dependent variables are the transition probability densities and satisfy the so called Kolmogorov-Fokker-Planck equations. In particular, the leading order transition probability density P_0 of the random phase process is given by

$$\partial_\sigma P_0(\sigma, \psi_0; \tau, \psi) + L^* P_0(\sigma, \psi_0; \tau, \psi) = 0, \quad \tau > \sigma, \tag{15}$$

with an inital condition $\lim_{\tau\downarrow\sigma} P_0(\sigma, \psi_0; \tau, \psi) = \delta(\psi - \psi_0)$. Here L^* is the adjoint operator of a limiting generator L.

In general, it is difficult to express the explicit solution representation of these equations in the closed form because of the dependence of coefficients on the variables τ and ψ. Here a pseudodifferential operator theory[7] is used and it is combined with an infinite dimensional functional construction to approximate the solution of (15). Then the operator symbol Λ_{L^*} of the generator L^* contains the complete spectral information for the transition probability density. To account for the τ-dependence in the operator symbol, the interval $[\sigma, \tau]$ is divided into N number of subintervals such that $\sigma = \tau_0 < \tau_1 < \cdots < \tau_N = \tau$ with the corresponding values ψ_i evaluated at τ_i and then repeatedly the Chapman-Kolmogorov equation is used to take the transition probability density function as a time-ordered product. In combination with the pseudodifferential operator calculus, the solution of equation (15) then takes the following approximate form:

$$P_0(\sigma, \psi_0; \tau, \psi) = \lim_{N \to \infty} (2\pi)^{-N} \int_{R^{2N-1}} d\psi_1 d\psi_2 \cdots d\psi_{N-1} \, dp_1 dp_2 \cdots dp_N \qquad (16)$$

$$\cdot \exp\left(i \sum_{j=1}^{N} \left(p_j(\psi_j - \psi_{j-1}) + ((\tau - \sigma)/N)\Lambda_{L^*}(\tau_j; \psi_j, p_j))\right)\right).$$

5 Conclusion

The random half space, on which the time harmonic acoustic wave is obliquely incident, was assumed to have a turning point. The two different scattering formulations were made corresponding to the two different physical wave phenomena. In terms of the random phase of the reflection coefficient, the reflection process above the turning point undergoes rapid phase wrapping whereas the reflection process below the turning point decays exponentially to leading order. Asymptotic diffusion limit theory of stochastic differential equations with three scales of variables was used to characterize both significant diffusion effect above the turning point and rapid variation effect with random noise below the turning point. A unified limit law of the phase process was obtained by putting both results together. The marching limit process was shown to be realized by solving the corresponding Kolmogorov-Fokker-Planck equations of the transition probability densities.

Acknowledgments

This work was supported by the interdisciplinary research program of the KOSEF 1999-2-103-001-5 and by the Brain Korea 21 Project

References

1. R. Z. Khasminskii, "A limit theorem for the solutions of differential equations with random right hand sides", Theory Prob. Appl., **11**(1966), 390–406.
2. S. Helgason, *Differential Geometry, Lie Groups and Symmetric Spaces*, Academic Press, New York, NY, 1978.
3. W. E. Kohler, "Reflection from a one-dimensional, totally refracting random multilayer II, Internal field statistics", SIAM J. Appl. Math. **48**(1988), 652–661
4. J.-H. Kim, "A uniform diffusion limit for random wave propagation with a turning point", J. Math. Phys., **37**(1996), 752–768.
5. J.-H. Kim, "An asymptotic limit law with a singularly perturbed drift and a random noise", J. Math. Phys., **37**(1997), 2660–2675
6. R. Y. S. Lynn and J. B. Keller, "Uniform asymptotic solutions of second order linear ordinary differential equations with turning points", Comm. Pure Appl. Math., **23**(1970), 379–408
7. G. B. Folland, *Harmonic Analysis in Phase Space*, Princeton University Press, Princeton, NJ, 1989.

Theoretical and Computational Acoustics 2001
E.-C. Shang, Qihu Li and T. F. Gao (Editors)
© 2002 World Scientific Publishing Co.

Finite Element Analysis on Waves Propagation in Cylinders

Xianmei Wu, Menghu Qian

Institute of Acoustics, Tongji University, Shanghai, 200092, P. R. China

Abstract   The application of finite element method (FEM) in analyzing the waves propagation in cylinders is presented in this paper. The finite element equation is obtained by the virtual displacement theorem and solved by the central difference method. The spatial and temporal steps of discretization of the finite element computation are discussed. In order to compute high-frequency problems, a new sub-matrix method was set up and applied in the computations. The displacement fields, excited by a pulsed line-source with 5 MHz central frequency, are computed and displayed by gray plots. The waveforms of receiving points on the surface of the cylinders, identified as A-scan results, are also shown. The comparison between the FEM and experimental results shows that the FEM is a good way for analyzing the waves propagation in cylinders.

Keywords: finite element method, waves propagation

1   Introduction

Lots of research works have been done in the theoretical analysis of wave-propagation problems in elastic solids. But most of them are restricted to infinite or semi-infinite space. If the arbitrary boundary or complex shaped defects exist, it will be very difficult to analyze the propagation of waves in elastic solids by the theoretical methods. Several numerical methods have been applied to solve this problem, but the finite element method (FEM) is better because it is easier to deal with the awkward geometries than other methods.

2   Problem statement

The finite element solution of a problem includes the discretization of the computed region into a series of simple elements, the approximation of the interior field values in terms of its nodal values through the shape function, and the determination of the nodal values through the minimization of energy by the virtual displacement theorem. The problem we solve with the finite element method is to simulate the propagation of the waves generated by a pulsed line-source on the curved surface of cylinders. It can be simplified into a two-dimensional plane problem (Fig. 1). The cross section of a cylinder is chosen to be discretized. Only half of the geometry is discretized by triangular elements because of the symmetry. The governing equation for an isotropic medium in the absence of body forces is as following:

$$(\lambda + \mu)\nabla(\nabla \cdot \vec{u}) + \mu\nabla^2\vec{u} = \rho\ddot{\vec{u}} \tag{1}$$

where $\bar{u}$ is the displacement vector, λ and μ are the Lame constants of the material, ρ is the density of the material. After discretizing the problem domain with triangular elements, the following matrix equation is derived through the minimization of energy by the virtual displacement theorem[1][2].

$$\mathbf{M\ddot{U} + KU = F} \tag{2}$$

where $\mathbf{K}$ is the global stiffness matrix determined by the elastic properties of the solid, $\mathbf{M}$ is the global mass matrix determined by the density of the cylinders, $\mathbf{F}$ is the surface traction vector, and $\mathbf{U}$ is the displacement vector.

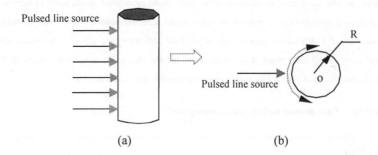

Pulsed line source

Pulsed line source

(a)   (b)

Fig. 1   Schematic diagram of the FEM computation

Eq. 2 may be solved by several methods. Central difference method is suitable for computing wave propagation and can be easily programmed on a personal computer. Therefore, it is applied to solve our problem. In this method, the acceleration can be expressed as

$$\mathbf{\ddot{U}}_t = (\mathbf{U}_{t+\Delta t} - 2\mathbf{U}_t + \mathbf{U}_{t-\Delta t})/\Delta t^2 \tag{3}$$

Substituting Eq. 3 to Eq. 2, yields

$$\frac{\mathbf{M}}{\Delta t^2}\mathbf{U}_t = \mathbf{F}_{t-\Delta t} - (\mathbf{K} - \frac{2\mathbf{M}}{\Delta t^2})\mathbf{U}_{t-\Delta t} - \frac{\mathbf{M}}{\Delta t^2}\mathbf{U}_{t-2\Delta t} \tag{4}$$

If the mass matrix is diagonalized through a mass lumping technique[3][4] , then

$$\mathbf{U}_t = \Delta t^2 [\mathbf{M}]^{-1}\{\mathbf{F}_{t-\Delta t} - (\mathbf{K} - \frac{2\mathbf{M}}{\Delta t^2})\mathbf{U}_{t-\Delta t} - \frac{\mathbf{M}}{\Delta t^2}\mathbf{U}_{t-2\Delta t}\} \tag{5}$$

Eq. 5 is an iteration formula and can be started with the initial conditions

$$\mathbf{U}_0 = 0, \quad \mathbf{U}_{-\Delta t} = 0 \tag{6}$$

Just as in any propagation problem, the step size of temporal and spatial discretization is important in solving the wave equation. In the discretization of FEM, at least eight nodes per shortest wavelength is required in order to simulate all the wave types (head, surface, longitudinal and shear waves). Hence, the step size of the spatial discretization depends on the frequency of the exciting source. Sixteen nodes per shortest wavelength are needed for computing the A-scan data[4]. Additionally, the numerical stability of Eq. 5 requires that the temoral step, Δt, and the spatial step, Δh, meet the following criterion[5]: $\Delta t \leq \Delta h / C_{max}$ where C_{max} is the maximum wave velocity in the material. In isotropic materials, C_{max} is the longitudinal velocity. The relationship between Δt and Δh used in our computation is about:

$$\Delta t \approx 0.68 \Delta h / C_{max} \tag{7}$$

3 Submatrix method to solve the large scale FEM problem

Although many researchers have studied the application of finite element method in the simulation of waves propagation, most people use the frequency of 1MHz to simulate the transducer source, which is not suitable for simulating the laser ultrasonic source with the 5MHz or higher central frequency. In fact, the higher the frequency of the exciting source, the more difficult the FEM computation because the stiff matrix will be huge and often hard to be handled. From the following example, the size of the matrix can give us a clear idea of how much memory will be necessary for the computation in high frequency.

In order to simulate a laser pulse source with 5MHz center frequency, a 5MHz sine source of one period is chosen as the load vector. In the previous section, we know that the spatial step is decided by the frequency of the exciting source. To achieve good accuracy, the 0.034mm spatial step was used in our computation. From the relationship in Eq. 7, we know that the temporal step also has to be very small. 4ns is applied in our computation. If the diameter of the cylinder is 10mm, the total nodes is 32782, and hence the stiff matrix is a 65564×65564 sparse matrix, really a very large matrix. Moreover, if the diameter increases slowly, the matrix increase quickly. Although such a matrix could be handled on some computational machines, it usually causes an error of out of memory in personal computer, especially when the exciting source has a higher center frequency. In order to solve this problem, a sub-matrix method is set up by separating the stiffness matrix and mass matrix into a series of submatrixes, which is described below.

When there are n nodes in the computation, nodal displacements in Eq. 5 can be expressed as

$$[U_t]_{1 \times 2n} = \Delta t^2 [M]^{-1}_{2n \times 2n} \bullet \{[F_{t-\Delta t}]_{1 \times 2n} - [K]_{2n \times 2n} - \frac{2}{\Delta t^2}[M]_{2n \times 2n}\}[U_{t-\Delta t}]_{1 \times 2n} - \frac{1}{\Delta t^2}[M]_{2n \times 2n}[U_{t-2\Delta t}]_{1 \times 2n}$$

If **K** and **M** is too large to compute, we can divide them into submatrixes:

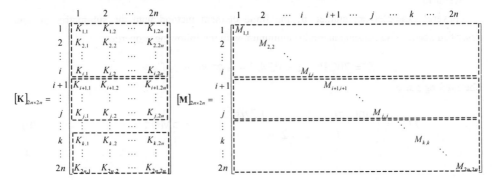

If defining
$$\hat{\mathbf{F}}_{t-\Delta t} = \mathbf{F}_{t-\Delta t} - \left[\mathbf{K} - \frac{2}{\Delta t^2}\mathbf{M}\right]\mathbf{U}_{t-\Delta t} - \frac{1}{\Delta t^2}\mathbf{M}\mathbf{U}_{t-2\Delta t}$$

Eq.5 can be rewritten as

$$\mathbf{U}_t = \Delta t^2 \mathbf{M}^{-1}\hat{F}_{t-\Delta t} \tag{8}$$

After **K** and **M** are divided into submatrixes, Eq.8 takes the following form:

$$
\begin{bmatrix}\left(\hat{\mathbf{F}}_{t-\Delta t}\right)_1 \\ \vdots \\ \left(\hat{\mathbf{F}}_{t-\Delta t}\right)_i\end{bmatrix} = \begin{bmatrix}\left(\mathbf{F}_{t-\Delta t}\right)_1 \\ \vdots \\ \left(\mathbf{F}_{t-\Delta t}\right)_i\end{bmatrix} - \left\{\begin{bmatrix}K_{1,1} & \cdots & K_{1,2n} \\ \vdots & \ddots & \vdots \\ K_{i,1} & \cdots & K_{i,2n}\end{bmatrix} - \frac{2}{\Delta t^2}\begin{bmatrix}M_{1,1} & \cdots & M_{1,2n} \\ \vdots & \ddots & \vdots \\ M_{i,1} & \cdots & M_{i,2n}\end{bmatrix}\right\}\begin{bmatrix}(U_{t-\Delta t})_1 \\ \vdots \\ (U_{t-\Delta t})_{2n}\end{bmatrix} - \frac{1}{\Delta t^2}\begin{bmatrix}M_{1,1} & \cdots & M_{1,2n} \\ \vdots & \ddots & \vdots \\ M_{i,1} & \cdots & M_{i,2n}\end{bmatrix}\begin{bmatrix}(U_{t-2\Delta t})_1 \\ \vdots \\ (U_{t-2\Delta t})_{2n}\end{bmatrix}
$$

$$
\begin{bmatrix}\left(\hat{\mathbf{F}}_{t-\Delta t}\right)_{i+1} \\ \vdots \\ \left(\hat{\mathbf{F}}_{t-\Delta t}\right)_j\end{bmatrix} = \begin{bmatrix}\left(\mathbf{F}_{t-\Delta t}\right)_{i+1} \\ \vdots \\ \left(\mathbf{F}_{t-\Delta t}\right)_j\end{bmatrix} - \left\{\begin{bmatrix}K_{i+1,1} & \cdots & K_{i+1,2n} \\ \vdots & \ddots & \vdots \\ K_{j,1} & \cdots & K_{j,2n}\end{bmatrix} - \frac{2}{\Delta t^2}\begin{bmatrix}M_{i+1,1} & \cdots & M_{i+1,2n} \\ \vdots & \ddots & \vdots \\ M_{j,1} & \cdots & M_{j,2n}\end{bmatrix}\right\}\begin{bmatrix}(U_{t-\Delta t})_1 \\ \vdots \\ (U_{t-\Delta t})_{2n}\end{bmatrix} - \frac{1}{\Delta t^2}\begin{bmatrix}M_{i+1,1} & \cdots & M_{i+1,2n} \\ \vdots & \ddots & \vdots \\ M_{j,1} & \cdots & M_{j,2n}\end{bmatrix}\begin{bmatrix}(U_{t-2\Delta t})_1 \\ \vdots \\ (U_{t-2\Delta t})_{2n}\end{bmatrix}
$$

$$\vdots$$

$$
\begin{bmatrix}\left(\hat{\mathbf{F}}_{t-\Delta t}\right)_k \\ \vdots \\ \left(\hat{\mathbf{F}}_{t-\Delta t}\right)_{2n}\end{bmatrix} = \begin{bmatrix}\left(\mathbf{F}_{t-\Delta t}\right)_k \\ \vdots \\ \left(\mathbf{F}_{t-\Delta t}\right)_{2n}\end{bmatrix} - \left\{\begin{bmatrix}K_{k,1} & \cdots & K_{k,2n} \\ \vdots & \ddots & \vdots \\ K_{2n,1} & \cdots & K_{2n,2n}\end{bmatrix} - \frac{2}{\Delta t^2}\begin{bmatrix}M_{k,1} & \cdots & M_{k,2n} \\ \vdots & \ddots & \vdots \\ M_{2n,1} & \cdots & M_{2n,2n}\end{bmatrix}\right\}\begin{bmatrix}(U_{t-\Delta t})_1 \\ \vdots \\ (U_{t-\Delta t})_{2n}\end{bmatrix} - \frac{1}{\Delta t^2}\begin{bmatrix}M_{k,1} & \cdots & M_{k,2n} \\ \vdots & \ddots & \vdots \\ M_{2n,1} & \cdots & M_{2n,2n}\end{bmatrix}\begin{bmatrix}(U_{t-2\Delta t})_1 \\ \vdots \\ (U_{t-2\Delta t})_{2n}\end{bmatrix}
$$

Substituting the above equations into Eq.8,

$$
[\mathbf{U}_t]_{1\times 2n} = \Delta t^2 [\mathbf{M}]_{2n}^{-1}\begin{bmatrix}\left(\hat{\mathbf{F}}_{t-\Delta t}\right)_1 \\ \left(\hat{\mathbf{F}}_{t-\Delta t}\right)_2 \\ \vdots \\ \left(\hat{\mathbf{F}}_{t-\Delta t}\right)_i \\ \left(\hat{\mathbf{F}}_{t-\Delta t}\right)_{i+1} \\ \vdots \\ \left(\hat{\mathbf{F}}_{t-\Delta t}\right)_j \\ \vdots \\ \left(\hat{\mathbf{F}}_{t-\Delta t}\right)_k \\ \vdots \\ \left(\hat{\mathbf{F}}_{t-\Delta t}\right)_{2n}\end{bmatrix}_{1\times 2n}
$$

which is just the displacement field at time t. If all the displacement fields in the studied period of time are obtained, the problem of the waves propagation in cylinder will be solved by this numerical way.

4 Results

Several examples are studied by the finite element method with the sub-matrix method described above. The chosen material is aluminum with the following properties:

$$E = 70\text{GPa}, \quad \upsilon = 0.34, \quad \rho = 2700 \ kg \ / \ m^3$$

The exciting pulse is

$$f(t) = \begin{cases} \sin(2\pi ft) & 0 < t \le 0.2\mu s) \\ 0 & t > 0.2\mu s \end{cases} \quad f = 5\text{MHz} \tag{9}$$

In order to compare with experimental results, the diameters of the cylinders are chosen as 10.18mm and 16.05mm. The following images display the normal displacement fields in different instants when diameter is 10.18mm. The deepness of the color represents the numerical value of the displacement. White corresponds to the maximum of the displacement and black corresponds to the minimum.

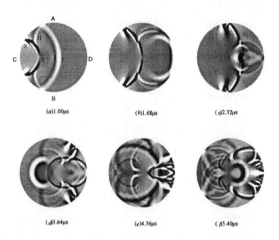

(a)1.00μs   (b)1.68μs   (c)2.32μs

(d)3.64μs   (e)4.56μs   (f)5.40μs

Fig. 2: The normal displacement fields at different instants

For demonstration purpose, several letters are inserted in Fig.2(a). The letter P, S, H and R represent the longitudinal, transverse, head and Rayleigh waves, respectively. The exciting source and receiving point are located at two terminals of a diameter CD, and the A-scan waveform at receiving point D will be given later. The diameter CD and AB are perpendicular to each other.

These images clearly show that the longitudinal, shear, surface, and head waves all start at the line-source, and separate when traveling further into the material because of different velocities. The energy of longitudinal wave mainly focuses along CD direction. But almost no energy of shear wave exists along this direction. Its energy mainly focuses in a direction near CA and CB. The energy of two surface waves mainly confines near the surface of the cylinder. Surface wave also owns the highest energy among all wave types. Head waves have curved wavefronts, quite different from the linear wavefront in semi-infinite space. Before the longitudinal wave passes through points A and B, only one head wave exists in either semicircle. After the longitudinal wave passes through A and B, another head wave appears in either semicircle and exists until the P wave arrives at D (see Fig.2 (b)). Other three wave pulses arrive at point D in Fig.2 (c), (d) and (e), respectively. Cylindrical Rayleigh waves arrive at point D in Fig.2 (f).

The A-scan results at point D are also given in Fig. 3 for two cylinders with diameters 10.18mm and 16.05mm, respectively. Laser ultrasonic experiments were carried on these two samples. The laser pulse was generated by Nd:YAG laser with energy about 16mJ, the generated ultrasound was detected by a commercial interferometer SH130. The detected signal was averaged

592

by 128 times. The experimental results are also shown in the figure. The unit of the y scale is voltage, which is for the experimental signal. Before displayed in the figure, the data of the FEM results are all multiplied by 10^{10} and added by a shift in y scale, which make the experimental and FEM results in the same order and easy to be compared in the same figure. From Fig.3, we can find that almost every corresponding pulse arrive at the same time for FEM and experimental results. The good agreements between FEM and experimental results show that FEM is valid in simulating the propagation of elastic waves in cylinders.

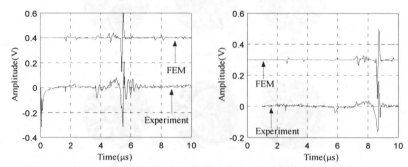

Fig. 3: Elastic wave pulses in aluminum cylinders generated by a line source (laser pulse energy E=16mJ, averaging times n=128). (a) d=10.18mm; (b) d=16.05mm.

4 Conclusion

In this work, the finite element method is applied to simulate the wave propagation in cylinders. Sub-matrix method is applied to solve the large-scale problem. The displacement fields of cylinders generated by a pulsed line-source are computed and displayed by gray images respectively. The numerical A-scan results and experimental results at receiving point are displayed in the same figure. The good agreement between the numerical and experimental results shows that the FEM is a good way for simulating the propagation of elastic waves in cylinders.

References

1   R. Ludwig and W.Lord, *IEEE Trans. UFFC*,   1988; 35: 809-820.

2   R. Ludwig and W.Lord, *IEEE Trans. UFFC*,   1989; 36: 342-250.

3   D. Moore and R. Ludwig, *Review of Progress in Quantitative NDE*,   1989; 8A:103-108.

4   Z. You and W. Lord, *Review of Progress in Quantitative NDE*,   1989; 8A:109-116.

5   Mary Sansalone, Nicholas J. Garino, and Nelson N. Hu, *Review of Progress in QNDE*, 1987; 6A: 125-133

Theoretical and Computational Acoustics 2001
E.-C. Shang, Qihu Li and T. F. Gao (Editors)
© 2002 World Scientific Publishing Co.

Solution of 2-D Wave Reverse-Time Propagation Problem by the Finite Element-Finite Difference Method (FE-FDM)

Yuan Dong, Huizhu Yang

Department of Engineering Mechanics, Tsinghua University, Beijing 100084, P. R. China

Lizhen Gu

Department of Mathematical Sciences, Tsinghua University, Beijing 100084, P. R. China

Aimed to time relay partial differential equation (PDE), 2-D wave equation, a method of joint using Finite Element Method (FEM) and Finite Difference (FD) Method in spatial domain, which is named as Finite Element – Finite Difference Method (FE-FDM), has been proposed in this paper. By using the semi-discretization technique of FEM in spatial domain, the origin problem can be written as a coupled system of lower dimensions PDEs that continuously depend upon time and part of space. FDM is employed to solve these lower dimensions PDEs. The concept and theory of this method have been discussed in this paper. A numerical example of 2-D wave reverse-time propagation shows the excellent performance and potential of it.

1. Introduction

Finite Element – Finite Difference Method (FE-FDM) proposed in this paper is a numerical method of joint using FEM and FDM in spatial domain for time relay partial differential equation. For the spatial multi-dimensional problem, FE-FDM uses FFM for some spatial dimensions and FD for remaining spatial dimensions and time dimension.

The FE-FDM has strong resemblance to a number of numerical methods such as the classical Kantorovich method, method of lines (MOL), finite element method of line (FEMOL), finite difference method and finite element method. A brief comparison on the basic differences between FE-FDM and the above-mentioned methods is given below.

1. Semi-analytical method. Most of semi-analytical method, such as Kantorovich method, MOL and FEMOL, semi-discretizes the partial differential equation into a group of ordinary differential equations (ODE). Then, analytical method is used to solve the ODEs. The basic difference of them is Kantorovich method uses analytic function or global polynomials, MOL uses FD and FEMOL uses FEM in semi-discretization (Ames, 1992; Yuan, 1993). The FE-FDM semi-discretizes PDE using FEM into a system of PDEs. Then, FD method is employed to solve the PDEs. For these reasons, semi-analytical method can be employed for elliptic model equation and FE-FDM for parabolic and hyperbolic model equation that described the time relay problems.

2. FEM and FDM. FEM fully discretizes a static problem into a system of algebraic equations with discrete nodal values as the basic unknowns. For the time relay problem, FEM fully discretizes it in spatial domain into ODEs and solve them with FD method (Hughes, 1987), whereas the FE-FDM semi-discretizes the PDE using FEM in spatial domain into a coupled system of PDEs. These PDEs still continuously depend upon both time and space (although not all the space dimension). They will be solved with FD method. Thus, the advantages of FEM such as the adaptation to arbitrary domain, boundary, material and loading are remained. The shortcoming of FEM such as large demand of computer memory and high computation costs is reduced because of semi-discretization. Comparing with FD method, the computation precision is increased for the reason of FEM semi-discretization. The technique of FD for solving PDEs in lower dimensions can decrease frequency dispersion in space and has looser conditions of stability for explicit FD schemes.

In this paper, the basic concept and theory of the finite element – finite difference method are described through 2-D wave equation. A numerical example of wave reverse-time propagation to demonstrate the tremendous performances of this method is given too. It is encouraged that the result is accurate and effective.

2. Principle

Considering the hyperbola model problem, the 2-D scalar wave equation is

$$\frac{\partial^2 u}{\partial x^2} + \frac{\partial^2 u}{\partial z^2} = \frac{1}{a^2(x,z)} \frac{\partial^2 u}{\partial t^2}, \quad \text{in } \Omega \tag{2.1a}$$

here $u(x, z, t)$ denotes the wave disturbance at horizontal (lateral) coordinate x, vertical (depth) coordinate z, where the z axis point downward ant time t, respectively, and $a(x, z)$ is the medium velocity.

Boundary condition (B. C.) is:

$$u(x,z,t) = \begin{cases} \varphi(x,t) & z=0 \\ 0 & z \neq 0 \end{cases}, \quad \text{in } \partial\Omega \tag{2.1b}$$

Initial condition is:

$$u(x,z,t=T) = \phi(x,z), \quad \& \quad u(x,z,t=T) = 0, \text{ in } \Omega \tag{2.1c}$$

where the two-dimensional rectangle domain Ω is bounded by piecewise smooth boundary $\partial\Omega$ (z=0, z=z0, x=0, x=x0). Reverse-time wave propagation problem is to solve the above equation so that the recorded wave field at t=T can propagates reversely back to t=0, hence the reflect wave lies at the reflection interface (Yilmaz, 1987). FE-FDM discretizes (2.1a) in x-coordinate using FEM, and solves the remaining equations in z-t coordinates using FD method.

2.1 FEM Semi-Discretization in X-Coordinate

P1 denotes equation (2.1). P2 denotes the corresponding Galerkin method of P1. P2 is:
Find $u \in S_\varphi^1$ such that for all $v \in S_0^1$

$$D(u,v) - F(v) = 0, \tag{2.2}$$

here

$$S_\varphi^1 = \left\{ u \mid \int \left[u^2 + \left(\frac{\partial u}{\partial x} \right)^2 \right] dx < \infty, \ u(0,z,t) = \begin{cases} \varphi(0,t) & z=0 \\ 0 & z \neq 0 \end{cases}, \ u(x0,z,t) = \begin{cases} \varphi(x0,t) & z=0 \\ 0 & z \neq 0 \end{cases}, \ \text{in} \partial\Omega \right\},$$

$$S_0^1 = \left\{ v \mid \int \left[v^2 + \left(\frac{\partial v}{\partial x} \right)^2 \right] dx < \infty, \ v(0,z,t) = 0, v(x0,z,t) = 0, \text{in} \partial\Omega \right\},$$

$$D(u,v) = \int_x \left[\frac{\partial^2 u}{\partial x^2} + \frac{\partial^2 u}{\partial z^2} - \frac{1}{a^2(x,z)} \frac{\partial^2 u}{\partial t^2} \right] v dx, \quad F(v) = 0,$$

$D(u,v)$ can be rewritten as

$$D(u,v) = \int_x \left[\frac{\partial u}{\partial x} \frac{\partial v}{\partial x} - \frac{\partial^2 u}{\partial z^2} v + \frac{1}{a^2(x,z)} \frac{\partial^2 u}{\partial t^2} v \right] dx. \tag{2.3}$$

Semi-discretizing the horizontal coordinate (x) in the region of [0, x0], the element number is NE. One constructs finite element function space as

$$u_h(x,z,t) = \sum_{i=1}^{NE} u_i(t,z) N_i(x), \tag{2.4a}$$

$$\frac{\partial}{\partial x} u_h(x,z,t) = \sum_{i=1}^{NE} u_i(t,z) \frac{d}{dx} N_i(x) = \sum_{i=1}^{n} u_i(t,z) B_i(x). \tag{2.4b}$$

Here, $u_i(t,z)$, $N_i(x)$, and $B_i(x)$ are the nodal values, the interpolation function and the derivative of interpolation function of element i. By substituting equation (2.3) and (2.4) into (2.2), one gets the discrete style description of P2.

$$D(u_h, v_h) = \sum_{n=1}^{NE} \int_{\Delta z} \left[v_e^T B^T B u_e - v_e^T N^T N \frac{\partial^2 u_e}{\partial z^2} + \frac{1}{a^2(x,z)} v_e^T N^T N \frac{\partial^2 u_e}{\partial t^2} \right] dx = 0$$
,

For the reason of function v is arbitrary, one obtains semi-discretized PDEs as

$$M \frac{\partial^2 u}{\partial t^2} + Ku = H \frac{\partial^2 u}{\partial z^2}$$
,

(2.5a)

boundary condition (B. C.) is:

$u(z = 0, t) = g(t)$, $u(z = z0, t) = 0$,

(2.5b)

initial condition is:

$u(z, t = T) = f(z)$, & $u(z, t = T) = 0$,

(2.5c)

where $g(t)$ and $f(z)$ are the discretization of $\varphi(x,t)$ and $\phi(x,z)$ respectively.

$$M = \sum_{n=1}^{NE} M_e \quad K = \sum_{n=1}^{NE} K_e \quad H = \sum_{n=1}^{NE} H_e$$
,

(2.5d)

$$M_e = \int_e \frac{1}{a(x,z)} N^T N dx \quad K_e = \int_e B^T B dx \quad H_e = \int_e N^T N dx$$
.

(2.5e)

It can be seen that the matrix M, K and H are all symmetric. M and H are positive-definite, and K is positive-semidefinite. The PDEs (2.5) are hyperbolic model equations when the velocity is constant because the matrix M and H can be diagonalized at the same time under this condition. It should be emphasized that only the matrix M varies with the depth.

2.2 FDM Solution of Matrix PDEs

One of the explicit schemes, five-point central scheme is selected to solve this problem. The difference equation has the form

$$Mu[i]_j^{n-1} = Au[i]_j^n + B(u[i]_{j+1}^n + u[i]_{j-1}^n) - Mu[i]_j^{n+1}$$
,

(2.6)

here

$$A = 2M - \tau^2 K - \frac{2\tau^2}{h^2} H \quad B = \frac{2\tau^2}{h^2} H$$
.

τ, h are the constant time and space step, i, j, k are the discrete denotation of lateral direction, depth direction and time, respectively. The local truncation error of this scheme has the form of $O(\tau^2 + h^2)$ (Durran, 1999).

The stability of wave equation with BC and IC is much complicated. For this problem, the stability is much difficult because it is related to the FEM semi-discrete scheme and the form of interpolation function.

The scheme stability condition analyses of the simplest condition that has only one element using piecewise linear interpolation function is discussed hear. The element length is l and the velocity is a. The interpolation function is

$N(x) = (\xi, 1 - \xi)$, here $\xi = \frac{x_i - x}{l}$.

The coefficient matrix of the equation (2.5a) is

$$M = \frac{l}{6a^2} \begin{bmatrix} 2 & 1 \\ 1 & 2 \end{bmatrix} \quad K = \frac{1}{l} \begin{bmatrix} 1 & -1 \\ -1 & 1 \end{bmatrix} \quad H = \frac{l}{6} \begin{bmatrix} 2 & 1 \\ 1 & 2 \end{bmatrix}$$
,

By diagonalzing these equations, one gets two individual equations

$$\frac{\partial^2 u_1}{\partial t^2} + \frac{12}{l^2} u_1 = a^2 \frac{\partial^2 u_1}{\partial z^2}$$
,

(2.7a)

$$\frac{\partial^2 u_2}{\partial t^2} = a^2 \frac{\partial^2 u_2}{\partial z^2}$$
.

(2.7b)

By using central scheme in both time and space, the stability condition of equation (2.7b) is $a^2\lambda^2 < 1$, here τ, h are the time and space steps, $\lambda = \tau/h$. Consider equation (2.7a) only, one uses the Fourier analysis method. The amplification matrix has the form as

$$G = \begin{bmatrix} 2 - 4a^2\lambda^2\sin^2\dfrac{kh}{2} - 12\dfrac{\tau^2}{l^2} & -1 \\ 1 & 0 \end{bmatrix} = \begin{bmatrix} b & c \\ 1 & 0 \end{bmatrix},$$

Because of $|c| = 1$ and $|b| \le 1 - c$, the eigenvalue of amplification matrix $|\mu(G)| \le 1$ meet the sufficient condition of scheme stability. From $|b| \le 1 - c$, it can be obtained that

$$a^2\lambda^2 + 3\dfrac{\tau^2}{l^2} \le 1.$$

(2.8)

It can be verified that equation (2.8) is the sufficient and necessary stable condition of equation (2.7a). Especially, when the element length is equal to the depth step ($l = h$), the stability condition become $(a^2 + 3)\lambda^2 < 1$. When the velocity of wave equation is much large than 3 as in real rock (It is about several hundred meter/s), the stability condition of $a^2\lambda^2 < 1$ for equation (2.7b) is still valid for (2.7a). This condition is much looser than that ($a^2\lambda^2 < 1/2$) of 2-D space and time central scheme FD method (Lu and Guan, 1987). It means that the time-step restriction imposed in FD-FDM is often much smaller than that needed in FDM.

2.3 The Skills of Program Realization

The formula given by the Galerkin formulation for the mass matrix is sometimes referred to as consistent mass matrix. It leads to the optimal error estimates. In practice, diagonal or lumped mass matrices are often employed due to their general economy and because they lead to some especially attractive explicit FD schemes. The accuracy of consistent mass can often be achieved by a much simpler lumped mass (Hughes, 1987, Wang and Shao, 1997). There are several ways of going about the construction of lumped-mass matrices. We use the way that is to employ nodal quadrature rules.

The initial condition of wave reverse-time propagation problem (equation (2.1c)) is the value of function and one-order derivative of this function on time t. A scheme of two-order difference is adopted for initial derivative condition to fit the two-order accuracy scheme of PDEs.

3. Numerical examples

Wave reverse-time propagation problem, also called reverse-time migration, is one of the key problems in seismic exploration data processing. Some migration algorithms such as Kirchoff migration, frequency-wavenumber (F-K) migration, finite difference (FD) migration and finite element method (FEM) migration have been represented in the past 30 years.

The best reason for using FD in migration is that it adapts to complicated structure (including transverse varying of velocity) and can be performed efficiently. But its drawbacks are also obvious: the problem of flow computation accuracy and frequency dispersion (Wang, 1998).

Comparing with FD, FEM has advantages of higher precision and adaptation to arbitrary boundary. Researches of acoustic and elastic wave equation show that FEM is effective in migration of real seismic data (Teng and Dai, 1989). But it's not widely used in seismic exploration for the problem of large demand of computation costs and computer memory. Here, finite element – finite difference method (FE-FDM) migration is carrying out.

3.1 Impulse Response

Impulse response, the fundamental model of migration (Robinson, 1983) is selected to test FE-FDM migration algorithm. The poststack profile in a constant-velocity medium (velocity a=2000m/s) shows as Fig. 1. An impulse locates at x=1000m (lateral) and t=0.8s (double travel time). This means that the reflection interface is a semicircle with a radius of 800m and center at x=1000 on the ground. The migration steps of time, lateral and depth are 4ms, 10m and 10m.

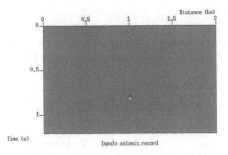

Fig. 1. Impulse record (The impulse locats at 1000m, 0.8s)

The ideal result shows in Fig. 2a, one can get it by placing the impulse to the reflection interface uniformly. The result of FE-FDM, Gazdag's phase-shift method and omig-x domain FD method (90 degree) show as Fig. 2b, 2c and 2d, respectively. The program which we got the result is Fig.2c and 2d is part of the free software package Seismic Unix (SU) coming from Center for Wave Phenomena (CWP), Colorado School of Mines.

It can be seen from Fig. 2 that all the three method can image the semicircle reflection interface correctly. The steeply dipping part of the interface (near ground) is well imaged by FE-FDM migration (Fig. 2b) and phase-shift migration (Fig. 2c), but not well by omig-x domain FD migration (Fig.2d). The result of FD migration (Fig. 2d) has much noise because of the spatial dispersion. The result of phase-shift and FE-FDM migration is much better than it. Especially, the noise is so small in Fig. 2b that it can not be seen with bare eyes.

Selecting the center traces (x=1000) of Fig.2, and translating into temporal domain and Drawing together, we get Fig.3. The impulse shapes of the three migration results are all differ from the original one. The obvious drawback of the result of FE-FDM migration is the shape of impulse becoming low and fat because of the numerical dissipation. The dissipation of FE-FDM migration leads to decrease the spatial distinguishes ability. Changing scheme is one of the possible methods to improve accuracy. Massive pseudo wave with high amplitude appear such as at the time of t=0.1 and t=0.25 in the result of omig-x domain FD method because of the numerical dispersion. The dispersion in omig-x domain FD migration maybe leads to the wrong interpretation results.

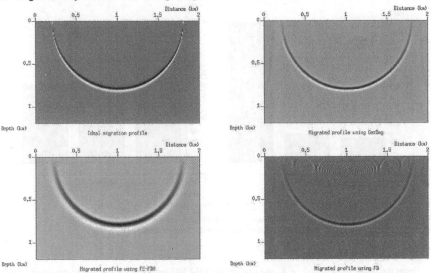

Fig. 2. Results of impuls depth migration.

(a) ideal result; (b) FE-FDM in this paper;

(c) Gazdag's phase shift method; (d) omig-x domain FD method

3.2 Efficiency Comparison

Adopting the impulse response model, we compare the expenditure of FE-FDM, phase-shift and omig-x domain FD migration in PC. The configuration of PC shows in table. 1. The result can be seen in Fig. 4.

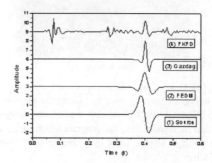

Fig. 3. Impuls record at 1000m in Fig. 2(a, b, c, d).

Table 1: The configuration of PC we used.

Content	Parameters
CPU	Intel MMX-200
MEM	64M (PC-100)
HARDDISK	IMPRIMIS 94601-15
OS	Linux RedHat 6.2

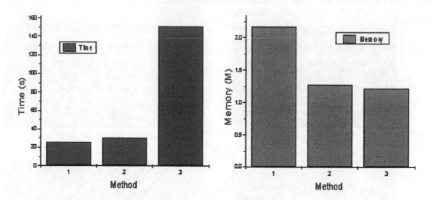

Fig. 4. The efficiencies of these three methods.

(Left) elapsed time; (Right) used memory.
1. FE-FDM; 2. Gazdag's phase-shift; 3. omig-x domain FD

From the result showed in Fig. 4, one can see the FE-FDM migration is the fastest one in the three methods, but unfortunately occupies much larger memory (about 2.2M) than the others.

3.3 Discussions

Gazdag's phase-shift algorithm is for lateral constant media (The velocity is only the function of depth z. i.e. $a(x, z)=a(z)$). Original poststack data is translated into frequency-wavenumber domain by 2-D Fourier transform (FT) of lateral coordination x and time t. Downward extrapolation of the wavefield is achieved by applying a series of phase-shifts to the Fourier transformed data. The migration result is obtained by inverse Fourier transform (IFT) (Gazdag, 1978). It elapses little time and occupies few memories. It can use for complex interface, but not for large lateral velocity variation condition.

Omig-x domain FD migration is a method for the one-way approximation wave equation. The eqation used in this paper is accurate for propagation directions to 90 degrees. FT and IFT are applied for time only. FD method is used in omig-x domain for wavefield extrapolation (Lee and Suh, 1985). This algorithm fits lateral velocity variation and complex interface so well that it becomes the most popular migration method today. The drawback of it is the computation costs, spatial dispersion and the worse image ability for steeply dipping interface.

FEM migration is one of the much accurate method for the problem of arbitrary shaped domain, lateral velocity, complex and dipping interface (Teng and Dai, 1989). But it is not widely used in seismic exploration for the reason of large demand of computation costs and computer memory. FE-FDM migration inherits most the advantages of FEM migration presented above. The computational efficiency is improved through spatial domain semi-discretization and FD method solution. As shown above, it is acceptable to use the FE-FDM migration to field data.

4. Conclusions

A numerical method named finite element – finite difference method (FE-FDM) for the solution of time relay partial differential equations such as parabolic and hyperbolic model equations is presented in this paper.

As the numerical example, 2-D scalar wave equation reverse-time propagation problem (2-D wave equation reverse-time depth migration), has been showed above, it is encouraged that the result is accurate and effective enough for steeply dipping interface imaging.

This method combines FEM and FDM based upon the semi-discretization of spatial domain. The most strongpoint of FEM such as adaptation to arbitrary domain and accuracy and that of FEM such as computation efficiency are inherited. It can be used to solve time relay problems that are described by parabolic and hyperbolic mode PDEs more accurate than FD method and much faster than FEM.

It is one of the obvious shortcoming that it is much complex to realize a FE-FDM program. The semi-discretization result is a coupled system of lower dimensions PDEs that continuously depend upon time and part of space. Appropriate schemes of FD method should be selected to solve the lower dimensions PDEs. The explicit scheme used in this paper is not the best choice because of the complicated stability and much numerical dissipation. Am implicit scheme should get much better results but expend more time. Velocity-stress finit-difference approach is one of the much faster and accuracy schemes to study wave propagation in multilayered media having irregular interfaces (Yang, 1993) and in anisotropic media (Liang and Yang, 1995). It would be one of the appropriate finit difference schemes.

It can be said that it would be a useful and promising numerical method to study wave propagation problem, especially the reverse-time problem.

Acknowledgments

The authors thank Drs. Zhu Yaping of Institute of Seismic Exploration (ISE) of Tsinghua University for his helpful advices about mathematics problem. We also thank Professor Wang Yong (JiangSu Oil Field, China) and Professor Du Qizhen (ISE, Tsinghua University) for useful discussions about migration. At last, we express sincere thanks to Center for Wave Phenomena (CWP) of Colorado School of Mines. Their free software Seismic Unix (SU) provides availability and credibility test for our method.

References

1. W. F. Ames, *Numerical methods for partial differential equations, -3rd edit* (Academic Press, 1992).
2. S. Yuan, *The finite element method of lines: theory and applications* (Science Press, Beijing-New York, 1993).
3. J. R. Thomas Hughes, *The finite element method: linear static and dynamic finite element analysis* (Prentice-Hall, 1987).
4. O. Yilmaz, *Seismic data processing* (Society of Exploration Geophysicists, Tulsa, 1987).
5. D. R. Durran, *Numerical methods for wave equations in geophysical fluid dynamics* (Spinger-Verlag New York, 1999).
6. J. Lu and Z. Guan, *Numerical method of partial differential equations (In Chinese)* (Tsinghua University Press, Beijing, 1987).
7. X. Wang and M. Shao, *Fundamental theory and numerical method of finite element method (In Chinese), -2nd edit,* (Tsinghua University Press, Beijing, 1997).
8. Y. Wang, "Seismic wavefield migration in anisotropic media," Ph. D. thesis, University of Petroleum (Beijing), 1998.
9. Y. C. Teng and T. F. Dai, "Finite-element prestack reverse-time migration for elastic wave", Geophysics, Vol. 54, No. 9: 1204~1208 (1989).
10. E. A. Robinson, *Migration of geophysical data* (International Human Resources Development Corporation, 1983).
11. J. Gazdag, "Wave-equation migration by phase shift", Geophysics, 43: 1342~1351 (1978).
12. M. W. Lee and S. H. Suh, "Optimization of one-way wave equations". Geophysics, 50: 1634~1637 (1985).
13. H. J. Yang, "A velocity-stress finite-difference approach to study wave propagation in multilayered media having irregular interfaces", Proceedings of Int. Conf. on Theoretical and Computational Acoustics, 1993,
14. F. Liang, and H. J. Yang, "A numerical method of modeling elastic wave propagation in anisotropic media", *ACTA MECHANICA SOLIDA SINICA*, Vol. 8, No. 4, 327-336 (1995).

Theoretical and Computational Acoustics 2001
E.-C. Shang, Qihu Li and T. F. Gao (Editors)
© 2002 World Scientific Publishing Co.

Imaging by Common Converted Point (Common Reflection Point) Set

Guangming Zhu, Qingchun Li and Yugui Wang
Institute of Applied Geophysics, Chang'an University, Xian 710054, China

1. Problems and strategy

The multiple coverage technique has been introduced in reflection seismics since 50s of the 20th century. This technique has successfully suppressed those coherent noises such as surface wave, refraction wave, diffraction wave and especially multiples. The data of multiple coverage are also satisfactory for eliminating random noises. The signal noise ratio (S/N) of the N fold data can be improved $\sqrt{N}$ times if the random noises are stable nornal distribution with zero mean. Moreover, the redundant information from multiple coverage data may be used to velocity analysis, residual static correction, migration and parameters inversion. Therefore, the application effect and exploration fields have been widely enhanced. The multiple coverage acquision has completely replaced one time coverage and it is appraised as a revolution in seismic industry.

However, the basic and most important processing for multiple coverage data, such as velocity analysis, normal moveout(NMO) correction and horizontal stacking, are all done by means of the common midpoint(CMP) gathers on the supposition that CMP are just the common reflection point(CRP) gathers. So the multiple coverage theory is based on the homogeneously horizontal layered model. Fig.1 shows the relationship between the CMP and CRP of this model. It can be seen that the downward wave rays from sources to reflection point and the upward wave rays from reflection point to geophones are symmetrical on the reflection point or common depth point(CDP) which is fixed and below the common midpoint. So CMP gathers are same as the CDP ones.

That is to say, velocity analysis, NMO correction and horizontal stacking should theoretically be done by CRP gathers, but they may be practically processed by CMP gathers since CMP is equivalent to CRP for same waves in horizontal layered media. This assumption is obviously correct when strata are nearly horizontal or with small dips. If dip is big as

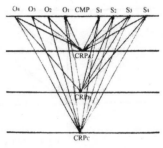

Fig.1. Common midpoint and common reflection point of the homogeneous horizontal layered model.

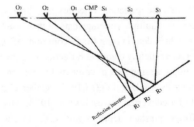

Fig.2. Common midpoint and common reflection point of the homogeneous inclined layered model.

shown in Fig.2, the downward rays are asymmetric with the upward ones. The reflection point is not below the CMP. It deviates from CMP with ocation of sources and geophones, dip and depth of the interface. The reflection points are scattering and there is no CRP. So we cannot simply take CMP gathers as CRP ones, the CRP gathers cannot be sorted out in fact.

Situation for converted waves is more complicated. Fig.3 (a) is CMP gathers of P-wave and (b) is those of PS-wave. Fig.4 shows the situation of reflection point of P-wave and converted point of PS-wave varying with depth. We may notice that downward rays are asymmetrical with upward ones on converted point not below CMP even though the media are homogeneously horizontal layered. The converted point deviates towards side of S-wave rays with location of sources and geophones. So the converted points are scattering and there are no common converted point(CCP) gathers. In this media, reflection trajectory of P-P wave depends on only depth while that of P-S wave depends both depth z and horizontal coordinate x.

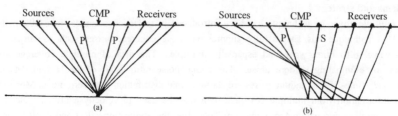

Fig.3　Comparison of CMP gathers.　(a) P-P;　(b) P-S

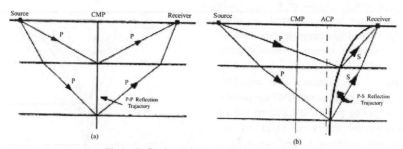

Fig.4　Reflection trajectory.　(a) P-P,　(b) P-S

2. Imaging by Common Reflection Point Set(CRPS) based on models

It has been shown from analyses above that CRP gathers cannot be sorted out for reflections from big dip interface and those gathers of converted waves cannot be acquired even in horizontal layered media because reflection trajectory is a function of both depth and horizontal coordinate. How should we process the multiple coverage data in this case? We put forward a concept of the common reflection point set, in which determining firstly common reflection or converted point set that is not same as gathers and then reconstructing common reflection or converted point gathers according to the reflection trajectory.

In processing data of converted waves, many scholars found difficulties for sorting the CCP gathers, building the CCL sorting(Li,1997), gather classification by asymptotic approximation (Thomsen,1999), DMO for converted waves(Tessmer,1988; Alfaraj, 1992; Sun,1999), PSVCCP mapping(Eaton,Slotboom, 1990) and prestack depth migration(PSDM) which are all trying to solve the 2-dimensional function of the converted point location. For example, sev..al layers with different depths are selected in CCL sorting and a few CCP gathers are sorted according to location of the converted points, the gathers are then processed respectively and put together because every CCP gather is accurate only with its corresponding depth. Asymptotic approximation adopts analytic formula of the horizontal layered media to calculate coordinate of the converted points. DMO for converted waves converts the P-S CMP gathers into CCP gathers by conversion of travel time into corresponding echo time t_0. PSVCCP mapping and PSDM are those of sorting no CCP gathers. We here emphasize that it is impossible for sorting CCP gathers directly and only common converted point set(CCPS) may be acquired, or reconstruct CCP gathers is necessary. We determine models by modeling and evaluate reflection trajectory based on the models.

Imaging by CCPS based on models will be accomplished as follows:

(1) Model building

The strategies include: ☐ determining model parameters by P-P data, ☐ determining velocity of S-wave by analyzing velocities to P-S data, ☐ making synthetic records by calculating travel time using ray tracing technique, ☐ determining the model parameters of P-S data by stacking the records with CCPS, ☐ comparing the parameters of P-P and P-S model and modifying the velocity of S-wave according their errors, ☐ repeating processes ☐ to ☐ until the error is small enough(See flow chart in fig.5).

(2) Normal Moveout(NMO) correction

Converting travel time of every sample to be depth or two way time corresponding to that converted point on common shot gathers.

(3) Stacking

Reconstructing CCPS to data after NMO according to the x-coordinate of the converted points and then stacking to output P-S stacked section.

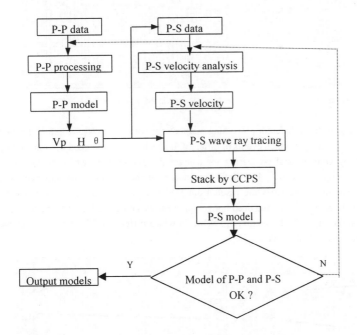

Fig.5. Flow chart of model building by Common Converted Point Set

3. Example

Table 1 includes the theoretical model parameters by imitating real geological situation in a district.

The synthetic records of P-P and P-S from these parameters are shown in fig.6 and fig.7 respectively.

Table 1 Theoretical parameters of inclined layered media

Layer No.	1	2	3
Interface depth at x=0(m)	420.0	2799.0	3200.0
Interface dip(deg.)	-15.0	20.0	0.0
Vp(m/s)	1800.0	3000.0	2400.0
Vs(m/s)	1000.0	1600.0	1200.0

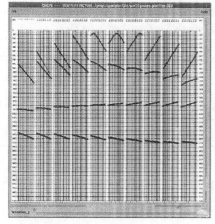
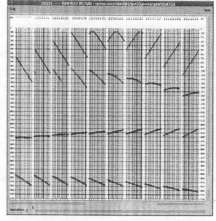

Fig.6. Synthetic P-P records   Fig.7. Synthetic P-SV records

We compare the imaging from different processing methods for converted wave. Fig.8 is the depth section by CCPS based on models, in which events of different depths have been imaged accurately. Fig.9 demonstrates the result by conventional processing for converted wave data, in which shallow event appears to be pseudo-pinching and the deep horizontal event has been changed into a dip one because of the horizontal velocity variation. Fig.10 is the result from DMO for converted wave, in which low frequencies appear on inclined events and the deep horizontal event was changed into dip because of the horizontal velocity variation. Fig.11 is the section of PSDM, in which different events have been imaged accurately but with quite large calculation cost.

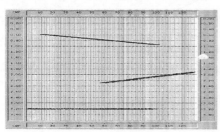
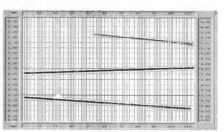

Fig.8. Imaging from CCPS based on models   Fig.9. Imaging from conventional P-S processing

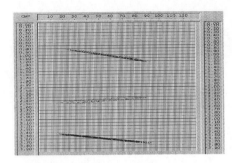
Fig.10. Imaging from P-S DMO

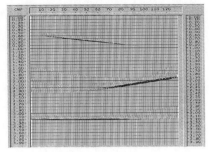
Fig.11. Imaging from P-S PSDM

4. Discussions

CRP or CDP gathers are same as the CMP ones for P-P data in homogeneous horizontal layered media, so we may sort out CDP and determine the velocities, separate some wave fields, enhance S/N ratio, or extract parameters. But CRP gathers do not exist for P-S data even in the simplest media above and P-S reflection trajectory depends on offset, depth and acoustic impedance of the interface.

CRP gathers do not exist for both P-P and P-S data in homogeneous inclined layered media. The reflection point is not below the CMP. The reflection points are scattering. Deviations from CMP change with location of sources and geophones, dip and depth of the interface and acoustic impedance of the media.

We should build the concept of Common Converted Point Set (CRPS), and determine model parameters by model building and evaluate reflection trajectory, stacking by reconstruct CCPS. Numerical calculation shows that CCPS method is fast, robust and accurate compared with the PSDM.

References

[1]. Li Luming and Luo Shenxian, Principle of multicomponent seismic exploration, Publishing House of Chengdu Institute of Technology in Chinese, 1997

[2]. Sun Peiyong and Li Chengchu, Integration DMO and applications, Petroleum Geophysical Prospecting in Chinese, Vol.37, No.1, 1999。

[3]. Mohammed Alfaraj, Conversion of Converted Wave into Zero-offset by Fourier Transform, Translation of Petroleum Geophysical Prospecting in Chinese, No.4, PP53-59, 1993。

[4]. D.W.S.Eaton，R.T.Slotboom，R.R.Stewart and D.C.Lawton， Stacking of converted waves by depth variation, 1990.

[5].Tessmer G. and Behle O.A, Common reflection point data-stacking technique for converted wave, Geophysical prospecting, 36(7) :671-688, 1988.

[6].Leon Thomsen，Converted wave reflection seismology over inhomogenous anisotropic media，Geophysics, Vol.64，No.3， PP 678-690，1999.

Theoretical and Computational Acoustics 2001
E.-C. Shang, Qihu Li and T. F. Gao (Editors)
© 2002 World Scientific Publishing Co.

High Power and High Resolution Ultrasonic Sources by Combining Width-Adjustable Pulses

Yongguang Mu, Bangrang Di
Key Lab. of Geophysical Exploration, CNPC
The University of Petroleum (Beijing)

Introduction

In the applied field of ultrasonic, the demands for ultrasonic sources become more and more strict, especially strict for the transmitting energy of ultrasonic sources and vibration continuing time of ultrasonic transducers.

In general, we demand that the transmitting energy of ultrasonic sources is as strong as possible, the vibration continuing time of ultrasonic transducer is as short as possible. So that in various science and technology applied field of ultrasonic, the detecting distance is as far as possible and the resolution is as high as possible.

However, due to the limitation of the vibration character of piezoelectric crystal itself and the techniques of the manufacture of sensor, the ultrasonic generated by the ultrasonic transducer is not suitable for the science and technology field stated above in the aspects of energy and vibration continuing time under the generating of single sharp pulse.

To improve the transmitting energy of ultrasonic sources and shorten the vibration continuing time of ultrasonic transducer, we put forward combining width-adjustable pulses method to improve the method of sharp pulse now used

608

Basic Theory

Suppose we apply a pulse voltage δt(figure 1a) to a ultrasonic transducer. The resulting vibration f(t) is shown in figure 1b.

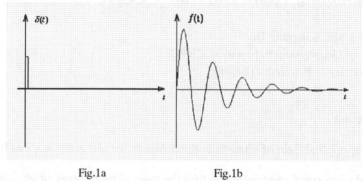

Fig.1a   Fig.1b

Figure 1

When widen the pusle width to a width τ ,we get a width-adjustable pulse Δt. It can be seen as a wide pulse stacked with some single pulses delayed $i\Delta\tau$ as shown in figure 2a. Then we have

$$\Delta(t) = \sum_{i=0}^{N-1} \delta(t - i\Delta\tau)$$

Where $N\Delta\tau = \tau$ i=0,1,2,••••••N-1

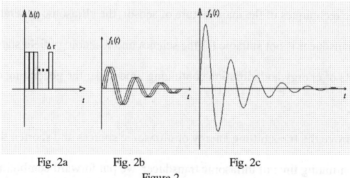

Fig. 2a   Fig. 2b   Fig. 2c

Figure 2

The response f1(t) is also stacked with some single pulse responses f(t) delayed $i\Delta\tau$,as shown in figure 2b an fig.2c. Then

$$f_1(t) = \sum_{i=0}^{N-1} f(t - i\Delta\tau)$$

From figure 2c we can see that the stacked response f1(t) is significantly increased, and in some range of pulse width, the greater the pulse width the more powerful pulse energe. Therefore, we can significantly increase the transmitting energe of ultrasonic sources.

When we combine some width-adjustable pulse, according to a certain mode as shown in figure 3a, the resulting response f2(t) is stacked with some responses of width-adjustable pulse as shown in figure 3b. Then

$$f_2(t) = \sum_{j=1}^{M} A_j f_1^j(t - t_j) = \sum_{j=1}^{M} \sum_{i=0}^{N-1} A_j f^j(t - i\Delta\tau - t_j)$$

Where M is the number of width-adjustable pulse ; tj is the time delay of the jth width- adjustable pulse ;Aj is the amplitude of the jth width-adjustable pulse.

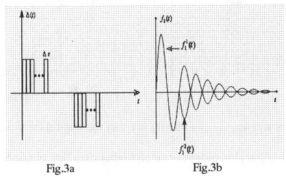

Fig.3a   Fig.3b

Figure 3

The response is changed by changing the amplitude Aj and the time delay of the jth width-adjustable pulse. So long as the amplitude Aj and the time delay are adjusted properly, we can cancel out the tail of the response of the first width-adjustable pulse. Therefore we can greatly improve the resolution of ultrasonic sources.

Results

We made a combined width-adjustable pulses ultrasonic source according to the basic theory as mentioned above. Using the new ultrasonic sources, we received the tramsmitted wave through a piece of plexiglass. We obtained a record shown in Figure 4b. For comparison, we got a record by using a ultrasonic source produced by Panametrics Company in the same way (shown in figure 4a).

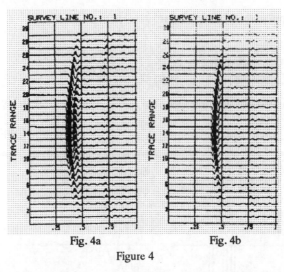

Fig. 4a   Fig. 4b

Figure 4

Form figures 4a and 4b, we can clearly see that the resolution obtained by using the combined width-adjustable pulses is much higher than that obtained by using single pulse ultrasonic source.

Theoretical and Computational Acoustics 2001
E.-C. Shang, Qihu Li and T. F. Gao (Editors)
© 2002 World Scientific Publishing Co.

Investigation on Sound Source in Cascades by Computational Aero-Acoustics

Yongbo Zhu, Jing Tian, Genshan Jiang
Institute of Acoustics, Chinese Academy of Sciences,
17 Zhongguancun Street. P. O. Box 2712, Beijing 100080, P. R. China

ABSTRACT   The wake/stator interaction has been predicted using an N-S solver, in order to obtain unsteady pressure fluctuations on blade surface and the variety of unsteady vortex. So the noise can be computed. The equations are solved using a finite volume time-integration Ni scheme. For improving the ability to simulate the turbulent flow , K- ε turbulence models are utilized for the turbulent closure. Then the numerical calculations on a turbine and an axial compressor cascades are carried out in this paper.

Key Words: Unsteady flow, N-S equation, Computational aero-acoustics

1. INTRODUCTION

The computational aero-acoustics (CAA) technique is developed in world widely. First, computational fluid dynamics (CFD) is used for analyzing flow structures in time domain and calculating the unsteady pressure fluctuations. Flow induced noise is then determined in frequency domain using fast Fourier transform for calculating the sound pressure level based on unsteady pressure fluctuations on domain boundary.

The main research of this paper is focused on the first step, to get unsteady pressure and vortex which are the sources of sound in the cascades. The results of an axial turbine blade show how the wake moves toward downstream and it is cut and migrates from suction side to the pressure side.

The results of calculating the unsteady flow in an axial compressor agree with the experimental data quite well. Then a study is carried out to determine the influences of wake parameters— wake velocity defect, wake width, wake traverse speed and axial gap on the surface pressure.

1. NUMERICAL SIMULATION OF THE UNSTEADY FLOW IN AN AXIAL TURBINE

The unsteady prediction described below is obtained at the mid-span section of a low speed turbine blade[1].

In order to study unsteady phenomena, the mean velocity is subtracted from the instantaneous velocity so that the unsteady parameter can be seen more clearly. The unsteady velocity and distribution of entropy are shown in Fig1. The wake motion and how wake is chopped and sheared can be seen clearly in the figures. The behavior of wake is as "negative jets", and it has high entropy and low total pressure, migrating from pressure surface to suction surface. The further result of the wake migration is the movement of main flow towards the pressure surface, which replaces the wake fluid. This recirculating motion then results in the formation of a pair of large vortex, on opposite sign, which lie on each side of the wake.

As a wake first moves into a blade passage, it becomes bowed because the mid-passage velocity is much higher than that near the leading edge. Once the wake is chopped by the downstream blade, the newly formed wake becomes distorted. There are three reasons for this distortion. The first, which results in a broadening of the wake near the suction surface and a thinning near the pressure surface, occurs because the "negative jet" draws wake fluid away from the pressure surface and onto the suction surface. The second form of distortion results in a shearing and therefore lengthening of the wake segment. It arises as a result of the different between suction and pressure velocities. The third form of wake distortion begins as the wake first meets the leading edge of a blade, when the downstream side of the wake is accelerated over the blade surfaces and away from the upstream side, which remains in the stagnation point.

2. WAKE/BLADE INTERACTION IN AN AXIAL COMPRESSOR CASCADES

The results of calculating the unsteady flow in an axial compressor agree with the experimental data quite well. Then a study is carried out to determine the influences of wake parameters— wake velocity defect, wake width, wake traverse speed and axial gap on the surface pressure. The simulation of axial compressor cascade is carried at 50% blade high.

(1) Effects of wake velocity defect on pressure distribution on blade surfaces

The wake velocity defect indicates the difference between the main flow velocity and wake velocity, and has significant influence on the unsteady pressure. At higher wake velocity defect, larger difference between main flow and wake appears and the unsteady effect is more obviously, which results in large variation of pressure on blade surface.

(2) Effects of wake width

In order to study the effect of the wake width, three wake widths are considered. A smoother variation of velocity in the wake due to larger wake width results in decrease of unsteady effect in flow fluid.

(3) Effects of wake traverse speed

The increase of wake transverse speed provides different wake inflow angles. Increase of wake transverse speed results in higher reduced frequency and wake inflow angle. The more unsteady pressure variation will appear on the blade surface at higher reduced frequency and wake inflow angle.

(4) Effects of axial gap

The wake from the upstream decays as it progresses through axial gap. Wake dissipation is stronger with the increase of axial gap. So the energy transfer between velocity defect. Then the increase of axial gap will reduce the unsteady effect and the variation of unsteady pressure.

CONCLUSIONS

The results of an axial turbine blade show how the wake moves toward downstream and it is cut and transform, and it has high entropy, vortex, low total pressure. Wakes act as "negative jet", move the main flow fluid towards pressure surface to replace the wake fluid. Three mechanisms are found to be responsible for the generation of wake distortion.

The parameters such as wake velocity defect, wake width, wake traverse speed, axial gap have significant influence on the unsteady pressure.

REFERENCES

[1] Hodson, H.P., and WN Dawes ,"On The Interaction of Measured Profile Losses in Unsteady Wake-Turbine Blade Interaction Studies",96-GT-494

[2] Fan,S. and B.Lakshminarayana "Computation and Simulation of Wake-Generated Unsteady Pressure and Boundary Layers in Cascadea" (Part 1-2) 94-GT-140   and   94-GT-141

[3] Hodson, H.P., R.G.Dominy "Three-Dimensional Flow in a Low-Pressure Turbine Cascade at its Design Condition",86-GT-106

[4] Hodson, H.P.,"Boundary Layer and Loss Measurements on the Rotor of an Axial Turbine", Journal   of Engineering for Gas Turbines and Power.   Vol.106 pp.391

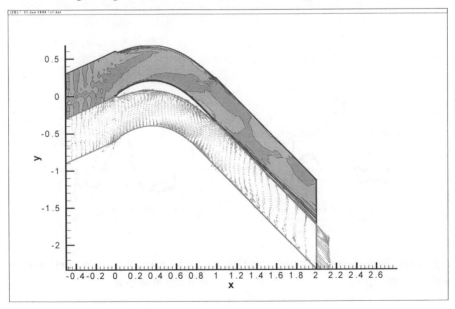

Fig1   t/T=0 unsteady velocity and entropy distribution

614

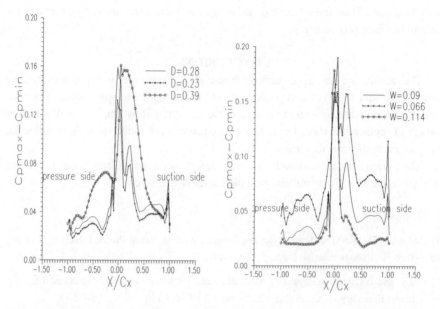

Fig2 unsteady pressure coefficient in different wake velocity defect

Fig3 unsteady pressure coefficient in different wake width

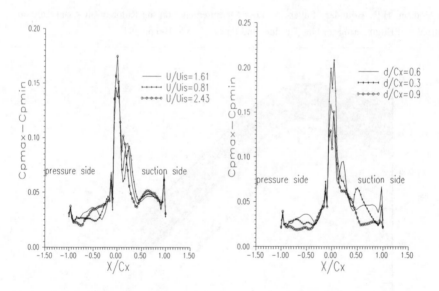

Fig4 unsteady pressure coefficient in different wake traverse speed

Fig5 unsteady pressure coefficient in different axial gap

Theoretical and Computational Acoustics 2001
E.-C. Shang, Qihu Li and T. F. Gao (Editors)
© 2002 World Scientific Publishing Co.

On Detecting the Source of Acoustical Noise

Victor Isakov
Department of Mathematics and Statistics,
Wichita State University, Wichita, KS 67260-0033,
USA
E-mail: *victor.isakov@wichita.edu*

1   Introduction

We consider the problem of identifying the source of the acoustical noise and the normal velocity of the sound on the surface Γ of a three-dimensional domain Ω. The acoustical field u of frequency k in Ω satisfies the Helmholtz equation

$$\Delta u + k^2 u = 0 \quad \text{in} \quad \Omega(\text{ or in } \Omega_e = \mathbf{R}^3 \setminus \overline{\Omega}) \tag{1}$$

The application we are interested in deals with Ω which is a cabin of an aircraft or of a car. A solutions to (1) in Ω_e are assumed to be satisfying the Sommerfeld radiation condition

$$limr(\partial_r u - iku)(x) = 0, \qquad \text{as } r = |x| \to \infty \tag{2}$$

Such u are called radiating solutions. The acoustical sensors are located on a surface Γ_0 inside or outside the cabin. They can measure the field u and the problem is to recover from these measurements u inside Ω and in particular the so-called normal velocity

$$v = \partial_\nu u \quad \text{on} \quad \Gamma = \partial\Omega \tag{3}$$

which is the surface enclosing Ω. Here ν is the unit exterior normal to $\partial\Omega$. We will show that there is a unique representation of u in Ω by the single

---

*Supported by the NSF grants DMS-9803816 and ITR/ACS 0081270

layer potential

$$u(x) = S_\Gamma \varphi(x) = \int_\Gamma K(x,y)\varphi(y)d\Gamma(y), \quad x \in \Omega \quad (\text{ or } x \in \Omega_e) \qquad (4)$$

where

$$K(x,y) = \frac{e^{ik|x-y|}}{4\pi|x-y|} \qquad (5)$$

is the free space radiating fundamental solution to the Helmholtz equation. Now our problem is reduced to solving the linear integral equation

$$\int_\Gamma K(x,y)\varphi(y)d\Gamma(y) = u(x), \quad x \in \Gamma_0 \qquad (6)$$

Given φ in the interior problem one can find the normal velocity from the formula

$$v(x) = \varphi(x)/2 + \int_\Gamma \nabla_x K(x,y) \cdot \nu(x)\varphi(y)d\Gamma(y), \quad x \in \Gamma \qquad (7)$$

which follows from (4) and the jump relations for the normal derivative of single layer potentials. This approach in principle allows to handle general domains Ω while most of existing methods are applicable to very special (rotationally symmetric Ω) when the Green's function of the Neumann problem for the Helmholtz equation can be found quite explicitly. In the paper ([3]) we considered a two-dimensional version of the interior problem. In the forthcoming paper ([4]) we handle the complete three-dimensional case.

In a popular approach one uses the represenation by single and double layer potentials

$$D_\Gamma \varphi(x) = \int_\Gamma \partial_{\nu(y)} K(x,y)\varphi(y)d\Gamma(y) \qquad (8)$$

coupled with the additional boundary integral equation (the Helmholtz-Kirchhoff systems (19),(20) or (21),(22)). We will show that for the interior problem the corresponding representation by the sum of these potentials is possible and unique, while for the exterior problems density for the single layer may be not unique.

In section 2 of this paper we show that any $(H_{(1)}(\Omega)\text{- })$ solution to the Helmholtz equation (disregard of possible Dirichlet or Neumann eigenvalues) in a bounded Lipschitz domain Ω with connected complement can be uniquely represented by a single layer potential, while for exterior domains this reprentation is not unique. A similar result holds for the representation by the sum of single and double layer potentials related to the systems (19),(20) or (21),(22). Section 3 contains uniqueness results and some stability estimates for recovery of u on Ω from u on Γ_0. These conditional estimates are of Hoelder type inside of Ω and of logarithmic type in Ω. We attempt to find explicit formulae for constants in the estimates for the particular case when Ω is the ball $|x| < r_1$ and Γ_0 is the sphere $|x| = r_0$. In section 4 we show completeness of some families of particular solutions of the Helmholtz equations, which justifies mathematically the HELS method of S. Wu and his collaborators ([12]), working very well in practice.

We will remind that $H_{(k)}(\Omega)$ denotes the Sobolev space of functions on Ω whose partial derivatives up to order k are square integrable and $\|\|_{(k)}$ denotes the standard norm in this space. We let $\|\|_2 = \|\|_{(0)}$ to be the norm in the space $L^2(\Omega)$. Correspondingly, $\|\|_{l+\lambda}, 0 < \lambda < 1$ is the norm in the Hoelder space $C^{l+\lambda}$ of functions whose partial derivatives up to order l are Hoelder continuous of exponent λ.

2   Representation by potentials

In the first approach we will characterize the source of an acoustical field $u(x)$ in $\Omega \subset \mathbb{R}^3$ by its surface density φ on $\partial\Omega$. In other words, for a solution u to (1) we would like to find a function φ such that (4) holds. We will assume that $\partial\Omega \in Lip$ which means that $\partial\Omega$ is locally the graph of a Lipschitz function. We will consider only Ω with connected $\mathbb{R}^3 \setminus \overline{\Omega}$ which is denoted by Ω_e. In the following theorem we will use traces of functions in Sobolev spaces. These traces are well defined for C^∞-functions, moreover the corresponding trace operators ($u \to u$ on $\partial\Omega$ from $H_{(1)}(\Omega)$ into $H_{(\frac{1}{2})}(\partial\Omega)$ and $u \to \partial_\nu u$ on $\partial\Omega$ from $H_{(1)}(\Omega)$ into $H_{(-\frac{1}{2})}(\partial\Omega)$) are continuous, so the traces can be understood using approximation by smooth functions. The continuity of the mentioned operators follows from the known results on the solutions of the Dirichlet problem in $H_{(1)}(\Omega)$ ([8], p.226).

A possibility of the representation (4) is guaranteed by the following

Theorem 2.1 *For any solution $u \in H_{(1)}(\Omega)$ to the Helmholtz equation (1) there is an unique function $\varphi \in H_{(-\frac{1}{2})}(\Gamma)$ such that (4) holds. Moreover, for*

some constant C depending only on Ω, we have

$$C^{-1}\|u\|_{(1)}(\Omega) \leq \|\varphi\|_{(-\frac{1}{2})}(\Gamma) \leq C\|u\|_{(1)}(\Omega) \qquad (9)$$

and if in addition $u \in C^{1+\lambda}(\overline{\Omega})$ and $\partial\Omega \in C^{1+\lambda}$, then

$$C^{-1}|u|_{1+\lambda}(\Omega) \leq |\varphi|_\lambda \leq |u|_{1+\lambda}(\Omega) \qquad (10)$$

Proof:
We first prove uniqueness of the representation (4). By the known proper-ties of the single layer potential S with $H_{(-\frac{1}{2})}(\Gamma)$-density ([8], pp.203, Chapter 9) we have

$$-\Delta S_- - k^2 S_- = 0 \qquad \text{in } \Omega, \quad -\Delta S_+ - k^2 S_+ = 0 \qquad \text{in } \mathbf{R}^3 \setminus \overline{\Omega},$$

$$S_- - S_+ = 0, \ \partial_\nu S_- - \partial_\nu S_+ = \varphi \qquad \text{on } \Gamma \qquad (11)$$

and the radiation condition at infinity. Here S_- is S on Ω and S_+ is S outside Ω. To show uniqueness it suffices to assume that $S_- = 0$ on Ω and to derive that $\varphi = 0$. If $S_- = 0$ on Ω, from the first jump relation we have $S_+ = 0$ on Γ. Then $S_+ \in H_{(1)}(B \setminus \Omega)$ (for any ball B) solves the homogeneous Helmholtz equation outside Ω, satisfies the homogeneous Dirichlet condition on Γ and the radiation condition at infinity. It is known ([8], pp. 286-289) that such $S_+ = 0$. Hence from the second of the jump relations (11) we conclude that $\varphi = 0$.

Now we similarly will prove existence of φ. As mentioned above, there is the trace $g \in H_{(\frac{1}{2})}(\Gamma)$ of the function u on Γ and the trace operator from $H_{(1)}(\Omega)$ into $H_{(\frac{1}{2})}(\Gamma)$ is bounded. According to the cited elliptic theory ([8]) there is a unique radiating solution $u^+ \in H_{(1)}(B \setminus \Omega)$ to the Helmholtz equation in $\mathbf{R}^3 \setminus \overline{\Omega}$ with the Dirichlet data g on Γ. Here B is any ball in $\mathbf{R}^3$. As known ([8]) for the solutions u, u^+ of the interior and of the exterior Dirichlet problems in Ω with the Dirichlet data g there are traces $\partial_\nu u, \partial_\nu u^+ \in H_{(-\frac{1}{2})}(\Gamma)$ and the trace operator is continuous from $H_{(\frac{1}{2})}(\Gamma)$ into $H_{(-\frac{1}{2})}(\Gamma)$. Let $\varphi = \partial_\nu u - \partial_\nu u^+$ on Γ. We claim that (4) holds.

Indeed, let S be the single layer potential $S\varphi$ and U be the function defined as u on Ω and as the solution to the exterior Dirichlet problem with the data g on Γ outside Ω and with the radiation condition. Then S and U solve the same transmission problem (11) with the radiation condition at infinity. A solution of this transmission problem is unique. This follows from the fact that a solution to a homogeneous problem ($\varphi = 0$) solves the homogeneous Helmholtz equations in $\mathbf{R}^3$ and it satisfies the radiation condition. Hence it is zero. Indeed, when $\varphi = 0$ by using the definition of a weak solution one can be convinced that S also solves the Helmholtz equation in $\mathbf{R}^3$ and satisfies the radiation condition. It is well known that then $S = 0$.

The bounds (9) follow from the known results ([8]) on solvability of the Dirichlet problem in Lipschitz domains and from the continuity properties of layer potentials in these domains. In particular, we use that the Dirichlet-to-Neumann operator mapping the Dirichlet data u on $\partial\Omega$ for the Helmholtz equation into the Neumann data $\partial_\nu u$ on $\partial\Omega$ is bounded from $H_{(\frac{1}{2})}(\partial\Omega)$ into $H_{(-\frac{1}{2})}(\partial\Omega)$.

The bounds (10) can be proven similarly, when we replace Sobolev space by Hoelder space and use known regularity properties of potentials and solutions to elliptic boundary value problems in these spaces collected for example in ([6]), section 1.6.

The proof is complete.

The above proof is a combination of our original proof in ([3]) in the two-dimensional case, of an idea of John Sylvester, and of the available elliptic theory ([8]) for Lipschitz domains.

A similar result holds for exterior problems, however under the condition that k^2 is not a Dirichlet eigenvalue in Ω.

To formulate results on exterior problems it is convenienst to fix a sufficiently large ball B (containing $\overline{\Omega}$).

Theorem 2.2 *Let k^2 be not a Dirichlet eigenvalue in Ω.*

For any radiating solution $u \in H_{(1)}(B \setminus \Omega)$ to the Helmholtz equation (1) there is an unique function $\varphi \in H_{(-\frac{1}{2})}(\Gamma)$ such that (4) holds in Ω_e.

Moreover, for some constant C depending only on Ω we have

$$C^{-1}\|u\|_{(1)}(B \setminus \Omega) \le \|\varphi\|_{(-\frac{1}{2})}(\Gamma) \le C\|u\|_{(1)}(B \setminus \Omega) \qquad (12)$$

and if in addition $u \in C^{1+\lambda}(\overline{\Omega_e})$ and $\partial\Omega \in C^{1+\lambda}$, then

$$C^{-1}|u|_{1+\lambda}(\Omega_e) \le |\varphi|_\lambda \le |u|_{1+\lambda}(\Omega_e) \qquad (13)$$

This result is known. Its proof is similar to the proof of Theorem 2.1.

We will observe that if k^2 is a Dirichlet eigenvalue in Ω then the representation (4) is not unique.

Indeed, the well known Green's formula for radiating solutions u to the Helmholtz equation outside Ω yields

$$
D_\Gamma u(x) - S_\Gamma \partial_\nu u(x) \quad
\begin{aligned}
&= u(x), \ x \in \Omega_e \\[2mm]
&= 0, \ x \in \Omega
\end{aligned}
\tag{14}
$$

Also for any solution u to (1) in Ω one has

$$
S_\Gamma \partial_\nu u(x) - D_\Gamma u(x) \quad
\begin{aligned}
&= u(x), \ x \in \Omega \\[2mm]
&= 0, \ x \in \Omega_e
\end{aligned}
\tag{15}
$$

Let u_0 be a Dirichlet eigenfunction in Ω. Then the formula (15) with $u = u_0$ ($u_0 = 0$ on Γ) and $x \in \Omega_e$ yields

$$
0 = \int_\Gamma \partial_\nu u_0(y) K(x,y) d\Gamma(y), \ x \in \Omega_e
\tag{16}
$$

Finally we will discuss a popular representation by the sum of single and double layer potentials (4), (8). We will remind the known jump relations for double layer potentials

$$
D_{e\Gamma}\varphi(z) = \frac{\varphi(z)}{2} + D_\Gamma\varphi(z), \quad D_{i\Gamma}\varphi(z) = -\frac{\varphi(z)}{2} + D_\Gamma\varphi(z),
\tag{17}
$$

where e denotes the limit from outside Ω and i denotes the limit from inside Ω.

Lemma 2.1 *Let $u \in H_{(1)}(\Omega)$ be a solution to (1).*

Then there are unique functions $\varphi_j \in H_{(\frac{1}{2}-j)}(\Gamma), j = 0, 1$, such that

$$
u(x) = -S_\Gamma\varphi_1(x) + D_\Gamma\varphi_0(x), \ x \in \Omega,
$$

$$
\frac{1}{2}\varphi_0(z) = -S_\Gamma\varphi_1(z) + D_\Gamma\varphi_0(z), \ z \in \Gamma,
$$

Proof:

To show existence of φ_0, φ_1, let $\varphi_0 = u$ on Γ, $\varphi_1 = \partial_\nu u$ on Γ and use the relation (15) in Ω. To obtain the relation (20) we let in (19) $x \to z$ and use the jump relations (17) to pass to the limit. Then

$$u(z) = S_\Gamma \varphi_1(z) + \frac{1}{2}\varphi_0(z) - D_\Gamma \varphi_0(z), \tag{18}$$

and observing that $\varphi_0 = u$ on Γ we obtain (20).

To show uniqueness of φ_j it suffices to let $u = 0$ and to deduce that $\varphi_j = 0$. Letting $x \to z$ in (19) with $u = 0$ and using the jump relations (17) we obtain (18) with $u = 0$. Now from (20) it follows that $\varphi_0 = 0$. So from (19) we have $S_\Gamma \varphi_1 = 0$ on Ω. Repeating the argument from the proof of Theorem 2.1 we conclude that $\varphi_1 = 0$.

Lemma 2.2 *Let $u \in H_{(1)}(B \setminus \Omega)$ (for any B) be a solution to (1) in Ω_e. Then there are functions $\varphi_j \in H_{(\frac{1}{2}-j)}(\Gamma), j = 0, 1$, such that*

$$u(x) = S_\Gamma \varphi_1(x) - D_\Gamma \varphi_0(x), \ x \in \Omega_e$$

$$\frac{1}{2}\varphi_0(z) = S_\Gamma \varphi_1(z) - D_\Gamma \varphi_0(z), \ z \in \Gamma$$

The function φ_0 is unique. If k^2 is not a Dirichlet eigenvalue in Ω, then φ_1 is unique.

A proof is similar to the proof of Lemma 2.3.

Now we can write the Helmholtz-Kirchhoff integral equations for our interior inverse problem

$$u(x) = -S_\Gamma \varphi_1(x) + D_\Gamma \varphi_0(x), \ x \in \Gamma_0 \subset \Omega \tag{19}$$

$$\frac{1}{2}\varphi_0(z) = -S_\Gamma \varphi_1(z) + D_\Gamma \varphi_0(z), \ z \in \Gamma \tag{20}$$

and for the exterior problem

$$u(x) = S_\Gamma \varphi_1(x) - D_\Gamma \varphi_0(x), \ x \in \Gamma_0 \subset \Omega_e \tag{21}$$

$$\frac{1}{2}\varphi_0(z) = S_\Gamma \varphi_1(z) - D_\Gamma \varphi_0(z), \ z \in \Gamma \tag{22}$$

3 Uniqueness and stability for the inverse source problem

In this section we will discuss uniqueness and stability for the inverse problem and for the related integral equations. For the interior problem we will denote by Ω_0 a Lipschitz subdomain of Ω. For the exterior problem, Ω_0 is a Lipschitz domain with connected Ω_{0e} containing $\overline{\Omega}$.

The interior problem of detecting of the source of the acoustical noise or (the interior inverse problem) is to find u on Ω given u on Γ_0. The exterior problem seeks for u in Ω_e from the same data.

Due to results of section 2 the interior inverse problem is equivalent to the integral equation (6) for density of a single layer distribution on Γ or to the Helmholtz-Kirchhoff system (19), (20). For exterior problems this equivalence is valid only when k^2 is not a Dirichlet eigenvalue in Ω.

Lemma 3.1 *(Interior problem)*
Let k^2 be not a Dirichlet eigenvalue for Ω_0.

If $\Gamma_0 = \partial\Omega_0$, then $u \in H_{(1)}(\Omega)$ solving (1) is uniquely determined by u on Γ_0.

If $\partial\Omega_0$ is analytic and Γ_0 is a nonvoid open part of $\partial\Omega_0$, then u is also uniquely determined.

Lemma 3.1 immediately follows from the uniqueness of the continuation for elliptic equations ([7], section 3.3). To prove uniqueness we assume that $u = 0$ on Γ_0 and we have to conclude that $u = 0$ in Ω. Since k^2 is not a Dirichlet eigenvalue, a solution to the Dirichlet problem in Ω_0 is unique, so $u = 0$ in Ω_0. Then by uniqueness of the continuation from Ω_0 onto Ω we conclude that $u = 0$ in Ω.

If $u = 0$ on Γ_0 which is a part of analytic $\partial\Omega_0$, then by uniqueness of analytic continuation on $\partial\Omega_0$ we conclude that $u = 0$ on $\partial\Omega_0$. After that we repeat the previous argument.

Lemma 3.2 *If $\Gamma_0 = \partial\Omega_0$, then a radiating solution $u \in H_{(1)}(B \setminus \Omega)$ to the Helmholtz equation is uniquely determined by u on Γ_0.*

If $\partial\Omega_0$ is analytic and Γ_0 is a nonvoid open part of $\partial\Omega_0$, then u is uniquely determined by u on Γ_0.

A proof of this lemma is similar to the proof of Lemma 3.1. We observe that a solution of the (exterior) Dirichlet problem with the data on $\partial\Omega_0$ is always unique.

Now we will discuss uniqueness for integral equations (6),(19)-(22).

Lemma 3.3 *(Interior problem)*

If k^2 is not a Dirichlet eigenvalue for Ω_0, then under the conditions of Lemma 3.1 a solution $\varphi \in H_{(-\frac{1}{2})}(\Gamma)$ to the equation (6) is unique and a solution $(\varphi_0, \varphi_1) \in H_{(\frac{1}{2})}(\Gamma) \times H_{(-\frac{1}{2})}(\Gamma)$ to the Helmholtz-Kirchhoff system (19), (20) is unique.

This result follows directly from Theorem 2.1, Lemmas 2.1 and 3.1.

Lemma 3.4 *(Exterior problem)*

If k^2 is not a Dirichlet eigenvalue for Ω, then under the conditions of Lemma 3.2 on Γ_0 a solution $\varphi \in H_{(\frac{1}{2})}(\Gamma)$ to the equation (6) is unique and a solution $(\varphi_0, \varphi_1) \in H_{(\frac{1}{2})} \times H_{(-\frac{1}{2})}(\Gamma)$ to the Helmholtz-Kirchhoff system (21),(22) is unique.

This result follows directly from Theorem 2.2, Lemmas 2.2 and 3.2.

Now we will discuss stability considering for brevity the interior problem.

Nonuniqueness caused by Dirichlet eigenvalues in Ω_0 can not be avoided, but we can get rid of eigenvalues by changing the measurement surface Γ_0. Away from them the Dirichlet problem in Ω_0 is stable in classical Sobolev (or Hoelder) spaces. The problem of the continuation of solutions of elliptic equtions (from Ω_0 onto Ω) is notoriously unstable, however assuming that $\|u\|_2(\Omega) < M_0$ one can control exponentially growing solutions and to obtain the conditional Hoelder type estimate

$$\|u\|_{(k)}(\Omega_1) < CM^{1-\theta}\|u\|_2^\theta(\Omega_0) \tag{23}$$

where C, θ depend on $\Omega, \Omega_0, k, 0 < \theta < 1$ and on the distance from $\Omega_1 \subset \Omega$ to $\partial\Omega$. We refer for proofs to ([7]), sections 3.2,3.3. In general situation it is hard to get expicit bounds for C, θ.

Now we give a stability estimate for spheres $\Omega = \{|x| < r_1\}, \Omega_0 = \{|x| < r_0\}$ with almost explicit constants. As known, a solution u to the Helmholtz equation (1) in Ω admits the expansion

$$u(x) = \sum u_{m,n} j_n(kr) Y_{m,n}(\sigma)$$

where j_n is the n-th Bessel function, $r = |x|$, $\sigma = \frac{x}{r}$, and $Y_{m,n}$ are the spherical harmonics orthonormal on the unit sphere. We denote by u_0 the sum of the terms with $n \leq n_1 = \frac{kr_1^2 - 3}{2}$ and by u_1 the sum of the remaining terms of the series.

Theorem 3.1 *Let u be a solution to the Helmholtz equation (1) in the ball $\Omega(r_1)$ and*

$$\|u\|_2(\Omega(r_1)) \leq M_0, \|\nabla u\|_2(\Omega(r_1)) \leq M_1 \qquad (24)$$

Then

$$\frac{r_0^3}{r^2}\|u\|_2^2(\partial\Omega(r)) \leq C_1(r)\epsilon_0^2 + C_2\epsilon_1^{2\theta}, \quad r < r_1 \qquad (25)$$

where $\epsilon_j^2 = \|u_j\|_2^2(\partial\Omega(r_0))$, $C_1(r) = max|\frac{j_n(kr)}{j_n(kr_0)}|^2$ over $n \leq n_1$, $C_2 = 5M^2$, $M^2 = 4M_0^2 + r_1^2 M_1^2$, $\theta = \frac{lnr_1 - lnr}{lnr_1 - lnr_0}$.
   In addition,

$$\|u\|_2^2(\partial\Omega(r_1)) \leq C_1(r_1)\varepsilon_0^2 + C_3\varepsilon_2(C_4 - ln\varepsilon_2) \qquad (26)$$

where $C_3 = \frac{1}{r_0}max(k^2 M_0^2, M_1^2)ln\frac{r_1}{r_0}$, $\varepsilon_2 = -\frac{1}{ln\frac{\varepsilon_1}{M}}$, and $C_4 = 1 - ln\frac{r_0^3 C_3}{5M^2}$.

A proof of this stability estimate is given in ([4]). It is based on the explicit bounds for Hankel functions and on the logarithmic convexity method. We observe that C_1 can be easilily calculated and it is relatively small. Since with increasing k the "stable" component u_0 of the solution is increasing we have increased stability in this problem for larger k. We observed this increased stability in our numerical experiments in ([3]).

4   Approximation and completeness results

Now we discuss approximation of general solutions u by simplest solutions. Its purpose is to interpolate our data on Γ_0 for further solution of integral equations. This can be crucial for larger k.

Theorem 4.1 *Let Ω be a bounded Lipschitz domain in $\mathbf{R}^3$. Let $u \in H_{(1)}(B \setminus \Omega)$ be a radiating solution to the Helmholtz equation $\Delta u + k^2 u = 0$ in Ω_e. Let a ball $B_0 \subset \overline{B_0} \subset \Omega$.*

Then for any positive ε there is a radiating solution u_ε to the Helmholtz equation outside $\overline{B_0}$ such that

$$\|u - u_\varepsilon\|_{(1)}(\Omega_e) < \varepsilon \tag{27}$$

Proof:

By extension theorems for Sobolev spaces in Lipschitz domains ([7]) there is an extension $u^* \in H_{(1)}(B)$ of u onto $\mathbf{R}^3$. Let $f^* = \Delta u^* + k^2 u^*$. Then $f^* \in H_{(-1)}(\mathbf{R}^3)$ and $supp f^* \subset \overline{\Omega}$. It is known that $f^* = f_0 + \sum \partial_j f_j$ for some $f_0, ..., f_3 \in L^2(\mathbf{R}^3$ supported in $\overline{\Omega}$. Let χ_n be a sequence of measurable functions with values 0 or 1 supported in Ω and convergent to 1 a.e. on Ω. Then f_n^* defined as $f_0 \chi_n + \sum \partial_j (f_j \chi_n)$ are convergent to f in $H_{(-1)}(\mathbf{R}^3)$ and $supp f_n^* \subset \Omega$. From the theory of elliptic equations ([7], section 4.1) and scattering theory ([10]) it follows that radiating solutions to the Helmholtz equition $(\Delta + k^2) u_n = f_n^*$ in $\mathbf{R}^3$ converge to u in $H_{(1)}(B \setminus \Omega)$ for any ball B. So one can find u_n such that

$$\|u - u_n\|_{(1)}(B \setminus \Omega)) < \frac{\varepsilon}{2} \tag{28}$$

By the Runge property for scattering solutions in $\mathbf{R}^3 \setminus \overline{\mathbf{B_0}}$ there is a radiating solution u_ε to the Helmholtz equation outside $\overline{B_0}$ such that

$$\|u_n - u_\varepsilon\|_{(1)}(B \setminus \Omega) < \frac{\varepsilon}{2} \tag{29}$$

From (28) and the last inequality we obtain (27).

The proof is complete.

In applications it is very helpful to use the special family of radiating solutions to the Helmholtz equation

$$e_{n,m}(x) = h_n^{(1)}(kr) Y_{n,m}(\sigma) \tag{30}$$

where $h_n^{(1)}$ is the spherical Hankel function of the first kind and $Y_{n,m}$ are spherical harmonics orthonormal in $L^2(S^2)$ on the unit sphere S^2. It is convenient to approximate solutions u to the Helmholtz equation by linear combinations

$$u_e(x) = \sum u_{n,m} e_{n,m}(x), \ m = 0, ..., 2n+1, n = 0, ..., N \tag{31}$$

Corollary 4.1 *Let $0 \in \Omega$.*

For any positive ε there is u_e such that

$$\|u - u_e\|_{(1)}(B \setminus \Omega) < \varepsilon$$

Proof:

Let B_0 be a ball centered at 0. By Theorem 2.1 there is a radiating solution u_ε to the Helmholtz equation in $\mathbf{R}^3 \setminus \overline{B_0}$ such that $\|u - u_\varepsilon\|_{(1)}(B \setminus \Omega) < \frac{\varepsilon}{2}$. Let B_1 be a ball of radius r_1 (greater than the radius of B_0) centered at the center of B_0 so that $\overline{B_1} \subset \Omega$. The spherical harmonics $Y_{n,m}$ form an orthonormal basis in $L^2(S^2)$. Expanding the function $u_\varepsilon(r_1\sigma)$ with respect to this basis we conclude that partial sums of the corresponding series are convergent in $L^2(\partial B_1)$ and hence due to known results about these series (e.g. ([2],Theorem 2.14) these partial sums are convergent to u_ε on $B \setminus \Omega$. Hence one can find a partial sum u_e such that $\|u_\varepsilon - u_e\|_{(1)} < \frac{\varepsilon}{2}$. Now the claim follows from the triangle inequality.

A similar result is valid for interior problems.

Theorem 4.2 *Let $u \in H_{(1)}(\Omega)$ be a solution to the Helmholtz equation $\Delta u + k^2 u = 0$ in Ω.*

Then for any positive ε there is a solution u_ε to the Helmholtz equation in $\mathbf{R}^3$ such that

$$\|u - u_\varepsilon\|_{(1)}(\Omega) < \varepsilon \tag{32}$$

For interior problems a particular family of useful solutions is spanned by the functions

$$E_{n,m}(x) = j_n(kr)Y_{n,m}(\sigma) \tag{33}$$

Now we will discuss how to use these results for approximation of u by u_e.

Letting in (32) $\varepsilon = 1$ we conclude that for $\varepsilon < 1$ there are approximating functions u_e such that

$$\|u_e\|_{(1)}(B \setminus \Omega) \le M_1 = \|u\|_{(1)}(B \setminus \Omega) + 1 \tag{34}$$

Slightly weaking this constraint we have that $\|u_e\|_{(0)}(B \setminus \Omega) < M_0 \le M_1$, or

$$\sum |u_{n,m}|^2 \le M_0^2, \; m = 0, ..., 2n + 1, n = 0, ..., N \tag{35}$$

Since the data are given on Γ_0 one can try to approximate u by u_e by solving the minimization problem

$$min\|u - u_e\|_{(0)}(\Gamma_0) \tag{36}$$

subject to the constraint (35). By solving this problem for sufficiently large $N = N(\delta)$ one can find $u_e(;\delta)$ such that

$$\|u - u_e(;\delta)\|_{(0)}(\Gamma_0) < \delta \tag{37}$$

and that the constraint (35) holds.

Lemma 4.1 *Let Ω_0 be a bounded domain, $\overline{\Omega} \subset \Omega_0$. Let either i) $\Gamma_0 = \partial\Omega_0$ or ii) $\partial\Omega_0$ be analytic and Γ_0 be a nonvoid open part of $\partial\Omega_0$.*
   Then there is a function $\omega(\delta) \to 0$ as $\delta \to 0$ such that $\|u - u_e\|_{(0)}(B\backslash\Omega_1) < \omega(\delta)$ and $\|u_e\|_{(0)}(B \setminus \Omega) < M_0$.
   In addition, if $\overline{\Omega} \subset \Omega_1, \overline{\Omega_1} \subset \Omega_0$ then $\omega(\delta) = C\delta^\theta$ ($\theta \in (0,1)$ and depends on distances from $\partial\Omega$ to $\partial\Omega_0$ and to $\partial\Omega_1$); and if $\Omega_1 = \Omega$, then $\omega(\delta) = -C\log\delta$.

A proof of this result can be obtained by using stability estimates in the Cauchy problem for the Helmholtz equation (see [7], section 3.3).
   Numerical experiments (in part reported in([3]), ([4]) based on the conjugate gradient method and stopping rule motivated by our stability analysis show good resolution and fast convergence. In the geometry simulating the cabin of a midsize aircraft at frequencies $k = 1, 2, 3$ one percent of random error in the data resulted in 3-4 percent in solution error

The results of this paper, except of section 4, were obtained together with Tom DeLillo, Nicolas Valdivia, and Lianju Wang. The research is in part supported by the NSF grants DMS-9803816 and ITR/ACS 0081270.

References

[1] K.E. Atkinson, *The Numerical Solution of Integral Equations of the Second Kind*, Cambridge Univ. Press, 1997.

[2] D. Colton, R. Kress, *Inverse acoustic and electromagnetic scattering theory*, Springer-Verlag,New York, 1992

[3] T. DeLillo, V. Isakov, N. Valdivia, L. Wang, *The detection of the source of acoustical noise in two dimensions*, SIAM J. Appl. Math. (to appear).

[4] T. DeLillo, V. Isakov, N. Valdivia, L. Wang, *The detecting of the source of interior acoustical noise* (in preparation).

[5] H. W. Engl, M. Hanke, A. Neubauer, *Regularization of inverse problems*, Kluwer, Dordrech, 1996.

[6] V. Isakov, *Inverse Source Problems*, AMS, Providence, R.I., 1990.

[7] V. Isakov, *Inverse Problems for Partial Differential Equations*, Springer-Verlag, New York, 1998.

[8] W. McLean, *Strongly Elliptic Systems and Boundary Integral Euqations*, Cambridge Univ. Press, 2000.

[9] P. Morse, K. Ingard, *Theoretical Acoustics*,McGraw-Hill, New York, 1968.

[10] *Scattering*, Academic Press, 2001.

[11] E.G. Williams, *Fourier Acoustics*, Academic Press,New York, 1999

[12] S. Wu, *On reconstruction of radiated acoustic pressure fields by using the HELS method* , J. Acoust. Soc. Am., 2000

Theoretical and Computational Acoustics 2001
E.-C. Shang, Qihu Li and T. F. Gao (Editors)
© 2002 World Scientific Publishing Co.

Stability, Resolution and Error Study for Implementation of Beamforming in Seismic Data Processing

Tianyue Hu
Peking University, Beijing 100871, China

Renqiu Wang
Petroleum University, Beijing 102200, China

Abstracts

The basic MVU (minimum variance unbiased) beamforming designs are defined to satisfy the general design constraints of (1) no signal distortion (boresight constraint) and (2) minimizing output noise power. The aim of applying MVU beamforming in seismic data processing is to extract seismic reflection signals. For this purpose, additional criteria in MVU beamforming for coherent noise, such as multiples, rejection are (1) zero or minimize the response to coherent (multiple) and (2) constrain random noise amplification. The solution of this problem can be obtained by using the method of Lagrange multiplier. The main challenger for the implementation of beamforming is the way to calculate the inversion of an ill conditional matrix to obtain the solution. This paper will show the stability, resolution and error study of different inverse operators for the ill conditional matrix of implementation of beamforming in seismic data processing.

1. Introduction

Multichannel (moveout) filters are effective moveout-based techniques in wavefield decomposition to attenuate coherent noise (multiples) in seismic data processing. Straight stacking is the simplest multi-channel method. But it is a very primitive one. For a long time, it was felt that more elaborate methods should outperform stacking. Beamforming, based on the coherent components model, has significant advantages for amplitude/phase versus offset analysis when attenuating multiples. The potential for using beamforming to improve the performance of multi-channel methods was recognized in the early 1960's in the fields of sonar and radar as well as seismic data processing, and research began to be published on improved methods of beamforming, see Shumway and Dean (1968).

For the purpose of applying robust beamforming to seismic data processing, the first job is to examine the implementation and response characteristics of beamformers. Moveout and amplitude variation is directly related to the incidence angle of the seismic wave, and may be associated with the change in lithology and/or fluid fill. In adaptive beamforming design, time, amplitude and phase variations are accounted for in the decomposition of the seismic data wavefields.

Cox et al (1987) and White (1988) concentrated on robust adaptive beamforming which is a critical issue in the development of beamforming. Robustness and its associated trade-offs raise important points of technique in applying beamforming to seismic data processing. The most evident trade-off is between coherent noise (multiple) attenuation and random noise attenuation.

This paper discusses (1) the implementation of basic beamforming, (2) the stability of the matrix inversion in beamforming design, (3) the beamformer responses, particularly the trade-off between coherent noise attenuation and random noise attenuation, and (4) the application to marine seismic data processing of using beamforming.

2. Basic MVU beamforming theory

In a common mid point (CMP) gather, the seismic data model can be described in frequency domain by

$$\mathbf{x} = \mathbf{As} + \mathbf{u} \tag{1}$$

where $\mathbf{x}$ is the recorded data, $\mathbf{s}$ the coherent components to be estimated and $\mathbf{u}$ weakly stationary zero mean noise ; and

$$\mathbf{A} = \{A_{kj}(f)\} \;, \; A_{kj}(f) = a_{kj}\exp(-2\pi i f\tau_{kj} - i\theta_{kj}) \tag{2}$$

where a_{kj}, θ_{kj}, and τ_{kj} are the amplitude, phase and the time delays of the jth coherent component at trace k. The coherent components $\mathbf{s}$ include signals and coherent noise, i.e. primaries and multiples. They are normalized to have unit power.

The basic MVU (minimum variance unbiased) designs are defined to satisfy the general design constraints of: (i) no signal distortion (boresight constraint), and (ii) minimising output noise power. In frequency domain, if the inverse matrix $(\mathbf{A}^H\mathbf{Q}^{-1}\mathbf{A})^{-1}$ exists, the multichannel filter can be constructed by

$$\mathbf{H} = (\mathbf{A}^H\mathbf{Q}^{-1}\mathbf{A})^{-1}\mathbf{A}^H\mathbf{Q}^{-1} \tag{3}$$

where $\mathbf{Q} = \mathrm{E}[\mathbf{uu}^H]$, the superscript H denotes conjugate transpose and E[ ] denotes expectation. If the noise is random and has the same power on every channel, then the spectral matrix is just

$$\mathbf{Q} \Rightarrow \mathbf{Q}_R = \mathrm{E}[\mathbf{uu}^H] = \sigma^2\mathbf{I} \tag{4}$$

where σ^2 is the noise variance or power on each channel. Then the equation (3) becomes

$$\mathbf{H} = (\mathbf{A}^H\mathbf{A})^{-1}\mathbf{A}^H \tag{5}$$

The estimated coherent components are given by

$$\hat{\mathbf{s}} = (\mathbf{A}^H\mathbf{A})^{-1}\mathbf{A}^H\mathbf{x} \tag{6}$$

3. Stability of the matrix inversion

The main problem of calculating the estimated signals by equation (6) is the inversion of the matrix $\mathbf{A}^H \mathbf{A}$. Aki and Richards (1980) and Hatton et al (1986) gave some discussion of the problems of such matrix inversion in terms of its eigenvalues and eigenvectors. Fortunately, the symmetry properties of matrix $\mathbf{A}^H \mathbf{A}$ allow the matrix decomposition

$$\mathbf{A}^H \mathbf{A} = \mathbf{V} \boldsymbol{\Lambda} \mathbf{V}^H \tag{7}$$

where $\boldsymbol{\Lambda}$ is a diagonal matrix with the eigenvalues of $\mathbf{A}^H \mathbf{A}$ as the elements of the main diagonal, and $\mathbf{V}$ is a matrix whose columns are the normalised eigenvectors of $\mathbf{A}^H \mathbf{A}$.

The main objective of the present discussion is to obtain an acceptable inverse operator $\boldsymbol{\Omega}$. If there are no zero eigenvalues of $\mathbf{A}^H \mathbf{A}$, $\boldsymbol{\Omega}$ can be obtained by:

$$\boldsymbol{\Omega} = (\mathbf{V} \boldsymbol{\Lambda} \mathbf{V}^H)^{-1} = \mathbf{V} \boldsymbol{\Lambda}^{-1} \mathbf{V}^H \tag{8}$$

where $\boldsymbol{\Lambda}^{-1}$ is also a diagonal matrix with elements $\{\lambda_j^{-1}\}$, and $\{\lambda_j\}$ are the elements of diagonal matrix $\boldsymbol{\Lambda}$.

However, from equation (8) it may be seen that any very small values of λ_j (very large values of λ_j^{-1}) will dominate the estimate of the signals. Actually, at low frequencies, the change of the factor $\exp(-2\pi i f \tau_{kj})$ is very slow with respect to the offset, which corresponds to one or more small eigenvalues in the matrix $\mathbf{A}^H \mathbf{A}$. Setting a threshold on the eigenvalues overcomes the condition of near singularity. Then the generalised inverse operator $\boldsymbol{\Omega}_p$ is defined to replace the previous inverse operator $\boldsymbol{\Omega}$:

$$\boldsymbol{\Omega} \Rightarrow \boldsymbol{\Omega}_p = \mathbf{V}_p \boldsymbol{\Lambda}_p^{-1} \mathbf{V}_p^H \tag{9}$$

where $\boldsymbol{\Lambda}_p$ and $\mathbf{V}_p$ are associated with the uppermost eigenvalues.

Figure 1 shows all the eigenvalues at different frequencies for different signal and noise models. The more pass bands and reject bands and the lower the frequencies, the bigger the differences in the eigenvalues. At 15 Hz, for a signal pass band of 100ms (5 components) and a coherent noise reject band of 100ms (5 components), the relative eigenvalue ratio of minimum to maximum is less than 10^{-9}!

There are other approaches to calculate equation (6). One is simply to add white noise to the diagonal of $\mathbf{A}^H \mathbf{A}$; but this leads to sacrificing the resolution; see Aki and Richards (1980). Singular value decomposition (SVD) is a well known way to deal with the problem of the inversion of a singular matrix. Searle (1982) gave an important and useful consequence of SVD. For a matrix of rank r

632

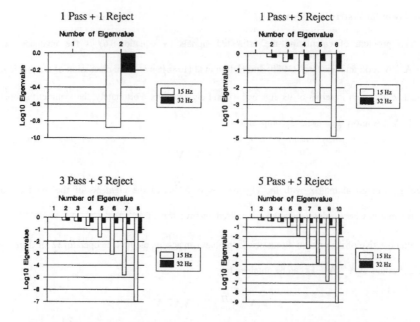

Figure 1: All the eigenvalues at different frequencies from different signal pass band and coherent noise reject band designs; the design components are spaced at 25 ms increments of maximum moveout.

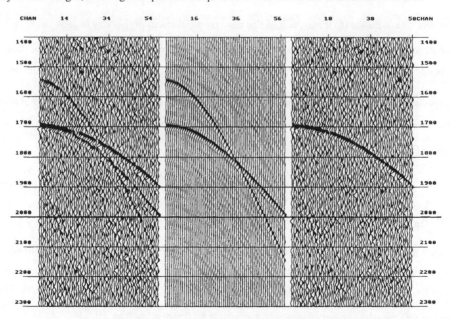

Figure 2: An example: the input data containing signal and coherent noise (multiples) and the output reconstructed signal without multiples. The left is the synthetic input data, the mid reconstructed coherent noise, and the right residuals of signal and random noise. Here the passband for signals is 150---250 ms and the reject band for coherent noise is 300---600 ms at maximum moveout.

$$A_{K \times J} = U \begin{pmatrix} \Delta_r & 0 \\ 0 & 0 \end{pmatrix} V^H , \qquad (A^+)_{J \times K} = V \begin{pmatrix} \Delta_r^{-1} & 0 \\ 0 & 0 \end{pmatrix} U^H \qquad (10)$$

is the Moore-Penrose inverse, where U and V are unitary and where Δ_r^2 is the diagonal matrix of the nonzero eigenvalues δ_j^2 of AA^H and of $A^H A$; and

$$U^H AA^H U = \begin{pmatrix} \Delta_r^2 & 0 \\ 0 & 0 \end{pmatrix}_{K \times K} , \quad \text{and} \quad V^H AA^H V = \begin{pmatrix} \Delta_r^2 & 0 \\ 0 & 0 \end{pmatrix}_{J \times J} \qquad (11)$$

By applying SVD and the Moore-Penrose inverse to equation (6), the estimated components can be calculated by

$$\hat{s} = V \begin{pmatrix} \Delta_r^{-1} & 0 \\ 0 & 0 \end{pmatrix} U^H x \qquad (12)$$

Comparison between equation (12) and equation (7) shows that $\lambda_j = \delta_j^2$. This means that the condition number of matrix $A^H A$ is the square of the condition number of matrix A. Therefore SVD could lead to more stable estimation.

In practice, for a typical seismic CMP gather with 60 traces and 10 coherent components, the matrix AA^H has 60×60 elements and the matrix $A^H A$ has 10×10 elements. Then the cost of calculating eigenvectors of matrix AA^H is over 50 times more than that of calculating the eigenvectors of matrix $A^H A$. Stewart (1979) developed an algorithm for the SVD of complex matrix by using matrix perturbation theory to save some computing. In general, the cost of applying SVD is more expensive. Later examples will show that estimation using the generalised inverse operator is good enough in seismic data applications.

4. Resolution and error for the different inverse operators

This section discusses the effects of the generalised inverse operator of equation (9) for the different models.

4.1 Over determined case

In a recorded seismic CMP gather, usually the number of channels, K, is bigger than the number of coherent components, J, and each coherent component is independent of every other. Then,

$$\text{rank}[A_{K \times J}] = J \quad \Rightarrow \quad \text{rank}[(A^H)_{J \times K} A_{K \times J}] = J \qquad (13)$$

Therefore, there are no zero eigenvalues of $A^H A$ and the matrix V of the eigenvectors can span the whole model space. The inverse operator of the matrix $A^H A$ can be obtained from equation (8) and the coherent components can be estimated by equation (6) as the least-squares solution, i.e.

$$\hat{s} = \mathbf{V}\boldsymbol{\Lambda}^{-1}\mathbf{V}^H\mathbf{A}^H\mathbf{x} \tag{14}$$

A special over-determined case is the 1-component model, i.e. $J=1$. In fact, after applying multiple attenuation to the seismic CMP gather, there is only one significant component left, the primaries. The 1-component model can be used in beamforming design over a very small time window in order to estimate the isolated primary waveform. Now, the matrix $\mathbf{A}^H\mathbf{A}$ reduces to a scalar $\sum_{k=1}^{K} a_{k1}^2$. From the solution of equation (6), the estimated primary $\hat{s}_1(f)$ is

$$\hat{s}_1(f) = \frac{\sum_{k=1}^{K} a_{k1}\exp(2\pi i f \tau_{k1} + i\theta_{k1})x_k(f)}{\sum_{k=1}^{K} a_{k1}^2} \tag{15}$$

Applying the inverse Fourier transform to equation (15) gives the estimated primary in the time domain

$$\hat{s}_{1,t} = \frac{\sum_{k=1}^{K} a_{k1}p(\theta_{k1})^* x_{k,t-\tau_{k1}}}{\sum_{k=1}^{K} a_{k1}^2} \tag{16}$$

where $p(\theta_{k1})$ is the phase shift operator. Obviously, the solution of 1-component beamforming of equation (16) is no more than a weighted stack and the solution is very stable.

4.2 Under determined case

In a seismic CMP gather, the amplitude and phase variations with offset (AVO and PVO) for different primaries are different. For the purpose of matching the AVO and PVO optimally for each primary, it is necessary to use a small time window in the beamforming design. However, in real marine seismic data processing, the shallow part of a CMP gather is very often muted to get rid of the direct wave, refracted events and so on. As the result, the number of live data traces becomes much smaller than the total number of recorded traces. Sometimes the number of live data traces is smaller than the number of components in the model, i.e. $K < J$. Then,

$$\mathrm{rank}[\mathbf{A}_{K\times J}] < J \quad \Rightarrow \quad \mathrm{rank}[(\mathbf{A}^H)_{J\times K}\,\mathbf{A}_{K\times J}] < J \tag{17}$$

Therefore, $\mathbf{A}^H\mathbf{A}$ must have some zero eigenvalues and the matrix $\mathbf{V}_\mathrm{p}$ of the eigenvectors cannot span the whole model space. The generalised inverse operator of $\mathbf{A}^H\mathbf{A}$ can be obtained from equation (9). From equations (6), (9) and (1), the coherent components can be estimated by

$$\hat{s}_\mathrm{p} = \mathbf{V}_\mathrm{p}\boldsymbol{\Lambda}_\mathrm{p}^{-1}\mathbf{V}_\mathrm{p}^H\mathbf{A}^H\mathbf{x} = \mathbf{V}_\mathrm{p}\mathbf{V}_\mathrm{p}^H\mathbf{s} + \mathbf{V}_\mathrm{p}\boldsymbol{\Lambda}_\mathrm{p}^{-1}\mathbf{V}_\mathrm{p}^H\mathbf{A}^H\mathbf{u} \tag{18}$$

and

$$E[\hat{s}_p] = V_p V_p^H s \tag{19}$$

where the matrix $V_p V_p^H$ is the resolution matrix. If $V_p V_p^H$ is the identity matrix I, resolution is perfect and the particular solution is equal to the true solution. If the row vectors of $V_p V_p^H$ have components spread around the diagonal (with low values elsewhere), the particular solution represents a smoothed solution over the range of signals in s.

For the under-determined case, Aki and Richards (1980) also gave an equivalent result of the signal estimation in order to avoid the zero eigenvalue problem:

$$\hat{s} = A^H (AA^H)^{-1} x = A^H U \Lambda'^{-1} U^H x \tag{20}$$

where

$$AA^H = U \Lambda' U^H \tag{21}$$

Λ' is a diagonal matrix with the eigenvalues of AA^H as the elements of the main diagonal, and U is a matrix whose columns are the normalised eigenvectors of AA^H.

In practice, there are some options if the number of live data traces is smaller than the number of components in the model. The simplest is to leave this part of the data alone. Another way is to treat it as an under-determined problem, recognising the limited resolution of the results. Third, one can reduce the number of the coherent components in the model to avoid the under-determined case and get better results. However appropriate models need to be specified throughout the muting zone.

4.3 Near singularity

In most real seismic data cases, the matrix $A^H A$ is nearly singular at low frequencies. The generalised inverse operator of equation (9) has been introduced to overcome the singularity. As in the case of under-determination, the estimated coherent components can be obtained by equation (18). According to equations (3) and (18), the residuals δx_p between the recorded data x and the estimated modelled data $\hat{x}_p$ can be calculated by

$$\delta x_p = x_p - \hat{x}_p = A V_s V_s^H s + (I - U_p U_p^H) u \tag{22}$$

where V_s consists of the eigenvectors associated with the lowermost eigenvalues Λ_s, which are truncated by the threshold of the relative eigenvalues and where

$$A = U \Delta V^H \tag{23}$$

is the SVD of $\mathbf{A}$. $\mathbf{I} - \mathbf{U}_p \mathbf{U}_p^H$ is a projection operator which projects the noise onto the signal space. In an alternative representation, according to equation (14), the accurate solution should be

$$\hat{\mathbf{s}} = \mathbf{V}\Lambda^{-1}\mathbf{V}^H \mathbf{A}^H \mathbf{x} = \hat{\mathbf{s}}_p + \hat{\mathbf{s}}_s \tag{24}$$

where

$$\hat{\mathbf{s}}_s = \mathbf{V}_s \Lambda_s^{-1} \mathbf{V}_s^H \mathbf{A}^H \mathbf{x} \tag{25}$$

comprises the residuals of coherent components from truncating the operator $\mathbf{\Omega}$. Since $\mathbf{V}_p^H \mathbf{V}_s = 0$, which means that $\hat{\mathbf{s}}_p$ and $\hat{\mathbf{s}}_s$ are uncorrelated, I get

$$\left|\hat{\mathbf{s}}\right| = \left|\hat{\mathbf{s}}_p\right| + \left|\hat{\mathbf{s}}_s\right| \geq \left|\hat{\mathbf{s}}_p\right| \tag{26}$$

The generalised inverse gives the minimum norm of all possible solutions subject the number of signals being p.

5. Beamformer response

5.1 Signal and noise model

In order to separate the primary and the multiples, i.e. to reject the multiples and estimate parameters of the primaries, the data model may be rewritten in terms of signals $\mathbf{s}$, coherent noise (multiples) $\mathbf{v}$ and random $\mathbf{u}$ noise:

$$\mathbf{x} = \mathbf{B}\mathbf{s} + \mathbf{C}\mathbf{v} + \mathbf{u} \tag{27}$$

where

$$\mathbf{B} = \{B_{kj}\}, \quad B_{kj} = b_{kj} \exp(-2\pi i f \tau_{kj}^{(s)} - i\theta_{kj}^{(s)}) \tag{28}$$

$$\mathbf{C} = \{C_{kl}\}, \quad C_{kl} = c_{kl} \exp(-2\pi i f \tau_{kl}^{(v)} - i\theta_{kl}^{(v)}) \tag{29}$$

and, b_{kj}, $\tau_{kj}^{(s)}$ and $\theta_{kj}^{(s)}$ are the amplitude, moveout and phase of the jth signal; c_{kl}, $\tau_{kl}^{(v)}$ and $\theta_{kl}^{(v)}$ are the amplitude, moveout and phase of the lth coherent noise at the kth channel.

5.2 Additional constraints

Additional criteria in MVU beamforming for coherent noise rejection are: (i) zero or minimise the response to coherent noise (multiple); (ii) constrain random noise amplification. The columns of the matrix $\mathbf{B}$ are the steering vectors of the beamformer. The constraint of no signal distortion is written as

$$\mathbf{W}^H \mathbf{B} = \mathbf{I}_{J \times J} \tag{30}$$

where $\mathbf{I}_{J \times J}$ is the identity matrix of dimension $J \times J$ and the beamforming filter $\mathbf{W}^H$ is $\mathbf{H}$ of equation (5). Also a null response to coherent noise components along the correlated noise steering matrix $\mathbf{C}$ can be described by

$$\mathbf{W}^H \mathbf{C} = \mathbf{0}_{J \times L} \tag{31}$$

where $\mathbf{0}_{J \times L}$ is the null matrix of dimension $J \times L$.

5.3 Solution

The two constraints of equations (30) and (31) can be combined as:

$$\mathbf{A}^H \mathbf{W} = \mathbf{G}^H \tag{32}$$

where $\mathbf{G} = (\mathbf{I}_{J \times J}, \mathbf{0}_{J \times L})$ and $\mathbf{A} = (\mathbf{B}, \mathbf{C})$.

Frost (1972) obtained the solution of this problem, by using the method of Lagrange multipliers:

$$\mathbf{W} = \mathbf{Q}^{-1} \mathbf{A} (\mathbf{A}^H \mathbf{Q}^{-1} \mathbf{A})^{-1} \mathbf{G}^H \tag{33}$$

where $\mathbf{Q}$ is the estimation of the noise covariance matrix $\mathrm{E}[\mathbf{u}\mathbf{u}^H]$. Using the random noise model of equation (4) gives the filter

$$\mathbf{W}_R = \mathbf{A} (\mathbf{A}^H \mathbf{A})^{-1} \mathbf{G}^H \tag{34}$$

The signals can be estimated by the filter of equation (34),

$$\hat{\mathbf{s}} = \mathbf{W}_R^H \mathbf{x} = \mathbf{G} (\mathbf{A}^H \mathbf{A})^{-1} \mathbf{A}^H \mathbf{x} \tag{35}$$

Figure 2 shows an example: (a) the input data containing one signal and coherent noise (multiples) and (b) the output reconstructed signal without multiples and the residuals of random noise and the coherent noise.

P.Harris (personal communication) suggested another design to replace the constraints of no signal distortion, i.e, equation (30), and coherent noise rejection, i.e, equation (31), by

$$\mathbf{I}_{J \times J} + \varepsilon_1 \mathbf{1}_{J \times J} \geq \mathbf{W}^H \mathbf{B} \geq \mathbf{I}_{J \times J} - \varepsilon_1 \mathbf{1}_{J \times J} \tag{36}$$

$$\varepsilon_2 \mathbf{1}_{J \times J} \geq \mathbf{W}^H \mathbf{C} \geq -\varepsilon_2 \mathbf{1}_{J \times J} \tag{37}$$

respectively, where ε_1 and ε_2 are the small value parameters to control the signal distortion and coherent noise rejection; and where $\mathbf{1}_{J\times L}$ is the 1's matrix of dimension $J \times L$. Equation (36) says that signals can be slightly distorted and a small amount of cross-coupling between signal component responses is allowed. Equation (37) says that a small amount of coherent noise can leak through. Note that the inequalities should be interpreted element by element.

In order to compare with this design, the performance of the estimated filter $\mathbf{W}^H$ of equation (34) is checked by

$$\mathbf{W}^H \mathbf{A} = \mathbf{G}(\mathbf{A}^H \mathbf{A})^+ \mathbf{A}^H \mathbf{A} \tag{38}$$

As in section 3, for the matrix decomposition

$$\mathbf{A}^H \mathbf{A} = \mathbf{V} \mathbf{\Lambda} \mathbf{V}^H \tag{39}$$

its generalised inverse is

$$(\mathbf{A}^H \mathbf{A})^+ = \mathbf{V}_p \mathbf{\Lambda}_p^{-1} \mathbf{V}_p^H \tag{40}$$

where the subscript p associates with the uppermost eigenvalues for the diagonal matrix $\mathbf{\Lambda}$ and the matrix $\mathbf{V}$ of the normalised eigenvectors. Then equation (38) can be calculated as

$$\mathbf{W}^H \mathbf{A} = \mathbf{G} \mathbf{V}_p \mathbf{V}_p^H = (\mathbf{I}_{J\times J}, \mathbf{0}_{J\times L})(\mathbf{V}_p \mathbf{V}_p^H)_{(J+L)\times(J+L)} \tag{41}$$

For the comparison with equations (36) and (37), an operator $\mathbf{\Gamma}$ is introduced by

$$\mathbf{\Gamma}_{(J+L)\times(J+L)} = (\mathbf{V}_p \mathbf{V}_p^H)_{(J+L)\times(J+L)} - \mathbf{I}_{(J+L)\times(J+L)} \tag{42}$$

From equations (41) and (42), the design of equations (36) and (37) can be translated as

$$\left\| \mathbf{\Gamma}_{j,i} \right\| \le \varepsilon_1, \quad j=1,\cdots,J; i=1,\cdots,J \tag{43}$$

$$\left\| \mathbf{\Gamma}_{J+l,i} \right\| \le \varepsilon_2, \quad l=1,\cdots,L; i=1,\cdots,J \tag{44}$$

where $\mathbf{\Gamma}_{j,i}$ and $\mathbf{\Gamma}_{J+l,i}$ are the elements of matrix $\mathbf{\Gamma}$.

Therefore, the design of equations (36) and (37) just puts some constraints on the resolution matrix $\mathbf{V}_p \mathbf{V}_p^H$ and aims to increase stability. In the next section, it is shown that a threshold on the eigenvalues can control stability.

5.4 Definition of beamformer response

First consider a simple design aimed at enhancing one signal and rejecting a specific multiple. If I input a coherent component to this filter and vary its maximum moveout, the beamformer response can be defined as the estimated output as a function of maximum moveout.

Figure 3 shows the beamformer response of such a design. When the difference of the maximum moveout between the beamformer's steering vector and the signal direction vector is over 10ms the average response is less than -2dB. Furthermore, if the beamformer's steering rejection vector mismatches the multiple vector by more than 15ms the rejection is no better than the usual stack.

Generally, designing an operator for several signal components and several multiple components increases the width of the pass band and the reject band in terms of maximum moveout. The mismatching can be reduced by summing reconstructions of the signal components. The beamformer response is then defined as the average, over all channels of the summed reconstructions of the signal components:

$$R(f) = \frac{1}{K}\sum_{k=1}^{K}|\hat{x}_k(f)|^2 \tag{45}$$

where

$$\hat{x}_k(f) = \sum_{j=1}^{J}B_{kj}(f)\hat{s}_j(f) \tag{46}$$

and

$$\hat{s}_j(f) = \sum_{k=1}^{K}H_{jk}(f)x_k(f) \tag{47}$$

The mean bandpass beamformer response is defined by

$$R_B = \frac{\Delta f}{f_h - f_l}\sum_{f=f_l}^{f_h}R(f) \tag{48}$$

where f_l, f_h and Δf are the low-cut frequency, the high-cut frequency and the frequency increment.

For a given width of the pass band and/or reject band, the number of coherent components in the model is decided by the interval between two coherent components. A small interval has better response in both pass band and/or reject band, however it means more computation and problems from singularity.

Figure 4 shows the response for a model having 3-pass and 3-reject moveouts for a 100ms width of signal pass band and the coherent noise reject band, where the interval is 50ms. The response is less than -7dB between two signals in the pass band and bigger than -15dB in the reject band, which is no better than the

640

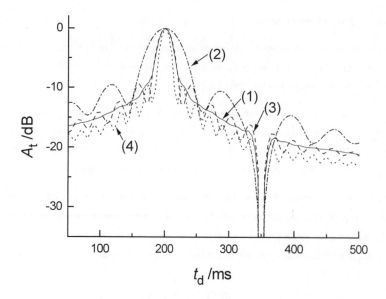

Figure 3: Beamformer responses for the design of a 1-signal pass band and 1-coherent noise reject band model at different frequencies and the mean over the frequency band. (1) average, (2) 15Hz, (3) 32.5Hz, (4) 50Hz

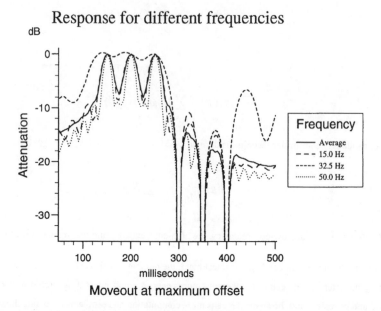

Figure 4: Beamformer responses for the design of a 3-signal pass band and 3-component reject band model at different frequencies and the mean over the frequency band.

usual stack. After reducing the interval to 20ms, the pass band becomes very flat without attenuation and the reject band has attenuation over 15dB higher than the usual stack; see figure 5 of the response for a model having 6-pass and 6-reject components.

5.5 Response for amplitude and time jitter and noise

Newman and Mahoney (1973) discussed in detail the effects of amplitude jitter and timing jitter on the array response function. The appendix of the paper gave the derivation of the results. Here their results are reproduced.

In the model of equation (28), if the amplitude jitter can be described by:

$$b'_{kj} = b_{kj}(1 + \varepsilon_{a,k}) \tag{49}$$

where the relative weight errors $\varepsilon_{a,k}$ are mutually independent and each have zero mean and standard deviation σ_a, then the response function in the reject band is:

$$R_a(f) = 20\log_{10}(\sigma_a) - 10\log_{10}(K) + \delta \tag{50}$$

where K is the number of channels and δ depends on the shape of the weight distribution but lies between 0 and 1.2 dB for most practical arrays.

Also, if the timing jitter can be described by:

$$\tau'_{kj} = \tau_{kj} + \varepsilon_{t,k} \tag{51}$$

where the timing errors $\varepsilon_{t,k}$ are mutually independent Gaussian distributed and each have zero mean and standard deviation σ_t, then the response function in the reject band is:

$$R_t(f) = 20\log_{10}(\sigma_t) - 10\log_{10}(K) + \delta \tag{52}$$

Figures 6 and 7 show mean (total frequency band) beamformer responses for the design of 6-signal pass band and 6-component reject band for different standard deviations σ_a of amplitude jitter and different standard deviations σ_t of time jitter, respectively. In the design, the number of traces is 60. There is an assumption of no systematic amplitude variation for signal or multiple with offset.

5.6 Trade-offs

If, as is usually assumed, signals, coherent noise, and random noise are independent then the cross spectral matrix of the data model of equation (27) is

$$\Phi_{xx} = E[xx^H] = BB^H + CC^H + \sigma_n^2 I \tag{53}$$

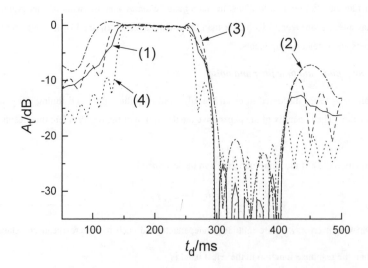

Figure 5: Beamformer responses for the design of a 6-signal pass band and 6-component reject band model at different frequencies and the mean over the frequency band. (1) average, (2) 15Hz, (3) 32.5Hz, (4) 50Hz

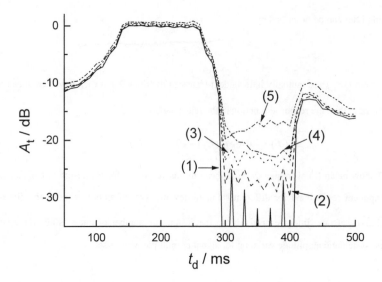

Figure 6: Mean (total frequency band) beamformer responses for the design of a 6-signal pass band and 6-component reject band model for different standard deviations σ_a of amplitude jitter. (1) 0.0, (2) 0.1, (3) 0.2, (4) 0.3, and (5) 0.4.

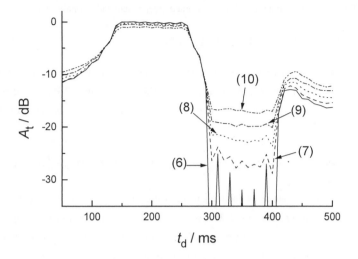

Figure 7: Mean (total frequency band) beamformer responses for the design of a 6-signal pass band and 6-component reject band model for different standard deviations σ_t of time jitter. (6) 0ms, (7) 1ms, (8) 2ms, (9) 3ms, and (10) 4ms.

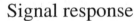

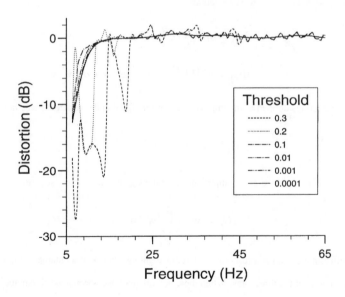

Figure 8: Signal distortion of the design of 1-signal pass band and 6-component reject band for different eigenvalue thresholds.

where the waveforms of signals and multiples are assumed independent and normalised:

$$E[ss^H] = I, \quad E[vv^H] = I \tag{54}$$

and the random noise model of equation (4) is applied. In practice it requires long time segments of independent random signals to approximate equation (54). This model is therefore a considerable over simplification of seismic beamforming. Similarly, the cross spectral matrix of the output of the beamforming filter W is:

$$\Phi_{yy} = W^H x(W^H x)^H = W^H BB^H W + W^H CC^H W + \sigma_n^2 W^H W \tag{55}$$

For analysing the gain of coherent noise and the gain of random noise, I consider a simple model of one signal, one coherent noise and random noise. That is, $J = 1$ and $L = 1$. Commonly there is only one signal of interest and c is chosen to be representative of the correlated noise that is to be rejected. For example, c is taken in the middle of the reject band. Alternatively an average response over the reject band could be used. For convenience, vectors b, c and w are used instead of the matrices B, C and W and are normalised:

$$b^H b = 1 \quad \text{and} \quad c^H c = 1 \tag{56}$$

Then, equations (53) and (55) can be rewritten as:

$$\Phi_{xx} = bb^H + cc^H + \sigma_n^2 I \tag{57}$$

$$\Phi_{yy} = w^H bb^H w + w^H cc^H w + \sigma_n^2 w^H w \tag{58}$$

The coherent noise gain is defined as the improvement of the signal to the coherent noise ratio:

$$G_v = \left| w^H b \right|^2 \Big/ \left| w^H c \right|^2 \tag{59}$$

The random noise gain is defined as the improvement of the signal to random noise ratio:

$$G_n = \left| w^H b \right|^2 \Big/ w^H w \tag{60}$$

By applying different eigenvalue thresholds, different weighting filters w are obtained. G_v and G_n may be plotted as functions of the thresholds. When there is more than one normalized signal and/or more than one normalized coherent noise, the coherent noise gain for each signal is about the same. The random noise gains for each signal are also close. The average of coherent noise gain for each signal is defined as the coherent noise gain for the signal pass band. The random noise gain for the signal pass band is defined in a similar way.

Figure 8 shows the signal distortion at different frequencies of the design of 1-signal pass band and 6-component reject band for different eigenvalue thresholds. Here, the test signal has been filtered by a band pass (10-50Hz) filter. Obviously, at frequencies lower than 10Hz, the signal distortion is not important. The threshold values between 0.2 to 0.3 result in the signal distortion up to 20Hz. The best values of the threshold to keep signal distortion free is in the range of 0.0001 to 0.01.

Figure 9 shows the mean response of the design of 1-signal pass band and 6-component reject band for different eigenvalue thresholds. It is very clear that the best eigenvalue threshold is in the range of 0.001 to 0.01. If the eigenvalue threshold is too big, the signal will be distorted and the coherent noise attenuation will not work perfectly in whole the reject band. However, if the eigenvalue threshold is too small, the singularity can not be compressed very well.

Figure 10 shows the trade-off between coherent noise gain and random noise gain associated with the threshold of eigenvalues of the singular matrix for the model of pass band with 6 signals and reject band with 6 coherent noise components. The thresholds are chosen from 10^{-7} to 0.3. If the value of the threshold is more than 0.3 the trade-off becomes unstable. This figure also confirms that the best value of the threshold is in the range of 0.001 to 0.01 for this design. Further studies also suggest that the best value of the threshold to keep signal distortion free is in the range of 0.001 to 0.01.

6 Application to seismic data

The adaptive beamformer has been installed on an interactive processing system. The input parameters are similar to those used in Radon transform multiple attenuation. There is an option to switch to a fast non-adaptive beamformer which assumes constant amplitude on all traces and therefore needs to be preceded by trace-to-trace amplitude equalisation.

So far, two multiple attenuation methods are generally used in practice, where f-k method is the simple one with relative less computation and can not efficiently attenuate multiples at small offset. The another popular method is Radon transform method. In comparison, Radon transform method attenuate multiples well on all traces including small offset. We compare the multiple attenuation results by using Radon transform method and adaptive beamforming method on real data examples. Figure 11 shows a CMP gather from a survey in the South China Sea after prestack processing including NMO and finally after beamforming. Figure 12 shows the comparison of the stacked sections after applying the conventional multiple attenuation method and after applying beamforming multiple attenuation technique for a survey from South China Sea.

The normal moveout correction applied before using beamformer will not lead signal distortion at far

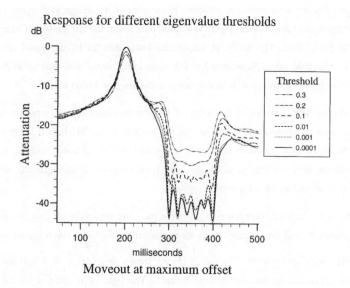

Figure 9: Mean (total frequency band) beamformer response of the design of a 1-signal pass band and 6-component reject band for different eigenvalue thresholds.

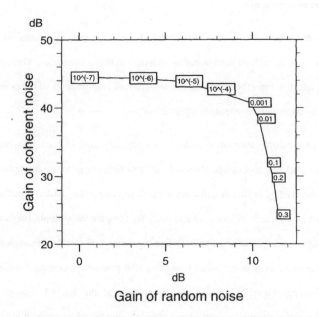

Figure 10: Trade-offs between the gain of the coherent noise and the gain of random noise corresponding to different thresholds of the eigenvalues for 6-signal pass band and 6-coherent noise reject band designs.

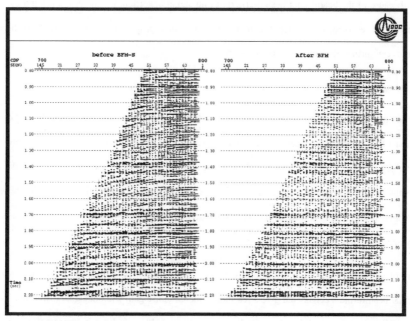

Figure 11: A CMP gather from a survey in the South China Sea after prestack processing including NMO (left) and finally after beamforming (right).

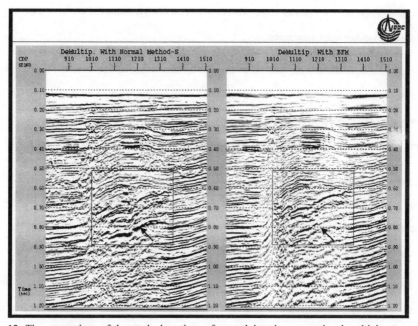

Figure 12: The comparison of the stacked sections after applying the conventional multiple attenuation method (left) and after applying beamforming multiple attenuation technique for a survey from South China Sea.

offset, because of the beamfomer design criteria of no signal distortion, i.e. the primaries and the nearby information is reserved. In practice, the application of a small processing time window focusing on the primaries can further guarantee the primaries unbias.

7. Conclusion

There are many aspects to the design and characteristics of MVU beamforming. The matrix inversion of equation (9) is handled by means of a generalised inversion. It led to the use of a threshold on the eigenvalues of the spectral matrix in beamformer design. Here the setting of the threshold of the eigenvalues of the singular matrix gives a reasonable result and also controls the main trade-off between coherent noise attenuation and random noise attenuation. However the constraint of no signal distortion is not maintained at low frequencies.

A way of characterising multiple rejection using a flexible model is developed. In the ideal case, the particular beamformers considered give over 40dB coherent noise gain without signal distortion. However, timing and amplitude jitter bring the gain down very rapidly, although the gain is generally much better than that of a stack. A standard deviation of time jitter over 2ms causes noticeable degradation of beamformer response. The coherent noise gain has to be controlled and checked to ensure the filter attains a satisfactory random noise attenuation.

The eigenvalue threshold also controls the quality of reconstructed signal. The threshold of eigenvalues of the singular matrix is defined as the ratio of the minimum accepted eigenvalue to the maximum eigenvalue of the matrix. If the threshold exceeds 0.3, the signal is distorted and there is an associated overall attenuation of 2.5dB. As the result, the best threshold should be in the range of 0.001 to 0.01.

One way to overcome the signal distortion is to treat all coherent components except the target signal as noise. Each coherent component can be analysed in turn in this way. This would involve more computation whereas an interactive process needs to be fast. However it may be necessary on occasion and is a design that needs to be studied. It could be very slow but might be more reliable.

The real data examples demonstrate that adaptive beamforming can attenuate multiples better than other popular techniques. This advantage is most clearly evident on good signal-to-noise data (such as marine data).

Acknowledgements

This project is partly supported by Foundation for University Key Teacher by the Ministry of Education and Foundation for University Excellent Young Teacher by the Ministry of Education of People's Republic of China. The real data used in this paper is provided by China National Offshore Oil Corporation.

References

Aki K and Richards P G, 1980, Quantitative seismology: W H Freeman & Co.

Cox H, Zeskind R M, 1987, Owen M K. Robust adaptive beamforming. *IEEE Trans. Acoust. Speech Signal Processing*, **ASSP-35** (10): 1365-1376

Hatton L, Worthington M H, Makin J, 1986, Seismic data processing. Oxford: Blackwell Scientific Publications.

Newman S R, Mahoney J T, 1973 Patterns—with a pinch of salt. *Geophysical Prospecting*, **21**, 197-219

Searle, S R, 1982, Matrix algebra useful for statistics: Wiley series in probability and mathematical statistics, John wiley & Sons, New York

Shumway R H, Dean W C, 1968, Best linear unbiased estimation for multivariate stationary processes. *Technometrics*, **10**, 523-534

Stewart G W, 1979, A note on the perturbation of singular values, Linear algebra and it applications, 23, 213-216

White R E. 1988, A multichannel method of multiple attenuation based on hyperbolic moveout curves. 50th EAEG Meeting, The Hague, Extended Abstracts, 1988, 1-22

Theoretical and Computational Acoustics 2001
E.-C. Shang, Qihu Li and T. F. Gao (Editors)
© 2002 World Scientific Publishing Co.

Finite Element Method Analysis of Sound Characteristics from Tiles Containing Cavities

Hongbo Tan, Hong Zhao, Haiting Xu

Qingdao Acoustic Laboratory, Institute of Acoustics, Chinese Academy of Sciences, Qiangdao 266023, E-mail: *Hb_tan@263.net*

In this paper, the finite element method (FEM) is applied to analyze the sound characteristic of the tiles with the periodic arraying spherical and cylindrical cavities. The reflection and transmission coefficients are obtained. Moreover, numerical results are obtained for tiles with different periodic structures and the effect of the cavities' size and shape is analyzed.

1. Introduction

More and more submarines are covered with tiles for silence, so it is very important to research sound characteristic of the tiles. The tiles mostly contain air cavities that are arrayed doubly periodically. The shapes of the air cavity may be different, sphere, cylinder, cone and so on. Most of them are not the standard shapes. It made the researching of tiles very complex. In the past, some methods were used to research the characteristics of tiles, like T-matrix method[1] and theory of wave in multi-layered media[2]. These methods were all based on the different assumption. These methods rely upon the simplify assumption for the structure and often ignore the resonance of the cavity. The finite element method (FEM) can tackle this problem. A.C. Hennion *et al*[3,4] have used the FEM to calculate the transmission coefficient of the gratings and D.J.W.Hardie has computed the transmission loss[5]. In this paper the FEM is used to analyze the effect of size and shape of the cavity.

The tile that will be researched is doubly periodic structure. It is infinite in the X and Y direction. The air cavities are arrayed periodically. The tile immersed in water. For using the FEM, the calculating region must be reduced into finite region. Bloch-Floqent relations are useful. By them, the calculating region can reduce into one unit that can be meshed into finite elements[6]. The calculating region includes three successive parts: top water, bottom water and tile's layer.

The code of the finite element is programmed. Above all, the theoretical formulation is given. Second, the calculating results and analysis are shown. Finally, the conclusions are given.

2. Theoretical formulation

2.1. Relation of doubly periodic structure

The tile is assumed infinite in the X direction and the Y direction. The structure of the tile is periodic, that is shown in Fig.2.1.

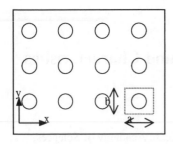

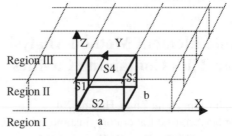

FIG 2.1 The structure from the top view   FIG.2.2 Three regions of one unit cell

Because the tile is assumed to immersed in water, the two semi-infinite water domains which around the tile are chosen. The domain is split into three regions by the plane S+ and plane S- that parallel to the X and Y plane in Fig. 2.2. One cell includes one cavity. It is shown in Fig.2.3. The tiles can be taken as this cell arraying periodically.   One cell includes two layers of water and one layer of the tile and they contact

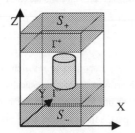

FIG 2.3 The structure of one cell   FIG2.4 The direction of the incident

with the plane Γ^+ and plane Γ^- (Fig.2.3). The cell is a wide in the x direction and b wide in the y direction. A plane, monochromatic wave whose direction of incidence is shown in the Fig.2.4 excites the tile.

If the time dependence $e^{j\omega t}$ is ignored and the amplitude of incident wave is unit, the incident wave can represent follow

$$p^{in}(x, y, z) = e^{j(k_x \cdot x + k_y \cdot y - k_z \cdot z)},\qquad(2.1)$$

$$k_x = k \sin\theta \cos\varphi\,, k_y = k \sin\theta \sin\varphi\,, k_z^2 = k^2 - k_x^2 - k_y^2 .(2.2)$$

The Bloch type relation is applied. Any space function F can represent follow relation.

$$F(x+a, y+b, z)$$
$$= F(x, y, z)e^{jak\sin\theta\cos\varphi}e^{jbk\sin\theta\sin\varphi}\qquad(2.3)$$

So the total pressure can be written in the region I as

$$P^{tol} = P^{in} + \sum_{m,n=-\infty}^{+\infty} V_{mn} P_{mn}^s \tag{2.4}$$

$$p_{mn}^s(x, y, z) = e^{j[(2m\pi/a+k_x)\cdot x+(2n\pi/b+k_y)\cdot y+k_{mn}\cdot z]} \tag{2.5}$$

$$k_{mn}^2 = k^2 - (2m\pi/a+k_x)^2 - (2n\pi/b+k_y)^2 \tag{2.6}$$

The first term is the incident wave and the second term is reflection wave, which is expanded as a doubly infinite series of waves. V_{mn} is the amplitude of the reflection wave (m, n).

It likes as above, the pressure in the region III is shown by follow.

$$P^t = \sum_{m,n=-\infty}^{+\infty} T_{mn} P_{mn}^t \tag{2.7}$$

$$p_{mn}^t(x, y, z) = e^{j[(2m\pi/a+k_x)\cdot x+(2n\pi/b+k_y)\cdot y-k_{mn}\cdot z]} \tag{2.8}$$

T_{mn} is the amplitude of the Transmission wave (m, n).

2.2 The basic functions

Due to the doubly periodic structure, the pressure or displacement (M) in the region II can represent as a doubly infinite series of basic functions.

$$M = \sum_{mn=-\infty}^{+\infty} (\alpha_{mn}^+ M_{mn}^+ + \alpha_{mn}^- M_{mn}^-) \tag{2.9}$$

The basic function M_{mn}^+ satisfies not only the Helmholtz equation but also the around boundary conditions.

$$\frac{d}{dn} M_{mn}^+(x, y)\bigg|_{S_+} = e^{j(k_x\cdot x+k_y\cdot y)} e^{j(2m\pi\, x/a+2n\pi\, y/b)} = f_{mn} \qquad \text{on S+} \tag{2.10}$$

$$\frac{d}{dn} M_{mn}^+(x, y)\bigg|_{S_-} = 0 \qquad \text{on S-} \tag{2.11}$$

$$M_{mn}^+(x, y)\bigg|_{S_1} = e^{-jak_x} M_{mn}^+(x, y)\bigg|_{S_3} \qquad \text{on } S_1 \text{ and } S_3 \tag{2.12}$$

$$M_{mn}^+(x, y)\bigg|_{S_2} = e^{-jbk_y} M_{mn}^+(x, y)\bigg|_{S_4} \qquad \text{on } S_2 \text{ and } S_4 \tag{2.13}$$

M_{mn}^- is defined as follow.

$$\left. \frac{d}{dn} M_{mn}^-(x, y)\right|_{S_+} = 0 \qquad\qquad \text{on S+} \qquad (2.14)$$

$$\left. \frac{d}{dn} M_{mn}^-(x, y)\right|_{S_-} = e^{j(k_x \cdot x + k_y \cdot y)} e^{j(2m\pi\, x/a + 2n\pi\, y/b)} = f_{mn} \qquad \text{on S-} \qquad (2.15)$$

$$\left. M_{mn}^-(x, y)\right|_{S_1} = e^{-jak_x} \left. M_{mn}^-(x, y)\right|_{S_3} \qquad\qquad \text{on } S_1 \text{ and } S_3 \qquad (2.16)$$

$$\left. M_{mn}^-(x, y)\right|_{S_2} = e^{-jbk_y} \left. M_{mn}^-(x, y)\right|_{S_4} \qquad\qquad \text{on } S_2 \text{ and } S_4 \qquad (2.17)$$

2.3 Finite element formulas for basic functions

Every function is gotten by finite element method. The cell is meshed into tetrahedral elements. The finite element method equations[7] are

$$\begin{pmatrix} [K] - \omega^2[M] & -[L] \\ -\rho^2 c^2 \omega^2 [L]' & [K_L] - \omega^2[M_L] \end{pmatrix} \begin{pmatrix} U \\ P \end{pmatrix} = \begin{pmatrix} F \\ \rho c^2 \Phi \end{pmatrix}. \qquad (2.18)$$

[K] and [M] are the stiffness matrix and the mass matrix. [L] is the coupling matrix between the solid and the fluid. $[K_L]$ and $[M_L]$ are the fluid compressibility matrix and the mass matrix. ρ and c are the fluid density and the sound speed. ω is the angular frequency. F contains the nodal values of the applied forces. Φ is the sound pressure normal gradient on the fluid domain. The pressure P in the fluid domain and the displacement U in the solid domain are unknown.

For using the around periodic conditions, the above equations must be modified. If a general limitative condition is $X_1 = d \cdot I \cdot X_2$, the equation (2.19) can be modified to equation (2.20).

$$\begin{pmatrix} A_{11} & A_{12} & A_{13} \\ A_{21} & A_{22} & A_{23} \\ A_{31} & A_{32} & A_{33} \end{pmatrix} \cdot \begin{pmatrix} X_1 \\ X_2 \\ X_3 \end{pmatrix} = \begin{pmatrix} B_1 \\ B_2 \\ B_3 \end{pmatrix} \qquad (2.19)$$

$$\begin{pmatrix} d \cdot \overline{d} \cdot A_{11} + \overline{d} \cdot A_{12} + d \cdot A_{21} + A_{22} & \overline{d} \cdot A_{13} + A_{23} \\ d \cdot A_{31} + A_{32} & A_{33} \end{pmatrix} \cdot \begin{pmatrix} X_2 \\ X_3 \end{pmatrix} = \begin{pmatrix} \overline{d} \cdot B_1 + B_2 \\ B_3 \end{pmatrix} \qquad (2.20)$$

By this way, the around periodic conditions: $g_{s3} = e^{j k_x \cdot a} \cdot g_{s1}$ and $g_{s4} = e^{j k_y \cdot b} \cdot g_{s2}$ are added into the FEM equations.

2.4 The matching on surface nodes for sound parameters

The coefficient of every basic function and the amplitude of the reflection and transmission waves will be computed by the matching method. The pressure is successive on the top and bottom surfaces.

$$\sum_{m,n=-\infty}^{+\infty} (\alpha_{mn}^{+} \cdot M_{mn}^{+}\{R^{+}\} + \alpha_{mn}^{-} \cdot M_{mn}^{-}\{R^{+}\}) = p^{t}\{R^{+}\}$$ (2.21)

$$\sum_{m,n=-\infty}^{+\infty} (\alpha_{mn}^{+} \cdot M_{mn}^{+}\{R^{-}\} + \alpha_{mn}^{-} \cdot M_{mn}^{-}\{R^{-}\}) = p^{s}\{R^{-}\} + p^{in}\{R^{-}\}$$ (2.22)

The pressure normal gradient is successive on the top and bottom surfaces.

$$\sum_{m,n=-\infty}^{+\infty} (\alpha_{mn}^{+} \cdot f) = \frac{d\, p^{t}\{R^{+}\}}{dz}$$ (2.23)

$$\sum_{m,n=-\infty}^{+\infty} (\alpha_{mn}^{-} \cdot f) = -\{\frac{d\, p^{s}\{R^{-}\}}{dz} + \frac{d\, p^{in}\{R^{-}\}}{dz}\}$$ (2.24)

Solve these equations to get α_{mn}^{+}, α_{mn}^{-}, V_{mn} and T_{mn}.

3 Calculating Results

The reflection coefficient in the region I and the transmission coefficient in the region III are defined as follow.

$$V = \left(\sum_{real(k_{mn})>0} V_{mn}^{2}\right)^{1/2}, T = \left(\sum_{real(k_{mn})>0} T_{mn}^{2}\right)^{1/2},$$ (3.1)

The transmission loss is defined like as

$$TL = 1 - V^{2} - T^{2}.$$ (3.2)

3.1 The validity of FEM

The sound characteristics of the tiles without cavity have analytical results by theory of layered media. To verify the FEM, there are comparison between analytical results and the results of FEM in Fig.3.1. There is good agreement between them. (The thickness of the tiles without cavity is 5cm. The widths of the cell in X and Y direction are 5cm. The physical constants are density $\rho = 1100 kg/m^{3}$, Young's modulus $E = 1.4 \times 10^{9} Pa$ and the Poisson's ratio $\gamma = 0.49$. The loss angle for the Young's modulus is 0.23.)

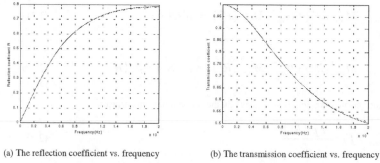

(a) The reflection coefficient vs. frequency   (b) The transmission coefficient vs. frequency

FIG3.1Comparsion of the FEM and analytical ·········· analytical ·········· FEM

3.2 The results of the tile containing spherical inclusions

The cell contains a spherical air cavity whose radius is 1cm and the sizes of the cell like above. The physical constants are density $\rho = 1100 kg/m^3$, Young's modulus $E = 7.1 \times 10^7 Pa$ and the Poisson's ratio $\gamma = 0.49$. The loss angle for the Young's modulus is 0.2.The cell is excited by a plane wave at normal incidence. There are partial reflection amplitude V_{mn} and transmission amplitude T_{mn} in Fig3.2 (a) and (b). The partial wave (m=0,n=0) is dominating and the others almost are zero.

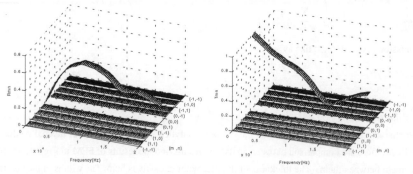

(a)   The reflection coefficient vs. frequency and terms   (b) The transmission coefficient vs. frequency and terms

FIG 3.2 The results containing spherical cavity

The reflection coefficient will increase and the transmission coefficient will decrease for the tile contains cavity in Fig.3.3(a). The minimum transmission coefficient corresponds to a resonance of the air spherical cavity[8,9] and the frequency of this point is 13.5 kHz in Fig.3.3(b). The Fig.3.3(c) show the comparison of the TL between none cavity and spherical cavity. The bigger TL is due to the resonance the spherical air cavity.

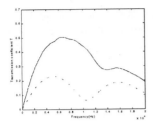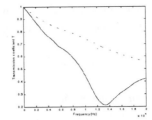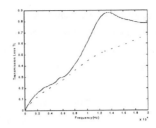

(a) The reflection coefficient vs. frequency(b)The transmission coefficient vs. frequency(c)The transmission loss vs. frequency

FIG3.3Comparsion between none cavity and spherical cavity ········ none cavity,——— spherical cavity(R =1cm)

The Fig.3.4 shows the results of tiles containing different spherical cavities whose radius are 1cm and 1.5cm. The frequency corresponding the minimum transmission coefficient is lower than result of the tiles containing smaller inclusion. The results show that the reflection coefficient is larger and the transmission coefficient is lower with the increasing of the radius of spherical cavity.

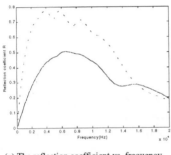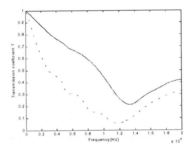

(a) The reflection coefficient vs. frequency　　　(b) The transmission coefficient vs. frequency

FIG3.4 Comparison of results of two kinds of spherical cavity;······ R =1.5cm,——— R =1cm

3.4 Cylindrical cavity

The physical constants are like as above. The radius of the containing cylindrical cavity is 1.5cm and its height is 3cm and the thickness of the covered layer is 1cm.

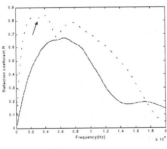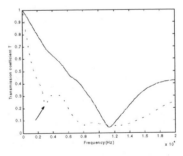

(a) The reflection coefficient vs. frequency　　　(b) The transmission coefficient vs. frequency

FIG3.5 Comparison of results of two kinds of cylindrical cavity, ········ R =1.5cm, H =3cm, ———R =1cm, H=2cm

The effect of the flexure wave is shown at the arrow in Fig3.5. At the 3kHz, the resonance is mainly due to a flexural motion of the cover layer[1,10]. The full lines are results of another cylindrical cavity, the radius of which is 1cm and the height of which is 2cm. The thickness of the covered layer is 1.5cm and the resonance of the flexural motion can't be found from the results.

3.5 The comparison of the tiles containing different shape cavities

The Fig.3.6 show the results of the tiles with the different cavities. These results show the cylindrical or cone-shaped inclusion may excite more effect than the spherical inclusion. Their positions of the minimum transmission coefficient are different for their shapes.

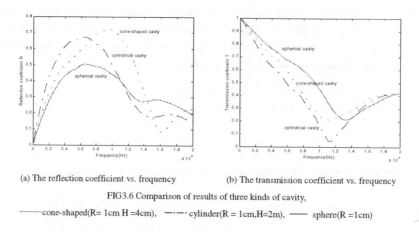

(a) The reflection coefficient vs. frequency (b) The transmission coefficient vs. frequency

FIG3.6 Comparison of results of three kinds of cavity,

·············cone-shaped(R= 1cm H =4cm), —·—·cylinder(R = 1cm,H=2m), ——— sphere(R =1cm)

3.6 The results of the tile covered with semi-infinite air domain

The region III is changed to semi-infinite air domain. Now, only the reflection coefficient interests us. The Fig.3.7 represents the results of the different tiles in this environment. The sizes of the cavities and the physical constants are like above.

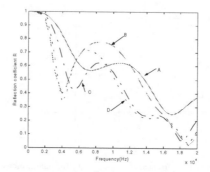

FIG. 3.7 The reflection coefficient vs. frequency.

Curve A: none inclusion.   Curve B: cone-shaped inclusion. Curve C: spherical inclusion. Curve D: cylindrical inclusion

These results show the cavities decrease the reflection coefficient at most frequency. There is extremum at the low frequency band in every result of the tiles containing cavity. But at some frequency, the reflection coefficient of the tile containing air inclusion is larger than that of tile containing none inclusion.

4 Conclusions

This paper presented analysis of the sound characteristics of the tiles containing doubly periodic cavities by the finite element method. There is good agreement between the FEM results and analytical results when the tiles has none inclusion. The results show that the tile immerse in water, the air cavity increase the reflection coefficient but it decrease the transmission coefficient because of the resonance of the air cavity. The reflection coefficient of the tiles containing bigger air inclusion is larger and the transmission coefficient is smaller. The minimums of the transmission of the tiles containing cavities are varying versus the shape of the cavity. When the tile is immersed between the semi-infinite air domain and semi-infinite water domain, the inclusion will decrease the reflection coefficient.

References

1   A. Lakhtakia, V.V.Varadan, and V.K.Varadan, "Reflection characteristics of an elastic slab containing a periodic array of elastic cylinders: SH wave analysis," *J.Acoust.Soc.Am.* 80,331 -316 1986

2   He Zuoyong, and Wang Man, "Investigation of the sound absorption of non-homogeneous composite multiple-layer structures in water" *Journal of Applied Acoustics*,15 May. 12-20 1996

3   A.C.Hennion, R.bossut and J.N.Decarpigny   "Analysis of the scattering of a plane acoustic wave by a periodic elastic structure using the finite element method: Application to compliant tube gratings," *J.Acoust.Soc.Am.* 87(5),1861-1870 1990

4   Anne-Christine Hladky-Hennion and Jean-Noel Decarpigny "Analysis of the scattering of a plane acoustic wave by a periodic elastic structure using the finite element method: Application to Alberich anechoic coatings," *J.Acoust.Soc.Am.* 90(6),3356-3367 1991

5   D.J.W.Hardie "Finite element analysis of decoupling tile performance," *UDT 1995*, 187-190 Conference of Connes, France 4-6 July 1995

6   T.C.Ma, R.A. Scott, and W.H.Yang, "Harmonic wave propagation in an infinite elastic medium with a periodic array of cylindrical pores," *J.Sound Vib.*71, 473-482 1980

7   Gordon C. Everstine and Francis M. Henderson "Coupled finite element/boundary element approach for fluid-structure interaction,"   *J.Acoust.Soc.Am.* 87(5) 1938-1947   1990

8   D.Bai and J.B.Kitahara, "sound waves in a periodic medium containing rigid spheres," *J.Acoust.Soc.Am* 78,43-48 1986

9   J.D.Achenban and M.Kitahara "Harmonic waves in a solid with a periodic distribution of spherical cavities.," *J.Acoust.Soc.Am* 81,595-598   1987

10 A. Lakhtakia, V.V.Varadan, and V.K.Varadan, "Reflection characteristics of an elastic slab containing a periodic array of circular elastic cylinders: P and SV wave analysis,"   *J.Acoust.Soc.Am.* 83,1267-1275 1988

Theoretical and Computational Acoustics 2001
E.-C. Shang, Qihu Li and T. F. Gao (Editors)
© 2002 World Scientific Publishing Co.

Wave Propagation and Imaging Using Gabor-Daubechies Beamlets

Ru-Shan Wu and Ling Chen
Modeling and Imaging Laboratory, IGPP, University of California, Santa Cruz,
CA 95064, USA

Abstract

Beamlet propagation and imaging using Gabor-Daubechies frame (G-D frame) decomposition with local perturbation theory is developed and tested. The method is formulated with a local background velocity and local perturbations for each window of the wave field decomposition. The propagators and phase-correction operators are obtained analytically or semi-analytically by one-way operator decomposition and screen approximation in beamlet domain. The numerical tests on point-spreading responses in homogeneous and laterally varying media and the imaging example for the SEG-EAGE salt model demonstrate the validity and the great potential of this approach applied to seismic wave propagation and imaging.

Keywords: wave propagation and imaging, Gabor-Daubechies frame, beamlets

Introduction

Steinberg (1993), Steinberg and Birman (1995) derived the localized propagators using the WFT (windowed Fourier transform) and a perturbation approach for wave propagation and imaging. These studies represent the effort of developing localized propagators instead of the traditional global propagator methods. The localized propagators are mainly controlled by the local properties of the heterogeneous media and therefore are much easier to make good approximations than the global propagators. However, the decomposition and reconstruction using WFT and inverse WFT are very expensive, so that the method is difficult to adopt for practical use. In addition, the method of Steinberg and Birman is formulated with a global perturbation to the partial differential wave equation. It may be inaccurate at sharply varying velocity areas. Wu et al. (2000a) developed a formulation with local background and local perturbations and derived the localized propagators in discrete representations. In this work we use a tight or nearly tight frame with the Gabor elementary function (Gabor-Daubechies frame) to decompose both the propagator and wave field. Beamlet propagator in heterogeneous media is split into a local free-propagator and a local phase-correction operator. The free propagator and the perturbation operator are obtained analytically or semi-analytically by one-way operator decomposition and screen approximation in beamlet domain. Point spreading responses (impulse responses) are tested for both homogeneous and laterally varying media. Finally, the SEG-EAGE salt model post-stack data are used to test the validity of the approach. The obtained image is shown in comparison with that of the traditional Kirchhoff migration and the global hybrid pseudo-screen method. The excellent results demonstrate the feasibility of the approach.

Windowed Fourier Transform (WFT)

For a function $f(x)$, its windowed Fourier transform (WFT) can be defined as

$$f(\bar{x}, \bar{\xi}) = \int dx f(x) w^*(x - \bar{x}) e^{-i\bar{\xi}x} \tag{1}$$

where $\bar{x}$ and ξ are the localized space and wavenumber parameters, respectively. $w(x)$ is a square integratable window function centered at $x = 0$, and "*" denotes the complex conjugate. We call the phase-space parameterized by $(\bar{x}, \bar{\xi})$ the beamlet domain. The inverse WFT (reconstruction) is given by

$$f(x) = \frac{1}{2\pi N^2} \iint d\bar{\xi} d\bar{x} f(\bar{x}, \bar{\xi}) w(x - \bar{x}) e^{i\bar{\xi}x} \tag{2}$$

where N is the L^2 norm of the window,

$$N^2 = \|w\|^2 = \int |w(x)|^2 dx \tag{3}$$

For a Gaussian window,

$$g(x) = \pi^{-1/4} \exp(-x^2/2) \tag{4}$$

$N^2 = \|g\|^2 = 1$.

Gabor-Daubechies Frames

Since the WFT reconstruction is very time-consuming, many studies have been performed to obtain sparsely sampled, but still accurate reconstruction schemes. For the Gaussian window function, which is closely related to the canonical coherent states in physics, Gabor (1946) proposed a phase-space decomposition using the critical sampling $\Delta_x \Delta_\xi = 2\pi$ in the phase-space (time-frequency) domain, where Δ_x and Δ_ξ are the space and wavenumber sampling intervals, respectively. However, Daubechies (1990) has proved that the reconstruction using critical sampling is unstable. Daubechies showed that for stable reconstruction oversampling $\Delta_x \Delta_\xi < 2\pi$ must hold, $\Delta_x \Delta_\xi$ measures the size of windowed Fourier atoms (Latticed coherent states). Daubechies gave the necessary and sufficient conditions for the stable reconstruction based on the frame theory. The Windowed Fourier Frames with Gaussian window is called by Daubechies the Weyl-Heisenberg coherent states frame, since it is associated with the Weyl-Heisenberg Group.

If the windowed Fourier atoms

$$g_{mn}(x) = e^{im\Delta_\xi x} g(x - n\Delta_x) \tag{5}$$

constitute a frame, where $g(x)$ is a window function and $\Delta_x \Delta_\xi < 2\pi$, then frame coefficients of a function $f(x)$ can be calculated as

$$\langle f, g_{mn} \rangle = \int dx f(x) e^{-im\Delta_\xi x} g^*(x - n\Delta_x) \tag{6}$$

The function $f(x)$ can be reconstructed from the so-obtained frame coefficients:

$$f = \sum_m \sum_n \langle f, g_{mn} \rangle \tilde{g}_{mn} \tag{7}$$

where $\tilde{g}_{mn}$ is the dual frame vectors,

$$\tilde{g}_{mn}(x) = e^{im\Delta_\xi x} \tilde{g}(x - n\Delta_x) \tag{8}$$

It has been proved that the dual frame is also a windowed Fourier frame. That means that the dual frame vectors are time and frequency translations of a new window. In the case of the dual frame equal to the original frame, the frame is then called tight frame. There have been various methods and algorithms developed to construct dual frames, such as the conjugate gradient iterations or pseudo inverse method (Daubechies, 1992; Mallat, 1998, Qian and Chen, 1996).

Compared to the critical sampling case ($\Delta_x\Delta_\xi = 2\pi$), frame decomposition is overcomplete. The representation by its coefficients contains redundant information. The moderate redundancy results in a robust and stable reconstruction. From wave propagation point of view, the tight frame representation with moderate redundancy leads to good localizations in both time and frequency (or space and wavenumber), a very desirable feature for efficient extrapolation of wavefield. However, even if the dual frame is very close to the original frame in the nearly tight case, the error accumulation during wave propagation is still noticeable if the dual frame is approximated by the original frame. Moreover, at oversampling cases, frame decomposition is redundant, therefore the dual frame is not unique. In this work, we select the dual frame window whose shape is the closest to the original frame in the sense of the least square error and precompute it for each application using the method of Qian and Chen (Qian and Chen, 1996).

For simplicity, we call the above-described Windowed Fourier Frames with Gaussian window as Gabor-Daubechies (G-D) frame. Gaussian window and its dual frame window with different redundant ratios are plotted in Fig.1. The higher the redundant ratio, the closer the dual frame window function is to the original window function. G-D frame has many desirable properties compared to other orthogonal decompositions, especially for wave field related problems:

1. Good time-frequency localization as we mentioned before.
2. Commutabilities for x and ξ translations. This leads to the convenience in theoretical treatment, and some analytical or semi-analytical expressions for wave propagation can be obtained.
3. Familiarity and similarity. Applications of windowed Fourier transform in wave phenomena have been well studied and the wave behavior under such a representation bears more similarities than other orthogonal wavelet representations.
4. Simple construction of frame vectors.

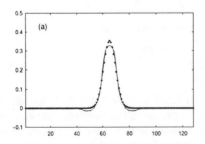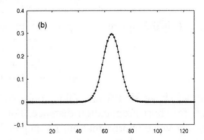

Fig.1 Gaussian window (dotted lines) and its dual frame window (solid lines)
(a) redundant ratio = 2; (b) redundant ratio = 4

Local Perturbation Theory and the beamlet migration

For the purpose of demonstration, we limit our derivation to the 2-D (x, z) case. The generalization to the 3-D case should be straightforward. In frequency-space $(f - x)$ domain, the scalar wave equation can be written as

$$\left[\partial_x^2 + \partial_z^2 + \frac{\omega^2}{v^2(x,z)}\right]u(x,z) = -f(\omega)\delta(x - x_0) \tag{9}$$

where $v(x,z)$ is the velocity field in space domain, and $f(\omega)$ is the source function.

The field at depth z can be decomposed into beamlets (basis vectors or frame vectors) with windows along the x-axis:

$$u(x,z) = u_z(x) = \sum_n \sum_m u_z(\overline{x}_n, \overline{\xi}_m)\widetilde{g}_{mn}(x,z)$$

$$= \sum_n \sum_m \langle u, g_{mn}\rangle\widetilde{g}_{mn}(x,z) \tag{10}$$

where $\widetilde{g}_{mn}$ are the synthesis (reconstruction) vectors (atoms) which are the dual vectors of the decomposition vectors g_{mn}, $u_z(\overline{x}_n, \overline{\xi}_m)$ are the coefficients of the decomposition atoms located at $\overline{x}_n$ (space locus) and $\overline{\xi}_m$ (wavenumber locus):

$$\overline{\xi}_m = m\Delta_\xi, \qquad \overline{x}_n = n\Delta_x \tag{11}$$

For a local beam

$$g_n(x,z) = \sum_m \langle u, g_{mn}\rangle\widetilde{g}_{mn}(x,z) \tag{12}$$

which is the summation of all the local Fourier components.

We can introduce a local perturbation to reduce the strength of perturbation. Let

$$k^2(x,z) = k_0^2(\overline{x}_n, z) + \left[k^2(x,z) - k_0^2(\overline{x}_n, z)\right] \tag{13}$$

where the first term in the right-hand-side is the local background (reference) wavenumber, and the second term is the local perturbations. Upon substituting the field decomposition (12) and local perturbation (15) into equation (11), we obtain the wave equation in beamlet domain (dropping the force term):

$$\sum_n (\partial_z^2 + A_n^2)\sum_m u_z(\overline{x}_n, \overline{\xi}_m)\widetilde{g}_{mn}(x,z) = 0 \tag{14}$$

where A_n is the square-root operators (pseudo-differential operators)

$$A_n \equiv \sqrt{\partial_x^2 + k_0^2(\overline{x}_n, z) + \left[k^2(x,z) - k_0^2(\overline{x}_n, z)\right]}$$

$$= \sqrt{\partial_x^2 + k_0^2(\overline{x}_n, z) + k_d^2(x, \overline{x}_n, z)} \tag{15}$$

Beamlets have significant values only within their local windows and their spreads are small for short propagation distance. For a local beam evolution problem, invoking the one-way wave approximation, i.e. neglecting interactions between the forward-scattered and backscattered waves, we can write a formal solution for the evolution of beamlets

$$G(x, z + \Delta z; \overline{x}_n, \overline{\xi}_m) = G_{mn} = e^{\pm iA_n\Delta z}\widetilde{g}_{mn}(x,z) \tag{16}$$

Here, G_{mn} is the Green's function in the mixed domain describing the evolution of $\widetilde{g}_{mn}(x)$ in the heterogeneous medium. The $\exp(\pm iA_n\Delta z)$ is a thin-slab propagator for the beamlets with an assumption of vertical homogeneity of the thin slab. The propagator matrix in beamlet domain is

$$P_{jl,mn} = P(\bar{x}_l, \bar{\xi}_j; \bar{x}_n, \bar{\xi}_m) = \langle G_{mn}(x), \tilde{g}_{jl}(x) \rangle \tag{17}$$

The field at $z+\Delta z$ can be obtained as the superposition of contribution from all the beamlets:

$$u_{z+\Delta z}(x) = \sum_n \sum_m u_z(\bar{x}_n, \bar{\xi}_m) G_{mn}(x)$$
$$= \sum_l \sum_j u_{z+\Delta z}(\bar{x}_l, \bar{\xi}_j) \tilde{g}_{jl}(x) \tag{18}$$

where

$$u_{z+\Delta z}(\bar{x}_l, \bar{\xi}_j) = \sum_n \sum_m P(\bar{x}_l, \bar{\xi}_j; \bar{x}_n, \bar{\xi}_m) u_z(\bar{x}_n, \bar{\xi}_m) \tag{19}$$

The beamlet propagator matrix represents the beamlet propagation and cross-coupling.

The square-root operator A_n, which is a pseudo-differential operator, can be expanded into perturbation series (De Hoop et al., 2000) or Padé series (Xie and Wu, 1998). Because a local perturbation scheme is adopted in which the reference velocities vary with each window, $k^2(x) - k_0^2(\bar{x}_n)$ can be treated as small perturbations. In Hamilton path integral formulation, $e^{iA_n\Delta z}$ can be considered as a thin-slab Green's function (Wu and De Hoop, 1996; De Hoop et al., 2000). The left symbol for the square-root operator A_n can be decomposed into background and perturbation parts and different approximations can be applied to the operator (see Wu et al., 2000b). In this way the beamlet propagator can be decomposed into a free-propagatior for the local homogeneous media and a perturbation operator to account for the phase-correction by local perturbations.

The G-D (Gabor-Daubechies) frame free-propagator can be calculated as

$$P_{jl,mn}^0 = \frac{1}{2\pi} e^{-i(\bar{\xi}_j\bar{x}_l - \bar{\xi}_m\bar{x}_n)} \int d\xi g(\xi - \bar{\xi}_m) \tilde{g}(\xi - \bar{\xi}_j) e^{i\xi(\bar{x}_l - \bar{x}_n)} e^{i\zeta_n\Delta z} \tag{20}$$

where ζ_n is the local vertical wavenumber, $\zeta_n = \sqrt{(\omega/c(\bar{x}_n))^2 - \xi^2}$. Fig.2 shows the free-propagators with different redundancies in case of low frequency and high frequency, respectively. We can see that they are all highly sparse matrices. For calculation of perturbation operator, two approaches are tested. One is the approximation of weak coupling between windows; the other is the beam summation in the mixed domain. In the following we present the results of beam summation approach.

Numerical tests on point spreading responses of beamlet propagators

Considering a point source located on the surface of the media with a ricker wavelet time function, we calculated the point spreading responses for a homogeneous medium (Fig.3) and a three-layered heterogeneous medium (Fig.4). The latter is a model with a laterally varying layer (Gaussian profile) embedded in a homogeneous medium (Fig.4a). In Fig.3, as expected, the wavefront of the point spreading response is a semi-circle; while in the layered medium, the wavefront is distorted after passing through the second layer as shown in Fig.4b.

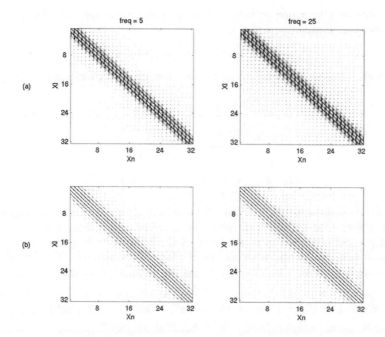

Fig.2 Free propagators in beamlet domain with Gabor-Daubechies frame in case of low
and high frequencies, respectively. (a) redundant ratio = 2; (b) redundant ratio = 4

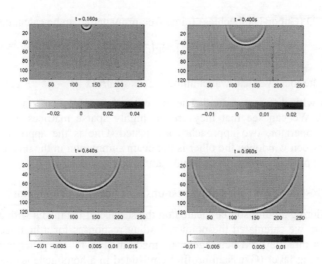

Fig.3 Point spreading response for homogeneous medium

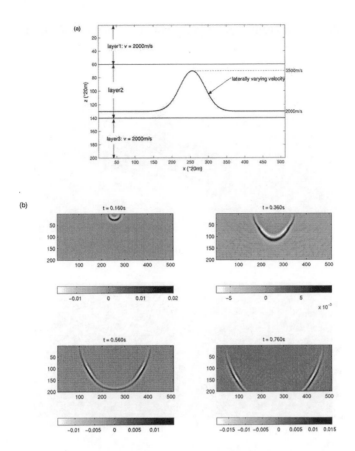

Fig.4 (a) A three-layered medium; (b) Point spreading response of the layered medium

Numerical Tests Using SEG-EAGE Salt Model Data

In this section we show a preliminary test result using the poststack data of the SEG-EAGE model. Fig.5a shows the velocity model of the A-A' profile from the salt model. The salt body has a strong velocity contrast with the surrounding medium and an erratic top surface. Fig.5b is the image obtained by beamlet domain G-D frame propagator. In comparison, we also plot the same part of the image migrated by the traditional Kirchhoff migration (Fig.5c). Apparently, the former has a superior quality compared with the latter, especially for the sharp irregular boundary of the salt body and the subsalt faults. In Fig.6, we show the comparison between the image obtained by the G-D beamlet propagator (Fig.6a) and that by the hybrid pseudo-screen propagator (Jin et al., 1999) (Fig.6b). We can see that all the main features including the boundaries of the salt body, the steep faults and sharp edges are correctly imaged with beamlet method. The overall image quality is comparable with that of the screen method. This demonstrates the

668

potential of application of beamlet propagators with local perturbation theory to seismic imaging.

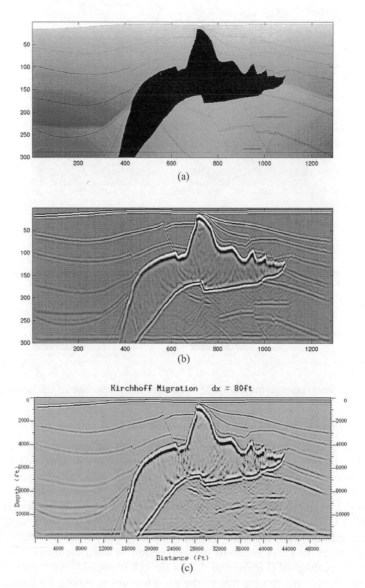

Fig.5 (a) velocity model of A-A' profile of the SEG-EAGE salt model; (b) image by beamlet migration; (c) image by Kirchhoff migration

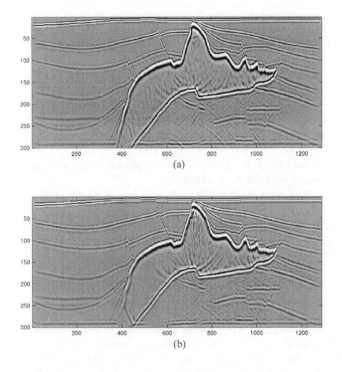

Fig.6 Images of the A-A' profile of the SEG-EAGE salt model.
(a) by beamlet migration; (b) by hybrid pseudo-screen migration

Conclusions

Beamlet propagation and imaging using G-D (Gabor-Daubechies) frame propagator with local perturbation theory has been developed and tested. The use of local background velocity and local perturbations allows to optimize the local beamlet propagators and easily to handle the strong lateral velocity variations. We have obtained the analytical form of G-D frame propagator and phase-correction operator based on the one-way operator decomposition and screen approximation. The examples shown here for the point spreading responses and SEG-EAGE salt model image demonstrated the potential of this approach. Further study is needed for improving both the accuracy and efficiency.

Acknowledgements

The support from the WTOPI (Wavelet Transform On Propagation and Imaging for seismic exploration) Project at University of California, Santa Cruz is acknowledged.

670

References

Daubechies, I., 1990, The wavelet transform, time-frequency localization and signal analysis, *IEEE Trans. Inform. Theory*, **36**, 961-1005.

Daubechies, I., 1992, Ten Lectures on Wavelets, Philadephia, Pennsylvania: Society for Industrial and Applied Mathematics.

De Hoop, M., Rousseau, J. and Wu, R.S., 2000, Generalization of the phase-screen approximation for the scattering of acoustic waves, *Wave Motion*, **31**, 43-70.

Gabor, D., 1946, Theory of Communication, *J. Inst. Electr. Eng.*, London, **93**(III), 429-457.

Jin, S., Wu, R.S., and Peng, C., 1999, Seismic depth migration with screen propagators: *Computational Geosciences*, **3**, 321-335.

Mallat, S., 1998, A Wavelet Tour of Signal Processing, Academic Press.

Qian, S. and Chen, D.P., 1996, Joint Time-Frequency Analysis, Methods and Applications, Prentice-Hall Inc..

Steinberg, B.Z., 1993, Evolution of local spectra in smoothly varying nonhomogeneous environments-Local canonization and marching algorithms, *J. Acoust. Soc. Am.* **93**, 2566-2580.

Steinberg, B.Z. and R. Birman, 1995, Phase-space marching algorithm in the presence of a planar wave velocity discontinuity-A Qualitative study, *J. Acoust. Soc. Am.* **98**, 484-494.

Wu, R.S., Wang, Y. and Gao, J.H., 2000a, Beamlet migration based on local perturbation theory, *Expanded abstracts, SEG 70th Annual Meeting*, 1008-1011.

Wu, R.S., Jin, S., Xie, X.B. and Moster, C.C., 2000b, 2D and 3D generalized screen migration, *Expanded Abstract, EAGE 62nd Annual Meeting*, B-15.

Xie, X.B. and Wu, R.S., 1998, Improve the wide angle accuracy of screen method under large contrast, *Expanded abstracts, SEG 68th Annual Meeting*, 1811-1814.

Theoretical and Computational Acoustics 2001
E.-C. Shang, Qihu Li and T. F. Gao (Editors)
© 2002 World Scientific Publishing Co.

Numerical Simulation Study on Mode-Coupling Caused by Solitary Waves in Shallow-Water Waveguide

Y. Y. Wang and E. C. Shang

CIRES, University of Colorado/NOAA/Environmental Technology Laboratory,

325 Broadway, Boulder, CO 80303, USA

Abstract

Mode-coupling induced by internal solitary waves in shallow-water is investigated by numerical simulation. It has been found that mode-coupling is very sensitive to frequency, mode number as well as water depth.

Introduction

The non-linear internal waves in the ocean have been known for over a century, and extensive investigations have been made in the oceanographic community in the early 1970s. Much more attention has been paid in the ocean acoustic community during the 1990s, it was stimulated by Zhou's paper [1], which as the first time took the internal solitary wave (ISW) as a possible candidates to explain the strong frequency-selected acoustic wave transmission loss observed in the shallow-water of the Yellow Sea [2]. The recent progresses can be found in [3-5].

In this paper, the numerical simulations of the acoustic wave interacted with ISW are performed using the PE code [6] and Normal mode code [7]. Firstly, based on the real ISW data observed by NOAA conducted COPE experiment at Oregon coastal area in September 1995 [8]. Secondly, mode-coupling based on a single soliton of KdV theory is investigated.

Numerical simulation based on COPE data

NOAA/ETL conducted the "Coastal Probing Experiment " (COPE) in the Oregon Coastal area in Sept-Oct 1995. A lot of tide related ISW data were collected. One example of the isotherm record measured by FLIP is shown in Fig.1 (transformed to Space distribution). Acoustic wave propagating through this ISW has been performed at the water depth of 150 m . Acoustic wave transmission loss of a single source has been calculated in the frequency range of 50 – 1000 Hz. Modal repopulations of single mode incident are also calculated. The strong mode-coupling case of f=200 Hz , mode 6 incident is illustrated on Fig2 a), and the adiabatic case of f=60 , mode 2 incident is illustrated on Fig.2 b). The transmission loss of f=50 Hz and f=850 Hz are illustrated on Fig.3 and Fig.4 respectively. As we can see that the f=50 case is adiabatic , only a slight phase shift caused by ISW, but for the f=850 Hz case there is 5-10 dB loss increasing caused by ISW.

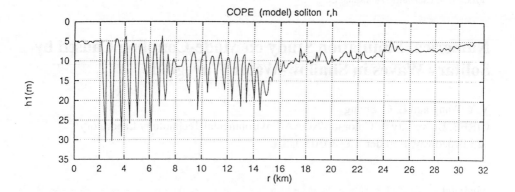

Fig.1 The isotherm (14^0c) of the solitary waves in Oregon coastal zone
measured by FLIP at water depth of 150 m. (COPE data).

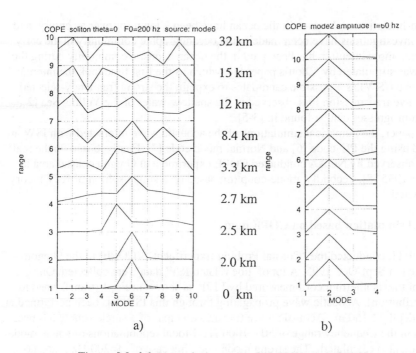

a)   b)

Fig.2 Modal repopulation caused by ISW, a) f=200 Hz, m=6 ;
b) f=60 Hz, m=2.

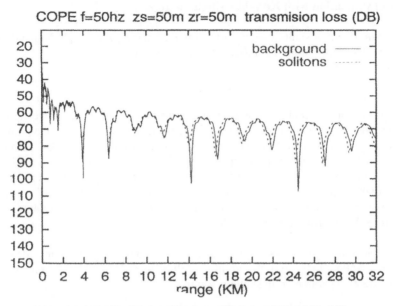

Fig.3   Transmission loss at f= 50 Hz, z_s =50m , z_r =50m, H=150m,
          Solid – background, dash- with ISW

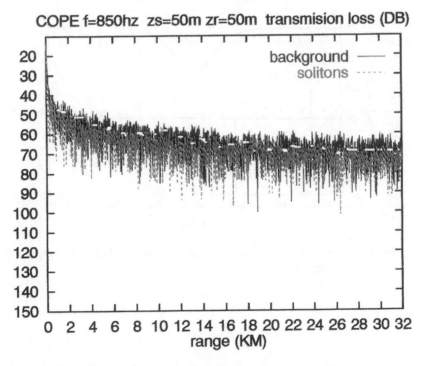

Fig.4   Transmission loss at 850 Hz, z_s=50m, z_r=50m, H=150m,
          Red- background, green-with ISW

Numerical simulation with KdV-like single soliton

The environment of shallow-water is the same as in the paper of Preisig and Duda[5], (see Fig. 5). The depression of the thermocline is given by

$$\eta(x) = A \operatorname{sech}^2 [(x - x_c)/L]$$ (1)

Different cases for different parameter A and parameter L, have been investigated. It has been found that mode-coupling is very sensitive to frequency and mode number, so the mode coupling effect has to be demonstrated on the $f - m$ space. For example, the Case of A=10 m, L = 50 m is illustrated in Fig. 6. One interesting thing is the L dependecy of the mode-coupling. Between the "cancellation regime" for very small L, and the "adiabatic regime" for large enough L, there is a "resonant regime" for moderate L for a fixed frequency and a fixed mode number. Fig.7 shows the "resonant" L is around 200 ~ 300 m for mode 3 at f=300 Hz.

We also found that for the same parameter A and L, but change the water-depth H from 50 m to 150 m, in general, the coupling will be reduced. Perhaps, parameter (A/H) is a more effective parameter in shallow-water.

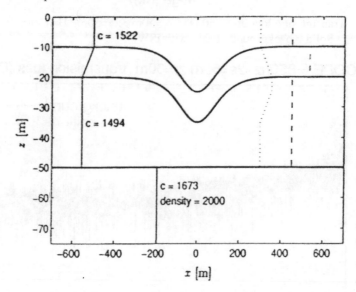

Fig.5   The basic shallow-water environment and the KdV like single soliton (from [9]).

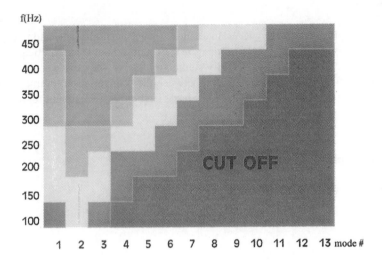

Fig.6 Modal adiabaticity in (f-m) space for a single KdV soliton in H=50m
Shallow-water (A=10 m, L=50 m).
Red-strong coupling, orange-weak coupling, green-adiabatic

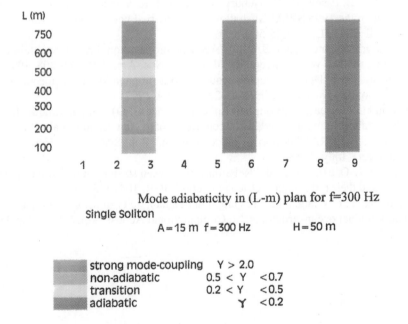

Mode adiabaticity in (L-m) plan for f=300 Hz
Single Soliton
A = 15 m f = 300 Hz   H = 50 m

strong mode-coupling	Y > 2.0	
non-adiabatic	0.5 < Y	<0.7
transition	0.2 < Y	<0.5
adiabatic	Y	<0.2

Fig.7 Modal adiabaticity in (L-m) space for f=300 Hz, A=15m, H=50m.

Summary

(1) Mode-coupling induced by ISW in shallow-water is significant, and is very sensitive to frequency f and mode-number m , as pointed out by theoretical analysis [9] , the detail features should be carefully studied on the $f-m$ space.

(2) The interesting L dependency is that between the "cancellation regime" for very small L and the "adiabatic regime" for very large L, there is a "resonant regime" for certain mode at a given frequency.

Acknowledgments

This work was supported by NOAA and ONR.

References

[1]. J.X. Zhou, X.Z. Zhang, and P.Rogers, "Resonant interaction of sound waves with internal solitons in the coastal zone", J. Acoust. Soc. Am., 90, 2042-2054, 1991.

[2]. E.C. Shang, "Some new challenges in shallow-water acoustics", Progress in Underwater Acoustics, Ed. H. Merklinger, (Plenum, New York), pp.461-471, 1987.

[3]. J.R. Apel, M. Badiey, C-S. Chiu, S. Finette, R. Headrick, J. Kemp, J.Lynch, A. Newhall, M.Orr, B. Pasework, D. Tieburger, A. Turgut, K. Vonde Heydt, and S. Wolf, "An overview of the 1995 SWARM shallow-water internal wave acoustics scattering experiment," IEEE, JOE, 22, 465-500, 1997.

[4]. D. Tieburger, S. Finette, and S. Wolf,"Acoustic propagation through an internal wave field in a shallow water waveguide," J. Acoust. Soc. Am., 101, 789-808, 1997.

[5]. J.C. Preisig and T. Duda, "Coupled acoustic mode propagation through continental shelf internal solitary waves," IEEE, JOE, 22, 256-269, 1997.

[6].D. Lee and G. Botseas, " An implicit finite difference (IFD) computer model for solving the parabolic equation," NUSC, Tech. Rep. 6659, New London Lab. 1982.

[7]. M. Porter, "The KRAKEN normal mode program," Rep. SACLANT Undersea Center, La Specia, Italy, 1988.

[8]. R. Kropfli, L. Ostrovsky, et al, "Relationship between strong internal waves in the coastal zone and their radar radiometric signatures," JGR, 104, 3133, 1999.

[9]. E.C. Shang, Y.Y. Wang, and T.F. Gao, 'On th adiabaticity of acoustic propagation through non-gradual ocean structures," J. Computational Acoust. Vol.9, No. 2, pp.359-365, 2001.